Carbon Free

Carbon Free

The Way Forward

Edited by
Jörg Böttcher

DE GRUYTER

This book is published in an open access edition and released on University Press Library Open (UPLOpen.com) as part of the 'UPLOpen Climate Change Collection.' UPLOpen is an initiative of Paradigm Publishing and the De Gruyter eBound Foundation.

ISBN 978-3-11-078417-6
e-ISBN (PDF) 978-3-11-078429-9
e-ISBN (EPUB) 978-3-11-078436-7
DOI https://doi.org/10.1515/9783110784299

Library of Congress Control Number: 2024949276

Bibliographic information published by the Deutsche Nationalbibliothek
The Deutsche Nationalbibliothek lists this publication in the Deutsche Nationalbibliografie;
detailed bibliographic data are available on the internet at http://dnb.dnb.de.

Typesetting: Integra Software Services Pvt. Ltd.

www.degruyter.com
Questions about General Product Safety Regulation:
productsafety@degruyterbrill.com

Contents

Jörg Böttcher

1 Introduction

Nunc aut nuncquam – it's now or never: this motto may serve as a leitmotif for the tasks we face as a society with regard to the upcoming transformation process of our economy and living environments.

The goal of achieving global climate neutrality is essential for ecological, economic and social reasons, is well documented scientifically and requires immediate and decisive action from politicians and society. Clarity, determination and unity are the principles on which action must now be based. The climate crisis not only endangers the health and lives of individuals, but also threatens the livelihoods of humanity as a whole.

We have all been aware for decades that we must act to avert catastrophic consequences for ourselves, our descendants and the environment: The Club of Rome named the "Limits to Growth" as early as 1972, the environmental study "Global 2000" informed the US Congress in 1980 about the foreseeable, growing environmental problems and climate change, and since the Earth Summit in Rio in 1992 at the latest, the issue of climate protection has reached the global public. It would be unfair to say that nothing has happened in these three decades. The fact is, however, that too little has happened and we are now faced with the situation of having to take far-reaching measures very quickly in order to be able to stop even more catastrophic developments on a global level.

Let's be clear: it's not "just" about the earth or emissions targets, it's about us humans. According to the European Environment Agency, Europe is the fastest warming continent. Since the 1980s, warming in Europe has been around twice as fast as the global average. Hundreds of thousands of people could die from heatwaves alone if politicians do not act immediately.[1] And this is just looking at Europe – some of the most populated regions are getting too hot to live.[2]

The climate turnaround is certainly not something that can be achieved by a single measure. On the contrary, it requires measures at many levels that are interlinked and interact with each other. In addition to ecological issues, suitable technologies and energy infrastructures, these include questions of economic efficiency, the legal framework and social acceptance. When weighing up the measures and side effects, our society has so far opted for a slower transition, certainly also in order to keep the economic burden and the impact on production sites to a minimum.

1 European Environment Agency (2023): https://www.eea.europa.eu/publications/european-climate-risk-assessment.

2 Kerry Emanuel (ed.) 2023, https://doi.org/10.1073/pnas.2305427120.

https://doi.org/10.1515/9783110784299-001

Russia's war of aggression against Ukraine, with all its consequences for the use of gas and oil as a bridging technology, then came as a surprise to some and the plans for a slow transition were shattered. The magic triangle of security of supply, ecological compatibility and economic efficiency had to be rebalanced virtually overnight. We are now faced with the difficult task of developing a climate-neutral society and expanding renewable energies in a very short space of time. At the same time, we need to achieve security of supply without overburdening people and companies. This also means that we will have to continue using fossil fuels in the coming years, which we would have liked to do without from an environmental perspective. However, this is sensible in the short term and will have to be compensated for later.

The current crisis is acting as a catalyst. It is important to involve all consumers in the restructuring of the energy and economic landscape and to consider environmental protection and other uses in addition to the appropriate legal and infrastructure measures. In particular, the framework conditions for investments in renewable energies must be improved so that the necessary investment volume can be raised and the measures can be implemented quickly. The aim over the next few years must be to restore security of supply and pursue a faster transformation to climate neutrality. A fossil-free, climate-neutral energy system offers security of supply, frees us from price risks and is ecologically indispensable.

Around 80% of final energy demand worldwide is based on fossil fuels. Although, depending on the country, electricity generation from renewable energies is significant – in Germany it will account for over 50% of gross electricity consumption in 2023 – the consumption sectors of buildings, industry and transportation in particular are heavily dependent on oil and gas. It is crucial that we achieve a fundamental transformation within the next two decades that excludes climate-damaging fossil fuels from our lives. The time for half-measures and hesitating steps is long gone. We need bold and decisive action to transition away from fossil fuels and towards a sustainable future for all.

Let us take a brief look back: the "Limits to Growth" and "Global 2000" mark two milestones on the way to social awareness that "business as usual" cannot work. In fact, the world's governments have been reluctant to act on this awareness for a very long time. Sociologists would say that the political planning horizon of politicians ends with the next re-election, but does not necessarily cover a period of several decades. After all, a change in living conditions is punished by the electorate – at least that is the fear, and so it is postponed first once and then repeatedly. So it was only shortly before twelve that the realization dawned that something had to be done now – and with force: At the World Climate Conference in Glasgow in November 2021, the EU strengthened its climate protection target to make Europe climate-neutral by 2050. One key element is a CO_2 price on natural gas and oil, which signals to us all that climate-damaging behavior will cost more and more money.

The last global crisis, the coronavirus pandemic, has taught us several things: we can deal with uncertainty and dramatic changes and, if necessary, we can adapt within a few days. We have also learned that there are two sides to change and that we want to keep it that way: Video conferences save travel time and, conversely, we are happy to meet colleagues in person in the office again. There will also be something similar in the fight against climate collapse: be it greater independence thanks to our own PV system on the roof or a better quality of life in cities thanks to less air pollution.

The negative issues of recent years should only be mentioned briefly, as they have had more than enough of a stage: firstly, a poorly communicated Building Energy Act, followed by a disinformation campaign against heat pumps and in favor of hydrogen heating, a short-lived political discussion about e-fuels and the repeated promotion of nuclear energy as an energy option. The hijacking of COP 28 by the oil industry is certainly also part of this.

What is perhaps particularly remarkable is what did not happen, despite the severity and depth of the energy crisis: blackouts, gas shutdowns, mass insolvencies, serious recession – none of this happened. The cost pressure on companies and people was great and in the hectic rush of crisis management, some things were done right, but some were also done wrong. The social balance was and is inadequate. Nevertheless, overall, politics and the markets have led us through the crisis much better than many had predicted.

Despite all the media noise, the good news is that the positive list is longer than the negative list. This includes breathtaking industry momentum in photovoltaics, the identically emerging development in battery storage, increasing wind energy approvals, the start of the heating transition, global investment records in renewable energies and many starting shots for climate-neutral production from steel to glass. In short, decarbonization has made its way into the political and economic arena.

Which direction should we take in the future? Perhaps a division into The Good, The Bad, The Ugly and Elinor Ostrom can help.

The question we have to ask ourselves is: How can we do with electricity what was previously based on the use of coal and gas? Is our mobility electric – both on land and in the air? Can we use electricity to generate heat? Can we use electricity to maintain the production of a wide range of goods? A cross-reading of the various articles in this anthology reveals a core message: Electricity from renewable energy sources – especially photovoltaics, onshore wind and offshore wind – is the central key to the transformation towards climate neutrality. This is due to the fact that the electricity production costs of these generation classes have continued to fall in recent years and are cheaper than those of conventional power plants. The use of renewable electricity is now possible in areas where this would have been unthinkable a few years ago for economic reasons.

The transformation requires an unprecedented pace in the expansion of wind and solar energy, storage systems and other renewable energy technologies. Industrial plants and processes need to be redesigned in order to be able to use climate-neutral alternatives.

Energy efficiency must be significantly increased and energy consumption must be reduced across all sectors. Every unit of fossil energy saved reduces dependency on imports and creates scope for new investments. In the heating sector, it is all about saving energy and efficiency: if we look at the efficiency aspect, heat pump technology and the use of geothermal energy are of paramount importance. In most cases, the use of heat pumps enables an economically favorable heat supply in the individual sector. The use of geothermal energy enables a base load-capable, ecologically advantageous and also economically attractive heat supply.

Experience from many industrial sectors, most recently photovoltaics, has shown that economies of scale are essential in order to reduce production and operating costs and thus make a new technology economically viable.

What needs to be stopped as quickly as possible is the burning of fossil fuels. The vast majority of combustion processes release CO_2. This is something we absolutely have to stop. Given that these energies account for around 80% of primary energy consumption worldwide, we will not be able to avoid using these undesirable energies in a narrowly defined transition phase, but we must reduce this area as quickly as possible. The pricing of CO_2 is a key instrument here. A second issue requires a concerted political solution: our goal must be to ensure that fossil fuels remain in the ground. The only problem is that large companies, interest groups and even states have an immense interest in continuing their previously profitable business model, even if this is at the expense of the general public.

Individual political players and companies are putting forward the idea that we should develop new and innovative technologies that would then solve our energy problems in an ecologically compatible way when they are ready for the market. As much as I believe in technological progress, I am convinced that this approach is misleading. A wait-and-see strategy lulls us to sleep for the next decades, solves none of the problems we face and suggests that we can continue our lifestyle as we have done for the last few decades. The opinion sometimes expressed that the use of nuclear energy is a sensible option must also be clearly contradicted from an ecological, economic and social point of view.

Building dream castles or pointing to a past that is – perhaps – perceived as unencumbered does not solve any of our existing problems, but rather exacerbates them by waiting and is therefore socially irresponsible.

The Ugly has yet another facet: key physical knowledge on the subject of energy, which is necessary to support a transformation process, is not available to decision-makers, is not understood or is deliberately ignored. Fortunately, scientific institutions are working to make it clear that physical facts cannot be ignored and that this results in clear political recommendations for action.

The good news is that we have the tools at our disposal to implement the necessary changes.

Together with Oliver E. Williamson, Ms. Ostrom was awarded the Nobel Prize in Economics in 2012 for her research into the successful management of common property. Her thoughts on how people can act sustainably in and with ecosystems are more relevant today than ever. In essence, this is about incentive systems that are partly participatory and partly determined by legal norms. We cover this topic in several articles.

Our actions on the path to climate neutrality must be swift and decisive. At the same time, it is necessary to be aware of the values that underpin our actions. The principles of sustainability relate to technical, economic and social aspects. The moral basis of our actions is adequately substantiated by the 17 sustainability criteria, which specify differentiated individual goals and at the same time provide us with guidelines for action in the event of conflict.

The issue of *acceptance of the energy transition* is of paramount importance: this project can only succeed if the majority of the population can be convinced that the energy transition is necessary and that the social consequences of the disruption to the system can be absorbed. This requires political decision-makers to communicate the necessary change of direction very well: By continuing as we are, we are not only damaging future generations, but also ourselves directly. What gives us hope is that even from a purely economic perspective, the switch to renewable energies such as wind, photovoltaics, battery storage and the use of geothermal energy makes sense. And every dollar we invest here avoids many times over in follow-up costs. This is a fundamental difference to climate impact adaptation measures, as these represent an economic good that cannot be provided for profit and must be financed by the general public.

Communication is a political key to the success of the energy transition, which is all the more important the more concrete and far-reaching the transformation measures are. The population will project the failure of previous governments to implement reforms onto the current government. However, there is no sensible debate about what should have happened in the last 30 years to make society and the economy fit for the changes that are definitely on the cards. The effects of climate change were already known at least 30 years ago and we all knew that we had to move towards renewable energies. In the electricity sector, the picture of expansion looks quite positive, but little has happened in the development of primary energy consumption, both nationally and internationally.

And now democratic governments are confronted with the war in Ukraine and the conflicts in the Middle East. These are very difficult conditions for democratic governments. Added to this is the constitutional court ruling in Germany that – if the debt brake remains in place – will continue to have an impact over the next decade and restrict the ability of future governments to act.

The book is structured as follows: The first main section deals with the topics that have an overarching character and are necessary for understanding the individual measures that follow.

In his article "Investments and Regulation", Tibor Fischer describes the measures that need to be initiated and implemented now and in the coming years so that we can advance the expansion of renewable energies as quickly as possible. This requires an adjustment of the legal and economic framework with the aim of mobilizing private capital.

Heinz Bontrup looks at the socio-economic framework conditions of the energy transition. The transformation process results in significant macroeconomic effects that pose considerable challenges for policymakers. Essentially, it is about taking into account the distributional effects of the energy transition and not overburdening broad sections of the population.

The debate on climate neutrality recognizes that our world is being pushed to or beyond its limits by anthropogenic influence. In order to leave us and future generations a planet worth living on, we are making efforts to limit greenhouse gas emissions and thus stabilize the rise in temperature. These efforts can easily be thwarted if ethically justifiable efforts are not also made to influence population growth. An exponential increase in the world's population is not compatible with – at best – linear growth in resources. Jane O'Sullivan's article provides information on the background and shows ways of dealing with the situation in an ethically acceptable way.

In their article "A Road to Sustainability for Economy and Society", Jennifer Kowallik and Alexandra Oboda present the sustainability goals and describe the concepts of efficiency, consistency and sufficiency using examples.

In "The Political Will – The European Path to a Sustainable Future", Andreas Gabler deals with the "Green Deal" and its main contents, as well as with the Investment Plan for the European Green Deal and the Just Transition Fund. The Action Plan for Financing Sustainable Growth is explained with its core elements: The taxonomy, the disclosure framework for non-financial and financial companies, and investment tools including benchmarks, standards and labels.

The widespread acceptance of the goals of the energy transition and the associated adjustments is one of the central tasks facing politicians. Sebastian Helgenberger and Roman Buss present the concept of co-benefits, which relate to economic, social and environmental aspects.

It is often the case that the questions that arise are the same, and in many cases the answers are also the same. However, this is not always the case: especially in the area of climate protection, these are issues where national solutions are being developed. Thinking outside the box is regularly helpful to get ideas for our own actions. This is the reason why we have included a few examples from different countries: Farid Mohamadi presents the very comprehensive measures that Colombia has initiated to decarbonize its economy ("CO_2 Reduction Measures in Colombia"). In his contribution ("Ghana's Example of De-Carbonization"), Kwesi Annan-Takyi focuses not only on the government decarbonization measures but also on the difficulties in implementation, which are mainly due to the limits of the availability of suitable project financing.

The second main section of the book looks at the key sectors affected by the transformation to a climate-neutral economy: On the one hand, this is the energy sector, which the majority of experts believe has the most important role to play. In this respect, we are looking at several sub-aspects: Firstly, we need to take stock of the current situation and develop a target picture of how renewable energies can develop over the next few years.

Jelto Lange, Michael Schulthoff and Martin Kaltschmitt describe the status of the various renewable energies in the electricity, heating and mobility sectors and the steps that need to be taken as part of the transformation process. It is clear that the expansion of renewable energies must go hand in hand with a transformation of the energy infrastructure structure.

The following is a larger section on the topic of *heat*, which is of paramount importance for the energy transition.

Oliver Opel et al. deals with the topic of heat pumps. This is the technology that will play an absolutely outstanding role in the heating transition due to its efficiency. First of all, this applies to individual households if a connection to a local heating network is not an option. Furthermore, the use of modern large heat pumps enables efficient and economical usage options for supplying heat to neighborhoods.

The use of geothermal energy is a second central building block for the heat transition. The efficiency of heat generation is extremely high in many cases and the use of heat pumps can significantly expand the field of application. In the article "Use of Geothermal Energy", I describe the status and prospects of geothermal energy internationally. I also present possible strategies for hedging the exploration risk and the economic viability of a project in Germany. In "Indonesia's Deep Geothermal Drilling Program", Attakan Janpidok describes Indonesia's experience in developing geothermal energy. The findings on the mechanisms that can be used to improve the probability of successful drilling are interesting.

In his article, Jens Kühne describes the role of solar thermal energy for local supply networks. The presentation of the technical options for the use of solar thermal energy is followed by an economic analysis. Based on the heat production costs, there are good opportunities – also in Germany – to economically integrate solar thermal energy as a heat source into the heat mix.

In his article "Levelized Cost of Electricity", Jochen Weilepp points out the importance of electricity production costs. They are an indicator of the competitiveness of a particular form of energy generation. However, it is also important to understand the innovation process in order to reduce costs. Financial learning and technical learning are of central importance in this respect in order to reduce production costs. The chapter ends with recommendations for practitioners and political decision-makers on how to implement innovations successfully and promote them appropriately.

In the article "Project Finance", I explain why this financing method is so important for the further expansion of renewable energies. In this chapter, my main aim is to raise awareness of the various requirements of project financing so that this method can be used in the best possible way.

For many countries, including Germany, it is important to position themselves within the decarbonization value chain. In her article, Kathrin Langewald explains how the topic of green hydrogen is integrated into German-South African energy relations. The focus here is on the balance between three geostrategic goals of market development, security of supply and climate protection, with conflicting implications for German foreign energy policy.

Thies Goldner, Christoph Torwegge and Felix Jäger examine the question of what legal steps need to be taken in order to establish a suitable regulatory system for a sustainable energy supply. First, they look at the existing regulatory systems. They then describe the need for further systems, which essentially stem from the political objectives, but must also support the transition to a green energy supply. In this context, various instruments are outlined, such as capacity markets, power purchase agreements and the role of the EU in the design of support systems.

In the third part of the book, we look at individual technologies, instruments and concepts and examine the role they can play in the transformation process.

Andreas Luczak kicks things off by taking a closer look at the topic of green hydrogen. The picture here is mixed: on the one hand, there is the political will of the EU and Germany to help the topic achieve a breakthrough; on the other hand, there are efficiency issues that will probably lead to green hydrogen being able to play a role in the no-regret areas, but not beyond.

In his article "Pricing and Trading CO_2", Nicolai Herrmann describes the outstanding importance of taxing fossil energy through a CO_2 tax. First, he places the topic in the economic concept of market failure. He then examines the practical consequences of introducing and applying a CO_2 tax, in particular the effects on competition and trade between countries.

Essential renewable energies are volatile in nature, so that there are special requirements for the networking of the various sectors, but also for their control.

Jasmine Ramsebner deals with the networking of the various sectors ("The Concept of Sector Coupling"). The changing generation landscape for electricity and heat is giving rise to requirements and opportunities to connect the previously separate sectors and the resulting usage options – such as heat, transportation and industrial uses – and thus open up new business opportunities. In her article, she provides a comprehensive description of the objectives, implementation steps and requirements for sector coupling.

In their article "Improvement of Grid Infrastructure", Tomas Haug, Philipp Hiemann and Lorenz Wieshammer deal with the outstanding importance of electricity grids for the energy transition. In addition to the expansion of the grids, the focus is also on smart solutions for supply and demand management and adapting the regulatory system

The networking of energy and heating systems requires a significant expansion and upgrading of information technology, as Torsten Cziesla, Alexander Stuckenholz and Jens Thorn emphasize in their article "Digitalization and Climate Change". Decentralized and sustainable energy generation can only be achieved with comprehensive digitalization – this includes flexibility options, the concept of a smart grid and the development of business models. The authors outline the necessary steps and challenges.

One element of a sustainable way of life is a core element, as described in Jan Lüken's concept of a "circular economy". Jan Lüken describes the importance of a circular economy, a possible target image and the requirements that arise for companies. He takes a political, an economic and an entrepreneurial perspective.

Certain political representatives are relatively persistent in spreading the myth that nuclear energy is an option worth considering in order to advance the energy transition. Stefan Wenzel explains why this is not the case. He uses the history of the development and use of nuclear energy to demonstrate this. And no matter how you look at the issue: There are no ecological, economic or social benefits to nuclear energy.

Particularly in view of the considerable adjustment processes, it is necessary for political decision-makers to be aware that market failures can arise that may justify state intervention. Without this economic understanding, markets may not be able to coordinate efficiently or at all, which we should avoid if possible. I discuss these issues in the chapter "Market Failure and Consequences for the Financing of the Energy Transition".

In the final chapter, "Markets and Markets Behavior", I discuss various aspects that are central to understanding how markets work. This includes a review of LCOE, a description of the interaction between project developers, investors and banks and an appeal to political decision-makers to ensure a stable and reliable regulatory environment.

In summary, the following picture emerges for me:

There is no one predetermined path to climate neutrality, although we would probably think it would be great if all our climate problems were solved in one fell swoop by green hydrogen or nuclear fusion. The truth is, however, that we will have to take individual steps in many places, most of which will – hopefully – prove to be the right ones, but we may also have to admit along the way that individual measures were not the right ones. In a tense social and political climate, it may be frustrating, but it is unavoidable.

The aim is to tackle the climate problem against the backdrop of complex,[3] networked,[4] non-transparent[5] and dynamic[6] situations. The systems consist of many

3 Complexity is not an objective quantity, but a subjective one. This can be compared to riding a bicycle, where the beginner not only has to coordinate their riding but also process their environment and its changes. This is not a problem for the experienced cyclist. He evaluates a traffic situation as a shape resulting from experience. Such "super signs" reduce complexity. A system is therefore complex with regard to an actor with his or her set of gestalts and thus very different from individual to individual.

4 An intervention that affects one part of a system also affects other parts of the system. This phenomenon is called "interconnectedness". Interconnectedness means that the influence of a variable does not remain isolated, but has secondary and distant effects.

5 Non-transparency means that not everything that you actually want to capture is visible. This means that many features of the situations are not or not directly accessible to the planner.

6 The reality is not passive, but active: this creates time pressure, you cannot wait forever until you decide to act. Time pressure also means that information gathering and planning cannot be carried out in fine stages, but you have to be satisfied
with "approximate" solutions. And finally, you have to deal with development trends: You have to try to work out where the whole thing is heading.

influencing variables that influence each other to a greater or lesser extent. In addition, the systems are at least partially non-transparent and ultimately the systems continue to develop on their own, they have their own dynamics. In addition, the actors do not know all the system properties and may even have incorrect assumptions.

Now the question of transformation has become even bigger: What will the world look like in the next 20 years? The recommendations for action are very clear: and they say that we must consistently invest in renewable energies, develop the potential of renewable heat and pragmatically adapt the regulatory framework. We must follow this path consistently, absorb social hardship and will then benefit from significantly lower electricity costs. We also need to think about economic security and thus become more robust in terms of economic development.

Nevertheless, I am optimistic about the future. Optimistic, because I can see what has happened in the last years, but also in the decades before that. How many crises, how many threats, how many conflicts have been analyzed and discussed, and how often have we succeeded, sometimes late, sometimes with serious consequences, in freeing ourselves from them and working our way out of them. And I believe, especially in view of the last few years, that this is the view we need to look to the future.

There are many changes. There is a completely different urgency to the energy transition. Nevertheless, we have every reason to believe in success and to work to ensure that we can emerge from this time stronger as a liberal community of values.

For the sake of good order, it should be noted that the authors represent their individual opinions. Their statements and evaluations do not necessarily reflect the opinion of the companies or institutions for which the authors work, nor do they necessarily reflect the opinion of the other authors. Any errors are of course my own responsibility.

My sincere thanks go to the authors of this book, whose great enthusiasm and commitment made its realization possible in the first place.

Part 1: **Overview**

Tibor Fischer

2 Investments and Regulation – What Needs to be Done Now: A Systemic Investment Policy for the Further Expansion of Renewable Energies

The energy transition is rightly one of the key transformation processes of this century. At the same time, the energy transition as a process is itself subject to change. As the integration of renewable energies in the electricity and heating sectors progresses, new technical, legal, economic and social issues arise. The overarching goal of climate neutrality creates a new frame of reference for investments in the energy sector. The economic and legal framework needs to be adapted to link the dimensions of the energy transition even more closely with those of decarbonization and climate protection in order to create a resilient framework for the urgently needed investments up to 2045 and 2050. The aim must be to incentivize private investment alongside public funding. The EU can be the driving force behind this necessary development.

2.1 Introduction

The expansion of renewable energies is a central key of the energy transition. Germany is one of the countries that began promoting renewable energies over 30 years ago. The Renewable Energy Sources Act (EEG) in particular has led to a strong increase in renewable capacities in the electricity sector.

2.1.1 Challenge: From 50 Percent to 100 Percent in Just Under Two Decades

In 2023, more than half of the electricity consumed in Germany came from renewable energies for the first time, with a share of around 52%. In the heating sector, expansion was comparatively slow and amounted to just 18.8% by comparison.

By 2030, the share of renewable energies in the electricity sector is set to rise to a total of 80%, and by 2045 to 100% across all sectors as part of the goal of climate neutrality. The expansion targets in the heating sector appear to be even more ambitious. The official plans envisage a total of 50% climate-neutral heat (including waste heat) by 2030. The target for 2045 is 100 percent.

https://doi.org/10.1515/9783110784299-002

Even if the successes of the energy transition in Germany are obvious, especially in the electricity sector, it is also clear that the great need for expansion in the current phase of the energy transition means that it is important to accelerate the speed of expansion even more than before.

2.1.2 High Investment Requirements and Uncertain Economic Situation

This becomes all the clearer when the targets for the expansion of renewable energies are compared with the current status. It is estimated that a total investment of EUR 351 billion will be required to achieve the official expansion target in the electricity sector alone.[1]

For the energy transition, full renewable energy generation across all consumption sectors means enormous private and public investment. For Germany, it is estimated that around EUR 721 billion will be required for the necessary expansion and adaptation of infrastructure by 2030 alone.[2] At the same time, the current economic policy environment is difficult in many European countries, not least due to the Russian war of aggression in Ukraine. Rising interest rates are increasing the cost of capital (CoC) in particular and making new renewable energy projects more expensive. At the same time, budgets are coming under further pressure, not least due to the increased relevance of other policy areas such as defense.

2.1.3 Wanted: a Resilient Framework for Accelerated Expansion

In view of the challenges ahead, the legal and economic framework also needs to be adapted in order to achieve the necessary investments in the energy transition by 2045. This is because the economic and legal requirements for the energy transition itself have changed with the increasing integration of renewable energies, a sharp fall in production costs on the one hand and a stronger focus on sector coupling and a sharp rise in demand for green energy sources on the other.

1 BDEW (2023): Capital for the energy transition. https://www.bdew.de/media/documents/Bdew-Vku-Deloitte-Kapital-fuer-die-Energiewende_ZtGblNH.pdf

2 BDEW (2024): Energy transition progress monitor. https://www.bdew.de/energie/fortschrittsmonitor-energiewende-2024/

2.1.4 A Systemic Investment Policy for Renewable Energies as a New Approach

In this article, I would like to use the example of the German and European energy transition to show the extent to which climate neutrality creates a new frame of reference for a systemic investment policy and at the same time work out which central aspects should be taken into account in the future in order to accelerate the expansion of renewable energies in both the electricity and heating sectors.

My central thesis: The transformation of the energy transition into a comprehensive transformation and decarbonization project is a key success of the past two decades. In the second phase, however, we must now move away from the previous path dependency of energy policy and think of the energy transition as a central innovation and investment project. It is not enough to simply slightly adjust the legal and regulatory framework for the desired expansion. Rather, a comprehensive mix of market-based and subsidy-based instruments must enable large-scale investments in production capacities, generation technologies and infrastructure. At the same time, various areas such as economic, environmental and energy policy must be more closely interlinked in order to leverage investments on a broad basis.

This applies all the more in light of the high investment requirements, but also with regard to the necessary robustness when it comes to making the transformation resilient and sustainable. A stable framework is needed that provides a clear perspective for the next two decades.

2.2 Changed Frame of Reference: From Energy Transition to Climate Neutrality

What began over 30 years ago as energy transition with percentage expansion targets in the electricity sector is now part of a broader concept: climate neutrality.

This presents opportunities for the financing of renewable energies, particularly in the area of financing. This is because the climate neutrality requirement has the potential to leverage additional private investment and value creation. However, this also requires new instruments and a policy approach that takes a systemic and complementary view of different policy areas. The aim is to create a clear perspective for decarbonization through the use of renewable energies in order to achieve the goal of climate neutrality. Financing the transformation across all sectors has a key role to play.

2.2.1 Climate Neutrality as the Starting Point for a New Investment Policy

This is changing the frame of reference for the energy transition as a political transformation project. As the integration of renewable energies in the electricity and heating sectors progresses, new technical, economic and, with regard to the financing of the transformation, social issues are arising.

Whereas almost three decades ago, the focus was initially on developing the technology and making it ready for the market as well as expanding renewable capacities, particularly in the electricity sector, the aim now is to find a framework that promotes the further development of renewable business models and technologies across all sectors. This new framework should enable the large-scale investments that are necessary to actually achieve decarbonization in the three consumption sectors. This also involves the question of who else can be involved in the costs of building the necessary infrastructure. Public and private financing must be considered in a much more complementary way than before.

Energy policy is no longer the sole point of reference. Instead, the focus today is increasingly on interactions between climate protection, industrial, financial and technology policy. As a result, a systemic perspective that goes beyond traditional energy policy is also becoming increasingly important in the area of policy-making.

This highlights one of the key challenges for shaping energy policy in the coming years: the progressive and necessary integration of renewable energies into all sectors and the connection with other policy areas leads to the need to continuously review the current economic and legal framework for the transformation and adapt it where necessary. This makes the energy transition a transformation project in two senses: in relation to the target dimension of decarbonization and at the same time in relation to the political requirement to accompany the transformation through a continuously evolving framework.

2.2.2 First Phase of The Expansion of Renewable Energies in Germany

Alongside Denmark, Germany is one of the European industrialized countries that promoted the expansion of renewable energies at a comparatively early stage.

With market liberalization and the targeted promotion of renewable energies, Germany has achieved significant successes in the electricity sector in particular, while the integration of renewable energies has achieved comparatively low growth in other areas. This applies in particular to the heating and transport sectors.

2.2.3 Germany as an Early Mover in the European Electricity Market

In the first phase of the expansion, the main aim was to ensure access to the electricity market for renewable energies in the context of the parallel liberalization of the electricity market and to get the first onshore wind and PV projects onto the market as part of an innovation and technology policy. From the 1000 Roofs Program and the Electricity Feed-In Act to the early EEG remuneration: the success was visible. The first significant generation capacities were added. In the early years of the energy transition, this was mainly through regional community energy projects. With the emergence of a German and European market, a new sector and a new branch of industry developed in Germany.

The German renewable energy sector grew strongly, production capacities were created and "renewables made in Germany" also became a sought-after product and service abroad. The first wave of renewable energy expansion was characterized by a growing interest in renewable energies in Europe and globally. In Europe and worldwide, the legal framework for support programs was increasingly established, new players became involved in the energy market and established players also increasingly turned their attention to the energy transition.

Parallel to the development of a German and European domestic market, however, countries such as China also recognized the potential of the renewable energy industry, particularly the PV and wind industry, and are focusing massively on building up their own production capacities.

2.2.4 Limitation of Growth and Loss of Own Production Capacities

As the expansion of renewable energies increased, so did the costs for the subsidy-based expansion of renewable energies. Unauthorized, retroactive market interventions, such as in Spain or the Czech Republic, marked the beginning of a turning point in Europe, which not only massively impaired the expansion of renewable energies, but also had a negative impact on the established industrial value creation stages and chains. In addition, market confidence in the framework conditions was lost in many markets.

In Germany, too, the market ramp-up for PV and onshore wind in particular, but also for biogas, was limited by policymakers in 2008 and 2012. The support framework was not developed further in an evolutionary manner, but instead the future growth corridor was severely limited in order to limit higher electricity prices due to the rising EEG surcharge for the promotion of renewable energies. Although renewable energies also lead to falling market prices on the electricity exchange due to the merit order effect, further dynamic expansion was seen by politicians as a locational disadvantage in global competition, alternative models of support or applicable surcharges were not widely discussed and further expansion was severely limited as a result of political decisions.

The now weakening sales in the European markets had its price: European manufacturers came under increased pressure and were only able to partially compensate for the loss of sales in the emerging non-European markets. At the same time, China was able to use the production capacities it had built up in the meantime and put pressure on European manufacturers with its state-supported industry.

As a result, remaining companies increasingly relocated their production or parts thereof to China or other countries with markets with rising growth rates for renewable energies. As a result, Europe lost a large part of its PV production. Only individual stages of the value chain, such as the glass industry, have remained. The lack of expansion corridors also led to a loss of production capacity in the wind industry. Many of the former pioneers of the global wind industry relocated parts of their production to other countries. The sluggish domestic market is still having an impact today: The last rotor plant in Germany was closed in 2022. With negative consequences for the resilience of their own value chains.

2.2.5 Increased Market Integration of Renewable Energies

In 2015, the EU Commission pushed for a change in the promotion of renewable power plants with an initial pilot tender in the PV sector. This was followed by the general introduction in the electricity sector in 2017. On the one hand, this increased the cost pressure for larger plants and funding rates for the respective technology were no longer defined in the political process, but via the market. The definition of the basic award mechanism also harmonized funding across the EU.

The introduction of the market premium in 2012 already provided an impetus for direct marketing and represented a first step towards greater market integration of renewable energies.

Around seven years later, market-driven expansion via long-term purchase agreements also gained significant importance for the first time. Today, the financing of new plants through direct supply contracts via Power Purchase Agreements (PPAs) to electricity consumers in industry and commerce is playing an increasingly important role.

2.2.6 Interim Result

Even if the framework conditions have so far been changed by politicians, sometimes significantly and not always to the benefit of steady and sustainable market development, the expansion of renewable energies in Germany has been a success, particularly in the electricity sector: the expansion of renewable energies in the electricity

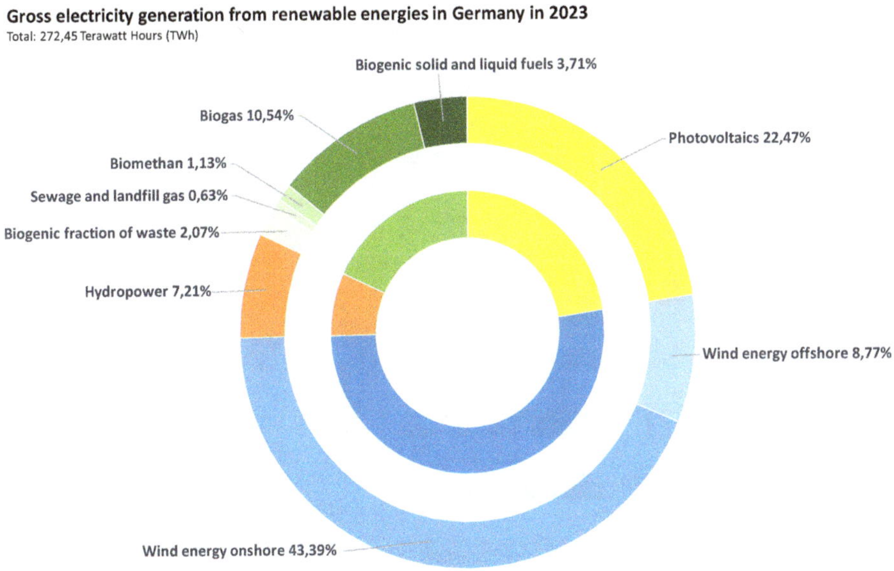

Figure 2.1: Gross Electricity Generation from Renewable Energy in Germany in 2023; figure adapted from AGEE-Stat 2024.

market has increased from 46.8 TWh in 2003 to 272.4 TWh in 2023.[3] This represents a more than five-fold increase. Last year, more than 50 percent of the electricity used in Germany was generated from renewable energies for the first time (see Figure 2.1).

2.2.7 White Spots: Expansion of Renewable Energies in the Heating Sector

In the first phase of the energy transition, the heating sector was not at the center of the expansion of renewable energies in Germany. Although the supply of heat from renewable energies also increased continuously, it was at a relatively low level. In contrast to countries such as Denmark, no fundamental transformation of the heating sector was initiated in which market access on the one hand and the expansion of new technology on the other was accelerated and the heating network was prepared for the use of renewable energies.

Growth was achieved above all in the areas of bioenergy, solar thermal energy and geothermal energy, albeit with much lower growth rates than in the electricity sector (see Figure 2.2). Overall, the share of renewable energies in heating and cooling generation amounted to 89 TWh_{th} in 2003. Twenty years later, this figure has more

3 AGEE-Stat (2024).

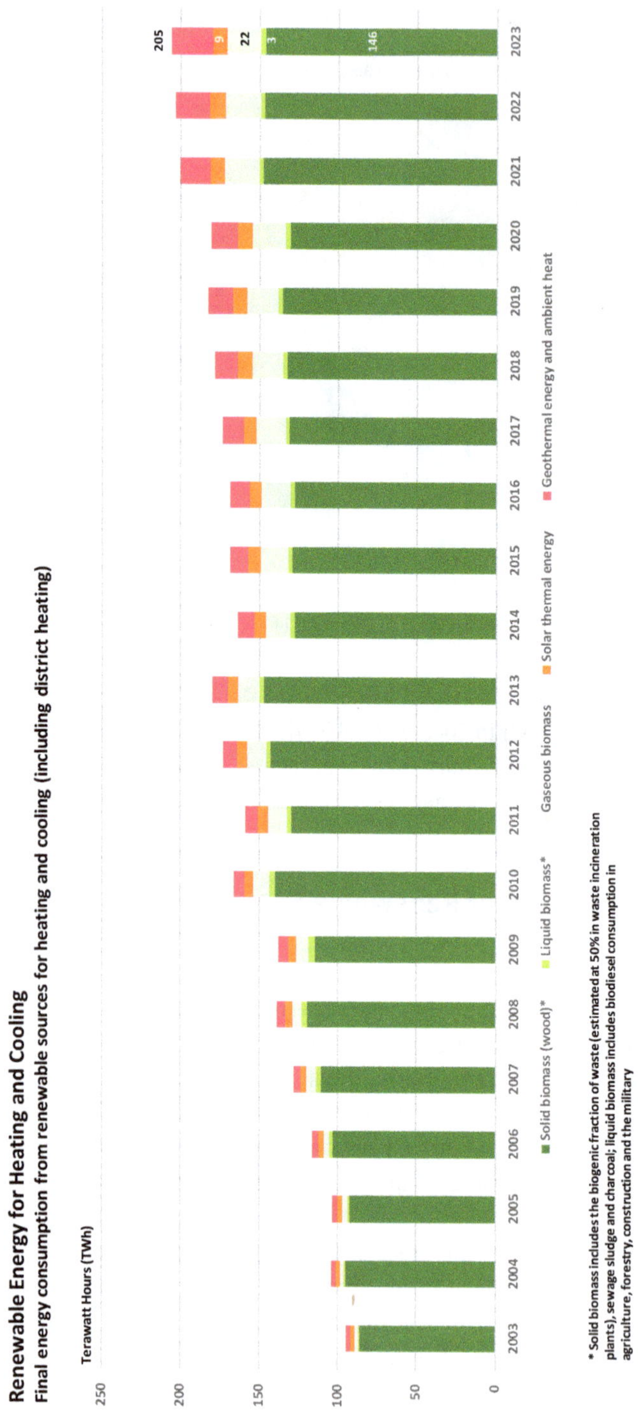

Figure 2.2: Renewable Energy for Heating and Cooling; figure adapted from Federal Environment Agency 2024.

than doubled to 205 TWh_{th}.[4] Compared to the electricity sector, however, it is clear that the relative growth has been much lower.

As the heating market is the largest final energy sector in Germany at around 59%, it is clear how central the expansion of renewable energies will be in this area in the coming years. The adoption of municipal heating planning and mandatory requirements for the use of renewable energies in the heating sector as well as a rising CO_2 price should lead to increased use of renewable energies in the heating sector in the coming years.

At the same time, however, it is also clear that market development must be far more dynamic if the target of 50 percent climate-neutral grid-based heating by 2030 is to be achieved as an interim target for climate neutrality in 2045.

2.2.8 Where do we Stand Today

What began as an energy transition in some countries over twenty years ago has now become a global industry. The focus of energy policy on strengthening the framework conditions for renewable energies in many countries, falling levelized costs of electricity (LCOE), particularly in the PV and wind sectors, have led to the emergence of a global market in two decades.

According to an analysis by the International Renewable Energy Agency (IRENA), 473 GW of renewable electricity capacity was added worldwide in 2023 alone – a new record.[5] Renewable energies accounted for 86% of the new power generation capacity built globally and conventional energy sources for just 14%.

As one of the pioneers, the example of Germany shows how great the influence of politics was on the initial emergence of the market in both a positive and negative sense.

In terms of the absolute size of its electricity market and the penetration of new generation technologies beyond hydropower, Germany is now one of the world's most developed markets for PV, wind and bioenergy.

At the same time, however, some harsh and selective interventions in the expansion paths have led to the loss of a large proportion of strategic production capacities in Germany and Europe, with negative consequences for the value creation and resilience of the European industry. Around ten years ago, the legal and regulatory framework for renewable energies was only able to develop partially and with little evolution in the face of the tension between an emerging market for renewable energies and political discussions and explicit decisions aimed at slowing down further expansion.

4 Federal Environment Agency (2024): Energy consumption for fossil and renewable heat | Federal Environment Agency

5 IRENA (2024): Renewable capacity highlights. Tracking COP28 outcomes: Tripling renewable power capacity by 2030 (irena.org)

The heating transition is one of the current Achilles' heels of the German energy transition. At the latest since the Russian war of aggression on Ukraine and the resulting energy crisis, the focus has been on developing the future infrastructure for generation based on renewable heat and waste heat. Local and district heating networks and the integration of new heat or electricity-based generators in industry play a central role in this.

In the European context, the EU has also extended its influence to strengthen renewable energies: From the liberalization of electricity markets, to European emissions trading, from the Electricity Market Design Directive (EMD) to the Renewable Energy Directive (RED III) and the Delegated Act for the production of green hydrogen.

The EU not only sets a framework for the policies of the member states, but also develops joint strategies and goals. Ultimately, this is also what the EU Commission's Green Deal stands for, with the aim of developing Europe into the first climate-neutral continent by 2050 through various measures.

Systemic issues such as sector coupling and leveraging flexibility play just as much a role here as the increased integration of renewable heat generators such as geothermal or solar thermal energy. The overarching question is how to create a market design that fully reflects the reality of 100% generation based on renewable energies and at the same time ensures the financing of existing and new plants.

2.3 What Matters Now

With the market development of renewable energies achieved in the past two decades, even higher annual investments in renewable energies are needed in view of the international and national climate protection targets and the goal of climate neutrality. At the same time, the long-term nature of the investments and the high initial investments have shown that a stable framework is necessary to ensure that the expansion targets are achieved. Both in terms of the level of investment required and the stability of the economic and legal framework, a systemic combination of different policy areas such as energy, economic and industrial policy and climate protection can lead to a framework that guarantees the necessary stability and at the same time promotes additional investment in renewable energies.

2.3.1 Global Investment Requirements: Triple the Billions

With a view to achieving the 1.5 degree target, IRENA also draws attention to the continued need for additional financing in the context of the recent record investments briefly described above: In 2023, a total of USD 570 billion was invested worldwide in the expansion of renewable energies in the electricity sector. However, annual investments of over USD 1.5 trillion will be required to achieve the Paris Climate Agreement target.[6] On top of this, USD 720 billion must be invested annually in grid expansion and strengthening flexibility options.

In short, these figures mean: The current investment record must be exceeded threefold – "Triple the billions" must be the slogan for the expansion of renewable energies worldwide in the coming years.

2.3.2 Investments in Germany Need to be Doubled

But even for a comparatively established market like Germany, current calculations for achieving the 80 percent target by 2023 also assume a not insignificant investment requirement for renewable energies in the electricity market of over 351 billion euros.[7] With a view to achieving the decarbonization of heat via renewables, a further 100 billion euros will be added. If grid expansion and e-mobility are also included, the total investment requirement increases to a total of EUR 721 billion.[8]

If we only compare these estimates of investment requirements from the electricity and heating sectors with the actual investments in renewable energies in 2023, it becomes clear how large the gap is. According to the latest statistics from the Federal Ministry for Economic Affairs and Climate Protection (BMWK), a total of EUR 36.6 billion was invested in renewable energies in both the electricity and heating markets in 2023.[9] This compares to an estimated investment requirement of EUR 451 billion without the necessary infrastructure by 2030. This means that around EUR 64.43 billion would have to be invested annually in the remaining seven years. In purely mathematical terms, this represents a shortfall of around 43 percent of the annual investment required in an environment that is currently characterized by rising interest rates and a general reluctance to invest.

So even for one of the strongest economies in the world, the motto is at least "double the billions" in order to secure the expansion targets by 2030.

---

6 IRENA (2024): Tracking COP28 outcomes: Tripling renewable power capacity by 2030. https://mc-cd8320d4-36a1-40ac-83cc-3389-cdn-endpoint.azureedge.net/-/media/Files/IRENA/Agency/Publication/2024/Mar/IRENA_Tracking_COP28_outcomes_2024.pdf?rev=6a40bf8184744e209283c159ab779603

7 BDEW (2023): Capital for the energy transition. https://www.bdew.de/media/documents/Bdew-Vku-Deloitte-Kapital-fuer-die-Energiewende_ZtGblNH.pdf

8 BDEW (2024): Progress report on the energy transition. https://www.bdew.de/media/original_images/2024/04/24/fortschrittsmonitor_2024_zCu1QX7.pdf

9 BMWK (2024) according to the Working Group on Renewable Energy Statistics (AGEE-Stat): https://www.bmwk.de/Redaktion/DE/Downloads/Energie/erneuerbare-energien-in-de-tischvorlage.pdf?__blob=publicationFile&v=6

These estimates make it clear that, despite all the technological progress and solutions, the investment framework in particular needs to be strengthened in order to achieve the energy transition targets.

2.3.3 Rising Capital Costs and Limited Budget Funds are Putting Pressure on the Expansion of Renewable Energies

The cost of capital is one of the key factors influencing the cost of a renewable energy project. They express the expected financial return and the minimum economic rate required for investments in a company or project. They are the average of borrowing costs (interest/interest rate) and equity costs (expected return) and are subject to different specific technology and project risks as well as the general economic risk factors typical of the country in question.

Due to very high initial investments and comparatively low operating costs, they have a high impact on the average levelized cost of electricity (LCOE). If they increase from two to ten percent for an onshore wind project, this leads to an 80 percent increase in the LCOE, according to IRENA[10] .

A study by the Swiss Federal Institute of Technology in Zurich (ETH Zurich) also comes to the conclusion with regard to the cost of capital in Germany that if interest rates were to return to the level before the outbreak of the financial crisis in 2008, the LCOE for PV systems would rise by 11 percent and for onshore wind systems by as much as 25 percent.[11]

Due to the current geopolitical and economic situation, the cost of borrowing has risen sharply in recent months. The ECB has already raised the key interest rate in 2022 after more than 11 years and has made repeated adjustments since then. For Germany, a cost of capital study confirms rising costs across all sectors as early as 2022. For the technology sector, the weighted average cost of capital (WACC) is 8%.[12] There are already increasing signs that rising capital costs are limiting the expansion of renewable energies, particularly in the area of projects financed directly on the market.

Rising costs for the procurement of capital and the associated higher costs for the construction of new plants have further negative economic effects and lead to higher costs and government expenditure in the case of state-subsidized plants or, in the

---

10 IRENA (2023): The-cost-of-financing-for-renewable-power. https://www.irena.org/Publications/2023/May/The-cost-of-financing-for-renewable-power

11 ETH (2019): Interest for competitive renewables is crucial: https://ethz.ch/de/news-und-veranstaltungen/eth-news/news/2019/09/zinsen-entscheidend-fuer-die-wettbewerbsfaehigkeit-erneuerbarer-energie.html

12 KPMG (2022): Cost of capital study 2022. https://kpmg.com/de/de/home/themen/2022/10/kapitalkostenstudie-2022.html

case of unsubsidized plants, to higher electricity costs on the part of the electricity-consuming companies.

At the same time, falling market values for renewable energies in some segments lead to higher differential costs for subsidized plants, which must be compensated for via the fixed feed-in tariff and thus via the state budget. Due to the interactions described here only very briefly, the future market design must also find answers to the interaction between the market and subsidy framework and address interactions and ensure complementarity to a greater extent than before. Ultimately, it is important to incentivize investment in both areas.

With the increasing relevance of further government tasks and the resulting shortage of public funds to finance the transformation, the question arises as to how the aforementioned investments can be secured at the necessary level. After all, experience has shown that a renewed cut in the expansion of renewable energies, as in the early 2010s, would set back the current growth of renewable energies significantly and sustainably. This would be associated with a further loss of value chains and added value.

At the same time, a self-imposed austerity policy is already jeopardizing a forward-looking industrial and technology policy aimed at building a sustainable energy infrastructure.

However, in view of the high and short-term investment requirements, the aim must always be to raise additional capital for the energy transition on the market alongside state funds. To this end, the instruments of traditional direct monetary support must be supplemented by additional instruments and the frame of reference must be expanded to include different policy areas.

2.3.4 Europe as an Opportunity for Value Creation 2.0

The goal of climate neutrality and the necessary economic and technical development are accompanied by considerable opportunities for innovation for existing and new value chains. For example, the "Net Zero Industry Act", which is also part of the "Green Deal Industrial Plan", aims to strengthen European industries in this area in view of the existing dependencies of strategically relevant stages of the value chain.

Increasing investment in relevant key areas of the energy transition can strengthen the resilience of the expansion and at the same time increase tax revenue in the medium term.

2.3.5 Climate Neutrality as a New Paradigm for a Systemic Investment Policy

If climate neutrality is seen more strongly as a paradigm and frame of reference for a new investment policy, effects can be achieved for the energy transition and climate protection. Although the frame of reference of the energy transition as part of an overarching climate neutrality concept has already been established in politics, too little consideration has been given to the interactions between the two policy areas and their instruments. An infrastructure based on renewable energies is the foundation of a climate-neutral society and a climate-neutral economic system.

The Green Deal adopted by the EU Commission in 2019 can be used as an example of such an expanded frame of reference. As an integral component, the EU taxonomy adopted in 2022 defines when investments are considered sustainable and establishes a uniform reporting standard for reporting companies as a basis for assessment. The taxonomy is intended to strengthen sustainable investments in the future. In future, companies with poor sustainability scores will only be able to obtain funding on the capital market on very poor terms. Due to the ever-increasing number of companies subject to reporting requirements, sustainability will thus become a central reference system in the area of financing. "Climate protection" and "adaptation to climate change" are already two of the mandatory criteria to be addressed today (see Figure 2.3).

The explicit definition of climate protection as a measurement criterion also indirectly creates an incentive for increased investment in renewable energies. This is because the purchase of electricity from their own plants (on-site or off-site) is already a key lever for many companies to reduce CO_2 emissions. The link to future conditions for access to debt capital therefore represents an indirect systemic incentive to expand renewable energies.

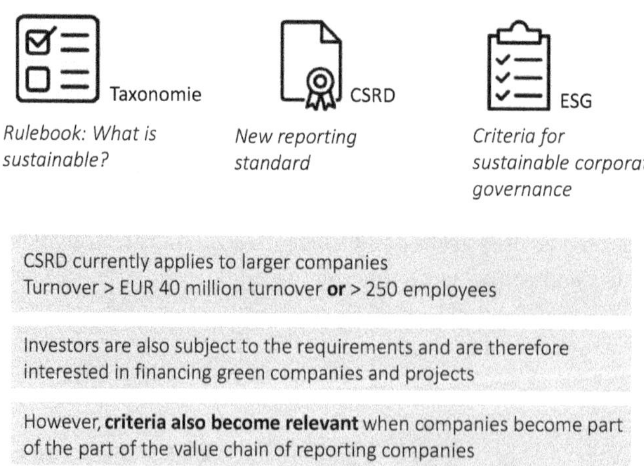

Figure 2.3: Relevance of climate credit rating; figure created by author.

The relevance of climate credit rating is increasing for companies. For policymakers, this represents a starting point for strengthening the expansion of renewables.

It is not only requirements from other areas such as financial policy that can strengthen investment in the energy transition. With the goal of climate neutrality, the expansion of renewable energies ultimately also becomes an overarching task for society as a whole. Companies with their local production are also required to make corresponding contributions locally. One example is the increased relevance of the regional availability of renewable energy for companies as a location factor. This is because the direct procurement of renewable energy enables industry to drive forward the decarbonization of its production and at the same time hedge against rising electricity prices.

For policymakers, this means that they can use the entrepreneurial perspective on the energy transition and decarbonization to raise additional funds on the market and accelerate expansion. In addition to state funding, it is important to strengthen corresponding business models by setting the right framework and thus boosting direct investment in renewable energies from the private sector.

2.3.6 The Financing of Renewable Energies Needs a Broad and Stable Foundation: Importance of Integrated and Complementary Approaches for a Systemic Investment Policy

The examples mentioned above show how an integrated approach can strengthen investments in renewable energies. In this context, it is increasingly important to understand the starting points in different areas and to identify and use the corresponding levers. This makes the promotion of expansion in terms of a systemic investment policy more complex. At the same time, however, effectiveness is increased as direct state funding is considered in a complementary manner with opportunities in the market and requirements are formulated at the same time.

In this way, additional financial resources can often be raised beyond traditional CAPEX and OPEX funding and reconciled with direct monetary funding. At the same time, other, previously less utilized instruments such as new business models (1), state obligations in the context of subsidies or exemptions (2), market and state risk hedging (3), investment incentives (4), corporate financing instruments (5), capital market instruments (6) or various participation models (7) offer the opportunity to strengthen the expansion of renewable energies. Climate neutrality combines different policy areas and increases the leverage for policymakers to generate more funding for the expansion of renewable energies.

Market-driven business models can serve as another concrete example of a field of action. For example, there is great interest in the market for green electricity supply contracts (green PPAs). Industry and commerce are very interested in corresponding supply contracts and are very willing to invest in new plants, not least in order to hedge against rising electricity prices and at the same time drive forward the decarbonization of their own production.

Green energy therefore has a value in the economy which, due to its high relevance for achieving climate neutrality, is also accompanied by correspondingly higher willingness to pay. In this context, policymakers must make strategic use of this to accelerate the expansion of renewable energies and strengthen direct investment in renewable energies as a complementary instrument.

In the recently adopted Electricity Market Directive (EMD), the EU also stipulated the strengthening of PPAs alongside the promotion of renewable energies via contract for difference. Member states are now called upon to remove barriers to the business model and introduce instruments to hedge risks.

With this approach, the EU is indirectly strengthening the inflow of private capital from the private sector for the expansion of renewable energies. The directive makes direct reference to strengthening the complementarity between state support via CfD and a business model.

Another example is the linking of access to subsidies and exemptions to requirements for the purchase of renewable energies. One example of this is the amended directive on electricity price compensation in 2022. Two years earlier, the EU had already made it possible for energy-intensive companies to receive it if they purchase green electricity via PPAs. By transposing the directive into national law, the German government has taken into account the fact that industry will only invest in direct electricity supply contracts if green electricity purchases are no less favorable than electricity purchases from fossil sources and they do not have to forego monetary compensation. At the same time, by making the use of electricity from specific regions in Europe mandatory, politicians have set a target for increasing the use of electricity from new plants in a regional context.

This example shows how important it is for politicians to open up opportunities for the expansion of renewable energies and at the same time address obstacles directly (see Table 2.1).

At the same time, electricity price compensation is also an example of the possibility of using the purchase of green energy as an access criterion for subsidies or exemptions and thus indirectly obliging companies to invest in renewable energies.

Another example of this interdependency is the special equalization scheme for companies. Here, applicants must obtain 30 percent of their electricity consumption from unsubsidized renewable energy plants. Here, too, the legislator explicitly relies on additional investment on the part of companies and only then grants them an economic advantage via the equalization scheme.

Another approach is the increasing importance of risk hedging for the expansion of renewable energies. Due to falling production costs and the associated competitiveness compared to fossil fuels, the assumption of specific risks such as default risks is

Table 2.1: Examples of qualitative criteria as access criteria for funding and exceptions (compiled by author).

Electricity price compensation	Covering at least 30% of total electricity requirements from CO_2-free sources. 80% of the plants are in Central Western Europe.
Special equalization scheme	30% of total electricity consumption from unsubsidized renewable electricity.
Federal funding for energy and resource efficiency	Renewable electricity purchased via PPAs can be taken into account with an emission factor of 0 when determining the CO_2 support cap.
Green hydrogen delegated act	Conclude a direct electricity supply contract for the quantity used to produce the green hydrogen, including time and location correlation

becoming increasingly important. At the same time, direct monetary support is becoming less important in some technology and market segments. This is because the minimization of various project risks can prevent higher premiums for the procurement of debt capital. In addition to state hedging instruments, the market also offers opportunities to address risks directly via corresponding projects.

For example, a mix of corresponding instruments could reduce the capital costs for the necessary ramp-up of geothermal energy in the heating sector with its high investment costs and exploration risk. Costs are then only incurred if a loss event occurs. There are already international examples of corresponding fund solutions, for example in Central America. With such a mechanism, which goes beyond monetary support for heating networks, the initial project risks can be addressed directly, financing costs can be reduced and municipalities can be put in a better position to finance such projects. Such investments also become more attractive for the additional private capital required. If you consider that geothermal energy could account for around a quarter of heat supply in Germany in the future,[13] it becomes clear how much leverage there could be for investment. This is because deep geothermal energy currently accounts for just 0.9 percent of final consumption.

The risks of market-driven business models such as PPAs could also be addressed in future using appropriate instruments. In this way, the risk premiums of the financiers on the capital provided can be minimized. This reduces the costs of the projects and the costs of electricity generation. Financing plants on the market would become even more important than before. Additional capital could be made available for the energy transition and the available state funds could be used for areas where direct monetary support is needed even more. This would take account of the economic viability of many renewable projects in the electricity sector by refinancing them on the market and at the same time address specific risks.

13 Fraunhofer (2022): Roadmap Deep Geothermal Energy for Germany.

In terms of investment incentives, the US Inflation Reduction Act (IRA) is certainly one of the most frequently mentioned instruments in recent months. It is an impressive example of how tax incentives can be used to boost investment in the future and strengthen competitiveness by combining the energy transition with green industrial policy.

Other instruments of state investment promotion are the low-interest loans or (repayable) grants often already used in the renewable energy sector.

Approaches that are more effective at company level are various forms of equity capital or other instruments for market or state risk hedging. Particularly in the area of foreign trade promotion, Germany has a broad portfolio of instruments that could also be used for the expansion of renewable energies in Germany.

Green bonds are one of the capital market instruments whose relevance has increased significantly in the past. As debt instruments, they offer any type of investor the opportunity to participate in the expansion of renewable energies. Issuers can be government agencies or companies. They undertake to invest the money raised in predetermined projects with a positive environmental impact.

In addition to strengthening the debt capital side, green bonds also offer the possibility of financial participation by the population and thus also have the potential to increase acceptance for the further expansion of renewable energies.

With regard to this aspect, Germany in particular already has a wealth of experience with various forms of economic participation in the expansion of renewable energies. These include financial participation models, or special forms such as cooperatives offer forms of financial participation in a regional context and strengthen local acceptance. Energy cooperatives are a special form. They offer shareholders the opportunity to generate and use energy. At the latest with the adoption of the European Renewable Energy Directive (REDII), the EU member states are called upon to strengthen corresponding models in the context of energy sharing. Some countries have already implemented successful forms, thereby strengthening the financing and acceptance of renewables. Strengthening these models is therefore in the dual interest of politics.

2.3.7 Conclusion

These selected and non-exhaustive examples make it clear that there are many different interdependencies and starting points for the design of financial instruments to strengthen the expansion of renewable energies. At the same time, it is now more important than ever to develop the promotion of renewable energies across different policy areas and interdependencies. A mix of regulations, market-based instruments and state support can enable the necessary tripling of the investment speed at a global level. In view of the necessary speed, overly complex bureaucratic processes should be avoided, particularly when designing state instruments.

For energy policy, this means engaging more closely than before with related policy areas. Investment in renewable infrastructure must be seen as a comprehensive task that must be tackled not only by the state, but also by the market. This often involves combining economic perspectives with business perspectives.

With this systemic perspective, a new financial framework and an investment policy can be developed that focuses on and finances the goals of expanding renewable energies and climate protection.

The goal of climate neutrality represents the guard rails for sustainable investments.

Heinz-J. Bontrup

3 Socio-Economic Framework Conditions – The Macroeconomic Effects of the Energy Transition and the Resulting Political Challenges

3.1 Brief Summary

The energy transition is about extensive technical transformation processes and not about minor socio-economic challenges. There is an interdependence. In the economic context alone, highly complex microeconomic relationships in companies and private households as well as macroeconomic implications must be taken into account. However, neither in the political nor in the scientific discussion is the order-theoretical market economy-capitalist system character, and thus the basis for everything, sufficiently discussed and taken into account. As a result, the findings on the energy transition have so far only resulted in a superficial and/or interest-driven discussion that is not scientifically tenable or productive. With its ruling of November 15, 2023 on the so-called "debt brake" in politics, the Federal Constitutional Court of Germany has now shown that the previously believed financing of the energy transition cannot be implemented so easily. Economic laws must also be taken into account in the energy transition.

3.2 Introduction

The target image of a neoliberal market constitution underlying the energy transition systematically ignores the fact that there are multiple concentrations of power on both the supply and demand sides of the markets, which do not lead to an optimal allocation and also contribute to redistribution effects in the value chain between and within the individual sectors. The market economy (neoliberal) approach is also one-sidedly profit-driven and deeply contradictory between the interests of capital and employees. This leads to constant distributional conflicts, which in reality are only insufficiently (suboptimally) taken into account when implementing the energy transition.

No one should believe for a moment that the energy transition would be feasible without a redistribution from primary value-added income to labor income. A purely secondary redistribution of income will not be sufficient. In addition, the wealth that is currently appropriated and accumulated by only a few must be redistributed through state intervention. The following article will demonstrate this.

https://doi.org/10.1515/9783110784299-003

3.3 Social Starting Conditions for the Energy Transition

The energy transition is not only about ecological issues, but also about economic and social issues. This raises questions: How can a good 13 million impoverished people in Germany, including almost 4 million children, and around 7.4 million precariously employed people in a low-wage sector and a good 3 million unemployed people cope with the rising costs of the energy transition? And how can an energy transition work when Germany, one of the richest countries in the world, has a poor population and a boom in the so-called "food bank movement"?[1] Furthermore, how can the approximately 2.5 million small and medium-sized enterprises with up to 20 employees cope with the burdens of the energy transition in Germany?

If these issues and the complex socio-economic implications and problems associated with them are not resolved and the people affected are not presented with credible and socially holistic concepts that are linked to development prospects that secure their livelihoods, the energy transition in Germany will not only fail economically and socially, but will also cause maximum political and, above all, economic damage.

3.4 Political Failure and the Question of Economic Power

Parliamentary (indirect) democracy is already under serious threat, as many recent sociological and political studies have shown.[2] How could Germany come to have such a divided society, a "torn republic", as described by the renowned political scientist Christian Butterwegge, among others?[3] In an indirect (parliamentary) democracy, the state, politics, is deeply permeated by conflicting party interests, including the personal interests of members of parliament, and is therefore burdened. There is no "neutral state" here that focuses on the overall interests of the people. This is one of

1 Cf. Selke, S., Die neue Armenspeisung, in: Blätter für deutsche und internationale Politik, Heft 1/2009, p. 95 et. seq., see also Grabka, M. M., Schupp, J., Etwa 1,1 Millionen Menschen in Deutschland besuchen Tafeln – vor allem Alleinerziehende und Getrenntlebende überdurchschnittlich häufig, in: DIW-Wochenbericht, No. 39/2023, p. 50 et. seq.

2 Cf. here only the latest study by the Körber Foundation: "Vertrauen in die Demokratie schwindet rapide", in: WAZ of August 18, 2023 and the study by Brülle, J., Spannagel, D., Einkommensungleichheit als Gefahr für die Demokratie, WSI-Verteilungsbericht 2023, WSI-Report No. 90, November 2023.

3 See Butterwegge, C., Die zerrissene Republik. Wirtschaftliche, soziale und politische Ungleichheit in Deutschland, 2nd ed., Weinheim/Basel 2020.

the reasons why it is so easy for the private sector to repeatedly assert and satisfy its partial interests against the majority of the people's interests.[4]

The impression of benevolent and aligned political decision-makers, also suggested by Keynesianism and post-Keynesianism, has always been wrong and is rightly rejected today by the New Political Economy (NEP). Just as there are market failures, there are also manifold state and political failures.[5] "The Bundestag acts aloof and far removed from the reality of people's lives," criticizes sociology professor Christiane Bender from the Helmut Schmidt University of the Federal Armed Forces in Hamburg in an interview with the Frankfurter Rundschau.[6] Understandably, this is leading to increasing disenchantment with democracy among the people. In the last federal election in 2021, non-voters were the largest "party". In terms of eligible voters, the current "traffic light government" was only elected by 49.5%, or just under half, of citizens. Trust in democracy is rapidly dwindling, according to a recent representative survey by the Körber Foundation. In the survey, 71% also said that leading representatives in politics and the media lived in their "own world" from which they looked down on the rest of the population. Almost half of Germans (46%) think that things are less or not at all fair in the country."[7] And economics as a science also contributes little to objective enlightenment here, because it is not unambiguous, but is interspersed with many contradictory theories. Interest-driven politicians can always choose their "appropriate theory" and thus hide behind science. "We have many economic and scientific players who are pursuing their own interests and have lost trust as a result," says the well-known environmental and sustainability researcher Ortwin Renn.[8]

There are two main academic explanations for this. Firstly, the two philosophers Theodor Adorno and Max Horkheimer spoke of a "privileged complicity" between representatives from business and politics in this context in their studies back in the 1960s.[9] Secondly, the laws of motion inherent in capitalism tend towards a "state-

4 The alternative to indirect democracy is direct democracy, which is currently only practiced in Switzerland. Cf. in detail Sommer, A. U., Eine Demokratie für das 21. Jahrhundert. Warum die Volksvertretung überholt ist und die Zukunft der direkten Demokratie gehört, Freiburg i. Br. 2022. The former Federal Constitutional Court judge Gertrude Lübbe-Wolff has written an interesting book on direct democracy from a legal perspective: Demophobie. Muss man die direkte Demokratie fürchten?, Frankfurt a. M. 2023.

5 Cf. Bontrup, H.-J., Marquardt, R.-M., Volkswirtschaftslehre aus orthodoxer und heterodoxer Sicht, Berlin, Boston 2021, p. 589 et. seq.

6 Cf. Frankfurter Rundschau of September 4, 2023, p. 2 f., "Raus aus der Blase, ran an die Basis".

7 Cf. WAZ from August 18, 2023.

8 Ortwin Renn, in: Deutschlandfunk, Internet retrieval on February 23, 2023.

9 Cf. Horkheimer, M., Adorno, T., Dialektik der Aufklärung. Philosophische Fragmente, 23rd ed., Frankfurt a. M. 2017, Voigt. L., Stand der privilegierten Komplizenschaft in Deutschland und die Folgen der Demokratie, University of Siegen 2020, Lindemann, K., Finanzkapitalismus als Beutesystem. Der

monopolistic capitalism" or "state-monopolistic complexes".[10] Here, companies with market power ally themselves with politics without having achieved a monopoly position. Today, critical economists speak of a "dictatorship of the corporations".[11] Thilo Bode writes: "Since the fall of the Berlin Wall, a new quality of lobbying has emerged due to the dramatic increase in the market and financial power of corporations. This market and financial power has become a political power. An industrial-political complex has emerged in which corporations and politicians form a mutually beneficial partnership of convenience that no longer makes decisions against corporations. This has devastating effects on democracy and causes enormous damage."[12] One of the characteristics of this power is that corporate costs are socialized through subsidies and profits are privatized.[13] Here is just one current example in the context of the much demanded hydrogen projects in relation to the energy transition. "According to a media report, the energy company RWE only considers investments in green hydrogen technology to be justifiable if the government provides subsidies. The Essen-based company is currently unable to initiate any climate-friendly hydrogen projects without government aid 'because we will not approve any project that does not pay off', Sopna Sury, the board member responsible for the hydrogen business, told the Süddeutsche Zeitung newspaper."[14]

An essential, undiscussed premise for the success of the energy transition, which requires acceptance by society as a whole, is therefore the dismantling of corporate power.[15] However, the entire eco and climate movement does not have the necessary knowledge or awareness for this in the slightest. For the same reasons, the ruling politicians do not want to fight corporate power. The Federal Cartel Office verifies this

---

Neoliberalismus und die Aktualität des Racket-Begriffs, in: Blätter für deutsche und internationale Politik, Heft 9/2014.

10 Cf. team of authors: Der Staat im staatsmonopolistischen Kapitalismus der Bundesrepublik. Staatsdiskussion und Staatstheorie, Frankfurt a.M. 1981, Dolata, U., Gottschalk, A., Huffschmid, J. Staatsmonopolistische Komplexe als neue Organisationsform des Kapitals im staatsmonopolistischen Kapitalismus, in: IMSF, Köln 1986, Binus, G., Landefeld, B., Wehr, A., Staatsmonopolistischer Kapitalismus, 2nd ed., Köln 2015.

11 See Bode, T., Die Diktatur der Konzerne, Frankfurt a.M. 2018.

12 Ibid, p. 8.

13 This was also the case with the *"Act on the Reorganization of Responsibility for Nuclear Waste Disposal"* (BT printed matter 18/10469). Here, the taxpayer bears the costs of nuclear waste disposal, from which the nuclear power plant operators E.ON, RWE, Vattenfall and EnBW as well as the Munich municipal utilities have been relieved, but previously realized gigantic profits with nuclear power in the private sector. Cf. in detail Bontrup, H.-J., Written statement on the law, Hearing of experts in the Bundestag on December 1, 2016, Committee on Economic Affairs and Energy, Committee document 18(9)1057.

14 Frankfurter Allgemeine Zeitung, " RWE will nur mit Subventionen in grünen Wasserstoff investieren", August 17, 2023.

15 Cf. in detail Attac Austria (ed.), Konzernmacht Brechen! Von der Herrschaft des Kapitals zum Guten Leben für Alle, Vienna 2016.

sobering finding of a completely failing policy on a daily basis.[16] It must now even be acknowledged that international politics also supports the corporations through huge subsidies. "The International Monetary Fund (IMF) calculated that coal, oil and gas were subsidized worldwide with around 5,300 billion dollars in 2015 – more than all government spending on health. Almost 90% of these subsidies (around USD 4,600 billion) are what economists call externalized costs, meaning that the polluters cause damage that others have to pay for – for example, taxpayers whose money is used to resettle climate refugees and build dams against rising sea levels, as on the coasts of the USA."[17]

But how can a necessary energy transition succeed under such conditions? This would require a radical paradigm shift in politics that ends the neoliberal redistributive power that still exists today in favor of a small social class from business, politics, media and science. However, there is sufficient leeway in Germany in terms of income and wealth to make things fair, as will be shown in detail later. The cost burden of the energy transition must be distributed fairly. It should be possible for at least the ruling politicians and the scientific community to agree on this in unison. And we must stop supporting those who can use their power to enforce this against politicians, but in reality do not need any support. Help must be reduced and concentrated on those who really need it.

3.5 Problematic Targets of the Energy Transition

The targets for the energy transition must also be discussed. They must be realistic and therefore feasible. This applies to the technical, economic and social conditions. Targets that are not based on a realistic foundation will not be able to ensure forecasting or planning accuracy. As the implementation of targets depends on too many factors that influence each other, it is not surprising that most targets for an energy transition have not yet been achieved. Economists would not come up with the idea of setting targets for nominal and real gross domestic product (GDP). Economists merely forecast GDP under certain economic premises. In the same way, management plans the economic development of a company. And here too, as with GDP, there are constant deviations between planned and actual figures. Therefore, if there is no rethink in the energy transition with regard to the desired targets, there will also be considerable deviations from the planned/actual figures in the future. This is also due to the pressure for change formulated by politicians, which leads to excessive demands on citizens and companies and not least on politicians themselves. As already mentioned,

---

16 Cf. Bontrup, H.-J., Marquardt, R.-M., Volkswirtschaftslehre aus orthodoxer und heterodoxer Sicht, Berlin, Boston 2021, p. 272 et. seq.

17 Bode, T., Die Diktatur der Konzerne, op. cit. p. 63 f.

even in Germany, the EU country with the strongest economy, there are economic limits to be observed. However, if politicians ignore these limits on the basis of rational, causal, dialectical and, not least, holistic thinking, this will ultimately lead to a lack of understanding and rejection among the population. The debate on the amendment of the Building Energy Act (GEG) may serve as an example here and the Federal Constitutional Court ruling on the "debt brake" is also evident here.

3.6 Structural Change Processes and Distribution Contradictions

The energy transition is causing a variety of structural change processes along the entire economic value chain. This will result in complex material and value-related adjustments within the individual sectors and also between the sectors through to the end consumer. There will be winners and losers and thus redistribution effects. Embedded in today's perverted market mechanism are private-sector companies that are exclusively pursuing an interest in maximum profit. And private households want to achieve a maximum benefit level through their consumer spending. Households must offer their labor to companies in order to generate income, because dependent employees do not own the means of production and are therefore dependent on the "capitalist playing field". All resulting exchange processes are inherently contradictory. Companies want maximum profit and employees want maximum wages. "What the one gets, the other can no longer have" (Adam Smith). The cumulative wage bill across all economic sectors and stages of value creation is therefore not sufficient to buy back the entire value product of the production process. The employees only use the wage bill to purchase consumer goods and give the capital owners the capital goods via the surplus value they realize on the markets. In this way, the owners of capital become richer and richer and the employees remain have-nots (Oswald von Nell-Breuning).[18]

But it is not as simple as that, because even capital owners do not agree with each other. They are only "brothers" when it comes to holding down wages for their employees, while they become bitter "enemies" when it comes to dividing up the surplus value. Interest rates, basic rents and profits are also antagonistic to each other and, in addition, the individual sectors also have completely different interests along the value chain (see Table 3.1). The turnover (sales prices multiplied by sales volumes) of the individual companies in the respective supplying sector are the input costs of the demanding sectors.

18 von Nell-Breuning, O., Kapitalismus und gerechter Lohn, Freiburg i.Br. 1960, p. 140 f.

Table 3.1: Value Chain (compiled by author).

	Raw Materials Economy	Energy Industry	Steel Industry	Automotive Industry	End User
Preliminary Work	6	16	32	44	60
Salary	4	6	5	7	
Interest	2	3	4	3	
Rent	1	2	1	2	
Profit	3	5	2	4	
Added Value	6	10	7	9	
Value Creation	10	16	12	16	
Turnover	16	32	44	60	

$$\text{Equation 1: } Price = \frac{Preliminary\ Work}{q} + \frac{Salary}{q} + \frac{Interest, Rent, Profit}{q}$$

It is therefore not surprising that in the value chain, the energy supply companies with market power, above all E.ON, RWE, Vattenfall and EnBW, realize disproportionately high profits in their price equations,[19] and have an exclusive interest in selling the electricity they produce to their customer industries and private households at the highest possible prices. Any shortage of electricity is exploited to further increase prices in view of the completely price-inelastic demand and the state also makes money from this.[20] So why should politicians intervene and break up the corporate power? RWE, for example, is reporting flourishing business despite the general economic crisis. "In the first nine months of the current financial year (2023), net income almost doubled and reached around 3.8 billion euros, according to the company's latest interim report. In the same period of the previous year, RWE generated a net profit of around 2.1 billion euros. 'Our balance sheet after the first nine months of 2023 is excellent,' said RWE CFO Michael Müller. This should also benefit the Group's shareholders, including coal-mining towns such as Dortmund, Essen and Mühlheim as well as the new major shareholder Qatar."[21] And even a coal-fired power producer[22] like STEAG, which had already been declared dead, achieved a profit of EUR 1.91 billion in 2022 despite the general economic crisis.[23]

19 Cf. Bontrup, H.-J., Marquardt, R.-M., Die Zukunft der großen Energieversorger, Konstanz, Munich 2015.

20 Cf. Kungl, G., Die grossen Stromkonzerne und die Energiewende, (PhD Thesis), Frankfurt a.M./ New York 2018.

21 Meinke, U., RWE kann Gewinn nahezu verdoppeln, in: WAZ of November 15, 2023.

22 See Bontrup, H.-J., Marquardt, R.-M., Perspektiven der STEAG GmbH als kommunales Energieunternehmen im Kontext der Energiewende, Hanover, Lüdinghausen 2012.

23 Schulte, S., Steag meldet Rekordgewinn von 1,9 Milliarden, in: WAZ of November 24, 2023.

However, the steel industry, which relies on electricity and is set to become even more dependent on renewable electricity production in the future as part of a "green hydrogen strategy", is suffering from the good results of the energy industry in the value chain. The market leader in the German steel industry, Thyssenkrupp, reported a loss of EUR 2 billion for the past financial year 2022/23. This was mainly due to value adjustments in the balance sheet of the steel division, which the company has wanted to divest for some time due to the high losses;[24] but has also just received EUR 2 billion in government subsidies for a hydrogen strategy. And even Deutsche Edelstahlwerke (DEW), already a market leader in "green steel", is posting high losses.[25] The steel industry can no longer pass on its massively increased energy bill to the sales markets, e.g. the automotive industry, which in turn is generating excellent results. The headlines here read: "Mercedes reports high profits – thanks to luxury cars". The Group is building fewer cars, but is still earning splendidly because customers primarily want expensive cars, and its competitor BMW has also been reporting record profits for years. In 2022 alone, this amounted to EUR 18.6 billion – almost half more than in the previous year. However, the costs and considerable profits, which are not market-based in the sense of performance and innovation gains, must ultimately always be paid for and absorbed by the end consumer in the value chain, and industries and companies that are unable to pass this on to their customers along the value chain will also face major problems. This applies in particular to companies that are in international competition. At the end of the value chain, however, all costs and added value factors (interest, basic rents and profits) in the price equation always end up with the end consumers, the approximately 40 million private households, including 21 million pensioner households.

In socio-economic terms, end consumers are a highly differentiated group. Many are unable to pay rising energy prices as a result of the energy transition and also all the price increases that are ultimately involved in many end products via the price equation in the value chain. The state has already had to help private households with massive transfer payments and this will continue to be necessary. This also applies to many small and medium-sized companies that are unable to pass on their increased electricity bills in the value chain via their sales market prices and may suffer losses as a result or only make such low profits that they can no longer invest sufficiently. Many are already and will continue to be forced out of the market as submarginal suppliers due to the energy transition. However, neither the political nor the scientific debate on the energy transition sufficiently addresses and takes into account the contradictions between the interests of the market economy and capitalism, as shown here, in addition to the dismantling of corporate power. As a result, the findings on the energy transition to date have only produced incomplete results that are

24 Cf. Meinke, U., Thyssenkrupp mit Milliarden-Verlust, in: WAZ of November 23, 2023.

25 Cf. Helmecke, J., DEW: Rote Zahlen trotz grünen Stahls, in: WAZ of September 11, 2023.

neither expedient nor tenable for the further process of implementing an energy transition. It is therefore urgently necessary to draw up a master plan for the energy transition that logically takes up and adequately considers the contradictions and power perverting that have been highlighted.

3.7 Market-based Primary Distribution and the Energy Transition

In order for the "price tag for nature" to be affordable for all, a fairer market-based primary distribution must be brought about. The associated distribution issue is regulated by trade unions and employers' associations within the framework of constitutional collective bargaining autonomy (Art. 9 (3) GG) and the Collective Bargaining Act (TVG). However, with collective bargaining coverage of only around 50%,[26] the trade unions are hardly in a position today to enforce at least the necessary distribution-neutral scope for all dependent employees, in which both gross wages and salaries and capital income (interest, basic pensions and profits) increase at the same rate in line with productivity and inflation. If nominal wages do not increase at the same rate as productivity and inflation, there will be a redistribution to the value-added ratio, and vice versa. In the past, this has led to an economically and socially massively counterproductive redistribution in favor of capital or value-added income and thus at the expense of wages. This does not yet take into account the different levels of income in eastern and western Germany,[27] between women and men[28] and the generally unequal distribution of income.[29] On the basis of the wage share (with depreciation offset in the value-added share) from 1991-2022, based on the year 1993, EUR 4.070 trillion was redistributed from employee compensation to value-added compensation in absolute terms. Excluding depreciation and amortization, the figure was still EUR 2.423 trillion (see Figure 3.1).

If we compare the different wage and value-added ratios with and without depreciation (see Table 3.2), the wage ratio without depreciation rose by just 0.2 percentage points over the entire period from 1991 (71.0%) to 2022 (71.2%). However, there was a sharp fall in the wage share in the years 1991-2007, down to 64.5%, by 6.5 percentage

26 In 2000, collective bargaining coverage was still at 68%. See Böckler Impuls, 16/2021.

27 Böckler Impuls, No. 15/2022, "Die innerdeutsche Lohnlücke".

28 Cf. Hobler, D., Lott, Y., Phahl, S., Unrau, E., Stand der Gleichstellung von Frauen und Männern, WSI Report, No. 72, February 2022.

29 Cf. in detail Mittelbach, H., Lohn- und Kapitaleinkommen in Deutschland von 1990 bis 2010, Cologne 2013, Görgens, H., Sind die Löhne in Deutschland zu hoch? Zahlen, Fakten, Argumente, Marburg 2007, Anselmann, C., Spitzeneinkommen und Ungleichheit. Die Entwicklung der personellen Einkommensverteilung in Deutschland, Marburg 2013.

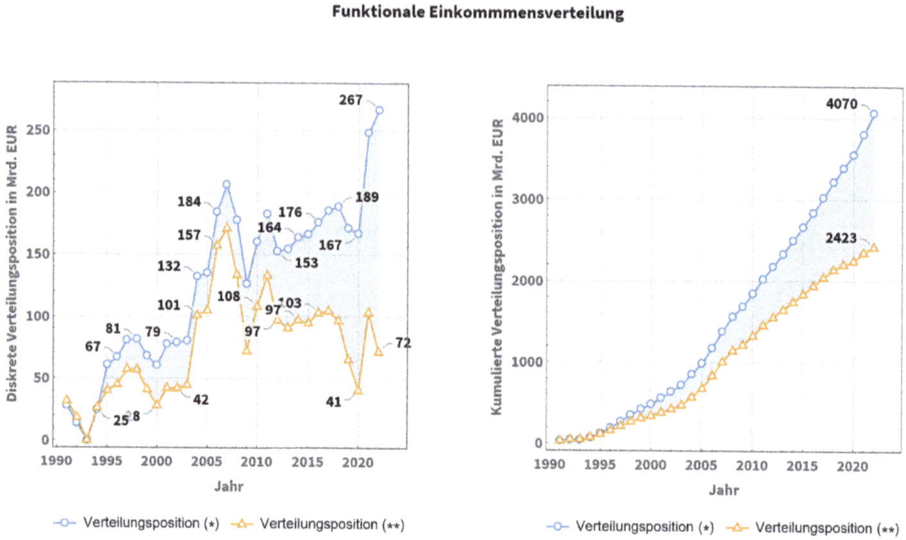

Figure 3.1: Functional Income Distribution (figure created by author).

points. The politically desired and neoliberally implemented redistribution policy from labor income to value-added income had a noticeable effect. From 2008, during the severe financial and economic crisis until 2009, the main cause of which was global redistribution,[30] the wage share rose again by 6.7 percentage points to 71.2% by 2022. However, 5.2 percentage points of this was caused by the crisis between 2007 and 2009 alone, because during the crisis, with national income falling overall from EUR 1,867.2 billion to EUR 1,805.3 billion, i.e. by EUR 61.9 billion, contractually determined wages fell less sharply than residual profit income. The decline in value added was also dampened by the predominantly contractually determined interest and basic pension income. The further increase in the wage share from 2007-2022 can be explained by the sharp rise in the number of dependent employees by almost 5.8 million.

The wage ratio was similar when depreciation was included in the value-added ratio, but at a much lower level. In addition to taking into account depreciation, the employee ratio and imputed entrepreneurial wages, the wage ratio was 66.2% in 1991. By 2007, it had fallen by 5.6 percentage points to 60.6% and was relatively stable at 60.5% until 2022, with minor fluctuations. The development of the value-added ratio was correspondingly complementary to 100%.

30 Cf. Bontrup, H.-J., Zur größten Finanz- und Wirtschaftskrise seit achtzig Jahren. Ein kritischer Rück- und Ausblick mit Alternativen, Hannover 2011, Huffschmid, J., Politische Ökonomie der Finanzmärkte, 2nd ed., Hamburg 2002.

Table 3.2: Functional Income Distribution in Germany (period: 1991-2022).[31]

Functional Distribution of Income																
	National Income (in bn EUR)	Employee Compen-sation	Added Value	Depreciation	Imputed Entre-preneurial Wages (in bn EUR)	Salaried Employees (tsd.)	Self-Employed	Employed	Salaries (in EUR)	Official Wage rate	Official Added Value Quota	Corrected Wage Rate I	Corrected Wage Rate II	Corrected Wage Rate III	Distribution Position *	Distribution Position **
1991	1,192.6	847.0	345.6	214.4	84.7	35,144	3,520	38,664	24,072	71.0	29.0	78.1	78.1	66.2	28.1	31.7
1992	1,269.8	917.2	352.6	234.5	95.0	34,489	3,577	38,066	26,568	72.2	27.8	79.7	79.7	67.3	13.9	18.4
1993	1,287.7	948.8	338.9	250.1	100.2	33,916	3,625	37,541	27,648	73.7	26.3	81.5	81.5	68.2	0.0	0.0
1994	1,341.0	961.9	379.1	260.8	106.0	33,763	3,725	37,488	28,464	71.7	28.3	79.6	79.6	66.7	24.8	26.2
1995	1,426.2	1,010.7	415.5	309.6	112.6	34,088	3,797	37,885	29,652	70.9	29.1	78.8	78.8	64.7	60.8	40.1
1996	1,445.2	1,019.8	425.4	317.0	115.5	34,036	3,854	37,890	29,964	70.6	29.4	78.6	78.6	64.4	66.8	45.0
1997	1,467.7	1,024.2	443.5	324.8	118.0	33,950	3,911	37,861	30,168	69.8	30.2	77.8	77.8	63.7	80.6	57.2
1998	1,496.6	1,045.8	450.8	332.5	120.6	34,355	3,960	38,315	30,444	69.9	30.1	77.9	77.9	63.8	81.4	56.9
1999	1,516.1	1,075.9	440.2	340.7	122.7	34,942	3,985	38,927	30,792	71.0	29.0	79.1	79.1	64.6	68.0	41.2
2000	1,554.9	1,117.4	437.5	354.4	124.7	35,797	3,995	39,792	31,212	71.9	28.1	79.9	80.7	65.7	60.4	28.3
2001	1,596.8	1,134.3	462.5	366.3	127.5	35,655	4,012	39,667	31,788	71.0	29.0	79.0	79.7	64.8	77.3	42.3
2002	1,606.7	1,141.9	464.8	374.6	130.7	35,438	4,060	39,498	32,184	71.1	28.9	79.2	79.8	64.7	79.0	41.9
2003	1,612.7	1,143.6	469.1	378.7	134.7	34,953	4,122	39,075	32,676	70.9	29.1	79.3	79.8	64.6	80.2	44.7
2004	1,692.5	1,146.1	546.4	385.9	139.4	34,960	4,258	39,218	32,736	67.7	32.3	76.0	76.7	62.4	132.3	101.0
2005	1,701.8	1,148.9	552.9	393.1	144.9	34,824	4,410	39,234	32,856	67.5	32.5	76.0	77.9	63.2	135.3	105.0
2006	1,801.3	1,169.9	631.4	402.9	149.5	35,083	4,483	39,566	33,348	64.9	35.1	73.2	75.1	61.4	184.2	157.4
2007	1,867.2	1,204.4	662.7	423.3	151.8	35,737	4,527	40,264	33,540	64.5	35.5	72.6	74.4	60.6	206.2	171.3
2008	1,879.5	1,251.2	628.3	440.9	153.9	36,317	4,503	40,820	34,188	66.6	33.4	74.8	76.4	61.9	177.7	133.6

(continued)

31 Source: Federal Statistical Office, National Accounts, various years, own calculations; wage ratio I = official wage ratio plus imputed entrepreneurial wage, wage ratio II = wage ratio I plus adjusted employment ratio, wage ratio III = wage ratio II plus depreciation; * Based on wage ratio III from 1993, ** Based on official wage ratio from 1993.

Table 3.2 (continued)

Functional Distribution of Income																
	National Income (in bn EUR)	Employee Compen-sation	Added Value	Depreciation	Imputed Entre-preneurial Wages (in bn EUR)	Salaried Employees (tsd.)	Self-Employed	Employed	Salaries (in EUR)	Official Wage rate	Official Added Value Quota	Corrected Wage Rate I	Corrected Wage Rate II	Corrected Wage Rate III	Distribution Position *	Distribution Position **
2009	1,805.3	1,258.0	547.3	451.9	155.4	36,367	4,492	40,859	34,596	69.7	30.3	78.3	80.0	63.9	126.4	72.2
2010	1,905.1	1,295.4	609.7	462.0	158.9	36,495	4,487	40,982	35,412	68.0	32.0	76.3	77.9	62.7	160.4	108.3
2011	2,016.1	1,352.2	663.9	478.1	166.4	36,973	4,563	41,536	36,468	67.1	32.9	75.3	76.7	62.0	182.8	133.3
2012	2,039.8	1,405.9	633.9	495.3	170.7	37,440	4,560	42,000	37,428	68.9	31.1	77.3	78.3	63.0	152.8	97.0
2013	2,086.8	1,446.6	640.2	509.8	170.1	37,790	4,458	42,248	38,148	69.3	30.7	77.5	78.0	62.7	154.6	91.0
2014	2,173.3	1,503.9	669.4	524.9	172.8	38,192	4,402	42,594	39,252	69.2	30.8	77.2	77.4	62.4	163.9	97.4
2015	2,252.8	1,564.8	687.9	542.8	175.8	38,632	4,405	43,037	39,900	69.5	30.5	77.3	78.2	63.0	166.4	95.1
2016	2,345.5	1,625.1	720.5	558.7	179.9	39,218	4,341	43,559	41,436	69.3	30.7	77.0	77.7	62.7	176.2	103.2
2017	2,444.2	1,696.3	747.9	581.4	181.9	39,858	4,273	44,131	42,564	69.4	30.6	76.8	77.3	62.5	185.8	104.6
2018	2,539.2	1,774.0	765.1	609.9	185.1	40,502	4,225	44,727	43,800	69.9	30.1	77.2	77.4	62.4	189.1	96.9
2019	2,608.2	1,856.2	752.1	640.4	188.4	40,973	4,160	45,133	45,300	71.2	28.8	78.4	78.5	63.0	171.5	65.6
2020	2,571.6	1,853.9	717.7	661.7	184.5	40,765	4,056	44,821	45,480	72.1	27.9	79.3	79.2	63.0	167.2	40.9
2021	2,743.4	1,918.0	825.4	704.9	185.6	40,908	3,958	44,866	46,884	69.9	30.1	76.7	76.4	60.8	248.7	103.4
2022	2,843.5	2,023.4	820.2	793.4	190.2	41,525	3,904	45,429	48,732	71.2	28.8	77.8	77.4	60.5	267.3	71.8
															4,070.0	**2,422.8**

Quelle: Statistisches Bundesamt, Volkswirtschaftliche Gesamtrechnungen, diverse Jahrgänge, eigene Berechnungen

Lohnquote I = Amtliche Lohnquote plus kalkulatorischer Unternehmerlohn, Lohnquote II = Lohnquote I plus bereinigte Beschäftigtenquote, Lohnquote III = Lohnquote II plus Abschreibungen

* Auf Basis der Lohnquote III von 1993, Auf Basis der amtlichen Lohnquote von 1993

Against this overall background of wage and value-added ratios, the empirical results of primary distribution offer considerable redistribution potential from value-added to labor income. With an actual wage share of 60.5% in 2022 and gross private investment of EUR 769.4 billion (gross government investment amounted to EUR 102.4 billion), the surplus value added was gigantic at EUR 666.0 billion. In 2022 alone, with an assumed wage ratio III (which takes into account depreciation, the employee ratio and entrepreneurial wages) of only 70.0% instead of 60.5% in real terms, a calculated EUR 344.5 billion could have been primarily redistributed from value-added income to wages and salaries. Working capital owners would also have benefited from this with their entrepreneurial wages. The surplus value of EUR 1,091.1 billion (surplus value ratio of 30%) would then have been more than sufficient to finance the overall private gross investment of EUR 769.4 billion without any problems and the non-working capital owners would still have received a high surplus value of EUR 321.7 billion. There would therefore still be a much greater redistribution opportunity here, also in favor of investments (see Table 3.3).

Table 3.3: Wage Share and Surplus Value Added – Scope for Redistribution (compiled by author).

Alternative Distributions 2022					
	National Income plus Depreciation	**Employee Compensation plus Entrepreneurial Wages**	**Added Value minus Entrepreneurial Wages**	**Gross Investment***	**Added Value Surplus**
	3636.9	2201.3	1435.6	769.4	666.2
Wage Quota III		60.53%	39.47%		
	3636.9	2545.8	1091.1	769.4	321.7
Wage Quota III		70.00%	30.00%		

*Private Gross Fixed Capital Formation (construction and equipment Investment), own calculation

Such a redistribution of primary income would solve all the financing problems of the energy transition at a microeconomic level for private households. All economic subjects would be able to pay the costs (the "price tag") of the energy transition without any problems. In addition, the primary redistribution from the value-added ratio to the wage ratio would have further positive effects: Firstly, there would be more demand, thus initiating growth and employment as well as increased investment, and secondly, the state would have more tax and social security revenue as a result.

However, the functional wage ratios say nothing about the distribution of employee remuneration. As the 5.8 million increase in the number of employees was mainly due to part-time and precarious employment for women, the distribution of income in real terms has deteriorated even further. In a study, the German Institute for Economic Research (DIW) found that the development of unequal working hours has massively increased the inequality of earned income. "If employees had been able to work as much or as little as they wanted, the inequality of earned income would only have increased by half as much."[32] This means that there is still considerable and much-needed potential in terms of working hours, and not just in the course of the energy transition, which must be tapped into with an appropriate labor market and family policy.

But who is to implement the necessary market-related primary redistribution? The increasingly eroding and weakened trade unions will be even less able to do this in collective bargaining against the interests of employers' associations in the future than they have been in the past. Without state support (intervention), redistribution will not succeed.

This shows how difficult the implementation and success of the energy transition is in reality. There are three essential economic necessities here in the context of the value added created by the division of labor in society and its primary distribution.

- Firstly, in order to strengthen the trade unions, but also the employers' associations, there must be compulsory membership of all employees in the trade unions and all companies in the employers' associations.
- Secondly, the state must ensure final distribution neutrality on the basis of productivity and inflation rates.
- And thirdly, employees must participate in the profits and equity of the companies that employ them. There must be a legally binding genuine profit and/or capital participation.[33]

Implementation would then make the current declaration of the general validity of collective agreements in individual sectors superfluous, as would the current political embarrassment that it is not even possible to award state contracts only to companies that are bound by collective agreements. And a statutory minimum wage and its never-ending political and ideological debates about an annual adjustment could then also be dispensed with.[34] However, despite all the optimism surrounding the implementation of the three redistribution instruments, it must be borne in mind that this

---

32 Beckmannshagen, M., Schröder, C., Entwicklung der Arbeitszeiten fährt die Ungleichheit der Erwerbseinkommen, in: DIW-Wochenbericht, No. 33+34/2022, p. 427.

33 Cf. in detail Wagner, K.-R., Wagner (ed.), Mitarbeiterbeteiligung, Visionen für eine Gesellschaft von Teilhabern, Wiesbaden 2002, Voß, E., Wilke, P., Maack, K., Mitarbeiter am Unternehmen beteiligen, Modelle, Wirkungen, Praxisbeispiele, Wiesbaden 2003, Bontrup, H.-J., Springob, K., Gewinn- und Kapitalbeteiligung, Eine mikro- und makroökonomische Analyse, Wiesbaden 2002, Bontrup, H.-J., Gewinn- und/oder Kapitalbeteiligungen – ökonomische Utopie oder Notwendigkeit?, in: Intervention, Zeitschrift für Ökonomie, Heft 1/2005, p. 95 et. seq.

34 Cf. Krebs, T., Drechsel-Grau, M., Mindestlohn von 12 Euro: Auswirkungen auf Beschäftigung, Wachstum und öffentliche Finanzen, IMK-Study, No. 73, August 2021.

can hardly be considered realistic in an indirect parliamentary democracy. There would certainly be massive resistance from the business camp and from representatives of capital interests in the parliaments. It must therefore be assumed that the energy transition will either fail to meet the most important economic prerequisite, i.e. the necessary primary redistribution, or will at best only take a suboptimal course.

3.8 Profit Rate and Investments

In addition to the primary redistribution, a further restriction for the energy transition must be assessed: the profit rate, the return on the capital provided or invested by capital owners through loans. No company invests here without an expected (high) profit rate with the shortest possible amortization periods. This correlation determines investments. From 1999 to 2022, net investment in the overall economy in Germany was low. The annual average net investment ratio was just 3.3%. We must therefore speak of significant underinvestment. However, investment is the driver of economic development. There is also serious underinvestment in the public sector: From 1999 to 2022, the state (federal, state and local) invested a cumulative gross amount of just EUR 1,501.1 billion. On an annual average, this amounted to just EUR 62.5 billion. In the same period, depreciation amounted to EUR 1,482.6 billion, i.e. an annual average of EUR 61.7 billion, so that the decisive net investments in the last 24 years only amounted to a cumulative EUR 20.5 billion or an annual average of EUR 0.854 billion. In the years 2003 to 2008 and from 2012 to 2016, net government investment was even negative, resulting in a considerable erosion of the state's capital stock (as shown in Figure 3.2).

The macroeconomic investment behavior is dominated by the macroeconomic profit rate, which in turn results from the multiplicative link between capital productivity and the value added ratio (value added/value added) or 1 minus the wage ratio.

$$Equation 2: Profit Rate = \frac{Labour Productivity}{Capital Intensity} = Capital Productivity * \frac{Interest, Rent, Profit}{Value Creation}$$

As with investments, the empirical findings from 1999 to 2022 show a sharp decline in the profit rate in recent times. The overall economic rate of return fell by 5 percentage points from 11.6% (1999) to 6.6% (2022). The annual average profit rate was 9.4% (Figure 3.3).

The decline in the profit rate can be explained by the individual components. Between 1999 and 2022, labor productivity increased by an annual average of only 1.3%, while capital intensity rose by 4.0% and, as a result, capital productivity fell by 2.7% (see Figure 3.4). This reflects the overall poor performance of the German economy as a whole. More and more capital had to be used per hour worked in order to achieve even low and insufficient labor productivity per hour worked. And this against the backdrop of an economic structural change that is increasingly moving towards the tertiary sector of the economy, which is more labor-intensive than capital-intensive.

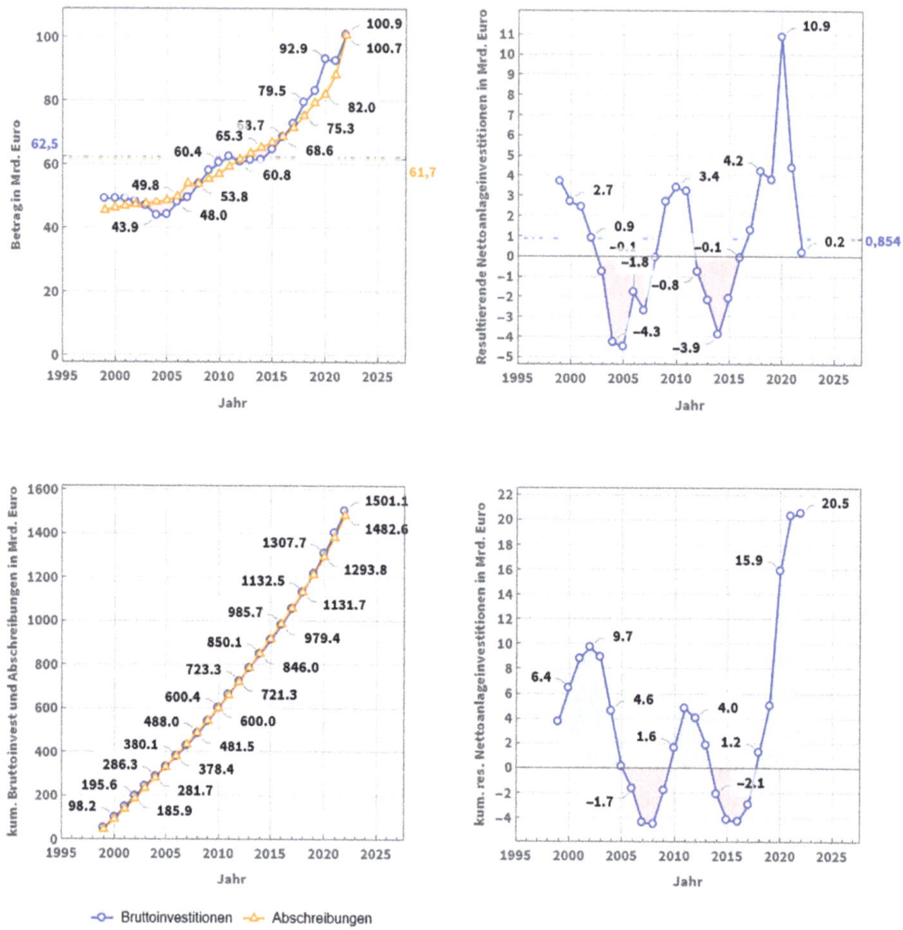

Figure 3.2: Gross Investment, Depreciation and Amortization in Germany (period: 1999-2022); figure created by author.

An increase in the profit rate in order to stimulate investment would have required a redistribution of the realized value added in the direction of the value added ratio. The wage share should have fallen. However, this was not the case. The actual wage share III shown stagnated at 60% between 1999 and 2022. However, it was previously concluded in 3.7 that a redistribution to wage share III is necessary to implement the energy transition and that this can and should be raised from a good 60% to 70%. This is obviously a contradiction, but one that is clarified by the fact that in a contradictory capitalist economy, it is not just a question of the distribution of value creation, although this is of course extremely important, but also, from the one-sided perspective of the owners of capital, of the profit rate, which is determined not only by distribution, but also by the ratio of labor productivity to capital intensity. If the quotient,

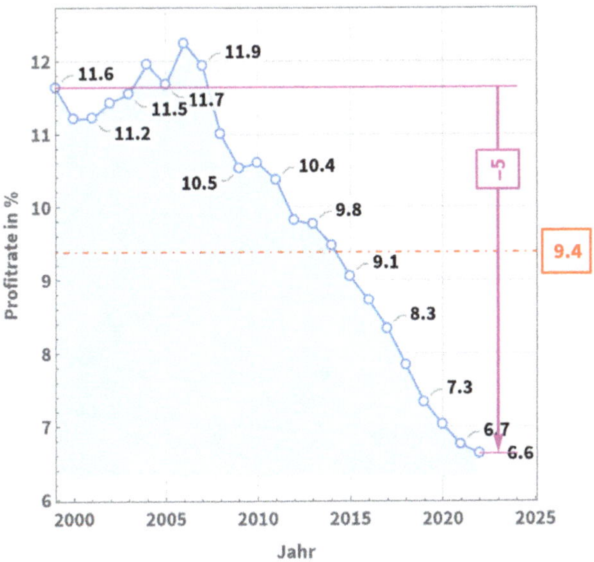

Figure 3.3: Development of the Profit Rate (period 1999 to 2022).

capital productivity, were to rise, there would also be scope for a higher wage ratio with a simultaneously rising profit rate. However, the growth rate of labor productivity would then have to be higher than the growth rate of capital intensity.

In addition to the derived redistribution findings, which are difficult to implement politically, the overall ex-post profitability and investment findings also show that the investments to be made as part of the energy transition will have a very difficult time being implemented. Every investment potentially considered here must pass through the "bottleneck" of the profit rate. If this appears too low to investors, there will be no investment in the energy transition. The higher the interest rate for borrowed capital, the higher the profit expectations on the equity capital invested must be.

Investments in the public sector must also be financed, just like investments in the private sector. In the case of the state, financing can take the form of taxes and levies as well as public debt. In this context, however, the state is dependent on the private sector. If this does not run well, the state also has a problem. It receives less state revenue and can therefore spend less if it does not want to get into debt. Alternatively, if the economy is doing well, the state's scope for political action will also increase.

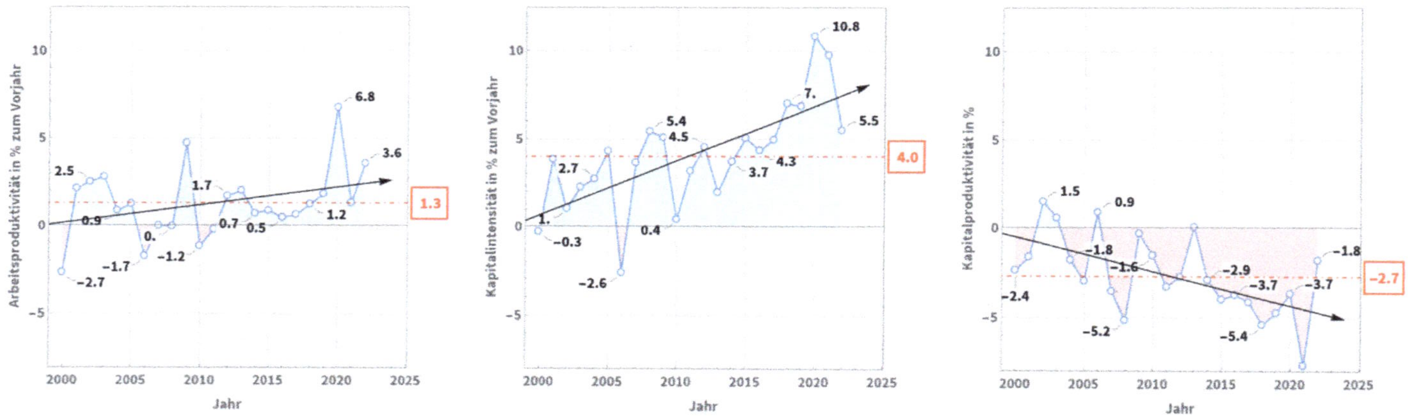

Figure 3.4: Labor Productivity and Capital Intensity in Germany (period: 1999-2022).[35]

35 Arrows based on a regression line determined using the least squares method.

3.9 High Volume of Public Investment is not Possible Without Public Debt

The investments to be made for the energy transition are enormous. According to Boston Consulting/Prognos, this will amount to around EUR 90 billion per year by 2050.[36] Added to this are investments for general government and private investments that have been neglected in recent decades.

- Infrastructure EUR 150 bn/a
  - Digitalization
  - Road and bridge construction
  - Rail, public transport
  - Social housing
  - School buildings, sports and cultural facilities
- Additional government spending EUR 100 bn/a
  - against unemployment
  - for health/nursing
  - for education
  - for pension
  - for general social benefits (incl. migration)

This means that in addition to the EUR 90 billion p.a. for the energy transition, there is a further EUR 250 billion p.a. for urgently needed future government investments and expenditure. Even if we deduct depreciation of 10% annually, i.e. EUR 34 bn/a, this leaves a net investment volume of EUR 306 bn/a to be financed from savings. The overall economic savings that can be generated in the future will not be sufficient for this. Neither by reducing private and public consumption, nor by reallocating corporate savings activities, nor by shifting savings activities from abroad to the domestic market.

This shows the macroeconomic ex-post equilibrium on an annual average for the years 1999 to 2022. According to the equation VerfE – C = S = $I_{Netto\ +/-ABeitrag}$, disposable income (VerfE) in Germany amounted to EUR 2,271.3 billion, of which EUR 2,024.5 billion was consumed (C) and thus EUR 246.8 billion was saved. The consumption rate amounted to 89.1% and the savings rate to 10.9%. Gross investments averaged EUR 584.6 billion for the year and EUR 493.3 billion was written off, meaning that overall net economic investment (I_{Netto}) amounted to just EUR 91.3 billion, of which EUR 20.5 billion was net government investment (22.4%). As a result, loans were granted to the rest of the world ($_{A\ contribution}$) in the amount of EUR 155.5 billion via capital exports.

36 See Boston Consulting/Prognos, Klimapfade für Deutschland, 2018, and Boston Consulting, Klimapfade 2.0: Ein Wirtschaftsprogramm für Klima und Zukunft, October 2021.

However, there is help with regard to macroeconomic savings because, in addition to savings, credit is created by the banking system and the volume of credit in circulation is therefore many times higher than cash and central bank balances. Loans are therefore not limited by social savings. "In macroeconomic terms, banks can make more money available for investment than was previously saved by society by forgoing consumption. This is one of the central statements of the Keynesian revolution in economic theory: it is not saving that is the prerequisite for investment, but the other way round: investment leads to an increase in national income and, since saving depends on the level of national income, also to an increase in the total amount saved in the economy as a whole."[37] Whether entrepreneurs therefore invest does not depend on an imaginary fund of available funds created by prior saving, but primarily on sales and return expectations.

Every cut in expenditure (austerity), wherever it takes place in the economy, whether in private households, companies or the state, or whether we are talking about the same players in the countries with which we trade, a cut in expenditure with no change in income always leads to a reduction in the profits of entrepreneurs, which are crucial in the system and which, according to the generally prevailing view, are important for the dynamics of the economy and therefore also for government debt.[38] In macroeconomic terms, this means that the income of entrepreneurs is always equal to the value of the investment, minus the value of the consumption of the entrepreneurs themselves, but minus the savings of all non-entrepreneurs, including the state. Government savings are directly reflected here as a reduction in the profits of entrepreneurs and government spending surpluses (debt) increase the profits of entrepreneurs.

It is therefore completely incomprehensible from a scientific point of view that entrepreneurs and politicians close to them oppose government debt and have obviously not understood that any deliberate creation of a spending surplus, whether by private households, the state or foreign countries, or by entrepreneurs themselves, directly improves the demand and profit situation of entrepreneurs. Entrepreneurs and politicians should therefore advocate debt. A look at the financial balances for Germany in the empirical findings should actually suffice here. Private households and the corporate sector would not have been able to save over the past 25 years without the systematic indebtedness of the state and, in particular, foreign countries, which clearly answers the question of the so-called state debt brake in the context of the ruling by the Federal Constitutional Court (BVerfG) on November 15, 2023. If the state does not succeed in forcing the entire corporate sector back into the role of debtor in the future, it will have to assume this role itself, unless Germany's economy continues to be geared towards using foreign countries as debtors and continuing to assume that Germany can remain "world export champion"

37 Huffschmid. J., Politische Ökonomie der Finanzmärkte, op. cit. p. 30.

38 If entrepreneurs then react to falling profits or even losses with further spending cuts, which is the normal case, they directly worsen the economic situation of all entrepreneurs, because their cuts mean nothing other than a loss of revenue for other companies.

until the end of time. However, if the state is to remain in the role of another debtor to finance the investments for the energy transition and other necessary investments and government spending, this would require the abolition of the debt brake. There would then be plenty of scope for debt for the necessary investments.

As long as the German economy does not incur net debt abroad, which requires current account deficits, debt-financed investments do not mean any additional burden for future generations. We cannot take anything away from a future generation that has not yet been produced. Their potential income is not reduced by the fact that high national debt is left behind. This only creates a distribution issue between the state's creditors and the remaining citizens, the resolution of which will be decided in the future. This is decided by the state, which can tax the winners of the distribution game appropriately. After all, if the state is indebted to its own citizens, the creditors are also taxpayers. Debt-financed investments only allow access to current economic output and can therefore only limit current consumption. Such investments can then only be of benefit to future generations.

The result can be seen in the macroeconomic creditor-debtor positions for Germany. The sum of all debts is always equal to the sum of all assets. The empirical findings reveal three creditor sectors: 1. private households (including all partnerships and sole proprietorships as well as non-profit organizations), 2. non-financial corporations and 3. the financial sector. From 1999 to 2022, these three sectors accumulated EUR 4,524 billion in asset growth. This amounted to an annual average of EUR 189 billion. This was offset by two debtor sectors: 1. the state and 2. foreign countries. They balanced out the assets with their debts. The balance was zero. However, the largest debtor by far was not the state, but foreign countries (cf. Figure 3.5).

3.10 State Taxation and Secondary Distribution

However, if you still do not want government debt, and even write it into the constitution with a debt brake, the only way to finance the energy transition and the additional government investment and spending described above is through higher taxation. Here, the current income of private households, the profits of companies and/or the wealth created from income in the past and further wealth from accumulated wealth could be siphoned off by the state through additional taxes. John Maynard Keynes refers to this as the absorption of "predatory savings". Indirect consumption taxes could also be raised.[39] For reasons of space, it is not possible in this article to

39 For the long history of German tax policy from 1871 to the present day, see the outstanding work by Marc Buggeln, Das Versprechen der Gleichheit, Berlin 2022.

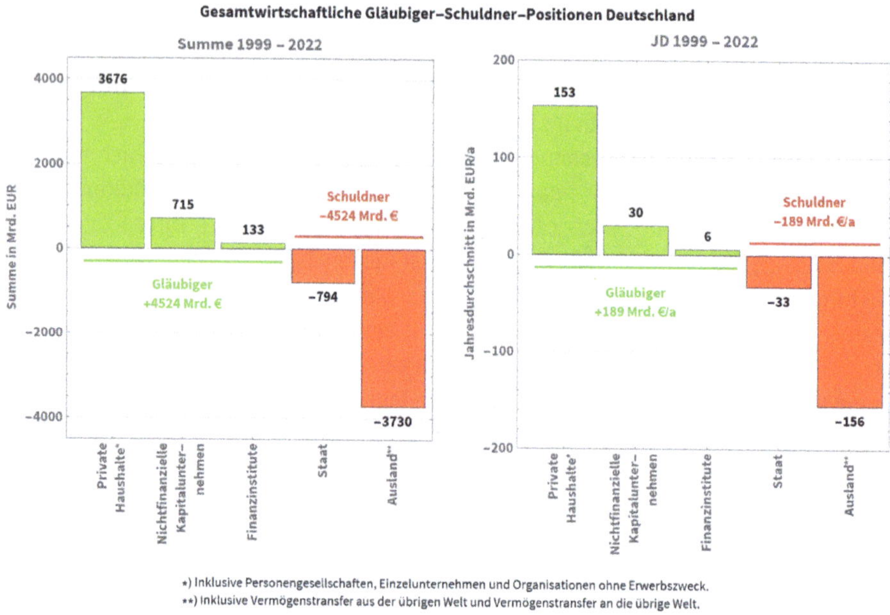

Figure 3.5: Germany's overall economic creditor/debtor positions; figure created by author.

show the complex effects that taxes (both positive and negative) have on economic performance, growth and employment, among other things.[40] However, it is an overarching and proven scientific finding that without the introduction of a progressive tax system, democratic communities would not have been able to develop over the last 100 years. "Since the publication of Thomas Piketty's book,[41] there has been no shortage of suggestions as to what a progressive tax system adapted to current circumstances might look like. In addition to Piketty, Anthony B. Atkinson, Emmanuel Saez and Gabriel Zucman in particular have made proposals in this regard, and for Germany, reference can also be made to the numerous contributions by Stefan Bach from DIW."[42]

There is widespread agreement that the income tax rate for the top one percent of incomes should be raised significantly. The values cited for this are marginal tax rates of between 60% and 80%. Income inequality in Germany has been a hotly debated topic for years. "While it had increased sharply between 2000 and 2005 during periods of high unemployment, the financial crisis of 2008/2009 briefly contributed to a slight reduction in inequality, mainly due to declining capital income and income

40 Cf. in detail Brümmerhoff, D., Finanzwissenschaft, 10th ed., Munich 2011, p. 423 et. seq.
41 See Piketty, T., Das Kapital im 21. Jahrhundert, Munich 2014.
42 Buggeln, M., Das Versprechen der Gleichheit, op. cit. p. 922.

from self-employment."[43] Since then, the development of the low-wage sector, which includes employees whose gross hourly wage is less than two thirds of the median wage, has fallen from just under 24% in 2007 to just over 20% in 2019. "Between 2000 and 2007, the proportion of employees in the low-wage sector grew significantly from around 19% to just under 24%. (. . .) After that, the proportion of low-wage employees stagnated until around 2012, after which their share gradually fell to 20.7% in 2019. According to the German Socio-Economic Panel (SOEP), however, a good 7.4 million employees in their main job were still in the low-wage sector."[44] It would be urgently necessary to introduce an immediate tax exemption for this group of people. The tax-free allowance here should be set at an annual taxable income of EUR 18,000.

The taxation of profits through corporation tax has been drastically reduced to a rate of 15%. Since the early 1990s, the neoliberal dogma has been preached here that states should compete like companies in an increasingly globalized world. Germany, the ruling and profiting elites agreed, would lose the competition if wages and taxes were too high. The Federation of German Industries (BDI) preached and labeled Germany as the "sick man" in Europe. But if you want the energy transition and no national debt, you have to demand at least a corporate tax rate of 30% and a financial market transaction tax as well as the abolition of the 25% flat tax on all private capital gains, which excludes them from the progressive income tax.

To compensate for the shortfall in direct income and profit taxes, indirect consumption taxes, above all VAT, have been increasingly increased. However, indirect taxes are regressive in nature, i.e. economic entities with a high savings rate are least affected by indirect taxes. The reduced VAT rate does little to change this, especially as this also only has to be paid by high-income earners. A reduction in indirect taxes would therefore make sense. This is demonstrated by the demand from individual sectors, such as the catering industry, which was able to calculate with the reduced VAT rate to ease the burden during the coronavirus pandemic. And when it comes to taxes, it should also be clear that there will have to be a significant tightening of tax law.

In view of the completely unequal distribution of wealth, there must also be adequate wealth taxation as part of the energy transition. The richest ten percent of private households own two thirds of all assets in Germany.[45] Pension assets are the only

43 Grabka, M. M., Einkommensungleichheit stagniert langfristig, sinkt aber während der Corona-Pandemie leicht, in: DIW-Wochenbericht, No. 18/2021, p. 308.

44 Ibid., p. 310.

45 See Bach, S., Grunderbe und Vermögensteuern können die Vermögensungleichheit verringern, in: DIW-Wochenbericht Nr. 50/2021, p. 801 ff., Bach, S., Vermögensabgabe, Aufkommen und Verteilungswirkungen, report by the German Institute for Economic Research (DIW) on behalf of the DIE LINKE parliamentary group in the German Bundestag and the Rosa Luxemburg Foundation Berlin 2020, Baresel, K., et al, Hälfte aller Erbschaften und Schenkungen geht an die reichsten zehn Prozent aller Begünstigten, in: DIW-Wochenbericht Nr. 5/2021, p. 63 ff., Heise, A., (Über-)Reichtum als gesellschaftliches Problem, in: spw, Heft 257, Ausgabe 4/2023, p. 53.

assets held by the poorer half of the population in Germany. As a result, the share of assets held by the bottom half of the overall distribution increases from two to nine percent. However, half of all inheritances and gifts go to the richest ten percent of all beneficiaries.[46] Even if a wealth tax is difficult to levy, a global wealth tax could be introduced here, "which would make the shifting of assets across national borders worthless. Inheritance tax is easier to levy. However, it has the disadvantage over wealth tax that it depends on the coincidence of death and is then levied all at once in a large sum. As a result, business assets are now hardly subject to inheritance tax in Germany. For this reason, a wealth tax that can be offset against inheritance tax in the event of an inheritance or gift could be an option. If a wealth tax is not levied, business assets should no longer be so comprehensively exempt from inheritance tax, although the taxes then due should be spread over a longer period of time so as not to jeopardize the existence of the business."[47] In summary, it can be said with regard to taxes and income distribution that income, as shown in 3.7, is first of all a matter of fair distribution of primary income and only then, in the course of secondary distribution, of fair taxation of income.

46 Cf. Bartels C., Bönke, T., Glaubitz, R., Grabka, M. M., Schröder, C., Rentenvermögen macht Großteil des Vermögens der ärmeren Bevölkerungshälfte in Deutschland aus, in: DIW-Wochenbericht, No. 45/ 2023, p. 626 et. seq.

47 Buggeln, M., The Promise of Equality, op. cit. p. 923.

Jane O'Sullivan

4 Population – The Emissions Multiplier

4.1 Introduction

Anthropogenic climate change, as a cumulative impact of human activity, is intimately entwined with the scale of the human population.[1] The more people there are, the bigger are our collective impacts on the environment. Yet none of the associations are simple, and many evoke sensitivities that cause population to be given little attention in climate change discourse.

Two key questions are: how much would future population growth contribute to future greenhouse gas emissions, and what can ethically be done to alter future population outcomes? These quite separate considerations are often conflated, when people who believe that it is unethical to intervene to reduce future population growth defend their position by arguing that it would make little difference to future emissions. To reach a well-informed position, we should not allow one of these questions to prejudice the other.

However, to pre-empt the objections that are so often cited to dismiss population discussion before it has begun, let it be said that the only actions advocated in this chapter involve voluntary birth control, including exploring factors influencing family size decisions and removing barriers to people (especially women and girls) exercising free and informed reproductive choices. Let it also be said that these actions would not detract in any way from actions to reduce emissions per person, but would increase the effectiveness of all other responses to climate change mitigation and adaptation.

For both of these key questions, the answers are different if we are considering developed countries or underdeveloped countries. This is another reason for controversy: since most future population growth will occur in underdeveloped countries, the topic is often viewed as an effort to shift blame for climate change onto the poor. Yet, as we will explore below, population is an important consideration in both settings, and the poor have the most to gain from addressing it. The "don't blame the poor" response is doing the poor a disservice by reducing political will for much-needed family planning programs.

4.2 Procreation Decisions in Developed Countries

For residents of industrialized countries, choosing to have one child fewer than they otherwise would is by far the most effective lifestyle choice available for reducing future carbon emissions (as well as for reducing all forms of human impacts on the envi-

1 Rees, W. E. (2022). The human eco-predicament: Overshoot and the population conundrum. *Vienna Yearbook of Population Research*, 21. https://doi.org/10.1553/p-eznb-ekgc.

https://doi.org/10.1515/9783110784299-004

ronment). Murtaugh and Schlax (2010) calculated that one avoided child in USA reduces future emissions by more than five times the lifetime emissions of each parent, on account of all the future descendants avoided.[2] This calculation attributes each parent with a one-half share of their offspring's emissions, a quarter share of grandchildren, and so on, and assumes as a baseline that future people will behave similarly to current residents of each country, in terms of emissions per capita and children per woman.

Wynes and Nicholas (2017) used these calculations to compare reproductive choices with other lifestyle changes individuals could undertake to reduce emissions.[3] Across a panel of developed countries, they found that an avoided child would save, on average, 59 tonnes of carbon dioxide equivalent per year of the parent's remaining life (bringing forward the anticipated emissions of later generations). This was more than 20 times the emissions saving from living without a car, and more than 70 times the impact of becoming vegan (Figure 4.1).

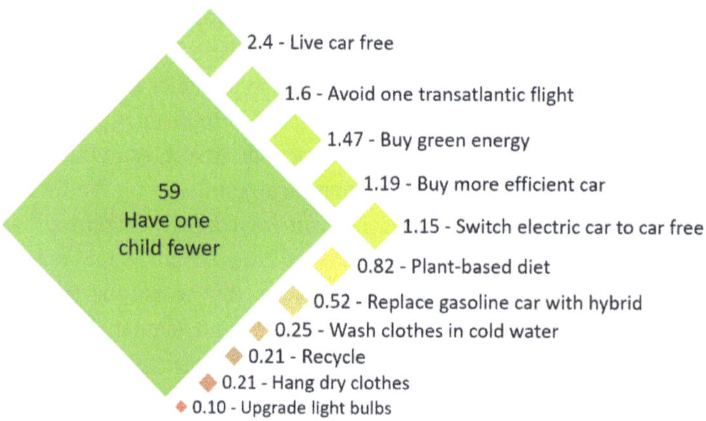

Figure 4.1: The emissions reduction (tonnes CO_2e per year) achievable from various individual actions.[4]

These results have been contested. Some people have argued that this calculation would count the same emissions multiple times: attributing responsibility to the individual and

2 Murtaugh P.A. and Schlax M.G. (2009). Reproduction and the carbon legacies of individuals Glob. Environ. Change 19: 14–20.

3 Wynes, S. & Nicholas, K.A. (2017). The climate mitigation gap: education and government recommendations miss the most effective individual actions. *Environ. Res. Lett.* 12, 074024. https://doi.org/10.1088/1748-9326/aa7541.

4 The Figure 4.1 shows the average value for a panel of developed countries. Emissions avoided by having one child fewer assumes each parent is responsible for half the lifetime emissions of their child, a quarter of a grandchild etc., divided among the parent's remaining years of life. Data from Wynes and Nicholas (2017).

their parents and their grandparents etc.[5,6] This is a misunderstanding, however, since we are referring to emissions *reductions*, not emissions. If an emissions reduction is attributed to a parent for deciding not to have a child, there is no child in the next generation to attribute the missing emissions to, so no double counting.

We should also avoid confusing the contentious issue of moral responsibility with the mathematical calculation of the impact of decisions. Of course, it is arguable that the children and grandchildren would have had smaller footprints than their parents, as their society transitions to renewable energy. Nevertheless, in accounting 'emissions reductions', what matters is the baseline. The baseline assumption is that current behaviours persist, including current fertility rates and current consumption behaviours. If a child has less impact than their parent, then those emissions reductions are attributable to the child, not the parent. But if there are fewer children than the baseline, those avoided emissions are legitimately attributable to the potential parents for choosing not to have them.

It is reasonable to question the usefulness of such a calculation if future emissions per person can be expected to decline rapidly. This raises the question of how much more rapidly the energy system can be converted away from fossil fuels if energy demand is also falling, and how much a reduction in population can contribute to that fall in energy demand. It is already a tall order to achieve net zero carbon emissions by 2050. If population increases by, for example, 50% in the same period, the rate of construction of new energy infrastructure must be similarly elevated. Shortages of mineral inputs would be exacerbated, adding to the cost for each megawatt of renewable capacity. If, on the other hand, the population were allowed to contract, all these challenges are lessened.

It is always true that reducing energy demand per person could negate the impacts of adding more people. However, it is equally true that population growth negates the efforts and sacrifices we make to reduce emissions per person. Without population growth, any reduction in energy demand per person serves to accelerate decarbonization. Emissions can never be blamed entirely on consumption nor entirely on population: it is the product of both.

The issue of timing of the avoided emissions is also a valid consideration: the single act today of choosing not to have a child spreads the avoided emissions over many decades, whereas choosing not to take a long-haul flight has a rapid effect (if we assume that airlines respond quickly to falling demand by scheduling fewer flights). But many other actions also have delayed effects: installing solar panels initially causes additional emissions in their manufacture and installation, and spreads the avoided emissions over the next 30 years or so. Planting trees sequesters carbon with a similar

5 van Basshuysen, P., & Brandstedt, E. (2018). Comment on 'The climate mitigation gap: education and government recommendations miss the most effective individual actions'. Environmental Research Letters, 13(4): 048001.). https://doi.org/10.1088/1748-9326/aab213.

6 Pinkert, F., & Sticker, M. (2021). Procreation, footprint and responsibility for climate change. *The Journal of Ethics*, *25*(3): 293–321. https://link.springer.com/article/10.1007/s10892-020-09345-z

time profile to not having a child: initially gradual, and peaking in the 20–100 year period, tapering off as tree generations turn over and total biomass plateaus. None of these timing issues are used to discredit these mitigation actions, so should not be used to dismiss the impact of avoided births.

An increasing number of young adults are choosing not to have children because of their environmental concerns. They are motivated by their desire to have less impact on the environment themselves, and/or their pessimism about the world their children would inherit.[7,8,9,10] Climate change is a dominant influence on both of these concerns. However, to date such sentiments are too rare to alter national fertility appreciably. Indeed, other cultural influences are operating in the opposite direction, including government promotion of births as a misconceived attempt to avoid population ageing[11,12] and an increasing tendency of celebrities to have large families.[13] Nevertheless, fertility has fallen in most developed countries in recent years, particularly since the Covid-19 pandemic. The trend is largely attributed to cost-of-living pressures.

4.2.1 The Dreaded 'Birth Dearth'

Nearly all developed countries already have 'below replacement' fertility, which means that people are choosing to have fewer children than would replace the parents' generation. Replacement fertility level is usually approximated at 2.1 children per woman, allowing for the slightly fewer girls than boys in natural birth ratios and for childhood mortality. Without a net influx of migrants, populations with below replacement fertility eventually contract. However, it takes several decades after fertility falls below 2.1 before population decline commences. This delay is referred to as 'demographic momentum', and is due to the youthful age profile of the population as a result of previously high fertility. Setting aside the impacts of migration (which we will

7 Overall, C. (2012) *Why Have Children?* Cambridge, MA: MIT Press. https://doi.org/10.7551/mitpress/8674.001.0001.

8 Conly, S. (2016) *One Child: Do We Have a Right to Have More?* Oxford: Oxford University Press, UK. https://doi.org/10.1093/acprof:oso/9780190203436.001.0001.

9 Hedberg, T. (2019) The Duty to Reduce Greenhouse Gas Emissions and the Limits of Permissible Procreation. *Essays in Philosophy* 20(1): eP1628. DOI: 10.7710/1526-0569.1629.

10 Cafaro, P. (2022) Just population policies for an overpopulated world. The Ecological Citizen 5(1): epub-046. https://www.ecologicalcitizen.net/article.php?t=just-population-policies-an-overpopulated-world.

11 Lee, R., Mason, A., & members of the NTA Network (2014). Is low fertility really a problem? Population aging, dependency, and consumption. *Science* 346(6206), 229–234. https://doi.org/10.1126/science.1250542.

12 Götmark, F., Cafaro, P. & O'Sullivan, J. (2018) Aging Human Populations: Good for Us, Good for the Earth. *Trends Ecol Evol.* 33(11):851–862. https://doi.org/10.1016/j.tree.2018.08.015.

13 The Guardian (2019) Is having five children really a middle-class status symbol? *The Guardian*, 9 January 2019. https://www.theguardian.com/lifeandstyle/shortcuts/2019/jan/08/five-kids-club-status-symbol-celebrity-sophie-ellis-bextor.

consider below), if fertility falls to 2.1 and then stays stable, the population would keep growing until the last generation that outnumbers its parents starts to die off, and then it would stabilize. If fertility falls well below 2.1, as it has in most developed and middle-income countries, the peak is reached earlier and is followed by a gradual decline.

Given the numerous environmental crises escalating around the world, population decline should be a very welcome prospect. Nevertheless, it has been met with alarm tending toward panic in mainstream media. Elon Musk has claimed, "the biggest problem the world will face in 20 years is population collapse."[14] This is hyperbolic exaggeration: even after the global population peak occurs, which is much more than 20 years away, the decline will be gentle and much easier to adjust to than the rapid growth rates of recent decades. Nevertheless, many governments have implemented measures hoping to raise fertility rather than lower it. These include parental leave payments, childcare benefits, tax incentives and in some cases lump-sum payments to parents. Some countries, like Iran and Poland, are withdrawing women's access to contraception and abortions in the hope of boosting the birth rate.[15] Such coerced motherhood unwinds many decades of women's emancipation and inevitably places more children into poverty.

The most substantive reasons for 'birth dearth' concerns relate to the increasing proportion of elderly people in low-fertility countries. This is the product of both increased lifespans and the smaller size of the younger generations. It should be seen as a symptom of our success: the triumph of survival and longevity, women's emancipation and responsible parenting that ensures sufficiency for the next generation. There is little doubt that services for the elderly, including healthcare and old-age care, will be growth industries, but changing the composition of the economy over the course of decades is nothing new. Pension systems will also require a greater distribution of public funds. However, older people are also becoming healthier and tending to retire later, so these shifts might not be as great as many models anticipate.[16,17] Also on the plus side is much less expenditure needed to expand infrastructure, and a little less spending on education as children make up a smaller share of the population. If a country implements policies to boost population growth, by boosting either birth rates or immigration, the extra costs for infrastructure and education can be greater than the savings in pensions, health and care sec-

14 Cox, H., 2022. 5 Reasons America's Birthrate Is Plummeting. *FEE Stories*, June 6 2022. https://fee.org/articles/5-reasons-america-s-birthrate-is-plummeting/.

15 Scigliano, M. (2021). *Welcome to Gilead: Pronatalism and the threat to reproductive rights.* A Population Matters report. https://populationmatters.org/news/2021/11/welcome-gilead-how-population-fears-drive-womens-rights-abuses.

16 Gratton, L. and Scott, A. (2021) The 100-Year Life: Living and Working in an Age of Longevity. Bloomsbury GB 432 pp. ISBN: 9781526622839.

17 Sanderson WC and Scherbov S (2010). Remeasuring ageing. Science 329: 1287–1288. https://doi.org/10.1126/science.1193647.

tors from having slightly smaller proportions of elderly.[18] The extra environmental impact from having more people is usually not counted in such economic studies.

People of 'working age' (typically defined as aged between 15 and 64 years) will also form a slightly lower proportion of the population as countries age. Many commentaries fear there will not be enough workers to fulfil all the required jobs. However, to date there is no evidence that ageing is causing any contraction of employment. Instead, unemployment and underemployment are lower, and workforce participation is higher, in countries with more elderly and fewer working-age people. This is how economic theory expects a tighter labour market to respond, with businesses working harder to recruit, train and retain workers. Societal benefits of a tight labour market are enormous. Disadvantaged job-seekers, such as those with mild disabilities or poor language skills, find it easier to gain employment; minimum wages tend to keep pace with productivity gains instead of stagnating; income inequality lessens and crime rates fall. Housing might also become increasingly affordable after the population peak has passed, although this depends on a range of other circumstances and policy settings.

Big businesses have an interest in keeping wages low and property developers gain from increasingly costly land and housing. Both groups are often advocates for policies to boost population growth. Their narrow interests should not be confused with the public interest. In this case, the public interest in minimizing future greenhouse gases is fully aligned with its interest in well-paid jobs, affordable housing and less income inequality.

4.2.2 The Prickly Issue of Immigration

Many developed countries still have growing populations despite having had below-replacement fertility rates for many decades. This is because they have more immigrants than emigrants. The issue of migration has become increasingly divisive in developed countries, with opposition to expanded immigration often presented as racist, nationalist and callous toward people in need. The discourse often presents a false binary choice between being 'pro-immigrant' or 'anti-immigrant'. The real situation is that immigration is, in all developed countries, a quantitative issue. None have open borders: all impose quotas and criteria to limit the inflow and to influence its composition.

No matter how welcoming countries might wish to be toward migrants, open borders is not a viable option. Gallup poll surveys show that nearly 900 million adults wished to migrate in 2021, including more than one in three adults in sub-Saharan

18 O'Sullivan, J.N. (2020) *Silver tsunami or silver lining? – Why we should not fear an ageing population.* Discussion paper, Sustainable Population Australia. ISBN: 978–0-6487082-3–0 https://population.org.au/discussion-papers/ageing/.

Africa and Latin America.[19] This contrasts with some 3–4 million who successfully migrate each year, and a total world population of migrants estimated at 281 million in 2020 (around 3.5% of global population).[20] The most preferred destination countries are United States, Canada, Germany, France, Australia, the United Kingdom, and Saudi Arabia. For five of these seven countries, would-be immigrants outnumber their current population.[21] Clearly, an open borders policy would overwhelm the capacity of these countries to employ, house and provide basic services for such numbers. Negative impacts on the existing population would cause a political maelstrom.

Given that satisfying the global demand for migration is not an option, countries have to consider a wider range of issues in deciding how and at what level to limit migration. This is a wicked problem with difficult trade-offs, but it is unhelpful to brand this necessary discussion of limits as "problematizing immigration" that "encourages hostility" toward migrants.[22]

There is clearly a trade-off between the quantity of migration and the quality of migrant experiences: where immigration is high, migrants are more likely to encounter exploitative employment and hostility from the public. In this sense, low-immigration is pro-migrant.

Although migration discourse focuses on humanitarian aspects, most developed countries prioritize highly skilled migrants, potentially doing more harm than good by causing a 'brain drain' in underdeveloped countries. One commentary reported, "There are more Ethiopian physicians practicing in Chicago today than in all of Ethiopia, a country of 80 million."[23] The refugee quota is often a small fraction of total immigration despite its high humanitarian value. However, very large intakes of asylum seekers who lack language, education and cultural familiarity with the host country can lead to poor integration and entrenched disadvantage. Given the cost of resettling refugees, greater good might be achieved by spending these funds to help more people in their own country. At least, the opposite situation has occurred: Sweden cut foreign aid to help pay for refugee resettlement, after influxes of both Syrian and Ukrainian refugees.[24]

---

19 Pugliese, A. & Ray, J. (2023) *Nearly 900 Million Worldwide Wanted to Migrate in 2021.* Gallup, published online 24 January 2023. https://news.gallup.com/poll/468218/nearly-900-million-worldwide-wanted-migrate-2021.aspx.

20 IOM (2022) *World Migration Report 2022.* https://worldmigrationreport.iom.int/wmr-2022-interactive/.

21 Chamie, J. (2022) An Open Borders World. *Inter Press Service*, 9 June 2022. https://www.ipsnews.net/2022/06/open-borders-world/.

22 Mulvey, G. (2010) When Policy Creates Politics: the Problematizing of Immigration and the Consequences for Refugee Integration in the UK. *Journal of Refugee Studies* 23(4): 437–462. https://doi.org/10.1093/jrs/feq045.

23 Tulenko, K. (2010) Countries without Doctors? Foreign Policy, June 11, 2010. https://foreignpolicy.com/2010/06/11/countries-without-doctors/.

24 Chadwich, V. (2022) Sweden pulls $1B in foreign aid for Ukrainian refugees at home. Devex 05 May 2022. https://www.devex.com/news/sweden-pulls-1b-in-foreign-aid-for-ukrainian-refugees-at-home-103164.

Within developed countries, the 'migration debate' is often framed as one of compassion versus xenophobia, with no acknowledgement of the potential impacts on national population size, which can be profound (see Figure 4.2). Achieving ecological sustainability should be an overriding imperative. Immigration should therefore be at levels that allow population stabilization or contraction.[25] Each country has a duty of custodianship toward the endemic species and ecosystems within its borders, as well as to future generations. Population stabilization is compatible with some net migration (more immigrants than emigrants), as long as fertility is well below replacement rate. For instance, a fertility rate around 1.5 children per woman (roughly the European average) would allow net migration up to a quarter of each generation. This translates into an immigration rate around 0.3% of population per year (more than most European countries are used to). In contrast, in a country like Australia, fertility around 1.8 children per woman and immigration approaching 1% of population annually results in population growth around 1.5%, doubling the population in little more than half a century.

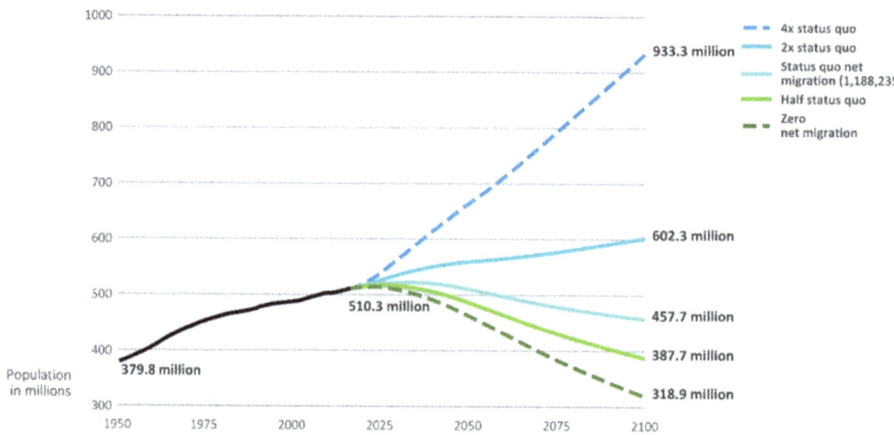

Figure 4.2: Projections of European population under varying levels of immigration. From Cafaro and Dérer (2018).[26]

It is often argued that migration doesn't change affect emissions because it doesn't change the number of people, just where they are. However, most migrants who move from poorer to richer countries increase their own greenhouse gas emissions as they adopt the

25 Cafaro, P. and O'Sullivan, J. 2019. How should ecological citizens think about immigration? *The Ecological Citizen* 3(1):85–92. https://www.ecologicalcitizen.net/article.php?t=how-should-ecological-citizens-think-about-immigration.

26 Cafaro P. and Dérer P. (2018) *New Policy-based Population Projections for the European Union, with a Consideration of their Environmental Implications*. Working paper, The Overpopulation Project, Gothenburg, Sweden. https://overpopulation-project.com/wp-content/uploads/2019/01/new-policy-based-population-projections-for-the-european-union.pdf.

lifestyle of the host country. By comparing per capita emissions in source and destination countries, studies in Canada[27] and Australia[28] estimated that migrants to those countries increased their emissions three- to fourfold. In addition, population growth itself can increase per capita emissions. The construction sector is particularly emissions-intensive, and as cities densify, timber and brick (relatively low emissions) give way to concrete and steel (higher embedded emissions). To provide services to ever-larger populations, municipalities resort to options that require vastly more energy and expense, such as road and rail tunnels, and water desalination.

Far from helping to ease population pressures in poor countries, support for high levels of migration to combat ageing can undermine political will for the reproductive health and family planning services in those countries. With ill-founded ageing catastrophizing constantly in media, emerging economies worry about "getting old before getting rich" and fear fertility falling too fast or too low.[29] A growing number of countries try to increase their birthrates, including some with high-fertility and very low proportions of elderly. In 2018, Tanzania's then-president Magufuli denounced contraception, saying women who don't want many children are lazy. He said "I have travelled to Europe and elsewhere and have seen the harmful effects of birth control. Some countries are now facing declining population."[30]

Such bizarrely misplaced concern is a product of the population taboo: although the negative effects of population growth vastly outweigh those of demographic ageing, it is deemed politically incorrect to attribute any harm to population growth. In this climate of self-censorship, most people are either unaware of, or have cultivated denial of, the connections between population growth, poverty, food and water insecurity, environmental deterioration and conflict. Tanzania's population has grown from 10 million at independence in 1961 to 66 million in 2022, and is forecast to reach 244 million by the end of this century. Ageing should be among the least of its concerns.

---

27 Grubel, H. and Grady, P. (2021) Immigration may make global net-zero harder. Financial Post, 23 Jan 2021. https://financialpost.com/opinion/opinion-immigration-may-make-global-net-zero-harder.

28 Bradshaw, C. J., & Brook, B. W. (2016). Implications of Australia's population policy for future greenhouse gas emissions targets. *Asia & the Pacific Policy Studies*, 3(2): 249–265, pp. 260, 261. https://doi.org/10.1002/app5.135.

29 Ziegenhain, P. (2021). Getting Old Before Getting Rich (and not Fully Realizing It): Premature Ageing and the Demographic Momentum in Southeast Asia. In: Goerres, A., Vanhuysse, P. (eds.) *Global Political Demography*. Palgrave Macmillan, Cham. https://doi.org/10.1007/978-3-030-73065-9_7.

30 BBC (2018) Tanzania's President Magufuli calls for end to birth control. *BBC News*, 10 September 2018. Available at https://www.bbc.co.uk/news/world-africa-45474408.

4.2.3 Climate Change Refugees or Overpopulation Refugees?

The term "climate change refugees" has been widely used in anticipation of many millions of people having to relocate from areas affected by sea level rise, drought or flood. The potential future scale of human displacement due to climate change is used as an argument for developed countries to be more open to migration in general, rather than as an argument to reserve some of their limited migrant-absorbing capacity for climate change refugees.

However, there is a tendency to overuse the term "climate change refugee", applying it to any migrant whose previous life has been impacted in any way by climate change, regardless of whether this was the main cause of their migration. It infers an obligation of developed countries to receive them, since developed countries contributed most to climate change. The pro-migration discourse rarely acknowledges population growth as a driver of out-migration, since this shifts responsibility for the migrants' plight from the receiving country to the sending country, and highlights that unwillingness to receive large numbers can be based on a rational desire to avoid population pressure rather than xenophobic nationalism.

Yet population growth is clearly the greater cause of relocation. A World Bank study estimated that, in sub-Saharan Africa, South Asia and Latin America, by 2050 more than 140 million people could relocate within their country due to effects of climate change.[31] The authors suggest that strong climate change mitigation measures could reduce this flow to around 50 million. However, these numbers represent a small proportion of the anticipated rural-to-urban migration due mainly to demographic pressure. UN projections suggest that well over 800 million people will move from rural to urban settings between 2020 and 2050, just in the three regions included in the World Bank study.[32]

Likewise, Lustgarden (2020) described a study using the same modelling framework as the World Bank report, which found that unmitigated climate change could increase international migration from Central America to USA by more than a million people between 2020 and 2050. But this represented less than four per cent of the total migrant flow of 30 million expected over this period. This migration, mainly driven by population pressure, was expected to increase each year regardless of climate change.[33]

---

31 Kumari Rigaud, K., de Sherbinin, A., Jones, B., Bergmann, J., Clement, V., Ober, K., Schewe, J., Adamo, S., McCusker, B., Heuser, S. and Midgley, A. (2018) *Groundswell: Preparing for Internal Climate Migration.* Washington, DC: The World Bank. https://www.worldbank.org/en/news/infographic/2018/03/19/groundswell—preparing-for-internal-climate-migration.

32 United Nations Population Division (2018) *World urbanisation prospects 2018.* United Nations Department of Economic and Social Affairs https://population.un.org/wup/.

33 Lustgarten, A. (2020) The Great Climate Migration. *New York Times Magazine*, 23 July 2020. https://www.nytimes.com/interactive/2020/07/23/magazine/climate-migration.html.

Kirezci et al. (2020) modelled the impact of climate change and sea level rise on land area exposed to coastal flooding events.[34] Only some of these areas would require permanent evacuation. They estimated that, by 2100 under the worst-case climate scenario, the population exposed to such flooding events would increase by 52%, from 148 million currently to around 225 million, based on current population distribution (not allowing for population growth). They emphasize that these figures could be lowered by the construction of protective infrastructure such as sea walls. Again, population growth on the most vulnerable islands and river deltas, particularly in South- and South-East Asia, will likely expose more extra people to this hazard than will climate change. Although coastal lowlands (less than 10 meters above sea level) occupy only 2% of global land area, they contain 10% of global population, and over 20% of the urban population of least developed countries – cities whose populations are doubling every few decades.[35]

These studies demonstrate the close interactions between climate change and other drivers of migration. Environmental, economic or security factors might be proximal drivers but, as a rule of thumb, if the population of the sending region is not permanently lowered as a result of emigration, the underlying driver is likely to be population pressure. In the case of small islands, out-migration has long been the strategy of surplus youth. If Pacific atolls are eventually evacuated due to sea level rise, the people who migrate at that time will be a relatively small addition to the diaspora already settled in Pacific Rim countries, evicted from their homeland due to population pressure.

Migration literature, particularly under the "new economics of labour migration" (NELM) theory, tends to ignore population growth as a driver of migration. Analyses typically present the decision of a household to send migrants as one of income diversification and self-insurance. Taylor (2002) sees rural-urban migration as a phenomenon driven by GDP growth and its implicit link with economic diversification, and suggests that constraints on local production and livelihoods are due to "market failures" such as inadequate market access, finance and insurance systems.[36] The presumption is that, without climate change or other exogenous factors undermining livelihoods, the economic situation would be stable or gradually improving due to development, and migration offers a means to enhance development. But nothing is stable where populations are growing. The climate migration literature does not discuss the common reality that the alternative to out-migration from rural areas is an ever-dwindling allocation of natural resources per household (arable land, water, or access

34 Kirezci, E., Young, I.R., Ranasinghe, R., Muis, S., Nicholls, R.J., Lincke, D. & Hinkel, J. (2020) Projections of global-scale extreme sea levels and resulting episodic coastal flooding over the 21st Century. *Nature Scientific Reports* (2020) 10:11629 | https://doi.org/10.1038/s41598-020-67736-6.

35 McGranahan, G., Balk, D., & Anderson, B. (2007). The rising tide: Assessing the risks of climate change and human settlements in low elevation coastal zones. *Environment and Urbanization*, 19, 17–37. https://doi.org/10.1177/0956247807076960.

36 Taylor J. E. (2002). The new economics of labour migration and the role of remittances in the migration process. *International Migration* 37(1): 63–88. https://doi.org/10.1111/1468-2435.00066.

to common forest, pasture or fishing resources),[37] and the inevitable degradation of those resources due to overuse.[38] Equally absent is any recognition that such subdivisions and degradations over the past two generations have contributed to the impoverishment of households and their vulnerability to climate change.

Often impacts of human overuse of resources are attributed to climate change, especially when they occur at a distance from the activity causing them. For instance, deforestation not only changes the hydrology of river catchments, increasing flooding and the seasonal variability of flows, but reduces rainfall downwind across whole continents.[39,40,41] Groundwater depletion contributed 13% of sea level rise between 2000 and 2008.[42] In addition, over-extraction of groundwater is causing widespread land subsidence, sometimes more than two meters per decade, affecting agricultural lands and coastal cities.[43] Climate change-related sea level rise is often cited as the cause of saltwater intrusion into groundwater of deltas and coastal plains,[44] where over-extraction of groundwater and expansion of aquaculture are mostly responsible.[45,46] This is not to belittle the vulnerability of atolls and coastal lowlands to inundation from sea level rise and related storm surge. But we should be mindful that climate change is not the only, and often not the biggest, cause of loss of livelihoods due to environmental change.

For migrants to be called climate change refugees, the sending region should be depopulated, reflecting its lower carrying capacity due to climate change. An absence

37 Garedew, E., Sandewall, M. & Soderberg, U. (2012) A Dynamic Simulation Model of Land-Use, Population, and Rural Livelihoods in the Central Rift Valley of Ethiopia. *Environmental Management* 49, 151–162. https://doi-org.ezproxy.library.uq.edu.au/10.1007/s00267-011-9783-4.

38 Taddese, G. (2001) Land degradation: a challenge to Ethiopia. *Environmental Management* 27(6):815–824. doi:10.1007/s002670010190.

39 Lawrence, D., & Vandecar, K. (2015) Effects of Tropical Deforestation on Climate and Agriculture. *Nature Climate Change* 5 (1): 27–36. https://doi.org/10.1038/NCLIMATE2430.

40 Ellison, D., Morris, C.E., Locatelli, B., et al. (2017) Trees, Forests and Water: Cool Insights for a Hot World. *Global Environmental Change* 43 (March): 51–61. https://doi.org/10.1016/j.gloenvcha.2017.01.002.

41 Pearce, F. (2018) Rivers in the Sky: How Deforestation Is Affecting Global Water Cycles. *Yale Environment 360*, 24 July 2018. https://e360.yale.edu/features/how-deforestation-affecting-global-water-cycles-climate-change.

42 Konikow, L. F. (2011), Contribution of global groundwater depletion since 1900 to sea-level rise, *Geophys. Res. Lett.* 38, L17401, https://doi.org/10.1029/2011GL048604.

43 Herrera-García, G., Ezquerro, P., Tomás, R. et al. (2021) Mapping the global threat of land subsidence. Science 371(6524): 34–36. https://doi.org/10.1126/science.abb8549.

44 Chen, J.J. & Mueller, V. (2018) Climate change is making soils saltier, forcing many farmers to find new livelihoods. *The Conversation*, 29 November 2018. https://theconversation.com/climate-change-is-making-soils-saltier-forcing-many-farmers-to-find-new-livelihoods-106048.

45 Mabrouk, M., Jonoski, A., Oude Essink, G.H.P. & Uhlenbrook, S. (2018) Impacts of Sea Level Rise and Groundwater Extraction Scenarios on Fresh Groundwater Resources in the Nile Delta Governorates, Egypt. *Water*, 10(11), 1690; https://doi.org/10.3390/w10111690.

46 Chang, S.W., Clement, P., Simpson, M.J. & Lee, K-K. (2011) Does sea-level rise have an impact on saltwater intrusion? *Advances in Water Resources* 34(10):1283-1291. https://doi.org/10.1016/j.advwatres.2011.06.006.

of population decline does not mean a community is unaffected by climate change, but it means the community has been able to adapt to live with that change. In the meantime, hardships caused by climate change and extreme weather events might have contributed toward many households' decisions to leave, but their place has been filled by local population growth. Without climate change, perhaps the region could have sustained an even bigger population thanks to other advances that increase opportunities for local livelihoods. But without the population growth, the same advances would increase incomes and climate resilience. We should not ignore impacts of population growth on migration, as migration literature tends to do.

We are left with the conclusion that climate change will be a contributing factor in net migration flows which are largely driven by population pressure. The total volume of international migrations will depend on the willingness of countries to receive migrants, more than on the factors motivating people to move. Climate change is likely to have some influence on who migrates, but little effect on the volume of international flows. Most people displaced due to climate change will move within their own country.

4.3 Family Size in underdeveloped Countries

Almost all the population growth projected to happen in the coming century will be in poor countries, especially in sub-Saharan Africa. These are the last countries to undertake what is known as the "demographic transition", in which high death rates and high birth rates are replaced with low death rates and low birth rates. Since WWII, death rates have plummeted everywhere, particularly for infants and mothers, but birth rates have fallen more slowly, barely at all in some countries, resulting in population growth rates never before experienced in human history. There are now six Nigerians for each one present in 1950; meanwhile Madagascar's population increased seven-fold and Somalia's nearly eight-fold.

Importantly, the rate at which fertility has fallen across the remaining high-fertility countries has slowed down since the 1990s. As a result, the United Nations (UN) have increased their estimate of global population growth this century. Virtually all the extra growth is happening in Africa: instead of peaking around two billion, as the UN anticipated in 2004, Africa looks on track to sail past four billion (Figure 4.3).

Since 2017, more recent UN projections have slightly reduced the expected growth to 2100. But this is despite revising upwards their estimates of growth up to now. Indeed, the 8 billion population milestone, estimated by the UN in mid-2022 to have occurred on 15 November 2022, was ten weeks earlier than they estimated in 2019. This was despite an extra 15 million deaths due to Covid-19. A shocking 177 million more people are present in mid-2022 than were expected by the UN in their 2010 projection. Given these repeated upward revisions of recent growth, it seems the downward revisions of future growth since 2017 are politically motivated rather than evidence-

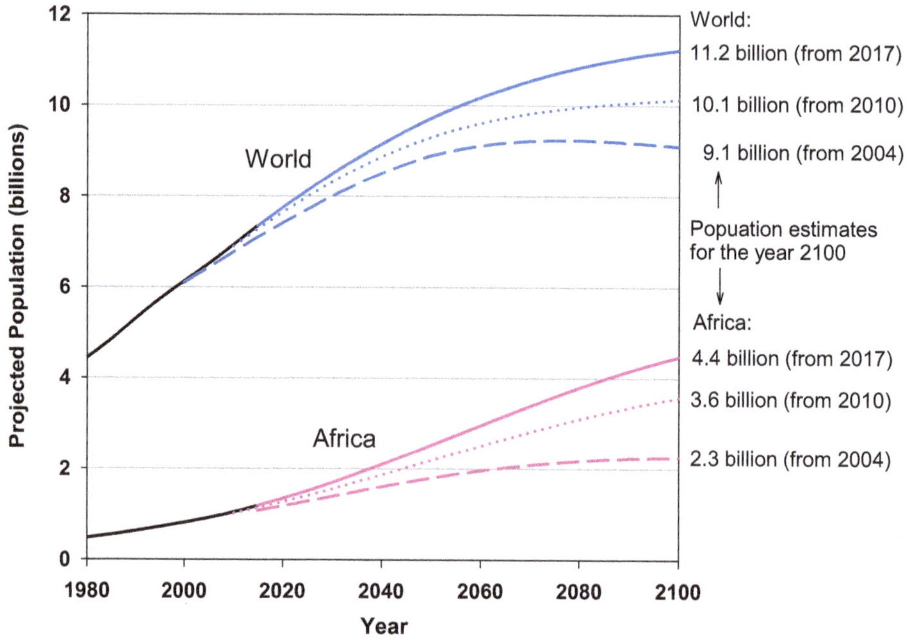

Figure 4.3: Population Projections from the United Nations 2004, 2010 and 2017 revisions.[47]

based.[48] The reassurance that world population growth will end in the 2080s has been used to dismiss population concerns, which the UN now calls "alarmist."[49] Yet the increasing food insecurity in many poor and increasingly common reports of near-famine conditions seem to justify a degree of alarm.

Apart from high-immigration countries (whose future immigration policies are hard to predict), almost all future population growth will happen in poor countries with very low greenhouse gas emissions per person. Hence, many people reject its relevance to climate change. They claim that including population growth in the climate change agenda serves only to "blame the poor" and deflect attention from consumption behaviors in rich countries. The inference is that poor countries would pay a penalty by lowering their population growth, and that "populationism" represents

47 Figure 4.3 shows the dramatic rise in expected outcomes since 2004 are mainly due to Africa's slower-than-anticipated fertility declines.

48 O'Sullivan, J.N. (2023). Demographic Delusions: World Population Growth Is Exceeding Most Projections and Jeopardising Scenarios for Sustainable Futures. *World* 4(3):545–568. https://doi.org/10.3390/world4030034.

49 O'Sullivan, J. (2022) The United Nations celebrates World Population Day by shaming population 'alarmists.' *The Overpopulation Project, 11 July 2022.* https://overpopulation-project.com/the-united-nations-celebrates-world-population-day-by-shaming-population-alarmists/.

an aggression against their interests and freedoms. This view entirely misrepresents the situation: high-fertility countries suffer from population growth and from the lack of support, from their own leaders and from the international community, for the programs and services that would slow it down. Women are burdened by frequent pregnancies they don't have the means to avoid. Poor families can't provide well for many children, and countries are unable to expand infrastructure fast enough to ensure basic services such as safe water, health services and school capacity.

In the context of climate change, reducing population growth is much more about climate change adaptation than mitigation. But it is more than that: in high-fertility countries, population growth is a far greater threat to food and water security and to peace and stability than is climate change.

4.3.1 Population Growth Heightens Vulnerability to Impacts of Climate Change

It is widely acknowledged that poor households in poor countries are most vulnerable to the impacts of climate change. This is partly because of their high reliance on the natural environment for their livelihood,[50] and partly because they live predominantly in equatorial zones likely to incur the greatest negative effects of climate change, from extreme heat[51] and disrupted rainfall patterns to intensified cyclones and flood events. But high population densities and growth rates also contribute to vulnerability.[52] With natural resources and human services stretched to their utmost to meet the needs of increasing numbers of people, the capacity to respond to crises is diminished and small disruptions to weather patterns can have large and cascading effects.

Much has been written on the threat of climate change to food and water security, but the role of population growth, and the potential to influence future population growth, often goes unmentioned,[53,54] even when the focus is on reducing food

50 Stern, N. (2006) *The Economics of Climate Change The Stern Review*. Cambridge UK, New York USA: Cambridge University Press.

51 Bathiany, S., Dakos, V., Scheffer, M. & Lenton, T.M. (2018) Climate models predict increasing temperature variability in poor countries. *Sci. Adv.* 4, eaar5809. https://doi.org/10.1126/sciadv.aar5809.

52 Das Gupta, M. (2013) *Population, Poverty, and Climate Change*. World Bank Policy Research Working Paper 6631. http://documents1.worldbank.org/curated/en/116181468163465130/pdf/WPS6631.pdf.

53 Beddington, J., Asaduzzaman, M., Fernandez, A., Clark, M., Guillou, M., Jahn, M., Erda, L., Mamo, T., Van Bo, N., Nobre, C.A., Scholes, R., Sharma, R., Wakhungu, J. (2011) *Achieving food security in the face of climate change: Summary for policy makers from the Commission on Sustainable Agriculture and Climate Change*. CGIAR Research Program on Climate Change, Agriculture and Food Security (CCAFS). Copenhagen, Denmark. https://ccafs.cgiar.org/publications/achieving-food-security-face-climate-change-summary-policy-makers-commission#.X03Yq8gzbIU.

54 Steiner, A., Aguilar, G., Bomba, K., et al. (2020) *Actions to transform food systems under climate change*. Wageningen, The Netherlands: CGIAR Research Program on Climate Change, Agriculture and

demand.[55,56] Indeed, the population taboo is evident in the disappearance of population as a focus in food security literature, from a central theme half a century ago to relatively rare today.[57] There is no doubt that the current food system is overreaching several planetary boundaries for sustainable impacts – by one estimate, current production and consumption patterns could sustainably provide a balanced diet for only 3.4 billion people.[58] Heroic shifts in production systems and dietary choices would be needed to allow sufficient food for 8 billion to be produced sustainably.[59] A study commissioned by The Lancet concluded that global food systems could provide healthy diets for up to 10 billion people by 2050 and remain within environmental boundaries, but it would take a global transformation of production systems, halving of food loss and waste, and red meat consumption reduced to about a third of current levels globally – all formidable challenges with low likelihood of achievement.[60] The authors conclude that, on current production systems and trends, emissions from food production will nearly double by 2050.

The regions most vulnerable to critical shortages of food and water tend to be those with high population densities and growth rates. In these regions, population growth is a much greater driver of water and food insufficiency than climate change. Modelling by Gunasekara and co-workers (2013) concluded that small reductions in population growth could have large effects on the numbers of people exposed to acute water stress.[61] Carter and Parker (2009, p. 676) evaluated threats to groundwater access in Africa, concluding, "The climate change impacts [on groundwater] are likely to be significant, though uncertain in direction and magnitude, while the direct and

Food Security (CCAFS). https://cgspace.cgiar.org/bitstream/handle/10568/108489/Actions%20to%20Transform%20Food%20Systems%20Under%20Climate%20Change.pdf.

55 Bajželj B, Richards KS, Allwood JM, Smith P, Dennis JS, Curmi E and Gilligan CA (2014) Importance of food-demand management for climate mitigation. *Nature Climate Change* 4:924–929. http://www.nature.com/nclimate/journal/v4/n10/full/nclimate2353.html.

56 Tilman, D., Balzer, C., Hill, J. and Befort, B.L. (2011) Global food demand and the sustainable intensification of agriculture. *PNAS* 108(50), 20260–20264. https://doi.org/10.1073/pnas.1116437108.

57 Tamburino, L., Bravo, G., Clough, Y. and Nicholas, K.A. (2020) From population to production: 50 years of scientific literature on how to feed the world. *Global Food Security* 24, 100346. https://doi.org/10.1016/j.gfs.2019.100346.

58 Gerten, D., Heck, V., Jägermeyr, J. et al. (2020) Feeding ten billion people is possible within four terrestrial planetary boundaries. *Nat Sustain* 3, 200–208. https://doi.org/10.1038/s41893-019-0465-1.

59 Conijn, J.G., Bindraban, P.S., Schröder, J.J., Jongschaap, R.E.E. (2018) Can our global food system meet food demand within planetary boundaries? *Agric. Ecosyst. Environ.* 251, 244–256. https://doi.org/10.1016/j.agee.2017.06.001.

60 Willett, W., Rockström, J., Loken, B., et al. (2019) Food in the Anthropocene: the EAT–Lancet Commission on healthy diets from sustainable food systems. *Lancet* 393, 447–492. https://doi.org/10.1016/S0140-6736(18)31788-4.

61 Gunasekara, N.K., Kazama, S., Yamazaki, D. & Oki, T. (2013) The effects of country-level population policy for enhancing adaptation to climate change. *Hydrol. Earth Syst. Sci.* 17, 4429–4440. https://doi.org/10.5194/hess-17-4429-2013.

indirect impacts of demographic change on both water resources and water demand are not only known with far greater certainty, but are also likely to be much larger. The combined effects of urban population growth, rising food demands and energy costs, and consequent demand for fresh water represent real cause for alarm, and these dwarf the likely impacts of climate change on groundwater resources, at least over the first half of the 21st century."[62]

Hall and co-workers (2017) modelled climate and population drivers of future food insufficiency, and concluded, "Very little to no difference in undernourishment projections were found when we examined future scenarios with and without the effects of climate change, suggesting population growth is the dominant driver of change."[63] Moreland and Smith (2012) found that even a modest increase in the rate of fertility decline in Ethiopia would negate the anticipated impacts of climate change on food insecurity in that country.[64] Thankfully, Ethiopia has since made considerable progress in extending and promoting family planning, as have Rwanda and Malawi, but most other tropical African countries are progressing more slowly.

In many areas, water scarcity will limit the further intensification of agriculture, leaving imports as the main source of increasing food supply. Food import dependence is growing in many high-fertility countries already. In terms of calories, internationally traded food has more than doubled in the past three decades.[65] An increasing number of countries depends on imports for more than 25% of their cereal needs.[66] In years with below-average yield, the price of internationally-traded cereals spikes up, placing severe stress on poor countries that depend on imported food. A flood or drought affecting major rice exporters such as Thailand and Vietnam could make food sufficiency unaffordable for hundreds of millions of West Africans: it was estimated that a 5% fall in the volume of traded rice could lead to prices rising 17%.[67]

62 Carter, R.C. & Parker, A. (2009) Climate change, population trends and groundwater in Africa. *Hydrological Sciences Journal* 54(4), 676–689. https://doi.org/10.1623/hysj.54.4.676.

63 Hall, C., Dawson, T.P., Macdiarmid, J.I., Matthews, R.B. & Smith, P. (2017) The impact of population growth and climate change on food security in Africa: looking ahead to 2050. *Int. J. Agric. Sustain.* 15, 124–135. https://doi.org/10.1080/14735903.2017.1293929.

64 Moreland, S. & Smith, E. (2012) *Modeling climate change, food security and population: pilot testing the model in Ethiopia.* Futures Group with MEASURE Evaluation PRH. https://www.measureevaluation.org/resources/publications/sr-12-69.

65 D'Odorico, P., Carr, J.A., Laio, F., Ridolfi, L. & Vandoni, S. (2014), Feeding humanity through global food trade. *Earth's Future* 2, 458–469. https://doi.org/10.1002/2014EF000250.

66 Gardner G. 2015. *Food Trade & Self-Sufficiency.* WorldWatch Institute – Vital Signs Report. https://farmlandgrab.org/24639.

67 Bren d'Amour, C., Wenz, L., Kalkuhl, M., Steckel, J.C. & Creutzig, F. (2016) Teleconnected food supply shocks. *Environ. Res. Lett.* 11 035007 https://doi.org/10.1088/1748-9326/11/3/035007.

A strong relationship exists between the global food price index and the incidence of violent unrest.[68] The 2022 food and energy price rises, largely triggered by the Ukraine war and its effects on supplies of fertilizers, wheat and vegetable oil, caused protests in many countries.[69] There is also the threat of countries banning exports in bad years, to ensure domestic needs are met first. Consequently, the adverse weather events caused by climate change, which might have caused ripples of hardship when most countries were self-sufficient, could turn into tidal waves of famine and violent conflict in overpopulated, import-dependent countries.

The relationship between natural resource scarcity and violent conflicts has been documented over many decades, in many countries.[70] Both the direct effect of population growth on the amount of land and water available per person, and the indirect effect of land degradation and water depletion due to overuse, contribute to this instability. Often the situation is enflamed by inequitable land titling and water access, with elites capturing the benefits of development such as irrigation infrastructure, disenfranchising communities who held traditional access. The Rwandan genocide was a case in point, escalating from severe land scarcity and power imbalances.[71] A 2003 study found that the risk of civil conflict was particularly elevated when land and water scarcity coincided with a youth bulge (a high proportion of those aged 15 to 29 in the adult population) and high rates of urban population growth.[72] Both of these stress factors are products of rapid population growth outpacing the growth in livelihood opportunities. These researchers found that "a decline in the annual birth rate of five births per thousand people corresponded to a decline of just over 5 percent in the likelihood of civil conflict." To put this in perspective, the birth rate in high-fertility countries is around 30 – 40 per thousand, while in low fertility countries it is around 10 per thousand.

This nexus between population growth and climate change has already been evident in recent unrest in the Middle East. High levels of population growth and unemployment

---

68 Lagi, M., Bertrand, K.Z., Bar-Yam, Y. (2011) *The food crises and political instability in North Africa and the Middle East.* New England Complex Systems Institute. http://arxiv.org/pdf/1108.2455.pdf.

69 Reuters (2022) Factbox: Surging food prices fuel protests across developing world. Reuters, 9 June 2022. https://www.reuters.com/markets/commodities/surging-food-prices-fuel-protests-across-developing-world-2022-05-18/.

70 Homer-Dixon, T. F., Boutwell, J. H., & Rathjens, G. W. (1993). Environmental change and violent conflict. Scientific American, 268(2), 38–45.

71 Gasana, J. 2002. Remember Rwanda? *World Watch Magazine* 15(5), 24–33. https://www.academia.edu/28702224/WORLD_at_BULLET_WATCH_WORLD_at_BULLET_WATCH_Remember_Rwanda_Working_for_a_Sustainable_Future.

72 Cincotta, R.P., Engelman, R. and Anastasion, D. (2003) *The security demographic: population and civil conflict after the Cold War.* Washington DC: Population Action International. https://pai.org/wp-content/uploads/2012/01/The_Security_Demographic_Population_and_Civil_Conflict_After_the_Cold_War-1.pdf.

heightened distrust between ethnic groups, laying the kindling for conflict.[73] Climate change played its role in a severe drought in Syria from 2007 to 2011,[74] which saw food imports rise steeply, and prices rose accordingly. This occurred at the same time as depleted groundwater was driving many farmers from their land, and as Syria's declining oil revenue was overtaken by its oil import needs. Oil revenues have enabled countries to increase their dependence on imported food, but they have a habit of running out. Middle East analyst Nafeez Ahmed argues that the converging effects of population growth, climate change, and fossil energy depletion in these chronically water-scarce countries are setting the stage for violent upheavals and failed states (see Table 4.1).[75]

The **Fragile States Index** scores countries on a range of political, social and economic indicators of vulnerability to political instability. The countries that scored highest for state fragility are all experiencing high rates of population growth.[76] Population density and growth rate have been described as challenge multipliers, as they exacerbate all environmental and social stresses. The stresses associated with climate change are only one dimension of this convergence of vulnerability in countries suffering high demographic pressure.

Greater resilience to the impacts of climate change is often argued as a co-benefit of measures intended to promote women's reproductive rights through family planning services.[77] Women are particularly vulnerable to climate change impacts, and the ability to regulate their childbearing enables their greater participation in livelihoods, natural resource management and community activities that build resilience.[78] Community-level effects of smaller families are also important, such as reducing pressure on natural resources, improving child nutrition and access to education, and simply adding fewer people to the numbers exposed to environmental crises.[79,80]

73 Friedman, T.L. (2013) Tell me how this ends. *New York Times*, 21 May 2013. http://www.nytimes.com/2013/05/22/opinion/friedman-tell-me-how-this-ends.html.

74 Kelley, C.P., Mohtadi, S., Cane, M.A., Seager, R. and Kushnir, Y. (2015) Climate change and the recent Syrian drought. *PNAS* 201421533; https://doi.org/10.1073/pnas.1421533112.

75 Ahmed, N.M. 2017. *Failing States, Collapsing Systems: Biophysical Triggers of Political Violence*. Springer. 94 pp. http://www.springer.com/us/book/9783319478142.

76 Population Institute (2015) *Demographic Vulnerability: Where Population Growth Poses the Greatest Challenges*. https://www.populationinstitute.org/demovulnerability/.

77 De Souza, R.-M.(2014) Resilience, integrated development and family planning: building long-term solutions. *Reproductive Health Matters*, 22(43), 75–83. https://doi.org/10.1016/S0968-8080(14)43773-X.

78 UNFPA (2009) *State of the world population 2009: Facing a changing world – women, population and climate*. United Nations Population Fund https://www.unfpa.org/sites/default/files/pub-pdf/state_of_world_population_2009.pdf.

79 Mogelgaard, K. (2018) *Challenges and Opportunities for Integrating Family Planning Into Adaptation Finance*. Population Reference Bureau, Washington DC. https://www.prb.org/wp-content/uploads/2018/03/Family_Planning_and_Adaptation_Finance_Full_Report_FINAL.pdf.

80 PRB and Worldwatch, (2014) *Making the connection: Population dynamics and climate compatible development recommendations from an expert working group*. Population Reference Bureau. http://www.prb.org/pdf15/population-climate-full-paper.pdf.

Table 4.1: Biophysical challenges (water scarcity, peak oil, population) for selected oil-producing nations.

Nation	x-Fold population increase 1960–2014	1962 Renewable freshwater per capita–m^3 yr^{-1}	2014 Renewable freshwater per capita–m^3 yr^{-1}	Peak oil year	2015 Barrels oil produced–1000s per day	2015 Barrels oil consumed–1000s per day	2015 Oil import dependence– %
Egypt	3.4	63	20	1995	493	824	63
Iraq	5.3	4,587	998		4,480	818	
Nigeria	4.2	4,690	1,245	2005	1,943	271	
Saudi Arabia	8.0	550	98		12,014	3,895	
Syria	4.1	1,456	380	1996	27	219	88
Yemen	5.5	393	86	2002	22	168	88

Source: Ahmed (2017), extra data from World Bank series EG.USE.PCAP.KG.OE. High water and energy insecurity are indicated in bold text. Moderate water shortage is deemed to occur when availability drops below 1700 m^3 cap−1 yr−1, and severe water shortage is deemed to be below 1000 m^3 cap−1 yr−1 (Kummu et al., 2016).

Kummu, M., Guillaume, J., de Moel, H. et al. (2016) The world's road to water scarcity: shortage and stress in the 20th century and pathways towards sustainability. *Sci Rep* 6, 38495. https://doi.org/10.1038/srep38495.

4.3.2 What Can be Done Ethically to Minimize Future Population Growth

Some people believe that falling infant mortality is the trigger that drives fertility decline, others believe it is increased economic development: urbanization, industrialization and reducing poverty. Others say it depends on the education and emancipation of women. Each of these factors contributes but, wherever birth rates fell rapidly, voluntary family planning programs have had by far the greatest influence.[81,82]

Countries as diverse as South Korea, Thailand, Mauritius, Iran and Costa Rica reduced fertility from around six children per woman to near two in the space of two decades. In Figure 4.4, the abrupt decline in fertility corresponds with the start of family planning efforts in each country. In each country, programs differed in response to local culture and specific barriers to family planning uptake.[83] They all made efforts to ensure contraception was accessible and affordable to people wherever they lived, but they also actively publicized the idea of contraception and the benefits of having fewer children through a range of media, from posters to television soap operas or enlisting religious leaders to endorse family planning.[84] Many also took measures to reduce child-bride marriages, promote wider spacing of children and increase girls' education and women's economic autonomy, acknowledging the synergy between women's empowerment and fertility reduction.

There are more than 200 million married women in underdeveloped countries who would like to avoid pregnancy but are not using effective contraception. However, a greater barrier to fertility decline is the persistent desire for large families, embedded by social norms (see Figure 4.5). In affluent countries, children are a significant financial burden; people are inclined to defer childbearing to establish their career and home, and limit births to ensure they can provide well for each child. In poor countries, children are often seen as both extra workers and a source of security for parents in old age. Women who are permitted no role in life other than raising children also gain prestige from a larger brood, particularly of sons. The advantage to children of having fewer siblings is generally not considered. But logical reasons are only a small part of the picture, since fatalism (it's in God's hands) and social norms (not wanting to be different to those around us) tend to make large families both ex-

81 de Silva, T. & Tenreyo, S. (2017) Population control policies and fertility convergence. *Journal of Economic Perspectives* 31(4) 205–228. https://pubs.aeaweb.org/doi/pdf/10.1257/jep.31.4.205.

82 Psaki, S.R., Chuang, E.K., Melnikas, A.J., Wilson, D.B. and Mensch, B.S. (2019) Causal effects of education on sexual and reproductive health in low and middle-income countries: A systematic review and meta-analysis. *SSM - Population Health* 8, 100386. https://doi.org/10.1016/j.ssmph.2019.100386.

83 Robinson, W.C. and Ross, J.A. (eds.) (2007) *The global family planning revolution*. World Bank, Washington D.C. 496 pp. ISBN-10: 0-8213-6951-2 https://openknowledge.worldbank.org/handle/10986/6788.

84 Parry, M. (2013) Broadcasting Birth Control: Mass Media and Family Planning. Rutgers University Press, 210 pp. ISBN-10: 0813561515.

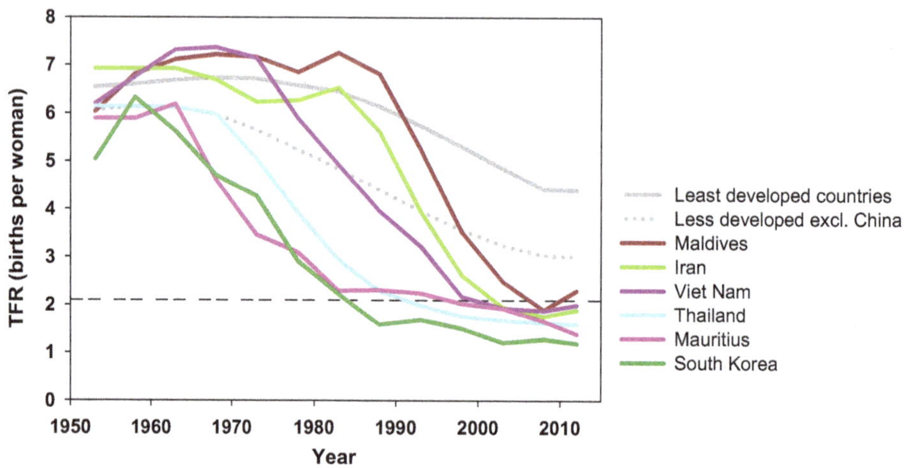

Figure 4.4: Time course of total fertility rate (TFR, average births per woman).[85]

pected and wanted.[86] The countries with the highest rates of population growth are generally those where tradition and poverty reinforce each other. Patriarchal traditions lead to large families, which result in resource scarcity and an inability to expand job opportunities, education and health services fast enough, leaving people poor and with neither access to family planning nor interest in using it.

Better education, changing attitudes to women's roles and rights, and urbanization all contribute to changing these social norms. But in countries that were most successful in reducing birth rates, the active promotion of small families and contraception methods greatly accelerated the change in attitudes. By reducing population growth, these countries were able to develop economically much faster than those where birth rates remain high. The correlation of fertility with education or incomes has causation in both directions, and the effect of family planning programs enabling better education access and poverty reduction is arguably stronger than the effect of education and enrichment on fertility.[87]

85 Figure 4.4 shows the TFR for selected countries which implemented population-focused voluntary family planning programs at differing times, showing rapid change in fertility, compared with slower decline across all less developed countries (excluding China), and for least developed countries.

86 Barrett, S., Dasgupta, A., Dasgupta, P., Adger, W.N., Anderies, J., van den Bergh, J., Bledsoe, C., Bongaarts, J., Carpenter, S., Chapin, F.S., Crépin, A.-S., Daily, G., Ehrlich, P., Folke, C., Kautsky, N., Lambin, E.F., Levin, S.A., Mäler, K.-G., Naylor, R., Nyborg, K., Polasky, S., Scheffer, M., Shogren, J., Søgaard Jørgensen, P., Walker, B., Wilen, J. (2020). Social dimensions of fertility behavior and consumption patterns in the Anthropocene. PNAS, 117 (12) 6300–6307; https://doi.org/10.1073/pnas.1909857117.

87 O'Sullivan, J.N. (2017). The contribution of reduced population growth rate to demographic dividend. 28th International Population Conference, Cape Town 30 Oct – 3 Nov 2017. https://iussp.confex.com/iussp/ipc2017/meetingapp.cgi/Paper/2521.

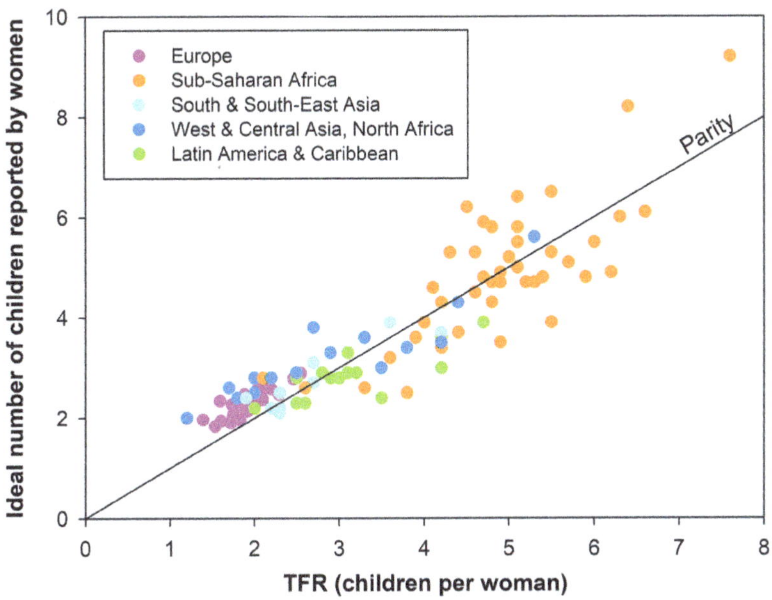

Figure 4.5: Family size is more about changing attitudes than changing access to contraception.[88]

The well-recognized relationship between poverty and high fertility led to the belief that economic development is a major driver of fertility decline. This belief is bolstered by the myth that population growth is economically neutral, since this myth dismisses the other explanation for the relationship: that population growth impoverishes, and easing it promotes economic advancement. To explore which direction of causation is more influential, Figure 4.6 contrasts the rate of fertility decline as a function of the prior level of wealth (Figure 4.6A), and the rate of income growth as a function of the prior level of fertility (Figure 4.6B). In Figure 4.6A, there is no evidence of fertility decline being enhanced by greater wealth: the poorest countries could reduce fertility as rapidly as any other, if they implemented appropriate programs. Conversely, in Figure 4.6B, we see that the level of fertility had a very strong influence on the likelihood of economic advancement. In countries with fertility above four children per woman, only a few oil-rich states sustained significant economic growth. Over a twenty-year period, all low-fertility countries increased GDP per capita, including those with shrinking populations.

An interesting phenomenon occurs when we compare countries that reduced fertility rapidly with those that didn't. In Figure 4.7, all the countries with fertility above five in 1950 are put into one of three groups according to the rate of their fertility tran-

88 The average family size countries achieve closely follows the number of children women say they want. Data from DHS surveys.

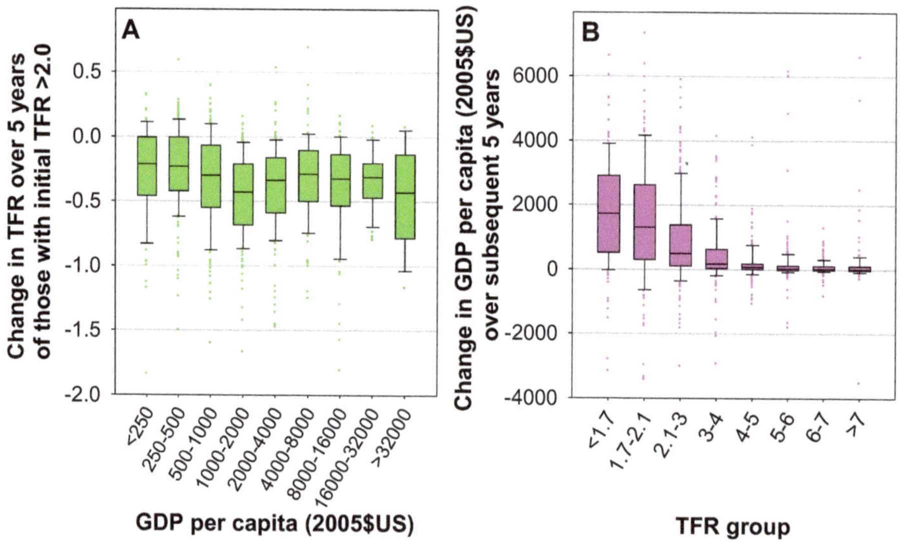

Figure 4.6: Exploring the direction of causation relating enrichment to fertility decline.[89]

sition. They are synchronized with respect to the start of their rapid fertility decline before averaging. Fast transition countries (achieving a fall of more than 3 units over a twenty-year period), all of which promoted family planning, are approaching a peak population around 2–2.5 times the population when they began fertility reduction efforts (Figure 4.7B). Those with intermediate transitions (1–3 units over two decades) have not reduced fertility as fast as the number of mothers has risen, so their population growth is not yet decelerating. The population of slow-transition countries has tripled over the same period, and still has at best another doubling in store due to demographic momentum, even if they adopt strong measures from now. In Figure 4.7C, we see a dramatic diversion of their economic progress. In Figure 4.4.7D, we find a remarkable commonality in the relationship between fertility and GDP per capita, all following the same path in which fertility falls first, and enrichment follows after. It appears that, barring exceptional mineral wealth, fertility decline is a prerequisite for sustained economic advancement. The faster fertility falls, the sooner countries move to middle income status.

89 Exploring the direction of causation relating enrichment to fertility decline: (A) the rate of fertility decline as a function of level of income, and (B) rate of economic development as a function of level of fertility. Data points represent each country in each five-year period between 1960 and 2010. All countries and time periods with available data are included. Box plots span 25 percentile, median and 75 percentile and whiskers extend to the 10th and 90th percentile. GDP per capita (inflation-adjusted 2005$US) are from the World Bank economic database, and fertility data from UNDESA (2015).

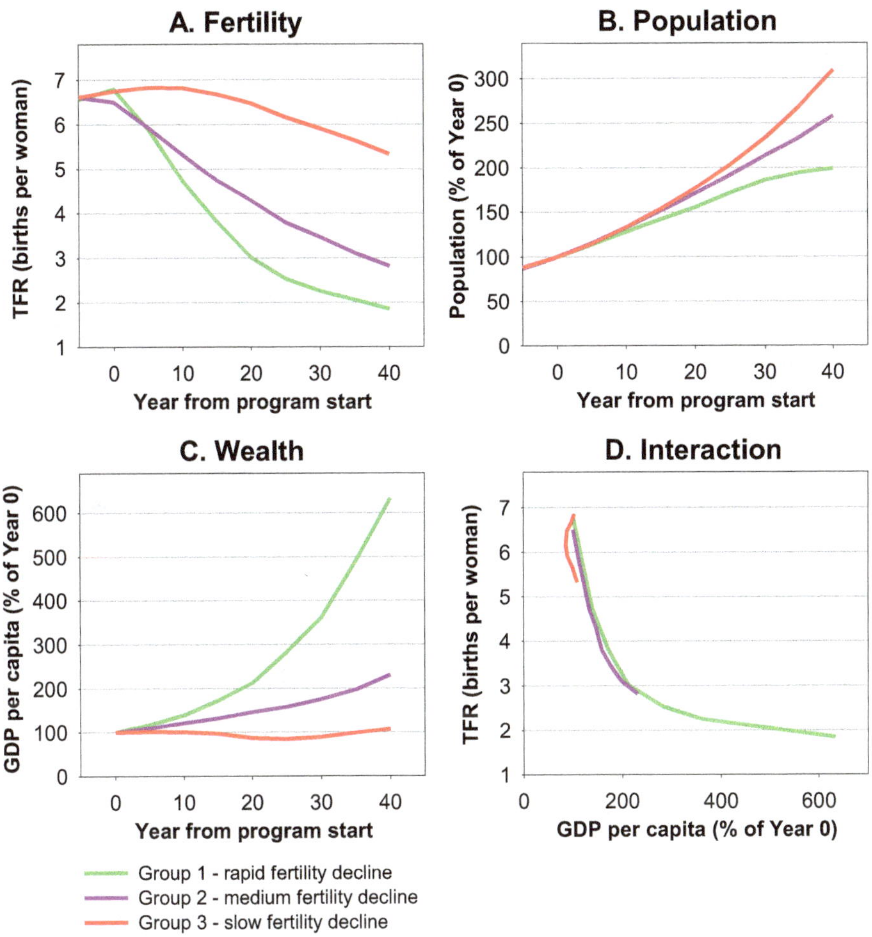

Figure 4.7: The average time-course for (A) fertility, (B) population, (C) GDP per capita (inflation-adjusted US$), and (D) the relationship between TFR and per capita GDP, for developing countries grouped according to the rate of their fertility transition.[90]

Voluntary family planning programs thus have a proven track-record for priming a virtuous cycle, in which smaller families lead to better household finances, education and gender equity, leading to smaller family preferences in the next generation. They are also much less costly than making significant direct impacts on poverty and education, and each dollar spent on family planning saves around three dollars in

90 Each of the rapid transition countries (Group 1) achieved a fertility decline greater than 3 units over twenty years under family planning programs. Group 2 countries had maximum falls between 1 and 3 units over twenty years. In Group 3 countries, fertility had not fallen by more than one unit in a 20 year period, up to 2010. Year 0 is the start of the fertility transition in each country, or 1970 for Group 3.

avoided health care for mothers and infants.[91] The economic stimulus from slowing population growth repays the investment more than a hundred-fold within a few years.[92] The same dollar liberates women from unwanted childbearing, saves lives of women and children, improves children's nutrition, education and employment prospects, enhances peace and security and eases pressure on natural resources and biodiversity. This makes family planning a 'best buy' for both development and the environment. The question should not be, "Is it ethical to intervene to lower population growth?" but "Is it ethical not to promote voluntary birth control to the best of our ability?"

Despite these immense benefits, reproductive health and family planning activities are chronically underfunded,[93] and shunned as too controversial by many aid agencies. In 2010 family planning received only 0.3% of European international aid (see Figure 4.8: 7% of 57% of 8% is 0.3%).

In 2012, the UK government and the Bill and Melinda Gates Foundation launched a campaign to rekindle international support for family planning, initiated through the 2012 London Family Planning Summit. The campaign aimed to halve the number of women with an unmet need for family planning by 2020, anticipating an extra 120 million contraception users in poor countries. The organization *Family Planning 2020* (now *Family Planning 2030*) was created to drive and monitor this agenda.[94] Some improvements in funding and service delivery were achieved, but by 2020 there were only 60 million additional contraception users, while the numbers of non-users increased due to population growth.

4.3.3 How Population Became Taboo

While never sufficiently funded to meet women's needs, the underfunding of family planning became dramatically worse after the 1994 United Nations Conference on Population and Development (ICPD).[95] Prior to 1994, the family planning agenda was regarded as an instrument enabling economic development. In most countries, services were entirely voluntary and focused on improving the health and rights of

91 Guttmacher Institute (2020). Adding it up: Investing in sexual and reproductive health in low- and middle-income countries. Fact Sheet. https://www.guttmacher.org/fact-sheet/investing-sexual-and-reproductive-health-low-and-middle-income-countries.

92 Kohler, H.-P. (2012). Copenhagen consensus 2012: Challenge paper on 'Population Growth.' PSC Working Paper Series. 34. https://repository.upenn.edu/cgi/viewcontent.cgi?article=1033&context=psc_working_papers.

93 Sully, E., Biddlecom, A., Darrock, J.E., Riley, T., Ashford, L.S., Lince-Deroche, N., Firestein, L. and Murro, R. (2020). Adding it up: Investing in sexual and reproductive health 2019. Guttmacher Institute. https://www.guttmacher.org/report/adding-it-up-investing-in-sexual-reproductive-health-2019.

94 The organization Family Planning 2020 has become FP2030. https://fp2030.org/.

95 Sinding, S.W. (2009). Population, poverty and economic development. Phil. Trans. R. Soc. B 364: 3023–3030. https://doi.org/10.1098/rstb.2009.0145.

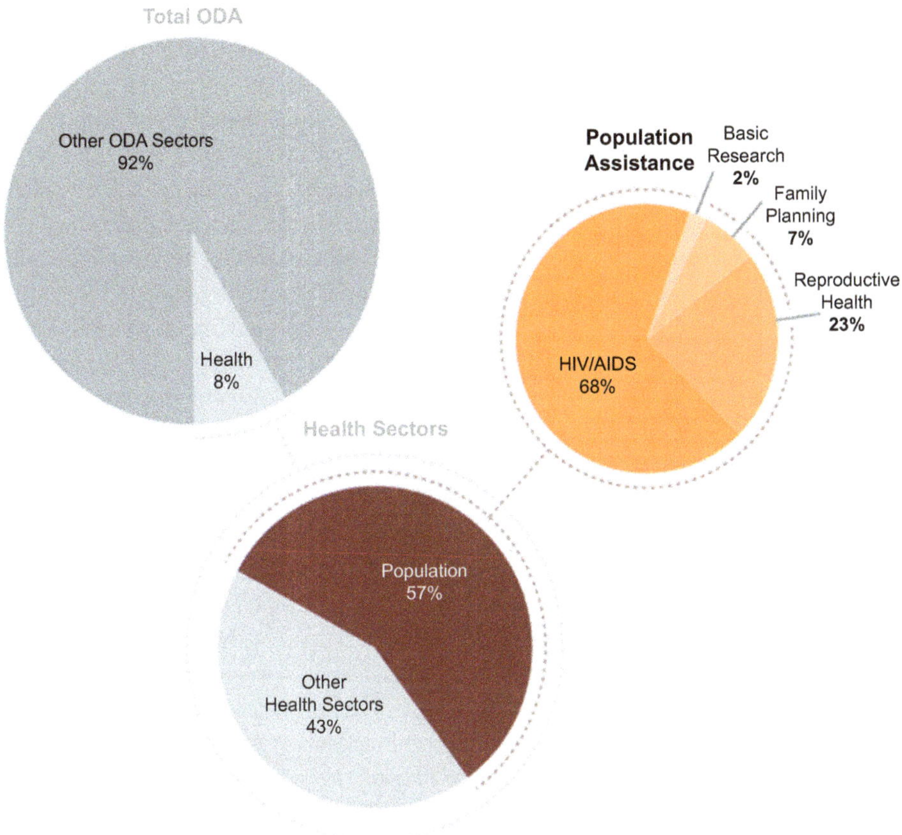

Figure 4.8: Distribution of European official development assistance to health sectors including family planning in 2010. In that year, the EU provided 63% of global development aid. Source: Pavao & Ongil (2011).[96]

women and infants, but the intention was to persuade couples, in their own interests, to limit childbearing with the aim of limiting population growth, knowing this was essential for reducing poverty and ensuring food security. However, in some countries, coercive measures were taken, including forced vasectomies in India in the late 1970s, and China's "one child policy" from 1979. A campaign leading up to the ICPD advanced the idea that any program explicitly aiming to reduce population growth would abet coercion and was incompatible with advancing women's rights.

96 Pavao, A. & Ongil, M. (2011). Euromapping 2011: Mapping European Development Aid & Population Assistance. German Foundation for World Population (DSW) and European Parliamentary Forum on Population & Development (EPF). https://www.europarl.europa.eu/meetdocs/2009_2014/documents/femm/dv/euromapping_2011_/euromapping_2011_en.pdf.

The campaign succeeded in reframing the UN's family planning agenda as serving women's reproductive health and rights exclusively, rather than ensuring (as the global family planning movement had always intended) that health, rights and population goals should be simultaneously and synergistically pursued. This shift demoted family planning from a central component of national development strategies to a minor activity of health departments. Funding and political interest in the provision of family planning services shrank,[97] and efforts to promote small family norms were abandoned.

The Vatican has been instrumental in the silencing of family planning and population discourse, from the very beginning of the World Health Organization in 1947[98] through its influence over USA policy from the 1970s[99] to its current privileged access and influence within UN headquarters.[100] The US Conference of Catholic Bishops has also been instrumental in orchestrating the anti-abortion campaign there.[101] A masterstroke has been to tar family planning with the same brush, despite contraception being the most effective method of reducing the incidence of abortion and criminalizing abortion the least effective. Indeed, abortion rates are highest in countries that ban abortion, because the same countries tend to make contraception difficult also.[102] But the Vatican has a lot at stake, since changing their stance on contraception would undermine the doctrine of papal infallibility on which much of the church's authority depends.[103]

The Vatican's influence has been evident in the gradual shift in rhetoric of the UNFPA, the UN agency created in 1969 to facilitate activities to end population growth. Since 1994, it has adopted increasingly euphemistic references to the benefits of reducing birth rates and the "link between population dynamics and development issues," eventually expunging all negative references to population growth and, more recently, actively denouncing "population alarmism."[104] UNFPA's 2022 World Population Day announce-

---

97 Sinding, S.W. (2009). Population, poverty and economic development. *Phil. Trans. R. Soc. B* 364: 3023–3030. https://doi.org/10.1098/rstb.2009.0145.

98 Mumford, S. (2012) Vatican Control of World Health Organization Policy: An Interview with Milton P. Siegel. Church and State. http://churchandstate.org.uk/2012/02/vatican-control-of-world-health-organization-policy/.

99 Mumford (1996) The Life and Death of NSSM 200: How the Destruction of Political Will Doomed a U.S. Population Policy. Church and State Press.

100 Rogers, B. (2021) Pro-natalism: The role of the Vatican. The Overpopulation Project Blog, 2 Nov 2011. https://overpopulation-project.com/pro-natalism-the-role-of-the-vatican/.

101 Bassett, L. (2011) The men behind the war on women. Huffington Post, 11 Feb 2011. https://www.huffpost.com/entry/the-men-behind-the-war-on_n_1069406.

102 Sedgh G, Bearak J, Singh S, Bankole A, Popinchalk A, Ganatra B, Rossier C, Gerdts C, Tunçalp Ö, Johnson BR Jr, Johnston HB, Alkema L. (2016) Abortion incidence between 1990 and 2014: global, regional, and subregional levels and trends. Lancet 388(10041):258–67. https://doi.org/10.1016/S0140-6736(16)30380-4.

103 Mumford, S. (2016) Infallibility and the Population Problem. Church and State. http://churchandstate.org.uk/2016/03/infallibility-and-the-population-problem-2/.

104 O'Sullivan, J. (2022) The United Nations celebrates World Population Day by shaming population 'alarmists'. The Overpopulation Project, 11 July 2022 https://overpopulation-project.com/the-united-nations-celebrates-world-population-day-by-shaming-population-alarmists/.

ment claimed "engineering population numbers has not proven successful in the past. Rather, it only serves to undermine human rights." This statement denies the many examples of successful voluntary family planning programs that vastly improved health, rights and economic conditions for women, children and whole countries.

The treaty agreed at the Cairo ICPD[105] did not belittle population concerns: it reaffirmed "interrelationships between population, resources, the environment and development." It advised, "To achieve sustainable development and a higher quality of life for all people, States should reduce and eliminate unsustainable patterns of production and consumption and promote appropriate policies, including population-related policies" and "Explicitly integrating population into economic and development strategies will both speed up the pace of sustainable development and poverty alleviation and contribute to the achievement of population objectives and an improved quality of life of the population." Family planning advocates welcomed the new Cairo text, to make more explicit opposition to coercion, commitment to women's health and rights, and more client-focused delivery of family planning. Few who signed onto the Cairo Agenda anticipated the subsequent deletion and delegitimization of all population focus from its implementation.

Cairo was the fifth decadal population conference convened by the UN. To prevent any challenge to this new agenda, no conference was held in 2004, 2014 or 2024. In 2014, a one-day session of the UN General Assembly was held "to assess the status of implementation of the Programme of Action and to renew political support for actions required for the full achievement of its goals and objectives." Despite the intended 20-year term of the Cairo treaty having expired, no renegotiation of the agenda was permitted. This session, on 22 September 2014, was sandwiched between UN Secretary-General Ban Ki Moon's Climate Summit on 23 September and the People's Climate March, the largest climate rally ever held, on 21 September. Needless to say, the population session was poorly attended and almost unreported.

The 2014 *Report of the Operational Review of the Implementation of the Programme of Action of the ICPD* contained one paragraph on population in the last chapter, declaring that earlier concerns about population growth

> "prioritized population control without heed to people's reproductive aspirations, their health, or the health of their children" and asserting "The Programme of Action reflected a remarkable consensus among diverse countries that increasing access to health and education, and greater human rights for women, including their reproductive health and rights, would ultimately secure a better social and economic future – **and also lead to lower population growth** – than targeted efforts for birth control. The evidence of 2014 overwhelmingly supports the accuracy of that consensus." (p. 223, original emphasis)

Yet the review contained no analysis of trends in population growth, nor of the impact of the UN's agenda on it. If it had, it would have revealed a marked slowing of

105 UNFPA (1994) Programme of Action adopted at the International Conference on Population and Development, Cairo, 5–13 September 1994. https://www.unfpa.org/sites/default/files/event-pdf/PoA_en.pdf.

fertility decline and a resurgence in global population growth since the mid-1990s.[106] This same document styled itself as the "Framework of Actions for the follow-up to the Programme of Action of the International Conference on Population and Development Beyond 2014", a framework that eliminated population entirely from its agenda. This population-free framework was subsequently endorsed at the 2019 *Nairobi Summit on ICPD25*, marking the 25th anniversary of the ICPD.

Various myths were promoted to help embed this perspective. Some denied the efficacy of voluntary family planning programs in reducing population growth. They maintained that poverty reduction and girls' education are the most effective ways to reduce fertility, citing studies that ignored the existence of family planning efforts. In fact, family planning program effort explains almost all of the variation among countries in their rate of fertility decline.[107]

Instrumental in undermining the humanitarian case for family planning in the 1980s was the economist Julian Simon, who argued that humans were "the ultimate resource," as our capacity for innovation would always find alternatives for scarce resources.[108] Simon's work was funded and promoted in the halls of power by the Catholic Church.[109] Despite transparently irrational arguments, he gained an almost cult-like following even among academic economists. Prior to Simon's influence,[110] there was near-consensus among economists that population growth would impede development and deepen poverty. As Lee (2009) articulates, "[it would seem] so obvious: Larger, more rapidly growing populations have fewer natural resources per person, less physical capital per worker, more dependents, and greater needs for new social infrastructure. Of course they must be economically worse off."[111] Population restraint as a development imperative was particularly championed within the UN by the USA government, until it backflipped spectacularly in the position it took to the 1984 UN Population Conference in Mexico City,[112]

---

106 Bongaarts (2008) Fertility Transitions in Developing Countries: Progress or Stagnation? *Studies in Family Planning* 39(2):105–10. https://doi.org/10.1111/j.1728-4465.2008.00157.x.

107 de Silva, T. & Tenreyo, S. (2017). Population control policies and fertility convergence. Journal of Economic Perspectives 31(4):) 205–228. https://pubs.aeaweb.org/doi/pdf/10.1257/jep.31.4.205.

108 Simon J.L. (1981) *The Ultimate Resource*. Princeton University Press.

109 Mumford, S.D. (2016) How the Catholic Church Undermined the Population Growth Control Movement. Church and State. http://churchandstate.org.uk/2021/03/how-the-catholic-church-undermined-the-population-growth-control-movement/.

110 Daly, H. (2003) Ultimate Confusion: The Economics of Julian Simon. *The Social Contract* 13(3): 194–197. https://www.thesocialcontract.com/artman2/publish/tsc1303/article_1144.shtml.

111 Lee R.D. (2009) *New perspectives on population growth and economic development*. Working Paper, University of California at Berkeley, Center on the Economics and Demography of Aging, http://www.ceda.berkeley.edu/Publications/pdfs/rlee/UNFPANewPerspectives09.pdf.

112 Finkle, J.L. and Crane, B.B. (1985) Ideology and Politics at Mexico City: The United States at the 1984 International Conference on Population. *Population and Development Review* 11(1): 1–28. https://doi.org/10.2307/1973376.

under the influence of Julian Simon and Catholic lobbying. The new adage became "with every mouth God sends a pair of hands" inferring production would always keep pace with population growth and dismissing the role of increasingly scarce natural resources.

Contention around the humanitarian motive and claims that family planning programs were "ineffective and even dangerous"[113] opened the opportunity to cast "population controllers" as misanthropic, racist, eugenicist or neo-colonial. Hence, it has become taboo to speak of population growth as a problem. Journalists, academic writers and NGOs avoided population for fear of being called racist or "blaming the poor". The myths have prevailed almost unopposed. Interest in the reasons for this taboo has generated a growing research literature.[114] Because of the taboo, the population issue has almost disappeared from development, environmental, climate change and food security literature.[115] Due to lower levels of funding and less attention to family planning, this reframing has not had the intended effect of elevating women's reproductive health and rights; it has done the exact opposite.

---

113 Davies, L. (2022) UN warns against alarmism as world's population reaches 8bn milestone. The Guardian, 18 Oct 2022. https://www.theguardian.com/global-development/2022/oct/18/global-population-growth-8-billion-unfdp-united-nations-warning-alarmism.

114 For examples, see:Campbell, M. (2007). Why the silence on population? *Population and Environment*, 28(4): 237–246. https://doi.org/10.1007/s11111-007-0054-5;Coole, D. (2013). Too many bodies? The return and disavowal of the population question. *Environmental Politics*, 22(2): 195–215. https://doi.org/10.1080/09644016.2012.730268;Coole, D. (2016). Population, environmental discourse, and sustainability. *The Oxford Handbook of Environmental Political Theory*, 274. http://dx.doi.org/10.1093/oxfordhb/9780199685271.013.35;Kopnina, H., & Washington, H. (2016). Discussing why population growth is still ignored or denied. *Chinese Journal of Population Resources and Environment, 14*(2): 133–143. https://doi.org/10.1080/10042857.2016.1149296;Betts, K. (2018). *Immigration and public opinion in Australia: how public concerns about high migration are suppressed.* Research report, The Australian Population Research Institute. https://tapri.org.au/wp-content/uploads/2016/04/Immigration-public-opinion-2018.pdf; Kuhlemann, K. (2019). The elephant in the room: the role of interest groups in creating and sustaining the population taboo. In *Climate Change Denial and Public Relations* (pp. 74–99). Routledge. https://www.taylorfrancis.com/chapters/oa-edit/10.4324/9781351121798-6/elephant-room-karin-kuhlemannWashington, H., Lowe, I., & Kopnina, H. (2020). Why do society and academia ignore the 'Scientists Warning to Humanity' on population?' *Journal of Futures Studies*, 25(1): 93–106. http://dx.doi.org/10.6531/JFS.202009_25(1).0009;Coole, D. (2021). The toxification of population discourse: A genealogical study. *The Journal of Development Studies*, 57(9): 1454–1469. https://doi.org/10.1080/00220388.2021.1915479.

115 Wolf, C., Ripple, W.J. & Crist, E. (2021). Human population, social justice, and climate policy. Sustainability Science 16: 1753–1756. https://doi.org/10.1007/s11625-021-00951-w;

Tamburino, L., Bravo, G., Clough, Y. & Nicholas, K.A. (2020). From population to production: 50 years of scientific literature on how to feed the world. Global Food Security 24: 100346. https://doi.org/10.1016/j.gfs.2019.100346.

4.4 How effective is the population lever?

If we were to treat population growth as a component of the response to climate change, how much could feasible, rights-based initiatives reduce future global population and how much would that change emissions, given that most of the reduction would be in countries with low per capita use of fossil fuels?

4.4.1 How Much Could We Bend the Curve of Future Population Growth?

Although most studies modelling the future assume population growth is a given, quite a wide range of plausible outcomes exist. Those depicted in Figure 4.9 are discussed below.

For many decades, the UN Population Division has been collating demographic data and formulating projections of future population growth. The UN's "medium fertility" projection is its best guess at future trends, and it publishes what it calculates to be the probable variation around this projection, as well as the "high fertility" and "low fertility" projections, which merely add or subtract half a child per woman to the fertility in each country, and are not considered probable. The UN's medium projection is based largely on the past relationships between the level of fertility in individual countries and the rate at which it has fallen. However, the UN's model doesn't seem to allow for either recalcitrant countries not yet embarking on steady fertility declines, or for mid-transition stalls.[116] As a result, over the past two decades of slower fertility decline, the UN has frequently overestimated the fertility fall and underestimated population growth.[117]

Modelling of future climate change mitigation scenarios has been coordinated internationally by the Intergovernmental Panel on Climate Change (IPCC) through the use of a set of "shared socioeconomic scenarios" (SSPs), which imagine different potential socio-political futures as contexts within which varying adaptation and mitigation actions can be applied.[118] The population projections used in the SSPs were supplied by the Wittgenstein Centre for Demography and Global Human Capital at Austria's International Institute for Applied Systems Analysis (IIASA). In contrast with

116 O'Sullivan, J. 2016. Population Projections: Recipes for Action, or Inaction? *Population and Sustainability* 1(1), 45–57. ISSN 2398-5496 https://jpopsus.org/full_articles/population-projections-recipes-for-action-or-inaction/.

117 O'Sullivan, Jane (2022) World population is growing faster than we thought. *The Overpopulation Project*, 4 August 2022 https://overpopulation-project.com/world-population-is-growing-faster-than-we-thought/.

118 O'Neill, B.C., Kriegler, E., Riahi, K. et al. (2014) A new scenario framework for climate change research: the concept of shared socioeconomic pathways. *Climatic Change* (2014) 122:387–400. https://doi.org/10.1007/s10584-013-0905-2.

the UN, they emphasize the role of human capital (particularly education attainment) in driving fertility decline.[119] The theory behind these projections is that future fertility declines will be more rapid than in the past, because the communities, and particularly the women, are better educated. Hence, the central projection (SSP2) lies well below the lower bound of the UN's range of probable outcomes, peaking around 9.4 billion in 2070 (see Figure 4.9). Because fertility is assumed to be dependent on human capital, and human capital development depends on economic progress and equity, the scenarios have higher population projections (SSP3) or lower (SSP1 and SSP5) according to whether economic development is, respectively, slower or faster than the central scenario.[120] While the SSP's demographers claim inclusion of the education factor makes their projections superior to those of the UN, the UN's demographers argue (correctly) that the impacts of education on fertility are incorporated in their model. If the proof of the pudding lies in the eating, we should note that the UN's medium fertility projection has substantially underestimated population growth over the past decade, making SSP2 even more wrong.[121] It would therefore be rash to put faith in these projections for the purpose of estimating future greenhouse gas emissions, climate change impacts and adaptation needs.

None of these projections examine the role of family planning efforts as drivers of fertility decline. In Figure 4.9, an attempt is made to do so. Two scenarios are modelled: a business-as-usual path in which failure to promote smaller families means that fertility in high-fertility countries continues on its current slow decline, and a proactive path in which all remaining high-fertility countries provide and promote voluntary family planning, and achieve the average path of fertility decline that was achieved by voluntary family planning countries in the past. The latter leads to a path similar to that of the UN's "low fertility" projection and IIASA's SSP1/SSP5 projection, without requiring fertility to fall as unrealistically low. (SSP1 expects Africa's fertility to be below 1.3 children per woman in 2100, whereas the UN's low projection assumes the average woman has half a child fewer than the medium projection in all countries, including those that already have very low fertility.)

The key to bending the population curve is greater political will to implement voluntary family planning programs. These programs should combine providing contraception and reproductive health services with communication campaigns to change social norms about family size and contraception acceptance. These activities should

119 KC, S. and Lutz, W. (2017) The human core of the shared socioeconomic pathways: Population scenarios by age, sex and level of education for all countries to 2100. *Global Environmental Change* 42, 181–192. https://doi.org/10.1016/j.gloenvcha.2014.06.004.

120 Jiang, L. (2014) Internal consistency of demographic assumptions in the shared socioeconomic pathways. *Popul Environ* 35:261–285. https://doi.org/10.1007/s11111-014-0206-3.

121 O'Sullivan, J. (2019) World Population Prospects, 2019 – good news or bad? *The Overpopulation Project*, 26 June 2019. https://overpopulation-project.com/world-population-prospects-2019-good-news-or-bad/.

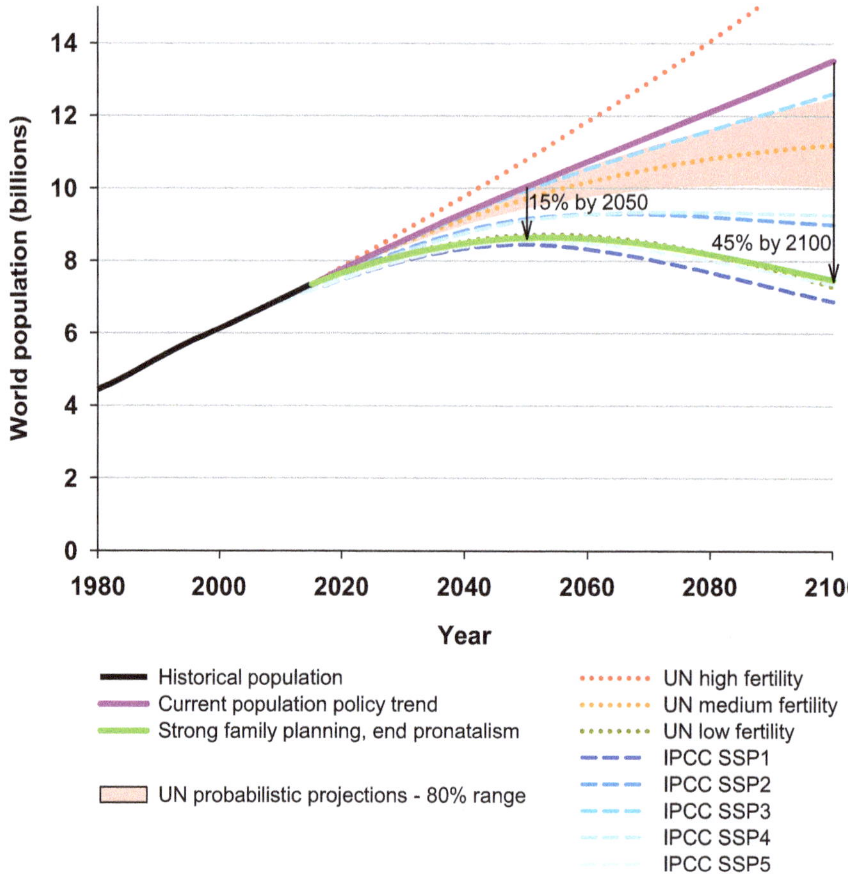

Figure 4.9: Policy-based projections of future global population.[122]

be an integrated part of national development plans (including climate adaptation plans). For this political will to emerge, governments and aid donors would need to acknowledge the profound benefits of minimizing further growth in human numbers, and the efficacy of direct, non-coercive approaches. Recent history suggests that waiting for the indirect drivers of education, urbanization and cultural globalization to shift social norms will be too slow.

122 Figure 4.9 compares outcomes if countries continue their recent trends —, or if remaining high fertility countries adopt strong family planning, achieving the average path that past family planning countries achieved, and assuming low-fertility countries abandon attempts to increase births —. These outcomes are compared with the UN projections (UNDESA 2015) and the IPCC's Shared Socioeconomic Pathways.

4.4.2 How Much Could Population Measures Contribute to Climate Change Mitigation

A study by the Global Center for Development asked whether providing much-needed additional funding for female education and family planning would be competitive with other options for climate change mitigation. Based on country-by-country estimates of emissions per person and costs per averted birth from family planning and female education, they found that "the population policy options are much less costly than almost all of the options for low-carbon energy development, including solar, wind, and nuclear power, second-generation biofuels, and carbon capture and storage. They are also cost-competitive with forest conservation and other improvements in forestry and agricultural practices."[123] In most of the 106 developing countries studied, the cost per tonne of carbon abated was less than $ 10. Interestingly, although expenditure on family planning averted more births per dollar than expenditure on education, the strong synergy between these two activities meant that dividing funding between them produced the lowest cost emissions reductions in their simulations.

Similarly, Wire (2009) estimated that providing family planning services to all women with an unmet need (then estimated at 215 million women globally), and thereby avoiding 75% of unintended births, would avoid 34 billion tonnes of carbon dioxide between 2010 and 2050 at a cost of less than $ 7 per tonne.[124]

Project Drawdown seeks to identify and promote the most impactful emissions reduction options available and model the scale of implementation needed to reverse the net accumulation to a net drawdown. Among the options it has analyzed is shifting global population growth from the UN's medium fertility projection to its low fertility projection. Its initial analysis, published in 2017, estimated the potential emissions avoidance at 119 Gt, the single biggest contribution among all the options assessed.[125] However, this shift in population was divided between two areas of intervention, family planning and education for girls, which consequently ranked 6th and 7th most impactful. In more recent updates, evidently responding to push-back against explicitly discussing population, the population effect is included under "health and

123 Wheeler, D., & Hammer, D. (2010). *The economics of population policy for carbon emissions reduction in developing countries.* Center for Global Development, Working Paper 229, November 2010. http://www.cgdev.org/publication/economics-population-policy-carbon-emissions-reduction-developing-countries-working.

124 Wire, T. (2009) *Fewer emitters, lower emissions, less cost: reducing future carbon emissions by investing in family planning – a cost/benefit analysis.* London School of Economics – Masters paper. https://www.srhr-ask-us.org/publication/fewer-emitters-lower-emissions-less-cost-reducing-future-carbon-emissions-investing-family-planning/.

125 Hawken, P. (ed.) (2017) *Drawdown: the most comprehensive plan ever proposed to reverse global warming.* Penguin Putnam Inc. 255 pp. ISBN: 9780143130444 https://drawdown.org/the-book.

education."[126] In Drawdown's scenarios integrating multiple interventions, as with SSP scenarios, the effect of a lower population is attributed to other interventions. For example, the deforestation avoided as a direct result of less population growth in tropical forest countries is attributed to 'Forest Protection'. Likewise, the lower demand for food reported under 'Reduced food waste' incorporates the lower demand for food resulting from a lower population than the baseline.[127]

Casey and Galor (2017) estimated that moving from the medium to the low variant of the UN global population projection could reduce emissions from fossil fuels and industry by 10% by 2050 and 35% by 2100, despite increasing income per capita.[128] For a similar shift in population trajectory, O'Neill et al. (2010) estimated emissions reduction around 15% by 2050 and 35–42% by 2100, factoring in emissions from the food system as well as effects of urbanization, household size and age structure, on lifestyle impacts, on a country-by-country basis.[129]

There is a fly in the ointment for these calculations, and that is the economic stimulus that reduced fertility levels are likely to promote. Extra consumption per person, particularly of fossil fuels and the products of their use, could counteract the emissions reductions from fewer births. If we compare successful family planning countries with similar countries that had slower fertility decline, we find that lower population growth does not mean lower total emissions. Thailand and the Philippines both had populations around 35 million in 1968. Due to Thailand's family planning program, Thailand's population will soon peak around 72 million. In contrast, the Philippines has already 116 million people and is expected to reach 180 million. Despite being initially poorer, less educated and less westernized than the Philippines, Thailand's GDP per capita is now almost twice that of its island neighbour, and its emissions per capita, at 4 t, is more than twice as great. Hence, despite the smaller population, the country's annual contribution of greenhouse gases was 276 Mt in 2018 compared with Philippines' 140 Mt.

The same pattern can be seen elsewhere: in 1977, when Bangladesh's family planning efforts started to reduce fertility, both Bangladesh and Pakistan had around 77 million people, and Pakistan was twice as rich. Now Bangladesh's population, at 170 million, trails Pakistan's by nearly 70 million and Bangladesh's GDP per capita overtook Pakistan's in 2016. However, while Bangladesh's emissions per capita have

126 Project Drawdown (2020) The Drawdown Review: Climate Solutions for a New Decade. https://www.drawdown.org/drawdown-review.

127 O'Sullivan, Jane (2020) Drawdown: a review of the Review. The Overpopulation Project blog, 06/05/2020. https://overpopulation-project.com/drawdown-a-review-of-the-review.

128 Casey, G., & Galor, O. (2017). Is faster economic growth compatible with reductions in carbon emissions? The role of diminished population growth. *Environmental Research Letters* 12, 014003. https://doi.org/10.1088/1748-9326/12/1/014003.

129 O'Neill, B. C., Dalton, M., Fuchs, R., Jiang, L., Pachaui, S., & Zigova, K. (2010). Global demographic trends and future carbon emissions. *PNAS* 107, 17521–17526.

grown faster than Pakistan's in relative terms, it is still only 0.5 t per person compared with Pakistan's 1 t. So the structure of the economy also matters, and GDP is not a good indicator of emissions. Nevertheless, Bangladesh's total emissions are probably more than they would have been, if it had not promoted family planning and now had a bigger but poorer population.

Given that those who argue against including population claim to be defending the poor, it is ironic that the only sound reason for dismissing the role of birth reduction efforts in climate mitigation is that they so powerfully enrich poor countries. If we don't support these measures for climate reasons, we should do so for equity reasons. We must strive to ensure that clean development strategies break this link between enrichment and emissions in the future. It is a positive sign that Thailand's emissions per capita peaked several years ago, but its GDP per capita keeps rising. Regardless of the effect on emissions, proactive birth reduction efforts should be supported for their impacts on food security, gender equity, poverty reduction, environmental protection and climate adaptation.

Climate mitigation modelling using the SSPs has shed only indirect light on the contribution of population growth to greenhouse gas emissions. As we saw above, each of the five "shared socioeconomic pathways" (SSPs) has a different population projection, but it is not possible to isolate the effect of lower or higher population because many other parameters also differ between the scenarios.[130] Within the model, fertility decline is assumed to be driven by investments in education and health. They assume that economic development causes fertility decline but not the other way around. Hence, the SSP framework does not lend itself to exploring the impacts of more direct investments in family planning.

Nevertheless, integrated assessment models (IAMs) using the SSPs do provide strong evidence that population is an essential component of successful climate change mitigation. One study applying the SSPs in six separate climate mitigation models found that even exceptionally high carbon prices and the most optimistic applications of emissions reduction options could not achieve less than 2°C of global warming using SSP3.[131] This was not due to high per capita consumption: SSP3 has by far the smallest global economy and per capita GDP, but it has the highest population. It should be noted that, since publication of the SSPs in 2014, global population growth has most closely followed the SSP3 path. Unless we make additional efforts to bend the population curve, according to these models, no feasible level of emissions reduction or carbon sequestration will suffice to prevent greater than 2°C of warming.

130 KC, S. & Lutz, W. (2017) The human core of the shared socioeconomic pathways: Population scenarios by age, sex and level of education for all countries to 2100. *Global Environmental Change* 42, 181–192. https://doi.org/10.1016/j.gloenvcha.2014.06.004.

131 Riahi, K., van Vuuren, D.P., Kriegler, E. et al. (2017). The shared socioeconomic pathways and their energy, land use and greenhouse gas emissions implications: An overview. *Global Environmental Change* 42, 153–168. http://dx.doi.org/10.1016/j.gloenvcha.2016.05.009.

4.4.3 Population Growth as A Driver of Land Use Change

The reason no climate mitigation model could contain climate change using SSP3 was found to be the inevitability of further deforestation to feed its larger population (Figure 4.10). Under scenarios that successfully transition away from fossil fuel use, land use change becomes a dominant determinant of climate outcomes. Agricultural expansion is the largest driver of deforestation,[132] and forest loss is closely correlated with the increase in rural populations.[133] In Africa, patterns of forest loss reveal a steady intensification from shifting to permanent cultivation. Under traditional management practices, small areas of forest were cleared and cropped for one or two years and allowed to regenerate forest during a fallow period of twenty or thirty years. Due to population pressure, fallow periods have progressively reduced so that forest remnants shrink and become degraded before being permanently cleared.[134] African researchers are increasingly forthright in identifying rapid population growth as "a major underlying force of environmental degradation and a threat to sustainable use of natural resources" as well as a driver of poverty, food insecurity, and colonization of forests and other natural habitats.[135] Elsewhere, "commodity agriculture" predominates, but this too is largely undertaken by rural smallholders migrating into the forest frontier to produce cattle, rubber, palm oil, rice and other cash crops, as well as illegal logging and charcoal production.[136] Globally, population growth has been the greatest driver of expansion of agriculture, although diet change in emerging economies (through more meat consumption) is also significant.[137,138] Thus, although only a tiny proportion of the world's population is directly involved in deforestation, regional population pressure is a push-factor, and global growth in food demand is a pull-factor, both acting to incentivize land clearing.

---

132 Carter, S., Herold, M., Avitabile, V., De Bruin, S., De Sy, V., Kooistra, L., & Rufino, M.C. (2018) Agriculture-driven deforestation in the tropics from 1990-2015: Emissions, trends and uncertainties. *Environmental Research Letters* 13: 1–13. https://doi.org/10.1088/1748-9326/aa9ea4.

133 FAO (2016) *State of the world's forests 2016. Forests and agriculture: Land-use challenges and opportunities.* Rome: FAO. www.fao.org/publications/sofo/en/.

134 Curtis, P. G., Slay, C. M., Harris, N. L., Tyukavina, A. and Hansen, M. C. 2018. Classifying drivers of global forest loss. *Science* 361:1108–11. https://doi.org/10.1126/science.aau3445.

135 Maja, M.M. and Ayano, S.F. (2021). The impact of population growth on natural resources and farmers' capacity to adapt to climate change in low-income countries. *Earth Systems and Environment* 5:271–283. https://doi.org/10.1007/s41748-021-00209-6.

136 Carr, D. 2009. Population and Deforestation: why rural migration matters. *Progress in Human Geography* 33(3), pp.355–378. https://doi.org/10.1177/0309132508096031.

137 Alexander, P., Rounsevell, M.D.A., Dislich, C., et al. (2015) Drivers for global agricultural land use change: The nexus of diet, population, yield and bioenergy. *Global Environmental Change* 35, 138–147. http://dx.doi.org/10.1016/j.gloenvcha.2015.08.011.

138 Henders, S., Ostwald, M., Verendel, V., & Ibisch, P. (2018) Do national strategies under the UN biodiversity and climate conventions address agricultural commodity consumption as deforestation driver? *Land Use Policy* 70: 580–590. https://doi.org/10.1016/j.landusepol.2017.10.043.

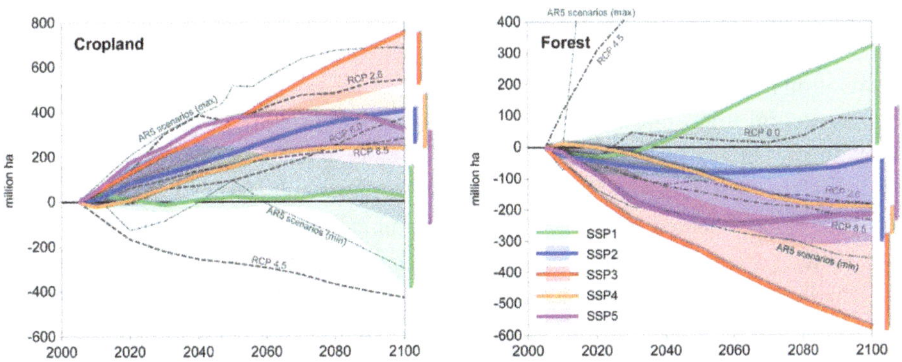

Figure 4.10: Changes in global cropland and forest areas for the SSP marker baseline scenarios (thick lines) and the range of other climate mitigation scenarios (coloured areas).[139]

Designation of protected areas is less effective at preventing forest encroachment by smallholders than by industrial projects.[140] In regions of acute land scarcity due to population growth, such as Kenya's Chyulu Hills National Park and Tanzania's Southern Highlands, it is not politically feasible to evict squatters from conservation areas.[141] In the north-western uplands of Cambodia, a 60% increase in agricultural area occurred between 2006 and 2016.[142] In Latin America, forest frontier migrants tend to have particularly high fertility, and their children are much more likely than other rural people to become frontier migrants themselves.[143] The construction of new roads into forests, often associated with mining or dam projects, provides access facilitating colonization by farmers.[144] Such infrastructure development has accelerated rapidly in recent years.[145]

---

139 Changes are shown relative to the base year of 2010 = 0. Dashed lines show projections from other studies for comparison. Note that cropland includes energy crops. Reproduced from Riahi, K., van Vuuren, D.P., Kriegler, E. et al. (2017). The shared socioeconomic pathways and their energy, land use and greenhouse gas emissions implications: An overview. *Global Environmental Change* 42, 153–168. http://dx.doi.org/10.1016/j.gloenvcha.2016.05.009.

140 Jayathilake, H.M., Prescott, G.W., Carrasco, L.R. et al. (2020) Drivers of deforestation and degradation for 28 tropical conservation landscapes. *Ambio*. https://doi.org/10.1007/s13280-020-01325-9.

141 Muriuki, G. (2016) Chyulu Hills burning reveals Kenya's squatter dilemma. *The Conversation*, 10/10/2016. https://theconversation.com/chyulu-hills-burning-reveals-kenyas-squatter-dilemma-65169.

142 Konga, R., Diepart, J.-C., Castella, J.-C., Lestrelin, G., Tivet, F., Belmain, E. & Bégué, A. (2019) Understanding the drivers of deforestation and agricultural transformations in the Northwestern uplands of Cambodia. *Applied Geography* 102, 84–98. https://doi.org/10.1016/j.apgeog.2018.12.006.

143 Lopez-Carr, D. & Burgdorfer, J. (2013) Deforestation drivers: Population, migration, and tropical land use. *Environment* 55(1) pp.3–11. https://doi.org/10.1080/00139157.2013.748385.

144 Laurance, W.F., Sayer, J., and Cassman, K.G. (2014) Agricultural expansion and its impacts on tropical nature. *Trends in Ecology & Evolution* 29: 107–116. https://doi.org/10.1016/j.tree.2013.12.001.

145 Laurance, W.F., Peletier-Jellema, A., Geenen, B. et al. (2015) Reducing the global environmental impacts of rapid infrastructure expansion. *Current Biology* 25: R259–R262. https://doi.org/10.1016/j.cub.2015.02.050.

Greenhouse gas emissions from agriculture-related deforestation were around 1.6 Gt CO_2 yr^{-1} in 2010–2015, and increasing rapidly in Africa.[88] Rapid growth of urban populations in Africa has also greatly increased demand for charcoal, which consumes several times more wood for the same cooking fuel than direct use of wood fuel, and generates significant amounts of methane, nitrous oxide and black carbon emissions in its production.[146] In addition, the exposed soil continues to lose carbon, and this is exacerbated by more frequent cropping and over-grazing resulting from population growth. The combination of soil carbon loss and deforestation contributed to a net loss of 1.65 Gt of carbon from Africa's tropical zone in 2016.[147] An estimated 116 Gt of carbon (425 Gt CO_2) have been released from soils over the history of agriculture.[148] Although increasing soil carbon has been widely promoted as a means of climate change mitigation, there are formidable technical, social and economic challenges to reversing soil carbon loss even in developed countries, and the prospects for net gains on a global scale are severely undermined by the growth in food demand.[101]

As mentioned earlier, deforestation and other reductions in vegetation also directly increase regional temperatures, and alter rainfall patterns locally and over substantial distances, as well as contributing to the seasonality of river flows and the severity of flooding.[31,32,33,149] Deforestation in the Amazon could reduce rainfall in the Midwest of USA, a vital grain exporter to the world. The drying already experienced in Africa's Sahel region has been exacerbated by deforestation in east and central Africa.[150,151]

The World Resources Institute estimated the extent to which various actions could reduce emissions from the global food system by 2050. Among the options considered was enhancing family planning to achieve replacement-rate fertility throughout Africa by 2050. This measure was estimated to contribute almost 1 Gt CO_2e yr^{-1} of emissions reductions from reduced deforestation, while saving an area of forest the

146 FAO (2017) *The charcoal transition: greening the charcoal value chain to mitigate climate change and improve local livelihoods.* Rome, Food and Agriculture Organization of the United Nations. http://www.fao.org/3/a-i6935e.pdf.

147 Palmer, P.I., Feng, L., Baker, D. et al. (2019) Net carbon emissions from African biosphere dominate pan-tropical atmospheric CO_2 signal. *Nat Commun* 10, 3344. https://doi.org/10.1038/s41467-019-11097-w.

148 Amundson, R. and Biardeau, L. (2018) Opinion: Soil carbon sequestration is an elusive climate mitigation tool. *PNAS* 115(46), 11652-11656. https://doi.org/10.1073/pnas.1815901115.

149 Mahmood, R., Pielke, R.A.Sr., Hubbard, K.G., et al. (2014) Land cover changes and their biogeophysical effects on climate. *International Journal of Climatology* 34(4), 929–953. https://doi.org/10.1002/joc.3736.

150 Zeng, N., Neelin, J.D., Lau, K.M. & Tucker, C.J. (1999) Enhancement of interdecadal climate variability in the Sahel by vegetation interaction. *Science* 286:1537–1540. https://doi.org/10.1126/science.286.5444.1537.

151 Zheng, X. & Eltahir, E. A. B. (1998) The Role of Vegetation in the Dynamics of West African Monsoons. *Journal of Climate* 11(8): 2078-2096. https://doi.org/10.1175/1520-0442(1998)011<2078:TROVIT>2.0.CO;2.

size of Germany.[152] While this contributed less than a tenth of the emissions avoided by 2050 from their full global menu of options, it was the only one offering much greater reductions, for no additional investment, in the latter half of the century. Other options, such as reducing meat consumption, or lowering methane emissions from cattle and rice paddies, are one-off changes that can't be repeated. In contrast, population contraction, if it occurs faster than our natural resource base is degrading, offers an ongoing dividend of increasing abundance for both people and wild nature.

4.5 Conclusions

Because population growth is occurring mainly in the world's poorest countries, reducing birth rates is far from being a remedy for climate change on its own. It is nevertheless an essential component of successful climate change mitigation. Choosing to have one child fewer remains the most powerful option available to people in developed countries to reduce future emissions.

It is a common fallacy to believe that reducing population growth must involve involuntary controls and human rights abuses. With around half of all pregnancies still unintended, there is considerable scope to improve reproductive rights by filling the unmet need for family planning services. Nevertheless, in many countries it is necessary to reduce desired family size, not through restrictions but by raising awareness of its advantages.

Given that a dollar spent on family planning services and promotion achieves cheaper greenhouse gas emission reductions than almost all other options, and the same dollar simultaneously empowers women, saves lives, improves child nutrition and survival, reduces environmental degradation, stimulates economic development in the world's poorest communities, reduces the extent to which other climate change responses are needed, and directly saves several dollars in unneeded health services, this surely represents the climate response's low-hanging fruit.

Climate change is likely to have only modest impacts on the size and distribution of human populations, other than in specific locations that are severely affected. In contrast, the extent of future population growth will have large impacts on both the vulnerability of communities to climate change and on the emission of greenhouse gases. With or without climate change, the security and prosperity of the world's poorest countries depend on minimizing further growth in their populations.

---

152 Searchinger, T., Waite, R., Hanson, C., Ranganathan, J., Dumas, P. & Matthews, E. (2018) *Creating a sustainable food future: A menu of solutions to feed nearly 10 billion people by 2050*. World Resources Institute, Synthesis Report, December 2018. https://www.wri.org/publication/creating-sustainable-food-future.

Policies and programs are available that would accelerate the reduction in fertility in those communities where it is still high, through voluntary measures that empower women and couples to take greater control over their lives and their children's wellbeing. If widely supported, they could lower global population by several billion people by the end of this century. The cost is modest: a doubling of current resourcing, from 1% to 2% of international aid, would make a substantial difference.[153] That such programs are, to date, absent from national and international climate change responses can best be explained by the prevalence of an irrational taboo against identifying population growth as a problem.

Ending population growth is only one of many important dimensions of a successful climate change response, but it is nevertheless essential for success. If global population growth meets or exceeds the UN's current medium projection, integrated assessment models suggest no feasibility of avoiding greater than 2°C warming. Lower projections, such as the IPCC's SSP2 projection, let alone SSP1, are very unlikely to be realized without substantially greater efforts to reduce desired family size and extend reliable and affordable access to contraception. The moral hazard of maintaining the population taboo has never been greater.

153 Bongaarts, J. (2016) Slow down population growth. *Nature* 530, 409–412. https://www.nature.com/news/development-slow-down-population-growth-1.19415.

Jennifer Kowallik and Alexandra Oboda

5 A Road to Sustainability for Economy and Society

5.1 Introduction

With the agreement reached at the UN Climate Change Conference in Paris in 2015 to limit global warming, the adoption of the 17 United Nations Sustainable Development Goals (UN SDGs), and the increasing and ever more ambitious commitments to climate neutrality by numerous economies, communities, and companies worldwide, it can be observed today that the global community is increasingly facing up to environmental and social challenges, but an enormous effort is still required to master the road to a resilient, livable, and more equitable future.

5.2 Executive Summary

This chapter explains the goal of a sustainable economy and society and the key milestones that have already been passed on the road there at global, European and national level. The context of sustainable development is discussed, and the sustainability strategies efficiency, consistency and sufficiency are presented as approaches to achieving the sustainability goals. Possible applications of the three sustainability strategies are exemplified using the sectors of energy, industry, transport, buildings as well as agriculture and forestry.

5.2.1 Main Challenges on the Road

Climate change is accelerating worldwide. Sea level rise, floods and droughts, biodiversity loss or damage to ecosystems because of climate change are on the rise. With the sixth assessment report of the Intergovernmental Panel on Climate Change (IPCC), it is unmistakably clear that the more global warming progresses, the greater the likelihood that these negative impacts will expand.[1]

Faster and increased actions are needed to limit global warming to 1.5 to 2 degrees – as agreed by the global community in the 2015 Paris Climate Agreement – according to the IPCC's Sixth Assessment Report on Climate Change Mitigation. Technologically and economically, it is still possible to limit long-term global warming to

1 UBA 2022.

https://doi.org/10.1515/9783110784299-005

1.5 degrees, it said. However, this requires an immediate global trend reversal and radical greenhouse gas reductions in all world regions and all sectors.[2]

In addition to ecological challenges, societal challenges must also be addressed. International political agendas, such as the 2030 Agenda with the 17 Sustainable Development Goals (SDGs) of the United Nations adopted by the global community of states in 2015, provide substantive orientation with regard to current societal challenges: these relate, for example, to global poverty, food security, health, education, gender equality, water and sanitation, the supply of sustainable energy and decent work.[3]

In the global risk report of the World Economic Forum 2022, growing inequality and social risks, currently further exacerbated by the COVID-19 pandemic and the war in Ukraine, are named as the main risks. Over the next five to ten years, however, the experts at the World Economic Forum expect ecological risks in the area of climate and the environment, such as extreme weather events or the loss of biodiversity, to dominate.[4]

Since fast action is now required, virtually all sectors are affected by a climate policy geared to the 1.5 to 2 degree limit. Either way, climate change will have significant consequences for companies' business activities in the coming years and decades: on the one hand, these will result from political, technological and social developments that will accompany the transition to a low-carbon economy that has been initiated. On the other hand, they result from the 'physical' effects of ongoing climate change.[5]

Like ecological challenges, social challenges are characterized by a high degree of complexity. They interact with the ecological challenges and in some cases reinforce each other. To achieve the goal of sustainable development, integrated approaches to solutions are therefore important. This article explains sustainability strategies and application examples for various economic sectors that can be used to implement the trend reversal (see Interrelationship of Sustainable Development and Examples of Application for the Sectors 5.3 and 5.6).

5.2.2 Goals for the Road to Sustainability

Today's understanding of the term 'sustainability' was significantly shaped by the final report of the UN Commission on Environment and Development, published in 1987, which includes what has become known as the Brundtland definition of sustainable development:

2 IPCC 2022.
3 UN 2015.
4 World Economic Forum 2022.
5 DGCN 2018, p. 2.

> Sustainable development is development that meets the needs of the present without compromising the ability of future generations to meet their own needs. It contains within it two key concepts:[6]
>
> the concept of needs, in particular the essential needs of the world's poor, to which overriding priority should be given; and
>
> the idea of limitations imposed by the state of technology and social organization on the environment's ability to meet present and future needs.

After many international conferences and mandatory targets on sustainability, such as the publication of the UN Millennium Development Goals (MDGs) in 2000, the global community adopted two landmark targets in 2015 that are valid today: the UN's 17 SDGs under the 2030 Agenda and the Paris Climate Agreement (see Figure 5.1).

Figure 5.1: The 17 Sustainable Development Goals (UN 2022).

The Agenda includes an action plan entitled 'Transforming our World: The 2030 Agenda for Sustainable Development', which takes up the UN MDGs. The 2030 Agenda provides guidelines for transforming the world in ways that enable economic progress consistent with social justice and within the Earth's ecological limits. The preamble states the following:

> The 17 Sustainable Development Goals and 169 targets which we are announcing today demonstrate the scale and ambition of this new universal Agenda. They seek to build on the Millennium Development Goals and complete what they did not achieve. They seek to realize the human rights of all and to achieve gender equality and the empowerment of all women and girls. They

6 WCED 1987, p. 41.

> are integrated and indivisible and balance the three dimensions of sustainable development: the economic, social and environmental.[7]

The 17 SDGs with their 169 subgoals are based on the three-pillar principle of sustainability and encompass all three dimensions: economic, social and environmental. A particular focus of the SDGs is on disadvantaged and discriminated population groups, and a key message is to 'leave no one behind' on the path to sustainable development.

5.2.3 Framework: Developments and Standards for the Topic of Sustainability

Since the 1980s, there have been many measures and activities at the global, European and national levels that have been dedicated to creating sustainable development. Major milestones as well as accompanying events on the global level are presented with this chapter (see Figure 5.2). The activities of the German government are used as an example for the national level.

5.2.4 Global Level

The UN Conference on Environment and Development in Rio de Janeiro in 1992 recognized sustainability or sustainable development as a normative, international guiding principle for the international community, the world economy, world civil society and politics. In principle, the conference focused on all areas of life, in particular the reorientation of production and consumption towards sustainability in the industrialized countries, and the fight against poverty in the developing countries. With the adopted Agenda 21, the states were called upon to follow words with actions and to develop national sustainability strategies.[8]

The Kyoto Protocol, signed in 1997, contains for the first time legally binding limitation and reduction obligations for CO_2 emissions for the industrialized countries and is a milestone for the implementation of the UN Framework Convention on Climate Change (UNFCCC). The protocol was to enter into force as soon as at least 55 countries, which together accounted for more than 55% of CO_2 emissions in 1990, had ratified the agreement. For a long time, the second condition prevented the agreement from coming into force, as major emitters refused to ratify it – for example, the

7 UN 2015, p. 1.
8 UNCED 1992.

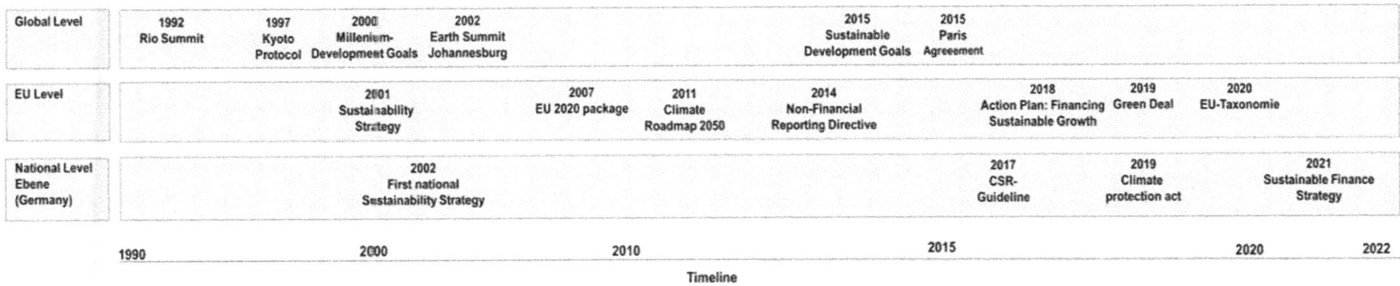

Figure 5.2: Timeline – Global, European and German Developments (figure created by author, sources are noted in the chapter).

USA since Republican President George W. Bush took office in 2000. After ratification by the Russian Duma, the Kyoto Protocol finally came into force in 2005.[9]

In 2000, the Millennium Summit of the United Nations adopted the Millennium Declaration, a catalogue of fundamental, binding goals – the MDGs – for all member states. The eight international development goals were primarily aimed at reducing poverty in developing countries, to be achieved by 2015. During the period of the MDGs, progress was made and the living conditions of many people improved significantly, but not all goals were achieved. In order to build on successes, learn lessons and continue collective efforts, the MDGs served as the basis for the SDGs developed with the 2030 Agenda.

Two years later, at the World Summit on Sustainable Development in Johannesburg in 2002, Agenda 21 was supplemented by the Johannesburg Action Plan, which contains further implementation-oriented recommendations for action and targets on the entire spectrum of sustainable development issues, in particular globalization and the environment, biodiversity, chemicals, water, basic sanitation and energy. Above all, however, the Johannesburg Summit provided the impetus for the rapid expansion of renewable energies.

In 2015, more than 20 years after the Rio Summit, the global community agreed on the 17 SDGs (see also "Goals for the Road to Sustainability") and was able to achieve even more concrete results in the same year with the Paris Climate Agreement.

With the decision of the UN climate agreement in Paris on December 15, 2015, the involved agreed to limit the increase in global mean temperature to below 2 degrees above pre-industrial levels and to make efforts to minimize this increase to as low as 1.5 degrees. To achieve this, each party is to prepare, communicate and comply with its targeted 'nationally determined contributions'. In this way, the international community has set the course for initiating a transformation of today's predominantly fossil fuel-based economy to net zero emissions in the second half of the century. The Paris Climate Agreement also linked the issue of sustainability to the financial system. For example, Article 2, paragraph 1c states: 'Making finance flows consistent with a pathway towards low greenhouse gas emissions and climate-resilient development.'

As part of this discussion, however, it was also stated in Paris that the redirection of financial flows cannot be reduced to climate issues alone, but must also be extended to the entire issue of sustainable development and take into account the 17 UN SDGs.[10]

5.2.5 EU Level

In order to implement sustainable development successfully, it must also be regulated internationally. Within Europe, the European Union has an important role to play

9 UN 1998.
10 Thimann 2019, p. 65.

here. Through programmes and treaties it sets a uniform direction for sustainable policy in the European countries.

On 15.05.2001, the first EU Sustainable Development Strategy (EU-SDS) was signed. The strategy contains a number of concrete propositions on how the European Union can improve its policies to make them more coherent and longer-term, as well as a number of important objectives and specific measures to achieve them. In particular, the strategy addresses the following issues: limiting climate change and increasing the use of clean energy; dealing with public health threats; using natural resources more responsibly; improving the transport system and optimize land use. The EU's sustainability strategy represents a cross-sectoral policy approach in which economic, social and environmental policies go hand in hand.[11]

The financial sector is seen as having a strategic role to play in achieving the global climate change and sustainability goals, and for this reason the EU published the Financing Sustainable Growth Action Plan in March 2018. The plan's measures aim to increase the involvement of the financial sector in raising investment for action to meet the goals of the Paris Climate Agreement and the UN 2030 Agenda. To meet the EU's climate and energy targets by 2030 alone, Europe will need to make up an annual investment shortfall of nearly € 180 billion.[12] The financial sector is expected to successfully support the sustainability transformation of the economy and society.

One of the fundamental measures from the EU Action Plan is the Sustainability Taxonomy Regulation ([EU] 2020/852), which entered into force on 18 June 2020.[13] In a classification system, the EU Commission thus defines in technical detail and in a binding manner which activities may be considered sustainable. This taxonomy thus provides clarity on what is meant by sustainable and climate-friendly financing. The taxonomy is intended to help distinguish sustainable activities from 'non-sustainable' ones. To date, the taxonomy has focused solely on the environmental pillar of sustainability, thus mapping almost exclusively environmental goals and climate change issues. However, the extension of the taxonomy to include social issues is already under development.[14]

Another measure from the EU action plan is the Corporate Sustainability Reporting Directive (CSRD). This is meant to replace the previously applicable Non-Financial Reporting Directive (NFRD) and tighten up sustainability reporting requirements. The CSRD obliges all large companies and stock-listed small and medium-sized enterprises in the EU to publish a comprehensive sustainability report. Large companies are defined as those that exceed two of the three size criteria – 250 employees, 40 million euros in annual sales and 20 million euros in total assets. This gives sustainability reporting a more significant role.

11 European Commission 2019.

12 European Commission 2018, p. 1.

13 Bassen and Lopatta 2020, p. 4.

14 Zhuang et al. 2020, p. 8.

5.2.6 National Level – Germany

In 2002, about 10 years after the Rio Conference, the German Federal Government published its National Sustainability Strategy, which has been regularly updated and revised since then. With the strategy, the German government sets out to make sustainability the fundamental principle of its policies.[15] Since the strategy was published, a number of measures have been taken in recent years to work toward improving its effectiveness. With the new edition in 2016, the sustainability strategy refers to the UN's 17 Sustainable Development Goals (SDGs) and breaks them down to national level.[16]

On November 15, 2019, the German Bundestag passed the Federal Climate Protection Act. The Act makes Germany's climate protection targets for 2030 legally binding for the first time. This has created the legal framework for climate protection in Germany that has long been called for. It is the first climate protection law to be enacted at the federal level. The purpose of the law is to ensure that national climate protection targets are met and that European targets are met. However, in May 2021, the Federal Constitutional Court declared parts of the Climate Protection Act unconstitutional following a lawsuit by various environmental associations – paving the way for more ambitious climate targets at the highest level. The German government immediately announced improvements.

In response to the EU's Sustainable Finance activities, the German Federal Cabinet adopted the German Sustainable Finance Strategy in May 2021. It focuses on financial market policy and is an important component of German sustainability policy. The aim of the strategy is to develop Germany into a leading location for sustainable finance.[17]

5.3 Interrelationships of Sustainable Development

The definition of sustainable development, which was coined by the Brundtland Report from 1987 (cf. previous section "Goals for the Road to Sustainability") and adopted by the 'Agenda 21' action programme at the 1992 environmental conference in Rio de Janeiro, involves all social groups (including society, politics, the economy) in the decision-making processes for implementing the guiding principle of sustainable development. Figure 5.3 shows a selection of the key challenges (cf. previous section 5.2.1) the interplay of the three-pillar model based on the 17 SDGs for the areas of society (social), environment/nature (ecology) and economy (economy) (cf. previous

15 Die Bundesregierung 2022.
16 Die Bundesregierung 2017.
17 Die Bundesregierung 2021b.

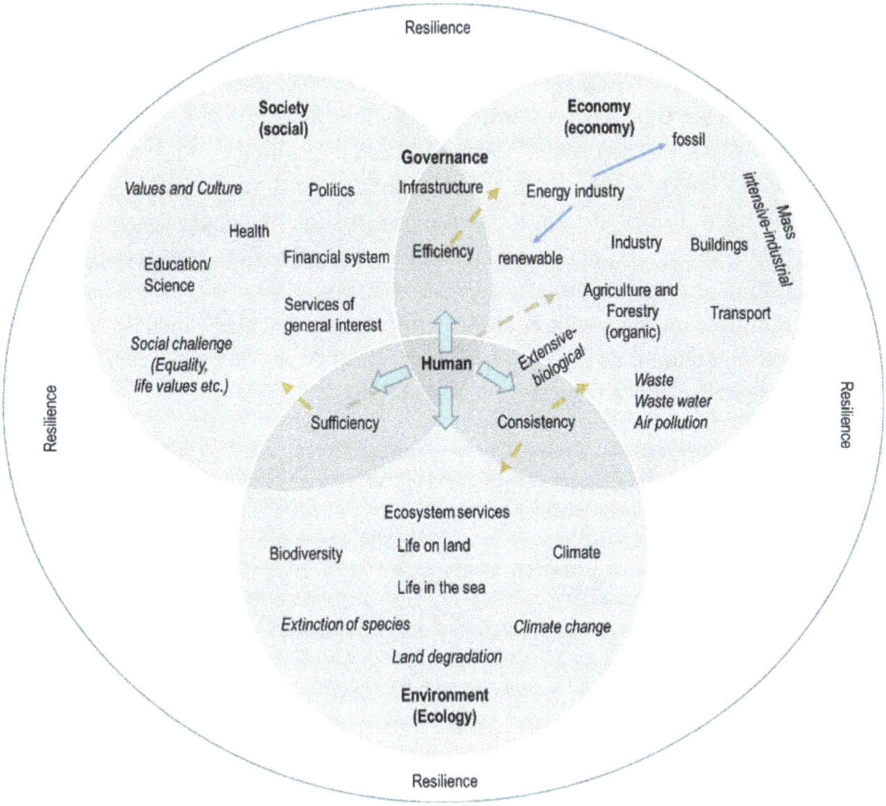

Figure 5.3: Interrelationships and interdependencies on the path to sustainable development (figure created by author, based on Haase, 2021[18]).

section 5.2.2) as well as the sustainability strategies efficiency, consistency and sufficiency that will follow in section 5.5.

It is important to emphasize here that this diagram only depicts one part of a complex system and that further influences and interrelationships can be added, for example, on a country- or sector-specific basis. Sustainable development requires the broad participation of various stakeholders. Thus, at the centre is the individual human being, from whom various positions in society and the economy are perceived that further affect the associated influences/effects. In the context of transformative development processes, the guiding principle of sustainable development thus presents interactions or conflicting goals.[19,20] The German Sustainability Strategy (based on the Sustainable

18 H. Haase 2021, p. 15.
19 H.G. Kastenholz et al. 2013, p. 5 et. seq.
20 K. Tomislav 2018, p. 68 et. seq.

Development Goals[21]), the Climate Protection Programme 2030 of the Federal Government of Germany for the implementation of the Climate Protection Plan 2050 and the Coalition Agreement (2021) are the fundament for some of the terms selected in the Figure 5.3, which are used similarly or identically in other countries. The demand sectors (also termed as 'transformation fields of action') include the industry, the energy sector as well as the transport and building sector. In the German sustainability strategy, the wording 'transformation areas' is also used for combinations of sectors, such as the construction (buildings) and transport sectors, etc. as well as other areas, such as energy transition and climate protection or social justice.

A paragraph from the coalition agreement 2021 is particularly concise, which includes almost all Figure 5.3's demonstrated terms in the area of future-oriented transformative research that are regarded as central future fields:

> (. . .) Modern technologies for a competitive and climate-neutral industry (such as steel and basic industries) (. . .), ensuring clean energy production and supply as well as (. . .) sustainable mobility of the future; (. . .) climate, climate impacts, biodiversity, sustainability, earth system and corresponding adaptation strategies, as well as sustainable agriculture and food system; (. . .) precautionary, crisis-proof and modern health care system, which uses the opportunities of biotechnological and medical procedures (. . .); technological sovereignty and the potentials of digitalization, e.g. in artificial intelligence and quartz technology, (. . .) e.g. in artificial intelligence and quantum technology, for data-based solutions across all sectors; (. . .) exploration of space and oceans and creation of sustainable uses; (. . .) societal resilience, gender equality, cohesion, democracy and peace.[22]

In this context, sustainable management means that profits should be generated ecologically and socially, and not in order to use them afterwards for socially responsible environmental concerns. The activities of the sectors have a circular impact on the environment and society.[23] In addition to the interdependent sustainability strategies (see following section "Sustainability Strategies" 5.5), a corresponding resilience strategy has to be additionally implemented. A system is resilient as long as it does not 'tip over' in the face of changes in environmental conditions due to exceedance of predetermined limits of the prevailing regime.[24] Climate change is a key example of this. The financial system has a crucial role in financing and promoting sustainable development and is reflected in international and national sustainable finance strategies.

21 J. Mensah 2019.

22 Die Bundesregierung 2021a.

23 I. Pufé 2012, p. 7.

24 Walker and Salt 2012.

5.4 Management of Sustainable Development

The path to sustainable development is complex. In order to manage this complexity, integrated sustainability management is required, which includes the economic, ecological and social dimensions.[25] Management systems are based on defining concrete factual and formal goals for public, private companies, NGOs, politics, etc. and implementing concrete concepts, measures, instruments, systems or tools for achieving them. Based on the SDGs, the equivalent sustainability strategy (cf. following section 5.5) can be selected on the basis of these 17 sustainability goals, impact management can be introduced and the results can be presented within the framework of sustainability reporting. To achieve the energy transition, a suitable measure in terms of a sustainability strategy is the expansion of renewable energies.

Broken down to the implementation level, concrete measures are derived from this in order to achieve the strategic target definition. These measures (see also 5.6) can include concrete concepts such as the circular economy or green economy and are supported, for example, by instruments or tools such as a sustainable balanced scorecard, environmental or energy management system. The German Federal Ministry for the Environment, Nature Conservation, Nuclear Safety and Consumer Protection provides the following definition:

> Core elements of comprehensive sustainability management are environmental management, quality management, risk management, addressing employee concerns, socially and environmentally responsible supply chain alignment, operational information systems and anti-corruption.[26]

The management approaches mentioned, such as environmental and quality management, but also systems of health management or occupational safety, have already been established since the mid-1990s in many of the sectors described in section 5.3. For example, data management can be based on existing working time models, corporate environmental information systems or supply chain management.

Sustainable corporate management is primarily used as a concept in Germany. Internationally, the term '(corporate) social responsibility' (CSR) is well established. In this context, sustainable economics and management aims to ensure that limited resources are handled responsibly in a company-related context. The terms CSR and sustainability, sustainability strategy and CSR strategy, and sustainability report and CSR report are often used synonymously, although the CSR term is more narrowly defined in the corporate sense.[27]

---

25 I. Pufé 2012, p. 8.

26 Bundesministerium für Umwelt und Naturschutz 2022.

27 Bundesministerium für Arbeit und Soziales 2022.

5.5 Sustainability Strategies

Mankind's ecological footprint is far beyond sustainability and urgently needs to be reduced. This is particularly difficult in view of the growing population and the pursuit of steady economic growth.[28] A sustainability strategy is needed, i.e. methods and instruments for the strategic implementation of sustainable development.[29] Sustainability strategies that can be applied to this transformation are the efficiency, consistency and sufficiency strategies. Efficient technologies and methods are often presented as the solution to all challenges, but efficiency alone will not be enough because the rebound effect will overcompensate for the successes. Consistency to make material flows more natural is important, but again will not be enough on its own. The reduction of resource consumption through behavioural change, so-called 'sufficiency', is also important, but politically difficult to implement.[30] To achieve the goals, the three strategies explained below must be combined.

5.5.1 Efficiency Strategy: More Productive use of Materials and Energy

The term 'efficiency' describes the relationship between effort and benefit required to achieve a certain result.[31] The efficiency strategy aims to create an economic output with the lowest possible use of materials and energy by improving the input–output ratio. Specifically, this means increasing material, raw material and energy efficiency. The effect is a relative reduction in resource consumption. The goal of achieving as much as possible with as few resources as possible can be achieved through new or improved technologies, processes and/or products.[32] More efficient engines, for example, lead to lower fuel consumption and thus pave the way for lower-emission mobility.

> The efficiency strategy is the closest thing to economic thinking: the maximization principle, according to which the yield is to be maximized with constant effort, and the minimization principle, according to which the effort is to be minimized with constant yield.[33]

The efficiency strategy therefore does not contradict economic goals, but actually supports them.

Unfortunately, technical efficiency gains are often offset by changes in behaviour and increased comfort, which is known as the rebound effect. For example, the inter-

28 See Chapter 1.5 (with regard to population issues) and Chapter 1.4 (Economic Growth and Distribution).
29 WCED 1987, p. 183 et. seq.
30 C. Berg, 2020, p. 314.
31 Irrek and Thomas, 2008, p. 1.
32 S. Behrendt (et al.), 2018, p. 8.
33 M. Schmidt, 2008, p. 6.

nal combustion engine is now more efficient and consumes less fuel; on the other hand, passenger cars have more and more power and comfort. Therefore, efficiency alone will not be enough to achieve the sustainability goals.[34]

5.5.2 Consistency Strategy: Environmentally Compatible Material Cycles, Recycling, Waste Avoidance

While efficiency addresses the quantity of resource consumption, consistency is about material flows and material cycles.[35] Ideally, with this strategy there is no waste, only products. Waste is avoided through recyclable resources. The prerequisite is that the technical processes are organized in such a way that in the technosphere, as in nature, there are only recyclable products and waste is avoided. If this is not possible, substances foreign to nature are to be recovered in closed cycles.[36]

In the course of the sustainability debate, a variety of approaches can be noted, for example the cradle-to-cradle approach by Braungart and McDonough.[37] The basic idea of this cradle-to-cradle approach is that there should actually be no waste in the process of industrial metabolism, because everything is fed back into the system and enriches either the natural or the technical cycles.[38]

In addition to the use of nature-friendly technologies, such as photovoltaics, wind turbines or green hydrogen for energy generation, the concept of consistency also includes the switch to electric drive systems in passenger and freight transport.

5.5.3 Sufficiency: Reducing Production and Consumption

The reduction of resource consumption through behavioural change is called sufficiency.[39] The sufficiency strategy calls for ecologically and socially compatible upper limits for the economy or economic growth in order to be able to comply with the ecological load limits of the ecological systems. This involves the view that a reduced consumption of resources and the environment also makes a satisfactory (sufficiency) life possible. The starting points for this are a corresponding change in people's awareness and the resulting change in lifestyle.[40]

---

34 C. Berg, 2020, p. 317.
35 See also Chapter 3.5.
36 J. Bauer, 2008, p. 64.
37 S. Behrendt et al., 2018, p. 8.
38 Braungart and McDonough, 2002.
39 C. Berg, 2020, p. 317.
40 Fischer and Grießhammer, 2013, p. 7 ff.

The use of car sharing instead of owning a car, is an example of the sufficiency strategy and can make a significant contribution to sustainable mobility.

5.6 Examples of Applications for the Sectors

In the following subsections, the sectors presented as examples in 5.3 are set in relation to the respective sustainability strategies with brief examples of their application (measures, concepts, tools). The examples presented for the individual strategies – efficiency, consistency and sufficiency – do not claim to be exhaustive and, as already described, can only pave the way for sustainable development in close interaction. Chapters 19–24 present selected measures, concepts and tools in greater detail.

5.6.1 Energy industry

The expansion of renewable energies and the associated phase-out of fossil energies will be accelerated in many countries since the Russian invasion of Ukraine (for the current status of the expansion of renewable energies, see section 10). The use of nature-compatible technologies, such as photovoltaics, wind turbines or green hydrogen (see sections 6, 10, 12, 18 and 19) for energy production embraces the concept of consistency (substitution). By applying this concept, our ecosystems are not destroyed and the reduction of fossil energies such as coal or gas accompany this. Consistency innovations also '(. . .) regularly require considerable sufficiency concessions (. . .) as feasible consistency solutions' which is why we also talk about'(. . .) qualified combinations of efficiency and sufficiency solutions (. . .)'.[41,42] Since renewable energy solutions often receive financial or institutional support from the government, this assurance must be reflected in a sustainable consistency policy. Unexpected regulatory changes or discontinuities, for example, are followed by larger risk margins in the context of future-based investment decisions and financing models.[43]

The reduction of final energy demands on the user side in the areas of electricity demand, transport, housing heat and industrial process heat, among others, is unavoidable, e.g. due to the requirements of the Green Deal, in order to achieve the reduction of CO_2 and other greenhouse gases.[44,45] The interconnection of the demand sectors should be coupled with the generation sector. This means efficient integration

41 cf. E. Seidel 2004, p. 431.
42 Seidel and Göllinger 2004, p. 402.
43 W. White et al. 2013, p. 97.
44 cf. Fraunhofer I.S.E. et al. 2020, p. 20 ff.
45 C. Schmitt et al. 2015.

of the fluctuating (variable) renewable energies and the linkage between the power sector with the building, transport and industrial sectors (sector coupling) (see sections 21 and 23).,[46],[47] In addition to grid integration, the electricity grids in particular must be expanded and their transmission strengthened.[48] Digital technologies can optimize grid operation and ensure efficient use through built-in sensors (smart metering and energy management systems). Central control of the grid system by artificial intelligence could analyze large volumes of data and flexibly adapt grid operation (decarbonization through digitalization). Furthermore, public acceptance for grid expansion should be created and information about the necessity of smart grids should be provided in order to be able to exploit the cross-sector potential of digitization (cf. 23).[49]

5.6.2 Industry

The different industrial sectors such as automotive or mechanical engineering, chemical-pharmaceutical industry, food, electrical engineering, metal production and processing as well as the plastics or paper industry have different requirements when it comes to establishing a sustainable production value chain (cf. e.g. 24). The transformation of the industry to a Green Economy requires different measures (cf. 24), which result in a value chain according to the sustainability strategies. The aim of the concept of a circular economy (cf. 24) is to create a production system which, among other things, implements the minimization of waste and emissions, the efficient use of resources and the reduction of material and energy losses in a standardized manner.[50]

Starting with the product design, the focus should be on a long service life as well as reusability and recyclability. The manufacturing process should be energy- and material-efficient. In sales aspects, the sufficiency idea of company tool sharing can be applied, as can the option of leasing machines and thus taking advantage of the possibility of always being able to use efficiency-optimized technology for one's own production. This pattern can also be applied to the sale of products so that they are lent or shared, thus ensuring sustainable use. Both waste prevention and reuse should be a focus to avoid resource loss. Bringing in regenerative and recycled raw materials closes the sustainable production cycle.[51,52]

46 cf. J. Ramsebner et al. 2021.
47 cf. J. Antoni et al. 2021.
48 cf. T. Brown et al. 2018, p. 721 ff.
49 cf. J. Strücker et al. 2021.
50 C. Dornack 2021, p. 28 et. seq.
51 S. Geisendorf 2018, p. 772 ff.
52 N. Gunarathne, K.-H. Lee 2021.

The aspects of sustainability should be anchored as a target definition of ecological responsibility in companies. As an example of waste management, the recycling of glass packaging in Austria illustrates that an established environmental and integrated sustainability management can involve all stakeholders in the region in order to optimize the creation and preservation of resources. Here, the focus is on saving raw materials, land consumption for primary raw materials, energy savings and CO_2 reduction, as well as social responsibility towards employees and all reference persons within the value chain.[53]

Another important indicator for a sustainable industry is not only to optimize and sustainably adapt the supply side, but also to solidify the awareness for sustainable consumption on the consumer and thus the demand side. Often, consumers do not want to consume less, but more sustainably. The sufficiency strategies of sharing, lending or second-hand consumption are still niche offerings that need to be expanded with innovative concepts from the industry. Digital support is essential here so that product services can be offered on a platform basis.[54]

5.6.3 Transport

For the transport sector, it is also important to apply a triad of coordinated efficiency, consistency and sufficiency strategies so that CO_2 emissions in this segment can be significantly reduced. New fuel technologies for lower-emission mobility, adapted combustion engines and the switch to electric and hybrid technologies are key future areas for passenger and freight transport that need to be implemented. Rising demand for goods and air travel will also increase heavy goods, shipping and air traffic. Hydrogen could be used in fuel cells or as a basis for synthetic fuels in the future. Biomass- or electricity-based fuels must be identified and further developed.[55,56]

The switch to electric driven systems is seen as one of the crucial consistency strategies, especially in passenger but also in freight transport. Battery technology and charging infrastructure must be efficiently optimized and expanded. Particularly in the extraction of non-renewable resources (water, raw materials, etc.), which are required for production and operation, a sustainable value creation system must be established. The expansion of the charging infrastructure poses different challenges for the respective local energy system.[57] Already during the process of subsequent planning, attention should be paid to resource-saving grid utilization. Automated and connected driving requires an intelligent load and energy management system that

53 H. Hauke 2014, p. 4 ff.

54 J. Camacho-Otero et al. 2018.

55 P. Sauermann, I. Bochum 2019, p. 903 et. seq.

56 K. Seidel 2020, p. 62 et. seq.

57 A. Kampker et al. 2020, p. 662 et. seq.

enables the integration of renewable energies on a demand-oriented basis.[58] A key driver component is that an integrated transport policy should transform the overall system:

> for a transport system geared towards sustainability, it is essential to take into account the complex interrelationships of economic, financial, fiscal, transport, land use, health, environmental, social and tourism policies. The aim of integrated transport policy is to ensure the mobility of people and goods, but to minimize the associated burdens on people, the environment and financial budgets.[59]

As an integrated sufficiency strategy, car sharing can make a significant contribution to sustainable mobility. Thus, it is important to reduce individual traffic and to create opportunities for cooperative service offerings. In this context, the different approaches of station-based car sharing should be adapted to the specific region with those of free-floating car sharing and should also be connected to the public transport system (bus, train, etc.).[60] The development of utilization varies socio-economically depending on the demographic situation and differs significantly between rural and urban areas in terms of offers and implementation possibilities. Comprehensive urban and regional development is essential in this context.[61]

5.6.4 Building Sector

The building sector can and should always be developed and optimized in harmony with the transport sector, among others. An integrated neighborhood development approach (village or urban) should be lived in this context and focus on sustainable buildings (green buildings). Here, the focus lies on environmentally friendly and resource-efficient criteria, which are reflected in established sustainability certificates for sustainable construction and renovation.[62] The world's first sustainability assessment method, BREEAM (Building Research Establishment Environmental Assessment Methodology), can be used alongside the LEED (Leadership in Energy and Environmental Design) green building certificate for master plan projects, infrastructure and buildings. The German counterpart DGNB (German Sustainable Building Council) is used to assess the sustainability of buildings and neighborhoods and integrates the entire building life cycle.[63]

58 M. Litzlbauer et al. 2020, p. 147 et. seq.
59 T. Bracher et al. 2014, p. 13.
60 B. Nansubuga, C. Kowalkowski 2021.
61 Doll and Krauss 2022.
62 C. He et al. 2021.
63 S. Ferrari et al. 2021.

The phases of a sustainable building life cycle comprise coordinated socio-cultural, ecological and economic dimensions, which should include all three sustainability strategy approaches:
1. Selection of project developers including consideration of sustainable business goal definition,
2. project concept phase with integration of public welfare aspects, life cycle costs and an environmental concept,
3. socially and environmentally compatible site (selection), which requires an integrated transport and regional concept,
4. sustainable financing and investment strategy,
5. architectural and construction specifications including life cycle assessment, energy concept (see for heat chapter 12, 13, 11 and 14), sustainable use of land and space as well as sustainable use of resources, compliance with sustainable building material and supply chain criteria including waste separation and recycling,
6. operating phase of the building with an integrated utilization and operating concept, energy and resource management, etc., and
7. the deconstruction phase (recycling and disposal of building materials), if applicable.[64,65]

In addition to the creation of socially acceptable housing, the focus must be on space-saving housing (sufficiency) and, above all, the optimization of existing buildings. A renovation concept originating in the Netherlands is the so-called 'Energiesprong' concept, which applies the energetic renovation of existing buildings by standardizing processes, production using digital technology and serial production technology. The conversion of the energy supply and the implementation of the NetZero standard (sufficiency) are as much part of this as the use of PV modules, heat pumps and heat recovery through room air technology.[66,67]

5.6.5 Agriculture and Forestry

The agriculture and forestry sector is also in the spotlight due to climate change and the associated adaptation strategies. If the transformation process is to succeed in this segment, the concept of the bioeconomy[68] is a possible answer to meeting '(. . .) current societal challenges such as climate change, energy and resource efficiency, health

64 I. Deden et al. 2018.
65 S. Plesser et al. 2015.
66 D. Brown et al. 2019.
67 N. Schäfstoß 2017, p. 162 et. seq.
68 McCormick and Kautto 2013.

and demographic change'.[69] As an example, the European Union has extended the European Agricultural Fund for Rural Development until 2027. In addition to establishing, maintaining and improving ecosystems associated with agriculture and forestry, another component is to promote resource efficiency and support the agriculture, food and forestry sectors in their transition to a low-carbon and climate-resilient economy. Efficiency improvements will be made in both water and energy usage, as well as in promoting carbon storage.[70]

The conversion to organic farming should include all stages –

1. production,
2. raw material trade,
3. processing and further processing,
4. logistics and
5. distribution channels of an organic value chain, ideally

– of an organic value chain that is, in the best case, fully regional.[71] The production of food and other agricultural products is carried out by environmentally friendly and self-contained material cycles, which include species-appropriate animal livestock maintenance. Here the renunciation of pesticide fertilizers, green genetic engineering or artificial preservatives is a consensus. As part of the German sustainability strategy, the goal is to achieve 20% organic farming by 2030. This target is not sufficient to achieve the sectoral climate protection goals. Rather, conventional production methods must be adapted in parallel and dovetailed with organic orientation.

Another sufficiency strategy to be expanded and established in the future is that of solidarity-based agriculture. Here, producers and consumers often cooperate in a living and working community. One possible approach is to choose the form of a cooperative model and to optimally balance resource use and consumption by expanding the regional value chain, such as the parallel operation of a citizens' energy cooperative. Interlocking with the regional (also conventional) structures on-site is essential in order to implement the concept of a cycle-oriented agriculture.[72,73]

69 A. Tschannen et al. 2021, p. 25.
70 S. Kummer et al. 2021.
71 Europäische Union 2022.
72 H. Klemisch 2021, S. 307 et. seq.
73 S. Gruber 2020.

5.7 Overview

The final tabular overview in Table 5.1 shows the sustainability strategies of efficiency, consistency and sufficiency for each of the above covered sectors separately.

Table 5.1: Overview of the Sustainability Strategies (compiled by author).

	Energy industry	Industry	Transport	Building area	Agriculture and forestry
Efficiency	Sector coupling	Optimize production value chain	Low-emission vehicles	Densification, building insulation, serial refurbishment	Resource efficiency (e.g. water utilization)
Consistency	Use of renewable energies	Use more environmentally friendly materials, cycle of production and consumption	E-Mobility	Sustainable construction and renovation (circular economy)	Conversion to organic farming
Sufficiency	Reduction of final energy consumption	Tool sharing, limit material consumption	Car sharing	Space-saving housing, zero-energy houses	Solidarity agriculture

It should be emphasized that the sectors cannot be considered separately from each other, but are always mutually dependent on concepts such as the circular economy and the path to a green economy. The respective value chains must be established and strengthened in a regionally sustainable manner and optimized in accordance with the SDGs. In addition to considering supply, production and service chains, a socio-economic–ecological balance should be established.

5.8 Discussion and Outlook

The global challenges of global warming, climate and demographic change, the scarcity of natural resources, the loss of biodiversity, poverty and conflicts must be met in a united effort by all communities of states. Past and present crises, such as the oil crisis in the 1970s, environmental catastrophes like Chernobyl in the 1980s or Fukushima in 2011, as well as smaller and larger financial crises in the last decades, demand a sustainable cohesion of all different actors, especially in current times, which are marked by the Ukraine war. Profit and economic growth find their natural limits in an economic order based on mass production, animal husbandry and 'exploitation', which we have only realized in recent years as not (any more) trendsetting. In particular countries with lower income and disadvantaged population groups bear the effects of the industrialized and emerging countries, which are responsible for the consumption but especially the long-term damage of limited resources.

In order to counteract climate change alone, immediate emission reductions are required in all of the sectors described, namely energy, industry, transport, buildings, agriculture and forestry. The measures to achieve this are largely in place. However, committed goals, future scenarios, technology development and individual instruments and concepts taken individually are no longer sufficient to advance the social, economic and ecological fields of action. The triad of sustainable efficiency, consistency and sufficiency strategies should be reflected in all areas of life. Transformation, the process of change in society as a whole, sustainable production and sustainable consumption behaviour – has the time come for a social contract of the kind proposed by the German Advisory Council on Global Change (WBGU) back in 2011?[74]

What does it take? A stable and reliable funding framework and the associated legal requirements and regulations from the political side. Environmental policy management with a wide range of instruments should be coordinated with one another and always adapted to the regional and supraregional policy context of each country. Clear, extended environmental and human rights rules should be established, e.g. for companies, with which compliance is mandatory. A socio-ecological transformation does not require individual criteria to be defined, but rather a uniform standard to which consumers, service providers, public administration, etc. can orient themselves. In this context, an important step has been taken with the EU Taxonomy Regulation, which must be refined in the coming years for application in practice.

For a Green Economy and ultimately a macroeconomic transformation, the redirection of financial flows is required, as is being initiated with the measures of the EU Action Plan. In this context, it will not only be necessary to define the side of investment and lending opportunities and the associated supply chain-specific dependencies in value chains, but also the individual position of public, private companies and NGOs. In their own internal and external financial flows and their own corporate strategy, these companies will have to specify which formal objectives, but above all which substantive objectives (integration of sustainability goals) they pursue in their (economic) activities.

As an essential driving force and transformer, practice-oriented scientific research on different subject areas is indispensable. Whether it is about technology, sustainable business models in the various industrial and service sectors or sustainable lifestyles as well as cultural sustainability, educational concepts of lifelong learning for the topic of sustainability must be further developed and socially integrated and implemented.

For sustainable development, the transformative interlocking of sectors (industries) and public participation, digitization must be advanced further. In this context,

74 WBGU 2011.

the IT infrastructures themselves must develop into green IT that pursues resource-conserving business models. Let's take the energy sector as an example, which will have to deal with increased fluctuating energy in the future. In addition to the expansion of storage capacities, virtual power plants will have to control and balance decentralized power generation units in a network, both regionally and nationally. This can include photovoltaic, hydroelectric, biogas, wind energy plants or combined heat and power plants. In addition to the efficiency strategy to save even more energy in the future and to identify and expand savings options, on-demand services for control and monitoring are to be introduced. For integrated sustainability management, which thrives on the collection and processing of data for the derivation of appropriate measures, technical progress is an essential component of sustainable development.

Participation and transparency are the cornerstones here and must go hand in hand with modern government and administrative action for a sustainable society (New Public Governance and Open Government). Strengthening community concerns can and must go beyond individual public relations campaigns. Sustainability goals must not only be discussed and defined at board, corporate and administrative management level or in political bodies, but must be developed in a participatory process and communicated both internally and externally. Sustainable awareness is created through repetition, continuation, practice-oriented participation, educational campaigns at all age levels and, finally, with the conscience of legal requirements in order to shape our future and cushion the negative effects.

With the agreement reached at the UN Climate Change Conference in Paris in 2015 to limit global warming, the adoption of the 17 United Nations Sustainable Development Goals (UN SDGs), and the increasing and ever more ambitious commitments to climate neutrality by numerous economies, communities, and companies worldwide, it can be observed today that the global community is increasingly facing up to environmental and social challenges, but an enormous effort is still required to master the road to a resilient, livable, and more equitable future.

Bibliography

Antoni, Johannes, Knoll, Fanny, Bieschke, Nils: Regulatorischer Rahmen für eine optimale Sektorenkopplung – Einführung in die Problemstellung und systematischer Überblick, Greifswald 2021.

Bassen, A., Lopatta, K.: Sustainable Finance: Neue Strategie im Finanzsektor trotz Coronakrise? ifo Schnelldienst, 2020, pp. 3–6., Online: https://www.ifo.de/publikationen/2020/aufsatz-zeitschrift/sustainable-finance-neue-strategie-im-finanzsektor-trotz.

Bauer, J.: Industrielle Ökologie – Theoretische Annäherung an ein Konzept nachhaltiger Produktionsweisen, Stuttgart 2008.

Behrendt, Siegfried; Göll, Edgar; Korte, Friederike: Effizienz, Konsistenz, Suffizienz Strategieanalytische Betrachtung für eine Green Economy, 2018, Online: https://www.izt.de/fileadmin/publikationen/IZT_Text_1-2018_EKS.pdf.

Berg, Christian: Ist Nachhaltigkeit Utopisch? Wie wir Barrieren überwinden und zukunftsfähig handeln, München 2020.

Bracher, Tilman,Gies, Jürgen,Thiemann-Linden, Jörg,Beckmann, Klaus J.: Umweltverträglicher Verkehr 2050: Argumente für eine Mobilitätsstrategie für Deutschland, Umweltbundesamt (editor), Dessau-Roßlau 2014.

Braungart, Michael; McDonough, William: *Cradle to Cradle: Remaking the Way We Make Things.* New York 2002.

Brown, Tom, et al.: Synergies of sector coupling and transmission reinforcement in a cost-optimised, highly renewable European energy system, in *Energy*, no. 160, 2018, pp. 720–739.

Brown, Donal; Kivimaa, Paula; Sorrell, Steven: An energy leap? Business model innovation and intermediation in the 'Energiesprong'retrofit initiative, in *Energy Research & Social Science*, 58. Jg., 2019, p. 101253.

Bundesministerium für Ernährung und Landwirtschaft: Zukunftsstrategie ökologischer Landbau Impulse für mehr Nachhaltigkeit in Deutschland, 2019(a), Online: https://www.bmel.de/SharedDocs/Downloads/DE/Broschueren/ZukunftsstrategieOekologischerLandbau2019.pdf?__blob=publicationFile&v=5.

Bundesministerium für Ernährung und Landwirtschaft: Richtlinie zur Förderung von Biowertschöpfungsketten im Rahmen des Bundesprogramms Ökologischer Landbau und andere Formen nachhaltiger Landwirtschaft (BÖLN), Bundesministerium der Justiz für Verbraucherschutz (editor), BAnz AT 06.08.2019(b) B1.

Bundesministerium für Arbeit und Soziales (BMAS): CSR und Nachhaltigkeit, 2022, Online https://www.csr-in-deutschland.de/DE/CSR-Allgemein/CSR-Grundlagen/Nachhaltigkeit-und-CSR/nachhaltigkeit-und-csr.html.

Bundesministerium für Umwelt, Naturschutz, nukleare Sicherheit und Verbraucherschutz: Nachhaltigkeitsmanagement, 2022, Online https://www.bmuv.de/themen/nachhaltigkeit-digitalisierung/wirtschaft/unternehmensverantwortung-/-zertifizierung/nachhaltigkeitsmanagement.

Camacho-Otero, Juana, Boks, Casper, Pettersen, Ida Nilstad: Consumption in the circular economy: A literature review, *Sustainability*, 10. Jg., no. 8, 2018, p. 2758.

Deden, Imke, Doe, Novi Johanna, Tacke, Fabian: Leitfaden für ethisch-nachhaltige Immobilieninvestments – Ein Überblick zu Kriterien und Bewertungsinstrumenten für Deutschland, Cric e.V., KlimaGut Immobilien AG (editor), 2018, Online https://www.trendreport.de/49-kriterien-fuer-ethisch-nachhaltige-immobilieninvestments/.

DGCN – GlobalCompact Netzwerk Deutschland, econsense: Neuer Impuls für die Berichterstattung zur Nachhaltigkeit?, 2018, Online: https://econsense.de/wp-content/uploads/2018/06/Studie-CSR-RUG_econsense-DGCN_2018.pdf.

Die Bundesregierung: Deutsche Nachhaltigkeitsstrategie. Neuauflage 2016. Berlin 2017.

Die Bundesregierung: Perspektiven für Deutschland. Unsere Strategie für eine nachhaltige Entwicklung. Berlin 2022.

Die Bundesregierung: Deutsche Nachhaltigkeitsstrategie – Weiterentwicklung 2021, 2020, Online www.bundesregierung.de/publikationen.

Die Bundesregierung: Koalitionsvertrag zwischen SPD, Bündnis 90/Die Grünen und FDP, 2021a, Online: https://www.bundesregierung.de/breg-de/service/gesetzesvorhaben/koalitionsvertrag-2021-1990800.

Die Bundesregierung: Deutsche Sustainable Finance-Strategie, 2021b, Online: https://www.bundesfinanzministerium.de/Content/DE/Downloads/Broschueren_Bestellservice/deutsche-sustainable-finance-strategie.pdf?__blob=publicationFile&v=8.

Doll, Claus, Krauss, Konstantin: Nachhaltige Mobilität und innovative Geschäftsmodelle, Expertenkommission Forschung und Innovation (EFI) (editor), no. 10-2022, Studien zum deutschen Innovationssystem, Berlin 2022.

Dornack, Christina: Welchen Beitrag kann die Kreislaufwirtschaft zur Ressourceneffizienz leisten?, in Arnhild Gerecke, Gilian Gerke, Hartwig Haase (editor), 25. Tagung Siedlungsabfallwirtschaft Magdeburg – Abfallwirtschaft ohne Grenzen? Magdeburg 2021, pp. 27–30.

European Commission: European Commission Action Plan: Financing Sustainable Growth – Factsheet, 2018, Online: https://ec.europa.eu/info/sites/info/files/180308-action-plan-sustainable-growth-factsheet_en.pdf.

European Commission: Der europäische Grüne Deal, 2019, Online: https://ec.europa.eu/info/sites/default/files/european-green-deal-communication_de.pdf.

Europäische Union: Gemeinsame Agrarpolitik (GAP) – Europäischer Landwirtschaftsfonds für die Entwicklung des ländlichen Raums, 2022, Online https://ec.europa.eu/info/food-farming-fisheries/key-policies/common-agricultural-policy/rural-development_de.

Ferrari, Simone, Zoghi, M., Blázquez, T., Dall'O, G.: New Level (s) framework: Assessing the affinity between the main international Green Building Rating Systems and the European scheme. Renewable and Sustainable Energy Reviews, 2021, p. 111924.

Fischer, Corinna, Grießhammer, Rainer: Mehr als nur weniger. Suffizienz: Begriff, Begründung und Potenziale, Institut für angewandte Ökologie, Freiburg 2013, Online: https://www.oeko.de/oekodoc/1836/2013-505-de.pdf.

Fraunhofer (I. S. E), Sterchele, Philipp, Brandes, Julian, Heilig, Judith, Wrede, Daniel, Kost, Christoph, Schlegl, Thomas, Bett, Andreas, Henning, Hans-Martin: Wege zu einem klimaneutralen Energiesystem. Die deutsche Energiewende im Kontext gesellschaftlicher Verhaltensweisen, Fraunhofer-Institut für solare Energiesysteme, Freiburg (editor) 2020, Online: https://www.ise.fraunhofer.de/content/dam/ise/de/documents/publications/studies/Fraunhofer-ISE-Studie-Wege-zu-einem-klimaneutralen-Energiesystem.pdf.

Geisendorf, Sylvie, Pietrulla, Felicitas: The circular economy and circular economic concepts: A literature analysis and redefinition, in *Thunderbird International Business Review*, 60. Jg., no. 5, 2018, pp. 771–782.

Gruber, Sabine: Bewältigungsstrategien alternativen Wirtschaftens: Wertrationalität und soziale Einbettung am Beispiel Solidarischer Landwirtschaft (Vol. 6). Baden-Baden 2020.

Gunarathne, Nuwan, Lee, Ki-Hoon: The link between corporate energy management and environmental strategy implementation: Efficiency, sufficiency and consistency strategy perspectives, in *Journal of Cleaner Production*, no. 293, 2021, p. 126082.

Haase, Hartwig: Genug, für alle, für immer – Nachhaltigkeit ist einfach . . . komplex, in: Arnhild Gerecke, Gilian Gerke, Hartwig Haase (editor): 25. Tagung Siedlungsabfallwirtschaft Magdeburg – Abfallwirtschaft ohne Grenzen?, Magdeburg 2021, pp. 9–20.

Hauke, Harald, Schöner, Haymo: Grünbuch Nachhaltiges Recycling von Glasverpackungen in Österreich: Handbuch zur zukunftsfähigen Gestaltung des Materialkreislaufs von Glasverpackungen, Österreich 2014.

He, Chenchen, Hou, Yuanyuan, Ding, Liping, Li, Peng: Visualized literature review on sustainable building renovation, in *Journal of Building Engineering*, 44, 2021, pp. 102622.

Intergovernmental Panel on Climate Change (IPCC): Climate Change 2022. Mitigation on Climate Change. Summary for Policymakers, 2022, Online: https://www.ipcc.ch/report/ar6/wg3/downloads/report/IPCC_AR6_WGIII_SPM.pdf.

Irrek, Wolfgang; Thomas, Stefan: Definition Energieeffizienz, 2008, Online: https://wupperinst.org/uploads/tx_wupperinst/energieeffizienz_definition.pdf.

Kampker, Achim, Kreisköther, Kai, Treichel, Patrick, Möller, Tom, Boelsen, Yannick: Elektromobilität – Trends und Herausforderungen der zukünftigen Großserienproduktion, in Handbuch Industrie 4.0: Recht, Technik, Gesellschaft, Heidelberg, 2020, pp. 661–680.

Kastenholz, Hans G., Erdmann, Karl-Heinz, Wolff, Manfred: Nachhaltige Entwicklung: Zukunftschancen für Mensch und Umwelt, Wiesbaden 2013.

Klemisch, Herbert: Genossenschaften und Non Profit Initiativen im Bereich Solidarische Landwirtschaft und Bürgerenergie, in Zeitschrift für das gesamte Genossenschaftswesen, 2021, 71. Jg., no. 4, pp. 306–317.

Kummer, Susanne, Klingbacher, Elisabeth, Petrasek, Richard, Bartel-Kratochvil, Ruth, Eichinger, Anja, Lindenthal, Thomas, Kranzler, Andreas, Niggli, Urs et al.: Stärkung der biologischen Landwirtschaft in Österreich bis 2030, 2021, Online unter: https://orgprints.org/id/eprint/39623/1/kummer-etal-2021-Bio2030_Projektbericht_FiBLat_final.pdf.

Litzlbauer, Markus, Philipp Rechberger, Schod, Franzl: Integration des eMobility-Demands in das lokale Energiesystem – Ansätze und Realisierungen, in e & i Elektrotechnik und Informationstechnik, no. 137.4, 2020, pp. 147–155.

McCormick, Kes, Kautto, Niina: The bioeconomy in Europe: An overview, in *Sustainability*, 5. Jg., no. 6, 2013, pp. 2589–2608.

Mensah, Justice: Sustainable development: Meaning, history, principles, pillars, and implications for human action: Literature review, in *Cogent Social Sciences*, 5. Jg., no. 1, 2019, pp. 1653531.

Nansubuga, Brenda, Kowalkowski, Christian: Carsharing: A systematic literature review and research agenda, in *Journal of Service Management*, Vol. 32 no. 6, 2021, pp. 55–91.

Plesser, Stefan, Koch, Marvin, Fisch, Norbert, Altendorf, Lars, Mühlbach, Ann-Kristin, Wilken, Thomas: Entwicklung einer Methodik zur Integralen Qualitätssicherung über den gesamten Gebäude-Lebenszyklus auf Basis der DIN V 18599, Stuttgart 2015.

Pufé, Iris. Nachhaltigkeitsmanagement, München 2012.

Ramsebner, Jasmine, Haas, Reinhard, Ajanovic, Amela, Wietschel, Martin: The sector coupling concept: A critical review. Wiley Interdisciplinary Reviews: Energy and Environment, 10. Jg., 2021, no. 4, p. e396.

Sauermann, Peter, Bochum, Ian Sharp: Kraftstofftechnologien für emissionsarme Mobilität, in Zukünftige Kraftstoffe, Berlin, Heidelberg 2019, pp. 902–917.

Schäfstoß, Nicolas: Flächenverbrauch und Wohnraum bedarf: Statistische Informationen zur politischen Gestaltung, in Meinel, G.; Schumacher, U.; Schwarz, S.; Richter, B. (editor.): Flächennutzungsmonitoring IX. Nachhaltigkeit der Siedlungs- und Verkehrsentwicklung?, Berlin, IÖR Schriften 73, 2017, pp. 161–169.

Schmidt, M.: Die Bedeutung der Effizienz für Nachhaltigkeit – Chancen und Grenzen, in Hartard, Schaffer, Giegrich (editor.): Ressourceneffizienz im Kontext der Nachhaltigkeitsdebatte, Nomos-Verlag: Baden-Baden, 2008, S. 31–46.

Schmitt, Corinna, Leuser, Leon, Brischke, Lars-Arvid, Duscha, Markus, Jacobsen, Sirkka: Suffizienz-Maßnahmen und -Politiken in kommunalen Klimaschutzkonzepten und Masterplänen – ein Überblick, in ifeu Institut für Energie- und Umweltforschung (editor), Heidelberg 2015. Online: https://energiesuffizienz.files.wordpress.com/2015/09/suffizienz_kommunalfinal_1509281.pdf.

Seidel, Eberhard: Brauchen wir eine nachhaltigkeitsorientierte Betriebswirtschaftslehre? in Hülsmann, Michael, Müller-Christ, Georg, Haasis, Hans-Dietrich (editor): Betriebswirtschaftslehre und Nachhaltigkeit, Wiesbaden 2004, pp. 369–435.

Seidel, Eberhard, Göllinger, Thomas: Konsistenz-Innovationen, in Schwarz, Erich (editor) Nachhaltiges Innovationsmanagement, Wiesbaden 2004, pp. 429–449.

Seidel, Kristian (et al.): Nachhaltigkeitsanalyse CO_2-emissionsarmer Aluminiumwerkstoffe, in ATZ-Automobiltechnische Zeitschrift, no. 122.11, 2020, pp. 62–67.

Strücker, Jens, Weibelzahl, Martin, Körner, Marc-Fabian, Kießling, Axel, Franke-Sluijk, Ariette, Hermann, Mike: Dekarbonisierung durch Digitalisierung: Thesen zur Transformation der Energiewirtschaft in Universität Bayreuth, Projektgruppe Wirtschaftsinformatik des Fraunhofer-Instituts für Angewandte Informationstechnik FIT und TenneT. Bayreuth (editor), Bayreuth 2021.

Thimann, C.: Finanzen und Nachhaltigkeit: das Ende des "Weiter so". BaFin Perspektiven, Volume 2, 2019, pp. 63–77.

Tomislav, Klarin: The concept of sustainable development: From its beginning to the contemporary issues, in Zagreb *International Review of Economics & Business*, 21. Jg., no. 1, 2018, pp. 67–94.

Tschannen, Amadea, Walker, David, Kammerhofer, Alfred W., Richter, Klaus, Sinabell, Franz: Strategien zur Bioökonomie in Deutschland, Österreich und der Schweiz, in Schweizerische Zeitschrift für Forstwesen, 172. Jg., no. 1, 2021, pp. 25–31.

Umweltbundesamt (UBA): IPCC-Bericht: Sofortige globale Trendwende nötig, 2022, Online: https://www.umweltbundesamt.de/themen/ipcc-bericht-sofortige-globale-trendwende-noetig.

United Nations (UN): Kyoto Protocol to the United Nations Framework Convention on Climate Change 1998, Online: https://unfccc.int/resource/docs/convkp/kpeng.pdf.

United Nations (UN): Sustainable Development Goals. Communication Material, 2022, Online: Communications materials – United Nations Sustainable Development.

United Nations (UN): Transforming our World: The 2030 Agenda for Sustainable Development 2015, Online: https://sdgs.un.org/sites/default/files/publications/21252030%20Agenda%20for%20Sustainable%20Development%20web.pdf.

United Nations (UN): United Nations Conference on Environment and Development (UNCED), Agenda 21, 1992, § 8.7.

United Nations: The Paris Agreement, 2015, Online: https://unfccc.int/sites/default/files/english_paris_agreement.pdf.

Walker, Brian; Salt, David: Resilience thinking: Sustaining ecosystems and people in a changing world. Island Press, Washington 2012.

WCED (World Commission on Environment and Development): Our Common Future. Oxford 1987.

White, William, Lunnan, Anders, Nybakk, Erlend, Kulisic, Biljana: The role of governments, in renewable energy: The importance of policy consistency. *Biomass and Bioenergy*, no. 57, 2013, pp. 97–105.

Wissenschaftlicher Beirat der Bundesregierung Globale Umweltveränderungen (WBGU): Welt im Wandel – Gesellschaftsvertrag für eine Große Transformation, 2011, Online https://www.wbgu.de/de/publikationen/publikation/welt-im-wandel-gesellschaftsvertrag-fuer-eine-grosse-transformation.

World Economic Forum: The Global Risks Report, 2022, Online: WEF_The_Global_Risks_Report_2022.pdf (weforum.org).

Zhuang, M., Löffler, K., & Rink, S.: Get ready: Regulatorische Entwicklungen hin zu einer nachhaltigen Finanzwirtschaft – Historie und Ausblick, in M. Granzow: Nachhaltige Finanzwirtschaft. Grundlagen und Konzepte für die Praxis, Berlin 2020, pp. 1–18.

Andreas Gabler

6 The Political Will – The European Path to a Sustainable Future

6.1 Overview

This article is about the Green Deal – the European answer to one of the most crucial issues of our time: combating the climate change.

First, it starts with a presentation of the Green Deal and its main contents, as well as the Investment Plan for the European Green Deal and the Just Transition Fund.

Then, the European Climate Law and the “Fit for 55” package, an essential part of the legislative implementation of the European Green Deal, will be presented.

Finally, the Action Plan for Financing Sustainable Growth is explained in more detail. Within this, the three key elements on which the Action Plan is based are examined in more detail: The taxonomy, the disclosure framework for non-financial and financial companies, and investment tools including benchmarks, standards and labels.

6.2 The Green Deal

With the Green Deal[1] as the central instrument, the European climate protection regime will be placed in the centre of EU policy. Based on the requirements of the Paris Agreement, the EU is to be climate-neutral by 2050 through measures in line with the Green Deal roadmap. To achieve this, greenhouse gas emissions must be reduced by at least 55% net by 2030 compared to 1990. The intention is to achieve the requirements by stimulating sustainable economic growth. In this respect, the Green Deal is to be understood as a growth strategy that enables a fair transition to an environmentally friendly economy whose growth is decoupled from resource use.[2] The “Fit for 55” package is intended to incorporate the climate targets set out in the Green Deal into legislation and thus revise all EU climate and energy regulations by summer 2022.[3]

The investments required for the integration of sustainability into all political and economic sectors are to be raised from the Union budget, national budgets and private sources. Therefore, the support of green investments as well as the mobilisa-

1 COM (2019) 640 final.

2 Pieper, Bergmann, Handlexikon der Europäischen Union, 6. Ed. 2021, Green Deal.

3 Frenz, Klimaschutz Gesamtkommentar, p. 27 marginal no. 1.

https://doi.org/10.1515/9783110784299-006

tion of research and the promotion of innovation are important aspects of the Sustainable Europe Investment Plan to finance the implementation of the Green Deal.[4]

6.2.1 Path to the Green Deal

The Green Deal is not Europe's first step towards an economic model in which natural resources are understood and recognised as a constraint on economic activity and, as a result, their conservation is integrated as an integral part of economic development. Green economic concepts were first discussed in the sustainability debate at the United Nations Conference on the Human Environment in Stockholm in 1972. Fleshing out this debate, the Brundtland Report 1987 provided a green economic concept based on three key elements of a sustainable economy: the social, an environmental and an economic one.[5] As a result of the Brundtland Report, which formulated the idea of an ecologically sustainable world economy as a common concern of mankind, the first UN Conference on Sustainable Development was held in Rio de Janeiro in 1992. The conference resulted in various environmental agreements, such as the United Nations Framework Convention on Climate Change ("UNFCCC") and the "Rio Declaration" as well as Agenda 21, which contained fundamental considerations on how to concretise sustainable development.[6] In 1997, the "Kyoto Protocol" was adopted, which provided for legally binding emission reduction targets with regard to the six gases of the Kyoto basket of at least 5% below the 1990 level for the participating states.[7] In 2015, the Kyoto Protocol was finally replaced by the Paris Agreement, which prescribes an internationally binding regulatory framework for a global energy transition. This framework aims to limit global warming to 1.5 degrees Celsius, if possible, and also to achieve greenhouse gas neutrality by 2050.[8]

6.2.2 Key Elements of the Green Deal

The Green Deal envisages a transformation of the European economy. The overarching goal is to achieve climate neutrality by 2050. In order to make this goal achievable and binding, the ambitious interim target of reducing greenhouse gases by at least

4 Becker, EuZW 2020, 441.
5 UN, Report of the World Commission on Environment and Development. Our Common Future, New York 1987.
6 UN, Rio Declaration on Environment and Development, Rio de Janeiro 1992, https://www.bpb.de/shop/-zeitschriften/apuz/green-new deals-2022/345729/der-europaeische-green-deal/#footnote-target-15.
7 Frenz, Grundzüge, p. 27 marginal no. 5.
8 Frenz, Grundzüge, p. 27 marginal no. 6.

55% by 2030 compared to 1990 was set in the European Green Deal.[9] The further decarbonisation of the energy system is also an elaborate project of the Green Deal. On the one hand, decarbonisation is necessary to achieve the envisaged climate targets of 2030, 2040 and 2050, and on the other hand, to be able to guarantee a supply of clean, affordable and secure energy for the European Union.[10] In addition, a mobilisation of industry towards climate neutrality is foreseen within the next 25 years.[11] Although the European industry has already started to change, it is still responsible for around 20% of greenhouse gas emissions. Implementation is expected to succeed through the European Industrial Strategy and the Circular Economy Action Plan, as well as the further pursuit of the Plastics Strategy and the 2018 Strategic Action Plan for Batteries. Reformed regulation is also planned for the building sector in the areas of construction, use and renovation, as this sector currently accounts for 40% of energy consumption.[12] This is clearly reflected, for example, in the Energy Performance of Buildings Directive, which was recast in 2018 with Directive (EU) 2018/844.[13] The rapid conversion of the transport sector to sustainable and intelligent mobility is also a key point in the Green Deal roadmap. Thus, in order to achieve the climate neutrality target, a 90% reduction in transport-related emissions by 2050 is unavoidable.[14] Another key component of the regulatory framework is the "Farm to Fork" strategy, which aims to make food systems fair, healthy and environmentally friendly.[15] In addition, a biodiversity strategy and an EU Forest Strategy aim to preserve or restore biodiversity.[16] Finally, the Zero Pollution Action Plan for air, water and soil is another element of the Green Deal. It aims to prevent further pollution by pollutants and to enable the elimination of existing pollution.[17]

---

9 COM (2019) 640 final, p. 5.

10 COM (2019) 640 final, p. 6.

11 COM (2019) 640 final, p. 8.

12 COM (2019) 640 final, p. 11.

13 Further amendments to the Energy Performance of Buildings Directive are pending. On 15.12.2021, the Commission submitted a new proposal to revise the Energy Performance of Buildings Directive (COM (2021) 802 final). On 18.05.2022, the Commission again proposed amendments to its own proposal of 15.12.2021 with the REPowerEU plan (COM (2022) 222 final). The REPowerEU plan is based on the "Fit for 55" package and supplements the measures on security of energy supply and storage (COM (2022) 230 final). The legislative process for both Commission proposals has not yet been completed.

14 COM (2019) 640 final, p. 12.

15 COM (2019) 640 final, p. 14.

16 COM (2019) 640 final, p. 15.

17 COM (2019) 640 final, p. 17.

Transition to a Green Economy

However, the Green Deal goes beyond pure climate protection. It aims at nothing less than the transition to a sustainable economy.[18] The measures laid out in the Green Deal Roadmap are intended to promote the efficient use of resources by moving to a clean and circular economy, halting climate change, promoting biodiversity and reducing pollution. It also shows how the planned transition can be ensured in an equitable and inclusive manner.

In order to realise the economic policy goals, immense investments are necessary. It is estimated that an additional 260 billion Euros will have to be invested every year.[19] Regions that are particularly dependent on CO_2-intensive activities are to receive financial and administrative support through a mechanism for a just transition. This Just Transition Mechanism, which accompanies the Green Deal, is expected to mobilise at least 100 billion Euro between 2021 and 2027.[20]

Green Minimum Criteria and Climate Neutral State Aid Framework

Within the framework of the Green Deal, the Commission proposes mandatory minimum “green” criteria or targets for public procurement in sectoral initiatives.[21] It also aims to create the conditions for sustainable investments by establishing an appropriate aid framework. This is intended to promote a cost-effective and socially inclusive transition to climate neutrality by 2050.[22] This is because, in addition to competition policy, state aid rules in particular represent an important aspect in terms of supporting the European Union in achieving its committed climate policy goals. Already in the Communication on the European Green Deal, attention was drawn to the need to revise state aid rules in order to promote a cost-effective and equitable transition to climate neutrality and also to facilitate the phase-out of fossil fuels. At the same time, fair conditions of competition in the internal market must continue to be preserved.[23]

The revised Guidelines on State Aid for Climate, Environmental and Energy Protection (CEEAG) provide information on how the Commission will in future examine whether state aid measures to promote environmental protection, including climate protection and the energy sector, are notifiable and compatible with the internal market under Art. 107 para. 3 point c of the Treaty on the Functioning of the European

18 Frenz, Grundzüge des Klimaschutzrechts, marginal no. 31 et. seq.
19 COM (2019) 640 final, p. 1; Becker, EuZW 2020, 441.
20 Frenz, Grundzüge des Klimaschutzrechts, marginal no. 36 et. seq.
21 COM (2020) 21 final, p. 13.
22 COM (2020) 21 final, p. 13.
23 COM (2020) 21 final, p. 13.

Union (TFEU).[24] In principle, state aid is inadmissible if it affects competition and thus trade between Member States. However, this does not apply if the aid measure promotes the development of an economic sector and at the same time does not alter trading conditions in a way that is contrary to the common European interest.[25] The CEEAG 2022 take this idea into account.[26]

6.2.3 European Green Deal Investment Plan

On 14.01.2020, the Commission published the Communication on an Investment Plan for a Sustainable Europe and the European Green Deal.[27] The core of the Communication is a strategy to combat climate change.

It is envisaged that by decoupling economic growth from resource use, clean economic growth and thereby sustainable economic development should be established in the EU. This would automatically be accompanied by a considerable saving of CO_2 emissions. The resource efficiency mentioned in Art. 191 para. 1 3rd indent TFEU as an environmental policy objective is a key element for clean economic growth and greenhouse gas neutrality.[28]

The Investment Plan envisages mobilising at least one trillion Euros in both private and sustainable public investment through the EU budget and its instruments for the new decade.[29] For climate and environment spending, the EU budget is expected to allocate 503 billion Euros for the period 2021-2030, which is in line with the spending target of at least 25% proposed for the Multiannual Financial Frameworks (MFF) 2021-2027 to achieve the climate targets. This spending target includes the environmental spending of all programmes.[30] In addition, national co-financing of 114 billion Euros is to be mobilised for climate and environmental purposes over this period. Furthermore, by providing an EU budget guarantee to reduce the risk of financing and investment, the "InvestEU" fund is expected to initiate around 279 billion Euros of private and public climate and environment-related investments from 2021-2030. The Just Transition Mechanism aims to simultaneously ensure a just transition through the provision of funds from the EU budget, Member State co-financing and contributions from InvestEU and the European Investment Bank (EIB). From 2021 to 2027,

24 (2022/C 80/01), Introduction no. 7.

25 (2022/C 80/01), Introduction no. 5.

26 (2022/C 80/01), Introduction no. 4.

27 COM (2020) 21 final.

28 Frenz, Grundzüge des Klimaschutzrechts, marginal no. 39 ff.

29 COM (2020) 21 final, Introduction no. 1; Frenz, Grundzüge des Klimaschutzrechts, marginal no. 41 ff.

30 European Agricultural Fund for Rural Development (EAFRD), European Agricultural Guarantee Fund (EAGF), Cohesion Fund, Horizon Europe and LIFE Programme (see COM (2020) 21 final, Funding no. 3).

this is expected to raise 100 billion Euros in investment, reaching 143 billion Euros over the next ten years. The Innovation Fund and the Modernisation Fund, which are financed with part of the revenues from the auctioning of allowances within the Emissions Trading System and are thus not part of the EU budget, are to contribute at least 25 billion Euros to achieve climate neutrality.[31]

The Investment Plan establishes the comprehensive framework for the desired sustainability shift in the EU, which translates into climate and environmental investments as well as social investments.[32] However, these public investments are not sufficient to raise the necessary volume. In addition, investments by private parties are needed to be able to achieve the targets of the Green Deal by contributing additional resources. To this end, the European Commission wants to combine new policy initiatives with increases in existing financial instruments within a relevant framework in order to set a new impulse for sustainable investment.[33]

The European Commission identifies three cornerstones of the Investment Plan for a sustainable Europe with a climate-neutral, green economy: First, mobilising at least one trillion Euros for sustainable investments in the coming decade. Secondly, the creation of suitable framework conditions for private investors and the public sector. And thirdly, the focused support of public authorities and project promoters in the selection, structuring and implementation of sustainable projects.[34]

6.2.4 Fund for a Just Transition

The European Commission adopted its legislative proposal to establish the Just Transition Fund (JTF)[35] on 14.01.2020. The fund aims to enable a just transition by mitigating the socio-economic costs of the areas most affected by the decarbonisation of the economy. In terms of content, the fund's support is mainly focused on economic transition, retraining of affected workers and job search assistance.[36]

The Just Transition Fund is endowed with a total of 17.5 billion Euros (in 2018 prices). This total is made up of 7.5 billion Euros[37] for commitments within the period 2021-2027 and 10 billion Euros[38] from the Next Generation EU Development Instrument for 2021-2023. Member States also contribute to the JTF programmes. In addition, they can

31 COM (2020) 21 final, Funding no. 3.
32 Frenz, Grundzüge des Klimaschutzrechts, marginal no. 41.
33 COM (2020) 21 final, Introduction, no. 1.
34 COM (2020) 21 final, Introduction, no. 1.
35 COM (2020) 460 final.
36 COM (2020) 460 final, p. 1.
37 Art. 3 para. 2 Regulation (EU) 2021/1056.
38 Art. 4 para. 1, 2 Regulation (EU) 2021/1056.

transfer funds from the European Regional Development Fund and the European Social Fund Plus. This enables the additional mobilisation of nearly 30 billion Euros.[39]

Art. 11 of Regulation (EU) 2021/1056 provides for a territorial approach for the support of affected regions and areas. Accordingly, the starting point for the territorial plans for a just transition to be drawn up by the Member States and the competent local and regional authorities are areas at NUTS 3 level according to Regulation (EU) 1059/2003.[40]

6.3 The European Climate Law

An essential element of the European Green Deal is the "European Climate Law",[41] which entered into force on 29 July 2021. With this regulation, a regulatory framework for the irreversible, step-by-step reduction of greenhouse gas emissions and the increase of greenhouse gas removals by sinks was created (Art. 1 subpara. 1 Regulation (EU) 2021/1119). To this end, the Regulation sets an interim target of reducing emissions by at least 55% by 2030 and a further interim target for 2040, to be determined no later than six months after the first stocktaking under the Paris Agreement planned for 2023. After achieving climate neutrality for Europe in 2050, negative emission levels are to be reached. In addition, the regulation provides a regulatory framework for the realisation of the global goal of adaptation provided for in Art. 7 of the Paris Agreement.[42]

With regard to the setting of an interim climate target for 2040 in accordance with Art. 4 of the European Climate Law, stakeholders who may be affected by a possible climate target for 2040 were asked about challenges and opportunities in their economic sectors as part of a public consultation from March to June 2023. The results of the consultation will then be used to inform the Commission's assessment of an appropriate climate target for 2040.[43]

---

39 Press release of the Council of the EU, dated 7.6.2021, Climate neutrality: Council adopts fund for a just transition, https://www.consilium.europa.eu/de/press/press-releases/2021/06/07/climate-neutrality-council-adopts-the-just-transition-fund/.

40 OJ 2014 L 241, p. 1.

41 Regulation (EU) 2021/1119 ("European Climate Law ").

42 https://eur-lex.europa.eu/legal-content/DE/LSU/?uri=CELEX:32021R1119, see Art. 1 subpara. 2, Art. 2 para. 1 Regulation (EU) 2021/1119.

43 Public consultation to set EU climate targets for 2040 and Impact Assessment Report as of 06.02.2024 (COM(2024) 63 final), https://ec.europa.eu/info/law/better-regulation/have-your-say/initiatives/13793-EU-climate-target-for-2040_en.

6.3.1 System and Arrangement of the Regulation

Based on the environmental legislative competence in Art. 192 para. 1 TFEU, the "European Climate Law"[44] consists of only fourteen articles. Art. 1 declares the establishment of a framework for irreversible, progressive reduction of greenhouse gas emissions and the enhancement of greenhouse gas removals as the object of the European Climate Law. In addition to the binding anchoring of the climate protection goals of the European Green Deal, the long-term goal of greenhouse gas neutrality by 2050 and the 2030 target, the competent organs of the Union and the member states are to be authorised to take necessary measures to make the goal of climate neutrality possible (Art. 2 para. 2).

In achieving the targets, the principles of solidarity and fairness as well as cost-effectiveness that apply between the Member States are to be taken into account above all. This implies that stronger Member States – especially as a result of the corona crisis – take on heavier burdens than weaker ones.[45]

Art. 3 regulates the tasks of the Scientific Advisory Panel on Climate Change, which was established by Art. 10 point a of Regulation (EC) No 401/2009 and which serves the EU as a reference point for scientific findings on climate change due to its independence and technical expertise. In order to strengthen the role of science in climate policy, each Member State is invited to establish a national climate advisory body responsible for providing expert advice to national authorities on climate policy (Art. 3 para. 4). Art. 4 para. 1 subpara. 1 sets the achievement of climate neutrality and the reduction of net greenhouse gas emissions within the EU by at least 55% compared to 1990 levels by 2030, as a binding climate target. In order to achieve the 2030 interim target, the competent bodies of the EU and the Member States must bring about rapid and predictable emission reductions and at the same time improve the removal of greenhouse gases by natural sinks (Art. 3 para. 1 subpara. 1). In order to ensure that sufficient mitigation action is taken by 2030, the contribution of net removals of greenhouse gases to the Union's 2030 climate target is limited to 225 million tonnes of CO_2 equivalent (Art. 4 para. 1 subpara. 3). With regard to the achievement of the climate neutrality target pursuant to Art. 2 para. 1 of the Regulation, a Union-wide climate target is set for 2040 (Art. 4 para. 3). When presenting its legislative proposal for the Union's 2040 climate target pursuant to Art. 4 para. 3 of the Regulation, the Commission shall at the same time publish in a separate report the projected indicative greenhouse gas budget of the Union for the period from 2030 to 2050, defined as the indicative total amount of net greenhouse gas emissions (in CO_2 equivalent and with separate information on emissions and removals) projected to be emitted during that

44 Regulation (EU) 2021/1119 ("European Climate Law").

45 Frenz, KSR, Introduction A marginal no. 30.

period, without thereby jeopardising the Union's commitment under the Paris Agreement (Art. 4 para. 4 sentence 1).

According to Art. 4 para. 5 of the Regulation, the Commission shall take into account the following when proposing the Union's climate target for 2040:
- the best available and most recent scientific evidence, including the latest reports of the IPCC and the Advisory Council,
- the social, economic and environmental impacts, including the costs of inaction,
- the need for a fair and socially equitable transition for all,
- the cost-effectiveness and economic efficiency,
- the competitiveness of the Union's economy, in particular small and medium-sized enterprises and those sectors of the economy at greatest risk of carbon leakage,
- the best available, cost-effective, safe and scalable technologies,
- energy efficiency and the principle of "energy efficiency first", energy affordability and security of supply,
- the fairness and solidarity between and within Member States,
- the need to ensure environmental effectiveness and progress over time,
- the need to conserve, manage and enhance natural sinks in the long term and to protect and restore biodiversity,
- the investment needs and opportunities,
- international developments and international efforts undertaken to achieve the long-term objectives of the Paris Agreement and the ultimate objective of the UNFCCC,
- the existing information on the projected indicative Union greenhouse gas budget for the period from 2030 to 2050 referred to in para. 4.

Art. 5 of the Regulation deals with measures for adaptation to climate change. Member States shall adopt and implement national adaptation strategies. In doing so, care must be taken to ensure that the measures taken by the Union and the Member States are coherent and mutually supportive and have positive spin-off effects for sector-specific policies (Art. 5 para. 3). By 30 July 2022, the Commission shall adopt guidelines setting out common principles and procedures for the identification, classification and regulatory management of significant physical climate risks in the planning, development, implementation and monitoring of projects and programmes (Art. 5 para. 5). In order to implement the requirement of Art. 5 para. 5, the "Technical guidance on the climate proofing of infrastructure in the period 2021-2027" were published in the Official Journal on 16 September 2021.[46] The purpose of the climate proofing process is to provide European institutional and private investors with a basis for making informed decisions on projects that are compatible with the Paris Agreement.

46 2021/C 373/01, ABl. C 373, dated 16.9.2021; COM (2022) 514 final, see footnote 316.

The process includes measures to mitigate climate change and adapt to its consequences in the development of infrastructure projects.[47]

Art. 6 lays down rules regarding the evaluation of the progress of the Union's measures. By 30 September 2023 and every five years thereafter, the Commission shall, together with the evaluation provided for in Art. 29 para. 5 of Regulation (EU) 2018/1999 (Governance Regulation), assess, on the one hand, the collective progress made by all Member States towards achieving the climate neutrality objective set out in Art. 2 para. 1 of this Regulation and, on the other hand, the collective progress made by all Member States in adapting towards climate neutrality (Art. 6 para. 1).[48]

Pursuant to Art. 7 para. 1, the Commission shall assess, at the same frequency, the compatibility of national measures to achieve the climate neutrality objective and the compatibility of the relevant national measures with ensuring progress towards adaptation to climate change.[49] If the Commission, taking into account all progress made jointly by the Member States, finds out that the measures of a Member State are not compatible with the climate neutrality objective, it shall issue a recommendation to the Member State, which shall be made publicly available (Art. 7 para. 2).[50]

In addition to the national measures mentioned in Art. 7 para. 1 point a, the Commission also bases its assessment under Art. 6 and 7 on the following criteria:

- Information submitted and reported in accordance with Regulation (EU) 2018/1999,
- reports from the EEA, the Advisory Council and the Commission's Joint Research Centre,
- European and global statistics and data, including statistics and data from the European Earth Observation Programme Copernicus, data on reported and projected losses due to adverse climate impacts and estimates of the costs of inaction and delayed action, where available,
- the best available and most recent scientific evidence, including the latest reports of the IPCC, IPBES and other international bodies; and
- any other information on environmentally sustainable investments made by the Union or Member States, including, where available, investments consistent with Regulation (EU) 2020/852 (Art. 8 para. 3).

47 2021/C 373/01, ABl. C 373, dated 16.9.2021, see summary under no. 1.

48 See "State of the Energy Union Report 2023", COM (2023) 650 final; reports relating to the state of the energy union for the years 2015 to 2022, https://energy.ec.europa.eu/topics/energy-strategy/energy-union_en#state-of-the-energy-union-reports.

49 COM (2020) 564 final: "An EU-wide assessment of National Energy and Climate Plans".

50 The National Energy and Climate Plans 2021–2030 including the necessary revision in 2023 and any recommendations by the Commission can be found under: https://commission.europa.eu/energy-climate-change-environment/implementation-eu-countries/energy-and-climate-governance-and-reporting/national-energy-and-climate-plans_en#national-energy-and-climate-plans-2021–2030.

In order to enable a fair and socially just transition to a climate-neutral and climate-smart society for all parts of society, the Commission shall address all of them through public participation. To this end, the Commission shall promote an inclusive, accessible process at all levels, including national, regional and local levels, to achieve the goal of climate neutrality (Art. 9 para. 1 and 2).

Furthermore, the Commission shall work with economic sectors that develop indicative roadmaps for achieving the climate neutrality objective and facilitate dialogue at Union level and exchange of best practices among stakeholders (Art. 10).

In summary, the EU Climate Change Act is more formal and procedural in nature. Apart from setting the central goal of achieving climate neutrality, it does not specify any other goals. Nevertheless, the EU Climate Act establishes a mechanism that enables the EU to intensively influence the member states to achieve the goal of greenhouse gas neutrality by 2050.[51]

6.3.2 Relation to the Governance Regulation

As the centrepiece of the legislative package "Clean Energy for All Europeans",[52] the Regulation on the Governance of the Energy Union and Climate Action[53] entered into force on 24 December 2018.[54] The Governance Regulation is intended to enable the mutual governance of energy and climate policy for the period from 2021 to 2030.[55] It constitutes a compromise and compensation for the EU's lack of competences in the area of energy supply, especially with regard to determining the energy mix in the Member States.[56]

In order to create a uniform European legal framework, Art. 13 of the European Climate Law provides for the adaptation of the Governance Regulation to the increased objectives of the European Climate Law.[57]

First of all, the objectives of the Governance Regulation are adapted to the objectives of the European Climate Law. According to Art. 13 no. 1 point a of the European Climate Law, the goals and targets already contained in the Governance Regulation are supplemented by the EU's climate neutrality target for 2050 with reference to Art. 2 para. 1 of the European Climate Law. The particularly serious importance of this target is emphasised by the wording "in particular". However, no further adjustments are made with regard to the increased target for the year 2030.[58]

51 Frenz, KSR, Introduction. A marginal no. 38; Schlacke/Köster/Thierjung, EuZW 2021, 620, 622.
52 COM (2016) 860 final.
53 Regulation (EU) 2018/1999.
54 Schlacke, EnWZ 2020, 355.
55 Schlacke/Knodt, ZUR 2019, 404.
56 Schlacke/Knodt, ZUR 2019, 404.
57 Schlacke/Köster/Thierjung, EuZW 2021, 620, 624.
58 Schlacke/Köster/Thierjung, EuZW 2021, 620, 624.

In line with the aforementioned amendment, the climate neutrality target is inserted into the monitoring system of the Governance Regulation. To this end, the European Climate Law extends the analysis and reporting obligations within the National Energy and Climate Plans[59] – the central instrument of the Governance Regulation – to the climate target.[60] The progress reporting to be carried out every two years according to Art. 17 para. 1 of the Governance Regulation is also to include progress towards climate neutrality in future (Art. 13 no. 7 point a European Climate Law). According to Art. 13 no. 6 European Climate Law, the long-term strategies pursuant to Art. 15 para. 1 Governance Regulation are also to be explicitly prepared as a contribution to achieving the climate neutrality target.[61]

Overall, the European Climate Law aligns the Governance Regulation with the 2050 target and largely brings about coherence between the two framework legal acts.[62]

6.3.3 Effects on the Member States

The Governance Regulation obliges the Member States to draw up national energy and climate plans and to monitor the Member States' targets and measures.[63] Specifically, each Member State will be required to submit an integrated national energy and climate plan to the Commission by 1 January 2029 and thereafter at ten-year intervals. The first plan, to be submitted as early as 2019, should cover the period from 2021 to 2030, taking into account a longer-term planning perspective. The subsequent plans shall each cover the ten-year period following the preceding plan.

Art. 5 para. 4 of the European Climate Law also obliges Member States to adopt and regularly update adaptation strategies and plans and to report on the update in accordance with Art. 19 para. 1 of the Governance Regulation.[64]

6.3.4 Effects on the Legal Position of Private Individuals

According to Art. 9 para. 2 of the European Climate Law, the Commission shall, for the purpose of public participation and voluntary business involvement, promote a process of discourse on best practices and the development of measures to achieve the

59 Shortcut: "NECP" ("National Energy and Climate Plan").

60 Schlacke/Köster/Thierjung, EuZW 2021, 620, 625.

61 See also footnote 352; the national long-term strategies up to the year 2050 can be found under: https://commission.europa.eu/energy-climate-change-environment/implementation-eu-countries/energy-and-climate-governance-and-reporting/national-long-term-strategies_en.

62 Schlacke/Köster/Thierjung, EuZW 2021, 620, 624.

63 Cf. Art. 3, 9, 14, 31, 32, 34 Governance-Regulation; Schlacke/Köster/Thierjung, EuZW 2021, 620, 621.

64 Schlacke/Köster/Thierjung, EuZW 2021, 620, 622.

objectives of the European Climate Law. This explicitly includes the national, regional and local levels, social partners, academia, business and citizens and civil society. The Commission can also make use of the Member State climate and energy dialogues according to Art. 10 and 11 of the Governance Regulation (cf. Art. 9 para. 3 of the European Climate Law). According to Art. 10 of the European Climate Law, the Commission shall also enter into a dialogue with various sectors of the economy that choose to develop indicative voluntary roadmaps for achieving the goal of climate neutrality in accordance with Art. 2 para. 1 of the European Climate Law.[65]

6.3.5 Implementation: Fit for 55

The "Fit for 55" programme[66] is the legislative implementation package to concretise the challenging targets identified in the European Climate Law and the Green Deal.[67] In particular, it refers to the target of reducing net greenhouse gas emissions by at least 55% by 2030 and aligning EU legislation with this target. The package was presented by the European Commission on 14 July 2021 and contains concrete interlinked proposals to shape European policies in the economic sectors of climate, energy, land use, fuels, transport, buildings and forestry to achieve the EU's 2030 interim climate target on the path to climate neutrality. The package of measures contains suitable instruments for a fundamental restructuring of the economy and society towards a fair, green and prosperous future. In total, it contains eight proposals concerning the strengthening of existing legal acts and five new initiatives for the different economic sectors mentioned.[68] The European Parliament's website contains a legislative train schedule for the "Fit for 55" package under the European Green Deal, which shows which legislative projects have been announced, are still ongoing or have already been completed.[69]

Overview of Measures

The carefully coordinated and complementary legislative instruments proposed are designed to ensure the necessary pace of greenhouse gas emission reductions over the next decade and combine the following measures: Emissions trading for new sectors and stricter requirements under the EU's existing Emissions Trading System; in-

65 Schlacke/Köster/Thierjung, EuZW 2021, 620, 625.
66 COM (2021) 550 final.
67 Franzius, KlimR 2022, 2.
68 COM (2021) 550 final, p. 4.
69 European Parliament, Legislative Train Schedule, Fit for 55 package under the European Green Deal: https://www.europarl.europa.eu/legislative-train/package-fit-for-55.

creased use of renewable energy; more energy efficiency; faster deployment of low-emission transport modes and related infrastructure and fuels; alignment of tax policy with the objectives of the European Green Deal; measures to prevent carbon leakage; instruments to preserve and increase our natural carbon sinks.[70]

Specifically, the package includes the following modernisation measures: As an overarching measure, the creation of a CO_2 border adjustment mechanism (CBAM), the reform of European emissions trading, the creation of a Social Climate Fund, the reform of the EU Effort Sharing Regulation (ESR) and Guidelines on state aid for climate, environmental protection and energy are foreseen.[71]

In the energy sector, the plan also provides for a reform of the Renewable Energy Directive, the Energy Efficiency Directive, Energy Taxation Directive and the Energy Performance of Buildings Directive.[72]

In the transport sector, an amendment of the regulation setting CO_2 emission standards for passenger cars and light commercial vehicles, a reform and conversion of the Alternative Fuels Infrastructure Directive into a regulation for the rapid expansion of infrastructure for alternative fuels[73] is foreseen, as well as the reform of directives and the creation of action plans for alternative fuels.[74]

Development of Emissions Trading

The EU Emissions Trading System (EU ETS) has reduced CO_2 emissions from power generation and other energy-intensive industries by 42.8% over the last 16 years through cap and trade.[75] However, an analysis by the Commission showed that if emissions reductions remained the same, the sectors currently covered by the EU ETS would only reduce their emissions by 51% by 2030 compared to 2005.[76] This means that the target of at least 55% compared to 1990 levels would be missed. In order to

70 Falke, ZUR 2021, 567, 569.

71 https://www.csr-in-deutschland.de/DE/CSR-Allgemein/CSR-Politik/CSR-in-der-EU/EU-Green-Deal/eu-green-deal.html;jsessionid=98DD4C246FAE6C44485A14C7798E2AC1.delivery1-replication.

72 https://www.csr-in-deutschland.de/DE/CSR-Allgemein/CSR-Politik/CSR-in-der-EU/EU-Green-Deal/eu-green-deal.html;jsessionid=98DD4C246FAE6C44485A14C7798E2AC1.delivery1-replication.

73 Regulation (EU) 2023/1804 published in the ABl. L 234/1 from 22.9.2023; Council of the EU, Press release from 25.7.2023, https://www.consilium.europa.eu/de/press/press-releases/2023/07/25/alternative-fuels-infrastructure-council-adopts-new-law-for-more-recharging-and-refuelling-stations-across-europe/.

74 https://www.csr-in-deutschland.de/DE/CSR-Allgemein/CSR-Politik/CSR-in-der-EU/EU-Green-Deal/eu-green-deal.html;jsessionid=98DD4C246FAE6C44485A14C7798E2AC1.delivery1-replication.

75 European Commission, Press release from 14.7.2021, p. 1.

76 COM (2021) 551 final, p. 2.

increase CO_2 reductions by means of the EU ETS, the cap on allocated climate allowances must be adjusted above all.[77]

Against this backdrop, the EU Council adopted the following laws on 25.04.2023:

- Revision of the ETS-Directive[78]
- Amendment of the MRV-Regulation[79] for shipping
- Revision of the ETS Aviation Directive
- Regulation on the establishment of a Social Climate Fund
- Regulation on the creation of a Carbon Border Adjustment Mechanism.[80]

The aforementioned legislative procedures and amendments were all announced in the Official Journal of the EU on 16.05.2023.[81] In general, these amendments enable the EU to reduce greenhouse gas emissions in the most important economic sectors and to support the most financially vulnerable citizens and micro-enterprises exposed to carbon leakage in the climate transition.[82]

Revision of the ETS-Directive

The Directive (EU) 2023/959, which was announced on 16 May 2023, amends the original EHS-Directive in many respects.

Stronger Reduction of Greenhouse Gas Emissions

Firstly, the new Art. 1 para. 2 stipulates that the ETS Directive provides for a greater reduction in greenhouse gas emissions in order to contribute to the achievement of the Union's climate neutrality target and the Union's climate targets from the European Climate Law. Overall, emissions from the sectors covered by the EU ETS must be reduced by 62% compared to 2005 in order to achieve the Union's emissions reduction target for 2030.[83]

77 COM (2021) 551 final, p. 2.

78 Directive 2003/87/EC ("ETS-Directive").

79 Regulation (EU) 2015/757 ("Regulation on the monitoring, reporting and verification of carbon dioxide emissions from maritime transport, and amending Directive 2009/16/EU", shortcut: "MRV-Regulation").

80 Council of the EU, Press release from 25.4.2023, https://www.consilium.europa.eu/en/press/press-releases/2023/04/25/fit-for-55-council-adopts-key-pieces-of-legislation-delivering-on-2030-climate-targets/.

81 2023/L 130/1, ABl. L 130 from 16.5.2023.

82 Council of the EU, Press release from 25.4.2023, https://www.consilium.europa.eu/en/press/press-releases/2023/04/25/fit-for-55-council-adopts-key-pieces-of-legislation-delivering-on-2030-climate-targets/; 2023/L 130/1, ABl. L 130 v. 16.5.2023.

83 Directive (EU) 2023/959, recital 39.

Separate Emissions Trading System for Buildings, Road Transport and Other Sectors (EU ETS II)

In addition, Chapter Iva of the ETS-Directive now provides for an "Emissions Trading System for Buildings, Road Transport and additional Sectors", which primarily includes small businesses[84] (EU ETS II).[85] The reason for the introduction of a new emissions trading system for these sectors is that it can and should address the different CO_2 abatement costs and emission cost-effective reduction potential.[86] Unlike the EU ETS I, the EU ETS II is a so-called "upstream" trading system, i.e. unlike the EU ETS I,[87] the end consumers do not participate as emitters (so-called "downstream"). Instead, the participants are the fuel traders as distributors of the fuels and not the end consumers.[88]

According to Art. 30b of the ETS Directive, affected companies must have an emissions permit from 01.01.2025. Overall, the provisions for monitoring, reporting, verification and permitting for those sectors covered by EU ETS II will apply from 01.01.2025.[89] The Member States shall ensure that each regulated entity holding a permit in accordance with Art. 30b reports its historical emissions for the year 2024 by 30 April 2025; monitors their emissions for each calendar year from 2025 onwards and reports those emissions to the competent authority in the following year, starting in 2026, in accordance with the implementing acts referred to in Art. 14 para. 1, see Art. 30f para. 4 and 2 of the ETS Directive.

Allowances for the emissions will be auctioned from 2027, Art. 30d para. 1 of the ETS Directive,[90] whereby the Union-wide quantity of allowances issued under this chapter shall decrease in a linear manner beginning in 2024, Art. 30c para. 1 of the ETS Directive.[91]

Emissions from the Maritime Transport

According to Art. 3a, the Art. 3ga to 3gg of the ETS Directive shall apply in respect of the maritime transport activities listed in Annex I. The scope of application extends to ships with a gross tonnage of 5.000; no later than 31 December 2026, the Commission should present a report to the European Parliament and to the Council in which it should examine the feasibility and economic, environmental and social impacts of the

84 Directive (EU) 2023/959, recital 77 ("small emitters"); Council of the EU, Press release from 25.4.2023, https://www.consilium.europa.eu/en/press/press-releases/2023/04/25/fit-for-55-council-adopts-key-pieces-of-legislation-delivering-on-2030-climate-targets/.

85 Directive (EU) 2023/959, Art. 30a – 30k.

86 Telschow/Grebe, IR 2023, 98; see COM (2021) 551 final.

87 German Bundestag, Scientific Services, 26.1.2023, WD 8 - 3000 - 001/23, see remark under footnote 289.

88 Kreuter-Kirchhof, KlimR 2022, 70,72; Stätsche, KlimR 2023, 171, 175.

89 Directive (EU) 2023/959, recital 100.

90 Directive (EU) 2023/959, recital 86.

91 Directive (EU) 2023/959, recital 80.

inclusion in Directive 2003/87/EC of emissions from ships below 5.000 gross tonnage, including offshore ships.[92] Art. 3 ga contains regulations on which voyages of ships, between which ports are subject to the allocation of allowances and the application of the levy obligations. Emissions from ships performing voyages departing from a port of call under the jurisdiction of a Member State and arriving at a port of call outside the jurisdiction of a Member State and also from ships performing voyages departing from a port of call outside the jurisdiction of a Member State and arriving at a port of call under the jurisdiction of a Member State are taken into account, which is intended to ensure that shipping companies do not shift their activities to other ports.[93] According to Art. 3 gb, the requirements for the remittance of emission allowances for maritime transport will be introduced gradually: shipping companies must remit the corresponding number of allowances for 40% of their verified emissions reported for 2024, 70% for 2025 and 100% for 2026 and each year thereafter.[94]

Reduction of Allowances

In addition, Art. 9 of the ETS Directive provides that the Union-wide quantity of allowances shall be decreased by 90 million allowances in 2024 and then by a further 27 million allowances in 2026. At the same time, in 2024, the Union-wide quantity of allowances shall be increased by 78.4 million allowances for maritime transport. The linear factor shall be 4.3% from 2024 to 2027 and 4.4% from 2028. As a result of this increase, allowance prices shall continue to rise, intensifying the economic pressure on plant operators to reduce their greenhouse gas emissions.[95]

Innovations in Free Allocation

Three major changes are planned with regard to the free allocation of allowances:

- The minimum adjustment of the benchmark values should be increased from 0.2% to 0.3% per year, and the maximum adjustment should be increased from 1.6% to 2.5% per year. This will allow better account to be taken of technical progress and at the same time create incentives to reduce emissions and reward innovation appropriately. For the period from 2026 to 2030, the benchmark values should thus be adjusted within a range of 6% to 50% compared to the value applicable in the period from 2013 to 2020.[96] In the context of free allocation, bench-

92 Directive (EU) 2023/959, recital 30.
93 Directive (EU) 2023/959, Art. 3ga; Telschow/Grebe, IR 2023, 98, 99.
94 Telschow/Grebe, IR 2023, 98, 99.
95 Kreuter-Kirchhoff, KlimR 2022, 70, 71.
96 Directive (EU) 2023/959, recital 48.

marks orient the free allocation of emission allowances to how many tons of CO_2 per ton of product the 10% most efficient plants in the EU emit.[97]

- In addition, the free allocation of emission allowances may in future be linked to the provision of environmental services[98] in return. According to Art. 10a para. 1, the amount of free allocation shall be reduced by 20% if an installation falls under the obligation to carry out an energy audit or a certified energy management system[99] and the recommendations contained therein are not implemented. However, Art. 10a para. 1 also provides for exceptions to such a reduction. There is also a threat of a 20% reduction in cases where, by 1 May 2024, operators of installations whose greenhouse gas emission levels are higher than the 80th percentile of emission levels for the relevant product benchmarks have not established a climate-neutrality plan for each of those installations for its activities covered by this Directive.
- With the implementation of the Carbon Border Adjustment Mechanism[100] ("CBAM"), there will also be no free allocation for sectors and sub-sectors that fall under the CBAM. This is intended to counteract the risk of carbon leakage.[101]

Amendment of the MRV-Regulation for Shipping

The scope of the MRV-Regulation[102] is extended by the newly adopted Regulation (EU) 2023/957. For example, Art. 2 para. 1 covers ships with a gross tonnage of 5,000 and above in respect of the greenhouse gas emissions released and, according to Art. 2 para. 1a, general cargo ships and offshore ships with a gross tonnage of 400 to 5,000 from 1 January 2025. According to Art. 2 para. 1b, the Regulation also applies to offshore ships with a gross tonnage of 5,000 and above from 1 January 2025.

The scope of the regulation has also been changed so that it no longer only covers CO_2 emissions, but greenhouse gases as a whole, i.e. both CO_2 and, from 2024 onwards, methane and nitrous oxide, see Art. 2 para. 1c.[103] From 2026, these emissions are also to be included in the ETS-Directive and thus in the European emissions trading system.[104]

97 Telschow/Grebe, IR 2023, 98, 99.
98 Telschow/Grebe, IR 2023, 98, 100.
99 This refers to energy management system under Art. 8 of Directive 2012/27/EU.
100 Regulation (EU) 2023/956, for further details see following section "Regulation on the creation of a Carbon Border Adjustment Mechanism".
101 Directive (EU) 2023/959, recital 46.
102 Regulation (EU) 2015/757 ("MRV-Regulation").
103 Regulation (EU) 2023/957, recital 9.
104 Directive (EU) 2023/959 ("ETS-Directive"), recital 20.

Revision of the ETS Aviation Directive

In addition to the changes to the ETS-Directive already mentioned, Directive (EU) 2023/958 also amends the Directive by providing for the gradual abolition of the free allocation of allowances for the aviation sector, which is already covered by the EU ETS. Free allocation is to be phased out in 2024 and 2025, with full auctioning to take place from 2026 onwards.[105]

In accordance with Art. 3c para. 6, during the period from 1 January 2024 until 31 December 2030, 20 million allowances should be reserved in order to be allocated to cover part of the remaining price differential between fossil kerosene and the eligible aviation fuels for individual aircraft operators.[106]

The European emissions trading system applies to intra-European flights, flights within the European Economic Area (EEA) as well as flights to the United Kingdom and Switzerland.[107]

Regulation on the Establishment of a Social Climate Fund

According to Art. 1 on the Establishment of a Social Climate Fund (Regulation 2023/955),[108] the measures and investments supported by the Social Climate Fund shall benefit households, micro-enterprises and transport users, which are vulnerable and particularly affected by the new EU ETS II. It thus serves to combat energy poverty.[109] The specific objectives of the Social Climate Fund shall be to support vulnerable households, vulnerable micro-enterprises and vulnerable transport users, through temporary direct income support and through measures and investments intended to increase the energy efficiency of buildings, decarbonisation of heating and cooling of buildings, including through the integration in buildings of renewable energy generation and storage, and to grant improved access to zero- and low-emission mobility and transport, Art. 3 no. 2 Regulation (EU) 2023/955. Disadvantaged households, disadvantaged small businesses and disadvantaged transport users are to be supported through temporary direct income support and through measures and investments that increase the energy efficiency of buildings, decarbonize the heating and cooling of buildings

105 Art. 3d para. 1 of Directive (EU) 2023/958; Directive (EU) 2023/958, recital 14.

106 Directive (EU) 2023/958, recital 15.

107 https://eur-lex.europa.eu/DE/legal-content/summary/greenhouse-gas-emission-allowance-trading-system.html; on the relationship between the EU ETS and the "Carbon Offsetting and Reduction Scheme for International Aviation" (CORSIA) of the International Civil Aviation Organization (ICAO), see Directive (EU) 2023/958, recitals 22, 23, 30; on CORSIA, see also Council of the EU, press release of 19.12.2022 with references there: https://www.consilium.europa.eu/de/press/press-releases/2022/12/19/council-adopts-decision-on-offsetting-requirements-for-air-transport-emissions-corsia/.

108 Regulation (EU) 2023/955.

109 "Energy poverty" means a household's lack of access to essential energy services that underpin a decent standard of living and health, including adequate warmth, cooling, lighting, and energy to power appliances, in the relevant national context, existing social policy and other relevant policies, see Art. 2 no. 1 of Regulation (EU) 2023/955.

and improve access to zero- and low-emission mobility and corresponding means of transport, Art. 3 para. 2 of Regulation (EU) 2023/955.

With a term of seven years from 2026 to 2032, the fund has 65 billion Euros at its disposal, Art. 10 para. 1 of Regulation 2023/955.

Regulation on the Creation of a Carbon Border Adjustment Mechanism ("CBAM")

Due to the strict requirements of the EU Emissions Trading System (EU ETS I and EU ETS II), the purpose of the Carbon Border Adjustment Mechanism[110] is, on the one hand, to prevent the risk of carbon leakage from certain sectors and sub-sectors of industry to other countries with less stringent climate protection ambitions and, on the other hand, to prevent imports from non-EU countries from replacing equivalent products that cause fewer greenhouse gas emissions.[111] On the one hand, such a shift would contradict the European Union's efforts to reduce CO_2 emissions and, on the other, would put Europe at a competitive disadvantage as an industrial location.

In light of this, the CBAM is now intended to ensure the same level of CO_2 pricing for imports and domestic products.[112] In principle, the CBAM only applies to certain goods listed in Annex I to the Regulation, namely iron and steel, cement, fertilisers, aluminium, hydrogen products and electricity originated in a third country, see Art. 2 para. 1 of Regulation (EU) 2023/956. According to Art. 4 of Regulation (EU) 2023/956, goods shall be imported into the customs territory of the Union only by an authorised CBAM declarant.[113] The CBAM is then used to record the embedded emissions of imported goods and assign a value to these emissions, which is measured in CBAM certificates.[114] According to Art. 3 no. 24 of Regulation (EU) 2023/956, such a "CBAM certificate" means a certificate in electronic format corresponding to one tonne of CO_2e of embedded emissions in goods.

A Member State shall sell CBAM certificates on a common central platform to authorised CBAM declarants established in that Member State;[115] the Commission shall calculate the price of CBAM certificates as the average of the closing prices of EU ETS allowances on the auction platform for each calendar week, Art. 21 para. 1 of Regulation (EU) 2023/956. By 31 May of each year, and for the first time in 2027 for the year 2026, the authorised CBAM declarant shall surrender via the CBAM registry a number of CBAM certificates that corresponds to the embedded emissions declared and verified for the calendar year preceding the surrender, Art. 22 para. 1 of Regulation (EU) 2023/956.

110 See Art. 1 para. 1 of Regulation (EU) 2023/956: "CBAM" shortcut for Carbon Border Adjustment Mechanism; Regulation (EU) 2023/956, recital 10.

111 Art. 1 para. 1 of Regulation (EU) 2023/956; Regulation (EU) 2023/956, recital 9.

112 Regulation (EU) 2023/956, recital 12.

113 "Importation" means release for free circulation as provided for in Art. 201 of Regulation (EU) No 952/2013, Art. 3 no. 4 of Regulation (EU) 2023/956.

114 Trennt/Ulke, EuZW 2023, 452, 453 f.

115 Art. 20 para. 1 of Regulation (EU) 2023/956.

In summary, the system works by EU importers buying certificates that correspond to the CO_2 price that would have been incurred if the goods had been produced in accordance with EU climate standards. However, importers are not obliged to make such compensation payments if they can prove that they have already paid a price for the CO_2 produced abroad.[116]

With the gradual introduction of the CBAM, the current system of free allocation of allowances in the sectors covered by the CBAM will be gradually phased out from 2024 until 2034.[117] In the CBAM sectors, companies will still receive an allocation of 97.5% of the allocation volume in 2026, whereby the percentage will be gradually reduced in the following years and reduced to 0% in 2034.[118]

According to Art. 36 para. 2 of Regulation (EU) 2023/956, this applies from 01.01.2023; deviating from this, certain articles will apply gradually from 31.12.2024 or 01.01.2026. In the period from 01.01.2024 to 31.12.2025, a transitional phase applies in which the obligations of importers are limited to certain reporting obligations in accordance with Art. 33 ff. of Regulation (EU) 2023/956.

Market Stability Reserve

The current market stability reserve will be modified in such a way that the percentages and quantity of allowances of 100 million EUR mentioned in Art. 1 para. 5 subpara. 1 of Decision (EU) 2015/1814 will be doubled.[119] This is intended to prevent a significant increase in the surplus of allowances in the EU ETS.[120]

Amendment of the Renewable Energies Directive (RED II)

With the RED Directive (RED II)[121] based on the TFEU, the European legislator prescribes a common framework for the promotion of renewable energies. Unlike its predecessor, the RED II does not contain any targets for the individual Member States, but only the binding overall target of the European Union for 2030. The RED Directive was already to be implemented by 30 June 2021. However, due to the stricter reduction requirements in the "Fit for 55" package, the Directive was comprehensively revised again in the revision process ("RED III").[122] The RED III came into force on

116 QANDA_21_3661, q. 1.

117 See Art. 31 of Regulation (EU) 2023/956; Regulation (EU) 2023/956, recitals 11, 12; Art. 10a of Directive 2003/87/EC; Weber kompakt/Hakenberg, CO2-Grenzausgleichssystem; Telschow/Grebe, IR 2023, 98, 100.

118 Directive (EU) 2023/959, recital 46; Telschow/Grebe, IR 2023, 98, 100.

119 Decision (EU) 2023/852; 2023/L 110/21, ABl. L 110 dated 25.4.2023.

120 Decision (EU) 2023/852, recital 15.

121 Directive (EU) 2018/2001.

122 Directive (EU) 2023/2413 ("RED III"), published in ABl. L dated 31.10.2023; COM (2021) 557 final.

20.11.2023, after it was published in the Official Journal on 31.10.2023, see Art. 7. According to Art. 5 para. 1 of RED III, the Member States have 18 months, i.e. until 21.05.2025, to implement the legal and administrative provisions.

The overarching goal of the revision of RED II was to achieve an increase in the use of renewable energies by 2030. For the implementation of the climate target, a significantly higher share of renewable energy sources in an integrated energy system is necessary than currently envisaged by RED II. While the RED II foresees a share of at least 32%, the Climate Target Plan (CTP) requires an increase to 38–40%. The RED III now states that the Member States must ensure collectively that the share of energy from renewable sources in the Union's gross final consumption of energy in 2030 is at least 42.5%; besides that the Member States shall collectively endeavour to increase the share of energy from renewable sources in the Union's gross final consumption of energy in 2030 to 45%, see Art. 3 para 1. Within the framework of the integration of the energy system, the hydrogen strategy, the strategy for renewable offshore energy and the biodiversity strategy, complementary flanking measures in various sectors are necessary to achieve the above-mentioned goals. Another objective is to promote the achievement of climate and environmental goals, for example to protect biodiversity. The revision of RED II and its adaptation to the current circumstances was essential both for achieving the challenging climate target set out in the Green Deal and for reducing Europe's dependence on energy imports, protecting the environment and health, and contributing to Europe's technological and industrial leadership and economic growth.[123]

Against this background, Member States must ensure by 21.02.2024 that, until climate neutrality is achieved, in the permit-granting procedure, the planning, construction and operation of renewable energy plants, the connection of such plants to the grid, the related grid itself, and storage assets are presumed as being in the overriding public interest and serving public health and safety when balancing legal interests in individual case.[124]

In addition, the RED III provides for faster project approval procedures overall. The Member States must designate renewables acceleration areas[125] by 21 February 2026, see Art. 15c of RED III; the public must also be involved, see Art. 15d of RED III. The permit-granting procedure shall not exceed 12 months for renewable energy projects in renewables acceleration areas according to Art. 16a of RED III; in the case of offshore renewable energy projects, the permit-granting procedure shall not exceed two years. The permit-granting procedure for the installation of solar energy equipment and co-located energy storage, including building-integrated solar installations, in existing or future artificial structures, with the exclusion of artificial water surfaces, shall not exceed three

123 COM (2021) 557 final, p. 1.

124 Art. 16f of Directive (EU) 2023/2413.

125 "Renewables acceleration area" means a specific location or area, whether on land, sea or inland waters, which a Member State designated as particularly suitable for the installation of renewable energy plants, see Art. 2 para. 2 no. 9a of RED III.

months, provided that the primary aim of such artificial structures is not solar energy production or energy storage, see Art. 16d para. 1. For the installation of solar energy equipment with a capacity of 100 kW or less, including for renewable self-consumers and renewable energy communities, the duration of the permit-granting procedure should not exceed one month, see Art. 16d para. 2. Art. 16e also contains regulations on permit-granting procedures for the installation of heat pumps.

According to Art. 16a para. 3, renewable energy projects, new applications for renewable energy plants, including plants combining different types of renewable energy technology and the repowering of renewable energy power plants in designated renewables acceleration areas for the relevant technology and co-located energy storage, as well as the connection of such plants and storage to the grid, shall be exempt from the requirement to carry out a dedicated environmental impact assessment pursuant to Art. 2 para. 1 of Directive 2011/92/EU, provided that those projects comply with Art. 15c para. 1, point (b), of this Directive.

Energy Efficiency of Buildings

The energy transition is accompanied by the need to reduce the current final energy consumption. Buildings offer an elementary savings potential. The Directive 2018/844[126] issued for this purpose amended Directive 2010/31/EU[127] on the energy performance of buildings and Directive 2012/27/EU[128] on energy efficiency.[129]

The existing building stock, which is responsible for approx. 36% of the total CO_2 emissions in the European Union, is to be taken as a starting point.[130] In addition, 50% of energy consumption in the EU is used for heating or cooling – 80% of which is inside buildings.[131] The building stock must therefore be decarbonised in order to ensure a transition towards a climate-neutral economy and a secure and decarbonised energy system by 2050. With regard to the Union's medium-term targets for 2030, 2040 and ultimately climate neutrality in 2050, it is essential that Member States develop long-term renovation strategies for the public and private residential and non-residential building stock in order to reduce greenhouse gas production by 80–95% by 2050 compared to 1990. To this end, Member States shall define national progress indicators, which shall be individually adapted to national circumstances and developments.[132] Art. 2a point a to g of Directive (EU) 2010/31 contains requirements regarding

126 Regulation (EU) 2018/844.
127 Regulation (EU) 2010/31.
128 Regulation (EU) 2012/27.
129 Frenz, Grundzüge, marginal no. 141.
130 Frenz, Grundzüge, marginal no. 138.
131 Frenz, Grundzüge, marginal no. 141.
132 Recital 6, Regulation (EU) 2018/844.

the minimum content of the renovation strategy. The strategies should guarantee a highly energy-efficient and decarbonised national building stock and enable the cost-effective conversion of existing buildings into low-energy buildings.

On 07.12.2023, the Council and Parliament reached a provisional agreement on a proposal to revise the Energy Performance of Buildings Directive, with the main objective of the revision being that all new buildings should be zero-emission buildings by 2030 at the latest and that existing buildings should be transformed into zero-emission buildings by 2050.[133] The revised Directive (EU) 2024/1275 of the European Parliament and of the Council of 24.04.2024 on the energy performance of buildings entered into force meanwhile.[134]

Promotion of Low-emission Modes of Transport, Infrastructure, Fuels

The step towards the lowest possible emission mobility is to be accelerated by stricter CO_2 standards. Against this backdrop, a Regulation to tighten CO_2 emission performance standards for new passenger cars and new light commercial vehicles was published in the Official Journal of the EU on 25.04.2023.[135] The Regulation provides for the following EU fleet-wide targets: from 01.01.2030, CO_2 emissions of the new passenger car fleet are to be reduced by 55% and for new light commercial vehicles by 50% compared to 2021 levels.[136] From 01.01.2035, the target for the new passenger car fleet and for the light commercial vehicle fleet will be to reduce the average emissions by 100% compared to 2021.[137] Practically, this means the end of combustion engines from 2035.[138] This drastic turnaround is intended to make an unmistakable statement in favour of investments by the automotive sector in zero-emission drive technologies and also to support the production and sale of low-emission or zero-emission vehicles. For the practical implementation of the transition of transport to low-emission fuels, the rapid development of an infrastructure for alternative fuels is indispensable. In conjunction with the tightening of CO_2 emission standards for new passenger cars and new light commercial vehicles, the Regulation on the deployment of alternative fuels

133 Council of the EU, Press release dated 7.12.2023, https://www.consilium.europa.eu/en/press/press-releases/2023/12/07/fit-for-55-council-and-parliament-reach-deal-on-proposal-to-revise-energy-performance-of-buildings-directive/; Council of the EU, 13280/22, 25.10.2022: Proposal for a Directive on the energy performance of buildings (recast).

134 The revised Directive (EU) 2024/1275 of the European Parliament and of the Council of 24.04.2024 on the energy performance of buildings entered into force meanwhile.

135 Regulation (EU) 2023/851, published in ABl. L 110/5 dated 25.4.2023; Council of the EU, Press release dated 28.3.2023, https://www.consilium.europa.eu/en/press/press-releases/2023/03/28/fit-for-55-council-adopts-regulation-on-co2-emissions-for-new-cars-and-vans/.

136 Art. 1 no. 1 point a) of Directive (EU) 2023/851.

137 Art. 1 no. 1 point b) of Directive (EU) 2023/851.

138 Falke, ZUR 2021, 567, 570.

infrastructure was adopted.[139] The Regulation establishes mandatory national targets leading to the deployment of sufficient alternative fuels infrastructure in the Union for road vehicles, trains, vessels and stationary aircraft.[140]

The Commission's 2022 proposal for the so-called "Euro 7" Regulation[141] is also linked to the European provisions for more sustainable transport, which aims to achieve more appropriate requirements for emissions, but also for other environmentally harmful factors, while also addressing other issues such as tyre abrasion and battery durability.[142] The legislative process is still ongoing: on 25.09.2023, the Council adopted its "general approach" on the Euro 7 Regulation;[143] a first reading was then held in the European Parliament on 09.11.2023.[144]

In addition, the "ReFuelEU Aviation"[145] and "FuelEU Maritime"[146] Regulations also aim to increase the use of sustainable fuels by aircraft and ships, thereby reducing their ecological footprint.[147] As a result, these Regulations also serve the overarching goal of reducing the EU's greenhouse gas emissions in order to achieve the target of a 55% reduction by 2030.

Alignment of Tax Policy

For the successful implementation of the European goal of climate neutrality, it is also necessary to adapt the taxation of energy products to energy and climate policy.[148] This includes the abolition of tax incentives for the use of fossil fuels in the form of

---

139 Regulation (EU) 2023/1804, published in ABl. L 234/1 dated 22.9.2023; Council of the EU, Press release dated 25.7.2023, https://www.consilium.europa.eu/en/press/press-releases/2023/07/25/alternative-fuels-infrastructure-council-adopts-new-law-for-more-recharging-and-refuelling-stations-across-europe/.

140 Art. 1 para. 1 of Regulation (EU) 2023/1804.

141 COM (2022) 586 final, "Proposal for regulation on type-approval of motor vehicles and engines and of systems, components and separate technical units intended for such vehicles, with respect to their emissions and battery durability (Euro 7) and repealing Regulations (EC) No 715/2007 and (EC) No 595/2009".

142 Council of the EU, Press release dated 25.9.2023, https://www.consilium.europa.eu/en/press/press-releases/2023/12/18/euro-7-council-and-parliament-strike-provisional-deal-on-emissions-limits-for-road-vehicles/.

143 Council of the EU, Press release dated 25.9.2023, https://www.consilium.europa.eu/en/press/press-releases/2023/12/18/euro-7-council-and-parliament-strike-provisional-deal-on-emissions-limits-for-road-vehicles/.

144 https://www.europarl.europa.eu/doceo/document/TA-9-2023-0394_EN.html.

145 Regulation (EU) 2023/2405, published in ABl. L dated 31.10.2023.

146 Regulation (EU) 2023/1805, published in ABl. 234/48 dated 22.9.2023.

147 Council of the EU, Infographics, https://www.consilium.europa.eu/en/infographics/fit-for-55-refueleu-and-fueleu/.

148 COM (2021) 563 final p. 2; QANDA_21_3662.

tax exemptions, for example in air and maritime transport, which are diametrically opposed to the realisation of the goals of the European Green Deal.[149]

The revision of the Energy Taxation Directive aims to ensure the following objectives: First, to create an adapted tax framework that has a positive impact on the EU's 2030 targets and on climate neutrality by 2050. Furthermore, the EU internal market is to be preserved and improved by updating the scope and structure of tax rates and streamlining the application of tax exemptions and reductions by Member States. In addition, scope should be created to allow Member States to generate revenue for their budgets.[150]

The aforementioned goals are to be achieved by setting tax rates for fuels and energy products according to their environmental performance in the future. Accordingly, conventional fossil fuels such as gas oil and petrol will be assigned to the highest tax category. The next highest tax rate will cover fossil-based fuels and combustibles that are less harmful than those mentioned and can thus still contribute to decarbonisation, at least in the medium term. Examples of this group are natural gas, liquefied petroleum gas and hydrogen of fossil origin. During the transitional period of ten years, these will be taxed at a rate of two-thirds of the reference rate. After the ten-year period, the tax rate will increase to the full reference rate. Another tax category is sustainable, but still non-advanced biofuels, which are taxed at a rate equal to half the reference rate. Regardless of its use, electricity is taxed at the lowest rate. In addition, biofuels, biogases and hydrogen of renewable origin are also included in this category. The taxation is significantly lower than the reference rate because the above-mentioned energy sources contribute to the achievement of European climate goals and their use drives the EU's transition to clean energy.[151]

In order to cushion the economic and social consequences of the introduction of the described taxation for sectors in which a full tax exemption is currently possible – e.g. aviation or heating fuels for non-vulnerable households – transitional periods are provided for.[152]

The proposal also takes social aspects into account. Member States are free to exempt vulnerable households from the taxation of heating fuels for a period of ten years. Member States are free to grant tax reductions for heating fuels to private households as long as these reductions do not fall below the minimum rates.[153]

149 Falke, ZUR 2021, 567, 570.

150 COM (2021) 563 final p. 4; Council of the EU, Infographics: https://www.consilium.europa.eu/en/infographics/fit-for-55-energy-taxation/.

151 COM (2021) 563 final p. 4.

152 COM (2021) 563 final p. 4.

153 COM (2021) 563 final p. 5.

Preserving and Increasing Our Natural CO_2 Sinks

The revised Regulation on Land Use, Land Use Change, Forestry and Agriculture (LULUCF[154])[155] now provides for a new Union target for CO_2 reduction of 310 million tons of CO_2 equivalent as a sum of the values of the net greenhouse gas emissions by 2030.[156] This represents a planned increase of 15%. In addition to the Union target for 2030, the Regulation also contains national targets for the Member States for net greenhouse gas removals in the LULUCF sector for the period from 2026 to 2030.[157] For the period from 2021 to 2025, Member States must ensure that their own greenhouse gas emissions do not exceed greenhouse gas removals.[158] The national targets are based on the most recent levels of greenhouse gas emissions or reductions and the potential for further reductions.[159]

Revision of the Effort Sharing Regulation

The revised Effort Sharing Regulation[160] obliges Member States to set higher national targets for their minimum contributions for the period from 2021 to 2030 in order to fulfill the Union's target of reducing its greenhouse gas emissions by 40% below 2005 levels in 2030.[161] The revised regulation thus contributes to the long-term target of climate neutrality in the Union by 2050. The scope of the Effort Sharing Regulation covers road and domestic maritime transport, buildings, agriculture, waste management and small industrial operations.[162]

The Regulation also provides for the adoption of annual emission allocations for each Member State to reach the cap in 2030 and to progressively lead to the 2030 greenhouse gas emission reduction target of each Member State.[163] To this end, the Commission shall adopt implementing acts setting out the annual emission alloca-

154 "LULUCF" shortcut for: Land Use, Land Use Change, Forestry and Agriculture Regulation, Art. 1 point a of Regulation (EU) 2023/839.

155 Regulation (EU) 2023/839, published in ABl. L 107/1, dated 21.4.2023; COM (2021) 554 final.

156 Art. 4 para. 2 of Regulation (EU) 2023/839.

157 Art. 1 point d) of Regulation (EU) 2023/839 in connection with Annex IIa of Regulation (EU) 2023/839.

158 Art. 4 para. 1 of Regulation (EU) 2023/839.

159 Council of the EU, Infographics, https://www.consilium.europa.eu/en/infographics/fit-for-55-lulucf-land-use-land-use-change-and-forestry/.

160 Regulation (EU) 2023/857, published in ABl. L 111/1, dated 16.4.2023.

161 Art. 1 of Directive (EU) 2023/857.

162 Regulation (EU) 2023/857, recital 9; Art. 2 of Regulation (EU) 2018/842; Council of the EU, Press release dated 28.3.2023, https://www.consilium.europa.eu/en/press/press-releases/2023/03/28/fit-for-55-package-council-adopts-regulations-on-effort-sharing-and-land-use-and-forestry-sector/.

163 Regulation (EU) 2023/857, recital 15.

tions, expressed in tonnes of CO_2 equivalent, for each Member State for the years from 2021 to 2030 in accordance with the established linear reduction pathways.[164]

The flexibility provisions of Art. 5 ff. of Regulation (EU) 2018/842, which continues to apply, provide assistance in achieving these targets.[165]

Green State Aid Law

State aid law is related to climate protection in two respects. On the one hand, this consists of the obligation of the Member States to promote a transition to climate friendliness by 2050 under the aspects of cost efficiency and social inclusion. On the other hand, these national subsidies find a clearly defined limit through the prohibition of state aid according to Art. 107 TFEU.[166]

The revised Guidelines on State Aid for Climate, Environmental and Energy Protection (CEEAG) were adopted and published by the EU Commission on 27.01.2022.[167] They became immediately applicable. The revision aligned the existing 2014 guidelines with the objectives set out in the European Green Deal and amended energy and environmental legislation. The update and also the change of name of the guidelines are primarily intended to give weight to the fact that climate protection is currently of more central importance than ever.[168]

A significant innovation in the CEEAG is the extension of the scope of application of the guidelines. Thus, the chapter on the promotion of renewable energies now generally includes aid for the reduction and elimination of greenhouse gas emissions, as well as for the promotion of renewable energies and energy efficiency. In addition, there are new chapters in the areas of energy performance and the environmental performance of buildings, resource efficiency and the circular economy, and in the area of clean mobility. The latter section includes support for the purchase or leasing of new or used clean vehicles for air, road, rail, inland waterway and maritime transport, as well as for clean mobile service equipment. Under special conditions, aid for the retrofitting, conversion or adaptation of these vehicles or service equipment is also possible.

Furthermore, there are new aid schemes in the areas of remediation of environmental damage and the decommissioning of coal, peat or oil shale power plants as well as the termination of the mining of the respective input materials.[169]

164 Art. 4 para. 2 and 3 of Regulation (EU) 2023/857.

165 Council of the EU, Press release dated 28.3.2023, https://www.consilium.europa.eu/en/press/press-releases/2023/03/28/fit-for-55-package-council-adopts-regulations-on-effort-sharing-and-land-use-and-forestry-sector/.

166 Frenz, Grundzüge, marginal no. 201.

167 2022 (2022/C 80/01).

168 Mora, EuZW 2022,195.

169 Mora, EuZW 2022,195, 196.

Reductions of taxes or charges similar to environmental taxes in the context of aid for energy-intensive enterprises will be strictly limited in the future. In addition, the number of eligible sectors listed in the Annex has been reduced. However, it is possible for Member States to demonstrate a risk of relocation in individual cases even outside the sectors listed in the Annex to the CEEAG.[170]

Energy Efficiency Directive

The revised Energy Efficiency Directive[171] requires Member States to collectively ensure a reduction of energy consumption of at least 11.7% by 2030 compared to the projections of the 2020 EU Reference Scenario so that the Union's final energy consumption amounts to no more than 763 Mtoe.[172] In addition, Member States shall make efforts to collectively contribute to the indicative Union primary energy consumption target amounting to no more than 992.5 Mtoe by 2030.[173] For the Member States, the consumption limit for the final consumption is binding; the target for primary energy consumption is indicative.[174]

Future Hydrogen and Gas Market

The Council and the Parliament have reached a provisional political agreement on a Regulation establishing common internal market rules for renewable and natural gases as well as for hydrogen.[175] The Regulation is part of the package to decarbonize the hydrogen and gas markets, which also includes the proposal for a Directive on common rules for the internal markets in renewable and natural gases and in hydrogen.[176] The Council and Parliament also reached a provisional political agreement on

170 Abl. (EU) C 80/11, marginal no. 406.

171 Directive (EU) 2023/1791, published in ABl. L 231/1, dated 20.9.2023.

172 Art. 4 para. 1 of Directive (EU) 2023/1791.

173 Art. 4 para. 1 of Directive (EU) 2023/1791.

174 Council of the EU, Press release dated 25.7.2023, https://www.consilium.europa.eu/en/press/press-releases/2023/07/25/council-adopts-energy-efficiency-directive/.

175 COM (2021) 804 final/2; Council of the EU, Press release dated 8.12.2023, modified on 21.12.2023, https://www.consilium.europa.eu/en/press/press-releases/2023/12/08/gas-package-council-and-parliament-reach-deal-on-future-hydrogen-and-gas-market/.

176 COM (2021) 803 final; Council of the EU, Press release dated 28.11.2023, modified on 21.12.2023, https://www.consilium.europa.eu/de/press/press-releases/2023/11/28/internal-markets-in-renewable-and-natural-gases-and-in-hydrogen-council-and-parliament-reach-deal/.

the Directive on 28.11.2023.[177] One of the aims of the two planned provisions is to facilitate the use of renewable and low-carbon gases in the energy system, thereby promoting their use in the EU up to 2030 and beyond. In addition, the package should also help to increase the security of gas supply and reduce dependence on imported fossil fuels.[178]

6.3.6 Methane emissions

In November 2023, the Council and the European Parliament reached a provisional agreement on a Regulation to lay down new rules to reduce methane emissions in the energy sector.[179] The Regulation was adopted on 13.06.2024.[180] The aim of the Regulation is to reduce methane emissions in the energy sector across the Union and thus to serve the overarching goal of climate neutrality in the Union.[181] In addition, the Regulation includes provisions on routine and non-routine inspections, reporting obligations in relation to estimated and directly measured methane emissions and regular inspections to check that operators are complying with the requirements set out in the regulation.[182] Three implementation phases are planned for imports: The first phase will focus on data collection and monitoring, including the creation of a methane transparency database for gas, coal and oil on the EU market; in the second and third phases, exporters to the EU should apply equivalent monitoring, reporting and verification measures by 01.01.2027 and have to comply with maximum methane intensity values by 2030.[183]

177 Council of the EU, Press release dated 28.11.2023, modified on 21.12.2023, https://www.consilium.europa.eu/de/press/press-releases/2023/11/28/internal-markets-in-renewable-and-natural-gases-and-in-hydrogen-council-and-parliament-reach-deal/.

178 Council of the EU, Infographics, https://www.consilium.europa.eu/en/infographics/fit-for-55-hydrogen-and-decarbonised-gas-market-package-explained/.

179 COM (2021) 805 final; Council of the EU, Press release dated 15.11.2023, modified on 21.12.2023, https://www.consilium.europa.eu/en/press/press-releases/2023/11/15/climate-action-council-and-parliament-reach-deal-on-new-rules-to-cut-methane-emissions-in-the-energy-sector/.

180 Regulation (EU) 2024/1787 of 13.06.2024 on the reduction of methane emissions in the energy sector and amending Regulation (EU) 2019/942 (OJ L, 2024/1787, 15.7.2024).

181 Council of the EU, Text of the provisional agreement dated 7.12.2023, 15927/23, p. 2.

182 Council of the EU, Text of the provisional agreement dated 7.12.2023, 15927/23, p. 3.

183 Council of the EU, Press release dated 15.11.2023, modified on 21.12.2023, https://www.consilium.europa.eu/en/press/press-releases/2023/11/15/climate-action-council-and-parliament-reach-deal-on-new-rules-to-cut-methane-emissions-in-the-energy-sector/; Council of the EU, Text of the provisional agreement dated 7.12.2023, 15927/23, p. 3 f.

6.4 The Green Deal Industrial Plan for the Net-Zero Age

The EU Industrial Plan as part of the Green Deal aims to support the twin transition to a green and digital economy, make EU industry more competitive globally and strengthen Europe's open strategic autonomy.[184] The Green Deal Industrial Plan is based on four basic principles:

- predictable, coherent and simplified regulatory environment,
- faster access to sufficient funding,
- expansion of skills,
- open trade for resilient supply chains.[185]

6.4.1 Net Zero Industry Act

Both the European Parliament[186] and the Council[187] have already adopted their own positions on the Commission's proposal for a Regulation on establishing a framework of measures for strengthening Europe's net-zero technology products manufacturing ecosystem (Net Zero Industry Act).[188]

The Regulation aims to establish measures for the renewal and expansion of manufacturing capacity for net-zero technologies[189] in the Union, such as ensuring the Union's access to a secure and sustainable supply of net-zero technologies. To this end, the Regulation should contain measures to ensure that by 2030, manufacturing capacity in the Union of the strategic net-zero technologies listed in the Annex approaches or reaches a benchmark of at least 40% of the Union's annual deployment needs for the corresponding technologies necessary to achieve the Union's 2030 climate and energy targets.[190] In addition, the free movement of net-zero technologies placed on the Single market is to be ensured.[191]

---

184 COM (2023) 62 final; QANDA_23_511, p. 1; European Commission, Factsheet: The Green Deal Industrial Plan, 1.2.2023.

185 https://commission.europa.eu/strategy-and-policy/priorities-2019-2024/european-green-deal/green-deal-industrial-plan_en.

186 European Parliament, Press release dated 21.22.2023, https://www.europarl.europa.eu/news/en/press-room/20231117IPR12205/meps-back-plans-to-boost-europe-s-net-zero-technology-production.

187 Council of the EU, Outcome of proceedings, dated 7.12.2023, 16521/23; Council of the EU, Press release dated 7.12.2023, https://www.consilium.europa.eu/en/press/press-releases/2023/12/07/net-zero-industry-act-council-adopts-position-to-boost-technologies-for-the-green-transition/.

188 COM (2023) 161 final.

189 A definition of "net-zero technologies" can be found under Art. 3 no. 1 point a) of the Commission's proposal, COM (2023) 161 final.

190 COM (2023) 161 final, Art. 1 of the proposal for the Net Zero Industry Act.

191 COM (2023) 161 final, Art. 1 of the proposal for the Net Zero Industry Act.

The Commission's proposal also provides for an annual injection capacity of at least 50 million tonnes of CO_2 to be achieved in certain storage sites by 2030.[192]

6.4.2 Critical Raw Materials Act

Another part of the Industrial Plan is the proposal for a Regulation establishing a framework for ensuring a secure and sustainable supply of critical raw materials.[193] The new Regulation (EU) 2024/1252 establishing a framework for ensuring a secure and sustainable supply of critical raw materials was adopted on 11.04.2024.[194,195]

In order to ensure the Union's access to a secure and sustainable supply of critical raw materials, the Regulation aims to

- strengthen the different stages of the strategic raw materials[196] value chain to ensure that the Union's capacities for each strategic raw material are significantly increased by 2030,
- diversify the Union´s imports of strategic raw materials to ensure that, by 2030, the Union's annual consumption of each strategic raw material at any relevant stage of processing can rely on imports from several third countries, none of which provide more than 65% of the Union's annual consumption,
- improve the Union's ability to monitor and mitigate the supply risk related to critical raw materials,
- ensure the free movement of critical raw materials and products containing critical raw materials placed on the Union market, while ensuring a high level of environmental protection, by improving their circularity and sustainability.[197]

6.4.3 Reform of Electricity Market Design

In December 2023 European Parliament and the Council of the EU reached a provisional agreement in light of the Commission's proposal[198] to revise the rules on electricity market design and to improve EU protection against market manipulation on

192 COM (2023) 161 final, Art. 16 of the proposal for the Net Zero Industry Act.

193 COM (2023) 160 final.

194 European Parliament, Press release dated 12.12.2023, https://www.europarl.europa.eu/news/en/press-room/20231208IPR15763/critical-raw-materials-plans-to-secure-the-eu-s-supply.

195 OJ L, 2024/1252, 3.5.2024.

196 COM (2023) 160 final, Art. 3 of the proposal for the Critical Raw Materials Act.

197 COM (2023) 160 final, Art. 1 of the proposal for the Critical Raw Materials Act.

198 COM (2023) 148 final; Council of the EU, Text of the provisional agreement dated 19.12.2023, 16964/23

the wholesale energy market.[199] The respective regulation adopting the provisional agreement entered into force in summer 2024.[200]

The reason for the reform of the energy market is the energy crisis experienced in 2022 and the resulting severe spikes in energy prices. The price of electricity on the EU electricity market depends on the cost of the fossil fuels used to generate electricity (known as the merit order principle).[201] The reform aims to avoid similar situations in the future.

The new provisions are intended to better protect consumers, for example by strengthening the measures to be taken by the Member States to protect vulnerable customers and customers affected by energy poverty.[202] In addition, with regard to the measures to be taken in the event of a crisis, the Member States are to be given the option of further reducing electricity prices for vulnerable and disadvantaged customers on the basis of the current Electricity Directive.[203]

Funding

Financial resources for innovation, production and the introduction of clean technologies should become more readily available, with a particular focus on REPowerEU, InvestEU and the Innovation Fund.[204] In addition, the establishment of the European Sovereignty Fund will be examined as a medium-term solution for investment needs.[205]

In this respect, the Temporary Crisis and Transition Framework was also amended and the Green Deal General Block Exemption Regulation was revised.[206]

199 Council of the EU, Press release dated 14.12.2023, modified on 3.1.2024, https://www.consilium.europa.eu/en/press/press-releases/2023/12/14/reform-of-electricity-market-design-council-and-parliament-reach-deal/.

200 Regulation (EU) 2024/1747 of 13.06.2024 amending Regulations (EU) 2019/942 and (EU) 2019/943 as regards improving the Union's electricity market design, OJ L, 2024/1747, 26.6.2024.

201 https://www.consilium.europa.eu/en/policies/electricity-market-reform/.

202 Council of the EU, Press release dated 14.12.2023, modified on 3.1.2024, https://www.consilium.europa.eu/en/press/press-releases/2023/12/14/reform-of-electricity-market-design-council-and-parliament-reach-deal/; https://www.consilium.europa.eu/en/policies/electricity-market-reform/.

203 Council of the EU, Press release dated 14.12.2023, modified on 3.1.2024, https://www.consilium.europa.eu/en/press/press-releases/2023/12/14/reform-of-electricity-market-design-council-and-parliament-reach-deal/.

204 QANDA/23/511, p. 2 f.

205 QANDA/23/511, p. 3.

206 https://commission.europa.eu/strategy-and-policy/priorities-2019-2024/european-green-deal/green-deal-industrial-plan_en.

Temporary Crisis and Transition Framework

On 20 November 2023, the Commission has adopted an amendment to the State aid Temporary Crisis and Transitioning Framework, changing its timetable for phasing out certain crisis tools.[207]

Since the beginning of Russia's war against Ukraine, the State aid Temporary Crisis Framework, first adopted on 23.03.2022, and subsequently amended in July and October 2022 and replaced on 09.03.2023 by the Temporary Crisis and Transition Framework, enables Member States to provide timely, targeted and proportionate support to businesses in need.[208] The amendment extended the "Limited amounts of aid" (section 2.1 of the Framework) by six months until 30.06.2024; in addition, the ceilings set out for the limited amounts of aid were increased; furthermore, the "Aid to compensate for high energy prices" (section 2.4 of the Framework) was also extended by six months until 30.06.2024.[209]

The other provisions of the Temporary Crisis and Transition Framework remain unaffected by the changes.[210]

General Block Exemption Regulation

On 09.03.2023, the Commission approved amendments to the General Block Exemption Regulation (GBER).[211] These amendments aim to ensure that the GBER provisions support the green and digital transitions. Against this background, Member States are granted more flexibility in the design and implementation of support measures in sectors that are key for the transition to a climate-neutral economy.[212]

---

207 European Commission, Press release dated 20.11.2023, https://ec.europa.eu/commission/presscorner/detail/en/ip_23_5861.

208 European Commission, Press release dated 20.11.2023, https://ec.europa.eu/commission/presscorner/detail/en/ip_23_5861; European Commission, Communication, 2023/C 101/03, published in ABl. C 101/03 dated 17.3.2023.

209 European Commission, Communication, 2023/C 101/03, published in ABl. C 101/03 dated 17.3.2023.

210 European Commission, Press release dated 20.11.2023, https://ec.europa.eu/commission/presscorner/detail/en/ip_23_5861; European Commission, Communication, 2023/C 101/03, published in ABl. C 101/03 dated 17.3.2023.

211 Regulation (EU) 2023/1315 published in ABl. L 167/1 dated 30.6.2023; European Commission, Press release dated 9.3.2023, https://ec.europa.eu/commission/presscorner/detail/en/IP_23_1523.

212 European Commission, Press release dated 9.3.2023, https://ec.europa.eu/commission/presscorner/detail/en/IP_23_1523.

REPowerEU

The REPowerEU plan was launched back in May 2022.[213] It proposed measures to save energy, diversify supplies, quickly substitute fossil fuels by accelerating Europe´s clean energy transition and smartly combine investments and reforms.[214] The provisions of the REPowerEU plan are intended to enable the EU as a whole to save energy, produce clean energy and diversify its energy supply.[215]

InvestEU

The InvestEU programme supports sustainable investment, innovation and job creation in Europe.[216]

- In line with the objectives of the European Green Deal, at least 30% of the InvestEU programme will support financing for investments that contribute to the EU's climate objectives; in addition, 60% of the investments supported under the "Sustainable Infrastructure Window" of the InvestEU Fund will contribute to the EU's climate and environmental objectives;[217]
- In relation to the objectives of the Green Deal Industrial Plan, the InvestEU programme will support the Industrial Plan in investing and financing the production of clean technologies in Europe to accelerate this development.[218]

Innovation Fund

The Innovation Fund is the EU fund for climate policy with a focus on energy and industry and aims, among other things, to support the transition of European industry to climate neutrality while promoting its competitiveness. The fund is the central funding instrument for meeting the EU's commitments under the Paris Agreement and the requirements of the REPower-EU plan, the Green Deal Industrial Plan and the Net Zero Industry Act.[219]

---

213 COM (2022) 230 final.

214 COM (2022) 230 final, Introduction.

215 https://commission.europa.eu/strategy-and-policy/priorities-2019-2024/european-green-deal/repowereu-affordable-secure-and-sustainable-energy-europe_en.

216 https://investeu.europa.eu/investeu-programme_en?prefLang=de.

217 https://investeu.europa.eu/contribution-green-deal-and-just-transition-scheme_en?prefLang=de.

218 https://investeu.europa.eu/investeu-and-european-green-deal-industrial-plan_en?prefLang=de.

219 https://climate.ec.europa.eu/eu-action/eu-funding-climate-action/innovation-fund/what-innovation-fund_en?prefLang=de.

6.5 Action Plan for financing Sustainable Growth

In order to ensure that the Green Deal achieves its goals, the assistance of the financial sector is required. To this end, the Commission presented a strategy for sustainable finance on 08.03.2021, which should above all enable the targeted channelling of private financial flows into economic activities relevant to climate policy.[220] Building on the "Action Plan on Financing Sustainable Growth"[221] from 2018, the EU has developed three building blocks for a sustainable finance framework: a classification system of sustainable activities (so-called "taxonomy"), a disclosure framework for non-financial and financial companies, and investment tools including benchmarks, standards and labels. The strategy comprises six packages of measures:[222]

- Expand the existing toolkit for sustainable finance to facilitate access to finance for transition;
- Better involvement of small and medium-sized enterprises (SMEs) and consumers;
- Increasing the resilience of the economic and financial system to sustainability risks;
- Increasing the contribution of the financial sector to sustainability;
- Ensuring the integrity of the EU financial system and monitoring its orderly transition to sustainability;
- Develop international initiatives and standards for sustainable finance and provide support to EU partner countries.[223]

6.5.1 Taxonomy

One of the main focuses of the EU Commission's Action Plan for Financing Sustainable Growth is to redirect capital flows towards sustainable investments. This is to be achieved through a uniform classification system for ecologically sustainable economic activities, the so-called taxonomy. This should ensure a uniform understanding of the environmental sustainability of economic activities within the EU and the EEA and provide investors with an objective basis for their investment decisions.[224]

On 22.06.2020, the Taxonomy Regulation,[225] as adopted by the European Parliament on 18.06.2020, was published in the EU Official Journal. It entered into force on 12.07.2020. Together with the Sustainable Finance Disclosure Regulation (SFDR), the Taxonomy Regulation forms a comprehensive regulatory framework with regard to

220 Falke, ZUR 2021, 706.
221 COM (2018) 97 final.
222 Falke, ZUR 2021, 706.
223 Falke, ZUR 2021, 706.
224 COM (2021) 390 final.
225 Regulation (EU) 2020/852.

transparency obligations on the green capital market.[226] The content of the Taxonomy Regulation is determined by delegated acts.

6.5.2 Essential Content of the Taxonomy Regulation

The Taxonomy Regulation is intended to enable companies and investors to classify economic activities as ecologically sustainable on the basis of uniform criteria valid within the EU. In addition, the Taxonomy Regulation is intended to prevent greenwashing. It is intended to function as a transparency instrument and to direct funds specifically to economic activities that are beneficial to at least one of the six environmental objectives listed in Art. 9 of the Taxonomy Regulation; namely climate protection, adaptation to climate change, sustainable use and protection of water and marine resources, transition to a circular economy, prevention and reduction of pollution, protection and restoration of biodiversity and ecosystems.[227] To this end, the Taxonomy Regulation classifies environmentally sustainable economic activities on the basis of technical assessment criteria set out in Commission delegated acts to the Regulation.[228]

The Taxonomy Regulation applies to all financial market participants[229] whose range of products includes financial products, and also to large companies[230] that are obliged to report on sustainability in accordance with the CSR Directive, i.e. have to submit a non-financial statement.[231] Pursuant to Art. 8 of the Taxonomy Regulation, affected companies must submit a report for the first time for the 2021 financial year. Affected companies are obliged to publish key figures on their sustainable economic activities (taxonomy quotas). On the one hand, the share of products and services of sustainable economic activities in the turnover (Capital Expenditures, CapEx) and on the other hand the share in the current operating costs (Operating Expenditures, OpEx) must be reported.[232]

According to the Taxonomy Regulation, an economic activity must cumulatively fulfil four conditions in order to be classified as environmentally sustainable. These four conditions are derived from Art. 3 and Art. 9 of the Regulation.

First, the economic activity under review must have a positive impact on at least one of the six environmental objectives listed above, i.e. it must make a significant contribution to the achievement of that environmental objective.[233] Secondly, the ac-

226 Steuer, BKR, 2022, 296.
227 Falke, ZUR 2022, 207.
228 COM (2021) 4987 final p. 1.
229 Art. 5–7 of Taxonomy-Regulation.
230 Art. 8 of Taxonomy-Regulation, defined in Art. 3 para. 4, 7 CSR-Direction.
231 Art. 1 para. 2 of Taxonomy-Regulation; Rectial 14 of Taxonomy-Regulation.
232 Lanfermann, BB 2020, 1643, 1645.
233 Art. 10–15 of Taxonomy-Regulation (EU).

tivity must not have a significant negative impact on any of the six environmental objectives. Thirdly, the activity must meet the technical criteria set out in delegated acts to determine under which conditions a "significant" contribution to an environmental objective, or a "significant" impairment, exists. Fourthly, the activity must be compatible with a minimum level of protection for workers.

Art. 10 to 16 of the Regulation identify three distinct classes of economic activities, each of which can be classified as environmentally sustainable.

The first group is the inherently "sustainable activities", which naturally fulfil the four criteria already explained. These include, for example, the economic activity of producing or transmitting renewable energy, which contributes to the environmental goal of climate protection by helping to reduce greenhouse gases.

In a second group, "transitional activities" are defined.[234] These can be classified as activities for which there is no technologically and economically feasible low-carbon alternative, but which then make a significant contribution to climate protection if they support the transition to a climate-neutral economy.

A third group is formed by the "enabling activities".[235] These do not contribute directly to achieving the environmental goals, but support an economic activity that has a positive impact on one of the environmental goals.

Shaping through Delegated Legal Acts

With the help of delegated acts, European legal acts, such as the Taxonomy Regulation, can be adapted to technical and scientific progress. In the Taxonomy Regulation, reference is made to such delegated acts with regard to the specification of the "technical assessment criteria".[236] The European Commission thus receives the power to issue legal acts without legislative character, but with general validity, to supplement or amend provisions of the Taxonomy Regulation.

This process of concretisation is already underway, initially through Delegated Regulation (EU) 2021/2139 (so-called "Climate Delegated Act"), which was published in the Official Journal of the EU on 9 December 2021. It sets out technical assessment criteria for the first two environmental objectives ("climate change mitigation" and "adaptation to climate change") to assess whether an economic activity makes a significant contribution to climate change mitigation or adaptation. It also contains criteria to determine whether this economic activity may have a significant negative impact on one of the environmental objectives of the EU taxonomy.

234 Art. 10 para. 2 of Taxonomy-Regulation.

235 Art. 16 of Taxonomy-Regulation.

236 Art. 10 para. 3, Art. 19 para. 5 of Taxonomy-Regulation.

Furthermore, Delegated Regulation (EU) 2021/2178 was published in the Official Journal of the EU on 10.12.2021, which supplements the content of Art. 8 of the Taxonomy Regulation, according to which companies that are required to disclose non-financial information pursuant to Art. 19a or Art. 29a of Directive (EU) 2013/34 must include in their non-financial statement or consolidated non-financial statement information on whether and to what extent the company's activities are linked to economic activities that are considered environmentally sustainable pursuant to Art. 3 and Art. 9. The Delegated Regulation requires financial and non-financial companies to disclose information to investors on the environmental performance of their assets and economic activities. The regulation specifies which information regarding business, investment and borrowing activities must be made available, according to which methodology and in which way. This is intended to enable markets and investors to obtain clear and comparable sustainability information.[237]

On 27.06.2023, two further Delegated Regulations were published by the Commission. One is Delegated Regulation (EU) 2023/2486 (the "Environmental Delegated Act") and the other is Delegated Regulation (EU) 2023/2485. Both Delegated Regulations were published in the Official Journal of the EU on 21.11.2023. The Delegated Regulation (EU) 2023/2486 contains, among other things, the technical assessment criteria for the four missing environmental targets pursuant to Art. 9 of the Taxonomy Regulation (so-called "Taxo 4").[238] According to Art. 6 of the Delegated Regulation (EU) 2023/2486, the Delegated Regulation applies from 1 January 2024 onwards.

With the second Delegated Regulation 2023/2485, the Delegated Regulation (EU) 2021/2139 (the "Climate Delegated Act") was amended to add to the list of economic activities listed in Annex I, among other things. Specifically, this includes seven new economic activities for the environmental objective "climate protection" and five new economic activities for the environmental objective "adaption to climate change".[239] In addition, the annexes to the technical assessment criteria will also be modified.[240] According to Art. 2 of the Delegated Regulation (EU) 2023/2485, the amendments to the Delegated Regulation (EU) 2021/2139 apply from 1 January 2024; the amendments relating to the provisions in Annex I no. 28 and Annex II no. 26 apply from 1 January 2025.

Furthermore, on 16.06.2023, the EU Commission published a Notice in the Official Journal of the EU on the interpretation and implementation of certain legal provisions of the Taxonomy Regulation and links to the Sustainable Finance Disclosure Regulation.[241] With regard to the Taxonomy Regulation, the Notice focuses in particular on

237 https://www.drsc.de/news/delegierte-verordnungen-zur-eu-taxonomie-verordnung-veroeffentlicht/.

238 Baumüller/Hrinkow, IRZ 2023, 539, 540.

239 Baumüller/Hrinkow, IRZ 2023, 539, 542.

240 Baumüller/Hrinkow, IRZ 2023, 539, 542.

241 European Commission, Commission Notice, 2023/C 211/01 published in ABl. C 211/1 dated 16.6.2023.

the requirements that operators must meet in order to comply with the minimum safeguards set out in Art. 18 of the Taxonomy Regulation.

The design of the EU climate taxonomy has not been completed by the aforementioned delegated regulations. In this respect, the Commission has already announced that it will also extend the list of economic activities that fall under the environmental objective of “adaptation to climate change”.[242]

Discussion on Natural Gas and Nuclear Energy

Particular attention is given to the European Commission’s proposal for a supplementary delegated act to the EU taxonomy from December 2021 and February 2022, respectively, to include certain gas and nuclear energy activities in the EU classification system under specific conditions.[243] In addition, the complementary delegated act sets out new disclosure requirements for companies regarding economic activities in the nuclear and gas sectors. These are intended to ensure better information for investors as well as greater transparency.[244] The Delegated Regulation (EU) 2022/1214 was published in the Official Journal of the EU on 15.07.2022, after neither the Council nor the European Parliament raised an objection within the four-month review period.[245] The Delegated Regulation thus entered into force and has been applicable since 01.01.2023.[246]

The Delegated Regulation (EU) 2022/1214 caused a heated debate among EU states. While some states (e.g. France and Poland) supported the categorisation of nuclear energy as environmentally sustainable, other Member States (e.g. Germany, Austria and Luxembourg) were critical of the European Commission’s announced plans.[247] In particular, the question of whether the European Commission, within the framework of its de facto taxonomy monopoly, was given the right to categorise nuclear energy and natural gas as environmentally sustainable, i.e. whether there was generally a competence-related entitlement to do so, as well as the question of which economic activities should be appropriately categorised as environmentally sustainable, are highly controversial.[248]

From a competence point of view, the question arises whether the European legislator was allowed to authorise the European Commission under Art. 290 TFEU with the Taxonomy Regulation as a basic legal act to classify specific forms of energy, such

242 Baumüller/Hrinkow, IRZ 2023, 539, 544.
243 COM (2021) 2800 final.
244 QANDA_22_712 q. 2.
245 European Parliament, Press release dated 6.7.2022, https://www.europarl.europa.eu/news/en/press-room/20220701IPR34365/taxonomy-meps-do-not-object-to-inclusion-of-gas-and-nuclear-activities.
246 See Art. 3 of Delegated Regulation (EU) 2022/1214.
247 Handelsblatt, 3.1.2022, EU-Taxonomie: Welche EU-Länder wollen Gas und Atomkraft als nachhaltig einstufen? Die wichtigsten Antworten auf die Brüsseler Pläne.
248 Centrum für Europäische Politik, cepInput no. 2, 2022, p. 1, Europe in the Taxonomy Trap.

as nuclear energy or natural gas, as sustainable economic activities within the meaning of the taxonomy system by means of a delegated act.[249]

It is true that the purpose of Art. 290 TFEU is to relieve the legislative secondary legislation and thus the Union legislator with regard to technical details. This relief is achieved by replacing the cumbersome EU legislative procedure with fast and flexible executive EU tertiary acts in the form of delegated acts.[250] However, the delegated acts may only serve to "supplement or amend non-essential rules". The legislative formulation of essential provisions, on the other hand, shall continue to be the responsibility of the legislator (cf. Art. 290 para. 1 subpara. 2 sentence 2 TFEU).[251] This reservation of essentiality is intended to prevent an undermining of the original legislative competence of the EU legislator (Council and Parliament) by transferring competence to the European Commission, in order to maintain the balance between the EU institutions and to comply with the principles of separation of powers and democracy.[252]

The core of the dispute was whether the classification of nuclear energy and gas activities as environmentally sustainable is such a "non-essential matter" in the sense of Art. 290 TFEU and thus whether a transfer of this decision to the European Commission is permissible at all. Regulations that require a political decision through which a fundamental orientation of EU policy takes place are considered essential.[253] The "reservation of sovereignty in energy policy" anchored in primary law in Art. 194 para. 2 TFEU, according to which it is up to each Member State to make an independent choice about the energy sources to be used, and the resulting different approaches of the various Member States with regard to the handling of, for example, nuclear energy, suggest that this issue was in no case immaterial within the meaning of Art. 194 para. 2 TFEU.

The practical consequence is, however, that from January 2023 investments in the nuclear and gas industries can be marketed under the label of climate friendliness. To what extent this will affect the credibility and functionality of the European taxonomy remains to be seen.

Practical Implications of the Taxonomy

With regard to the practical implementation of the taxonomy, companies will have to deal with three central questions in the future: Firstly, what data is needed? Secondly: Where is the data available? Thirdly: How is the data processed and output? What is required for the practical implementation of the taxonomy is the targeted collection

249 Centrum für Europäische Politik, cepInput no. 2, 2022, p. 1, Europe in the Taxonomy Trap.
250 Gellermann in: Streinz, EUV/AEUV, 3. Ed. 2018, Art. 290 TFEU, marginal no. 1.
251 Centrum für Europäische Politik, cepInput Nr. 2, 2022, p. 9, Europe in the Taxonomy Trap.
252 Gellermann in: Streinz, EUV/AEUV, 3. Ed. 2018, Art. 290 TFEU, marginal no. 7.
253 Gellermann in: Streinz, EUV/AEUV, 3. Ed. 2018, Art. 290 TFEU, marginal no. 7.

and processing of the necessary data in the company as well as ensuring quick access. The data is then included in the reporting and can be placed by the stakeholder in the context of the existing sustainability strategy. The implementation activities described mean an immense administrative process of data collection for the companies concerned, for which the associated costs should be taken into account.[254]

For companies, products and services that comply with the ecological requirements of the taxonomy, this will have a positive impact in the form of competitive and reputational advantages. Taxonomy-compliant sales and investments could, for example, result in financing at favourable conditions on the capital market and with banks, as well as the possibility of resorting to taxonomy-compliant financial products, such as EU Green Bonds.

The extent to which classification within the framework of the taxonomy will actually contribute to a climate-neutral economy is certainly being critically questioned. For example, many different lobby representatives will be interested in having their economic sector classified as ecologically sustainable, even if in doubt it is not (cf. nuclear power). At this point, one could ask whether the expenditures made for the taxonomy could not be used more effectively elsewhere for climate protection.

6.5.3 Disclosure Framework for Non-Financial and Financial Undertakings

The second key element of the Sustainable Growth Financing Action Plan is formed by the obligation for non-financial and financial companies to disclose information that enables investors to make an informed decision on investments taking into account their sustainability. Under the concept of "double materiality", companies must disclose the environmental and social impacts of the company's activities, as well as business and financial risks arising from the company's sustainability risks.[255]

The Regulation on sustainability-related disclosure requirements in the financial services sector[256] and the final delegated act supplementing Art. 8 of the EU Taxonomy Regulation[257] provide the regulatory framework for affected financial and non-financial companies, obliging them to disclose data to investors on the sustainability of their assets and economic activities. This disclosure obligation is intended to enable markets and investors to have access to uniform and comparable information regarding the sustainability of economic activities, and in particular to prevent

254 https://www.independentcapital.at/die-taxonomie-verordnung-das-herzstueck-des-eu-aktionsplans-zur-finanzierung-nachhaltigen-wachstums/.

255 COM (2021) 390 final.

256 Regulation (EU) 2019/2088.

257 Delegated Regulation (EU) 2021/2178.

greenwashing.[258] This term of American origin has meanwhile been transferred to the financial sector. There, one speaks of greenwashing if a financial product is declared as environmentally friendly although it does not fulfil the corresponding ecological requirements.[259]

In terms of content, the delegated regulation obliges financial institutions to disclose information on the share of environmentally sustainable economic activities in the total assets in which they invest or which are financed by them. Non-financial companies are obliged to disclose the ratio of their turnover and their operating and investment costs (OPEX and CAPEX) associated with sustainable economic activities according to the Taxonomy Regulation and the delegated acts on EU climate taxonomy.[260]

The Delegated Regulation (EU) 2023/2486 deleted Art. 8 para. 5 of the Delegated Regulation (EU) 2021/2178. The provisions on the scope of the disclosure obligations previously regulated there are now replaced by the new Art. 10 para. 6 and 7 inserted by the Delegated Regulation (EU) 2023/2486, which contain provisions on the entry into force of the disclosure obligations with regard to the four environmental objectives regulated therein.[261]

Besides that Art. 10 of the Delegated Regulation (EU) 2021/2178 contains soft transitional provisions for the first-time reporting. According to these, non-financial companies had to disclose the share of taxonomy-eligible and non-taxonomy-eligible economic activities in their total turnover as well as investment and operating costs in their reports until 31.12.2022. Financial undertakings had to disclose sector-relevant key figures regarding the taxonomy compliance of their economic activities by 31.12.2022. Until 31.12.2023 reporting was also required on further quantitative information pursuant to Art. 10 para. 3 point b-d of Regulation (EU) 2021/2178. Only as of 1.1.2026 will the key performance indicators for credit institutions apply, which also relate to their trading book and to commissions and fees for commercial services and activities other than the provision of financial resources.[262]

6.5.4 Investment Tools including Benchmarks, Standards and Seals of Approval

The third and final key element of the Action Plan consists of various investment instruments, namely benchmarks, standards and labels. Standards and labels, like the

258 Falke, ZUR 2021, 706.
259 Möllers, ZHR 2021, 881, 896; Rectial 11 sentence 3 of Taxonomy-Regulation.
260 Falke, ZUR 2021, 706.
261 Baumüller/Hrinkow, IRZ 2023, 539, 541.
262 Baumüller/Haring/Merl, IRZ, 2022, 77, marginal no. 16.

taxonomy, are intended to help channel financial resources to companies, issuers and investors that engage in sustainable activities and business models.[263]

6.6 Summary

With the "Green Deal", the European Commission has given itself a roadmap for measures that will enable a transition to a green economy whose growth is decoupled from resource use. The goal is to achieve greenhouse gas neutrality by 2050. Already by 2030, net emissions are to be reduced by at least 55% compared to 1990. To this end, the Commission has adopted an investment plan for a sustainable Europe, along with a mechanism for a fair transition. Further substantial and far-reaching changes are being added as part of the "Fit for 55" programme.

The investment plan for a sustainable Europe aims to create appropriate framework conditions for private investment and the public sector and to provide targeted support to project promoters and public authorities in the selection, structuring and implementation of sustainable projects. At least 1 trillion Euros are earmarked for this purpose. The investment plan not only targets the energy sector, but also other areas of EU environmental policy, especially the circular economy and the buildings sector. This is accompanied by a differentiated further development of the individual targets, e.g. in the area of energy efficiency in the building sector (long-term renovation strategy).

This goes along with a far-reaching restructuring of European economic law. The fundamental economic and competition principles will remain intact. However, they will be supplemented by increased consideration of environmentally related measures and the consideration of climate protection-relevant concerns in weighing processes. This is particularly evident in public procurement, which is to be made more ecological through green minimum criteria, as is the law on state aid for sustainable investments.

These measures are flanked by the European Climate Law, which, although limited to formal regulations, grants the Commission powers vis-à-vis the Member States to ensure the transition to climate neutrality by 2050. Thus, while the basic responsibility for energy and climate policy remains with the Member States, the European Governance-Regulation regulates extensive reporting obligations and enables the Commission to evaluate them. Even though this does not lead to directly enforceable legal consequences, this mechanism generates considerable public pressure on national decision-makers and thus contributes to a uniform European climate policy.

To achieve the aforementioned goals, however, substantial private investment is required. The provision of appropriately sustainable financial products is an effective

---

263 COM (2021) 390 final.

means of channelling private investment into such activities. The EU taxonomy, consisting of the central European Taxonomy Regulation (EU) 2020/852 and delegated implementing acts based on it, is intended to develop the necessary steering function for this. The taxonomy classification does not interfere with the energy mix of a Member State. Rather, the aim is to harmonise the requirements for marketing financial products or corporate bonds as environmentally sustainable investments in the Member States in order to increase investor confidence and awareness of the environmental impacts of these financial products or corporate bonds, to enhance visibility and to address concerns about "greenwashing".

Sebastian Helgenberger and Roman Buss

7 Social and Economic Co-Benefits on Our Way to Carbon-Free Societies and Economies

7.1 Co-Benefits: Turning Challenges into Opportunities

The global transition to renewable sources of energy is in full swing. The social and economic opportunities presented by this transformation are spurring the growth of climate-friendly and renewable energy generation in many countries.[1] Climate and energy policies are currently at crossroads. While countries implementing the Paris Climate Agreement have committed to reducing their carbon emissions in the foreseeable future, they are also tempted to prolong their dependence on fossil energy sources to cover their growing energy needs in the longer term. This is even more so the case after the beginning of the Russian military aggression in Ukraine. For decades, climate mitigation efforts and renewable energy policies have been regarded as costly luxury measures. The co-benefits approach aims to reverse this outdated picture.

Over the past decades we could observe shifting paradigms in the new renewable energy and climate policy arena: from burden-sharing to opportunity-sharing. In order to promote climate mitigation and adaptation at all levels, we argue that translating, where possible, climate policy objectives into the language and thinking of co-benefits, particularly those that will mobilise economic interests while also protecting the natural basis of life.

This broader approach should be based on a coalition of government, business and civil society actors operating at all levels of the global multi-level system of climate governance. Tapping into relevant co-benefits could help countries turn their energy challenges into opportunities. There are manyfold social and economic opportunities, such as cleaner air and related health system benefits, new jobs, a secure and independent energy supply, improved energy access, and rural development are among the immediate co-benefits of the global transition towards climate-friendly and renewable energy. A smart climate policy can give rise to many possible co-benefits for society and the economy. They offer us the opportunity to translate climate policy objectives into a broad variety of interests. The IPCC reports contain numerous references to co-benefits and can be roughly grouped into economic, social and environmental co-benefits (as shown in Table 7.1).

1 A comprehensive analytical framework for the assessment of the socio-economic effects of renewable energies was presented by IRENA (2016) on the basis of Fraunhofer ISI et al. (2014).

https://doi.org/10.1515/9783110784299-007

Table 7.1: Overview of Co-Benefits mentioned in IPCC reports.

Economic co-benefits	Social co-benefits	Environmental co-benefits
Energy security	Health impact (air quality, noise)	Ecosystem impact (e.g., via air pollution)
Employment impact	Energy access	Land use competition
New business opportunity / economic activity	(Fuel) Poverty alleviation	Water use / water quality
Productivity / competitiveness	Food security	Biodiversity conservation
Technological spill over / innovation	Impact on local conflicts	Urban heat island effect
	Safety / disaster resilience	Resource / material use impact
	Gender impact	

The social and economic co-benefits of climate change mitigation and renewable energies have moved from the sideline to the centre of climate- and energy-related debates and have become key drivers of the global transition towards the new renewable energy world. These developments correspond to an observable paradigm shift – from 'burden sharing' to an increasing degree of 'opportunity sharing' – a shift that was ultimately reflected in the 2015 Paris Climate Agreement. The article briefly outlines some of the major co-benefits and presents empiric evidence and quantifications from India, Vietnam, South Africa, Turkey, Kenya and Mexico. The article is structured into three sections:

- **Framing the discourse – "benefits not costs"** Why co-benefits of climate action are important to achieve carbon neutrality: harnessing diverging interest and political agendas for climate action; using opportunity-oriented narratives to motivate for action
- **International evidence, key data and figures** on important co-benefits (jobs, health, local income) **in emerging economies** (India, Turkey, Vietnam, Mexico, South Africa)
- **Shaping the (just) transition** – the multiple co-benefits of renewables in climate mitigation (key recommendations to mobilise the co-benefits for speeding up the race towards climate neutrality)

7.2 Framing the Discourse

International climate policy and action from its very beginning in the late 1970 has been confronted with a central issue. Even if a country or a decision-maker was willing to act and to reduce harmful CO_2 emissions, how can these actions be financed? In every single conference of parties (COPs, the most recent number 29 just being finished in Azerbaijan) one of the hardest issues negotiated officially and behind closed doors were of a financial nature. Who will pay for "additional" costs? Who will pay the "burden" of "more expensive" renewable sources of energy? Who is responsible for global warming? Who will pay compensation for the ecological damage already done or to prevent even worse impacts of the climate crisis? The largest emitters of the present, or countries that emitted most gigatons of carbon to the atmosphere over the past centuries? Do the least developed countries that suffer most from the effects of climate change not only need financial assistance in setting of a clean economy, but also be compensated for the ecological damage created by industrialized countries? These are some of the hardest cleavages visible in climate negotiations over the past decades. What they all have in common is that climate action is seen as a burden, an additional cost, a dramatic obstacle to overcome, or even as a luxury that only wealthy countries can afford, but certainly not developing economies or least developed nations of the global South. Proponents of renewable energy have argued for decades that the new energy technologies such as wind energy and solar photovoltaic power are facing a stiff opposition from the existing energy system and its major players that provide "cheap" fossil power and fuels. Oil, coal, gas. Due to massive subsidies, there is no "level playing field" for renewable sources of energy. The figures to estimate the volume of global subsidies for fossil fuels vary widely, from a USD 400 billion per year as estimated by the International Energy Agency (IEA) to USD 6000 as published by the International Monetary Fund (IMF). No matter what definition is taken, there is a massive intervention to the market that prevented a quicker market uptake of renewables that seemed to be unaffordable to economies and low-income households. The argument of costs and burden to state budgets and households has stifled innovation and investments into clean energy and low-carbon technologies for decades.

Only in recent years has this discourse gradually changed with the observation that eco-innovations and new emerging technologies and policy interventions do not only cause costs, but also may yield beneficial side effects – or collateral benefits – or co-benefits. These positive impacts can be found implicitly in many policy papers or strategies but only in recent years have been observed, described and monitored more systematically by scholars and the scientific community world-wide. The positive side effects were termed co-benefits. The most recent IPCC reports published over the past decades contain numerous references to co-benefits. These can be roughly grouped into economic, social and environmental co-benefits (see Table 7.1).

There are many definitions found in the literature. The term 'co-benefits' here refers to "simultaneously meeting several interests or objectives resulting from a political intervention, private-sector investment, or a mix thereof."[2]

2 Sebastian Helgenberger, Martin Jänicke, & Konrad Gürtler (2019): Co-benefits of Climate Change Mitigation. Encyclopedia of the UN Sustainable Development Goals.

In other words, the co-benefits approach may be understood as studying, implementing, and replicating the positive externalities of a political action. Implementing this approach requires fostering an environment for problem solving, by encouraging the idea that the solution to global problems, such as climate change and development, have more synergies with each other than trade-offs. The key is to move this method of thinking away from burden sharing, and towards opportunity sharing.[3]

More recently the un-locking of these co-benefits and creating an enabling environment were not only seen in relation to climate policy and international agreements, or the reform of fossil energy systems and the ecologic transformation of industry, but in the light of recovery of economies from the global Covid 19 pandemic:

> By fostering an enabling policy environment that succeeds in unlocking these co-benefits, the government can provide important stimuli toward recovery from the impacts of the COVID-19 pandemic and revive the health system and the national economy. The Paris Climate Agreement and the 2030 Agenda on Sustainable Development offer important internationally agreed frameworks to ensure economic recovery in the shorter term and for building resilient economies and health systems in the long run.[4]

The idea of co-benefits is not entirely new. And the approach has also been institutionalized for some years. One example is the Asian Cobenefits Partnership (ACP). In June 2009 during the first International Forum for a Sustainable Asia and the Pacific (ISAP) in Hayama, Japan, policymakers and experts proposed creating an informal network to improve stakeholder cooperation and knowledge management on co-benefits in Asia.[5]

More recently a project, funded by the Germany International Climate Initiative[6] (IKI) has created a huge body of evidence on co-benefits in the major emerging economies India, South Africa, Vietnam, Turkey, Mexico and Kenya. Jointly with many local knowledge partners and in close cooperation with relevant national authorities, government ministries, research institutions, and think tanks, the COBENEFITS project[7] developed key insights that enabled policy makers and the civil society to mobilise these co-benefits in their countries and accelerate domestic processes aimed at

---

3 See Aayushi Awasty & Kavya Bajaj (2019): Implementing the co-benefits approach in India, in: IASS/UfU/TERI. 2020. Making the Paris Agreement a success for the planet and the people of India. Unlocking the co-benefits of decarbonizing India's power sector. COBENEFITS Policy Report. Potsdam/New Delhi. www.cobenefits.info, last accessed July 2022.

4 IASS/UfU/TERI. 2020. Making the Paris Agreement a success for the planet and the people of India. Unlocking the co-benefits of decarbonizing India's power sector. COBENEFITS Policy Report. Potsdam/New Delhi. www.cobenefits.info, last accessed July 2022.

5 See *Asian Cobenefits Partnership* (ACP), https://www.cobenefit.org/index.html.

6 See COBENEFITS, https://www.cobenefits.info. The Federal Ministry for the Environment, Nature Conservation, Nuclear Safety and Consumer Protection (BMUV) has funded the COBENFITS project via its International Climate Initiative (IKI). https://www.international-climate-initiative.com/en.

7 www.cobenefits.info.

achieving their international climate protection commitments. Substantiated by scientific rigor and key technical data, the co-benefits approach provides guidance to governmental departments and agencies on further shaping an enabling environment to for expanding the use of renewable sources of energy and finally reducing greenhouse gas emissions in line with the Paris agreement.

7.3 International Evidence, Key Data and Figures on Important Co-Benefits

The next paragraphs will give a cursory overview of the most widely cited co-benefits – the collateral benefit of a policy measure such as economic prosperity, business development, job creation and employment opportunities as well as health impacts, or issues related to the water–energy–food nexus or the gender dimension in energy and climate policy. We will also explore and briefly describe their nature and give examples of these beneficial side effects of climate action, including empirical data and figures that quantify the co-benefits in different country contexts.

7.3.1 The Employment Impact

The employment dimension of the energy transition is particularly relevant as governments and their representatives look to increase economy-wide job creation and manage misalignments in labour markets across sectors in transition. Several regular updates for job creation through renewable sources of energy can be found in the literature. The International Renewable Energy Agency (IRENA) and the REN21 policy network are monitoring the global scale of renewable energy induced employment in their Renewable Energy and Jobs and Global Status Reports.[8] The EurObserv'ER project is publishing estimations on job creation through renewables and avoided fossil fuel use in the European Union.[9] Also individual countries' statistical bodies keep track on their renewable job markets for example in Germany,[10] Austria,[11] the United

8 IRENA Renewable Energy and Jobs, https://www.irena.org/publications/2021/Oct/Renewable-Energy-and-Jobs-Annual-Review-2021.

9 EurObserv'ER, The State of renewable energies in Europe, 22nd edition 2024, https://www.eurobserv-er.org/category/all-annual-overview-barometers.

10 BMWK 2022: Studies by the Federal Ministry for Economic Affairs and Energy on the energy transition and its impact on investment, growth and jobs https://www.bmwk.de/Redaktion/DE/Dossier/erneuerbare-energien.html, last accessed Novemeber 2024.

11 Bundesministeriums für Klimaschutz, Umwelt, Energie, Mobilität, Innovation und Technologie (BMK), Innovative Energietechnologien in Österreich Marktentwicklung 2021, Biomasse, Photovoltaik, Photovoltaik-Batteriespeicher, Solarthermie, Wärmepumpen, Bauteilaktivierung in Gebäuden und

Kingdom,[12] or Australia.[13] Beyond, European or United States Renewable Energy Industry Associations are monitoring job impacts of single technologies such as solar photovoltaic or wind energy besides estimations of globally active consultancies.[14] This broad range of publications suggests that employment and job creation through renewable energy or other new emission mitigating technologies is important for decision makers to convince the general public of climate action. This is specifically true for economies with relatively higher rates of (youth) unemployment.

Usually job impacts are categorized as direct, indirect or induced employment. Employment is measured in Full Time Equivalents (FTE). Direct jobs are typically onsite construction workers for example on renewable energy installations. But these direct jobs may also mean job at equipment manufacturers, design services maintenance workers, or security personnel. Indirect jobs in turn comprise jobs in legal services, natural resource suppliers, construction equipment suppliers, or wholesale businesses for replacement parts. The final category of so-called induced jobs are the result of direct and indirect jobs in that they benefit from the wages spent by directly employed staff in other sectors of the economy. Examples may be housing for workers, work for retailers, jobs in restaurants, health care providers, agriculture, food providers, accounting services.

Many of these conducted analyses found that the energy transition has net positive effects on job creation. The International Renewable Energy Agency (IRENA) estimates that the number of worldwide jobs directly or indirectly related to the renewable energy sector passed 11.5 million in 2019, doubling its initial assessment of 5.7 million jobs in 2012. The employment prospects for renewables are positive, not least because renewable energy generation tends to be more work-intensive than in the fossil energy sector. In Vietnam, for example, an estimated 1.94 million job years could be created in the country by transitioning the power sector towards larger shares of renewable energy between 2015 and 2030. Over that 15-year period, solar and wind would create 3.5 jobs and 2.8 jobs, respectively, per average installed MW capacity, whereas coal would create only a maximum of 1.4 jobs per MW. In other words, only half of the employment but with the full range of negative environmental damage and impacts.

Similar effects are predicted for other Asian countries. Also India could significantly increase employment through its power sector by increasing the share of re-

---

Windkraft, thttps://nachhaltigwirtschaften.at/resources/iea_pdf/schriftenreihe-2022-21a-marktstatistik-2021-kf-web.pdf.

12 REA 2021, REview21, REA's state of the industry annual report, https://www.r-e-a.net/wp-content/uploads/2021/07/REview-2021-.pdf.

13 UTS 2020, Renewable Energy Jobs in Australia: Stage One Prepared for Clean Energy Council by UTS Institute for Sustainable Futures, https://www.uts.edu.au/sites/default/files/2020-06/Renewable-Jobs-Australia-ISF%20F.pdf, June 2020.

14 See for example WindEurope, Solar Power Europe (SPE), American Wind Energy Association.

newable energy. The renewable energy sector is poised to become the largest employer in the future Indian power sector. By 2050 the renewable energy sector could potentially employ more than 3.2 million people – five times more than the entire Indian fossil-fuel sector employs today. This could increase the total number of jobs in India's power sector to 3.6 million by 2050 (IASS, UfU and TERI, 2020).

A co-benefits study for South Africa found that job creation through renewables exceeds the anticipated job losses in the coal sector. In the Mpumalanga region in South Africa, even if not all job losses in the fossil fuel sect tor can be compensated by clean energy jobs, still, under an ambitious decarbonisation scenario, these net losses can be minimised: Almost 79,000 clean energy jobs can be created, three times more than under the current policy scenario (25,000 direct, 26,000 indirect, and 28,000 induced versus 27,000: 8,000 direct, 9,000 indirect, and 10,000 induced, by 2030).[15] These figures suggests the enormous positive side effects of an ambitious climate policy. Job creation is a powerful narrative for rising public acceptance of ambitious climate policies.

7.3.2 Increasing National Health and Wellbeing

Ambient air pollution has emerged as massive health risk factor for billions of people worldwide. Polluted air (largely caused by coal power plants and particles from combustion of fossil fuels) contributes significantly to cardiovascular diseases, chronic respiratory diseases and lower respiratory tract infections. Fossil fuelled air pollution has also many further negative impacts, of which those of greatest concern include heart disease, lung cancer, stroke and chronic obstructive pulmonary disease. The consequences of such diseases include increased levels of morbidity, which further result in elevated health costs and losses of productivity for local communities and the entire economy, besides the human tragedy of premature deaths. The co-benefits of climate action are possibly most clearly visibly in the health sector. A shift to a less carbon-intensive power sector can help to reduce negative impacts and contribute to reducing costs in many health systems around the world. In other words: scaling up renewables can cut health costs associated with the power sector considerably and reduce negative health impacts and related costs for millions of people and businesses.

Since electricity generation in many countries is still largely coal-based, the power sector is an important contributor to ambient air pollution. UN reports estimate that air pollution is causing millions of deaths per years with disastrous effects on entire economies, higher costs in the health sector, besides the individual tragic of premature deaths for families and communities.

The Cobenefits project investigated health and economic costs associated with the exposure to particulate matter (PM2.5/PM10) and assessed the health benefits of in-

15 IASS/IET/CSIR. 2021, p. 6.

creased share of renewables in the energy and power sector of emerging economies. For India the researchers found that adopting an accelerated decarbonisation pathway will avoid many premature deaths and that India can markedly improve the livelihoods of its citizens by reducing ambient air pollution. In the business-as-usual scenario, during 2020 almost 500,000 people are assumed to die prematurely due to exposure to particulate matter (PM10); this number would rise to 830,000 premature deaths during 2050. By moving to the accelerated decarbonisation pathway, more than 200,000 premature deaths can be avoided.

Up to 44 million people are exposed to air pollution from coal power plants in South Africa. Health costs related to coal emissions will peak in 2022, at up to R45 billion (€ 2.7 billion) in that year alone. As many as 2080 premature deaths annually were predicted due to air pollution from power plants in South Africa. Health cost externalities of power plants range from Rand 5 to 15 cents per kWh. By 2050, South Africa can almost completely eliminate its health costs from the power sector by following an ambitious decarbonisation pathway South Africa can save as much as R 141 billion (€ 8.3 billion) in health costs by 2050 by following a rapid decarbonisation pathway€

Such encouraging co-benefits figures were also found in other country contexts for example in Turkey. According to the Cobenefits Study,[16] which examined the co-benefits to people's health and health cost savings resulting from increased deployment of renewable energy, Turkey can significantly reduce the number of premature deaths related to air pollution emitted from fossil-fuelled power plants. Under the current policy, mortality can be expected to increase from 2,100 cases in 2017 to more than 2,300 cases in 2028. By following an ambitious decarbonisation pathway (Advanced Renewables Scenario B), estimated mortality would be reduced to less than 1,600 cases in 2028, thus avoiding more than 750 deaths in that year alone. Beyond, Turkey can significantly unburden its health system by decarbonising the power sector: Under the current policy, annual health-related costs can be expected to increase from USD 2.15 billion in 2017 to USD 2.5 billion in 2028. By following an ambitious decarbonisation pathway health cost savings in 2028 can amount to USD 800 million in that year alone.

7.3.3 Energy Access and Energy Security

Energy access is essential for economic and human development and is an important driver for the economic development and progress of a country. Providing access to energy and electricity is a vital development goal and also a co-benefit increasingly investigated.[17] Access to modern forms of energy, especially electricity, becomes even

16 IASS/IPC, 2020a: Improving air quality and reducing health costs through renewable energy in Turkey.
17 Cobenefits Vietnam 2019d: Electricity access and local value creation for the un-electrified population in VietnamAssessing the co-benefits of decarbonising the power sector.

more important for the socio-economic development of rural areas (which lag behind urban areas in terms of infrastructure development). The full electrification of all households and local enterprises, even in rural communities is vital to achieve the social and economic development goals (and SDGs). Also here we can find beneficial impacts of alternative development paths based on climatically sound measures.

Solar-powered mini-grids of high installed power capacity can remain economically viable and cost-competitive with the centralised grid in rural areas of India and Vietnam. Renewables create opportunities for electricity access for the remaining 2% of the population in Vietnam, predominantly located in regions with terrain unfavourable to grid expansion. Solar mini-grid systems greater than 100 kW with interest rates as low as 8% and a 15% return on equity can achieve grid parity and a low cost of electricity supply to the rural consumer. Decentralized solar mini-grids are also effective for improving rural education in India, as most schools in remote areas of India experience continuous power cuts which impede the quality of education that the students receive. The mini-grid can provide electricity at schools or education centres consistently during the teaching hours to help stimulate better educational outcomes for the students in rural India. Vietnam has tremendous potential for off-grid renewable energy systems, which are cost competitive against grid extension in rural areas with challenging terrain that hinders navigation and connection. Deploying low-wind-speed wind turbines to electrify clusters in rural areas with a levelized cost of 9087 VND/kWh is the cheapest means of providing low-cost energy access to remote areas.

In times of rising energy bills for fossil fuels (oil, gas, LNG, petrol) on international markets, local energy generation also increases the resilience to the recent market fluctuations. Functional decentralized mini grids may also reduce household expenditure for unsustainable modes of energy supply.

7.3.4 Promoting the National Economy and Empowering Local Communities and Citizens

The energy transition is inducing new investments in the electricity production and infrastructure sectors worldwide. Along with these investment stream go beneficial impacts for the national economy as well as for local or even marginalised communities. Encouraging figures can be found in this respect on national level, as well as for rural and marginalised communities in emerging countries. A case in point is Turkey: The increasing energy demand met mostly by fossil fuel resources, faces significant risk of an escalation of its dependency degree on energy imports in the future. In order to address this issue, Turkey's public policy framework includes not only strategies to increase the share of renewable energy resources in its energy mix but also aims to develop a local manufacturing industry and to enable technology transfer. The conducted co-benefits study and report found and assessed a range of additional co-benefits of renewable energy, besides reducing energy sector greenhouse gas (GHG) emissions, when compared to conventional energy systems and provided insights on regional trade opportunities. Turkey could significantly boost the value of production by increasing the share of renewables. Increasing solar PV capacity by 60% and more than double the wind capacity over the next 10 years, increase the value of production along the solar value chain fifteenfold, and over 31% along the wind value chain in the next ten years alone. By following more ambitious renewable pathways for Turkey, the expected increases in value of production can be more than doubled across the wind power value chain and increased eightfold along the solar value chain, pushing up the total value of production by more than USD 69 billion in the next ten years compared to 2016.

Fostering competitiveness in manufacturing and closing the technology gap between imports and exports in both the solar and wind sectors is crucial to further improving the trade balance in Turkey's renewable energy sector. By 2028 it is possible for the solar energy sector to increase its value by USD 9.9 billion above the expected USD 1.3 billion estimated under the current policy, if more ambitious solar capacity additions are achieved. Likewise, the wind sector could peak to a total value of USD 83.5 billion from the expected USD 33.32 billion in the next ten years should RE capacity additions are in place. Across the value chains, each additional MW capacity of energy increases industrial production by around 452.5 thousand USD in the solar energy sector, and around 3.6 million USD in the wind sector, on average.

Ambitious renewable energy pathways may also generate large impacts for beneficiaries in marginalised communities. In South Africa the integrated resource plan is thought to stimulate positive socioeconomic development in various ways. The IRP 2018 may support up to 3,000 local enterprises and up to 10,000 local jobs can be created in marginalised communities. Beyond the IRP in South Africa is thought to enable up to 30,000 individuals in marginalised communities can benefit from access to education-related programmes by the year 2050. Within the context of the sites assessed, the types of jobs created locally through SED and ED spend include non-core services offered to projects, such as cleaning and catering services. In communities with other significant opportunities for economic activity, job creation may not necessarily support renewable power generation. For example, supported enterprises may create retail jobs or service jobs for other industries, including the mining industry.[18]

A higher share of renewables can significantly reduce average wholesale electricity prices in Turkey. The industrial sector in particular, can benefit from cost reductions, which in turn will improve the sector's economic competitiveness. Compared with the currently planned scale-up of wind and solar PV until 2030 (BAU scenario:

18 COBENEFITS South Africa (2019): Economic prosperity for marginalised communities through renewable energy in South Africa. Assessing the co-benefits of decarbonising the power sector available on www.cobenefits.info.

business as usual), a scenario with a high share of renewables (Advanced Renewables Scenario) can reduce wholesale electricity prices by 2.4% on average. This would amount to total savings of TRY 1.96 billion (USD 477 million) in 2030 alone. Compared with a market lacking any renewables, a high-renewables scenario can reduce wholesale market prices by 12.6%, amounting to total cost savings of up to TRY 13.49 billion (USD 2.8 billion) in 2030 alone.[19]

7.3.5 Gender Impact

For too long, professional careers in the energy sector have been male-dominated. This is true for the energy sectors worldwide. Until recently, women accounted for a mere one per cent of top management positions and six per cent of technical staff in the fossil energy sector globally. Countries will fail to achieve the societal traction necessary for energy transitions if women remain unable to access such co-benefits in the same ways as men. The sustained masculinities – the prevalence of a greater proportion of men in the energy sector, especially in positions of power, and the ways in which policies on energy are framed – raise ethical issues related to gender justice, universal energy access and inclusiveness.[20]

In comparison, the renewable energy sector has seen an increase in women employment over the past decade. However, men still outnumber women in the sector's key functions in technical, managerial and policy making positions.[21] Nevertheless, the renewables sector holds out high hopes. Women are estimated to hold 32 per cent of jobs in the renewable energy sector. While this exceeds the 22 per cent across the oil and gas sectors, both are far below the 47 per cent average in the global Figure.[22] Although women currently account for more than 50 per cent of science, technology, engineering and mathematics (STEM) university students in 144 countries surveyed, they occupy only 28 per cent of STEM jobs (such as facility operations and maintenance; equipment manufacturing, construction and installation project planning) in the renewable energy sector. The share of women in non-STEM technical careers, such as finance, statistics and information technology, is slightly greater, at 35 per cent, but the bulk of these positions in the energy sector encompass administrative, non-leadership roles.

19 IASS/IPC/UfU/IET. 2022. Increasing industrial competitiveness and hedging against fossil price volatility with renewables in Turkey. Assessing the co-benefits of decarbonising the power sector. COBENEFITS Executive Report. Potsdam/Istanbul: IASS/IPC. www.cobenefits.info.

20 IASS. 2021. Green Employment for Women. Towards gender-inclusive renewable energy careers. COBENEFITS Impulse. Potsdam. www.cobenefits.info.

21 IRENA, 2019. Renewable Energy: A Gender Perspective, www.irena.org/publications/2019/Jan/Renewable-Energy-A-Gender-Perspective, See also IRENA, 2020. Wind Energy: A Gender Perspective www.irena.org/publications/2020/Jan/Wind-energy-A-gender-perspective.

22 See IASS, UfU and TERI, 2020, Employment co-benefits of decarbonising the power sector in India.

Gender imbalance not only fails to tap economic potential, but also has implications for gender justice, equity, and inclusiveness. Given the lower participation of women in STEM and administrative jobs, neglecting to skill a female workforce in the rapidly growing renewable energy industry can be a disadvantage, particularly because the industry demands a greater share of high-skilled jobs than the fossil energy industry. There is also a risk that qualified maintenance will be neglected if a female workforce is not fully mobilized, such as when project developers attempt to reduce costs by focusing on the installation of renewable energy but reduce budgets for the integration of female workers. At an organizational level, gender diversity tends to be associated with greater creativity, innovation and openness. There is also evidence that women-owned start-ups generate "seventy-eight cents for every $ 1 invested versus thirty-one cents from their male counterparts". In view of the urgent need for climate action and decarbonization of the energy sector, a study by CRB (2012) found that "companies are more likely to increase investment in renewable energy and to decrease carbon emissions throughout their value chain when there are more women on the board of directors". This conclusion may be highly context-dependent, but it is an interesting connection between gender diversity at top leadership levels and its impact on the ways that institutions function.

7.4 Conclusion: The Multiple Co-Benefits of Renewables in Climate Mitigation – Key Recommendations for Climate Neutrality

These examples and data indicated above not only suggest that there are various feasible technologies and instruments out there, already existing to address the urgently needed action on climate change, but beyond, that they provide numerous co-benefits and beneficial impacts for local communities and as well as for entire economies. Whereas specific national or regional contexts need to be considered for the successful implementation of climate policies, the COBENEFITS project found some universal recommendations for policy makers in order to harness the mentioned beneficial side effects of climate measures.

Countries relying heavily on coal power can significantly cut health costs by increasing the share of renewable energy. Health impacts and related costs can be reduced even further by following more rapid and ambitious decarbonisation pathways.

In some cases, socio-economic benefits could even be increased by scaling up the adoption of renewable energy (RE) in line with the more ambitious low-carbon energy pathways. A strong political guidance, to local actors may deliver the anticipated level of significant benefits for marginalised communities.

One lesson learned is that in order to effectively drive the adoption of low-cost off-grid renewable energy systems in remote areas, a close dialogue between the government, private sector and financial institutions at the national and provincial levels is needed concerning suitable financing mechanisms. In general, the private sector or organised community groups need to be encouraged to invest in the off-grid renewable energy sector and be made exempt from import taxes for supplying electricity to households or businesses in rural communities. If effectively implemented, this can stimulate the localisation of skills for the off-grid solar PV.

To sum up: the co-benefits approach is a reversal of former thinking that links climate action with higher cost. The opposite is the case. The prolongation of the existing fossil energy infrastructure will cause an enormous burden on future generations and state budgets, whereas an ambitious climate policy in line or beyond the Paris targets will yield societal benefits of various kinds: employment, clean air, secure and clean energy, local value creation, new technologies, innovation, increased national competitiveness, new business models and incomes for individuals, communities and national economies.

As a consequence, social and economic co-benefits are at the core of a just transition, in terms of a forward-looking, sustainability-oriented Just Transition narrative, which (i) provides guidance to the transition without blocking it, acknowledging interests of the current and next generations; (ii) harnesses the social benefits of the new energy world and maximise the number of beneficiaries; (iii) provides the next generation with capabilities to thrive in the new energy world of renewables; and (iv) honors the accomplishments of the current generation and ensures financial and social dignity for those not participating in employment opportunities emerging with the transition.

Bibliography

IASS. 2021. Green Employment for Women. Towards gender-inclusive renewable energy careers. COBENEFITS Impulse. Potsdam. www.cobenefits.info

IASS/IET/CSIR. 2022. From coal to renewables in Mpumalanga: Employment effects, opportunities for local value creation, skills requirements, and gender-inclusiveness. Assessing the co-benefits of decarbonising South Africa's power sector. COBENEFITS Executive Report. Potsdam/Pretoria. www.cobenefits.info

IRENA 2024. **World Energy Transitions Outlook 2024, International Renewable Energy Agency, Abu Dhabi,** https://www.irena.org/Publications/2024/Nov/World-Energy-Transitions-Outlook-2024, last accessed November 2024.

IRENA 2016. Renewable Energy Benefits – Measuring the Economics, International Renewable Energy Agency, Abu Dhabi, https://www.irena.org/publications/2016/jan/renewable-energy-benefits-measuring-the-economics, last accessed November 2024.

Fraunhofer ISI; Ecofys; EEG; Rütter + Partner; SEURECO: Employment and growth effects of the sustainable energies in the European Union, Final Report, Karlsruhe: Fraunhofer ISI, 2014.

Helgenberger, Sebastian, and Jaenicke, Martin: Mobilizing the co-benefits of climate change mitigation. Connecting opportunities with interests, in the new energy world of renewables. IASS Working

Paper, July 2017, Institute for Advanced Sustainability Studies (IASS), 2017, https://www.cobenefits.info/resources-results, last accessed July 2021.

Manish Ram, Ashish Gulagi, Arman Aghahosseini, Dmitrii Bogdanov, Christian Breyer (2022), Energy transition in megacities towards 100% renewable energy: A case for Delhi, Renewable Energy 195 (2022)587–589.

India

COBENEFITS India (2019a): Improving health and reducing costs through renewable energy in India. Assessing the co-benefits of decarbonising the power sector, available in: https://www.cobenefits.info/resources/improving-health-and-reducing-costs-through-renewable-energy-in-india/

COBENEFITS India (2019b): Secure and reliable electricity access with renewable energy mini-grids in rural India. Assessing the co-benefits of decarbonising the power sector, available in: https://www.cobenefits.info/resources/secure-and-reliable-electricity-access-with-renewable-energy-mini-grids-in-rural-india/

COBENEFITS India (2019c): Future skills and job creation with renewable energy in India. Assessing the co-benefits of decarbonising the power sector, available in: https://www.cobenefits.info/resources/future-skills-and-job-creation-with-renewable-energy-in-india/

South Africa

COBENEFITS South Africa (2019a): Consumer savings through solar PV self-consumption in South Africa. Assessing the co-benefits of decarbonising the power sector, available in: https://www.cobenefits.info/resources/cobenefits-south-africa-consumer-savings/

COBENEFITS South Africa (2019b): Improving health and reducing costs through renewable energy in South Africa. Assessing the co-benefits of decarbonising the power sector available in: https://www.cobenefits.info/resources/cobenefits-south-africa-health/

COBENEFITS South Africa (2019c): Future skills and job creation through renewable energy in South Africa, Assessing the co-benefits of decarbonising the power sector, available in: https://www.cobenefits.info/resources/cobenefits-south-africa-jobs-skills/

COBENEFITS South Africa (2019d): Economic prosperity for marginalised communities through renewable energy in South Africa. Assessing the co-benefits of decarbonising the power sector, available in: https://www.cobenefits.info/resources/cobenefits-south-africa-prosperity

Turkey

COBENEFITS Turkey (2019a): Industrial development, trade opportunities and innovation with renewable energy in Turkey. Assessing the co-benefits of decarbonising the power sector, available in: https://www.cobenefits.info/resources/industrial-development-trade-opportunities-and-innovation-with-renewable-energy-in-turkey/

COBENEFITS Turkey (2019b): Future skills and job creation with renewable energy in Turkey. Assessing the co-benefits of decarbonising the power sector, available in: https://www.cobenefits.info/resources/future-skills-and-job-creation-through-renewable-energy-in-turkey/

COBENEFITS Turkey (2019c): Improving air quality and reducing health costs through renewable energy in Turkey. Assessing the co-benefits of decarbonising the power sector, available in: https://www.cobenefits.info/resources/improving-air-quality-and-reducing-health-costs-through-renewable-energy-in-turkey/

Vietnam

COBENEFITS Vietnam (2019a): Electricity access and local value creation for the un-electrified population in Vietnam. Assessing the co-benefits of decarbonising the power sector, available in: https://www.cobenefits.info/resources/electricity-access-and-local-value-creation-for-the-un-electrified-population-in-vietnam/

COBENEFITS Vietnam (2019b): Future skills and job creation through renewable energy in Vietnam. Assessing the co-benefits of decarbonising the power sector, available in: https://www.cobenefits.info/resources/future-skills-and-job-creation-through-renewable-energy-in-vietnam, last accessed July 2022

Kenya

SD Strategies 2019: The role of geothermal and coal in Kenya's electricity sector and implications for sustainable development, available in: https://newclimate.org/resources/publications/the-role-of-geothermal-and-coal-in-kenyas-electricity-sector-and, last accessed July 2022

Ambition to Action 2022, The role of renewable energy mini-grids in Kenya's electricity sector, available in: https://ambitiontoaction.net/country-analysis-kenya, last accessed July 2022.

Mexico

Crunching numbers Quantifying the sustainable development co-benefits of Mexico's climate commitments, available in: http://sdghelpdesk.unescap.org/sites/default/files/2019-12/Crunching_Numbers_cobenefits_vf_reduc_0.pdf.

Farid Mohamadi

8 CO_2 Reduction Measures in Colombia: Seizing Economic Reactivation Opportunities to Overcome Climate Change Challenges

8.1 Introduction

The Republic of Colombia, like 193 other countries, has signed and ratified the 2015 Paris Agreement, whose objective is to limit global warming increase to 2 degrees Celsius compared to pre-industrial values by 2050 and even to 1.5 degrees if feasible, paving the way to a complete decarbonization of the country. In fact, Colombia, to comply with the article 4, paragraph 2 of the Agreement, has updated its Nationally Determined Contributions (NDCs) for the 2020-2030: a target of 51% less greenhouse gas (GHG) emissions by 2030, the commitment on transformation of several sectors of the Colombian economy as well as a strategy to a Net Zero by 2050.

Indeed, the world's anthropogenic emissions must be reduced by 45% by 2030 compared to 2010 levels to keep global warming under 1.5 dc. And this transformation is expected to both create new climate-friendly economic opportunities and to contribute to the attainment of Sustainable Development Goals (SDGs). Seizing economy reactivation opportunities, in a sustainable way, to overcome climate change and gender equality challenges. This is the task awaiting Colombia's political leaders of the next thirty years and beyond. Economic reactivation policies are deployed in the context of a global post Covid-19 crisis and are expected to enable growth while containing GHG emissions below a certain level.

In that sense, formulating a long-term decarbonization strategy (LTS) as committed under the United Nation Framework Convention on Climate Change (UNFCCC) requires looking not only at the environmental and climate situation in Colombia but also in a holistic way at the economy, its strengths and weaknesses. Then, adapting the regulatory framework considering social aspects from culture and tradition to gender and equality to ensure adoption and prioritize 'no-regrets' programs. In the process of adapting the regulation, a sweat spot between 'command and control' policies, consisting in imposing and enforcing caps, forbidding businesses or materials, and incentivizing policies is to be found as well as the perfect timing for the switch between the two approaches for a successful energy transition. Colombia has performed a tremendous amount of work in removing inadequate incentives and substituting them with more efficient and climate-friendly ones that should contribute to reaching carbon neutrality by 2050. What economic growth for what GHG emissions

https://doi.org/10.1515/9783110784299-008

reduction? Climate risks, as highlighted in several reports from the center for disaster risks (Sistema de Gestion de Riesgo de Desastres), show the urgency to act now in a climate-vulnerable country.

On this ground, the government – through several institutions in charge of the country's decarbonization – has been assessing these risks and their potential costs and subsequently communicated responding measures in the 2020 NDC. Dedicated resources have been allocated proportionally to the needs of each sector and region especially to the most critical sectors such as Agriculture, Forestry and other Land Use (AFOLU) and energy. For each sector, some opportunities and sustainable business models are identified as well as endemic barriers to be overcome in order to support the transformation. Communication is extremely relevant in a society with dozens of ethnics, armed conflicts, half of the land owned by a handful of persons and a high informal economy because it allows involving all parties and stakeholders starting with the society itself. Social adoption by households and industries is a key driver too because both tend to stick to 'business as usual' avoiding new capital-intensive technology unless mandatory regulation is put in place, which is known as insufficient to allow attainment of the objectives. When all tranches of the society and all stakeholders are not convinced by the long-term benefits of the transition and by the necessity to transform their behavior and business models, 'carbon lock in' slows down progress in adoption retaining leaders and end-consumers in business and consumption as usual.

Monitoring and evaluating is also a critical area of work and a precondition for the success of the proposed energy transition: it allows making informed steering decisions at national and local level based on reliable information. It also contributes to the understanding and assessment of the needs over the time. Needs in terms of funding, by confronting the costs and benefits with the available public resources and the potential international public and private funds. Needs in correction and adjustment along the way of the regulatory framework applicable for the carbon neutrality strategy. What resources for adaptation and mitigation per sector? Per region? Which cross-sector and trans-regional resources? What resources to support technological innovation and education to have the right capabilities at the right point of time?

Funding is more challenging in middle-income countries than in Europe for instance, where quantitative easing allows for large capital to be deployed for the energy transition and decarbonization in general. Therefore, private capital is much needed in Colombia and in a very short supply, because the threat would be that the completion of targets gets postponed and NDCs updated still including stranded infrastructure, leading to new targets that would be even more difficult to achieve. Yet, it has been demonstrated that the later actions are taken, the higher the costs to decarbonize the economy (Sanderson & O'Neil, 2020).[1]

1 Sanderson Benjamin M. and O'Neil Brian C., Assessing the costs of historical inaction on climate change, *Sci Rep* 10, 9173, 2020, https://doi.org/10.1038/s41598-020-66275-4.

Finally, transparency plays a major role in the Colombian strategy, because it sends an encouraging message of confidence to foreign investors and to the citizens. The fact, for instance, that Colombia designed a Road Map with clear targets gives visibility to investors to dedicate funds to disruptive industries, innovation, impactful initiatives and resilient infrastructure, despite the uncertainties related to the long-term nature of such strategies. The definition of green taxonomy is a first knowledgeable step, but many hurdles remain along the way to full decarbonization that must be addressed in order to increase the chances of success. Tensions between the political parties and the majority would require compromises and diplomacy to avoid falling in the trap of postponing targets and to ensure that despite changes in the government, today's vision remains intact. Given the complexity of the topic and the numerous reports and publications for the Colombian energy transition, this chapter should be considered as a simplified glimpse over the current strategy and plans of the country with a perspective on opportunities and challenges brought by the decarbonization.

8.2 Part 1: Climate Policies to Create Opportunities

In the context of the economic crisis post Covid-19 and of the soaring global inflation, it is of the highest relevance for leaders to combine climate change with economic stimulus, adopting CO_2 reduction measures and creating economic reactivation opportunities. Policymakers have been working with experts, scholars and societal associations searching for the appropriate policy that could ally climate protecting society with economic growth. Colombia, in cooperation with some international technical assistance providers, assessed its country in terms of needs for climate policies and for funding and initiated a strategy as well as a myriad of plans and programs, all aiming at reducing the GHG emissions while maintaining economic growth. To better understand the local situation, in the light of the complex climate policies and actions puzzle, the regulatory framework and the current CO_2 emissions will be presented, then the situation will be analyzed from different perspectives such as regional, sectoral and social. Every policy plan or program for adaptation and mitigation is thought of and designed to generate opportunities to balance the costs of decarbonization. But most of them face serious barriers that require efforts in terms of finance and measures to change the mindset of both citizens and business and industry leaders. Finally, an introduction to the Colombian long-term Net Zero strategy, E2050, will be made.

The complexity lies in the fact that climate change is the concern of every part of a country's economy, in all sectors, and it is therefore extremely hard, even inevitable, to face redundancies in the programs or plans. Especially when considered from a macro level, all sectors are interconnected and so are the institutional systems. Therefore, and for the sake of simplicity, Colombian climate policies and status quo are presented from different perspectives: with political, sectoral, regional and social approaches. This does not necessarily reflect the real structure of the Colombian context. On the other hand, opportunities and challenges presented hereafter are illustrative and not exhaustive. Also, because of their interconnection between sectors, these opportunities and challenges might fit within different categories. For instance, electro mobility offers opportunities in the sector itself with the manufacturing or sale of EVs but also for the (renewable) energy sector because power production will need a ramp up to cope with the new consumption demand of EVs. Finally, some opportunities might originate directly from the economic reactivation policies, and some would exist irrespective of them.

8.2.1 Regulatory Framework: Around LTS and NDCs

Awareness about the risks for the environment caused by climate change has increased among the Colombian citizens and politicians have been responding by the introduction of a set of laws aiming at mitigating the impact of climate change. Different laws have been voted, a myriad of institutions have been created, partnerships with international organizations initiated, carbon emissions measured or estimated, NDCs and LTSs, programs and projects implemented. Finally, a strategy for 2050, the so-called E2050, has been designed and presented to the stakeholders including the Colombian society.

National Entities and Decarbonization Instruments

Based on three axes established by the UNFCCC, which are education, capacity building and awareness (i), technology transfer (ii) and financing (iii), Colombia's implementation of the decarbonization strategy goes hand in hand with the strategy to reach the United Nations SDGs and imply therefore an interaction with education, investment capital raising and innovation (Figure 8.1). The Colombian government added two further components to its Means of Implementation (MoI) (Medios de Implementación), namely, planning and construction with SISCLIMA (iv) and strengthening of capabilities (v). The objective is to obtain a more effective coordination between all programs and actors as well as a global articulation. The main actors steering the transition are the Ministry of Environment and Sustainable Development (MinAmbiente), the National Planning Department (acronym in Spanish, DNP), the National Unit for Disaster Risks Management (acronym in Spanish, UNGRD) and the Institute for Hydrology, Meteorology and Environmental Studies (acronym in Spanish, IDEAM) all joining forces through the National Plan for Adaptation to Climate Change (Plan Nacional de Adaptación al Cambio Climático [PNACC]). In fact, there are many other institutions involved in dozens of programs and policies, at national and regional level, and per sector, but those mentioned earlier are among the most relevant in Colombia. And at regional level, territorial and urban plans must consider the national level objectives. The plans (Planes Integrales de Gestion del Cambio Climático Territorial – PIGCCT) define the instruments to be used according to each region GHG emissions and vulnerability levels (NDC Colombia, 2020).[2] They assess, define and prioritize measures for mitigation and adaptation actions. The complexity of the coordination and articulation between the different institutions having a role to play in the decarbonization could be, however, a hurdling factor in the path to decarbonization, after the cost, the adoption and further hurdles. The duplication of similar roles and institutions (Sistema Nacional de Cambio Climático, Sistema Nacional de Gestion de Riesgos de Desastres . . .) can also exacerbate frustration of the involved actors on the costs of the management and the speed of progress.

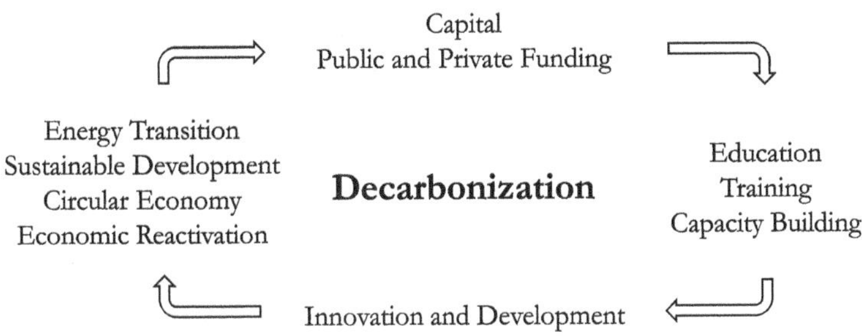

Figure 8.1: Decarbonization Strategy Components (based on Camara de Comercio de Bogotá, 2022 and UNFCCC).

The government obtained support from international organizations and governments such as Agence Française de Development, Expertise France, République Française, joined by World Resources Institute, UNPD, IADB and German BMK, a support that contributed to documenting through numerous reports, structuring, providing technical advisory and deploying resources for a sound climate action in the country. Together with the relevant entities they assess the particularities at macro level and define policies, measures and instruments needed to comply with the multilateral commitments of Colombia such as the pledge to establish a Net Zero by 2050.

Current Regulatory Framework

Colombia has initiated reflection on climate policy and started adopting laws for the protection of the environment in the early 90's, precisely in 1994 with the approval of

2 NDC Colombia, Actualización de la Contribución Determinada a Nivel Nacional de la República de Colombia para el periodo 2020-2030, Colombia: 2020, https://www.minambiente.gov.co/wp-content/uploads/2021/10/informe-actualizacion-contribucion-determinada-Colombia-ndc-2020.pdf.

the United Nations Framework Convention on Climate Change (UNFCCC) through the Law 164 (for an overview, see Figure 8 2 and Table 8.1). This triggered a series of laws and decrees to fight climate change while creating a sustainable and equal economy. But it was not until the Paris Agreement in 2015 that awareness among the political leaders and policymakers led to major change in legislation for climate change. In 2017, the law on the national climate policy (Política Nacional de Cambio Climático*)* was introduced. Then came the law of climate change (Ley 1931 de 2018, Ley de Cambio Climático*)*, adding to the establishment of the National System of Climate Change (Sistema Nacional de Cambio Climático -SISCLIMA) structured to use the regulatory framework and instruments and actors to institutionalize the climate change in the country. In 2020, the country's Long-Term Strategy (Estrategia de Largo Plazo E2050) is submitted to UNFCCC before the COP26 in Glasgow. Then came the law of climate action, (Ley 2169 del 22 de diciembre de 2021, Ley de Acción Climática). And recently, in 2022, the government created an intersectoral commission for climate action (Comisión Intersectorial del Gabinete Presidencial para la Accion Climática).

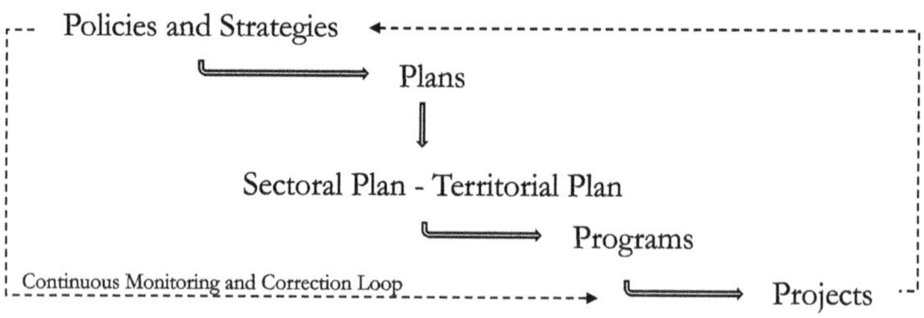

Figure 8.2: Climate Policy: From Strategy to Projects; figure created by author.

The design of strategies considered gender equality and inclusion, the SDG number 5, i.e. Gender and Climate Change among others. As a standard in climate policy design, one of the first steps was to identify regulatory barriers, the inappropriate or obsolete incentives such as subsidies for fossil fuel energy and the lacking ones such as subsidies for renewables or energy efficiency. Then to draft policies that would allow reaching the target position and go backwards up to today and set the measures. The strategy looked at short-term responses, using incentives, a proper taxonomy and adoption methods, the 'enabling phase' and at long-term objectives, the latter focused on obligations to be complied with such as carbon taxes and pricing for instance, the 'imposing phase'. In the first phase, there would be, as example, exemptions of import duties for EVs or tax incentives for renewable energy (RE) projects, while in the second phase, thermic vehicles emitting more than 100 g CO_2 per km would be forbidden for sale by 2040 or restrictions on mobility for thermic engines vehicles would be put in place in urban areas, and energy traders would be obliged have a minimum of 10% of the sold energy produced with clean sources. The National Determined Contributions (NDC) which are, in our context, composed of three main components: i) GHG emissions mitigation, ii) climate change adaptation, iii) implementation instruments of the political action for a low carbon development, adapted and resilient to the climate could be considered as the backbone of Colombian climate ambition for 2030-2050 and were communicated in 2020. In the case of Colombia, it considers human rights, gender equality and women empowerment, intergenerational equality, ethnic and vulnerable communities rights, ecosystems integrity, biodiversity protection, food security, poverty eradication, both sustainable production and consumption, education, peace and development (Acosta Giraldo et al., 2018).[3] All this in alignment with the 2030 Agenda for Sustainable Development Goals (SDGs).

But if there is a consensus about the effective need of NDCs, critics point to the poor tracking and monitoring of the measures, the under-estimated costs and funding efforts but also to the fact that NDCs get rapidly obsolete. Indeed, infrastructure being built now, for instance, lasts up to 50 years and its footprint might undermine decarbonization efforts in the long run, ending up in stranded assets. The risk that infrastructure built under current NDCs becomes obsolete once the NDCs are increased to appropriate levels, leading to abandonment with all the corresponding losses is high in Latin American context (BID and DDPLAC, 2019).[4] The latest communicated NDC for Colombia pledges for: not to exceed 169.44 Mt CO_2 eq in 2030 (51% of reduction compared to 2030 reference scenario) starting the reduction by 2027; to establish carbon budget for 2020-2030 latest by 2023; to reduce black carbon emissions by 40% compared with 2014 values and finally, to foster cooperation under article 6 of the Paris Agreement with a focus on environmental integrity as per the San José Principles. As mentioned earlier, Government level ambitions are then translated into sectoral and territorial plans, in Spanish Planes Integrales de Gestión de Cambio Climático (al nivel sectorial PIGCCS y territorial PIGCCT) in order to consider the local and sectoral specificities, such as strengths and weaknesses for example.

Greenhouse Gas Emissions

The Intergovernmental Panel on Climate Change (IPCC), a large group of experts, internationally recognized in the field, classified the GHG emissions in four categories: energy, industrial processes and product uses (IPPU), agriculture, forestry and other land use (AFOLU), and waste. It includes the following emissions: Carbon Dioxide (CO_2), Methane (CH_4), Nitrous Oxide (N_2O), Hydrofluorocarbon (HFC), Sulfur Hexa-

3 Acosta Giraldo Jenny A., Ovalle Zanabria Katherine, Arcila Burgos Katherine, eds.: Florian Buitrago Maritza and Cortés Ospina Erika, Dirección de Cambio Climático y Gestión del Riesgo Ministerio de Ambiente y Desarrollo Sostenible, Consideraciones de Cambio Climático para el Ordenamiento Territorial, Bogotá, D.C.: Colombia. 2018, ISBN: 978–958-8901-70–1.

4 BID and DDPLAC, Como Llegar a Cero Emisiones Netas: Lecciones de América Latina y el Caribe. Banco Interamericano de Desarrollo, Washington D.C: 2019.

Table 8.1: Relevant Climate Legislation (Mimambiente, 2022).

Year	Law, decree, resolution	Purpose
1994	*Ley 164*	Approval of the UN Framework Convention on Climate Change (UNFCCC)
2000	*Ley 629*	Approval of the Kyoto Protocol under the UNFCCC
2010	*Resoluciones 2733 y 2734*	Procedure to approve of Program of Activities (PoA) and Clean Development Mechanism (CDM) and authorized coordinating entities
2011	*Conpes 3700*	Strategy for the articulation of climate change policies and actions
2012	*Ley 1523*	Policy for the management of risk of disasters Creation of *Sistema Nacional de Gestion de Riesgos de Desastres*
2015	*Ley 1753*	Implementation of sectoral mitigation and adaptation plans for the climate change under the *Estrategia Colombiana de Desarrollo Bajo en Carbono (ECDBC)* Implementation of the national strategy for the reduction of emissions due to deforestation and forest degradation REDD+.
2016	*Decreto 298*	Creation of *Sistema Nacional de Cambio Climatico*
2016	*Ley 1819*	Creation of national carbon tax
2017		Set of the *Politica Nacional de Cambio Climatico*
2017	*Decreto 926*	Set the exemption to the carbon tax
2017	*Ley 1844*	Approval of the 2015 Paris Agreement
2018	*Ley 1931*	Set of directives for the management of climate change
2018	*Resolucion 1447*	Rules for the Monitoring, Reporting and Verification of the national mitigation actions
2020	*Decreto 446*	Rules applicable to organisms that measure GHG emissions reduction

fluoride (SF6) and Perfluorocarbon (PFC). They are considered as the relevant emissions to consider for inventories and diagnostic purposes since all gases do not have the same impact (intensity and lasting impact over time). Inventories of GHG emissions are bi-annual, the so-called Bi-annual Reports (BRs), and the latest inventory carried out in Colombia, is dated 2012, as stated in Table 8.2 based on Third National Communication on Climate Change (BR), (Pulido Guio et al., 2015).[5] In contrast with national communication (NatCom), which are more comprehensive reports and must

5 Pulido Guio Ana D., Jiménez Rodrigo, Turriago Juan D., Mendoza Javier E., IDEAM, PNUD, MADS, DNP, Cancillería, FMAM. Inventario Nacional de Gases de Efecto Invernadero (GEI) de Colombia: Tercera Comunicación Nacional de Cambio Climático de Colombia. Bogotá: 2015, ISBN: 978–958-8902-94–4; http://documentacion.ideam.gov.co/openbiblio/bvirtual/023421/cartilla_INGEI.pdf.

Table 8.2: GHG Emissions in Colombia (based on Pulido Guio et al., 2015).[6]

Sectors	GHG emissions (2010)	GHG emissions (2012)	Evolution 2010–2012
Transport	22.7 Mt CO2 eq	30.4 Mt CO2 eq	+7.7
Energy	22.8 Mt CO2 eq	24.5 Mt CO2 eq	+1.7
Industries	29.4 Mt CO2 eq	27.4 Mt CO2 eq	−2 (reduction)
Deforestation	78.9 Mt CO2 eq	27.3 Mt CO2 eq	−51.6 (reduction)
Residential	5.7 Mt CO2 eq	6.2 Mt CO2 eq	+0.5
Agriculture	27.7 Mt CO2 eq	27.9 Mt CO2 eq	+0.2
Livestock	26.2 Mt CO2 eq	23.8 Mt CO2 eq	−2.4 (reduction)
Waste	10.5 Mt CO2 eq	10.9 Mt CO2 eq	+0.4
TOTAL	223.9 Mt CO2 eq	178.4 Mt CO2 eq	−45.5 (reduction)

be submitted by countries every four years, according to the article 12 of the UNFCCC. The methodology used is based on the 2006 IPCC guidelines, on Tier 1 and Tier 2 data (Tier 3 being significantly detailed data collection, studies and correlation with international data). It is important to note that the methodology allows for deducting the absorption of emissions in forests, meaning that the results of AFOLU are net, i.e.; difference between emissions and absorptions.

In 2010, a Colombian, on average, originated 4.95 t CO_2 eq p.a. (based on a population of 45.2 Mio inhabitants) and 3.88 t CO_2 eq p.a. per capita in 2012 (on 46 Mio inhabitants). In comparison, an habitant from the European Union, had on average 9.3 t CO_2 eq per capita in 2012 (European Environment Agency, 2022).[7] The objective of 2030 is to reduce GHG emissions by 51% bringing them from 346 Mt CO_2 eq to 170 Mt CO_2 eq. assuming a growth scenario. And for the 2050 Net Zero objective, the emissions level per capita should be brought to around 0 to 1 t CO_2 eq. Forest and agriculture represent about half of Colombia's GHG emissions (Table 8.3). They, therefore, require special attention especially in a country with one of the world's highest biodiversity.

In 2020, although the figures have not been communicated to the UNFCCC, the total GHG emissions were estimated at 270 Mt CO_2 eq (Climate Watch, 2022).[8] For the sake of comparison, for Argentina, a country of similar size in population with different biodiversity, climate risks and economy model though, emissions are estimated at

6 Pulido Guio Ana D., Jiménez Rodrigo, Turriago Juan D., Mendoza Javier E., IDEAM, PNUD, MADS, DNP, Cancillería, FMAM. Inventario Nacional de Gases de Efecto Invernadero (GEI) de Colombia: Tercera Comunicación Nacional de Cambio Climático de Colombia. Bogotá: 2015, ISBN: 978–958-8902-94–4; http://documentacion.ideam.gov.co/openbiblio/bvirtual/023421/cartilla_INGEI.pdf.

7 European Environment Agency, 2022, GHG Emissions in the EU (accessed June 2022); https://ec.europa.eu/eurostat/databrowser/view/t2020_rd300/default/table.

8 Climate Watch, 2022, (accessed June 2022); https://www.climatewatchdata.org/ghg-emissions?end_year=2019&regions=COL%2CARG&start_year=1990.

Table 8.3: GHG Emissions Evolution (based on Colombia's LTS E2050, Colombian Government, 2021).[9]

Sources of Emissions in Mt CO_2 eq	1990	2010	2014	CAGR 1990-2014
Forest Land Management	113.93	84.21	79.23	−1.5% p.a.
Agriculture Activity	45.24	54.7	52	0.6% p.a.
Mining and Energy Industries	14.06	26.66	32.32	3.5% p.a.
Manufacturing Industries	15.40	25.10	27.63	2.5% p.a.
Transport	18.95	24.52	28.96	1.8% p.a.
Sanitation Activities	4.17	9.22	9.82	3.6% p.a.
Commercial Facilities and Residences	4.55	5.94	7.01	1.8% p.a.
TOTAL Emissions	216.29	230.36	236.97	0.4% p.a.
Sources of Removals in Mt CO_2 eq	1990	2010	2014	CAGR 1990-2014
Forest Land	−3.22	−17.15	−15.6	6.8% p.a.
Agriculture Activity	−3.68	−6.13	−7.06	2.7% p.a.
TOTAL Removals	−6.90	−23.28	−22.66	5.1% p.a.
NET BALANCE (Emissions – Removals)	209.10	207.08	214.31	0.1% p.a.

around 400 Mt CO_2 eq for the same year (Climate Watch, 2022).[10] Comparisons with developed countries are irrelevant given the obstacles Colombia faces to access large capital for climate change and the fact that as per today, the largest GHG emitters are developed countries such as the U.S.A. for instance. Furthermore, figures are difficult to verify, for example, according to Climate Watch, Colombia emitted 220 Mt CO_2 eq in 2012, whereas their 3rd BR communicated to the UNFCCC shows around 180 Mt.

Risk Assessment for Mitigation and Adaptation

The most common actions, programs and projects in climate policies are classified in three categories, mitigation, adaptation and cross-sectoral actions or synergies. The mitigation actions are mainly related to energy (generation and efficiency), transport, buildings, industries and the adaptation are actions that transform the existing infrastructure vulnerable to climate change such as extreme weather, floods or landslides

9 Colombian Government, Ministry of Environment, DNP, Foreign Affairs Ministry, AFD, Expertise France, E2050 Colombia's long-term climate strategy to meet the Paris Agreement, Full Report, WRI: Bogotá 2021.

10 Climate Watch, 2022, (accessed June 2022); https://www.climatewatchdata.org/ghg-emissions?end_year=2019&regions=COL%2CARG&start_year=1990.

into resilient and sustainable ones (Mohamadi, 2021).[11] The synergies have an impact on both mitigation and adaptation. These actions, programs and projects, grouped in Colombia under the Programs and Projects of Low Carbon Development (in Spanish acronyms PDBC), can also be classified by sectors such AFOLU, energy, infrastructure but also on country-specific sectors (Table 8.4).

These plans, programs and projects respond to criteria whose most relevant one is the risk behind climate change and the vulnerability of the associated population. And Colombia is considered as a high climate change risk country. These risks are classified per type of threat or natural event and their potential impact is assessed probabilistically per impacted sector and subsector by the Sistema Nacional de Gestion de Riesgos de Desastres the Colombian institutional disasters risk manager (Table 8.5). In a technical report commissioned for this institute, threats were assessed depending on their magnitude and their frequency of occurrence and classified per scenarios (Cardona et al., 2021).[12] Losses, yearly average and maximal potential losses are also quantified and estimated in percentage of GPD and in values. For instance, the yearly average losses due to floods and hurricanes are estimated to a significant 7.54% of the GDP (ca. 4.8 Bio COP).

Country's Low-Carbon Transition Road Map: E2050

Colombia's long-term strategy (LTS) is based on the conceptual model Resilience Thinking and proposes to attain climate socio-ecologic resilience by 2050 by working on 9 efforts in form of a pledge: the development of knowledge on climate change to make informed decisions on governance and policies (i), an integrated management of the biodiversity including the reduction of deforestation and ecosystems degradation, the maintenance of ocean and coastal ecosystems, and the security of access to water (ii), sustainable production and consumption, including an emerging circular economy, with less waste, less GHG emissions, efficient and environmental-friendly production systems, sustainable procurement criteria for the state, households and businesses, equality and innovation (iii), a fair workforce transition, including equality, improvement of conditions of life, inclusion, incentives and education offer to foster green jobs creation (iv), a rural development per region, promoting resilience to climate change, biodiversity and ecosystem services, forest and communities protec-

11 Mohamadi Farid, Introduction to Project Finance in Renewable Energy Infrastructure: Including Public-Private Investments and Non-Mature Markets, Springer Nature, Cham: 2021, ISBN: 978–3-030-68742-7 https://doi.org/10.1007/978-3-030-68740-3.

12 Cardona O.D., Bernal G., Pabón J.D., M. A., Marulanda M. C., Carreño M. L., González D., Villegas C., Marulanda P., Grajales S., Rincón D., Molina J.F., Estudio de Riesgo por Efectos del Cambio Climatico y Medidas de Adaptacion para la Estrategia a Largo Plazo E2050 de Colombia, Ingeniar Risk Intelligence, Bogota: 2021.

Table 8.4: Mitigation and Adaptation Programs in Colombia (compiled by author).

Sectors	Mitigation	Adaptation
All (Sub-sectorial)	Nationally Appropriate Mitigation Action (NAMAS)	
All	Economic Support Program: instruments to promote a transition into a competitive, sustainable and low carbon economy by reducing the population and ecosystems vulnerability to climate change effects. It includes mainly: – Carbon tax – Exemption of the carbon tax in case of offset through CERs – Emissions Trade Systems (ETS) will be allowed in a near future (through the *Programa Nacional de Cuotas de Emisiones de Gases de Efecto Invernadero* and in particular the *Programa Nacional de Cupos Transables Emision de Gases de Efecto Invernadero, PNCTE*).	
All		*Plan Nacional de Adaptacion al Cambio Climatico (PNACC):* to reduce the impact of climate change and/or build capacity to do so. This includes assessing threats to the vulnerable communities, preventing impacts to the territories, ecosystems and economies, mitigating damages and finally seizing opportunities related to the fight against these impacts. https://www.minambiente.gov.co/wp-content/uploads/2022/01/PNACC-2016-linea-accion-prioritarias.pdf
Ecosystems		*Adaptacion basada en Ecosistemas (AbE)* https://www.minambiente.gov.co/wp-content/uploads/2022/01/MADS_Guia_AbE_LIBRO_Digital-Cambio.pdf
Communities		*Adaptacion basada en Comunidades (AbC)*
Infrastructure		*Adaptacion basada en Infraestructuras (AbI)*
Technology		*Adaptacion basada en Teclogias (AbT)*
AFOLU	Reducing Emissions from Deforestation and Forest Degradation (REDD+) to handle emissions due to deforestation and forest degradation	
Energy	Clean Development Mechanism (CDM) to handle emissions through Certified Emissions Reductions (CERs)	
	Clean Development Mechanism – Program of Activities (CDM-PoA)	

Table 8 5: Climate Threats and probabilistic Impact on Sectors Matrix (Cordena et al., 2021 from E2050).

Threats and events vs. Sectors and subsectors	Construction	Infrastructure	Agriculture	Ecosystem Services
Floods	Residential Commercial Industry Government Education Health	Water and water treatment Energy Telecommunication Transport		
Droughts			Corn Rice	
Fires in vegetal area				Woods & forests Agriculture
Hurricanes	Residential Commercial Industry Government Education Health	Water and water treatment Energy Telecommunication Transport		
Landslides		Main roads network		

tion and security of access to food (v), an integrated urban development, including robust and inclusive urban governance, sustainable consumption, urban planning with efficient net-zero cycle buildings, optimized waste management and circular economies (vi), a diversified energy mix, including a target for full electrification, a decentralized power generation with smart management, digitization, renewable energies located according to their resources and connected to the national grid, an obligation by 2050 to capture carbon (CCUS) or to offset it for coal and gas power plants, and preserve the environment (vii), mobility and sustainable infrastructure, based on competitiveness for the users and the economy, including electrified mobility, clean-energy powered heavy transport, air transport with reduced emissions, ban by 2040 of diesel vehicles import, smart roads and rivers, climate-friendly and resilient infrastructure (viii), health improvement, including climate actions co-benefits on health (ix) (E2050 website, accessed: June 2022).

Beside the extensive list of efforts or bets (apuestas) to be considered, LTS, in general, must look at where to source the funding to enable the transition, approach the private sector and the institutional funds that catalyze financing. Understanding the vision of these actors is fundamental to design proper and efficient policies, to substitute obsolete incentives and remove regulation barriers. There are three major phases for the carbon neutrality of the country, with milestones and targets: 2020-2030, the phase of significant increase in ambition, 2030-2040, the phase of multidimensional transformation and 2040-2050 the consolidation phase of a climate resilient future, as described in Table 8.6 (Colombian Government, 2021).[13]

Table 8.6: Extract of E2050 ambitions vs. phase (Colombian Government, 2021).[14]

E2050 ambitions (out of 195) / Phases	2020–2030 Ambition	2030–2040 Transformation	2040–2050 Consolidation
Coverage of hydrometeorological networks	35% of the country	35% of the country	35% of the country
Technology trends in business	30–40% of companies	60–80% of companies	80–100% of municipalities
Knowledge within scope of territorial actors	20–50% of municipalities	50–80% of municipalities	80–100% of companies
Zero deforestation and ecosystem degradation	2550000 ha	3400000 ha	Min. 4250000 ha
Forest with sustainable management	>20%	>35%	>50%
Ecological restoration for connectivity	1000000 ha	1130000 ha	1300000 ha
Conservation *ex situ*	30% of threatened species	60% of threatened species	90% of threatened species
Bioeconomy and GDP Increase in circular economy companies	8–10% of GDP 20–40%	10–15% of GDP 40–60%	20% of GDP 80–90%
Sustainable forest	15–20%	20–45%	>45%
Increase in agriculture productivity	10-20%	40–60%	30–50%
Sustainable livestock	40–60%	60–70%	70–100%

The conceptual work has started triggered by a decision of the Intersectoral Climate Change Commission (Comisión Intersectorial de Cambio Climático [CICC]) in 2019 with an emphasis on public consultations and participation of societal associations, including a panel of international and national experts. After assessing the transformation

13 Colombian Government, Ministry of Environment, DNP, Foreign Affairs Ministry, AFD, Expertise France, E2050 Colombia's long-term climate strategy to meet the Paris Agreement, Full Report, WRI: Bogotá 2021.

14 Colombian Government, Ministry of Environment, DNP, Foreign Affairs Ministry, AFD, Expertise France, E2050 Colombia's long-term climate strategy to meet the Paris Agreement, Full Report, WRI: Bogotá 2021.

needed with priorities and comparing the impacts of action vs. inaction, among other analyses, the final document has been sent to the UNFCCC. Several models were used to estimate the different scenarios of decarbonization, and in this context, Universidad de los Andes, a major university in Colombia has analyzed the impact of the full electrification on the economy and displayed them in the form of trajectories. They used a comprehensive method considering all major drivers such as energy plans, carbon-free energy mix, electrification of mobility, use of energy efficiency, carbon capture (avoided emissions and carbon absorption) using different technology and even different levels of meat consumption by the Colombians.

8.2.2 Sector Approach: Energy, Industries, Transport and AFOLU

The complexity behind the energy transformation required by the commitment made in the framework of the Paris Agreement and by a Net Zero dynamic is that Colombia, like just like any jurisdictions in a similar context, must find a way to decarbonize its energy sources for all sectors, including energy production, transport, agriculture and industries, including the costs of externalities while ensuring a safe, secured, always available, competitive energy supply. This starts with greening the energy mix with more non-conventional renewable energy (with large-scale infrastructure and decentralized distributed production), reducing energy consumption through energy efficiency measures, introducing electric buses in large cities, promoting individual electrical vehicles purchase, transforming the agriculture and forestry sector, launching road maps for hydrogen production among others, enhancing circular economy and collaborating with the private sector. Sectorial guides for Gender and Climate Change were published in 2020 and 2021 for the transport, housing, industry, environment, mining and energy, and agriculture. In the energy sector, for example, the group of national and international experts that worked at the transformation of the energy sector, under the leadership of the MinEnergia, considered for their road map five key aspects: i) the electrical market structure (including participation and competitiveness), ii) the place of the gas in the transformation, iii) decentralization, digitalization and efficient management of the demand, iv) improving the incentives system and v) review of the regulatory framework (7) (Misión de la Transformación Energética, 2022).[15]

15 Mision de Transformacion, 2022, (accessed March 2022)https://energiaevoluciona.minenergia.gov.co/transformacion#.

Energy and Deployment of Non-conventional Renewable Energy

The country is rich in natural resources to produce energy: oil, coal, wind, rivers for hydropower and sun radiation. In fact, Colombia has enough oil and coal to be able to export which it does. Thus, today's energy mix in Colombia is dominated by hydropower with c. 11.1 GW about 65% of the total installed capacity, 17.6 GW. This energy is, however, considered a conventional renewable energy source due to the impact of large-scale hydropower plants and dams on the environment. This energy is relatively reliable and firm, compared to wind and solar energy because it is significantly less intermittent (capacity factors of 60 to 80% compared to 20 to 40% for solar power). It is the reason why Colombia maintained this energy mix for decades. In 1992, the country faced a blackout due to the El Niño phenomena, an event where droughts bring the water reserve down reducing the capacity of hydropower plants to supply sufficient power to respond to the consumption demand. This event triggered policies to develop new capacity using so-called "non-conventional renewable energy". Indeed, often, in these latitudes, the wind blows extensively in seasons when water reserves reduce. These policies, beyond the fact that Colombia needed to diversify its energy mix with clean energy, were meant to persevere the citizens from experimenting blackouts, with all the consequences that they could have on the economy. In 2014, a new regulation was established with the vote of the Law 1715, aiming at promoting non-conventional renewable energy, with a set of fiscal incentives among others. In 2022, only one small-scale (20MW) wind farm was erected, adding to a 18 MW pilot project built by Medellin public utility EPM back in 2007. Solar energy, in contrast, has developed more actively, going from almost none by the time the law is introduced to above 200 MW. The Law 1715 has been amended in its article 11 by the article 174 of the Law 1955 of 2019 (National Development Plan), clarifying and enhancing certain incentives, and further additions introduced to close regulation gaps. Still, compared to countries of similar size such as Argentina for instance, the pace is extremely slow. In 2019, following the Law 1955 of 2019 and after a first auction that failed, about 1300 MW of mainly wind energy were awarded long-term PPAs. Most of them being planned in La Guajira, a region with poor infrastructure and marked indigenous communities, technical and social hurdles have slowed down and delayed the implementation of these large-scale projects. These projects and further solar projects from a second auction, are expected to enter the national interconnection system by 2025, some of them subject to the commissioning of a high-voltage transmission line facing the same hurdles. Energy being the second largest producer of GHG emissions in Colombia after AFOLU, the government had to pay special attention on how to decarbonize this sector, and the recent adoption of the obligation for registered energy traders to have a minimum of 8–10% of their sold energy produced by non-conventional renewable energy, is a good start. This, combined with the willingness of major utilities to anticipate the need for Environmental, Social and Governance (ESG) compliance will contribute significantly to the energy transformation of the country's energy mix. Also, the government started initiatives to discuss with the stakeholders the most appropriate business environment to kick start large-scale hydrogen production and offshore wind. The road maps resulting from these actions led to a first license application for a wind farm off the shores of Cartagena and green hydrogen pilot projects. Still, the lack of articulation between the different institutions and general planning, for instance between ANLA and UPME to dedicate priority RE development zones or the management of points of grid connection by UPME through the recent CREG 075 decree, a decree with good intention of purging the 'paper projects' and having only bond guaranteed connections, created confusion instead within the RE developers' community.

Relevant Applying Regulation

One of the major pivotal regulations in favor of the development of renewable energy was introduced in 2014, the Law 1715. It sets the ground for the development of renewable energy, large- and small- scale generation and energy efficiency, including net-metering for instance, detailing the competent authorities, incentives etc. Relevant changes were added through the Law 2099 of 2021 to include newest technologies such as hydrogen energy and financing support schemes (see Table 8.7).

The energy sector encompasses several sub-sectors such as energy efficiency, use of energy for cooking in rural areas, energy for transport and heating but the energy production represents the most significant sub-sector to look at and with a focus on clean energy. According to the Vice Minister of Energy and Mines, Miguel Lotero, around 3.000 small-scale solar projects for self-production have been realized during their 4-year mandate under the President Ivan Duque, together with the 3.400 MW of large-scale wind and solar energy to be built and some 800 MW to be commissioned, renewables will save 1 Mt CO_2 per annum and take RE share in the energy mix to 16% by 2023. His government contributed to reduce the power interruptions by 30% going from around 41 hours yearly in 2019 to 29 hours in 2021, based on Asocodis figures. It has also provided first-time access to power to 80.000 families in remote areas, 30.000 from them with solar energy and first-time substitution of wood for cooking by natural gas, for a further 70.000 families (Vanguardia, 2022).[16] In the energy sector, electricity being a major energy sub-sector, the objective is to have around 76% from hydro, wind, solar, biomass and geothermal (cf. Table 8.8) and the remaining 24% being from fossil fuel whose emissions would be absorbed with CCUS and reforestation. The challenge will remain the 'full electrification' of the demand, which requires heavy transformation especially in the transport and mobility sector. Objectively, Colombia is on the right path: now with the latest auctions, with an average energy

16 Vanguardia, Columna Miguel Lotero, Política Energética Sostenible y Sostenida, (accessed May 2022) https://www.vanguardia.com/opinion/columnistas/miguel-lotero/politica-energetica-sostenible-y-sostenida-AM5248064.

Table 8.7: Relevant Regulation in Colombia (Minergia, 2022).[17]

Year	Law, decree, resolution	Purpose
2014	*Decreto 2469*	Manages excess of energy production
2014	*Decreto 2492*	Sets mechanisms for the response to demand
2015	*Decreto 1623*	Adds *Decreto 1073 de 2015* and sets political guidelines for the extension of the energy coverage in the connected grid system *SIN* and non-connected areas
2015	*Resolucion CREG 024*	Sets the rules for energy self-production at large scale in the grid system *SIN*
2015	*Resolucion UPME 0281*	Defines the maximum power capacity for self-production at small scale
2015	*Decreto 2143*	Adds *Decreto 1073 de 2015* and defines the rules of application of the incentives established in Chapter III of the Law 1715 of 2014.
2016	*Resolucion MinAmbiente 1283*	Sets the procedures and requirements to obtain certification of environmental benefits for new investments in renewable energy and thus obtain fiscal incentives under articles 11, 12, 13 and 14 of Law 1715 of 2014.
2016	*Resolucion MinAmbiente 1312*	Sets the terms and requirements to elaborate an Environmental Impact Assessment (EIA) required to obtain an environmental license for wind farm
2018	*Resolucion UPME 703*	Sets the procedures and requirements to obtain certification of renewable energy projects for the benefits of VAT exclusion and exemption of import duities under articles 12 and 13 of Law 1715 of 2014.
2019	*Ley 1955 de 25.05.2019*	Introduces the National Development Plan (*Plan de Desarrollo Nacional*), modifies the Law 1715
2021	*Ley 2099 de 10.07.2021*	Modifies the Law 1715 of 2014 in order to extend it to further technologies such as hydrogen and geothermal energy and carbon capture usage and storage (CCUS) and to a larger scope; sets as public interests certain projects *Proyectos de Interes Nacional y Estrategico (PINES)*; sets a fund (FENOGE) for the promoting and funding of these projects; modifies incentives such as accelerated depreciation among others; sets priorities for the economic reactivaton. Modifies the Law 143 of 1994 in particular in terms of the responsible institutions.

17 Minenergia, 2022, Fuentes No Convencionales de Energia Renovable, (accessed May 2022) https://www.minenergia.gov.co/energias-renovables-no-convencionales.

price of 100 COP per kWh in the 2019 auction and 130 COP per kWh in the second one, the demonstration has been made that RE is cheaper than fossil fueled energy production. Still the challenge of the intermittency of the renewables remains especially for the grid system in a scenario of high share of RE. In this context, Colombia, aware of the challenge of RE variability, initiated discussions with the relevant stakeholders about the need and potential of hydrogen energy and concluded in 2022 its national hydrogen road map.

Energy Reinforcement 'Very Long-Term' Plan by 2050

Today, in 2022, the total installed capacity is 17 GW with 70% of hydro power and some coal and gas power plants. The so-called very long-term scenario (muy largo plazo) established by UPME by 2050, includes 17.8 GW of wind and solar (with a contribution of 2 GW of offshore wind), representing 42% of the total energy mix (see Table 8.8). Adding the hydro power, clean energy will represent about 85% of the total in capacity, not in generation, different sources having different capacity factors. These figures are the ones of the least optimistic 'very long-term' scenario, the most optimistic one has about 2 GW more (44 GW instead of 42 GW). Other scenarios like scenario 10 for example foresees an additional of up to 7.8 GW of wind and solar, respectively 4.6 GW and 3.2 GW. In general, the scenario established by UPME considers recommendations from market operator XM and regulator CREG to ensure grid stability with a high share of intermittent energy. Furthermore, in Colombia, after the El Nino event, a concept of guaranteed firm capacity (so-called Cargo por Confiabilidad) has been set to always be able to rely on firm capacity. This makes the contribution of energy production for the E2050 plan relatively realistic. And although, often discussed especially in reports about the security of energy supply, the question of developing or not nuclear power, is still ongoing but is not included in most scenarios.

Opportunities Generated by the Energy Transition to Seize

The latest relevant regulatory change through the Law 2099 of 2021 considers the necessity to stimulate investments in the energy sector to reactivate the economy. And even before its ratification, some projects were accelerated to generate short-term job creation such as the large-scale solar projects that were awarded PPAs during the second long-term renewable energy auction in 2021. There are opportunities in the sector in both grid transmission reinforcement and electricity production, pushed by an impressive 10.6% of economic growth in 2021 (World Bank Data, 2022)[18] and thus energy demand growth as illustrated by the expansion plan (Plan Indicativo de

---

18 World Bank Data, Economic Growth per Country, (accessed March 2022); https://data.worldbank.org/indicator/NY.GDP.MKTP.KD.ZG?locations=CO.

Table 8.8: Very Long-Term Scenario 2 with 42% of wind and solar (figures: UPME, 2019)[19].

Energy Resources	Basis	Closed extension/in construction	Extension Scenario 2	Extension 2035–2050	Total in MW	Share per resource
Hydro	11122	2400	0	798	14320	34.01%
Gas	3726	762	0	350	4838	11.49%
Coal	1623	0	0	860	2483	5.90%
Wind	18	2042	1658	2744	6462	15.35%
Large-scale Solar	18	713	700	7080	8511	20.22%
Distributed Solar	15	594	0	179	788	1.87%
Offshore Wind	0	0	0	2000	2000	4.75%
Others (biomass, geo . . .)	1138	81	270	1209	2698	6.41%
TOTAL	17660	6592	2628	15220	42100	100%

Expansion de la Cubertura Electrica 2019-2023, PIEC and Plan de Expansion de Referencia Generacion-Transmision 2020-2034) from the planification entity (Unidad de Planificacion Minero Energetica, UPME). To facilitate the execution of clean energy projects, dedicated funds have been made available such as FENOGE and FINDETER. However, insiders from the industry have been criticizing the system for the complexity of access to these funds and the disproportion between the needs and the available funds. Offshore wind, green and blue hydrogen, for which Road Maps have been released, offer further opportunities, even if their implementation might present higher risks than more mature technology, especially in a Colombian green market that is just starting, and might also require more time. Local content requirement is a means to foster local manufacturing of renewable energy equipment, for example through solar panels made in Colombia. Although not much appreciated by developers because the requirement adds a further constraint to a non-mature market with all its challenges, experience has shown that the economy can benefit from it with part of the investment remaining in the country and significant jobs creation. Globally, in the path of attaining a 1.5 d. Celsius, the renewable energy sector is expected to reach 38 Mio jobs by 2030 and 43 Mio by 2050 (compared to 12 Mio in 2020), with the majority in the solar energy branch with around 20 Mio jobs, bioenergy with 13.7 Mio jobs, wind with 5.5 Mio and hydropower with 3.7 Mio jobs by 2050 (IRENA and ILO, 2021).[20] This distribution varies from country to country but the relevance and the number of new jobs could realistically apply to Colombia's energy transformation.

19 UPME, Plan de Expansión Referencia Generación Transmisión 2020-2034 Volumen 2. Generación, Colombia: 2019; http://www.upme.gov.co/Docs/Plan_Expansion/2020/Volumen2_Plan_Expansion_Generacion_Transmision_2020_2034_Final.pdf.

20 IRENA and ILO, Renewable Energy and Jobs: Annual Review 2021, International Renewable Energy Agency and International Labor Organization, Abu Dhabi, Geneva: 2021, ISBN: 978–92-9260-364–9.

Challenges to Overcome

The transmission lines that must be built, and the electric grid reinforced are infrastructure that are on the one hand business opportunities but also barriers for the energy projects developers who cannot easily find where to connect their solar or wind farms. The new transmission lines such as Colectora that goes from La Guajira to the center of the country have seen their date of commissioning delayed which caused major complications to the utilities that counted on them and even entered PPAs. A further challenge could be the complexity of the coordination between the different institutions having a role to play in the energy transformation such as Ministry of Energy and Mines, UPME, XM, Ministry of Environment with the ANLA, PDN and local authorities. In this context, developers have been struggling, for instance, to obtain eligibility of certain incentives such as tax reduction and credit under the Law 1715 which led the government to set working groups (Mesas de Trabajo) bringing the above-listed public institutions and the private developers and utilities to discuss ways to remedy these issues. Finally, the fact that gas and oil remain somehow subsidized with about 0.66 Bio USD in 2019 (Climate Transparency, 2020)[21] keeps affecting the competitiveness of RE against the energy sources (cf. Costs-benefits of the Energy Transition).

Integration of Energy Efficiency Measures

Policies to promote energy efficiency have been introduced in 2014, with regimes for public companies and private ones. They go hand in hand with the energy production sector. Indeed, one must assume, in order to set a target of closing fossil fuel-based energy production and of reinforcing non-conventional renewable energy, the potential contribution of energy efficiency (DNP, 2022).[22] Among the energy efficiency measures, mandatory energy or emissions tags and labels on household equipment (RETIQ) and for streets and public lighting (RETILAP) play an important role. International standards and labels such as the well-recognized LEED certification as well. But also, business models proposed by specialized companies, the so called Energy Services Companies (ESCos), to help large consumers in the industry to reduce their energy consumption by renewing their equipment.

21 Climate Transparency, Report on Colombia's Climate Action and Response to the Covid-19 Crisis, 2020, https://www.climate-transparency.org/wp-content/uploads/2021/01/Colombia-CT-2020.pdf.

22 DNP Departamento Nacional de Planeación, Eficiencia Energética, (accessed on May 2022); https://www.dnp.gov.co/Crecimiento-Verde/Ejes-estrategicos/Paginas/Eficiencia-energ%C3%A9tica.aspx.

Opportunities Generated by the Energy Transition to Seize

International ESCos such as French Greenyellow for instance are entering the market joined by national locally founded ones. They propose several schemes including the financing through long-term leases depending on their customer's profile. There is an important market to cover, so opportunities can be seized in this area. Furthermore, specialized funds such as FENOGE support initiatives that improve the efficient use of energy. In the commercial building and construction sector, opportunities for passive housing are also available as the market of energy efficiency is just starting to develop. Local manufacturing of solar panel-based lighting for streets is also a market niche given the low technology requirements, even though requirements set in public tenders might rule out local product offers due to lower standard or quality non-compliance.

Challenges to Overcome

The lack of local and international standardization as well as the missing clarity in the use of labels could be a hurdle since it prevents a good understanding of the efficiency levels by the lenders and thus could complicate the funding. The limited offer in and the access to funding is a further challenge, as well as the poor awareness of business and industry leaders of the benefits offered by energy efficiency. Finally, the high initial investment costs needed to get new efficient equipment could discourage potential adopters.

Electrifying the Transport Sector

The ambitious 2050 target requires an intensification of the use of electrical vehicles (EV), combined with more adoption of public transportation, also to be fully electrified. Around 16,000 hybrid EVs and only 1000 EVs were matriculated in 2021 in the country and new cities adopted electric buses in their fleet. In the case of Bogota, for instance, already a user of electric buses, 1002 electric buses are pending delivery from the last auction (Andemos, 2022), with which the capital will remain the leader among Latin American capital cities for the fifth year in a row. In Colombia, two third of bus services need to be electrified to reach Net Zero by 2050 and 4 times more public transport users (BID and DDPLAC, 2021).[23] The sector with an energy consumption of 28.96 Mt CO_2 eq (figures for 2014) represents about 8% the country's total GHG emis-

23 BID and DDPLAC, Como Llegar a Cero Emisiones Netas: Lecciones de América Latina y el Caribe. Banco Interamericano de Desarrollo, Washington D.C: 2019.

sions value (Colombian Government, 2021).[24] This is because not only urban centers contribute negatively to global warming but the transport of goods across the country. The latter requires adoption as well by logistics companies. Full electrification of the transport sector requires a pivotal change in the use of railways and fluvial channels, whose usage is at its infancy in Colombia, a significant addition of renewable energy, which might be underestimated even in the most optimistic scenario of energy production extension and the reinforcement of corresponding infrastructure such as charging stations among others. Electromobility requires measures to foster adoption as well as economical support to trade off the heavy impact of the initial investment in EVs for low- and middle-class households. Colombia, because of its complex orography, faces the issues of high investments for roads and railways, and subsequently developed a strong domestic flights market. A trip from Bogota to Medellin, 245 km, would last 8 to 9 hours on the local highway, when a flight would take only 30 min from takeoff to landing, and be even cheaper. Aviation is therefore a serious environmental issue in Colombia given the high carbon intensity of air traveling. Cycling is part of the urban culture in Colombia for leisure and to commute between home and workplace and very popular among all social classes: Bogota, for instance, has the world's longest bicycle lanes with over 500 km. The contribution of this CO_2 free transportation is also crucial to achieve Net Zero and must be developed and reinforced. For the sake of simplicity, maritime transport and 'imported' emissions are not analyzed here.

Opportunities Generated by the Energy Transition to Seize

Because electrifying the transport sector represents a major market, there are a myriad of opportunities for business and industry leaders and investors. Colombia disposes, for instance, of a history and know-how with car manufacturing: in the 80's French Renault had a major factory in Medellin and today Chevrolet from the USA is assembling several of its models. Transformation, if market signals were detected by local or foreign investors, could take place and lead to local manufacturing of EVs at national scale. Furthermore, existing train railway is not exploited: this could be retrofitted to allow electric trains for freight (example of Costa Rica with its 250 Mio USD soft loan and 21 Mio USD grant from Green Climate Fund), this would reduce by up to tenfold the CO_2 emissions per ton-km compared to road transport (Thirion and Geoffron, 2020).[25] Inter-modality of the transport is a vector of enhancement of sustainable transport, since it could connect railways with fluvial ways and thus allow a complete

24 Colombian Government, Ministry of Environment, DNP, Foreign Affairs Ministry, AFD, Expertise France, E2050 Colombia's long-term climate strategy to meet the Paris Agreement, Full Report, WRI: Bogotá 2021.

25 Thirion Benoit, Geoffron Patrice, Les Co-bénéfices du Fret Ferroviaire: Éléments d'évaluation et propositions, Altermind, Paris: 2020.

network and make feasible the transition from a conventional logistic offer to carbon-free offer (Minambiente and Ministerio de Transporte, 2020).[26]

Challenges to Overcome

One of the major challenges in electromobility is the absence of easy access to or availability of charging stations and auxiliary infrastructure. This must be intensively built in a short term to foster adoption by the citizens. The latter is also a further obstacle and electromobility must be promoted to both citizens and business and industry leaders, assuming that municipalities and regions are already committing to extend the use of EVs especially urban and inter-city buses. Maybe allowing more incentives than the current exemption of import duties, during an 'enabling phase' would help accelerate the path. Finally, the high purchase price of EVs is a strong barrier to massive adoption: same as investment in energy efficiency, the concept of 'spending more today and less over the time' is difficult to adopt by society and particularly in middle-income countries.

Agriculture, Forestry and Land Sector

This sector which is often referred to as Agriculture, Forestry and Other Land Use (AFOLU) is extremely relevant in the decarbonization path in Colombia, first, because, the country generates a significant part of its GDP through agriculture (8.5% of total GDP in Q1 2022), second, because it is considered as a "bio superpower" with its biodiverse natural resources, and third because it encompasses sub-sectors such as forestry which is a means to absorb carbon emissions, contributing to the national GHG emission reduction (DANE, 2022a).[27] The latter, to be efficient and 'net negative' requires the prevention of CO_2 rich ecosystems and of deforestation and even reforestation. And one of the biggest concerns of environmentalists is the fact that intensive agriculture is a major threat for biodiversity. Improvement in the management and use of soil is an urgent need, and the government made it its priority in the context of the decarbonization strategy.

Opportunities Generated by the Decarbonization to Seize

Colombia's abundant resources in water and food, which is a source of capital, generates benefits called 'ecosystem services', including the export of food. But not only. Developing innovation in bioeconomy could create niche markets such as bio-

26 Minambiente and Ministerio de Transporte, Estrategia de Movilidad Electrica, 2020, ISBN Medio Electronico: 978–958-5551-18–3 ISBN Impreso: 978–958-5551-17–6.

27 DANE, Estadísticas por Tema: Sociales, Gobernabilidad, Sector Agricultura (accessed in June 2022) quoted as 2022a; https://www.dane.gov.co/index.php/204.

chemicals and -pharmacy and -cosmetics as well as inputs for biomass energy adding up to 20% to the country's GDP by 2050 (Colombian Government, 2021).[28] According to the OECD benefits from ecosystem services such as in water, flood resilience or carbon capture represent 125 to 140 Bio USD in 2011 globally, when the cost of inaction (loss of forest area due to expansion of agriculture activities) represents several billions per year but significantly less (OECD, 2019;[29] Costanza et al. 2014).[30] Sustainable or 'climate smart' farming is an opportunity for the transition, since consumers are changing their standard and demanding biologic products enlarging this niche market but requires an initial investment cost and cost of changing the methods that small owners cannot afford without external support. Further, new techniques in the management of livestock with grazing and salvo-pastoral improvement could increase the density of head per ha (World Bank 2014).[31]

Challenges in the AFOLU Sector

Experts from the Interamerican Development Bank and the Deep Decarbonization Pathways Latin America and Caribbean highlighted three recommendations in one of their reports (BID and DDPLAC, 2019).[32] To preserve forestry, the assignment of rights for its use, the expansion of natural protected areas, the support to native and indigenous communities (financially and by building capacity) and sustainable management of concessions are key for the transition. On the other hand, the agriculture sub-sector requires improvement of the crop management, the reconversion of rice into permanent crops and the increase of intermittent dry system in rice cultivation. Finally, regarding livestock, Colombia should bring the animal density from 0.8 head/ha to 2 head/ha to allow freeing 12 Mio ha for other agriculture uses and thus end pressure on forest area.

---

28 Colombian Government, Ministry of Environment, DNP, Foreign Affairs Ministry, AFD, Expertise France, E2050 Colombia's long-term climate strategy to meet the Paris Agreement, Full Report, WRI: Bogotá 2021.

29 OECD, Finance and the Economic and Business Case for Action, Chapter 3 The Socio-economic case for biodiversity action, 2019, ISBN: 9789264597044, https://doi.org/10.1787/a3147942-en.

30 Costanza Robert, De Groot Rudolf, Sutton Paul, Van der Ploeg Sander, Anderson Sharolyn J., Kubiszewski Ida, Farber Stephen, Turner R. Kerry, Changes in the Global Value of Ecosystem Services, Global Environmental Change, Vol. 26 May 2014, 152–158 https://doi.org/10.1016/j.gloenvcha.2014.04.002.

31 Banco Mundial, Departamento Nacional de Planeación, Desarrollo de Bajo Carbono para Colombia, Washington D.C. 2014 https://colaboracion.dnp.gov.co/CDT/Ambiente/Desarrollo%20Bajo%20En%20Carbono%20Para%20Colombia.pdf.

32 BID and DDPLAC, Como Llegar a Cero Emisiones Netas: Lecciones de América Latina y el Caribe. Banco Interamericano de Desarrollo, Washington D.C: 2019.

Private Sector

The private sector, including services, banking and the industry, represents a potential strong ally for the government and the main actors involved in the decarbonization process in Colombia. Its contribution combined with the efforts of public companies such as Ecopetrol, the public oil and gas giant, is valuable and might require imposing less constraints. Indeed, large companies are now aware of the new expectations of their stakeholders (e.g. customers, suppliers, employees, investors and shareholders, the society in general) in terms of Environmental, Societal and Governance (ESG) considerations and corporate culture is subsequently changing. Standards such as ISO 5001 for sustainability in enterprises and efficient use of energy are widely adopted in Colombia for instance in industrial processes and product use (IPPU). Business and industry leaders owning large infrastructure are also aware that the costs of not changing (the status quo) will be higher than the costs of immediate action. They do not want to do 'business as usual' and bear risks related to stranded assets or even jeopardize their businesses. New infrastructure like coastal or along-the-mountain roads for instance is thought to be 'climate proofing' i.e. resilient, located far from the shores for example. In general, the Colombian carbon neutrality project cannot be pursued without the significant contribution of private investments and Voluntary (environmental) Agreements (VA) with the private sector are inevitable. These agreements involving government and polluters are often used as a means to enhance capacity building and awareness even though their initial objective was to preserve the environment (Blackman et al., 2012).[33] Heterogeneity of ESG rules, confusing labels and lack of standardization make it complicated to measure and control though. Same applies for investment and banking where the inconsistency or incomplete taxonomy leads to fewer impact investments than expected or in some cases to greenwashing. Though, Asobancaria, the Association of Colombian Banks has been a member of the Sustainable Banking Network (SBN) for already a decade. This occurs because returns from non-green infrastructure or investments are still higher than climate resilient investments, but also because banks and investments tend to underestimate the long-term climate risks (the IFC Green banking academy offers seminars and training to bankers in Latin America including Colombia). Also, because if any support, tax shield or others, is subject to conditionality policies, ESG are difficult to quantify or label in a standardized way.

33 Blackman Allen, Uribe Eduardo, van Hoof Bart, Lyon Thomas P., Voluntary Environmental Agreements in Developing Countries: The Colombian Experience, 2012; https://www.jstor.org/stable/resrep14958.

Success Stories

The Bogota City and Region Chamber of Commerce highlighted several success stories of companies that successfully embraced carbon neutrality policies and are making significant progress towards their corporate objectives contributing to the national decarbonization effort. These are the cases of two major companies in Colombia, one in the building sector and one in the service and transport sector (Camara de Comercio de Bogotá, 2022). Holcim, a global leading cement maker, has launched two product lines, EcoPlanet and EcoPact, that produce 30% less of CO_2 emissions than comparable products. Holcim has integrated quantified objectives into the corporate strategy and decided to have emissions values measured by externals such as Science Based Targets (SBTi). They also developed automation and digitization to further improve their carbon footprint. Avianca, the Colombian major airline, 120 years of international operation, has started initiatives to reduce its footprint: it has installed solar panels on the roof of its new maintenance center, allowing for 40% of its energy consumption, and it has managed to recycle 80% of the waste during the construction of the complex. Among other measures, it has set targets of reduction of fuel such as 5% of reduction of the CO2 emissions per 100 passengers. It has brought fuel consumption from 4.62 liters per 100 passengers per km in 2014 to 4.28 liters. These cases show that carbon neutral opportunities can be seized while generating businesses and thus economic growth.

Circular Economy and 'Prosuming'

There are, both at national and regional level, initiatives to foster responsible and sustainable production and consumption, including the use of so-called 'short circuits'. The notion of short circuits or proximity production and consumption refers to the planning and development of production sites close to urban centers where most of the consumption is located, leading to shorter supply chains and thus lower GHG emissions. In this sector, both sides demand and offer have to be rethought: business and industry leaders must think of pivotal changes such as transforming their product offer into a service offer (e.g. competitive rental of domestic tools instead of sale; stopping programmed obsolescence of products) and consumers must embrace changes in their consumption behaviors focusing on high-quality products that last longer for instance or depending on their purchase power on organic products and local farming (concept of 'prosumption'). Changes might be more difficult to adopt for the lowest social class that worry more about security of access to water and food, but a large part of the urban population in Colombia can economically afford these changes. Car and bicycle sharing could be further developed especially in large cities such as Bogota, Medellin or Cali, where less restrictions -thus incentives in a way-, are offered for electric cars and car sharing. On the other hand, innovative recycling and new waste management technologies are in development in the country pushed by municipalities aware of the impact of waste in their local GHG emissions. Citizens play an important role because recycling at home is far more cost-efficient than at central urban waste treatment plants. In that sense, plastic recycling rate in Colombia is still low and because the country still has a strong informal economy (about 40% of the GDP) and street collectors of steel and junk equipment and waste pickers in general play a role in the recycling activities across the country as well even if their contribution is undervalued (Wiego, 2019). Many issues remain in the value chain especially when it comes to PET bottles recycling (Zapata Bravo et al. 2021). Finally, monitoring and evaluation, as provided by Sistema de Informacion de Economia Circular, are primordial in the success of such policies.

8.2.3 Regional Approach: Sustainability without Degrowth

As relevant as the sectoral approach, the regional approach plays an important role in Colombia due to its variety of endemic particularities. Indeed, regions are different in terms of natural resources, people and culture and economic model. Urban centers attract more services and have a higher population density, while rural areas enjoy less people concentration and focus on agriculture and livestock and finally coasts developed some industries including fishery and capitalize on sea infrastructure such as ports and export distribution channels.

Smart Cities and Urban Development

Technological progress, for instance in energy efficient buildings, or digitization of electricity smart metering can contribute to decarbonization and particularly in the energy transition. Having a measurement and tracking of the electricity consumption helps to monitor and make savings. Newest standards in construction of housing and commercial buildings, i.e. buildings that go beyond the management of their sole own GHG emissions, pave the way to negative emissions. Serving this purpose, the Resolución 0549 of 2015 established a Sustainable Construction Guide that sets requirements in terms of energy efficiency among others. Once again objectives in climate policy go hand in hand with the SDGs, for instance, in the case of urbanization, development planning should integrate a perspective that bring housing closer in distance to working centers, by doing so, commuting will be shorter and public transport energy demand lower, and users' comfort and health significantly improved. Cities look at improving the life in metropolis creating new business models based on sustainability without affecting the economic growth. Urbanization being one of the biggest threats for biodiversity, with potential uncoordinated construction projects development affecting green areas, permit granting should be prioritized to abandoned old industrial buildings before projects with high impact on environment. Finally, public private partnerships (PPP) in the waste management sector can also allow bringing large investment and new technology to the country.

Region-Specific Resources

Both in coastal and inland areas, Colombia is characterized by a high biodiversity. Coastal regions are commonly distinguished depending on the seaside: the Atlantic Ocean (el Caribe) and Pacific Ocean (el Pacifico). For instance, in the inland, temperatures are expected to rise by 0.9 degrees between 2011 and 2040, by 1.6 between 2041 and 2070 and 2.14 between 2071 and 2100, when in the Atlantic, they are expected to increase by 0.5, 0.85 and 1.6 for the respective periods and finally in the Pacific by 0.7, 1.2 and 2 degrees (Acosta Giraldo et al., 2018 and Pulido Guio et al. 2015).[34,35] Therefore, climate risks generated by the warming are different and resources as well. On the coast the main risk is the increase of the sea level and in inland droughts and floods from heavy rain precipitations are the most relevant threats. Impacts of hurricanes and heavy forest fires cannot be underestimated though. Having droughts reduces the hydro power plants reserves leading to use of fossil fuel power generation; stresses the drinking and agriculture water reserves leading to less food reserves; challenges the remote communities environment leading to 'nomadification' toward less desertic or urban areas; and finally, droughts increase the consumption of energy through the increase utilization of air conditioning. Increasing sea level pushes communities to leave their original environment and living places disturbing local economies, habitats and lifestyles; it also significantly affects the living sea ecosystems. And the objectives per territory or regions are the same, even though adapted to the risks, resources and infrastructure of each region: reduce emissions, increase CO_2 capture, reduce vulnerability to climate change and manage the impact on economies at local level. Information collection and analysis must contribute not only to tracking emissions and further impacts of climate change in regions for national statistics but also to adapt the regional existing infrastructure (ports, roads, water and waste treatment, energy) and to build new resilient one and conserve the ecosystems.

34 Acosta Giraldo Jenny A., Ovalle Zanabria Katherine, Arcila Burgos Katherine, eds.: Florian Buitrago Maritza and Cortés Ospina Erika, Dirección de Cambio Climático y Gestión del Riesgo Ministerio de Ambiente y Desarrollo Sostenible, Consideraciones de Cambio Climático para el Ordenamiento Territorial, Bogotá, D.C.: Colombia. 2018, ISBN: 978–958-8901-70–1.

35 Pulido Guio Ana D., Jiménez Rodrigo, Turriago Juan D., Mendoza Javier E., IDEAM, PNUD, MADS, DNP, Cancillería, FMAM. Inventario Nacional de Gases de Efecto Invernadero (GEI) de Colombia: Tercera Comunicación Nacional de Cambio Climático de Colombia. Bogotá: 2015, ISBN: 978–958-8902-94–4; http://documentacion.ideam.gov.co/openbiblio/bvirtual/023421/cartilla_INGEI.pdf.

8.2.4 Social Approach: Redistribution, Inclusion and Equality

Colombia is a country of high people diversity: beside the 4.7 Mio Afro-Colombians, there are about 1.9 Mio indigenous people from 4 main communities (above 100.000 indigenous): the Wayuu in La Guajira, the Senu in Cordoba and Sucre, the Nasa in Cauca and Putumaya, the Pasto in Nariño among many others. They represent about 14% of the population though, the rest is mostly settled in urban centers (DANE, 2022b),[36] but their territory covers 53.4% of the 60 Mio ha of Colombian natural forest (IDEAM, 2022).[37] The objective of the Colombian long-term strategy considers these populations and the obligation to include them in the equation. Every citizen must be considered in terms of redistribution of wealth and chances to access business or professional opportunities, development etc. Scholars insist on the relevance of the dialogue with sector unions, power companies, academia, civil society representatives, associations and indigenous communities for the success of the strategy (Waismann 2019,[38] Bataille 2016).[39] Yet, dialogue with some communities is extremely challenging and requires gaining trust first before enhancing adoption. In that context, the government involved 5 national organizations representing indigenous communities including 23 representatives and 11 national organizations representing Afro-Colombian with 84 participants.

Deforestation, Agriculture vs. Habitat, Ecosystems Conservation

Ethnic communities in Colombia cover a significant geographic area where use of land is a debatable question given the tensions between government representatives, landowners with or without legitimate ownership titles, and even guerrillas and narcotraffic groups in some departments such as Nariño for example. If land is sacred for some indigenous and thus preserved from industrial activities and areas habited by indigenous people suffer less from deforestation (Ding et al., 2016),[40] it is less the

36 DANE, Estadísticas por Tema: Demografia y Poblacion, Grupos Etnicos, Grupo Etnicos Informacion Tecnica (accessed in June 2022) quoted as 2022b.

37 IDEAM, Instituto de Hidrología, Meteorología y Estudios Ambientales, Bosques y Recurso Forestal, Indicadores, Proporcion de la Superficie Cubierta por Bosque Natural, (accessed 2022) http://www.ideam.gov.co/web/ecosistemas/bosques-y-recurso-forestal.

38 Waisman Henri, Bataille Chris, Winkler Harald et al., A Pathway Design Framework for National Low Greenhouse Gas Emission Development Strategies, Nat. Clim. Chang. 9, 261–268 (2019) https://doi.org/10.1038/s41558-019-0442-8.

39 Bataille Chris, Waisman Henri, Colombier Michel, Segafredo Laura, Williams Jim, and Jotzo Frank, The Need for National Deep Decarbonization Pathways for Effective Climate Policy, Climate Policy Vol. 16, Issue sup. 1 (June, 2016): S7–S26 https://doi.org/10.1080/14693062.2016.1173005.

40 Ding Helen, Veit Peter, Gray Erin, Reytar Katie, Altamirano Juan-Carlos, Blackman Allen, Climate Benefits, Tenure Costs: The Economic Case for Securing Indigenous Land Rights in the Amazon. World Resources Institute, 2016, ISBN: 978–1-56973-894–8.

case for drug organizations. This makes conservation of ecosystems and habitat quite a challenge in Colombia. Yet, tackling climate change, according to the IPCC, is limiting the global warming, and developing in a resilient way, leading to human and planet health, equity, justice and requires a transition of human systems (societal, energy, industry, urban/rural infrastructure) by adapting and mitigating as well as a transition of ecosystems (land, freshwater, coastal, ocean, biodiversity) by restoring and conserving (Table 8.9). Finding the balance between implementing these measures and satisfying all involved parties has become almost impossible according to local negotiators. There are, for the use of soil and the protection of the biodiversity, compensations or offsets (in Spanish regalias) required by law, but their use is questioned and often communities prefer having a leisure facility before infrastructure that could improve conditions of life on the long term (e.g. access to drinking water, waste management, street lighting, roads, schools and care centers).

To improve the efficiency of the regalias system and to create a demand-offer market, Environmental Real Obligations (ERO) or 'compensation banks' used in other countries could be implemented to allow projects that preserve the biodiversity and soil to get a compensation from land users impacting the biodiversity (impacting users would be obliged to give compensation credits to preserving users). Though scholars are still analyzing if compensation by demand is better than the creation of an offer (Bureau and Schubert, 2020).[41] On the other hand, in urban areas, priority (through incentives for instance) should be given to the refurbishment and renovation of abandoned industrial sites before the use of new land for construction.

Externalities for the Citizens and Social Adoption

Externalities could be classified as positive and negative for the Colombian population. Distinction between rural and urban population, between gender, between ethnic population could also be made to rigorously analyze them. In general, positive externalities or co-benefits of the energy transition and decarbonization policy are assessed in terms of earned compensation, employment, health and social conditions, economic growth per territory or per community as well as further opportunities. Often associated with the amount of would-be-saved tons of CO_2. Some of them such as externalities directly related to decarbonization measures or programs could be quantified and associated with the transition; others are hardly connectable and thus even if measurable, their classification as a positive externality remains questionable. One further major obstacle in the benefits of decarbonization is the fact that they are

41 Bureau Dominique, Schubert Katheline, Compensation écologique: à la demande ou par le développement d'une offre ? Conseil d'analyse économique, Focus nr. 047-2020. Septembre 2020.

Table 8.9: Observed Impacts in Ecosystems: Confidence in Attribution to Climate Change (IPCC, 2022).

Ecosystem	Changes in ecosystem structure			Species range shifts			Changes in timing (phenology)		
	Terrestrial	**Freshwater**	**Ocean**	**Terrestrial**	**Freshwater**	**Ocean**	**Terrestrial**	**Freshwater**	**Ocean**
North America	High or very high								
Europe	High or very high								
Central and South America	High or very high			High or very high			Insufficient evidence		Low
Asia	High or very high	Medium	High or very high	Low	Medium	Insufficient evidence	Low		Medium
Africa	High or very high			High or very high	Insufficient evidence	High or very high	Insufficient evidence	Low	
Australasia	High or very high	Insufficient evidence	High or very high	High or very high	Insufficient evidence	High or very high	High or very high	Insufficient evidence	Low
Global	High or very high								

rarely distributed equally socially and geographically (BID and DDPLAC, 2019).[42] On the other hand, the negative externalities are related to the costs and their repercussions on the prices of the adaptation constraints brought by the energy transition. Indeed, even if new business models that include CO_2 reduction measures are economically viable on the long run, often the initial reconversion costs affect the end-users. These negative effects such as loss of jobs due to the shutdown of fossil fuel power plants or stranded infrastructure or business, higher energy, food and public transport prices could unequally affect the communities (Vogt-Schilb and Hallegatte 2017;[43] Trebilcock 2014).[44] Adoption by lower social class could result in a more challenging task due to bottom class's affordability limits.

8.2.5 Tracking Emissions and Monitoring Decarbonization

Monitoring and communication on economic, social and environmental results of the CO_2 reduction measures are key for their success: with the complex interactions and interdependencies between the policies, monitoring allows policy leaders to make sound steering and corrective decisions and communication increases adoption from all the stakeholders starting with the society. Furthermore, social adoption for households i.e.; accepting for instance that products are subject to a carbon tax depends significantly on the clarity of the communication. Because often generated tax revenues are used to reduce fiscal deficit, proper monitoring of the emissions but also of relevant further indicators such as externalities helps to equally relocate to and distribute benefits and co-benefits among the social classes instead. As mentioned earlier, involving all parts of the society through working groups and associations into the national strategy enhances adoption, even if opportunities and chances might remain unequal. Beside these reasons to implement a rigorous monitoring and communication on the progress made towards the decarbonization, there are obligations set by the pledge and commitment entered by Colombia in the framework of the UNFCCC, namely of Monitoring Reporting Verifying (MVR), a verification for mitigation actions and Monitoring and Evaluating (ME), an assessing for adaptation actions. Monitoring supports strategic methods which consist in setting new rules gradually: for instance, the stop of incentives for a sector or the introduction of a climate tax could jeopardize businesses, turn businesses or infrastructure rapidly obsolete with all the economic

42 BID and DDPLAC, Como Llegar a Cero Emisiones Netas: Lecciones de América Latina y el Caribe. Banco Interamericano de Desarrollo, Washington D.C: 2019.

43 Vogt-Schilb Adrien, Hallegatte Stephane, Climate Policies and Nationally Determined Contributions: Reconciling the Needed Ambition with the Political Economy, IDB Working Paper Series 818, June 2017.

44 Trebilcock, Michael J., Dealing with Losers: The Political Economy of Policy Transitions, New York: Oxford University Press, 2014. DOI: 10.1093/acprof:oso/9780199370658.001.0001.

and social consequences of an insufficiently prepared measure. Finally, for an optimized contribution of the instruments for territorial and environmental planning, national and regional policies must be properly articulated but also in a timely manner and in sequence (CIDER, 2021),[45] completed with a reinforcement of monitoring.

8.3 Part 2: Challenges to Reach the Carbon Neutrality

In a context of a serious global crisis that seems endless, political leaders must prioritize between reactivating the economy and pushing for ecological sustainability. They must ensure that their choices and decisions do not provoke regression and degrowth in the short term, and find reasonable ways to fund the transition, not only through the public fiscal budget which would be insufficient. To conciliate "end of the world with end of the month", an expression to refer to the complex equation of combining the care of the environment to avoid a calamity on earth and the affordability of the cost of this care. For policymakers and leaders, finding the right balance between imposing a high cost of carbon emissions (with a possible negative impact on competitiveness) and a low one (which does not cover enough for the negative externality of the anthropogenic activity) is a continuous challenge they are confronted with each time the NDCs must be updated. Sustainable economic reactivation is possible, in fact scholars showed that there is no empirical evidence that carbon tax has a negative impact on growth and employment (Metcalf and Stock, 2020;[46] Dechezleprêtre and Kruse, 2018).[47] But each jurisdiction has its constitution and legal system that allow with more or less room to set climate protection, each economy has strengths and weaknesses and middle-income countries like Colombia have less access to funds (carbon tax collection is not enough) and must think first of its most vulnerable people. The costs-benefits profiles of the energy transition are necessarily different from continent to continent, country to country and region to region. The challenges lie in the fact that the transformation, for instance, from classic agriculture into a sustainable or bio agriculture requires significant upfront investments for the owners and sometimes a change in the mindset to allow reengineering of the working methods. Efforts

45 CIDER Centro Interdisciplinario de Estudios sobre Desarrollo, Análisis de los Instrumentos de Planificación Territorial y Ambiental en el Marco de la Estrategia de Largo Plazo de Colombia para la Carbono-Neutralidad y la Adaptación E2050, Boletin 112, 2021.

46 Metcalf Gilbert E., Stock James H., Measuring the macroeconomic impact of carbon taxes. AEA Papers and Proceedings, Vol. 110, pp. 101–106. 2020.

47 Dechezleprêtre Antoine, Kruse Tobias, A Review of the Empirical Literature combining Economic and Environmental Performance Data at the Micro-level, OECD Economics Department Working Papers, Nr. 1514, Paris: 2018 https://doi.org/10.1787/45d269b2-en.

must be done to thoroughly explain the long-term benefits of the reconversion and financial support (e.g. through FINAGRO for agriculture, FINDETER for energy or FENOGE for energy efficiency) given to the vulnerable or small owners. And these challenges are also valid for any investment required to convert an old business model into a resilient climate friendly one, for instance the implementation of new equipment to obtain a more energy efficient building.

8.3.1 Costs-Benefits of the Energy Transition

The task of confronting the possible costs of climate change impact with the costs of adaptation and mitigation is an extremely demanding task for climate economy experts. The debate is passionate and opponents to carbon neutrality often argue that there is no scientifically proven global warming, instead only biased probabilistic models, and for the least 'climatoseptic' ones, that the only way to decarbonize the economy is through degrowth. There are no doubts that the energy transition has collateral costs. The cost impact of climate in GDP in the region could exceed 1% (Delgado et al., 2021).[48] According to the IADB, in Latin America and the Caribbean, between 7 and 19% of the GDP (up to 1.3 Trio USD) would be required to achieve the Paris Agreement goals (Galindo et al., 2022).[49] Policies that demand the shutdown of polluting factories or power plants lead to unemployment and depending on the sector this could represent a massive job loss, which at turn leads to less activities in the affected business ecosystem and less end-users consumption in general. Policies that de-incentivize or remove subsidies in certain sectors often lead to cuts in jobs too, which could be somehow considered as degrowth. Unless transformation into low carbon business models is considered and support is provided to the concerned sectors and businesses. In Colombia, the oil and gas sector represents 60% of total annual exportation and significantly contribute to the fiscal budget: a careful strategy must be established before the complete step out from this sector since a brutal stop of tax collection from oil, gas and coal would exacerbate an already tense situation of reduced fossil fuel production revenues and corresponding job losses. So does a GHG emissions reduction necessarily mean a drop in GDP? It has not been proven so far in the case of Colombia. All policy decisions for low carbon scenarios must consider the security of energy supply at competitive prices, the access to water and food for all etc. and cost-benefits analysis of each policy or mitigation and adaptation measure is biased by the transversality between the sectors, with cross-costs and collateral benefits. The list of benefits is long and include contribution

48 Delgado Raúl, Eguino Huáscar, Pereira Aloisio Lopes, Fiscal Policy and Climate Change: Recent Experiences of Finance Ministries in Latin America and the Caribbean, 2021, Inter-American Development Bank, IDB Monograph 941.

49 Galindo Luis Miguel, Hoffmann Bridget, Vogt-Schilb Adrien, How Much Will it Cost to Achieve the Climate Goals in Latin America and the Caribbean?, IDB Working Paper Series 01310, March 2022.

to the attainment of SDGs but their list of inherent costs is as long too, and their comprehensive analysis is proportionally complex. For instance, indirect positive externalities such as health improvement or others must be internalized in the equation just like negative externalities indirectly caused by GHG emissions. Scholars and experts developed tools and models to estimate the burden of implementing national low carbon policies. In Colombia, two leading universities Universidad de Los Andes and Universidad del Rosario recently used the Global Change Assessment Model. Costs-benefits analysis could be made using simplified and thus limited method such as the Marginal Abatement Cost (MAC), which when displayed in the curve, show for each opportunity the saving or reduction in tons equivalent of CO_2 in an axis and on the other, the cost of this abatement in USD per ton, as did experts of the Word Bank for Colombia in 2014 (Banco Mundial and Departamento Nacional de Planeación, 2014).[50] This is a method used in the energy sector but also in AFOLU. Another model used by experts is the Global Change Analysis Model (GCAM) an open-source tool that features demand-supply and inputs-outputs to simulate drivers such as land, energy, economy and CO_2 emissions from the Joint Global Change Research Institute. A more comprehensive and thus complex model is the MEG4C matrix (general equilibrium model for climate change) used in 2012 by Colombia consists in assessing the socio-economic drivers and search for an equilibrium between costs and benefits of adaptation and mitigation actions and tax measures (Loboguerro and DNP, 2012).[51] Or the Integrated Assessment Model (IAMs), that assumes technologic and economic pathways and baseline characteristics as variable to assess the impact of GHG emissions in a cost-benefit approach, considering estimated damages estimated with other tools (Sanderson & O'Neil, 2020).[52] In general, as pointed out by DDPLAC experts, cash transfer to substitute efficiently existing subsidies: 1 USD in subsidies is less worth for social value than 1 USD in cash (Feng et al., 2018;[53] Schaffitzel et al., 2020).[54] Dealing with the removal of subsidies and substituting them with more climate-friendly and efficient ones is crucial for the success of climate resilience policies and Colombia has been working on adapting its cli-

50 Banco Mundial, Departamento Nacional de Planeación, Desarrollo de Bajo Carbono para Colombia, Washington D.C. 2014 https://colaboracion.dnp.gov.co/CDT/Ambiente/Desarrollo%20Bajo%20En%20Carbono%20Para%20Colombia.pdf.

51 Loboguerro A.M., MEG4: A Computable General Equilibrium Model for Colombia, Presentation of DNP, 2012, http://www.globalchange.umd.edu/data/lamp/presentations/costarica2012/Day2/Loboguerrero_LAMP_Model_overview_colombia.pdf.

52 Sanderson Benjamin M. and O'Neil Brian C., Assessing the costs of historical inaction on climate change, Sci Rep 10, 9173, 2020, https://doi.org/10.1038/s41598-020-66275-4.

53 Feng Kuishuang, Hubacek Kaus, Liu Yu, Marchán Estefanía, Vogt-Schilb Adrien, Managing the Distributional Effects of Energy Taxes and Subsidy Removal in Latin America and the Caribbean, Applied Energy 225 (September, 2018): 424–436, 10.1016/j.apenergy.2018.04.116.

54 Schaffitzel Filip, Jakob Michael, Soria Rafael, Vogt-Schilb Adrien, Ward Hauke, Can Government Transfers Make Energy Subsidy Reform Socially Acceptable? A Case Study on Ecuador, Energy Policy, vol. 137 (February 2020) 111120, https://doi.org/10.1016/j.enpol.2019.111120.

mate-related tax regime for several decades now. There is a tax on use of water and environmental services and further air pollution, whose revenues are assigned to projects for the conservation of water sources and valleys, the Law 99 of 1993, though with debatable impact and efficiency. The Law 1715 of 2014 for renewable energy projects. For urban planning, a costs-benefits balancing instrument is established by the Law 338 of 1997 with the aim of redistributing by the developers in a fair way costs and benefits to affected parties. Also, in the area of inclusion and gender, the Law 1876 of 2017 allows for subsidies with priority target low social class population, victims of conflicts, women alone etc. Further law projects are under study to foster the bioeconomy, circular economy among others. On the other hand, other instruments are in process of being removed given their negative impact on the path to decarbonization: incentives for the intensive use of soil by livestock owners, the fund that stabilizes the gasoline prices (Fondo de Estabilizacion de Precios de los Combustibles) and the subsidies in gasoline prices for municipalities at country borders, the tax on polluting vehicles that is not adjustable according to the vehicle vintage or level of contamination and finally subsidies in the extractive sector. The latter being a significant hurdle in the way to carbon neutrality and at the same time a dilemma for the country leaders, because of the important revenues obtained by companies (large employers) in the mining, oil and gas and by the taxes collected by the government. These removals of old tax incentives could have an impact on employment and on tax recollection and require special care from policy makers if the economy is capable of coping with reduced tax revenues and sudden increase of unemployment in targeted sectors. In case of gasoline, opposition from the end-users might arise leading to roadblocks and riots such as in Ecuador in 2019 and 2022. Furthermore, even if incentives are in place for a green sector, for instance renewables, it must be revised regularly, because the market might have become mature in the meantime and those incentives are not necessary any longer. Modeling impacts and costs, from a social, economic and ecological perspective, of such tax policies is therefore of high relevance in order to monitor and correct them over the time and ensure an affordable and equal socio-ecological transition.

8.3.2 Financing the Sustainability Transformation

Financing is one the four pillars of the energy transition, beside innovation, education and the transition itself. And in that sense, it deserves special attention and proportionally dedicated resources. In the case of Colombia, there are three main instruments and sources of funding: carbon tax revenues thus the fiscal budget, the carbon trading market (Sistema de Comercio de Emisiones – SCE) and the different incentives listed previously. A list of 32 instruments formulated by AFD, Expertise France and Econometria Consultores is accessible in a technical report used for the formulation of the Colombian long-term strategy E2050 (AFD et al., 2021).[55] The carbon tax has been created following the Law 1819 of 2016, a fiscal reform, at a price of 15.000 COP per ton of CO_2 eq. for 2017 (in 2022, through indexation 18.891 COP, about 4.9 USD) applying on most fossil fuels with exception for biofuels among others. Several laws entered into force and reviewed the distribution of the collected tax revenues, including reforestation, coasts conservation, resolution of armed conflicts (Minambiente, 2020).[56] A pending challenge is the right carbon taxing per industry: usually the transport sector is subject to higher taxes than energy production, industries or agriculture. The carbon market, with the trade of Cupos Transables de Emision (CTE), sort of local ETS, works on a basis of 'cap-and-trade' where Colombia would set a cap of GHG emissions per year and allow trading of CTEs either during public auctions in the primary market or 'over-the-counter', i.e. directly between users or through an intermediary in the secondary market (Fuss et al. 2018).[57] The third main instrument is the collection and redistribution of taxes related to water or oil and mining compensation (Sistema General de Regalias-SGR), shared between central institutions and regions for social equality expenses. Finally, Colombia has access to climate finance, which includes instruments provided by development banks such as soft loans, provision of guarantees, grants and most important national and international funds specially dedicated to support climate-related projects and investments (e.g. Green Climate Fund-GCF, Global Environmental Facility-GEF etc., and national ones: besides FENOGE, FINDETER and others, Bancoldex and FDN). The UK PACT is a bilateral cooperation between Colombia and Great Britain, whose mission is to support the transition with financing instruments. The relevance of these funds lies in their role as catalysts to leverage more funds from the private sector, both investors and private banks. Globally, climate finance is supposed to generate 100 Bio USD p.a. under the UNFCCC for the fight against climate change. In addition, green bonds are at their infancy in Colombia, but more and more companies are structuring their climate-related activities and obtaining long-term funding through green bonds (e.g. major utility CELSIA with the 420 Bio COP green bonds issued to finance solar parks among others). Taxonomy plays an important role in the socio-ecological transition, and recently Colombia published its green taxonomy in compliance with its 2020 NDC commitment, being the first jurisdiction in all Latin America to have it. This will provide more transpar-

55 Agence Francaise de Développement, Expertise France, Econometria Consultores, E2050, Instrumentos Económicos y Financieros de la Estrategia Climática de Largo Plazo de Colombia para cumplir con el Acuerdo de París E2050, Informe Técnico Los Lineamientos para la Modificación o Creación de Instrumentos Econόmicos y/o Financieros en el Largo Plazo, 2021.

56 Minambiente and Ministerio de Transporte, Estrategia de Movilidad Electrica, 2020, ISBN Medio Electronico: 978–958-5551-18–3 ISBN Impreso: 978–958-5551-17–6.

57 Fuss Sabine, Flachsland Christian, Koch Nicolas, Kornek Ulrike, Knopf Brigit, Edenhofer Ottmar, A framework for assessing the performance of cap-and-trade systems: Insights from the European Union Emissions Trading System. Review of Environmental Economics and Policy, Vol. 12 nr. 2, 220–241, Chicago: 2018.

ency on which activities and actions are relevant for the country's development towards a sustainable transition, accelerating investment decisions and eligibility criteria for climate funding and enhancing monitoring. Creating political and business environments that enable climate and sustainability finance is key also through intangible or no-monetary actions. One tangible support that governments can offer is the political and regulatory stability because this would create more confidence in the private sector so that potential investors interested in impact investing in the Latin American region choose Colombia. Finally, the banking sector is promoting 'green finance': already in 2012 most large commercial and development Colombian banks signed the Green Protocol with the government, a voluntary agreement that fosters green finance in the country (Climate Transparency, 2018).[58]

8.3.3 Technological Innovation, R&D and Education

Disruptive technology in all areas of business, including nescent geo-engineering (also referred to as climate engineering), is extremely relevant because they are expected to bring enough efficiency or to create new business models that compensate for the high costs of low carbon producing and servicing. Decarbonization scenarios in the energy sector and in ecosystem services among others are considering the use of best available technologies (BAT), including but not limited to carbon capture for fossil fuel plants that remain necessary to balance the intermittency of RE, hydrogen and offshore wind energy. Road Maps have been launched to give visibility to the sector's players and investors. Furthermore, the Law 633 of 2000 allows exemptions on revenues taxes and tax credits when SMEs (up to 50% of the project costs) and large companies (25%) invest in science, technology and innovation (in Spanish CTI) projects related to climate. Innovation and education are key components of the MoI: for each aspect (apuesta) of the Colombian long-term strategy, proposals are made regarding both components: from cross-topics awareness of the society to improve home recycling or the daily use of the bicycle for example to region- and sector-specific education programs to prepare for tomorrow's industries. In this context, the German BMBWF and AHK together with further partners are establishing a RE academy to train future Colombian wind turbines and solar panels technicians. The offer in capacity building, in the energy transition, is thought, despite inter institutional complexity, from territorial and inclusion perspectives as well, considering local specificities, traditions, cultures, gender and human rights.

58 Climate Transparency, Report on Colombia's Climate Action and Response to the Covid-19 Crisis, 2020, https://www.climate-transparency.org/wp-content/uploads/2021/01/Colombia-CT-2020.pdf.

8.3.4 ESG, Green Washing and Climate Realpolitik

Environmental, Social and Governance (ESG) is gaining ground in all sectors: business and industry leaders embraced new needs from the market and the society: it is necessary to establish ESG policies in the organizations and communicate to internal and external stakeholders. The issue related to so-called 'green washing' is when communication is used imprecisely or even incorrectly, intentionally or not, to mislead the recipients about their level of measures to protect the environment. In Colombia, notorious cases like Ecopetrol with the 'clean diesel' (Diesel Limpio), or Cerrejon, with its communication towards sustainability in contradiction with the real impact of its activities on the environment and the communities are still in the memories (Tascon Choco and Roa Martiney, 2021). Yet the Decree 1369 of 2014 of Minambiente regulates the information and marketing communication of products or services including their level of sustainability. ESG strategies displayed in websites must be accurate and distinguish the objectives from the actually realized actions. With the growing need from end-users for information about the products and services 'level of green', it is important for the consistency of the transition policy to have all players on board starting with the private sector. Industry, agriculture, energy and services are at the forefront, but banking is also required to work on its image to create confidence for the consumers and give them the feeling that every player in Colombia is doing his part. Finally, communication is key: policies must survive changes of presidencies or governments and the challenges posed by day-to-day politics with possible conflicts between different political parties and not fall in the trap of the BaU and the so-called environment 'Realpolitik'. Strong industrial associations and lobbies should be heard but always maintained away from inappropriate influence on leaders and on their low carbon targets.

Bibliography

Acosta Giraldo Jenny A., Ovalle Zanabria Katherine, Arcila Burgos Katherine, eds.: Florian Buitrago Maritza and Cortés Ospina Erika, Dirección de Cambio Climático y Gestión del Riesgo Ministerio de Ambiente y Desarrollo Sostenible, Consideraciones de Cambio Climático para el Ordenamiento Territorial, Bogotá, D.C.: Colombia. 2018, ISBN: 978-958-8901-70-1

Agence Francaise de Développement, Expertise France, Econometria Consultores, E2050, Instrumentos Económicos y Financieros de la Estrategia Climática de Largo Plazo de Colombia para cumplir con el Acuerdo de París E2050, Informe Técnico Los Lineamientos para la Modificación o Creación de Instrumentos Económicos y/o Financieros en el Largo Plazo, 2021

Andemos, Durante el 2021, las matrículas de vehículos eléctricos e híbridos crecieron un 107% en América Latina, Articulo (accessed May 2022), https://www.andemos.org/index.php/2022/04/04/durante-el-2021-las-matriculas-de-vehiculos-electricos-e-hibridos-crecieron-un-107-en-america-latina/

Banco Mundial, Departamento Nacional de Planeación, Desarrollo de Bajo Carbono para Colombia, Washington D.C. 2014 https://colaboracion.dnp.gov.co/CDT/Ambiente/Desarrollo%20Bajo%20En%20Carbono%20Para%20Colombia.pdf

Bataille Chris, Waisman Henri, Colombier Michel, Segafredo Laura, Williams Jim, and Jotzo Frank, The Need for National Deep Decarbonization Pathways for Effective Climate Policy, Climate Policy Vol. 16, Issue sup. 1 (June, 2016): S7–S26 https://doi.org/10.1080/14693062.2016.1173005

BID and DDPLAC, Como Llegar a Cero Emisiones Netas: Lecciones de América Latina y el Caribe. Banco Interamericano de Desarrollo, Washington D.C: 2019

Blackman Allen, Uribe Eduardo, van Hoof Bart, Lyon Thomas P., Voluntary Environmental Agreements in Developing Countries: The Colombian Experience, 2012; https://www.jstor.org/stable/resrep14958

Bureau Dominique, Schubert Katheline, Compensation écologique: à la demande ou par le développement d'une offre ? Conseil d'analyse économique, Focus nr. 047-2020. Septembre 2020

Cámara de Comercio de Bogotá, Guia de Descarbonizacion y Accion Climatica para Bogota Cundinamarca, 2022, ISBN 978–958-688–514-0

Cambio Climático y Gestión del Riesgo, Política y Normativa, Ministerio de Ambiente y Desarrollo Sostenible-Mimambiente, (accessed May 2022), https://www.minambiente.gov.co/cambio-climatico-y-gestion-del-riesgo/politica-y-normativa/

Cardona O.D., Bernal G., Pabón J.D., M. A., Marulanda M. C., Carreño M. L., González D., Villegas C., Marulanda P., Grajales S., Rincón D., Molina J.F., Estudio de Riesgo por Efectos del Cambio Climatico y Medidas de Adaptacion para la Estrategia a Largo Plazo E2050 de Colombia, Ingeniar Risk Intelligence, Bogota: 2021

CIDER Centro Interdisciplinario de Estudios sobre Desarrollo, Análisis de los Instrumentos de Planificación Territorial y Ambiental en el Marco de la Estrategia de Largo Plazo de Colombia para la Carbono-Neutralidad y la Adaptación E2050, Boletin 112, 2021

Climate Transparency, Report on Colombia's Climate Action and Response to the Covid-19 Crisis, 2020, https://www.climate-transparency.org/wp-content/uploads/2021/01/Colombia-CT-2020.pdf

Climate Watch, 2022, (accessed June 2022); https://www.climatewatchdata.org/ghg-emissions?end_year=2019&ions=COL%2CARG&start_year=1990

Colombian Government, Ministry of Environment, DNP, Foreign Affairs Ministry, AFD, Expertise France, E2050 Colombia's long-term climate strategy to meet the Paris Agreement, Full Report, WRI: Bogotá 2021

Costanza Robert, De Groot Rudolf, Sutton Paul, Van der Ploeg Sander, Anderson Sharolyn J., Kubiszewski Ida, Farber Stephen, Turner R. Kerry, Changes in the Global Value of Ecosystem Services, Global Environmental Change, Vol. 26 May 2014, 152–158 https://doi.org/10.1016/j.gloenvcha.2014.04.002

DANE, Estadísticas por Tema: Demografia y Poblacion, Grupos Etnicos, Grupo Etnicos Informacion Tecnica (accessed in June 2022) quoted as 2022b

DANE, Estadísticas por Tema: Sociales, Gobernabilidad, Sector Agricultura (accessed in June 2022) quoted as 2022a; https://www.dane.gov.co/index.php/204

Dechezleprêtre Antoine, Kruse Tobias, A Review of the Empirical Literature combining Economic and Environmental Performance Data at the Micro-level, OECD Economics Department Working Papers, Nr. 1514, Paris: 2018 https://doi.org/10.1787/45d269b2-en

Delgado Raúl, Eguino Huáscar, Pereira Aloisio Lopes, Fiscal Policy and Climate Change: Recent Experiences of Finance Ministries in Latin America and the Caribbean, 2021, Inter-American Development Bank, IDB Monograph 941

Ding Helen, Veit Peter, Gray Erin, Reytar Katie, Altamirano Juan-Carlos, Blackman Allen, *Climate Benefits, Tenure Costs: The Economic Case for Securing Indigenous Land Rights in the Amazon*. World Resources Institute, 2016, ISBN: 978–1-56973-894–8

DNP Departamento Nacional de Planeación, Eficiencia Energética, (accessed on May 2022); https://www.dnp.gov.co/Crecimiento-Verde/Ejes-estrategicos/Paginas/Eficiencia-energ%C3%A9tica.aspx

European Environment Agency, 2022, GHG Emissions in the EU (accessed June 2022); https://ec.europa.eu/eurostat/databrowser/view/t2020_rd300/default/table

Feng Kuishuang, Hubacek Kaus, Liu Yu, Marchán Estefanía, Vogt-Schilb Adrien, Managing the Distributional Effects of Energy Taxes and Subsidy Removal in Latin America and the Caribbean, Applied Energy 225 (September, 2018): 424–436, 10.1016/j.apenergy.2018.04.116

Fuss Sabine, Flachsland Christian, Koch Nicolas, Kornek Ulrike, Knopf Brigit, Edenhofer Ottmar, A framework for assessing the performance of cap-and-trade systems: Insights from the European Union Emissions Trading System. *Review of Environmental Economics and Policy*, Vol. 12 nr. 2, 220–241, Chicago: 2018

Galindo Luis Miguel, Hoffmann Bridget, Vogt-Schilb Adrien, How Much Will it Cost to Achieve the Climate Goals in Latin America and the Caribbean?, IDB Working Paper Series 01310, March 2022

https://www.cae-eco.fr/Focus-no47-Compensation-ecologique-a-la-demande-ou-par-le-developpement-d-une

https://www.minambiente.gov.co/wp-content/uploads/2021/10/informe-actualizacion-contribucion-determinada-Colombia-ndc-2020.pdf.

https://www.minambiente.gov.co/wp-content/uploads/2022/01/Consideraciones_de_Cambio_Climatico_para_el_Ordenamiento_Territorial_VF.pdf

IDEAM, Instituto de Hidrología, Meteorología y Estudios Ambientales, Bosques y Recurso Forestal, Indicadores, Proporcion de la Superficie Cubierta por Bosque Natural, (accessed 2022) http://www.ideam.gov.co/web/ecosistemas/bosques-y-recurso-forestal

IPCC, 2022: Summary for Policymakers [H.-O. Pörtner, D.C. Roberts, E.S. Poloczanska, K. Mintenbeck, M. Tignor, A. Alegría, M. Craig, S. Langsdorf, S. Löschke, V. Möller, A. Okem (eds.)]. In: Climate Change 2022: Impacts, Adaptation, and Vulnerability. Contribution of Working Group II to the Sixth Assessment Report of the Intergovernmental Panel on Climate Change [H.-O. Pörtner, D.C. Roberts, M. Tignor, E.S. Poloczanska, K. Mintenbeck, A. Alegría, M. Craig, S. Langsdorf, S. Löschke, V. Möller, A. Okem, B. Rama (eds.)]. Cambridge University Press. In Press.

IRENA and ILO, Renewable Energy and Jobs: Annual Review 2021, International Renewable Energy Agency and International Labor Organization, Abu Dhabi, Geneva: 2021, ISBN: 978–92-9260-364–9

Loboguerro A.M., MEG4: A Computable General Equilibrium Model for Colombia, Presentation of DNP, 2012, http://www.globalchange.umd.edu/data/lamp/presentations/costarica2012/Day2/Loboguerrero_LAMP_Model_overview_colombia.pdf

Metcalf Gilbert E., Stock James H., Measuring the macroeconomic impact of carbon taxes. *AEA Papers and Proceedings*, Vol. 110, pp. 101–106. 2020

Minambiente and Ministerio de Transporte, Estrategia de Movilidad Electrica, 2020, ISBN Medio Electronico: 978–958-5551-18–3 ISBN Impreso: 978–958-5551-17–6

MinAmbiente, ABC Decreto 926 de 2016 Decreto 446 de 2020, Preguntas Frecuentes del Impuesto Nacional al Carbono y el Tratamiento Tributario de No Causación por Carbononeutralidad, 2020; https://www.minambiente.gov.co/wp-content/uploads/2022/01/ABC_DECRETO_926_de_2017.pdf

Minenergia, 2022, Fuentes No Convencionales de Energia Renovable, (accessed May 2022) https://www.minenergia.gov.co/energias-renovables-no-convencionales

Mision de Transformacion, 2022, (accessed March 2022)https://energiaevoluciona.minenergia.gov.co/transformacion#

Mohamadi Farid, Introduction to Project Finance in Renewable Energy Infrastructure: Including Public-Private Investments and Non-Mature Markets, Springer Nature, Cham: 2021, ISBN: 978–3-030-68742-7 https://doi.org/10.1007/978-3-030-68740-3

NDC Colombia, Actualización de la Contribución Determinada a Nivel Nacional de la República de Colombia para el periodo 2020-2030, Colombia: 2020

OECD, Finance and the Economic and Business Case for Action, Chapter 3 The Socio-economic case for biodiversity action, 2019, ISBN: 9789264597044, https://doi.org/10.1787/a3147942-en

Pulido Guio Ana D., Jiménez Rodrigo, Turriago Juan D., Mendoza Javier E., IDEAM, PNUD, MADS, DNP, Cancillería, FMAM. Inventario Nacional de Gases de Efecto Invernadero (GEI) de Colombia: Tercera Comunicación Nacional de Cambio Climático de Colombia. Bogotá: 2015, ISBN: 978–958-8902-94–4; http://documentacion.ideam.gov.co/openbiblio/bvirtual/023421/cartilla_INGEI.pdf

Sanderson Benjamin M. and O'Neil Brian C., Assessing the costs of historical inaction on climate change, Sci Rep 10, 9173, 2020, https://doi.org/10.1038/s41598-020-66275-4

Schaffitzel Filip, Jakob Michael, Soria Rafael, Vogt-Schilb Adrien, Ward Hauke, Can Government Transfers Make Energy Subsidy Reform Socially Acceptable? A Case Study on Ecuador, Energy Policy, vol. 137 (February 2020) 111120, https://doi.org/10.1016/j.enpol.2019.111120

Tascon Choco and Roa Martiney, Greenwashing en Colombia, Universidad ICESI, Santiago de Cali: 2021 https://repository.icesi.edu.co/biblioteca_digital/bitstream/10906/92483/1/TG03404.pdf

Thirion Benoit, Geoffron Patrice, Les Co-bénéfices du Fret Ferroviaire: Éléments d'évaluation et propositions, Altermind, Paris: 2020

Trebilcock, Michael J., *Dealing with Losers: The Political Economy of Policy Transitions*, New York: Oxford University Press, 2014. DOI: 10.1093/acprof:oso/9780199370658.001.0001

UPME, Plan de Expansión Referencia Generación Transmisión 2020-2034 Volumen 2. Generación, Colombia: 2019; http://www.upme.gov.co/Docs/Plan_Expansion/2020/Volumen2_Plan_Expansion_Generacion_Transmision_2020_2034_Final.pdf

Vanguardia, Columna Miguel Lotero, Política Energética Sostenible y Sostenida, (accessed May 2022) https://www.vanguardia.com/opinion/columnistas/miguel-lotero/politica-energetica-sostenible-y-sostenida-AM5248064

Vogt-Schilb Adrien, Hallegatte Stephane, Climate Policies and Nationally Determined Contributions: Reconciling the Needed Ambition with the Political Economy, IDB Working Paper Series 818, June 2017

Waisman Henri, Bataille Chris, Winkler Harald et al., A Pathway Design Framework for National Low Greenhouse Gas Emission Development Strategies, Nat. Clim. Chang. 9, 261–268 (2019) https://doi.org/10.1038/s41558-019-0442-8

Wiego, Parra Federico, In Colombia a Global Model for Inclusive Recycling is under Threat, 2019 (accessed June 2022) https://www.wiego.org/blog/colombia-global-model-inclusive-recycling-under-threat

World Bank Data, Economic Growth per Country, (accessed March 2022); https://data.worldbank.org/indicator/NY.GDP.MKTP.KD.ZG?locations=CO

Zapata Bravo Alvaro, Vieira Escobar Valentina, Zapata-Dominguez Alvaro, Rodriguey-Ramirez Alfonso, The Circular Economy of PET Bottles in Colombia, Universidad de Cali: 2021; https://doi.org/10.25100/cdea.v37i70.10912

Kwesi Annan-Takyi

9 Ghana's Example of De-Carbonization

9.1 Executive Summary

Ghana has since the 1990s introduced several policy actions (programs and projects) in fuel switching, energy efficiency, renewable energy promotion measures, afforestation and reforestation in the effort to decarbonize the economy, though precipitated by economic and energy security reasons. From Minimum Energy Performance Standards (MEPS) for lighting, refrigeration and air conditioners, as well as eighteen other appliances, to switching from LCO and diesel to imported natural gas as well as gas from its own oil & gas industry to generate electricity, and switching from biomass to LPG usage, Ghana's carbon reduction measures have contributed to reducing the carbon emissions significantly. The role of energy pricing in the achievement of decarbonization cannot be overemphasized due to the peculiar macroeconomic environment prevailing. This paper attempts an examination of these decarbonization measures, through their origination and institutional delivery, success stories and the outlook for decarbonization of the Ghanaian economy.

9.2 Introduction

Among the several measures that governments around the world, including Ghana, embarked on to reduce the carbon footprint of their economies are mitigation and adaptation programs such as energy efficiency, renewable energy, fuel switching, and others. With motivation from the nation's struggles with deforestation and power supply challenges, amidst demand side inefficiencies, Ghana embarked on a remarkable decarbonization journey since the 1990s. These measures did not only reduce electricity consumption but also carbon, protected carbon sequestration in remaining forests, a trend that has been sustained by sector institutions in Ghana.

The IEA (2021) reports that presently, energy efficiency is the leading contributor to decarbonization, a trend that is expected to continue till 2050, at least, and placing Ghana's efforts in lock-step with economies at the front of the decarbonization agenda.

9.3 Sources of Increase in Carbon Emissions

Ghana's gross domestic product is currently about 77.5 Billion USD (World Bank 2022), with much of the economic growth springing from services, agriculture and mining (Trading Economics 2022). Industry's (i.e., manufacturing) contribution has been un-

https://doi.org/10.1515/9783110784299-009

dulating though descent growth is has been registered over the past decade (Trading Economics 2022). Carbon emissions from these sectors have been the most significant besides the transport and the agro-forestry sector (Energy Commission 2020). The number of vehicles on Ghanaian roads keep increasing whiles the deforestation for farming, biomass harvesting, lumbering and related practices have resulted in loss of much of the forest cover and the decarbonization potential it carries. The power generation sector become one of the major sources of carbon emissions as Ghana introduced thermal power plants to complement the mainly hydro generation plants that existed post-independence (Energy Commission 2020). These factors presented a dire situation in Ghana's peculiar context, which created the need for decarbonization efforts (see Table 9.1).

Table 9.1: Grid Electricity Generation in Ghana (compiled by GRIDCO and ECG/PDS).

Year	Generation (GWh)				Share (%)		
	Hydro	Thermal	Renewables	Total	Hydro	Thermal	Renewables
2000	6.610	614	0	7.224	91,5%	8,5%	–
2001	6.609	1.250	0	7.859	84,1%	15,9%	–
2002	5.036	2.237	0	7.273	69,2%	30,8%	–
2003	3.885	1.996	0	5.881	66,1%	33,9%	–
2004	5.280	758	0	6.038	87,4%	12,6%	–
2005	5.629	1.159	0	6.788	82,9%	17,1%	–
2006	5.619	2.811	0	8.430	66,7%	33,3%	–
2007	3.727	3.251	0	6.978	53,4%	46,6%	–
2008	6.196	2.129	0	8.325	74,4%	25,6%	–
2009	6.877	2.081	0	8.958	76,8%	23,2%	–
2010	6.995	3.171	0	10.166	68,8%	31,2%	–
2011	7.561	3.639	0	11.200	67,5%	32,5%	–
2012	8.071	3.953	0	12.024	67,1%	32,9%	–
2013	8.233	4.635	3	12.871	64,0%	36,0%	0,0%
2014	8.387	4.572	4	12.963	64,7%	35,3%	0,0%
2015	5.844	5.644	3	11.491	50,9%	49,1%	0,0%
2016	5.561	7.435	27	13.023	42,7%	57,1%	0,2%
2017	5.616	8.424	28	14.068	39,9%	59,9%	0,2%
2018	6.017	10.195	33	16.245	37,0%	62,8%	0,2%
2019	7.252	10.885	52	18.189	39,9%	59,8%	0,3%

9.4 Planning and Implementation Institutions

A number of key governmental institutions take on the role of decarbonizing the economy. These include Ministry of Energy, Ministry of Environment, Ministry of Trade and Industry, and the Ministry of Finance. These Ministries have several institutions such as the Environmental Protection Agency, Energy Commission, Energy

Foundation, Ghana Standards Authority and a host of others who design and undertake specific programs, as well as co-ordinate with local and international bodies and organizations, to reduce carbon emissions. Several private sector organizations also advocate for and undertake projects in decarbonization activities. A clear cut budget for decarbonization activities in these institutions may not be easily identified. However, Ghana Government's contributions, GCF and other climate funding support, donor support through USAID, GIZ, AfDB, and other development bodies have been responsible for funding most of the activities undertaken in decarbonization.

9.5 De-Carbonization Activities

9.5.1 Overview of De-Carbonization Interventions

Through a combination of policies in the energy, transport, industry and agriculture, sectors Ghana has successfully been reducing its carbon footprint since the 1990s. Though thermal generation became necessary around the same time, the interventions served to significantly reduce the potential emissions, as summarized in Figure 9.1.

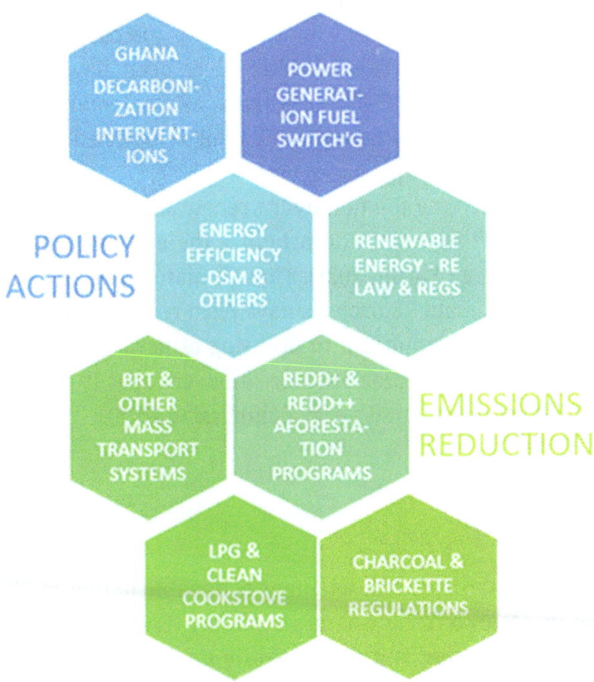

Figure 9.1: Policy Actions that have resulted in significant GHG emissions in Ghana since 1995; figure created by author.

9.5.2 Fuel Switching

The growth in Ghana's power generation sector has been rapid, as demand growth rate ballooned (Centre for Global Development 2017), necessitating the use of thermal generation units to complement the largely hydro generation at that time. Obviously, carbon emissions also increased, as cost of electricity also increased. Government subsidies to the power sector resulted in a lot of fiscal pressure due, in part, to the hydrocarbon feedstock (crude oil, diesel and natural gas) utilized (Volta River Authority [VRA] 2022). To ease the fiscal pressure (World Bank 2019) and to reduce the environmental impact, a policy of full cost recovery (Ministry of Energy, Ghana 2019) was adopted in 2010 and programs to switch the generation feedstock to natural gas and gas from Ghana's own oil fields was embarked upon (EIA 2018), resulting in significant reduction in carbon emissions from the national electricity grid. These policies and programs have continued to present days with carbon capture in the oil production fields being considered as an additional measure to reduce carbon emissions.

9.5.3 Renewable Energy

Ghana passed its Renewable Energy (RE) Law in 2011 (Act 832, amended in 2020). A program to scale up RE pilot projects that have been conducted over the past decade and beyond, Scaling-Up Renewable Energy Projects (SREP) has been developed (Aglanu 2015), awaiting completion of financial arrangements, as implementation preparations also proceed presently (Ministry of Energy 2015). This is expected to further reduce carbon emissions in the country.

Additionally, a national target to increase the non-hydro renewable energy sources picked up steam, growing the effort to further decarbonize the economy, as RE electricity became cheaper (World Energy Forum 2021). Though many pilot projects have been done and a number of utility scale RE projects have been realized and connected to the national electricity grid, a moratorium to licensing of new RE projects became necessary to sanitize the regulatory regime and cure an over-capacity in generation the nation was experiencing at that point (Energy Commission 2018).

9.5.4 Energy Efficiency

Ghana introduced a program of demand side management actions including MEPS which has played a major role in decarbonizing Ghana's economy (Energy Commission 2022). Started by the Energy Foundation and continued by the Energy Commission of Ghana, MEPS have been developed for about 21 home and industrial appliances, promulgated into law and public awareness built to ensure that the value chain of home and industrial appliances, from importers to end users, is well informed and equipped in the drive towards demand side energy efficiency and decarbonization (LI 1815 – 2005).

Major efforts in energy efficiency include the introduction of energy efficient lighting, starting with CFLs and later LEDs to replace energy inefficient incandescent light bulbs and T 12 fluorescent light to LED bulbs and T 5s. Another major initiative by the government of Ghana was to replace 20 million incandescent light with CFL bulbs 2007-2008, which yielded a reduction in peak demand of 124 MW and an accompanying carbon reduction of 112,320 tons per annum, according to the Energy Commission of Ghana (2009). Additional programmes for energy efficient cooling and food preservation have also been implemented.

The domestic consumption sector of the electricity market has also benefited from projects such as the Energy Efficient Refrigerator Program (Energy Commission 2022) project and supply of timer switches for freezers to reduce peak consumption between 6pm and midnight, and the Energy Efficiency for Households and SMEs (ECREEE 2022), in addition to recurring public education programs in energy efficiency. A project on Air Conditioners has been developed for implementation in addition to a Green Cooling Program to replace GHG gases in refrigeration (Green Cooling Initiative 2022), to further reduce peak demand and support the decarbonization drive. Industrial energy efficiency programs preparations under GCF Readiness program has been commenced with UNIDO as the main implementation partner, aiming to regulate industrial energy efficiency and reduce carbon emissions in the manufacturing and other industrial sectors of the economy (UNIDO 2022).

Though it may not be a silver lining, energy pricing played a significant role in these efficiency for decarbonization measures, as increasing generation cost raised the political expediency for action and the nation's fiscal circumstances necessitated energy cost reduction, as well as passing some of this cost to the end-user, which provided the motivation for consumers to employ efficiency methods.

9.5.5 Transportation

In terms of carbon emissions, the transportation sector is one of the most significant sectors due to the use of fossil fuel as the main source of energy to power vehicles (Energy Commission 2020). However, the petroleum and transport sector had been almost abandoned as far as decarbonization is concerned, until recently. Reducing the sulfur content in vehicular fuels in Ghana is one of the major initiatives to reduce carbon emissions, among other reasons behind the initiative (Daily Graphic 2017). Fuel conservation awareness building has also been one of the measures that has been undertaken to decarbonize the sector, as the motoring public is educated on the methods and benefits of fuel efficiency and fuel conservation periodically. From the revival of government owned public transport system, creating dedicated bus lanes and Bus Rapid Transit System (Business & Finance Times 2021), to surcharging over-aged (10 years and above) vehicles from being imported into the country (Ghana Revenue Authority 2002), the same economic motivation thread can be found across all these measures of decarbonization. Drive-Electric is another program that has been launched to promote the use of electric vehicles in Ghana (Energy Commission 2022). The future of this initiative holds promise as fuel prices keep rising due in part to macro-economic factors and the price of petroleum products.

9.5.6 Forestry and Biomass

Ghana commenced REDD, REDD+ and other programs a while back in a massive effort to reduce the rate of losing sequestered carbon from the ever reducing forest cover (Forestry Commission 2015). About 78% of Ghana's forest cover was lost to agricultural practices in a 29 year period (Acheampong 2019), as biomass harvesting (including charcoal burning), lumbering, surface mining and desertification also wreaked havoc. Wood fuel (biomass) accounts for 37.4% of the final energy consumption on the average in Ghana (Energy Commission 2020), hence the sector has attracted a number of initiatives to stem the tide on the rising trend of losing the forest cover. Charcoal production has been heavily regulated (Energy Commission 2022), and efficient 'Clean Cookstoves' initiatives (Ghacco 2022) have been introduced to reduce consumption of wood fuel. The pioneer program introduced to reduce deforestation was the introduction of Liquefied Petroleum Gas (LPG) as replacement fuel for cooking in households and by other outlets who needed wood fuel as energy source (Statistica 2022). This initiative implemented over three decades has been hugely successful as LPG usage has spread across the country, though wood fuel is still utilized in many homes and outlets for food preparation (Energy Commission 2015). The REDD and REDD + program has financed afforestation and reforestation programs to return the nation to higher levels of forest cover, with their attendant decarbonization benefits (Forestry Commission 2015).

9.6 Obstacles and Solutions

9.6.1 Overview

In effectively pursuing these policies, it quickly becomes obvious what the main drawbacks are in achieving the desired policy goals and results – finance, capacity, regulations, etc Developing or adopting a climate-friendly policy, signing up to an international treaty such as the Kyoto Protocol and the Paris Climate Agreement are relatively easy because there is always bi-partisan political support for protecting the environment. However, the major challenge that Ghana as a country encounters in the execution of policies and implementation of programs and projects towards decarbonization is the lack of financial resources. This part of the discussion is centered on understanding the nature of the financial obstacles and how solutions identified create a sense of direction for accelerated action to implement climate friendly policies sustainably.

Particularly evident from Table 9.1 is the very low contribution of renewable energy since 1990s, even to 2019, mainly due to the lack of access to finance for implementing decarbonization projects such as green energy generation.

9.6.2 Obstacles Encountered Through the Period

Out of the economic barriers faced, high interest rates and long term loan tenure needed by Green Energy projects emerged as the most important barriers, meaning that the discount factor that would be arrived at for discounting the cash flows is too high and renders most green energy/decarbonization projects economically not viable. In contrast to other regions of the world where lower interest rate and favorable financial market factors exist, they have discount rates low enough to keep projects viable over the lifetime of the projects as accounted by Esty (2014). These debilitating factors also indicate that the Local Financial Institutions may not be adequately capitalized to sustain long term projects, hence they cannot offer Green Energy projects the patient or long term capital needed to ensure successful implementation, such as seen under the Energiewende in Germany (Bohringer et al 2014).

9.6.3 Solutions and Sense of Direction

The most prominent solution to the economic barriers strewn across the board is that special interest rates for Green Energy projects is needed. To ensure more secure project cash flows, forex hedging instruments and long term credit lines from the MDBs hold the key to eliminating the barriers that prevent the local banks from using project finance mechanisms to fund Green Energy projects. Directly implicated in these solutions are the Multilateral Development Banks and the pivotal role they could play in eliminating the barriers through strengthening and focusing their activities towards the needs at hand to enable the LFIs engage more meaningfully and create the room for favorable downstream lending products targeting Green Energy projects using project finance.

9.7 Outlook

Ghana submitted its INDCs seven years ago (CTCN 2015) with moderate ambitions but upgraded the goals in the newly submitted NDCs for Ghana that is focused on further decarbonization in industry, forestry and other economic sectors of the nation (UNDP 2020). With several projects continuing and springing up, such as SREP, REDD+, Drive Electric, amongst others, the path to further decarbonization of the economy seems to have a sound spring board to progress from. What remains is a locally based comprehensive tracking system that combines the various subsets of the efforts into one tracking mechanism as the nation pursues a low carbon economy.

Bibliography

Acheampong, E. O. & Sayer, J. (2015). Deforestation is Driven by Agricultural Expansion in Ghana's Forest Reserves https://www.sciencedirect.com/science/article/pii/S2468227619307070

Croce, R.D., Paula, J. & Laboul, A. (2015). Infrastructure financing instruments and incentives. The Organization for Economic Co-operation and Development (OECD).

Aglanu, L. M. (2015). Diffusion of Renewable Energy Policy Innovations in Ghana. https://www.scirp.org/(S(oyulxb452alnt1aej1nfow45))/reference/ReferencesPapers.aspx?ReferenceID=1840538

Böhringer, C., Cuntz, A.N., Harhoff, D. & Asane-Otoo, E. (2014). The impacts of feed-in tariffs on innovation: Empirical evidence from Germany.

Böttcher, J. ed. (2017). Rechtliche Rahmenbedingungen von EE-Projekten: Band 2. BWV-Verlag.

Business & Financial Times (2021). Bus Rapid Transport. https://thebftonline.com/2021/08/09/bus-rapid-transport-brt-will-revolutionize-nations-urban-transport/

Centre for Global Development (2017). The Electricity Situation in Ghana: Challenges and Opportunities. https://www.cgdev.org/publication/electricity-situation-ghana-challenges-and-opportunities

Choudhury, T., Kamran, M., Djajadikerta, H.G. & Sarker, T., (2021). Can banks sustain the growth in renewable energy supply? An international evidence. *The European Journal of Development Research*, pp. 1–31

Croce, R.D., Paula, J. & Laboul, A. (2015). Infrastructure financing instruments and incentives. The Organization for Economic Co-operation and Development (OECD).

CTCN (2015) INDC of Ghana. https://www.ctc-n.org/content/indc-ghana

Daily Graphic (2017) Ghana Revises Fuel Sulphur Content. https://www.graphic.com.gh/business/oil-gas/ghana-revises-fuel-sulphur-content-from-3000ppm-to-50ppm-beginning-july-1.html

ECREEE (2022). Energy Efficiency for Households in Ghana. http://www.ecowrex.org/news/energy-efficiency-households-ghana

Energy Information Administration – USA (2018). Ghana. https://www.eia.gov/international/analysis/country/GHA

Energy Commission (2022). Drive Electric Initiative. http://www.energycom.gov.gh/efficiency/drive-electric-initiative

Energy Commission (2022). Licencing in the Renewable Energy Sector. http://www.energycom.gov.gh/licensing/licensing-renewable-energy-sector

Energy Commission (2022). Energy Efficient Refrigerator Programme. http://www.energycom.gov.gh/efficiency/the-success-story-of-the-eerp

Energy Commission (2020). Energy Statistics. https://energycom.gov.gh/planning/energy-statistics

Energy Commission (2009). Final Report – CFL Exchange Programme Impact Assessment

Energy Commission (2022). Ghana Appliance Energy Efficiency Standards and Labelling Programme. http://www.energycom.gov.gh/efficiency/standards-and-labelling

Energy Commission (2022). LI 1815. www.energycom.gov.gh%2Ffiles%2FLI_1815.pdf

Energy Commission (2018). Suspension of Issuance of Provisional Wholesale Electricity Supply Licenses. http://www.energycom.gov.gh/public-notices/93-suspension-of-issuance-of-provisional-wholesale-electricity-supply-licences

Esty, B. (2014). An overview of project finance and infrastructure finance-2014 update. HBS Case, (214083).

Forestry Commission Ghana (2015). National REDD+ Strategy. www.forestcarbonpartnership.org%2Fsites%2Ffcp%2Ffiles%2F2015%2FApril%2FGhana%2520National%2520REDD%252B%2520Strategy%2520Final.pdf&usg=AOvVaw3gkbjhBm2Fw_eS-ngx9MKx

Ghana Alliance for Clean Cookstoves (2022). www.cleancookingalliance.org%2Fbinary-data%2FRESOURCE%2Ffile%2F000%2F000%2F334-1.pdf

Ghana Revenue Authority (2002). Over-aged Vehicle Importers To Be Penalised. https://www.ghanaweb.com/GhanaHomePage/business/artikel.php?ID=27118

Green Cooling Initiative – Ghana (2022). https://www.green-cooling-initiative.org/news-media/news/news-detail/2022/06/13/kenya-and-ghana-take-a-step-towards-improving-training-in-the-area-of-hc-cooling-technology

International Energy Agency (2021). CO2 Emissions Reductions by Measure in the Sustainable Development Scenario Relative to the Stated Policies Scenario, 2010-2050. https://www.iea.org/data-and-statistics/charts/co2-emissions-reductions-by-measure-in-the-sustainable-development-scenario-relative-to-the-stated-policies-scenario-2010-2050

Ministry of Energy (2019). Energy Sector Recovery Program

Ministry of Energy (2015). Ghana SREP Investment Plan: https://www.climateinvestmentfunds.org%2Fsites%2Fcif_enc%2Ffiles%2Fmeeting-documents%2Fsrep_ip_presentation_ghana_may13_2015_0.pdf&usg=AOvVaw16JDOiWY_SRBLji8Rwkot3

Renewable Energy Act (2011). https://energycom.gov.gh/files/RENEWABLE ENERGY ACT 2011 (ACT 832 pdf).

Trading Economics (2022). Ghana GDP by Sectors) https://tradingeconomics.com/ghana/gdp

Trading Economics (2022). Ghana GDP – Manufacturing. https://tradingeconomics.com/ghana/gdp-from-manufacturing

Statistica – Share of population with primary reliance on Liquefied Petroleum Gas (LPG) as source for cooking in Ghana from 2013 to 2020. https://www.statista.com/statistics/1293943/population-using-lpg-for-cooking-in-ghana/

UNDP (2020). Ghana Review of National Emission Contributions in GH-NDC under the Paris Agreement. https://www.ndcs.undp.org/content/ndc-support-programme/en/home/impact-and-learning/library/ghana–review-of-national-emission-contributions-in-gh-ndc-under.html

UNIDO – Ghana Industrial Energy Efficiency Readiness. https://open.unido.org/projects/GH/projects/190289

Volta River Authority (2022). Power Generation: Facts and Figures. https://www.vra.com/resources/facts.php

World Bank (2022). Ghana Data. https://data.worldbank.org/indicator/NY.GDP.MKTP.CD?locations=GH

World Bank (2019). Ghana Joint World Bank-IMF Debt Sustainability Analysis December 2019. http%3A%2F%2Fdocuments1.worldbank.org%2Fcurated%2Fen%2F829241580327419447%2Fpdf%2FGhana-Joint-World-Bank-IMF-Debt-Sustainability-Analysis-December-2019.pdf&usg=AOvVaw3UbtTa8xnuLzXWMPHw1adJ

World Energy Forum (2021). Renewable Energy Is Cheaper Than Previously Thought. https://www.weforum.org/agenda/2021/10/how-cheap-can-renewable-energy-get/

Part 2: **Sector Specific Approach**

Jelto Lange, Michael Schulthoff and Martin Kaltschmitt

10 Status of Renewable Energies and the Way Forward

10.1 Introduction

Renewable energies offer a way to reduce greenhouse gas emissions and, thus, the environmental impact associated with global energy consumption. However, the transition towards a broad utilization of renewable energies poses various challenges in every energy sector. In the power sector, for example, the transition will likely be accompanied by a fundamental change in electricity supply characteristics, as power production will increasingly rely on fluctuating wind power and solar photovoltaics and thus become more weather-dependent. In the other energy sectors (e.g., heating, transportation), options for transitioning to renewable energies are limited, resulting in an intensified need for electrification, which amplifies the necessity of expanding the electricity generation based on renewables. Overall, this already ongoing development will lead to profound changes in energy systems globally.

Against this background, this chapter provides a brief overview of the status of renewable energies and the way forward. Therefore, the impact of our current energy consumption on global greenhouse gas emissions – and thus on the global climate – as well as the latest progress relating to the expansion of renewable energies and their further development trends in different parts of the global energy system, are discussed. This will also include a discussion of possible changes to the current energy demand (i.e., electrification) and additional challenges for the transition toward renewable energies and net greenhouse gas neutrality.

For this purpose, the elaboration is structured as follows: Section 10.2 will outline the status and development of renewable energies in the electricity, heating, and transportation sector and discuss technology-specific prospects for the future. The section will close with an intermediate conclusion summarizing the main development trends and necessities for an increasingly renewable energy supply. Then section 10.3 goes into more detail on the impacts associated with the development of renewable energies and discusses further requirements for the way toward a greenhouse gas-neutral global energy system. Finally, section 10.4 closes with a conclusion, summarizing the status and prospects of renewable energies and the way forward toward energy sustainability.

https://doi.org/10.1515/9783110784299-010

10.2 Status and Prospect of Renewable Energies

Average global atmospheric and oceanic temperatures at the end of the second decade of this century have been more than 1 °C higher compared to preindustrial times.[1] The temperature increase is thus approaching the 1.5 °C limit, defined in the Paris Agreement as a boundary that should not be exceeded to mitigate the likelihood of irreversible climate change.[2]

Exceeding warming of 1.5 or even 2.0 °C will most likely cause changes to the earth's climate that will strongly impact civilizations globally.[3] Therefore, to limit further warming of the planet, greenhouse gas emissions – especially CO_2 emissions, which contribute the strongest to anthropogenic climate change – must be reduced drastically and promptly. In this context, the greenhouse gas emissions associated with the use of fossil fuels are of particular importance.

Despite this necessity, overall and energy-related greenhouse gas emissions steadily increased throughout the first two decades of the 21st century (except for temporary reductions during the world economic crisis and the first year of the COVID-19 pandemic).[4] Figure 10.1 shows this development of greenhouse gas emission levels from 2000 to 2021. By now, overall annual CO_2-equivalent emissions amount to roughly 50 Gt_{CO2eq}/a globally, with approximately 75% of this stemming from energy utilization (i.e., roughly 37 Gt_{CO2eq}/a).[5]

The remaining carbon budget for limiting global warming to 1,5 °C with a probability of 66% set as a goal within the Paris Agreement is estimated to amount to roughly 400 $Gt_{CO2eq.}$[6] Supposing the current global greenhouse gas emission levels are not reduced significantly in the future, the available carbon budget will be exhausted by 2030. In about a decade, energy-related emissions alone would use up the remaining emission budget aligned with the Paris Agreement targets. Therefore, global efforts to reduce greenhouse gas emissions in the short term must be drastically stepped up.

1 World Meteorological Organization (WMO): State of the Global Climate 2020 - Unpacking the indicators, Genf, 2021

2 United Nations: The Paris Agreement, Paris, 2015.

3 IPCC: Climate Change 2021: The Physical Science Basis. Contribution of Working Group I to the Sixth Assessment Report of the Intergovernmental Panel on Climate Change – Summary for Policymakers, 2021.

4 IEA: Global Energy Review 2021, Paris, https://www.iea.org/reports/global-energy-review-2021, 2021. Zuletzt geprüft: 22.06.2021.

5 IEA: Global Energy Review 2021, Paris, https://www.iea.org/reports/global-energy-review-2021, 2021. Zuletzt geprüft: 22.06.2021. Ritchie, H., Roser, M., Rosado Pablo: CO_2 and Greenhouse Gas Emissions – Published online at OurWorldInData.org, https://ourworldindata.org/co2-and-other-greenhouse-gas-emissions, 2020. Zuletzt geprüft: 24.08.2022.

6 IPCC: Climate Change 2021: The Physical Science Basis. Contribution of Working Group I to the Sixth Assessment Report of the Intergovernmental Panel on Climate Change - Summary for Policymakers, 2021.

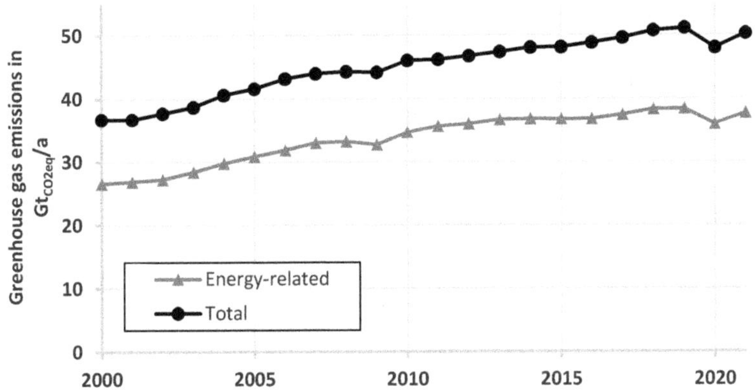

Figure 10.1: Global energy-related and total greenhouse gas emissions from 2000 to 2021 ([7,8]see above and own calculations).

Renewable energies can reduce energy-related greenhouse gas / CO_2 emissions by substituting energy from fossil fuels and thus avoid the emission of greenhouse gases otherwise released during their combustion. However, this replacement is particularly difficult because energy from fossil fuels is currently dominant in every energy sector. Thus, renewable sources of energy would have to substitute fossil primary energy as well as secondary energy provided from the use of fossil fuels within the power sector, the heat supply sector, the transportation sector, and various other industrial and non-industrial applications.

The following sections aim to provide a deeper quantitative understanding of the described challenging endeavor of transforming energy systems globally, thereby eliminating net greenhouse gas emissions. In that, the focus will lie on the status of renewable energies in the different energy sectors and their likely further development in the near future.

10.2.1 Electricity Based on Renewable Energies

In the past, electricity was almost exclusively generated based on the combustion of fossil fuels like, e.g., hard coal, lignite, natural gas, or heavy fuel oil. The only exception was hydropower as an early option for renewable power supply. Its development started in the late 19th century, and during the last decades, it has contributed a rela-

7 IEA: Global Energy Review 2021, Paris, https://www.iea.org/reports/global-energy-review-2021, 2021. Zuletzt geprüft: 22.06.2021 Ritchie, H.

8 Roser, M., Rosado Pablo: CO_2 and Greenhouse Gas Emissions - Published online at OurWorldInData.org, https://ourworldindata.org/co2-and-other-greenhouse-gas-emissions, 2020. Zuletzt geprüft: 24.08.2022.

tively stable share of about one-sixth to global electricity generation.[9] However, up until today, coal-fired power plants – and this is true for hard coal as well as for lignite – have been by far the most dominant power generation technology, supplying more than a third of electricity globally during recent history. Besides this, natural gas contributes significantly to covering the global electricity demand. As a result, more than half of the electricity generated globally results from these two fossil fuels contributing significantly to global greenhouse gas emissions.[10]

Despite this dominance of fossil fuels for providing electrical energy, power generation from wind and solar photovoltaics currently are the strongest growing technologies for supplying electricity. While at the beginning of the millennium, they contributed a combined share of less than 0.25% to the electricity supply globally, they now account for more than 10% of the total global electricity generation.[11]

Figure 10.2 shows the development of the different technologies for power generation between 2000 and 2021. The recent growth of electricity generation from wind and solar is clearly noticeable. However, due to a strongly growing global electricity demand, power generation from fossil fuels is also still considerably increasing. Consequently, the more substantial increase in global electricity demand currently overcompensates electricity generation growth from renewable energy sources.[12] Therefore, the expansion of renewables has to be further accelerated on a global scale if not just carbon intensity but also absolute greenhouse gas emissions from power generation are to be reduced.

Figure 10.3 shows the development of greenhouse gas emissions from electricity generation from 2000 to 2021 in more detail (note the different y-axis scales from 0 to 180 Mt-CO_{2eq}/a and from 500 to 8 500 Mt-CO_{2eq}/a). Again, fossil fuels are the primary source of electricity-related greenhouse gas emissions, with coal-based power generation emitting the highest quantities. Greenhouse gas emissions from this carbon-intensive energy source currently exceed 8 Gt_{CO2eq}/a. Wind power, however, while providing a little less than 20% of the electricity generated from coal-fired power plants, causes merely around 0.25% of the greenhouse gas emissions caused by coal-based power generation. Thereby, emissions that impact global climate are reduced by several orders of magnitude when generating electricity from renewable energies. Moreover, this difference will grow further as emissions from wind power and many other renewables mainly stem from manufacturing, which is becoming increasingly efficient and less dependent on carbon-intensive energy.

9 Ember: Global Electricity Review 2022, https://ember-climate.org/insights/research/global-electricity-review-2022/, 2022. Zuletzt geprüft: 13.06.2022.

10 Ember: Global Electricity Review 2022, https://ember-climate.org/insights/research/global-electricity-review-2022/, 2022. Zuletzt geprüft: 13.06.2022.

11 Ember: Global Electricity Review 2022, https://ember-climate.org/insights/research/global-electricity-review-2022/, 2022. Zuletzt geprüft: 13.06.2022.

12 International Energy Agency, IEA: Electricity Market Report, 2021.

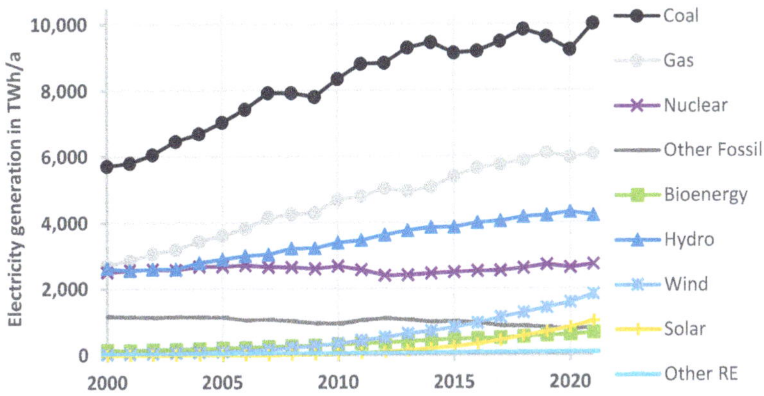

Figure 10.2: Development of electricity generation from fossil and renewable sources from 2000 to 2021[13] (RE: renewable energies).

Renewable energies can, therefore, effectively reduce greenhouse gas emissions from electricity generation. However, to reach a net reduction in emissions to achieve the goals defined within the Paris Agreement, renewable expansion has to be accelerated substantially so that renewable electricity not only covers growth in demand but also substitutes fossil fuels for the supply of the existing power demand.

Many different technology options for power generation based on renewable energies are available in various stages of their technological development and are most likely playing a particular role in a global energy transition. Nevertheless, the most promising technologies for accelerated growth in electricity supply from renewable energies are currently wind power and photovoltaics. Moreover, with regard to capacity additions, they are currently the strongest growing power generation technologies on a global scale, even compared to any conventional (fossil fuel-based) option.[14]

Furthermore, wind turbines and photovoltaic systems are generally expected to grow strongest when analyzing possible scenarios for reaching net zero greenhouse gas emissions in 2050.[15] Thus, according to current development trends and expectations, these two power generation options will contribute most to a sustainable energy supply by the middle of this century. Therefore, while power generation from fossil fuels – and especially coal-fired electricity generation – will need to decrease drastically to reach net zero greenhouse gas emissions, solar photovoltaics and wind capacities will need to grow by at least one order of magnitude (i.e., a factor of roughly 12 for wind and approximately

13 Ember: Global Electricity Review 2022, https://ember-climate.org/insights/research/global-electricity-review-2022/, 2022. Zuletzt geprüft: 13.06.2022.

14 International Energy Agency, IEA: Electricity Market Report, 2021

15 bp: Statistical Review of World Energy 2021 – 70th edition, 2021. IEA: World Energy Outlook 2021, 2021. Global Wind Energy Council, GWEC: Global Wind Report 2022, 2022.

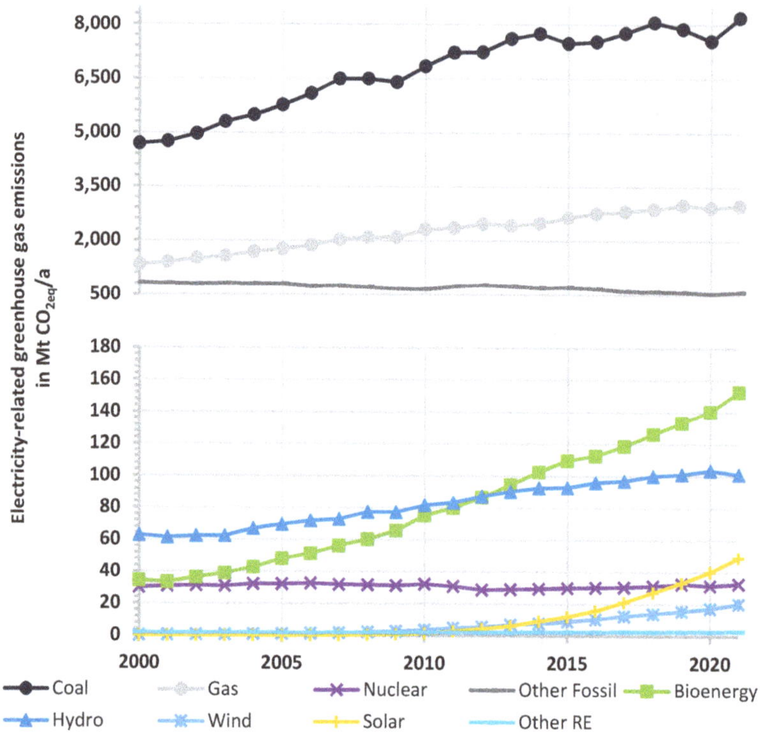

Figure 10.3: Development of greenhouse gas emissions from electricity generation based on fossil and renewable sources from 2000 to 2021 (note the different y-axis scales from 0 to 180 Mt-CO_2eq/a and from 500 to 8 500 Mt-CO_2eq/a) (RE: renewable energies).[16]

20 for photovoltaics.[17] This implies the need for an average annual growth of about 10% (i.e., roughly 8.5% for wind and approximately 10.5% for solar) if the transition to a greenhouse gas-neutral global energy system is seriously pursued within the envisaged timespan.

The following subsections go into further detail on the different options for electricity generation based on renewable energies.

Wind Power

The wind energy industry started developing earlier than the photovoltaic industry; substantial growth began in the early 2000s (Figure 10.4). During this development,

16 Ember: Data Explorer, https://ember-climate.org/data/data-explorer/, 2022. Zuletzt geprüft: 24.08.2022.
17 bp: Statistical Review of World Energy 2021 – 70th edition, 2021. IEA: World Energy Outlook 2021, 2021.

the total installed capacity grew considerably year on year. However, relative growth has decreased since the global wind energy expansion began. While during the first decade of the 21st century, total installed power increased by 20 to 30% annually, capacity additions currently amount to roughly 10% of the existing installed power. Recently, offshore wind power showed a strong development (especially in the Northern part of Europe and China) and is increasingly becoming an essential driver for the wind energy industry.

In order to achieve net zero greenhouse gas emissions by 2050, the overall growth rates for the use of wind energy should not decrease further. Therefore, global manufacturing and installation capacities need to increase tremendously over the coming decades to enable a stable growth rate of roughly 10% until the middle of the century.

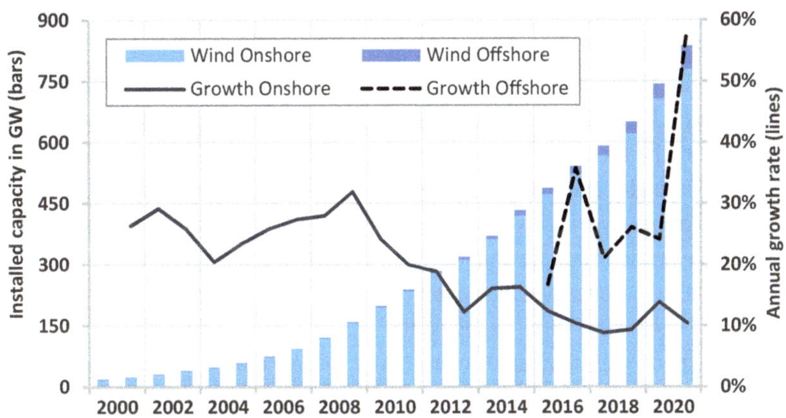

Figure 10.4: Development of installed capacities and annual growth rates of onshore and offshore wind energy from 2000 to 2021.[18]

In the course of wind energy expansion, the technology has been continuously refined and optimized. This development's primary focus was increasing rotor diameters as the essential design parameter for increasing rated turbine capacities.[19] At the beginning of commercial wind energy utilization, a single wind turbine's installed power laid in the two- to lower three-digit kW range, while nowadays, installed capacities of modern offshore wind turbines more commonly exceed 10 MW.[20] These giant wind turbines have rotor diameters of almost 200 m and reach heights exceeding 250 m. The next genera-

18 Global Wind Energy Council, GWEC: Global Wind Report 2022, 2022.

19 IEA-ETSAP and IRENA: Wind Power - Technology Brief, 2016.

20 IRENA: Future of Wind - Deployment, investment, technology, grid integration and socio-economic aspects, 2019.

tion of large-scale offshore wind turbines currently entering commercialization reaches capacities of roughly 15 MW.[21]

During wind power expansion and technological improvement during the last decades, the levelized cost of electricity generation from wind power experienced a substantial reduction. Figure 10.5 shows this development of the global weighted average utility-scale levelized cost of electricity from onshore and offshore wind power from 2010 to 2021.[22]

The average cost of electricity from onshore wind reduced by more than 50% in the past ten years. Meanwhile, the globally installed capacity roughly increased by a factor of 3.6 (i.e., a reduction of the levelized cost of electricity by roughly 35% for each time, globally installed capacity doubled during the depicted time span). On the one hand, this cost reduction can be attributed partly to developing more favorable locations (i.e., locations with higher average wind speeds). However, on the other hand, technological development contributed substantially to a drastic overall reduction in power generation costs from wind energy.

For offshore wind, the decline of the global weighted average utility-scale levelized cost of electricity started later. Nevertheless, after gaining global momentum in the middle of the last decade, the levelized costs of electricity were reduced by more than 50% from 2015 to 2020 while global capacities increased by a factor of 3 (i.e., a reduction of levelized cost of electricity by roughly 28% for each doubling of globally installed capacity). Moreover, if an electricity supply with net zero greenhouse gas emissions is reached (i.e., increasing global wind energy capacities by a factor of 12 by 2050 compared to today's capacities), levelized cost can be expected to further decline substantially.

Overall, wind power will be a significant driver for the global transition towards an energy system with net zero greenhouse gas emissions. In the course of this development, the power supply will become increasingly dependent on the prevalence of sufficiently high wind speeds and, thus, more weather-dependent and fluctuating. However, assuming a continuation of cost reduction trends, this volatile electricity supply will be available with diminishing costs.

Solar Photovoltaics

Photovoltaics is another technological option for electricity generation based on renewable energies characterized by considerable technological progress and signifi-

---

21 Siemens Gamesa Renewable Energy: SG 14–222 DD - Offshore wind turbine, https://www.siemensgamesa.com/products-and-services/offshore/wind-turbine-sg-14-222-dd, 2022. Zuletzt geprüft: 04.11.2022. Vestas Wind Systems A/S: V236-15.0 MW™, https://www.vestas.com/en/products/offshore/V236-15MW/V236-15MW, 2022. U. S. Department of Energy: Offshore Wind Market Report: 2021 Edition, 2021.

22 IRENA: Renewable Power Generation Costs in 2020, Abu Dhabi, 2021. IRENA: Renewable Power Generation Costs in 2021, Abu Dhabi, 2022.

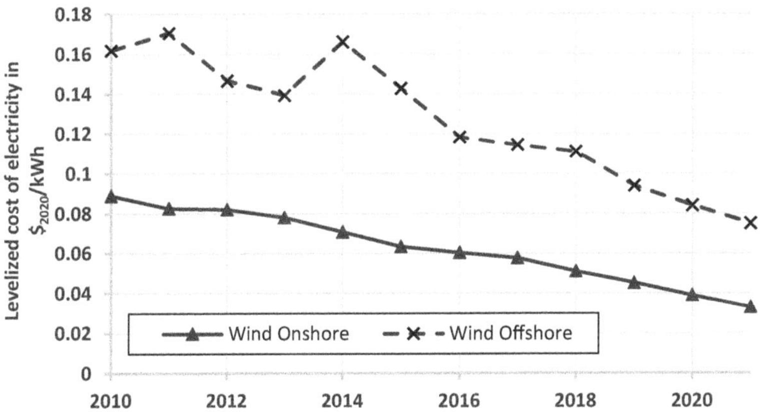

Figure 10.5: Development of global weighted-average utility-scale levelized cost of electricity for onshore and offshore wind energy from 2010 to 2021.[23]

cant growth, reaching even higher growth rates than wind energy in recent years. In terms of capacity growth, photovoltaics has been the fastest-growing power generation technology for more than 15 years, even though its commercial use started later than that of wind power.[24]

Global installed capacities of photovoltaics are approaching the one-terawatt mark. Based on recent annual capacity additions, it is likely that this mark will be reached within the next few years (Figure 10.6). Given that serious expansion of photovoltaics only began in the 2010s, this represents an unprecedented momentum in adding new power generation capacity for a single technology. During this expansion, average annual growth rates declined (whereby the overall low cumulative capacity strongly influenced earlier rates) but still amount to roughly 20%; however, a slight downward trend is discernible. Assuming growth rates will not decline considerably, globally installed capacities would roughly double every four years. Thus, the current growth rates significantly exceed the required growth for reaching the necessary capacities in 2050 to achieve the different net zero greenhouse gas emission scenarios (i.e., a growth rate for power generation from solar energy of roughly 10.5%.[25] However, as sustaining constant relative growth demands a substantial increase in absolute annual additions of installed power, manufacturing and installation capacities will also have to grow substantially to maintain current growth rates in the coming years.

23 IRENA: Renewable Power Generation Costs in 2020, Abu Dhabi, 2021. IRENA: Renewable Power Generation Costs in 2021, Abu Dhabi, 2022.

24 IRENA: Renewable Capacity Statistics 2022, Abu Dhabi, 2022.

25 bp: Statistical Review of World Energy 2021 - 70th edition, 2021. IEA: World Energy Outlook 2021, 2021.

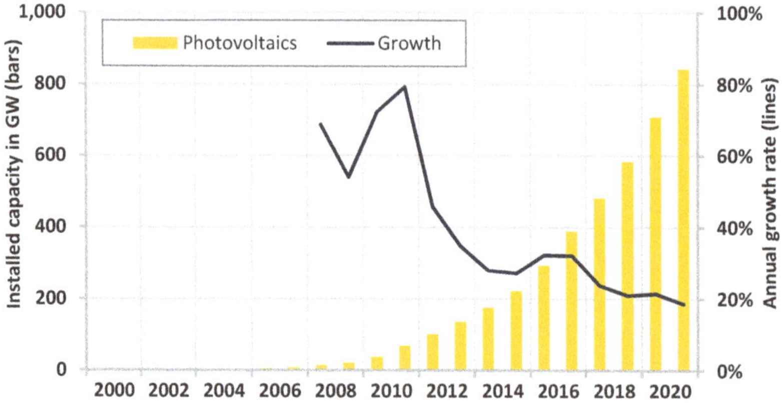

Figure 10.6: Development of installed capacities and annual growth rates of solar photovoltaics from 2000 until 2021.[26]

During the recent development of capacities for power generation based on solar energy, photovoltaic cells, modules, and systems have been continuously improved. One aspect of this development has been the increase in overall efficiencies.[27] An even more noticeable improvement has nevertheless been of economic nature; the amount of material used for cell and module production as well as manufacturing energy intensity were reduced substantially, leading to a drastic decline in manufacturing cost. Furthermore, the cost for peripheral systems (e.g. inverters) has also been reduced.

These improvements led to an unprecedented decline in the levelized cost of electricity generation from photovoltaics, which became the main driver for further global photovoltaics expansion. Figure 10.7 shows this continued development, depicting the global weighted average utility-scale levelized cost of electricity from photovoltaics over the last decade. While capacities increased by a factor of 18.3, the levelized cost of electricity declined by more than 85% (i.e., a reduction of the levelized cost of electricity by more than 35% for each doubling of globally installed capacity during the depicted time span). For photovoltaics, a significant influence on this reduction in the levelized cost of electricity has been that after initial development in solar-poor regions (e.g., Germany), photovoltaics has been adopted increasingly in solar-rich regions (e.g., North Africa or the Middle East). The reduction of global average cost can thus be attributed partially to this increase in projects in locations with higher solar irradiance. However, due to drastic cost reduction in, e.g., module prices,

26 IRENA: Renewable Capacity Statistics 2022, Abu Dhabi, 2022. IRENA: Renewable Capacity Statistics 2017, Abu Dhabi, 2017. Our World in Data: Installed solar energy capacity - Cumulative installed solar capacity, measured in gigawatts (GW)., https://ourworldindata.org/grapher/installed-solar-pv-capacity. Zuletzt geprüft: 16.06.2022.

27 Fraunhofer ISE: Photovoltaics Report, Freiburg, 2022.

the unprecedented decline in the levelized cost of electricity has been primarily achieved by technological advances in the modules themselves and by improving manufacturing (e.g., less material loss, greater automation).

Currently, the global weighted-average levelized costs of electricity from utility-scale photovoltaics amount to roughly 0.05 $\$_{2020}$/kWh with even lower values in favorable locations. In 2020, 5% of capacity additions were realized with a levelized cost below 0.039 $\$_{2020}$/kWh.[28]

Supposing, global photovoltaic expansion follows a path in line with net zero scenarios (i.e., a 20-fold increase of global capacities until 2050[29]), it can be assumed that the levelized cost of electricity for photovoltaics will continue to decline further considerably over the next few years.

Based on this cost development and the expected further improvement, photovoltaic power production will likely influence and shape the ongoing energy system transformation toward a more sustainable energy supply. In the future, photovoltaic power will increasingly influence electricity supply characteristics globally. Especially in the daytime, electricity will thereby be available almost without marginal cost.

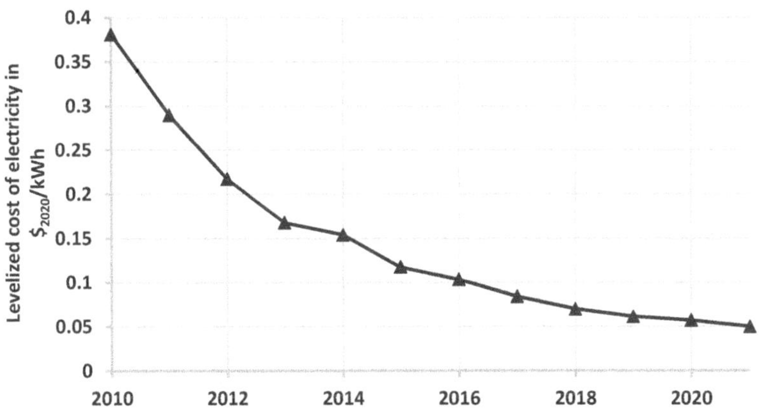

Figure 10.7: Development of global weighted-average utility-scale levelized cost of electricity for photovoltaics from 2010 to 2021.[30]

With wind power and photovoltaics becoming increasingly relevant, electricity supply characteristics will change considerably. A large part of electricity generation will be weather-dependent and thus fluctuate more strongly. In order to bridge times with

28 IRENA: Renewable Power Generation Costs in 2020, Abu Dhabi, 2021.

29 bp: Statistical Review of World Energy 2021 - 70th edition, 2021. IEA: World Energy Outlook 2021, 2021.

30 IRENA: Renewable Power Generation Costs in 2020, Abu Dhabi, 2021; IRENA: Renewable Power Generation Costs in 2021, Abu Dhabi, 2022.

low availability of electricity from wind and photovoltaics, various measures can be taken; a crucial aspect of this is the implementation and utilization of dispatchable renewable power generation.

Hydropower

Hydropower is the most important option for a more controllable electricity supply from renewable energies. Despite the recent strong growth of wind and solar, it is still the most important technology for renewable power production (Figure 10.2). Furthermore, during recent decades hydropower continued growing with remarkable absolute capacity additions (Figure 10.8), leading to global cumulated hydropower capacities of more than 1.3 TW.[31] China has driven more than 60% of this development during the last decade.[32]

Despite this remarkable absolute increase in electricity generation from hydropower, the relative growth declined continuously; year-on-year capacity additions amounted to less than 2% during recent years. However, to reach global net zero

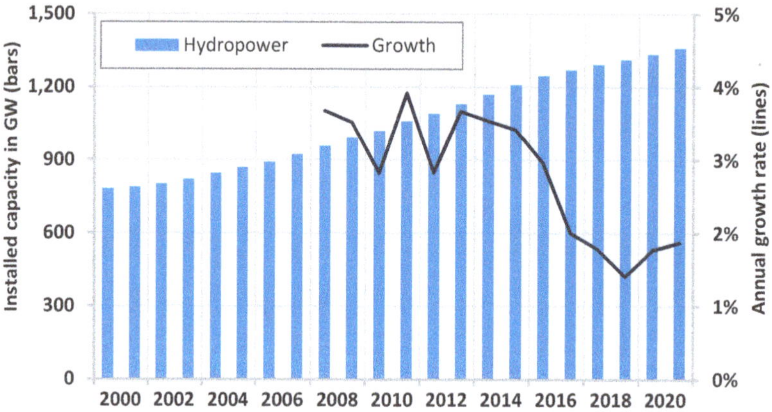

Figure 10.8: Development of installed capacities and annual growth rates of hydropower from 2000 until 2021.[33]

31 IRENA: Renewable Capacity Statistics 2022, Abu Dhabi, 2022.

32 Ember: Global Electricity Review 2022, https://ember-climate.org/insights/research/global-electricity-review-2022/, 2022. Zuletzt geprüft: 13.06.2022.

33 IRENA: Renewable Capacity Statistics 2022, Abu Dhabi, 2022. IRENA: Renewable Capacity Statistics 2017, Abu Dhabi, 2017. Our World in Data: Installed global renewable energy capacity by technology - Installed global renewable energy capacity in megawatts (MW) by technology (hydropower, solar, wind, biomass, marine and geothermal), https://ourworldindata.org/grapher/installed-global-renewable-energy-capacity-by-technology. Zuletzt geprüft: 05.10.2022.

greenhouse gas emissions, hydropower will need to grow substantially in the future.[34] Anyhow, considerable growth will be somewhat challenging due to foreseeably limited potential; many of the most promising sites are already exploited, and installing small-scale hydropower systems is generally time-consuming and costly. Furthermore, due to the ongoing climate change, the natural precipitation patterns will change in the coming years, potentially influencing economic viability due to unpredictable risks for hydropower plants.

With hydropower being a mature technology, further development is likely not accompanied by a cost decline comparable to the reductions in levelized cost achieved for solar photovoltaics and wind power. Moreover, it is not unlikely that electricity from hydropower might even become more expensive, as many of the most economical sites on large rivers are already utilized, and smaller hydropower plants do not profit as much from economies of scale. The fact that environmental concerns also gain more global importance reinforces the trend toward rising costs.

An increase in cost was already noticeable during the recent development (Figure 10.9). The levelized costs of electricity from hydropower have stagnated chiefly during the last decade, and they even showed a slight increase. Nevertheless, the achieved levelized costs already reach relatively low overall values, making hydropower a generally cheap option for electricity supply based on renewable energies. Furthermore, hydropower's ability to offer (partially) dispatchable/controllable power generation is additionally valuable. The fact that the technology is mature and highly reliable complements this positive adjustment. However, this might be partially compromised due to water availability changing because of ongoing climate change.

The limited availability of suitable sites can potentially be a restricting factor for further hydropower development. In some regions and countries, promising sites required for economic hydropower utilization (i.e., sufficient height difference and water volume flow) are either already exploited or nonexistent. On a global scale, however, suitable locations are still available in different areas, making the further expansion of hydropower highly likely – although not at growth rates comparable to those of wind power or photovoltaics.[35]

34 bp: Statistical Review of World Energy 2021 – 70th edition, 2021. IEA: World Energy Outlook 2021, 2021.

35 International Energy Agency: Hydropower Special Market Report - Analysis and forecast to 2030, 2021.

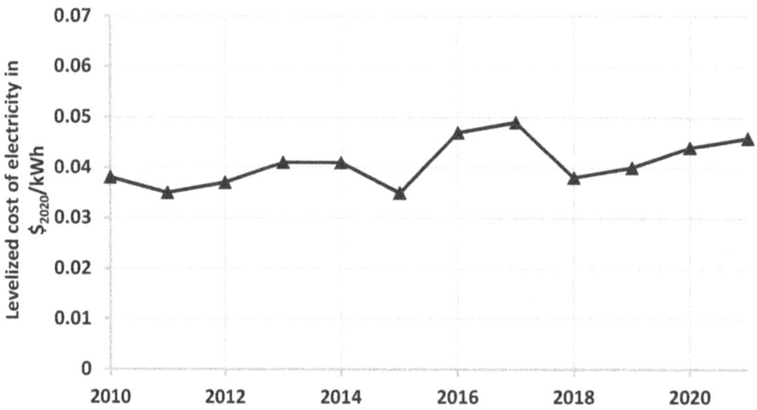

Figure 10.9: Development of global weighted-average utility-scale levelized cost of electricity for hydropower from 2010 to 2021.[36]

Biomass

Besides hydropower, power production based on biomass is another dispatchable and controllable form of electricity generation from renewable energies.

This is generally true for the power generation in biomass-based power plants (i.e., power plants burning lignocellulosic biomass driving a classical steam cycle) as well as the power generation based on the combustion of biogas (i.e., gas gained from the fermentation of moist biomass streams) in an internal combustion engine. Its controllability makes biomass-based electricity supply an additional, important technology for transitioning toward broad reliance on renewable energies. However, compared to hydropower, electricity generation based on biomass currently plays a much smaller role (Figure 10.10). Globally installed capacities recently surpassed 140 GW, hardly more than 10% of global hydropower capacities.

The relative growth of biomass-based electricity generation has been between 5 and 10% over the last decade. Generally, it is expectable that power production from biomass has to be expanded even faster than this in order to support the development toward net zero greenhouse gas emissions.[37]

36 IRENA: Renewable Power Generation Costs in 2020, Abu Dhabi, 2021. IRENA: Renewable Power Generation Costs in 2021, Abu Dhabi, 2022. Our World in Data: Installed global renewable energy capacity by technology - Installed global renewable energy capacity in megawatts (MW) by technology (hydropower, solar, wind, biomass, marine and geothermal), https://ourworldindata.org/grapher/installed-global-renewable-energy-capacity-by-technology. Zuletzt geprüft: 05.10.2022.

37 bp: Statistical Review of World Energy 2021 - 70th edition, 2021. IEA: World Energy Outlook 2021, 2021.

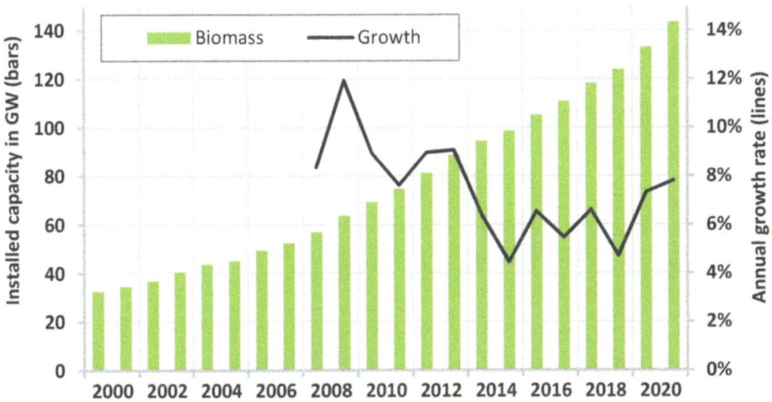

Figure 10.10: Development of installed capacities and annual growth rates for electricity generation from Biomass from 2000 until 2021 [19, 20, 23].

Much like hydropower, power production from biomass is based on mature power plant components, with little potential for (drastic) cost reduction. The stagnation of levelized cost of electricity of biomass-based power generation during the last decade reflects this (see Figure 10.11).[38] While the levelized generation cost can vary strongly from power plant to power plant – in some cases exceeding 0.2 $\$_{2020}$/kWh respectively falling below 0.05 $\$_{2020}$/kWh – global weighted average values consistently ranged between about 0.06 $\$_{2020}$/kWh and 0.08 $\$_{2020}$/kWh.

Despite concerns about the true emissions impact of biomass,[39] which potentially lead to a reduction or cancellation of political support,[40] further expansion of power generation from biomass can be considered likely. Especially where unavoidable and otherwise unusable organic waste streams are available, power generation from biomass can be an economically promising option supporting the further growth of biomass-based electricity supply.

Others

Besides wind power, solar photovoltaics, hydropower, and biomass, there are other options for electricity generation based on renewable energies (e.g., geothermal en-

38 IRENA: Renewable Power Generation Costs in 2020, Abu Dhabi, 2021. IRENA: Renewable Power Generation Costs in 2021, Abu Dhabi, 2022.

39 Ember: Global Electricity Review 2022, https://ember-climate.org/insights/research/global-electricity-review-2022/, 2022. Zuletzt geprüft: 13.06.2022.

40 Simon, F.: EU Parliament groups rally behind plans to end biomass subsidies, https://www.euractiv.com/section/biomass/news/eu-parliament-groups-rally-behind-plans-to-end-biomass-subsidies/, 2022. Zuletzt geprüft: 05.10.2022.

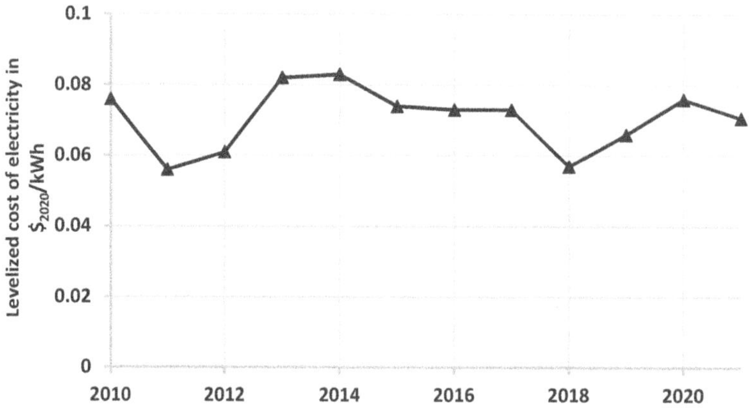

Figure 10.11: Development of global weighted-average utility-scale levelized cost of electricity from biomass from 2010 to 2020.[41]

ergy,[42] wave energy, concentrated solar power, and tidal energy). The respective technologies will potentially contribute to a more sustainable energy supply in the future as their expansion is relevant for the way toward net zero greenhouse gas emissions.[43] Especially in regions with high local potential (e.g., high enthalpy geothermal reservoirs in the case of geothermal energy), these technologies for utilizing renewable energies can be promising supply options. However, since growth has been moderate in these areas recently, it might be necessary to support this expansion further (e.g., via suitable policies and regulatory frameworks).

Summary

The overall need for reducing energy-related greenhouse gas emissions will push the development of renewable energies in the power sector. Besides the already expanded use of hydropower, this further development of renewables will likely be based primarily on a drastic increase in wind power and solar photovoltaic capacities, which will consequently claim the most significant market shares. Still, other options for renewable power generation will likely also grow during the transition to net zero greenhouse gas emissions in global energy systems. Nevertheless, due to the strong development of technologies with volatile supply characteristics (i.e., wind power and solar photovol-

41 IRENA: Renewable Power Generation Costs in 2020, Abu Dhabi, 2021. IRENA: Renewable Power Generation Costs in 2021, Abu Dhabi, 2022.

42 See sections 12 and 13.

43 bp: Statistical Review of World Energy 2021 - 70th edition, 2021. IEA: World Energy Outlook 2021, 2021.

taics), power generation will become more and more weather-dependent and thus show much more substantial fluctuations than today's electricity supply.

In conclusion, the rising share of non-dispatchable power production will substantially transform the operation of electricity supply systems. Additionally, renewable electricity will increasingly become a basis for reducing reliance on fossil fuels in other energy sectors, raising the value of power systems in general.

10.2.2 Heat Based on Renewable Energies

In contrast to the development of renewable energies in the electricity sector, heat supply from renewables has largely stagnated recently.[44] As shown in Figure 10.12, the overall share of renewable heat of total heat consumption did not show any distinct development trends between 2010 and 2020. The slight increase in modern renewable heat during the last decade can almost exclusively be attributed to replacing heat from traditional biomass through modernized technology (e.g., deployment of improved biomass cooking stoves). Non-renewable heat consumption stayed mostly constant despite some slight variations over the years.

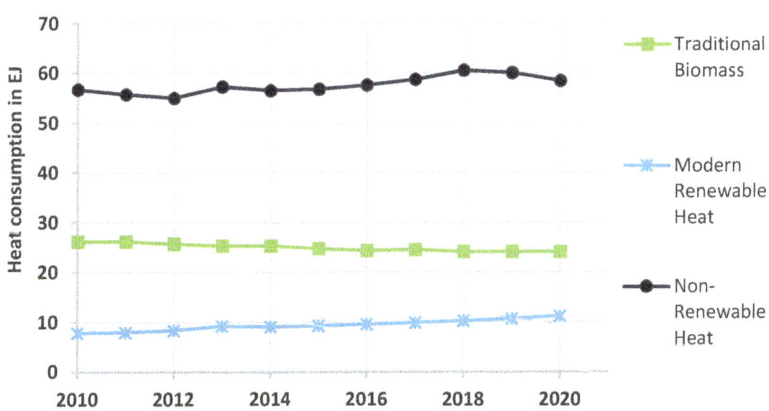

Figure 10.12: Development of heat consumption from fossil and renewable sources from 2010 to 2020 [26] (modern renewable heat covers indirect and direct final consumption of bioenergy, solar thermal energy, geothermal energy, as well as renewable electricity used for heat production).

Consequently, greenhouse gas emissions from heat supply have stagnated in recent history. Figure 10.13 shows that despite minor variations of heat-related greenhouse gas emissions, no overall reduction trend is visible. Carbon intensity for heat genera-

44 IEA: Renewables 2021 - Analysis and forecast to 2026, 2021.

tion thus – just like the amount of non-renewable heat – mainly stayed constant during the last decade.

Furthermore, future projections predict global heat consumption to grow substantially in the short- to medium-term.[45] Despite an even stronger expected relative growth of heat from renewable energies, the absolute increase of demand is likely to surpass the absolute expansion of the low carbon-intensity supply of thermal energy. Therefore, it is expectable that non-renewable heat – as well as greenhouse gas emissions from heat generation – will increase if no further measures are taken to expand the thermal energy supply based on renewable sources.

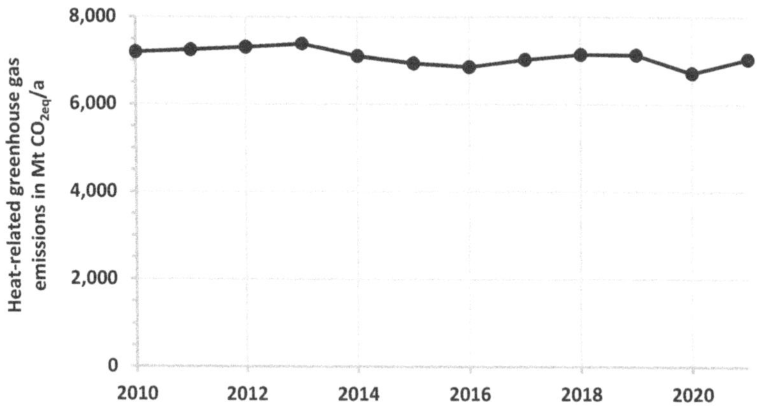

Figure 10.13: Development of greenhouse gas emissions from heat generation 2010 to 2021.[46]

Biomass

Increasing shares of heat generation from renewable energies will likely require various measures and technologies. The modern use of bioenergy is expected to play an important role. Shifting from traditional biomass to modern biomass utilization for heat generation is often accompanied by a strong increase in conversion efficiency (e.g., by replacing fireplaces with efficient stoves). Meanwhile, the overall amount of supplied usable energy can be increased considerably while using the same amounts of biomass resources; additionally, this shift generally improves the environmental performance. Therefore, this transition is generally not associated with further poten-

45 IEA: World Energy Outlook 2021, 2021. IEA: Renewables 2021 - Analysis and forecast to 2026, 2021

46 IEA: Global Energy Review 2021, Paris, https://www.iea.org/reports/global-energy-review-2021, 2021. Zuletzt geprüft: 22.06.2021. Ritchie, H., Roser, M., Rosado Pablo: CO_2 and Greenhouse Gas Emissions - Published online at OurWorldInData.org, https://ourworldindata.org/co2-and-other-greenhouse-gas-emissions, 2020. Zuletzt geprüft: 24.08.2022. Ember: Data Explorer, https://ember-climate.org/data/data-explorer/, 2022. Zuletzt geprüft: 24.08.2022. And authors' own calculations.

tially unsustainable biomass sourcing; if the transformation to modern biomass use is successful, less biomass is needed to provide the same amount of heat compared to the traditional way. As a result, modern biomass utilization can contribute to a sustainable reduction of greenhouse gas emissions from heat generation. However, this will not be sufficient to replace conventional non-renewable heat completely.

Solar Thermal Energy

Solar thermal systems are a promising option for increasing shares of renewable energies in the heating sector.[47] They showed considerable market development in recent years, especially for simple natural circulation systems providing hot water in areas with high solar irradiation. Especially for covering the demand for domestic hot water, these systems will likely gain more importance in the years to come.

Additionally, it might also be true for forced circulation systems that can partially support the heating of buildings and the domestic hot water supply. However, such systems are still of minor importance on a global scale. Furthermore, in many cases, it is unclear how solar thermal energy will perform in the competition with the strongly developing photovoltaic technology due to conflicting area usage of photovoltaic and solar thermal systems. The future of solar thermal energy is thus quite difficult to predict.

Geothermal Energy

Where geological layers within the earth's surface contain sufficiently hot water, this can be exploited based on existing technology for locally supplying heat. For promising geological conditions (e.g., Iceland, Turkey), such reservoirs are already used to a certain extent and might even be expanded. However, the economic risk of failure while unlocking such reservoirs is significant. The technology is demanding, and the distribution of geothermal heat typically needs a potentially expensive heat distribution network. For this reason, heat supply from geothermal energy will likely predominantly play a local role where geological conditions are promising. The extent to which geothermal energy will contribute to reducing energy-related greenhouse gas emissions on a global scale, however, remains unclear.

47 See also section 2.11.

Renewable Electricity

Besides a further intensification of the use of solar thermal and geothermal energy, a significant driver for reducing greenhouse gas emissions in the heating sector will likely be the utilization of renewable electricity for supplying heat predominantly via heat pumps.[48] By installing large amounts of efficient heat pump systems, electricity can effectively and efficiently substitute fossil fuels for the supply of thermal energy, thereby potentially reducing net greenhouse gas emissions in the heating sector. This emissions reduction will, however, in large parts rely on an increase of renewable energies within the electricity sector. As a result, the heating sector is becoming increasingly dependent on the power sector. At the same time, the power supply system is also increasingly influenced by the thermal energy system.

Summary

It can be expected that electricity will play an essential role in the reduction of greenhouse gas emissions in the heating sector. However, this will intensify the drive for a shift toward renewable power generation. In this context, the integrated assessment of the development of heat and power systems and their energy demands will likely gain importance.

Overall, there is a major gap between the current efforts (or lack thereof) to reduce greenhouse gas emissions in the heating sector and the requirements defined by a 1.5 or 2 °C boundary for global warming. Therefore, the expansion of renewables in both the thermal energy system and the power sector has to be accelerated.

10.2.3 Transport Based on Renewable Energies

Mobility based on renewable energies is commonly associated with biofuels produced from biomass (e.g., bioethanol, biodiesel). These fuels are usually blended with conventional fuels (e.g., gasoline, diesel); the amount of biofuel currently added to conventional fuels depends inter alia on the specific regulatory frameworks and the respective market conditions. On a global scale, the share of biofuels is almost negligible compared to total fuel consumption. As Figure 10.14 shows, the level of biofuel consumption is orders of magnitude lower than the consumption level for fossil fuels. Even though biofuel utilization has grown stronger than fossil fuel consumption over the last decade, biofuels still only make up slightly more than 2% of global liquid fuel consumption.[49]

48 See also section 2.9.

49 bp: Statistical Review of World Energy 2021 - 70th edition, 2021.

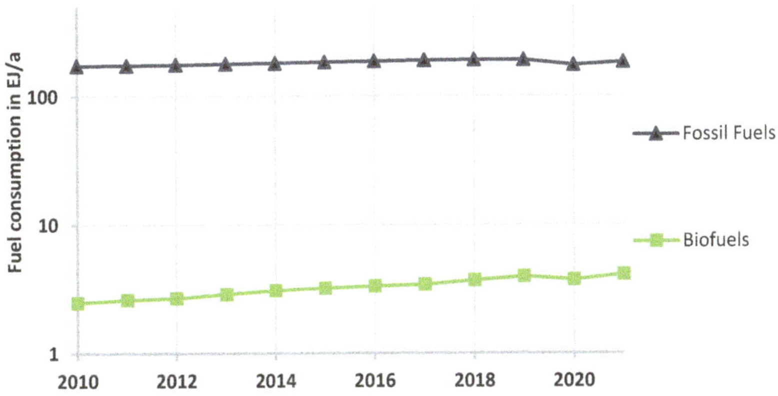

Figure 10.14: Development of fossil fuel and biofuel consumption from 2010 to 2021.[50]

As shown in Figure 10.15, the global increase in the utilization of conventional fuels led to an increase in greenhouse gas emissions within the transportation sector; this trend seems to be only temporarily interrupted by the COVID-19 pandemic. Meanwhile, the growth of biofuels had no visible impact on this global development. It is, therefore, apparent that even though greenhouse gas emissions need to be reduced drastically to meet the targets of the Paris Agreement, the transport sector is still not implementing appropriate measures on a global scale.

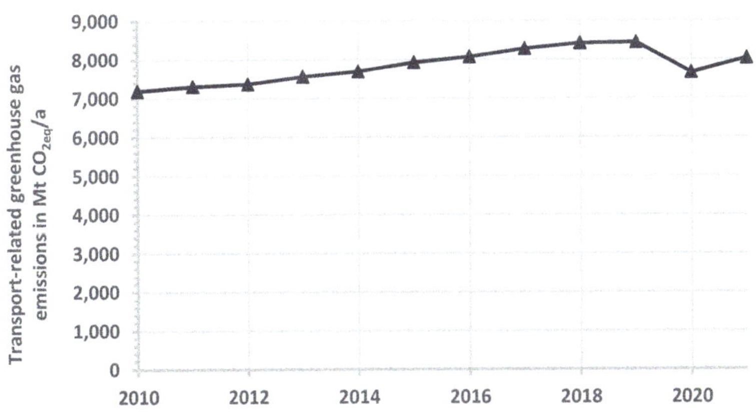

Figure 10.15: Development of greenhouse gas emissions from transportation 2010 to 2021.[51]

50 bp: Statistical Review of World Energy 2021 - 70th edition, 2021.

51 IEA: Global Energy Review 2021, Paris, https://www.iea.org/reports/global-energy-review-2021, 2021. Zuletzt geprüft: 22.06.2021. Ritchie, H., Roser, M., Rosado Pablo: CO_2 and Greenhouse Gas Emissions - Published online at OurWorldInData.org, https://ourworldindata.org/co2-and-other-greenhouse-gas-emissions, 2020. Zuletzt geprüft: 24.08.2022. And authors' own calculations.

Efforts to reduce CO_2 emissions within the mobility sector thus need to be substantially intensified to reach net zero greenhouse gas emissions. Besides potentially improving the effectiveness of mobility (e.g., strengthening public transport), a significant aspect of this reduction of greenhouse gas emissions will have to be to drastically increase the shares of renewable energies in the transportation sector.

Biofuels

Biofuels so far have had the most considerable impact on reducing greenhouse gas emissions in the mobility sector, even though this impact is limited (and overcompensated by an absolute increase in fuel consumption). Figure 10.16 shows the amounts of consumed bioethanol and biodiesel.[52] Bioethanol is thus the more widely used biofuel globally. However, biodiesel consumption showed a more robust growth rate, almost reaching bioethanol consumption recently. Nevertheless, with bioethanol and biodiesel accounting for roughly 1.25%, respectively 1% of global fuel consumption, their impact on reducing greenhouse gas emissions in the transportation sector is almost negligible. Especially as biofuel production also releases greenhouse gas emissions, reducing the specific mitigation potential.

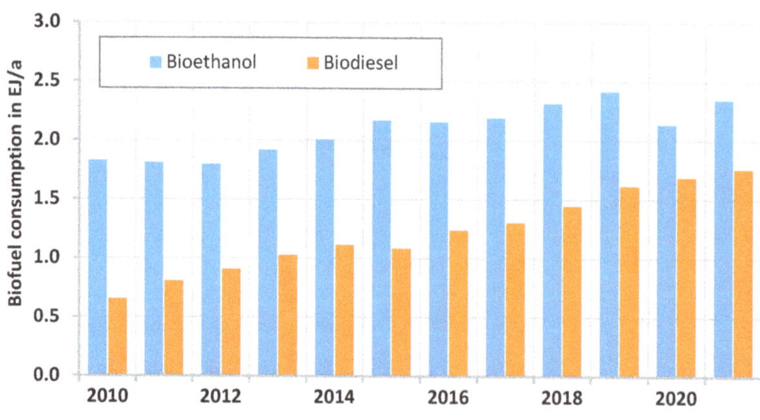

Figure 10.16: Development of bioethanol and biodiesel consumption from 2010 to 2021.[53]

However, it is expected that to reach higher shares of renewables in the mobility sector and enter a path toward reaching net zero greenhouse gas emissions in 2050, bio-

52 bp: Statistical Review of World Energy 2021 - 70th edition, 2021. bp: Statistical Review of World Energy 2022 - 71st edition, 2022.

53 bp: Statistical Review of World Energy 2021 - 70th edition, 2021. bp: Statistical Review of World Energy 2022 - 71st edition, 2022.

fuel production and consumption will need to increase considerably (e.g., expansion by more than 200% by 2030[54]).

Renewable Electricity

Beyond the utilization of biofuels, electrification of mobility currently appears to be the most promising driver for substituting fossil fuels within the transportation sector. Recently, battery electric vehicles (BEV) and plug-in hybrid vehicles (PEHV) experienced substantial growth in market shares in the automotive industry.[55] Figure 10.17 shows this development over the last decade. While in 2010, less than 20,000 (partially) electric vehicles were in stock globally, in 2021, the amount of BEVs and PHEVs exceeded 10 million and 5 million, respectively, with year-on-year growth of the global electric vehicle fleet commonly amounting to roughly 50%.

Moreover, the global electricity demand for electric passenger cars amounted to roughly 30 TWh in 2021, while it lay far below 0.1 TWh in 2010. The electrification of significant parts of transportation will, therefore, likely be a primary driver for the sector's transformation toward net zero greenhouse emissions. Already in 2035, major automotive markets (e.g., EU, US, China) are expected to be fully electric, equating to a global automotive industry disruption.[56]

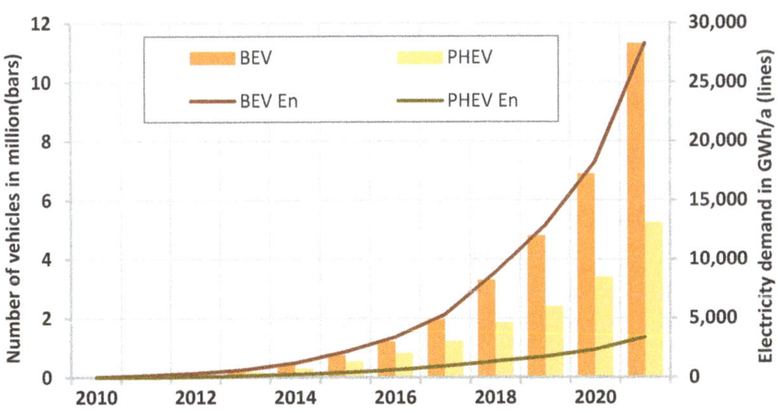

Figure 10.17: Development of BEV and PHEV stock as well as the associated electricity demand from 2010 to 2021.[57] (BEV: Battery electric vehicle, PHEV: Plug-in hybrid vehicle, En: Energy)

54 IEA: Transport Biofuels, Paris, https://www.iea.org/reports/transport-biofuels, 2021. Zuletzt geprüft: 21.06.2022.

55 IEA: Global EV Outlook 2022 - Securing supplies for an electric future, OECD.

56 IEA: Global EV Outlook 2022 - Securing supplies for an electric future, OECD.

57 IEA: Global EV Outlook 2022 - Securing supplies for an electric future, OECD.

It is expected that this disruptive transformation of electrification will support the future reduction of greenhouse gas emissions in the transportation sector. Currently, however, electricity plays an almost negligible role in transportation compared to fossil fuel consumption. In this regard, a disruptive transformation will likely be needed within the mobility sector to still achieve net zero greenhouse gas emissions by the middle of the 21st century.

Hydrogen and Synthetic Fuels

The utilization of hydrogen and other synthetic fuels can play a significant role in the future, especially for areas of the transportation sector, where direct electrification is hardly realizable, and available biofuel quantities are insufficient for a comprehensive substitution of fossil fuels (e.g., aviation, shipping).[58]

The practical demonstration of supplying and using hydrogen and other synthetic fuels for different mobility applications is currently the core of many research and industry projects. However, their impact on energy consumption within the transportation sector is currently negligible; neither the synthetic fuels nor the fleets able to use such fuels are available to create a relevant impact on the global energy and greenhouse gas balance. Additionally, from an energy system's perspective, the utilization of hydrogen and synthetic fuels is similar to electrification, as the basis for supplying these energy carriers environmentally friendly is, in general, renewable electricity.

Summary

As large parts of the transportation sector are likely to be based on electricity directly or indirectly, reducing greenhouse gas emissions within the mobility sector strongly depends on the carbon intensity of power generation. For electric vehicles, for example, the emissions related to power generation generally play an important role in the overall greenhouse gas emissions over their lifecycle.[59] Therefore, in addition to the heating sector, the transportation sector will increase the need for a substantial expansion of the utilization of renewable energies in the electricity sector. Furthermore, the exclusive use of biogenic fuels will not be an option as their potential is much too low to cover the global energy demand for mobility sustainably. Electricity (including electricity-based fuels) is, therefore, currently the only option for a comprehensive

58 See also section 19.

59 Conzade, J., Cornet, A., Hertzke, P., Hensley, R., Heuss, R., Möller, T., Schaufuss, P., Schenk, S., Tschiesner, A., Laufenberg, K. von: Why the automotive future is electric, 2021.

substitution of fossil fuels in the transportation sector, amplifying the need for an accelerated and substantial expansion of renewable electricity generation.

10.2.4 Intermediate Conclusion

The main conclusion that can be drawn from the status of energy supply and the immediate need for reducing greenhouse gas emissions on a global scale is the substantial importance of further developing electricity generation based on renewable energies. This is not solely true for the power sector but also for the overall energy system. Due to other energy sectors (i.e., heating and transportation) becoming increasingly entangled with and more reliant on the electricity grid, they enforce the accelerated development and increased expansion of renewables for electricity generation to fulfill their emission reduction necessities. The global energy system will thus likely become increasingly more reliant on electricity. Here, it can be strongly claimed that the underlying expansion of electricity generation will result in a drastic increase in power generation from wind and solar photovoltaics as they are cheaply and almost unrestrictedly expandable on a global scale.[60]

In conclusion, electricity will increasingly become the basis for the supply of the overall energy system. While the ongoing development of technology options for using electricity in heat or transportation (e.g., heat pumps, electric vehicles) is a critical aspect of the transformation toward an energy supply with net zero greenhouse gas emissions, the development and further expansion of renewable electricity in the power sector remains the most urgent and crucial objective.

10.3 Additional Necessities in an Increasingly Renewable Energy System – The way Forward

If the currently visible development trends and efforts are not intensified, reaching net zero greenhouse gas emissions will not be achievable until 2050.[61] Therefore, the most obvious necessity for the way forward toward a sustainable future is still the expansion of renewable energies throughout the entire energy system. In large parts, this will be driven by an expansion of renewable electricity generation, which in turn

60 Breyer, C., Khalili, S., Bogdanov, D., Ram, M., Oyewo, A.S., Aghahosseini, A., Gulagi, A., Solomon, A.A., Keiner, D., Lopez, G., Ostergaard, P.A., Lund, H., Mathiesen, B.V., Jacobson, M.Z., Victoria, M., Teske, S., Pregger, T., Fthenakis, V., Raugei, M., Holttinen, H., Bardi, U., Hoekstra, A., Sovacool, B.K.: On the History and Future of 100% Renewable Energy Systems Research, IEEE Access, 10, 2022, S. 78176–78218.

61 bp: Statistical Review of World Energy 2021 - 70th edition, 2021. IEA: World Energy Outlook 2021, 2021.

can enable the substitution of fossil fuels in different energy sectors and consequentially realize a reduction of greenhouse gas emissions.

The required and massive expansion of systems for a renewable energy supply – mainly for generating electricity – will fundamentally change energy systems globally. Some of the consequences emerging from this transformation, which partly demand further adaptations, will be briefly discussed in the following sections.

10.3.1 Electrification of Previously Non-Electric Energy Demands and Sectors

With solar and wind energy being market mature and still improving technologies, which enable harnessing renewable energies at reasonable prices, global energy systems will increasingly rely on electricity to fulfill existing energy demands. While combustion and thermal processes used to be the basis for providing energy in the past, this paradigm will shift during the ongoing energy transition. In the future, renewable electricity will become the basis for supplying energy to different energy sectors – and will thus become a new primary energy source. Moreover, this electrification will extend to numerous other sectors (e.g., industry, chemicals) and thus not only include the heat and transportation sectors discussed previously.

Guaranteeing the security of supply while enabling this transformation, electricity systems have to be strengthened considerably to secure the transmission of ever-larger amounts of electric energy. Besides the general expansion of transmission grids, this includes the improvement of interconnections to other energy systems to fulfill the electricity demands emerging in previously non-electric sectors. In the future, this so-called sector-coupling will lead to a more holistic interconnection of global energy systems, blurring the previously clear boundaries between the different energy demand sectors within an overall economy.

As discussed in previous sections, the utilization of electricity promises to support the reduction of greenhouse gas emissions in various fields of energy use that were previously operated almost exclusively without electricity. However, this demands new appliances since directly substituting fossil fuels with electricity is often inefficient or simply not possible. Therefore, it is generally essential that the transition begins as promptly as possible so that existing technical appliances are consistently replaced by modern technology, allowing further integration of (renewable) electricity.

On this basis, the development of the energy system is likely to lead to a “more electric world” in the coming years in which electricity from renewable sources evolves from a secondary energy carrier to a substantial primary energy source.

10.3.2 Decentralized Energy Systems

The increased reliance on electricity will intensify the electricity demand on every level of the energy system. Besides, e.g., industrial processes that shift from the utilization of fossil fuels to electrical energy from renewable sources on a larger scale, a considerable proportion of this electricity demand will be located in the residential sector and, thus, the lowest voltage level of the energy system. Consequently, this will lead to a decentralization of electricity demand as residential households and larger properties shift from, e.g., using gasoline and natural gas for mobility and heating purposes to local electricity utilization.

In parallel, the supply of electricity will become more small-scale. While large-scale power plants generated electricity in the past, in the future, these plants – generally powered by carbon-intensive fossil fuels – have to be decommissioned and replaced by renewable power generation. The latter – mainly solar photovoltaics and wind power systems – are generally much smaller in terms of single systems' installed capacities than conventional power plants. This development will likely result in a drastic increase in the number of energy facilities, which will be distributed much more broadly. Figure 10.18 shows this decentralization trend schematically.

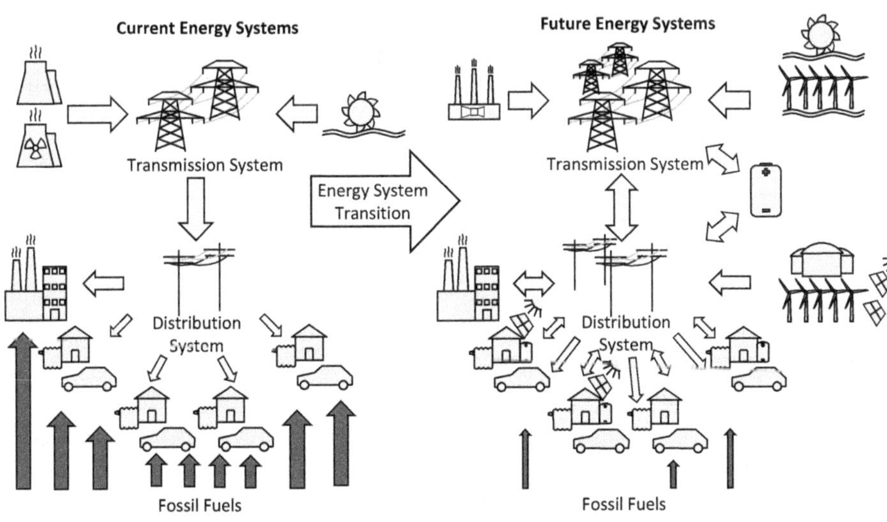

Figure 10.18: Decentralization of Electricity Systems (own illustration according to[62]).

62 Böttcher, J.: Green Banking - Realizing renewable energy projects, De Gruyter Oldenbourg, Berlin, Boston, 2020.

Overall, energy systems will become increasingly decentralized during the transition toward net neutral greenhouse gas emissions. Meanwhile, the operation of such energy systems is likely becoming ever more complex, as much more parts have to be controlled and organized. Consequently, this transition will intensify the reliance on different information and communication (ICT) systems that help organize system operation and energy flows (e.g., precisely forecasting power output from volatile renewables and flexibly adapting demand).

10.3.3 Volatile energy Supply and Demand Flexibilization

Renewable electricity will become an essential energy source for many sectors and applications. However, with wind and solar power being the most promising technologies for renewable electricity generation, energy will increasingly be available volatilely depending on the current availability of sufficiently strong winds and sufficient solar irradiation. This volatility calls for a change in the use of energy.

While in the past, consumers depended on an almost unrestricted availability of fuels and electricity, in the future they have to fulfill a more active part in global energy systems. As a result, demand has to act more flexibly and align with the fluctuating energy supply as much as possible to consume electricity when it is most abundantly available and thus the cheapest.

Examples of concepts for achieving such flexible demand (including flexible cross-sectoral demand) are the following:

- **Demand Side Management.** The active control of electricity demand is called demand-side management. Some appliances in the industry (e.g., electric arc blast furnaces) and private households (e.g., dishwashers, laundry machines) offer the potential to be operated more flexibly and realize a shift of demand to periods with higher availability of fluctuating renewable electricity. The consequentially improved match between supply and demand decreases the need for controllable power or storage capacities. Besides technical limitations, the overall potential of such demand flexibilization, however, strongly depends on the possible revenue respectively savings for shifting operations. Currently, in many markets, some regulations hinder such flexible operation (e.g., increased grid fees once demand patterns deviate from a base load profile). Nevertheless, if market frameworks are suitable, demand side management could help aligning demand with fluctuating power supply and thus ease the integration of volatile electricity generation.
- **Intelligent charging of Electric Vehicles.** Intelligently controlling the charging process of electric vehicles constitutes a particular form of (cross-sectoral) demand-side management. In contrast to an inflexible charging process at constant power levels, the intelligent charging of electric vehicles or vehicle fleets would aim at aligning the power demand of electric vehicles most optimally with the output of solar and wind power. Besides the availability of electricity, this charg-

ing management has to incorporate the expected demand for minimum charging states after defined periods (e.g., via forecasting the driver's mobility needs). Operationally, vehicle charging can be oriented on wholesale electricity prices, which are commonly low during high solar and wind feed-in periods. Customers can profit from low power prices for charging while simultaneously supporting the integration of volatile renewable electricity.

- **Flexible operation of heat pumps.** Similar to the charging of electric vehicles, the operation of heat pumps can be performed more flexibly to improve the alignment with the availability of renewable electricity. This flexibilization stems from exploiting (generally cheap) thermal storage capacities for a more power-system-oriented heat pump operation. Optimally managed thermal storage capacities (e.g., buffer storages, building masses) can bridge periods of low availability from renewable energies. Through this, power demand for heat pump operation can be oriented more strongly on volatile power supply while simultaneously meeting the local heat demand securely, e.g., guaranteeing sufficiently high room temperatures. This partially shifts the storage demand for balancing out volatile generation to the heating sector, where storing energy is generally feasible at much lower costs. However, this flexibilization is limited. For example, when solar and wind power are not available to a sufficient extent over longer periods with low ambient temperatures, heat pumps have to operate regardless of the availability of renewable electricity. Otherwise, the local heat supply could be compromised. Flexibilization is thus a secondary goal for heat pump operation, which is only available if the primary goal of residential heating (i.e., achieving certain room temperatures and providing hot water) is guaranteed. Nevertheless, despite these limitations, a flexible operation of heat pumps can improve the integration of renewable electricity compared to an inflexible heat pump operation while simultaneously reducing carbon emissions in the heat sector.[63]
- **Direct electric Power-to-Heat.** In addition to heat pumps, the direct electrical supply of heat (i.e., power-to-heat) can help integrate an electricity surplus during periods of especially high solar or wind feed-in. This can generally be realized on a larger scale (e.g., in district heating systems) and the residential level (e.g., in existing thermal storages). A benefit of such implementation is the complete additionality of the associated electricity consumption, exclusively when solar and wind power would otherwise stay unutilized. Another advantage of direct power-to-heat is the generally low specific investment cost for the required technical

63 Ruhnau, O., Hirth, L., Praktiknjo, A.: Heating with wind: Economics of heat pumps and variable renewables, Energy Economics, 92, 2020, S. 104967.

equipment, enabling an economic operation even in a few hours with a corresponding surplus of electricity per year.[64]

All of these and similar further concepts are generally in need of some form of ICT-system that manages supply and demand alignment via an intelligent control of the described flexibilization measures. Therefore, digitalization will likely be another driver for increased implementation and reliance on ICT systems.[65] Overall, however, this also increases the demand for cyber security so that corresponding ICT systems do not constitute a vulnerability in the energy system.

When renewables become ever more dominant, however, a flexible demand will likely not be sufficient to guarantee a secure energy supply based on volatile wind and solar power. Here, energy storage will become increasingly necessary to secure an energy supply even during low solar and wind power output.

10.3.4 Energy Storage and Hydrogen

Since solar irradiation and sufficiently strong winds – and thus electricity from solar and wind power – are sometimes absent simultaneously, an energy supply system increasingly reliant on these energy sources is in dire need of a backup supply. Guaranteeing energy security, therefore, demands (additional) energy storage.

Storage systems are categorized in various ways: for example, based on the realizable storage capacity (i.e., realizable with reasonable cost) or the duration over which energy can be stored, without losses rising excessively.[66] Batteries, for example, primarily address a diurnal shift of energy, complementing demand-side flexibility, which can potentially address a similar temporal range. These storage systems, combined with a more flexible demand can already enable a completely renewable energy supply in periods with periodical surplus and deficit of renewable power production (e.g. photovoltaics in the summer).

However, seasonal storage is needed for longer periodicity (e.g., seasonal differences). Due to its ability to store large amounts of energy over long times, hydrogen production based on renewable electricity is a promising option. Green hydrogen (i.e., hydrogen generated via electrolysis fed with renewable electricity) can be stored in large amounts (e.g., in underground salt caverns) and secure energy supply during periods of little immediate power production from renewables. As direct utilization of electricity is the most efficient option, the use of hydrogen for power generation should, however, be limited to

64 Lange, J.: Bereitstellung von Flexibilität durch am Strommarkt orientiert betriebene Heizpatronen in Endkundenheizsystemen, Dissertation, Verlag Dr. Kovač.

65 See also section 3.4.

66 Kaltschmitt, M., Streicher, W., Wiese, A.: Erneuerbare Energien - Systemtechnik, Wirtschaftlichkeit, Umweltaspekte, 6th ed., Springer Vieweg, Berlin, Heidelberg, 2020.

a necessary minimum. The production of hydrogen and a potentially needed generation of electricity in the power sector generally show rather high losses and thus increase the overall need for renewable electricity generation and the system costs.

Besides the storage systems that primarily address a temporal energy shift toward demand, an increasingly renewable energy system with fewer conventional power plants needs further storage systems that take over the tasks previously realized by fossil power plants. These system services include, inter alia, the provision of momentary reserves and the guarantee of voltage quality. The respective storage systems need to be highly dynamic and supply large amounts of power but generally do not need to store large amounts of energy.[67]

Hydrogen, as a particular form of energy storage, can also act as an enabler for substituting fossil fuels in other sectors (e.g., utilization as a feedstock in the industry). Despite that, it is commonly treated as a solution for especially hard-to-abate sectors and energy demands, which cannot easily be electrified (e.g., aviation, shipping). With renewable electricity increasingly becoming a primary energy source, the immediate use of electric power, without any further transformation processes and thus losses, generally enables the most effective substitution of fossil fuels. If valid alternatives are available (e.g., in individual passenger mobility), using hydrogen will increase overall energy losses and thereby call for higher overall installed capacities of renewable power generation in a net zero greenhouse gas energy system. Therefore, utilization of cost-intensive hydrogen (or hydrogen derivatives like power-to-liquid fuels) should be limited to areas where its use constitutes the most cost-effective option (resp. areas devoid of alternatives) and should still be treated as a valuable resource.[68]

Overall, it is likely that a comprehensive substitution of fossil fuels will ask for a multitude of different technical solutions. Thus, hydrogen will likely not be the only way to reduce greenhouse gas emissions of hard-to-abate sectors or applications. In addition, new technology options (e.g., for seasonal energy storage) might emerge during the complete transition to renewable energies.

10.3.5 International Trade of Renewable Energy

Renewable energy – in the form of hydrogen or its derivatives (e.g., ammonia, methanol) – will likely be traded internationally in the future. The basis for this is already being worked out, with various countries entering into trading partnerships for such

67 Kaltschmitt, M., Streicher, W., Wiese, A.: Erneuerbare Energien - Systemtechnik, Wirtschaftlichkeit, Umweltaspekte, 6th ed., Springer Vieweg, Berlin, Heidelberg, 2020

68 Ueckerdt, F., Bauer, C., Dirnaichner, A., Everall, J., Sacchi, R., Luderer, G.: Potential and risks of hydrogen-based e-fuels in climate change mitigation, Nature Climate Change, 11, 2021, 5, S. 384–393. Plötz, P.: Hydrogen technology is unlikely to play a major role in sustainable road transport, Nature Electronics, 5, 2022, 1, S. 8–10.

renewable energy sources. In this context, a heterogeneous development is expected, as the optimal transportation scheme often depends on the entire supply chain, from the synthesis of the energy sources to the specific mode of utilization.[69]

Figure 10.19 illustrates this diversity of possible supply chains. Besides the international transportation itself, the local distribution, as well as the way of reconditioning, play a major role when assessing overall cost-optimal configurations.

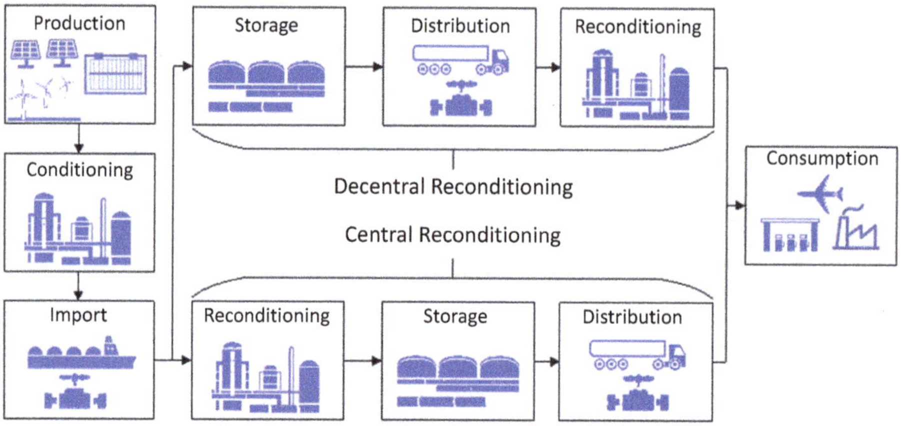

Figure 10.19: Supply chain for international trade of Renewable energy carriers (own illustration according to).

In conclusion, it is expected that besides direct electrification, renewable energy carriers will play an important role in various areas of the energy system in different forms (i.e., energy carriers like hydrogen, ammonia, and methanol as well as aggregate states like liquid or gaseous hydrogen). As a result, multiple different supply chains may emerge. Overall, this will likely make global energy systems more diverse and complex.

10.3.6 Market and Regulatory Framework

In order to support the profound transformation of global energy systems adequate market and regulatory framework conditions are required. These have to ensure that less carbon-intensive supply options increasingly and consistently become the more economically beneficial alternative. In this regard, the main instrument is the pricing of CO_2 emissions (e.g., via a carbon tax or cap-and-trade systems).[70] A globally imple-

69 Sens, L., Neuling, U., Wilbrand, K., Kaltschmitt, M.: Conditioned hydrogen for a green hydrogen supply for heavy duty-vehicles in 2030 and 2050 – A techno-economic well-to-tank assessment of various supply chains, International Journal of Hydrogen Energy, 2022.

70 See also section 3.2.

mented system to account for environmental and climate damages caused by greenhouse gas emissions would be most favorable, as it would eliminate the current market distortions caused by neglecting the consequential costs of CO_2 emissions. However, whether such a system will be realized globally must at least be questioned. Until the costs of greenhouse gas emissions are integrated into markets on a global scale, carbon leakage remains a serious issue. To address this, economically stronger regions could implement import tariffs based on carbon intensity to achieve carbon pricing, at least in their domestic market, without compromising global competitiveness (e.g., the EU's carbon border adjustment mechanism).

Besides the overarching goal to design less distorted markets, there are further market and regulatory aspects that should be overcome on a smaller scale. For example, with the energy system becoming more and more coupled respectively interconnected, it is becoming increasingly important to realize a level playing field between energy carriers, where this is not already the case. If, for example, there is different taxation for power consumption than for utilization of natural gas, it becomes more challenging to implement heat pumps in large quantities. This level playing field will be essential to boost the development of more environmentally friendly technology during the energy system transformation by enabling fair competition between different energy sources. Another example of unfavorable regulatory framework conditions is the coupling of grid fees to the evenness of power demand. It is currently realized in different countries (e.g., Germany) and hinders the flexibilization of demand by putting an indirect penalty on a flexible consumption oriented more strongly on the availability of fluctuating renewable electricity.

Furthermore, various industries are likely to transform profoundly (e.g. automotive industry[71]); here suitable policies are needed to start this transformation early on and give guidance and planning security to businesses directly affected by the change.

Overall, it is highly relevant that markets globally give stable and distinct incentives for further investment in the expansion of power generation from renewable energies, the flexibilization of demand, and the implementation of further storage capacities. Where these investment incentives are not reliably derivable from market prices (e.g., for power generation) due to volatile markets (i.e., markets with little foreseeability of long-term price development), further regulatory framework definitions have to ensure adequate development. If, for example, electrolyzers for the generation of green hydrogen are only built when hydrogen demand rises substantially, the supply will fall short of the demand and slow down the overall transition toward a comprehensive utilization of renewable energies. Here, adequate market and regula-

71 IEA: Global EV Outlook 2022 - Securing supplies for an electric future, OECD.

tory framework conditions must secure a steady and consistent transformation towards a net zero greenhouse gas emission energy supply.

10.4 Conclusion

Renewable energies experienced a tremendous development over the recent decades. Driven by increased awareness of the necessity to reduce greenhouse gas emissions, especially the power sector saw an immense expansion of electricity generation based on renewables. Initially, this development was strongly driven by the policy decisions of industrialized countries. However, based on cost reductions as a consequence of this expansion, the development is now largely economically driven, as power production from renewables often constitutes the cheapest option to supply electricity. In this course, photovoltaics and wind power emerged as the most promising technologies; their development has advanced to global proportions. Installed wind power capacities grew more than 11-fold over the last 15 years; an even stronger growth has been realized for solar photovoltaics, which increased 12-fold over the last decade.

Despite this unprecedented growth, renewable power generation still plays a subordinate role in electricity supply systems globally. While hydroelectric power generation made up a considerable share during the past, it still contributed much less to global power supply than fossil fuels like coal and natural gas – wind power and photovoltaics, however, only make up roughly 10% of global electricity generation even after the tremendous growth in recent years.

Looking at other energy sectors (e.g., heat, transportation), shares of renewables are even lower. For example, heat generation still relies on fossil sources for more than 60% of overall supplied amounts, and with traditional biomass (i.e., biomass burned for cooking and heating purposes often unsustainably and unsafely) making up roughly 25%, less than 15% of global thermal energy demand is supplied by sustainable, modern renewable energies. In transportation, the shares of renewable energies look even worse, with only around 2% of globally consumed fuels coming from renewable sources.

These relatively low contributions of renewable energies to global energy supply with ever-increasing demand levels result in increasing global greenhouse gas emissions. However, the latter have to decrease drastically to slow down and eventually stop global warming.

Increased awareness for the issue of climate change and the need for climate protection has, however, driven not only the development of renewable power generation but also the technology development for the substitution of fossil fuels in the heating and transportation sector. While battery electric vehicles grew the strongest in the automotive industry, the deployment of heat pumps is actively pushed in the building and heating sector. These technologies, however, need electricity to supply the demand previously supplied by fossil fuels – and thereby rely on reducing carbon intensity in the electricity sector to achieve a reduction of greenhouse gas emissions in their specific sector.

In order to achieve a carbon-neutral energy future, all of this results in a drastically intensified need to expand power generation based on renewable energies. Moreover, with wind power and photovoltaics being the most cost-effective options, their capacities will have to increase tremendously globally. As a result, net zero greenhouse gas emissions development scenarios foresee a further increase of wind power and photovoltaics by a factor of 12 and 20 respectively by 2050.

In the course of this development, global energy supply systems will change fundamentally. While fossil fuels have been the foundation of energy supply systems in the past, electricity from renewable energies will increasingly become a primary energy source in the future, which in turn will be converted into different forms of energy and transferred to different sectors. In addition to expanding power generation capacities, this requires a massive upgrade of the power supply system. This includes strengthening transport and distribution structures to ensure a reliable supply of both the power sector itself as well as other sectors connected to the electricity system.

Furthermore, new challenges will arise as power supply characteristics shift considerably due to an increased reliance on photovoltaics and wind power. As their electricity generation is weather-dependent, volatile, and fluctuating, current supply concepts must be reconsidered. While power supply was oriented on rather static and inflexible power consumption in the past, the demand side will have to play a much more significant role in the future. Flexible consumers will have to adapt to the availability of electricity from wind power and photovoltaics, and storage systems will need to be implemented to step in when demand flexibility is exhausted. In this regard, hydrogen will likely become an important option for seasonal storage – but the technological diversity will also increase, potentially offering new solutions in the future. Besides this technological development, market and regulatory framework conditions must be adapted to stimulate and secure the system transition.

The switch to renewable energies will comprehensively transform current energy supply systems globally. Through the development of renewables and the necessary accompanying changes, energy systems will become increasingly complex and more challenging to operate. However, if the existing hurdles are overcome, renewable energies will enable a sustainable future with net zero greenhouse gas emissions.

Bibliography

Böttcher, J.: *Green Banking – Realizing renewable energy projects*, De Gruyter Oldenbourg, Berlin, Boston, 2020

bp: Statistical Review of World Energy 2021 – 70th edition, 2021

bp: Statistical Review of World Energy 2022 – 71st edition, 2022

Breyer, C., Khalili, S., Bogdanov, D., Ram, M., Oyewo, A.S., Aghahosseini, A., Gulagi, A., Solomon, A.A., Keiner, D., Lopez, G., Ostergaard, P.A., Lund, H., Mathiesen, B.V., Jacobson, M.Z., Victoria, M., Teske, S., Pregger, T., Fthenakis, V., Raugei, M., Holttinen, H., Bardi, U., Hoekstra, A., Sovacool, B.K.: On the History and Future of 100% Renewable Energy Systems Research, IEEE Access, 10, 2022, S. 78176–78218

Conzade, J., Cornet, A., Hertzke, P., Hensley, R., Heuss, R., Möller, T., Schaufuss, P., Schenk, S., Tschiesner, A., Laufenberg, K. von: Why the automotive future is electric, 2021

Ember: Data Explorer, https://ember-climate.org/data/data-explorer/, 2022. Zuletzt geprüft: 24.08.2022

Ember: Global Electricity Review 2022, https://ember-climate.org/insights/research/global-electricity-review-2022/, 2022. Zuletzt geprüft: 13.06.2022

Global Wind Energy Council, GWEC: Global Wind Report 2022, 2022

IEA: Global Energy Review 2021, Paris, https://www.iea.org/reports/global-energy-review-2021, 2021. Zuletzt geprüft: 22.06.2021

IEA: Global EV Outlook 2022 – Securing supplies for an electric future, OECD

IEA: Renewables 2021 – Analysis and forecast to 2026, 2021

IEA: Transport Biofuels, Paris, https://www.iea.org/reports/transport-biofuels, 2021. Zuletzt geprüft: 21.06.2022

IEA: World Energy Outlook 2021, 2021

IEA-ETSAP and IRENA: Wind Power – Technology Brief, 2016

Fraunhofer ISE: Photovoltaics Report, Freiburg, 2022

International Energy Agency: Hydropower Special Market Report – Analysis and forecast to 2030, 2021

International Energy Agency, IEA: Electricity Market Report, 2021

IPCC: Climate Change 2021: The Physical Science Basis. Contribution of Working Group I to the Sixth Assessment Report of the Intergovernmental Panel on Climate Change – Summary for Policymakers, 2021

IRENA: Future of Wind – Deployment, investment, technology, grid integration and socio-economic aspects, 2019

IRENA: Renewable Power Generation Costs in 2020, Abu Dhabi, 2021

IRENA: Renewable Power Generation Costs in 2021, Abu Dhabi, 2022

IRENA: Renewable Capacity Statistics 2022, Abu Dhabi, 2022

IRENA: Renewable Capacity Statistics 2017, Abu Dhabi, 2017

Kaltschmitt, M., Streicher, W., Wiese, A.: Erneuerbare Energien – Systemtechnik, Wirtschaftlichkeit, Umweltaspekte, 6th ed., Springer Vieweg, Berlin, Heidelberg, 2020

Lange, J.: Bereitstellung von Flexibilität durch am Strommarkt orientiert betriebene Heizpatronen in Endkundenheizsystemen, Dissertation, Verlag Dr. Kovač

Our World in Data: Installed global renewable energy capacity by technology – Installed global renewable energy capacity in megawatts (MW) by technology (hydropower, solar, wind, biomass, marine and geothermal), https://ourworldindata.org/grapher/installed-global-renewable-energy-capacity-by-technology. Zuletzt geprüft: 05.10.2022

Our World in Data: Installed solar energy capacity – Cumulative installed solar capacity, measured in gigawatts (GW)., https://ourworldindata.org/grapher/installed-solar-pv-capacity. Zuletzt geprüft: 16.06.2022

Plötz, P.: Hydrogen technology is unlikely to play a major role in sustainable road transport, *Nature Electronics*, 5, 2022, 1, S. 8–10

Ritchie, H., Roser, M., Rosado Pablo: CO_2 and Greenhouse Gas Emissions – Published online at OurWorldInData.org, https://ourworldindata.org/co2-and-other-greenhouse-gas-emissions, 2020. Zuletzt geprüft: 24.08.2022

Ruhnau, O., Hirth, L., Praktiknjo, A.: Heating with wind: Economics of heat pumps and variable renewables, *Energy Economics*, 92, 2020, S. 104967

Sens, L., Neuling, U., Wilbrand, K., Kaltschmitt, M.: Conditioned hydrogen for a green hydrogen supply for heavy duty-vehicles in 2030 and 2050 – A techno-economic well-to-tank assessment of various supply chains, *International Journal of Hydrogen Energy*, 2022

Siemens Gamesa Renewable Energy: SG 14–222 DD – Offshore wind turbine, https://www.siemensgamesa.com/products-and-services/offshore/wind-turbine-sg-14-222-dd, 2022. Zuletzt geprüft: 04.11.2022

Simon, F.: EU Parliament groups rally behind plans to end biomass subsidies, https://www.euractiv.com/section/biomass/news/eu-parliament-groups-rally-behind-plans-to-end-biomass-subsidies/, 2022. Zuletzt geprüft: 05.10.2022

Ueckerdt, F., Bauer, C., Dirnaichner, A., Everall, J., Sacchi, R., Luderer, G.: Potential and risks of hydrogen-based e-fuels in climate change mitigation, *Nature Climate Change*, 11, 2021, 5, S. 384–393

United Nations: The Paris Agreement, Paris, 2015

U. S. Department of Energy: Offshore Wind Market Report: 2021 Edition, 2021

Vestas Wind Systems A/S: V236-15.0 MW™, https://www.vestas.com/en/products/offshore/V236-15MW/V236-15MW, 2022. Zuletzt geprüft:

World Meteorological Organization (WMO): State of the Global Climate 2020 – Unpacking the indicators, Genf, 2021

Oliver Opel, Jörg Böttcher, Darius Bonk, Jim Lukas Armbruster and Marlies Wiegand

11 Heat Pumps

11.1 Introduction

Heat pumps are the first option today for heating and cooling of new buildings and of growing importance for existing buildings. Air-water and ground-source heat pumps have been used for decades now for heating and cooling of buildings, especially in Scandinavia, the Netherlands and Belgium, but also Austria, Switzerland, Germany and of course Asia, the United States and everywhere where cooling equipment and air conditioning is widely abundant and used also for heating purposes. Market shares in Germany have been rising significantly over the past decade and heat pumps – besides e-mobility – are a key technology in lowering end energy demands and for electrification, thus enabling primary and final energy savings, the energy turnaround and decarbonisation of the heating and cooling sector.

However, existing district heating networks are still mainly based on fossil fuels and – due to relatively high temperatures – sometimes regarded more as a problem for decarbonisation than as a solution. So, how could these grids become decarbonised? Which technical solutions exist – and which constraints? This chapter covers heat pumps at the megawatt-scale – technology, coolants, heat sources and grid integration as well as regulatory aspects – to provide insight and practical information.

11.2 Technology

Heat pumps use mechanical compression of different media (left-handed Carnot process), leading to a higher temperature in the compressed medium compared to the expanded medium. Therefore, they are able to convert low-temperature heat into higher temperature heat, using exergy, most often electricity.

Most heat pumps furthermore use the evaporation of a working medium to absorb heat from the surrounding. The compressor compresses the vapor and the medium condensates at a higher pressure and therefore higher temperature, releasing heat. Thus, low-grade heat from a variety of environmental heat sources but also excess heat can be utilized to cover heat demands at different temperature levels.

As heat pumps use less energy for vapor compression compared to the amount of heat generated, a Coefficient of Performance (COP) is defined as:

$$COP = \frac{Q_{th.}\,(out)}{Q_{el.}\,(in)} \tag{1}$$

https://doi.org/10.1515/9783110784299-011

The higher the COP, the lower the demand of electrical energy given a certain heat demand covered: The electrical energy needed is:

$$P_{el.} = \frac{1}{COP} \cdot P_{th.} \tag{2}$$

As electrical energy demand rises with the pressure delta between evaporation and condensation due to higher forces during mechanical compression, the COP depends on the temperature delta between the (environmental) heat source and the heat sink in a non-linear relationship. The lower the temperature difference, the higher the COP rises. And the higher the temperature difference (temperature uplift), the lower the COP and thus higher electrical demand in relation to thermal power. Using heat pumps, therefore, temperatures of sources and sinks are a matter of high importance with regard to efficiency.

11.2.1 Basics of Heating Applications and District Heating

Thus, higher temperature at the environmental heat source and lower temperature of the heating network or application are favorable for heat supply using heat pumps. This fact, alongside the integration of thermal storage, which depends on the temperature difference at which the storage operates, explains the trend of lowering temperatures in heating networks (Lund et al. 2014):[1] The first generation of heating networks used steam. Second generation (2nd Generation) grids are using water, however during peak demand, the water is heated to over 100°C, often 120°C, to cover high heat consumption at limiting mass flow. Since

$$\dot{Q} = \dot{m} \cdot c_p \cdot \Delta T, \tag{3}$$

to maximize capacity of the heating network $\dot{Q}$, either the mass flow $\dot{m}$ has to be maximized or ΔT, the temperature difference between supply and return flow. As pressure loss for moving the water through the piping network depends on velocity squared, velocity is technically and economically limited to 1–2 m/s and therefore mass flow is limited depending on the piping dimensions. Higher velocities would result in high cost for electricity for pumping and could also result in higher corrosion rates through erosion of passive layers. Thus often, such temperatures are needed to cover peak loads in existing grids.

Lowering return flow temperature via cascaded use with decentralized lower temperature networks connected to the main network is also advantageous since ΔT

1 Lund, H., Mathiesen, B. V., Connolly, D., & Østergaard, P. A. (2014). Renewable energy systems—A smart energy systems approach to the choice and modelling of 100 % renewable solutions. *Chemical Engineering Transactions*, 39, 1–6. https://doi.org/10.3303/CET1439001.

and thus heating capacity of the grid $\dot{Q}$ is maximized and also condensation temperature of the heat pump is lowered. At a given supply temperature, lowering the return flow temperature is equivalent to lowering the overall temperature of the network, however less effective compared to lower supply temperatures in general.

Heat supply for 2nd generation heating networks with heat pumps is possible, but depends on the temperature of the heat source, the working medium and the technology. Some heat pumps provide a hot gas bypass to be able to heat up the water over the condensation temperature. Other heat pumps use the supercritical state of the medium to compress it to higher pressures and thus reach higher temperatures. A third option is to use electrical heaters or biomass / renewable gas vessels for peak demand. However, all of these technologies lower the efficiency and thus increase the demand for electrical energy or (in case of fossil oil and gas) emit CO_2. Thus, heat pumps are generally to be more suitable for lower temperatures.

Generation 3 networks use temperatures up to 95°C, which are possible to reach for two-stage heat pumps even from low temperature sources at $COP > 2$.

Generation 4 grids operate at temperatures of 55–65°C which are still suitable for hot water supply even in Germany, where hot tap water must be maintained above 55°C to ensure water hygiene. (Note that tap water is not chlorinated in Germany).

Most buildings do not need temperatures above 65°C for heating and production of hot tap water, so lower temperatures in heating grids are mostly limited by the overall heat demand of the buildings connected to the grid, following the limits given by equation 3. Economically, however, higher heat demand converts to higher income for the grid operator to cover operational cost and investment in the heating grid.

5th generation heating networks are “cold networks” or anergie networks. They do not supply heat at temperatures suitable for heating or tap water, but serve as a heat source for decentralized heat pumps. This technology is feasible for highly energy efficient buildings in connection with underground or ice thermal storage and can be used for integration of low-grade waste heat to provide better efficiency during cold winter times compared to air-water heat pump installations (Wirtz et al. 2022).

Most abundant in cities and districts and relevant for large scale heat pumps are generation 2–4 grids. This chapter will thus focus on heat pumps that are able to provide temperatures up to 95°C as a drop-in option for baseload in gen 2 networks and monovalent supply of gen 3 and 4 networks.

11.2.2 Working Media and Climate Change

Many gases can serve as working media for heat pumps. The first heat pumps used ammonia and propane. Later, chlorofluorocarbons (CFCs) were used due to their low risk of fire hazard and low toxicity, providing high levels of security in practice.

However, as it is widely known today, chlorofluorocarbons were later discovered to deplete stratospheric ozone and thus faded out and banned in 1989 in the UN by the Montreal Protocol. In 2009, all 197 united nations had agreed to the Montreal Protocol (Behringer et al. 2021).

After phase out, fluorinated gases replaced CFCs. Fluorinated gases (f-gases) also provided very high safety, low toxicity and low risk of fire, without depleting stratospheric ozone. CFCs and f-gases both provide lubrication for the compressors used in heat pumps.

However, like CFCs, fluorinated gases are very long-lived in the atmosphere and contribute to climate change. By now, already 10% of the radiative forcing attributes to synthetic fluorinated gases used in heat pumps and cooling equipment (IPCC 2014), as f-gases exhibit very high greenhouse warming potential of up to several thousand-fold of the warming potential of CO_2. Consequently, in 2014, the EU decided on the phase-out of f-gases. They are still in use, only f-gases with the highest global warming potential (GWP) were forbidden, but the market volumes are capped such that CO_2-equivalent emissions are reduced to 80% by 2030. To reduce the emissions, blends of different f-gases with similar thermodynamic properties are used as drop-in, "low-GWP" coolants with a GWP of < 500, besides natural and 4th-generation synthetic refrigerants.

The old natural refrigerants ammonia and propane, amongst others, provide a more sustainable alternative. Propane became common for air-water heat pumps for households recently. Risk of fire hazard is minimized by placement of the heat pump outside. Ammonia was still used for refrigeration in parallel to CFCs and f-gases and is still an option for refrigeration, but also large-scale heat pumps.

CO_2 is also used for refrigeration and high-temperature heat pumps using supercritical compression. Water can also be used for high temperatures. Also, synthetic alternatives with a lower lifetime of days instead of years were developed, for example HFO-1234ze and HFO-1234yf, two fluorinated olefines, with HFO-1234ze being the isomer with the lowest ecological risk associated. These working media are referred to as 4th-generation coolants. Those olefines provide $GWP < 1$, low risk of water contamination and low risk of fire. Fluorinated substances, however, even if exhibiting low lifetimes in the environment themselves, tend to build up breakdown-products and metabolites that could be very long-lived. A breakdown-product of high concern is trifluoroacetic acid (TFA) which is contaminating groundwater, surface and drinking water. Small changes in molecular structure can have significant effects on environmental fate and longevity of breakdown products. For example, HFO-1234yf breakdown alone (besides older coolants like R134a) produces TFA to a large extent, expected 2.5 kg/km^2 per year for Europe in 2050 (Luecken et al. 2010, Garry 2022), whereas HFO-1234ze breaks down almost without producing TFA (HFO-1234ze photolysis in atmosphere is regarded the main route of HFO-1234ze breakdown and does not produce TFA, however during hydrolysis, TFA could occur (Tian et al. 2023). Because of the very low water solubility and slow kinetics of HFO-1234ze hydrolysis compared to photolysis, the amount of TFA produced is regarded to be negligible by most authors (Behringer et al. 2021).

Nevertheless, per- and polyfluorinated alkyl substances (PFAS), the main class of fluorinated hydrocarbons and HFO, could be regulated in the future because of their longevity, leading to accumulation in the environment and food chain, and their build-up of potentially equally long-lived and harmful breakdown products.

11.2.3 Compressor Technology Used for Large Scale Heat Pumps

For compression, different types of compressors are feasible. Rotary piston compressors are common in refrigerators and smaller heat pumps. They are easily controlled by frequency modulation (EC drives) and thus very efficient during partial load conditions. As for heating and cooling purposes partial load is the prevalent condition, efficiency at partial load is most often critical.

For large scale heat pumps, three principles are most important: Turbo compressors, screw and scroll compressors, each having their advantages: Turbo compressors provide higher power density, reducing the space needed. Scroll and screw compressors, like rotary piston, perform better at partial loads and can be operated at speeds as low as 20%.

It is most important to notice that the choice of compressor technology also has implications on which working medium is feasible: Turbo compressors, a fluid machine, use centripetal forces accelerating the molecules and the kinetic energy of the compressed medium is converted into pressure. They are most efficient with gases with higher molecular weight, like fluorinated olefins.

Scroll and screw compressors do not rely on molecular weight. They reach higher efficiencies compared to turbo compressors, also with lower molecular weight gases like the natural coolants ammonia and propane. They also exhibit slightly higher cost. In 2024, for 5–10 Megawatts, costs were 450 €/kW for turbo and 550 €/kW for scroll compressor large scale heat pumps without electrical installation, foundation, housing and grid connection as well as connecting to the environmental heat source (Table 11.1). For a full installation, the cost will approximately double (for surface water heat pumps).

Table 11.1: Technology and cost of large scale heat pumps and direct electrical heaters (compiled by authors).

Technology	P	COP	Cost	Space	Coolant
Turbo-Compressor	Up to 25 MW	2,5–2,7	> 400 € / kW	100 m^2 / 10 MW	HFO-1234ze
Scroll-Compressor	Up to 13 MW	2,7–3	> 550 € / kWp	100 m^2 / 5 MW	R717, R290
Direct electrical heater	> 20 MW	1	Ca. 100 € / kW	10 m^2 / MW	–

11.3 Heat Sources for Large Scale Heat Pumps

As (environmental) heat sources, in principle any medium could possibly be used for extraction of heat to serve as a heat source for large-scale heat pumps.

Few options are common: Air, exhaust air, surface water, sewage water, ground water (including geothermal), borehole heat exchangers and drinking water (table 11.2). For most large-scale heat pumps, also if air is used, water (in some cases including antifreeze, depending on the application) is used as the primary heat source. In the case of air as a heat source, the water primary circuit is used to couple an air heat exchanger to the heat pump. Evaporation directly in the air heat exchanger, as it is often used with smaller heat pumps, is seldomly feasible for larger machines.

In the following sub-chapters, the environmental heat sources are described and discussed in detail.

11.3.1 Air

Air is universally abundant and easily accessible. However, the heat capacity of air (especially of cold and thus rather dry air in winter) with 1.0035 kJ/kgK and a low density of 1.293 kg/m^3 is rather low.

Therefore, for larger air source heat pumps, large volumetric flows are needed to extract enough heat. Also, temperature difference at the air heat exchanger is often higher compared with other environmental heat sources to keep the volumetric flows and energy for driving the electrical fans reasonable.

At 500 kW thermal power and a COP of 3, enough air has to be moved through the heat exchanger to provide 2/3 of thermal power, in this case 333,3 kW or $1.2 \cdot 10^6\ kJ/h$.

At a temperature difference of $\Delta T = 10K$ this resembles 119.5 t/h or 92.484 m^3/h of air per hour. Therefore, heat pumps with more than 750 kW thermal power seldomly use air as primary heat source.

As a consequence of the high temperature difference, evaporation temperature is low. At 5°C outside air temperature evaporation temperature is slightly above −5°C and the air exits the heat exchanger at −5°C.

It is a challenge to prevent a thermal short circuit during such operating conditions and keep noise emissions low. Often, air ducts and noise insulation are needed for larger scale air source heat pumps. Also, frequently at temperatures below 3–5°C, freezing of air moisture occurs, lowering thermal conductivity at the air heat exchanger. Thus, defrost cycles are needed to remove the ice and regenerate thermal coupling to the heat source medium. Both, the freezing of the air heat exchanger and the defrost cycles, lower the efficiency of the air-source heat pump.

11.3.2 Surface Water

Surface water includes rivers, lakes and sea water. Water has the advantage of much higher density $\rho = 1000\,kg/m^3$ and heat capacity c = 4.19 kJ/kgK compared to air, a roughly 4000-fold higher volumetric heat capacity. Therefore, volumetric flows of surface water needed for large-scale heat pumps are much lower in relation to thermal power needed for the installation. Also, a lower temperature difference is suitable.

The main disadvantage is the risk of freezing which limits the amount of heat extractable from surface water in winter, depending on the surface water temperature.

A 10 MW thermal power heat pump at COP=3 needs 6,667 kW or $24 * 10^6$ kJ/h from the primary heat source. At a temperature difference of $\Delta T = 3$ K, this resembles $1.9 * 10^6$ kg/h or 1,900 m^3/h of water, which is possible to transport using 700 mm diameter or 3 times 400 mm piping at 1.5 m/s velocity flow for supply and return, respectively. However, since the COP varies over the year due to lower source temperatures and higher supply and return temperatures of the heating grid in winter, the amount of heat extracted from the water also varies with regard to the heat supplied to the grid.

At a COP of 2 or lower, occurring in winter if high temperatures in the heating grid are needed and the temperatures of the sources are low, only 50% or even less of the supplied heat is extracted from the environmental heat source. Thus, at equal flow rates, ΔT can be lower at the heat source, allowing extraction of heat from surface water even at 2–3°C.

Nevertheless, whilst the volumetric flow needed for the primary heat source water even at low temperature differences of 1–3 K is much less compared to air, it is much higher than the mass flow in the district heating system, operating at 20 K temperature difference (spread) or above. Therefore, the heat pump should be located near to the source and the produced higher-exergy heat transported to the sink.

Surface water, like groundwater, although often not pumped and drained but reinfiltrated, most often could not be readily accessed and used because of competing ecological interests. Although extraction of heat is mostly regarded to be less adverse to the ecology of the respective water body compared to the very common use as a heat sink for cooling purposes permission is most often needed and risks of adverse ecological effects should be minimized.

Even the mere pumping of surface water could interfere with living organisms and should be carefully engineered to avoid suction of fish, insects and shellfish, for example. Permission is regularly needed according to regulations, however climate-friendly production of heat could be prioritized, as it is the case in Germany since 2023.

A more technical issue is the risk of corrosion and fouling in the heat exchanger, that needs to be built from resistant materials (most often stainless steel) and frequently cleaned by scraper techniques, for example.

To avoid leakage of coolant into the water, double wall heat exchangers should be used. An intermediate heat exchanger can be to avoid contact between the lake water and the evaporator of the heat pump, thus minimizing the build up of dirt and risk of contamination (Kammer 2018).

Surface water bodies in general regenerate incorporated amounts of heat via thermal inputs from the ground, groundwater inflow, inflow from surface water and solar irradiation. Even in winter, solar irradiation reaches 15–20 W/m^2 at a minimum (e.g. 17 W/m^2 in December in Germany), so large surface areas of lakes and rivers act as solar heat collectors. Heat flux from the ground can also be approximated in the range of 10 W/m^2, assuming temperatures of 10–15°C in 1 m depth in the sediment. Thus, besides flow and volume of the water body and inflow from surface water and groundwater inputs, surface area is an important factor in calculating capacities as environmental heat sources (Gaudard et al. 2017, Figure 11.1).

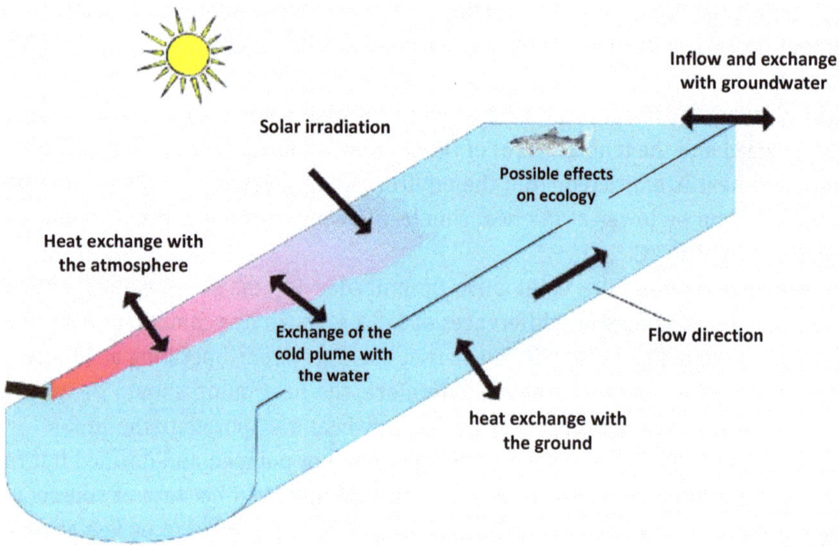

Figure 11.1: Thermal usage of rivers for extraction of heat (Gaudard et al. 2017).

11.3.3 Rivers

Rivers have the advantage of providing a more or less steady flow of water. The potential of extracted heat is directly proportional to the water mass flow. Typically, rivers provide thousands of m^3 per hour. They could potentially easily be cooled down to up to −1 K. However, with the limitation of the portion of total river water flow needed to be pumped through the heat exchanger.

Another concern is the cold plume, locally potentially affecting water and river bank ecology (Gaudard et al. 2017). However, at low ΔT in the heat exchanger, this effect should be negligible.

As a rough estimation, 10–50% of the heat corresponding to the river flow at 1 K temperature difference could easily be extracted from rivers, an equivalent of roughly 1 kW per 10 m^3 river flow per hour.

Generally, if surface water is used, the inflow should be designed carefully to avoid suction of organisms. For rivers it is advantageous if the inflow of the heat pump installation is slower than the river flow. This could be achieved by broadening of the inflow towards the river, if space is available. The inflow is then secured by a rake with ~ 1 cm spacing before the sand trap, narrowing of the inflow and pump sump.

The outflow can be designed as a diffuser to avoid crossflow and provide better mixing with the river water to avoid pronounced cold plumes.

11.3.4 Lakes

Lakes could potentially serve as primary heat sources for large scale heat pumps. Although many lakes have an inflow and outlet, the water body must more or less be understood as static. Therefore, the amount of heat that could be extracted during one year is limited by the volume of the water body. As a rough estimation, 1 kWh of heat could be extracted per m^3 of volume per year to keep the cooling below 1 K temperature difference, as heat is constantly supplied by irradiation and surroundings.

The water can be extracted by using an open- or closed-looped system. In an open-looped system the water is pumped from the lake and filtered. However, the filter needs to be cleaned regularly and the lake water could freeze at the intermediate heat exchanger in the winter, due to the fact that no antifreeze can be added (Schwinghammer 2012, Kammer 2018).

In contrast to this is the closed-loop system in which a heat exchanger is implemented directly into the lake. Thus, allowing the usage of antifreeze for the heat pump loop as well as removing the filter from the system. The direct implementation of the heat exchanger and any additional structures could negatively impact the lake (Kammer 2018).

The ecology of lakes depends on their stratigraphy that needs to be respected. The three layers of lakes are the epilimnion, the warmest and topmost layer, which is influenced by solar radiation and the wind, the metalimnion as separating layer and the hypolimnion as the coldest and most dense layer. In moderate temperature zones lakes go through a circulation in spring and fall. These layers and circulation should be considered if the lake is to be used as a heat source for large scale heat pumps, due to the fact that temperature fluctuations either shorten or prolong the time between the circulations, which could negatively affect the ecology (Gaudard et al. 2017).

Despite the aforementioned fact a reintroduction of cooled water, counteracts the higher temperature in the epilimnion layer caused by global warming. This could ideally be achieved by filling a heat reservoir in the summer and using the saved energy in the winter, for example.

11.3.5 Sea water

Sea water is highly available at the coast line, however accessibility could be an issue. Its use is not so much limited by the volume but by space available for the heat pump installation in the vicinity of district heating networks.

Often, water from a harbor basin would be used. In this case, the volume of the basin and exchange rates with the open sea should be taken into account to estimate heating potential. The tides often induce high water exchange to the harbor basin. However, induced crossflow from pumping water should be limited to not hinder original harbor operation (maximum 0.4 m/s).

A major difference to fresh water ecosystems is the salinity, which leads to higher risk of corrosion. Also, Microbially Induced Corrosion (MIC) can be expected especially in marine environments. Material properties and cleaning of the heat exchanger should reflect these higher requirements.

11.3.6 Sewage Water

Sewage water is very advantageous because of several reasons: Its use does not require extra permission, temperature is 10–15°C all year round like groundwater, but unlike groundwater, it is often easily accessible as effluent from water treatment plants in hundreds of m^3 per hour. The higher temperature is advantageous to the heat pump for better COP, also the water could be used with higher delta T up to 5 K, requiring lower amounts of pumped water (Pelda & Holler 2018).

The use of influent or heat exchangers in canalization often was less advantageous in first projects due to a number of reasons: Heat exchangers need a high surface area. Due to fouling, plate heat exchangers or tube heat exchangers that are common to the use of other water sources could not be used, but only flat heat exchangers at the bottom of the pipe or canal. Thus, heat extraction is limited and large installations are needed for comparable low heat extraction capability. A sufficient diameter for pipe or canal heat usage needs to be at least 600 mm. Lower than that the flow rate is insufficient. Such installations however can contribute to an ensemble of heat sources (Then et al. 2019).

Using treatment plant effluent would generally be a primary choice if located in vicinity of the heating network or not too far away, what could be the case often, however. As a very rough estimation, around 5–10% of the heating demand of a community could be covered by treatment plant effluent.

11.3.7 Drinking Water

Drinking water would resemble a readily available environmental heat source. Their potential capacity is comparable to sewage water. Heat exchangers could be added to raw water or drinking water pipelines, centralized at the drinking water plant or decentralized to supply heat for larger buildings or districts.

Since warming of drinking water is becoming a bigger and bigger problem, heat extracting from drinking water could exhibit positive effects on water hygiene, however, at least in Germany, the use as heating source is prohibited.

Care should be taken to use materials allowed for drinking water supply and heat exchangers that do not introduce large dead volume where biofilms could possibly form.

11.3.8 Ground Water and Geothermal

In absence of surface water, groundwater could be used. Its temperature is dependent on depth due to the geothermal gradient, such that higher temperatures could be extracted from deeper wells that however are more costly to develop (Ahrendts et al. 2023).

A single well could provide 50–200 m^3/h, enough for up to 1 MW of heating capacity for shallow aquifer use. Shallow groundwater provides temperatures of 10–15°C all year round making it possible to extract 5 K from the water.

Most often, a well duplet would be used for water extraction and reinfiltration. Thus, no permission is needed for groundwater extraction since the water is not drawn from the aquifer. However still, permission could be needed for exploration, opening, heat extraction and storage of heat, if the water is heated above the natural groundwater temperature at the site for integration of waste heat.

Deeper wells provide higher temperatures, around 30°C in 300–400 m depth, depending on the particular region. In higher depths of 1000-2000 m, around 60°C could be found and up to 120°C in even higher depths that allow direct heat use and sometimes even electricity production. However, permeability of the used aquifer is most important, since deeper aquifers are often more dense and provide lower yield (Holstenkamp et al. 2017). Also energy for pumping of the water needs to be taken into account, besides higher cost for drilling.

For use with large-scale heat pumps, shallow aquifers with high permeability and water yields are best suited. For high heat extraction rates, regeneration of the aquifer should be taken into account, to avoid cooling of the heat source over time. Low grade waste heat could be used for regeneration, for example from cooling applications, or low grade solar thermal, like from combined photovoltaic-thermal (PV-T) panels, water cooled PV, or low temperature solar heat exchangers.

At depths up to 100–500 m, also waste heat from higher temperature sources could be integrated and stored. Such regeneration of the heat source helps with efficiency of the heat pump and could also reduce needed pumping volume since the higher temperature source could be cooled down more than 5 K (Kuznik et al. 2021).

Care has to be taken with regard to water chemistry (iron and manganese content over 1–2 ppm) as well as high calcium carbonate and other possibly precipitating minerals, especially if heated significantly (Fleuchaus et al. 2020). Borehole and system maintenance is mandatory with regard to specific water chemical qualities present at the site.

11.3.9 Borehole Heat Exchangers

As an alternative to groundwater wells, for example in case of possible problems with water chemistry, borehole heat exchangers (BHE) are used. They provide a closed heat exchanger circuit made of crosslinked polyethylene (PE-X) that is cemented into 100–150 mm boreholes.

Dependent on the heat conductivity and water content of the ground, 20–80 W can be extracted per m. Such, a 100 m borehole is sufficient for 2–8 kW.

Thus for larger installations, typically hundreds of BHE have to be used. Regeneration of these shallow low temperature thermal storages is critical, providing similar means of potential waste heat usage compared to groundwater. Each borehole of 100 m depth is able to store around 5 MWh, at 5–20 K delta T.

Table 11.2: Environmental heat sources for large scale heat pumps – estimated flow rates (compiled by authors).

Heat Source	delta T	Flow for 1 MW thermal
Air	5–10 K	200.000 m^3/h
Surface Water	2–3 K	250 m^3/h
Groundwater	5 K	100–150 m^3/h
Sewage water	5 K	100–150 m^3/h
Drinking Water	5 K	100–150 m^3/h
borehole heat exchangers (BHE)	up to 10 K	125–250 BHE

11.4 Energy System Integration and Electricity Demand

11.4.1 Seasonality, Peak Load and Redundancy

The size of a heat pump installation needed to supply a given heating grid is highly dependent on seasonality of the heat demand. Seasonality means the difference between peak load (typically in cold winter times) and all-year-round baseload (typically due to hot water generation in buildings, swimming pools etc.). Obviously, seasonality is highly influenced by the climate. The colder the winter the higher the seasonality, but the longer the heating season, the lower the seasonality. Therefore, northern maritime influenced climates show lower seasonality than terrestrial climates with cold but comparably short winterly heating seasons.

The load is divided into weather-dependent and weather-independent load. The higher the weather-dependent load, which is directly connected to the size and insulation of the connected buildings, in relation to weather-independent base load, the higher the seasonality.

The seasonality can be expressed as yearly full-load-hours: The higher the full-load hours over the year, the lower the seasonality.

As with every asset, higher full-load hours over the year are advantageous in economic terms. Also, expenses for electrical grid connection and dimensions of the heating network needed in relation to heat consumption over the year are lower if seasonality is lower.

Typically, heating networks show yearly full-load hours of 2500 h/a. For planning of heat supply for heating networks, the load curve is typically sorted and displayed as yearly load duration. Figure 11.2 is showing a simplified example using monthly medium load curves (already including buffer storage).

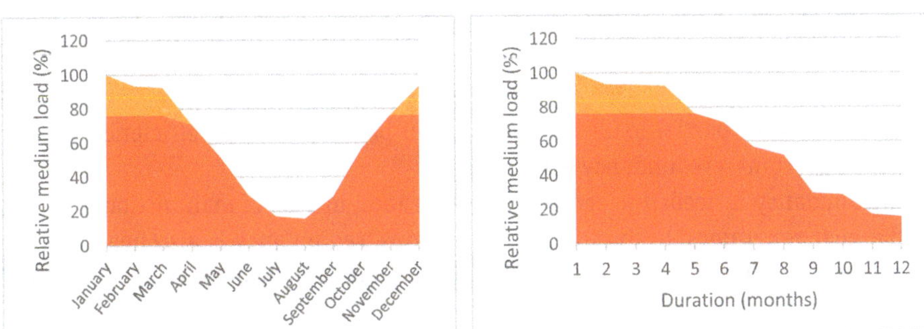

Figure 11.2: Load curve of a typical district heating network and load duration curve; figure created by author.

The supply of a heating network is thus frequently planned by dividing the load curve into base, medium and peak load. In Figure 11.2, baseload would be roughly 20% (hot drinking water), medium load is 20–75% and peak load is 75–100%, which is only used during the 3 winter months. Heat pumps in combination with buffer storages can be used to produce base and medium load at good efficiency. For peak load, biomass or (renewable) gas vessels or electrical direct heaters are frequently used.

Modelling of heating networks could be used to evaluate the optimal size of the heat pump. Figure 11.3 shows optimization of nominal thermal power output of the heat pump to reduce peak load demand. In this example, a designed thermal power of the heat pump of 75% of the monthly medium load (compare Figure 11.3), the peak load energy demand is reduced to < 5%. Even larger heat pumps, in this example, would not further reduce the peak load demand considerably.

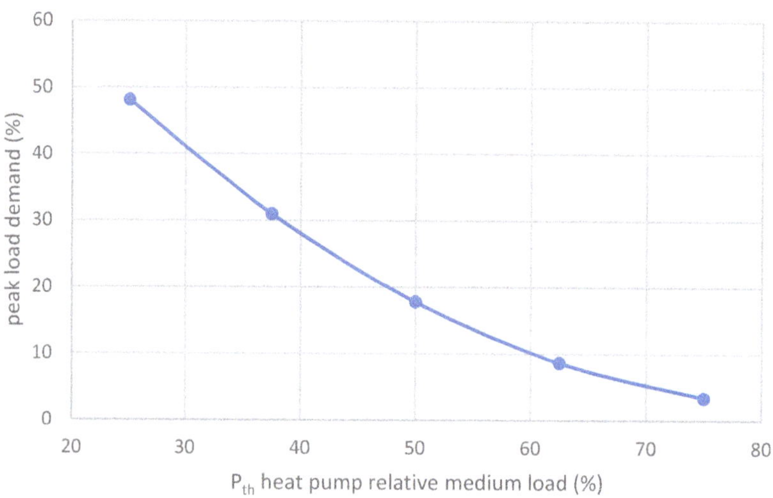

Figure 11.3: Optimal sizing of the heat pump.

It is often advantageous to divide the base, medium and peak load into 2 or more heat generators. Thus, partial load conditions could be met at better efficiencies and – equally important – redundancy is increased.

Redundancy is needed to secure heat generation in case of malfunction of a part of the heat generators. In the case of heat pumps, redundancy is also needed if the temperature of the environmental heat source is too low – especially in the case of surface water and air source heat pumps.

Depending on technology used for peak load and redundancy, these have also to be taken into account when calculating electrical grid connection needs. Direct electrical heaters, which could provide up to 100 MW heating capacity in small space, need equally powerful electrical grid connections to operate. As large heating networks of big cities could easily exceed gigawatts of peak power, appropriate solutions for peak load and redundancy have to be considered.

11.4.2 Buffer Storage

Buffer storage is an important part of every heating system and heating networks. It covers extreme peak loads, acts as hydraulic separator to enable efficient and controlled operation of heat pumps and peak load heaters, helps with redundancy and also covers extremely low load conditions that could not directly be met by partial load operation of the heat generators.

Buffer storages are designed to cover peak loads in the range of hours. At extreme peak load, the buffer storage alone should be able to meet the energy demand of the network for at least 15–30 min. In summer, the storage should be able to store heat production of the heat pumps of a few hours at lowest possible partial load to avoid switching and increased wear of the heat pumps.

Since temperatures and temperature spreads possibly used in the buffer storage are different in summer and winter, dimensioning should be evaluated in both cases.

As with the thermal power of the heat pumps, the size of the buffer storage could be optimized using modeling, as exemplarily shown in Figure 11.4. As can be seen, increasing buffer storage size will help reducing peak load demand to an extent.

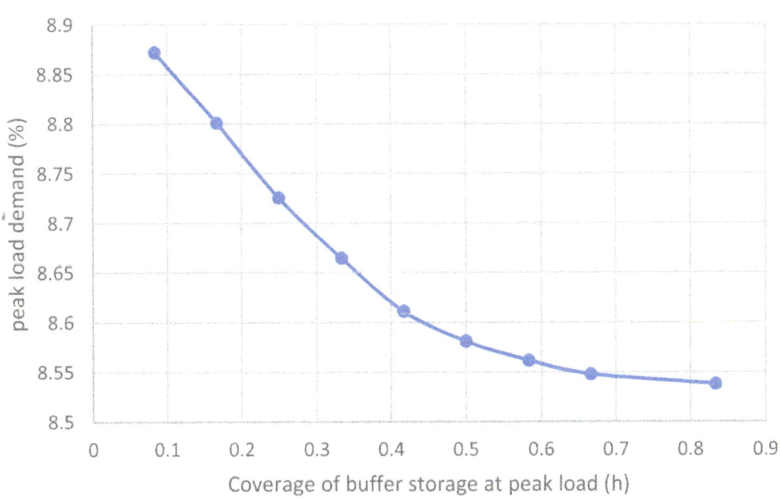

Figure 11.4: Optimization of buffer storage.

11.5 Electricity Demand of Heat Pumps

Whilst electricity demand is less critical compared to the appropriate environmental heat source, installations in multi-MW range also need high capacity electrical grid connections. However 10–20 MW is often possible at 10–20 kV grid level.

Besides the annual load curve and seasonal heating demand (compare Figure 11.2), the COP of the heat pump installation including electrical direct heating during peak load is relevant for the electricity demand. As the COP is dependent on temperatures of the environmental heat source as well as on supply and return flow temperatures of the sink or heating network, respectively, it also varies over the year.

Figure 11.5 shows the variable COP of a heat pump installation with an annual COP of 2.7, as can be achieved for 2nd or 3rd generation heating networks from surface water or 4th to 3rd generation network from air, using a two-stage large scale heat pump with scroll compressor and direct electrical heater as peak load supply.

Figure 11.5: Monthly COP of a surface water fed-two stage heat pump in a 2nd or 3rd generation heating network; figure created by author.

The load curve in Figure 11.2 could then be divided by the COP to derive the monthly electricity demand (compare Figure 11.6). The highly seasonal electricity demand can be compared to production profiles of renewable electricity production. Figure 11.6 shows the high simultaneity of electrical demand for heat pump operation and electricity production from wind turbines for northern Germany coastal regions.

Based on a zero annual balance with 90% coverage by wind and 10% coverage by solar power, the demand in winter is slightly higher, whereas there is little surplus from March to October. To cover electrical demand in winter on a monthly basis, more wind installation (+30%) or backup power generation is needed.

However, good simultaneity with wind power generation indicates the possibility of low-cost electricity usage in heat pump installations if coupled with the respective installation of renewable electricity generation from wind turbines, depending on regional circumstances.

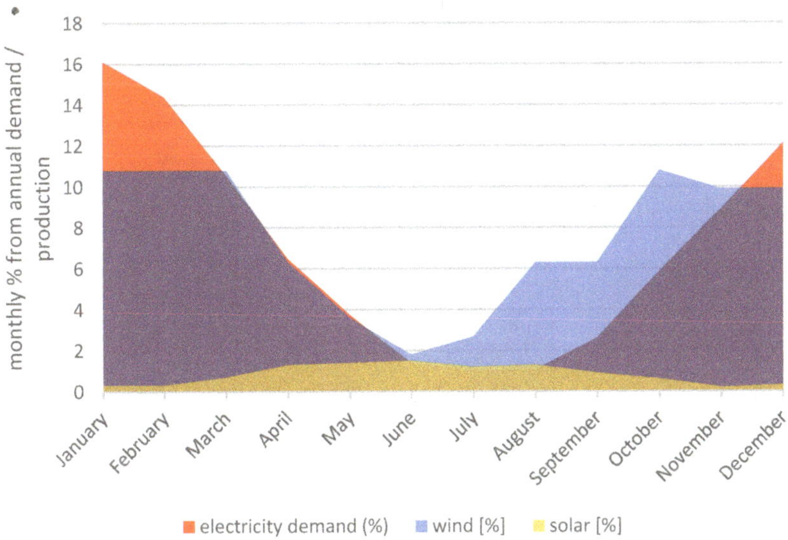

Figure 11.6: Seasonal electricity demand for a heat pump with annual COP = 2.7 (surface water, 2-stage heat pump, 2nd generation grid with peak load from direct electrical heater) in northern Germany and coverage with wind and solar at the Schleswig-Holstein north sea coast.

11.6 Economic Aspects of Heat Pumps

In the following, we would like to determine the economic efficiency of a large heat pump. Our aim is to look at the levelized costs of energy of a heat pump and then, in the following section 11.7 consequences for project financing.

The following Figure 11.7 shows the key data for assessing the economic efficiency of the heat pump:

The key parameters are the annual coefficient of performance (COP) and the electricity costs for heat production. In terms of their significance for the heat generation costs, the investment costs are less important on the time axis. This is due to the fact that the heat pump works very efficiently and – at the heat price assumed here – pays for itself in around four years. The ratio of electricity costs to the annual coefficient of performance therefore largely determines the economic efficiency; here it is 8 cents/kWh, see Table 11.3.

Project Description	Project Name:		Heat Pump Template		
Heat Pump	1			0	1: yes, 0: no
Annual Performance Factor	2,7				
Start of Operation	2025				
Full Load Hours p.a.	2.500				
Electric Power Consumption (MW)	3,704				
Electricity for heat pump (MWh)	9.259				
Installed Capacity (MW)	10				
Produced Heat (MWh)	25.000				
Investment / Financing (in Euro)					
Total Investment Costs	10.000.000		Debt Service Reserve Target Value:	50%	
Grid Connection			Initial Balance of Debt Service Reserve	0	
Project Costs					
Interest during Construction			Depreciation in years	16	
Costs for advisers			Rate of Assessment in %:	350%	
Reserve			Basic Federal Rate:	5%	
Sum	10.000.000		Allowance:	24.500	
	Amount in Euro		Interest Rate in % p.a.	Start of Repayment	Tenor
Equity	4.170.000				
Loan 1	5.830.000		5,30%	30.03.2026	15
Loan 2	0		5,40%	30.06.2027	19
Sum	10.000.000				
Decommissioning Bonds	0		1,00%		15
VAT-Facility	1.900.000		5,30%		1
				WACC	7,26%
Costs for Electricity and other costs (kWh)					
		BC		Inflation	
Electricity Costs Heat Pump p.a.	1.111.111			3,00%	
Initial Electricity Costs for Heat Pump (Cent/kWh)	12,00	12			
Sales Price for Heat (Cent/kWh)	10,50	10,5			
Income (Heat Sale)	2.625.000			3,00%	
Operative Costs (in € p.a.)					
			Inflation		
Maintenance	200.000		3,0%	approx. 2% of total investment costs	
Electricity Costs Heat Pump	1.111.111		3,0%		
Land lease	0		3,0%		
Direct Marketing	0		3,0%		
Insurance	0		3,0%		
Other Costs	0		3,0%		

Figure 11.7: Assumptions of a Heat Pump; figure created by author.

Two components are therefore relevant for the assessment of heat pumps, namely the development of the seasonal performance factor and the electricity costs. The **annual coefficient of performance** can be significantly higher due to the heat medium used (see above), and we can expect that technical progress will contribute to a gradual improvement in efficiency. The level of **electricity costs** depends largely on the source of supply and can also be considerably lower – e.g. if self-generated electricity is used.

Table 11.3: LCOE of Multi-MW Heat Pump (compiled by authors).

	1 2025	2 2026	3 2027	4 2028	5 2029	6 2030	7 2031
Operating Expenses	1.411.811	1.350.444	1.390.958	1.432.687	1.475.667	1.519.937	1.565.535
Annual Energy Production in kWh:	2.500.000	2.500.000	2.500.000	2.500.000	2.500.000	2.500.000	2.500.000
Discounted OPEX:	1.324.903	1.189.300	1.149.572	1.111.171	1.074.052	1.038.174	1.003.494
Descounted Annual Energy Production:	2.346.105	2.201.683	2.066.152	1.938.964	1.819.605	1.707.594	1.602.477
Total Investment Costs:	10.000.000						
Assumed Total Opex:	21.015.419						
Assumed Total Energy Yield:	40.251.454						
LCOE in Cent/kWh:	7,705						

In the example, we have assumed a gradual increase in electricity costs and thus calculate a heat production price of around 10.96 cents/kWh.

If we assume that electricity can be purchased at the production costs of a current renewable energy project in Germany (i.e. around 6.6 cents/kWh), the heat production costs are significantly reduced to just 5.72 cents/kWh. This does not yet include the transmission charges, but the example shows very impressively the possibilities that can arise from the combination of increasingly favorable electricity generation costs of renewable energies and the efficient use of a large heat pump. Economically, there is considerable potential here, which can also be realized in the short term.

11.7 Project Financing of a Large-scale Heat Pump – Key Aspects

In this section, we expand the above perspective of favorable heat generation costs to include that of bankability. In order for a heat pump to be financed, it must be capable of servicing the capital from a banking perspective. This means that the surpluses it generates from heat marketing must be so high that the electricity costs and debt servicing can be reliably serviced. Despite the advantages described in section 11.6 there are – for the time being – practically no examples of project financing for a large heat pump, which is mainly due to the fact that no need for the development of large heat pumps was seen. This section is intended to highlight the key aspects from a financing perspective.

We remain with the above example and now extend it to include the aspect of bankability. The essential extension to the previous calculation of the heat production costs is that the sales prices must now be included in the consideration, which are of central importance for the credit assessment of the banks. In our initial case, this results in the following cash flow consideration, which is central to the assessment of profitability:

Here (see Table 11.4) we see the cash flow waterfall for the project in question. The amount of heat produced is multiplied by the respective heat price to give the revenue.

As described above, the banks look at the CFADS and compare it to the debt service. In the initial case, this results in adequate DSCR values that could even allow the long-term loan to be extended. But more on this later. From the investor's point of view, there are substantial returns, which are shown above in the "Free CF" line. The example shows an internal rate of return of 17.98% over 20 years. The core data of the calculation is shown in Figure 11.8.

Table 11.4: Cashflow-Waterfall of a Heat Pump (compiled by authors).

Heat Pump Template	1	2	3	4	5	6	7
	2025	2026	2027	2028	2029	2030	2031
Energy Production	2.500.000	2.500.000	2.500.000	2.500.000	2.500.000	2.500.000	2.500.000
Income (Heat Sale)	2.625.000	2.703.750	2.784.863	2.868.408	2.954.461	3.043.094	3.134.387
Sum Income	2.625.000	2.703.750	2.784.863	2.868.408	2.954.461	3.043.094	3.134.387
Maintenance	200.000	206.000	212.180	218.545	225.102	231.855	238.810
Electricity Costs Heat Pump p.a.	1.111.111	1.144.444	1.178.778	1.214.141	1.250.565	1.288.082	1.326.725
VAT-Interests	100.700	0	0	0	0	0	0
Sum Operating Costs	1.411.811	1.350.444	1.390.958	1.432.687	1.475.667	1.519.937	1.565.535
EBITDA	1.213.189	1.353.306	1.393.905	1.435.722	1.478.794	1.523.157	1.568.852
Payable Business Income Tax	98.646	123.166	130.271	137.589	145.126	152.890	160.887
CFADS (after tax)	1.114.543	1.230.140	1.263.634	1.298.133	1.333.667	1.370.267	1.407.965
Interest	308.990	308.990	293.541	270.960	248.380	225.800	203.220
Repayment	0	291.500	426.038	426.038	426.038	426.038	426.038
CFADS (after debt service)	805.553	629.650	544.055	601.134	659.248	718.428	778.707
Target Value DSRA	300.245	359.789	348.499	337.209	325.919	314.629	303.339
Built-up of DSR	300.245	59.544	0	0	0	0	0
Decrease of DSR	0	0	11.290	11.290	11.290	11.290	11.290
Free CF	505.308	570.105	555.345	612.424	670.538	729.718	789.997
	2025	2026	2027	2028	2029	2030	2031
DSCR	3,61	2,05	1,76	1,86	1,98	2,10	2,24
DSCR (with DSRA)	3,61	2,55	2,26	2,38	2,51	2,65	2,81
DSRA	0	300.245	359.789	359.789	359.789	359.789	359.789

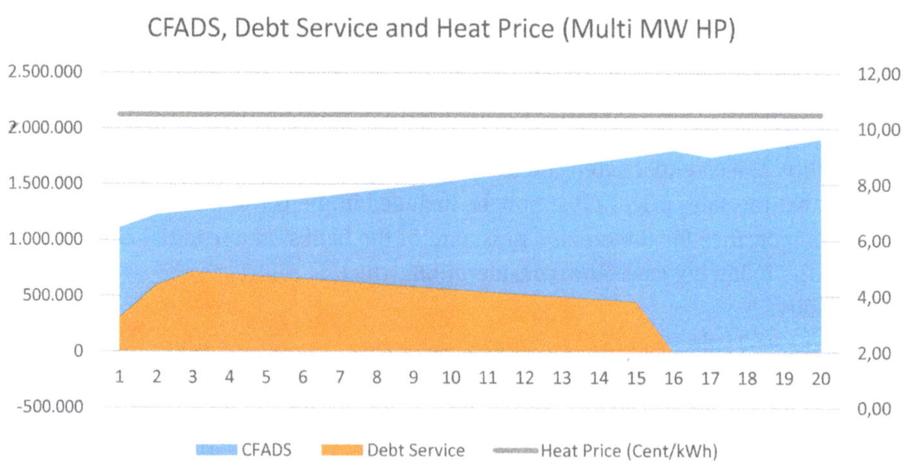

Figure 11.8: Multi-MW Heat Pump (Base Case Scenario).

The CFADS block increases gradually, as the heat revenues are higher than the electricity costs and both items are indexed to the same extent. The debt service is gradually reduced after the start of repayment, as the interest burden is slowly reduced by the ongoing repayment.

However, it is not the base case described above that is decisive for the bank, but rather a "realistic stress case" in which the debt service can still be provided. And this realistic stress case is at the heart of a bank's risk management for project financing. It is about finding out what the central risk drivers are that could jeopardize the servicing of the debt in the case of project financing.

Looking at the figures, there are two key risks – with different weightings: firstly, an adverse trend in revenue in relation to costs, and secondly, a decline in revenue, e.g. due to a technical problem.

If the cash flow falls short of expectations due to technical problems, it would make sense to conclude a long-term service and maintenance contract that provides for a certain technical performance. As the financing of heat pumps is a new market for banks, it can be assumed that banks will insist on the conclusion of a corresponding long-term maintenance contract, especially at the beginning of the financial learning process. This requirement will presumably be relaxed over time, as the technical operation of a large heat pump promises no particular surprises during operation and should be easy to manage.

From a risk perspective, however, the following aspect is key: the supply market and the sales market are not directly linked. This means that an increase in electricity costs (supply market) does not automatically lead to a corresponding increase in heating revenues (sales side). This is of particular importance here, as non-parallel changes in income and expenditure have a particularly dynamic effect on profitability.

From a banking perspective, there are basically two ways to counter this risk: The first option could be to agree with the supply side that cost increases on the supply side are passed on to the heat prices. This would transfer the risk of cost changes entirely to the heat consumers. This may be an option for commercial customers, but is largely out of the question for private customers. A second option would be for the legislator to take action and, for example, set a specific heat pump electricity price. In this case, any risk of a mismatch would be socialized by the general public. The fact that the risk of a mismatch is highly relevant for the financing bank is shown by the following consideration, which arises even with a very small deviation from our set of assumptions: The following calculation assumes that the prices for heat sales only increase by 1.0% p.a., while the electricity costs on the procurement side increase by 3.0% (Figure 11.9).

Even if the ability to service debt is still consistently given here, the change in CFADS compared to the previous figures is remarkable. A reduction in the price increase in the heat sector to 1% in the example would mean that the debt servicing capacity would no longer be given. For this sub-aspect, this means that when structuring project financing, care must be taken to ensure that a parallel rule is agreed with regard to inflation adjustment on both the purchase and sales side (back-to-back).

If the various partial risks are summarized, the assessment of the financing of a large heat pump can be divided into two parts from the bank's perspective: either it is

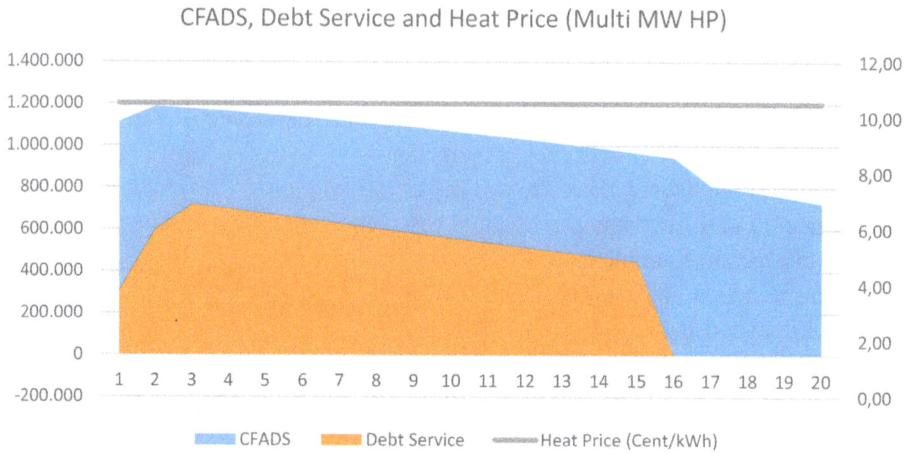

Figure 11.9: CFADS of a Multi MW Heat Pump (Stress Scenario).

possible to achieve a coupling of the supply and sales side or it is not. If this is not successful, project financing is indeed ruled out, as the bank would otherwise be taking a bet on a largely parallel development of the supply and sales sides with an uncertain outcome. And faith, love and hope are not categories of the credit business. In the following, we will therefore look at the case where suitable regulatory measures or the design of the project contracts succeed in hedging the risk of the project.

In this case, a DSCR value of 1.40 may be sufficient (see Figure 11.10). In the first step, this means a shift in financing to a higher proportion of borrowed capital, which now amounts to € 7.32 million (equity € 2.68 million).

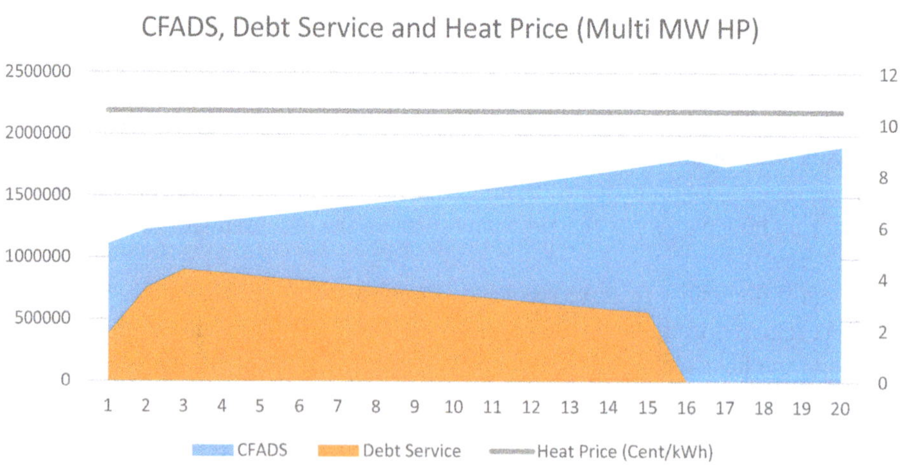

Figure 11.10: Multi-MW Heat Pump (Final Structure).

The slightly higher debt capitalization and the associated use of the leverage effect resulted in a slightly improved return on equity of 21.3%.

If the link between the procurement side and the sales side proves to be stable, financial learning is likely to begin among the banks, so that the required minimum DSCR values for later projects may also be lower or the terms of the loans may be longer. However, it will always be necessary to link the procurement side with the sales side.

Despite the very high efficiency of a large heat pump, it is necessary from a banking perspective that the purchase and sales contracts are structured to enable project financing. Otherwise, the financing of a heat pump can only be realized as corporate financing. The example also shows that it is imperative to consider the financing rules and economic facts in addition to the technical-physical laws.

11.8 Conclusion

As we have shown, large scale heat pumps are available as a drop-in option even for current 2nd and 3rd generation district heating networks. Electricity demand of such installations could be covered by wind turbines, which annual generation profiles show high simultaneity with the electricity demand of heat pumps. At least small to medium sized district heating networks with up to 100 MW thermal and temperatures of 95°C and above can be supplied using heat pumps and direct electrical heaters as peak load generators using conventional buffer storage.

However, two constrains arise, of which an appropriate environmental heat source being the most critical. Since air as a heat source is limited to a few hundred kilowatts as a maximum, typically water has to be used, either surface water, sewage water or ground water, including geothermal and waste heat stored in the underground, for example.

Typically, sewage water will be used primarily, as it is often readily available and a better heat source then surface water, at least if located not too far away from the district heating network. Drinking water is also potentially used, however according to regulations. Sewage and drinking water alone provide a potential of 5–20% of the heat demand of typical communities and cities.

Surface water, if available in abundance, is second best suited, especially from rivers. Rivers regenerate heat from solar irradiation, groundwater and surface water inflow and heat flux from the ground, which would typically enable extraction of heat equivalent to 1 K change in temperature. Thus, depending on the river flow, the heating capacity could easily be calculated.

The second constraint, especially for large installations, is the electrical grid connection. Whilst heat pumps even in large installations seldomly need up to 100 MW at a single site, direct electrical heaters used for peak load generation and redundancy could easily reach such power ratings, even for medium sized heating networks. In large cities, only a combination of heat sources, including waste heat, and more or less decentralized heat generation combined with heat from waste incineration could cover demands in the GW scale without usage of (renewable) gas vessels or cogeneration. However, depending on the availability of hydrogen or other renewable gases, combinations of heat pumps and cogeneration for base and medium load and renewable gas vessels as well as direct electrical heaters as peak load and redundancy may well be suited for decarbonized heat supply in district heating.

Even if the heat generation costs of heat pumps can be very low, it is essential from a financial perspective to ensure that there is always a sufficient gap between the heat yields and the electricity costs. This can be achieved by passing on increased electricity costs to the heat revenues or by subsidizing or capping the electricity tariff for heat pump electricity.

Jörg Böttcher

12 Use of Geothermal Energy

12.1 Geothermal Energy – Introduction

In the following, we would like to highlight the importance of using geothermal energy, which is particularly important for the future heat supply.

We would like to outline the technical and legal requirements that currently need to be met in order to realize a large-scale geothermal project using the project financing method. It is important to be aware that technology in particular is developing dynamically and that the legal framework data reacts to market conditions and obeys overriding energy policy requirements, so that geothermal projects must be managed dynamically and flexibly, especially during the development phase. This presentation is no substitute for project-specific support and advice from specialists in the respective fields – on the one hand, the projects are too specific for this and, on the other hand, legal, technical and economic aspects are also constantly evolving.

Geothermal projects have a unique risk profile within the renewable energy sector: While the completion phase harbors considerable risks, the potential risks that arise during long-term operation are easily manageable and generally enable ecologically and economically attractive operation.

We first present the basic technical principle of geothermal energy projects and then address two key issues relating to their implementation: First, the exploration risk, i.e. the risk that an exploratory well does not prove to be exploitable or not fully exploitable in such a way that it can be operated economically. Here we discuss the incentive effects and possible mechanisms for mitigating this risk. In the next step, we examine the economic viability of a geothermal project and look at the economic aspects if a project does not prove to be as economical as expected.

12.2 Geothermal Potential

Geothermal energy can be used to provide energy in the form of technically usable heat or electricity. Hot water-bearing rock strata, so-called hydrothermal systems, provide potential utilization horizons that are currently used in most large geothermal power plants worldwide.

The technologies for utilizing deep geothermal energy generally require at least one production well and one injection well each, which taps energy at a sufficient temperature from a deep geothermal reservoir as required. The thermal water cycle is closed above ground, the energy is usually transferred to the respective consumer

https://doi.org/10.1515/9783110784299-012

using a heat exchanger and the cooled water is returned to the reservoir via the injection well.

The geothermal reservoirs available in Germany are hot layers containing deep water (hydrothermal systems) and heat stored in deep rock (petrothermal systems) with no or limited water supply. For the most part, deposits exist with transitions from hydrothermal to petrothermal systems. Even with today's heat pump technology, efficient heat recovery can be realized even at lower temperatures, which significantly expands the field of application for the use of geothermal energy.

In particular, the possibility of being able to generate base-load electricity should make geothermal energy projects more financially viable, although the risks – depending on the technology and geological conditions – can be high, especially in the completion phase, and public acceptance must be achieved. Nevertheless, the use of geothermal energy should be of interest to policymakers as, in addition to base load-capable electricity generation, conventional combustion processes are also avoided, meaning that no direct CO_2 emissions are caused.

Statements on the economic viability of a geothermal system are crucially dependent on the hydraulic and thermal properties of the useful horizon and the composition of the water. The geothermal gradient is central to this: it describes the temperature rise towards the earth's core and in Germany averages 3 to 4°C per 100 meters.[1] Locations with higher temperature gradients can lead to cost savings because the drilling depth is lower. However, the production rate must always be taken into account.

For the economic operation of a geothermal system, the heat is regularly used all year round. Electricity generation is generally only possible at temperatures above 100°C with the appropriate technology. The higher the temperature level, the better the efficiency of electricity generation. The residual heat from this technology must also be marketed from an ecological and economic perspective. Similar considerations apply to the use of petrothermal systems.[2]

Germany's most important regions in terms of hydrogeothermal utilization are the South German Molasse Basin, the North German Basin and the Upper Rhine Graben.[3] In these regions, there are reservoirs with hot water in the deep subsurface, which, with temperatures of over 60°C, enable direct heat utilization and, at temperatures of over 100°C, also base load-capable electricity generation (see Figure 12.1).

Geothermal projects can be realized in the form of project financing so that the sponsors achieve a limitation of liability. However, this is only possible if the cash flows generated by the project can be regarded as so stable and predictable that joint liability on the part of the initiators can be waived for the entire duration of the proj-

1 Enerchange 2012, p. 4.

2 PK Deep Geothermal Energy 2007, p. 14.

3 The Geothermal Information System GeotIS (www.geotis.de) provides further information on the locations and potential of deep geothermal energy.

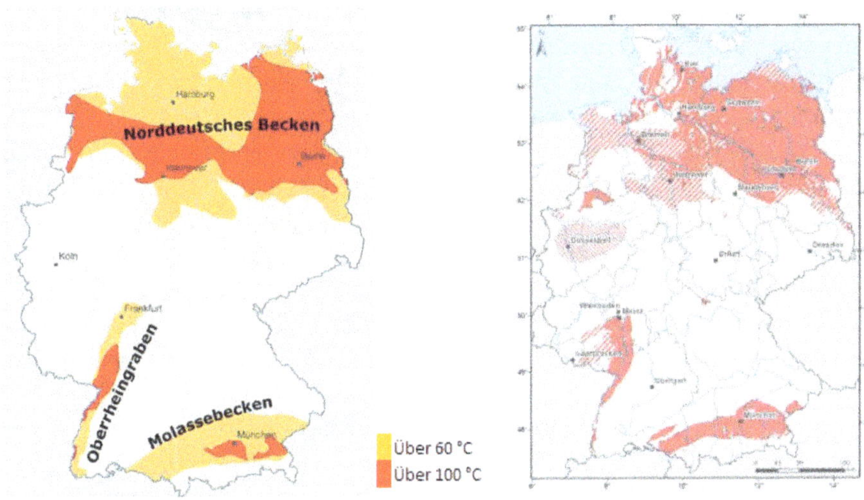

Figure 12.1: Deep geothermal utilization potential in Germany.[4]

ect. We explain the special methodological features that need to be taken into account in section 16 below.

The **direct utilisation of geothermal heat** covers a wide range of resource types and subsurface characteristics – geological setting, depth and temperature – as well as end-user heating applications. Geothermal resources, which are suitable for heating, are widespread globally, in the form of shallow geothermal resources, or deep-seated aquifers hosted in permeable formations in sedimentary basins, or volcanically and tectonically active zones. Geothermal energy for heating and cooling can be obtained at varying depths.

Geothermal energy sources are base-load capable and can therefore compensate for daily or seasonal fluctuations in the overall mix of renewable energies. The global reserves – theoretically exploitable with today's technology – are estimated to be thirty times greater than all fossil reserves. However, the large-scale use of geothermal energy is still in the early stages of development. The contribution of geothermal energy generation is likely to increase globally from approx. 20 PJ (5.5 TWh) in 2010 to 395 PJ in 2050. This also means a twenty-fold increase in the relative share of geothermal energy in renewable energies from 2.2% in 2010 to 13% in 2050 (DLR 2021). The theoretical potential is already enormous: 99% of our earth is hotter than 1,000°C and 99.9° is warmer than 100°C. For comparison: in Germany alone, 40% of primary energy input – across all energy uses – is used to generate heat below 100°C. The geoscientific and engineering challenges now consist of tapping these energies economi-

4 Left diagram: N.A. BMU 2011, P. 4. Right Diagram: https://www.geotis.de/homepage/sitecontent/info/publication_data/public_relations/public_relations_data/Positionspapier-Waermewende.pdf (2018).

cally and sustainably. However, whether this potential can be exploited in the long term also depends on the experience and lessons learned from implemented projects. Both the reliability of existing systems and the further reduction of costs – particularly drilling costs – play a key role here.

Large-scale plants for the use of geothermal energy to generate electricity can be found around the world, particularly in volcanically and tectonically active areas along plate boundaries of the earth's crust. The proximity to magmatic intrusions leads to particularly high heat flow densities and considerable geothermal gradients in such weak zones of the crust. This results in high-enthalpy reservoirs from which the geofluids can be extracted directly as superheated steam. This superheated steam from hydrothermal reservoirs can be converted directly into electricity using steam turbines.

Regions with particularly high potential and intensive use of geothermal energy are located on the edges of the Pacific, for example. On this so-called **"ring of fire"**, there are large geothermal deposits in the far west of the USA and along the subduction zones of Central and South America, as well as on the other side of the Pacific in the Philippines, Indonesia, Japan and New Zealand. Many countries are already making significant use of geothermal energy.

However, the development of geothermal reservoirs is not exclusively limited to high-enthalpy areas. On the contrary: these reserves are limited due to their tectonic position and can only be developed to a limited extent. In the remaining areas of the continental crust, i.e. outside active plate boundaries, are the low-enthalpy regions; they make up more than 95% of the earth's land-covered surface. This is where the real future of geothermal energy lies. Apart from comparatively few formations with a high natural thermal water flow, their development is largely tied to non-permeable or low-permeability sedimentary rocks or their crystalline basement.

Most of the geothermal energy available in Germany can only be exploited via hydraulic tapping of the deep rock; according to a potential study by the Technology Committee of the German Bundestag, this accounts for 85–90% of all available resources.[5]

5 TAB Working Report No. 84 (02/2003): Opportunities for geothermal power generation in Germany. German Bundestag, Committee on Education, Research and Technology Assessment. See also R. Bracke, E. Huenges 2022, https://doi.org/10.24406/ieg-n-645792.

12.3 Market Development in Geothermal Use

12.3.1 Electricity Generation

Situation in Germany

According to a study of Fraunhofer IEG et al.,[6] 26 deep geothermal plants are in operation in 2020. The installed geothermal capacity in Germany reached 47.6 MW_{el} end of 2021 and the electricity production amounted to 190.6 GWh in 2020.

Europe and the World

According to the International Renewable Energy Agency (IRENA),[7] geothermal energy provides electricity generation in more than 30 countries worldwide, reaching a total installed capacity of around 16 gigawatts (GW) in 2020. In the EU, the gross capacity for electricity was just over 1 gigawatts electric (GW_e) that year. Several other EU countries produce electricity from geothermal (Germany, Portugal, France, Croatia, Hungary and Austria), albeit with considerably smaller production.[8]

A look at the countries reveals that the focus is on North and Central America and the Far East. In contrast, usage in Europe falls significantly behind (see Figure 12.2):

If we look at the reasons for the restrained development of deep geothermal energy in Germany, we can cite the unique risk profile of geothermal projects as an important reason alongside the geological conditions. On the one hand, there is a pronounced exploration and borehole risk, which means a considerable risk of loss during the underground completion phase. This makes it rather unlikely that lenders will participate in this phase. On the other hand, once completed, geothermal power plants allow for base load energy production, which makes them generally well suited for project financing in this phase. This temporal difference in risk characteristics requires a corresponding adjustment of financing sources and financing structures: during the drilling phase and underground development, equity investors are in demand and must be in a position to make considerable advance payments from their own resources, while above-ground development also allows the use of debt capital.

6 https://publica-rest.fraunhofer.de/server/api/core/bitstreams/18358fcc-b833-4c98-a1df-047390520191/content, p. 6.

7 https://www.irena.org/-/media/Files/IRENA/Agency/Publication/2023/Feb/IRENA_Global_geothermal_market_technology_assessment_2023.pdf?rev=310b36d62bc749d6837b4460e050ed1b, p. 102.

8 Earlier sources: Technology Roadmap - Geothermal Heat and Power, OECD/IEA, Paris 2011. Special Report on Renewable Energy Sources and Climate Change Mitigation – Geothermal Energy, IPCC Working Group II, Final Release, Cambridge UniversityPress, Cambridge, United Kingdom and New York, NY, USA, 1075 pp., 2011.

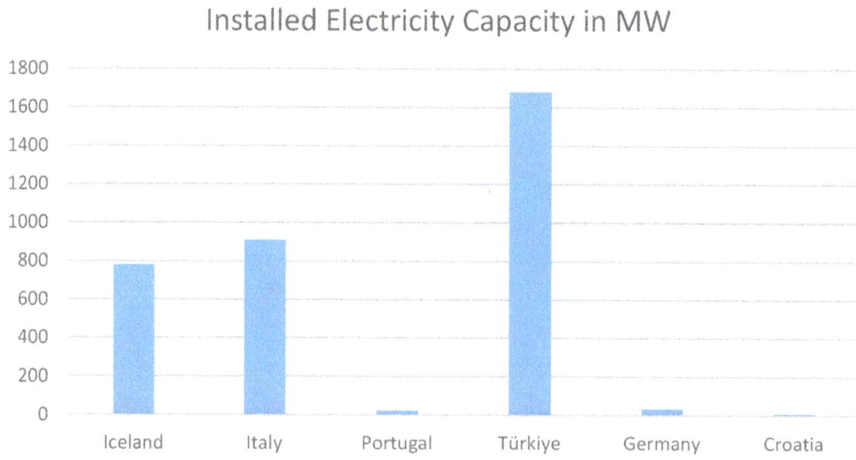

Figure 12.2: Installed geothermal electricity capacity by country.[9]

North America's previous strong position in geothermal energy is being replaced by developments in Asia – excluding China and India. Further expansion is expected here, particularly in Indonesia and the Philippines. Geothermal power generation in Europe, Africa, China and India will increase significantly. Expectations for development in Central and South America, which are subsumed under "Other" by the IEA, are particularly strong.

12.3.2 Heating Market and Deep Geothermal Energy

Geothermal energy has a far greater opportunity in the European heating sector than in the electricity market. By 2030, Europe (EU-25) aims to triple the global installed capacity of renewable energy sources to at least 11 TW.[10] The geothermal district heating and cooling sector has seen a 6% growth rate in installed capacity, reaching 2.2 GWth in 2021.

District heating networks already account for 10% of the total heating market in Europe. Of the 5,000 district heating networks currently in use in Europe, 395 are operated as geothermal district heating networks.[11] From the mid-1970s to the mid-1980s, there was the first major impetus for the construction of such networks. The market calmed down until 2009, then picked up again and has been developing

9 IGA 2012.

10 See https://energy.ec.europa.eu/topics/international-cooperation/eu-external-energy-engagements_en?etrans=de&prefLang=de.

11 EGEC Deep Geothermal Market Report, 12th edition, 2022, Brussels.

steadily ever since. In 2022, 142 geothermal power plants are in operation with an installed capacity of about 3.5 GWe and generating more than 22 TWh.

There are currently 41 geothermal district heating networks in Germany, most of which are located in Bavaria. A further 102 networks are in development. By then, Germany would be the leader in terms of numbers, but France (2022: 80, 2022+: +24), Netherlands (29, +22), Italy (30, +11) and Poland (7, +21) are also pursuing massive expansion. Production depths of 2,000 m to 3,500 m, mainly in sedimentary rock, are not uncommon. At lower depths and possibly lower extraction temperatures, heat pumps are regularly used. The aim must be to massively expand geothermal district heating as soon as possible in order to drive forward the energy transition in the area of renewable heat supply.

According to the BMU, more than 50% of the final energy demand in Germany is used for space heating and hot water; in Europe, this figure is around 50% for heating and cooling. The massive potential of geothermal energy for avoiding greenhouse gases and for expanding the energy transition on the thermal side is therefore becoming apparent.

12.4 Project Obstacles and Challenges

Due to the high investment costs, the technical risks involved in the drilling process itself and a general discovery risk of several million euros per drilling failure, it is still very difficult to secure project financing at this stage, both for local authorities and for municipal utilities as potential operators. Some industrial insurance brokers are now offering the first comprehensive insurance solutions, while at ministerial level, project funds are being considered to cover such risks for national and international projects.

There are positive expectations for the future development of geothermal energy, both in terms of electricity generation and in the heating and cooling market. The roadmap published by the IEA (International Energy Agency of the OECD) in 2011 assumes strong growth in the traditional OECD countries as well as in China and India. The International Geothermal Association (IGA), the International Energy Agency and other organizations see the development and expansion of EGS technology as a particular challenge for science and industry. This is the only way to help geothermal energy achieve a significant market share of renewable energy sources. To achieve this, five milestones must be reached over the next 20 years:

1. The development of EGS power plants or plants in different geological and tectonic situations, such as crystalline/sedimentary rocks or basin or orogenic structures within the next ten years. This also includes their further development from and overlap with hydrothermal systems.
2. The development of hydraulic and chemical stimulation technologies that are particularly geared to geothermal issues.

3. For reasons of environmental protection and social acceptance, the focus must be on environmentally and socially compatible stimulation processes. To this end, project management structures must first be defined and standardized with regard to health protection, safety and environmental protection. This also applies, for example, to aspects of induced seismicity.
4. Achieving long-term availability of the resource with the associated operational management and monitoring processes for the reservoir.
5. From 2025, the learning curve should lead to the expansion of power plant capacities up to 50 MW_{el} and then to an order of magnitude of 200 MW_{el}. Cascaded, modular systems must be developed for this.

The expansion of geothermal energy requires a series of measures, as shown in Table 12.1:

Table 12.1: Suggested Steps for the Promotion of Geothermal Energy.[12]

	Requirements
Clear Expansion Goals	Apart from the definition of clear installation targets, the regulator has to create a supportive regulatory regime. I.a., the process of approval has to streamlined.
Instruments for the Mitigation of the Drilling Risk	From an economic point of view we need risk mitigation instruments as well as geo-physical examinations in suited areas.
Investment in Key Technologies	These are technical procedures like Engineered Geothermal Systems, heat pumps, intercommunal heating network . . .
Capacity Building	Support for education along the value chain
Broad Public Relation	Promotion of a broad acceptance within the society (via information and participation)

The easiest and quickest way to increase the market share of heating networks is to increase the connection rate to existing heating networks in urban neighborhoods. This is why instruments for the expansion of heating networks and the use of geothermal energy should start here first. Where heating networks already exist, the infrastructure costs for the development of new district heating customers are significantly lower than for the development of new areas. Especially in high-density urban areas with existing heating networks, the densification and expansion of these heating networks offer a very good option for decarbonizing the urban building stock in a cost-efficient manner. This approach of harvesting the "low hanging fruits" first has the

12 See R.Bracke; E. Huenges 2021 (ed.), p. 7 and p. 19 (abbreviated version).

advantage of making the technology more efficient, which will benefit regions that are not the primary focus.

12.5 Relevant Individual Risks – Allocation of Responsibilities

In the following sections, the industry-specific characteristics of geothermal projects are dovetailed with the traditional risk management process of project financing. The presentation determines the respective risk profile for various forms of geothermal projects and describes suitable risk management measures. This section ends with an evaluative summary of the individual risks considered.

12.5.1 The Resource Risk – Estimation of the Energy Yield

A realistic forecast of the energy yield is of central importance for the economic viability of a geothermal project.

In the field of geothermal energy, the resource risk is referred to as exploration risk. The exploration risk is the risk of developing a geothermal reservoir in insufficient quantity or quality. The quantity is defined by the thermal output that can be achieved with the help of a borehole. This output P is proportional to the production rate Q and the temperature T:[13]

$$P = Q * T$$

Quality essentially refers to the composition of the water. There may be components in the water that make geothermal use impossible or difficult. However, all water found in geothermal boreholes in Germany to date has been considered manageable in terms of its composition for geothermal use, albeit with varying degrees of technical effort.

A geothermal borehole is considered successful if

- the thermal water discharge reaches more than a minimum production rate Q at a maximum drawdown Δs and
- a minimum temperature T is reached.

It is usually easier to estimate the reservoir temperature before construction begins than the achievable production rate.[14]

13 PK Deep Geothermal Energy 2007, p. 12.
14 Enerchange 2012, p. 19.

In hard rock aquifers, the permeability and thus the yield of the aquifer is based on the presence of open fractures or caverns, on sufficient flow-effective porosity and on other macroscopic cavities, such as those that can be encountered in fault zones. If the expected permeability is not encountered during the development, then upgrading or even stimulation measures are required. These measures include, for example, acidizing carbonate rock or hydraulic stimulation, possibly in combination with acidization.

From a project financing perspective, the exploration risk in geothermal projects is closely related to the completion risk and the technical risk. This is important because there are different risk carriers for the individual partial risks. It can often be difficult to prove in court whether a lower than expected discovery is a manifestation of the resource risk or is due to the fact that the injection well and production well were planned too close together.

There are some special features to the discovery risk: Equity providers and lenders have a largely equal interest in reliably assessing the quality of resources: The higher the coverage ratios are from the perspective of the lenders, the more economical the project is from the perspective of the equity providers. This assessment does not change much if the different starting points – sponsors case (owner) and banking case (lender) – are taken into account. An overestimation of the quality of the location tends to lead to a higher level of external funding than the project can bear. As a consequence, there is an increased probability that the project will not be able to service the debt as planned, leading to a restructuring that usually involves lengthy negotiations and concessions from both sides. The situation is significantly different if a project developer wants to sell a project to the equity providers before or on completion, as is typical for a high proportion of German geothermal projects. In this constellation, there is an incentive for the developer to overestimate the project quality, as this allows him to realize a higher sales price.[15]

In a risk assessment, geothermal projects are characterized by the fact that there is a certain probability that the well will not be found and that the well will therefore be lost from an economic perspective ("exploration risk"). The probability of loss is assumed to be between 20% and 35%.[16] This risk means that investors and banks are reluctant to finance a geothermal project. Nevertheless, the use of geothermal energy

---

15 There is a principal-agent problem that can lead to a market failure. The principal-agent problem describes the situation in which the contractor gains an information advantage in the course of its activities, which it can use to its advantage and to the disadvantage of the client. From a dynamic perspective, this can lead to a market failure, as supply and demand no longer match. See G. Akerlof 1970, pp. 488–500. See also section 4.1.

16 These values must be checked on a project-specific basis and can also be significantly reduced using suitable investigation methods and a good data situation. The exploration risk is considerably lower in the Molasse Basin. The values mentioned here are more about the basic principle.

appears to be advantageous from an economic, ecological and supply security perspective.

The question is whether and how the state can and wants to provide support here. Two state instruments are discussed below: Exploration insurance and geothermal funds. The perspective of investors and banks, but also of the state, is taken.

We look at a numerical example to illustrate the profit and loss situation of the operator and the bank, with and without a hedging instrument (see Figure 12.3).

			Equity	Debt	Total
	1	Total Cost 1st drilling	10.000.000	40.000.000	50.000.000
	2	Loss Probability	25%	25%	25%
	3	Calculated Loss	2.500.000	10.000.000	12.500.000
	4	Interest Rate Bank		4,50%	
	5	Interest Payments p.a.	1.800.000		
	6	Interest Yield p.a.		1.800.000	
		Margin p.a. (1,5 %)		600.000	
Variant 1	7	Loss Guarantee of Investor	10.000.000		
	8	Compensation by Investor		10.000.000	
Variant 2	9	No Loss Guarantee	0		
	10	Compensation by Investor		0	
Earnings Value					
	11	Total Loss, Compensation of Bank	14.300.000		
	12	Total Gain, Compensation of Bank		600.000	
	13	Total Loss, no Compensation of Bank	4.300.000	8.200.000	12.500.000

Figure 12.3: Geothermal Project: Consequences of Exploration Risk (no risk mitigation).

We consider 10 wells, each costing € 5 million. Each individual well is financed proportionately by the investor and the bank (1). Assuming a 25% discovery risk, € 12.5 m is lost across all 10 projects (2). The investor has to pay interest on the borrowed capital (€ 40 million) at 4.5% p.a., which amounts to an annual interest expense of € 1.8 million (4 and 5). If the investor absorbs the bank's loss, the bank earns from the margin income – that is € 0.6 million (6). For the investor, this results in a cumulative loss of € 14.3 million – he bears the risk of loss of the entire project (€ 12.5 million) and must also bear the interest costs (11). If the investor does not assume the bank's loss, the bank loses € 8.2 million (the interest income is already included, 13). For the investor, this results in a cumulative loss of € 4.3 million – he bears the risk of loss on his share (€ 2.5 million) and must also bear the interest costs (€ 1.8 million) (13).

If the investor is obliged to also assume the bank's risk of loss, the risk is not acceptable for him. If the bank risk is not assumed, the scenario is not attractive for the investor either, unless he has a lot of capital and can develop several projects. For the bank, only the scenario in which the investor assumes the risk is attractive. As a result, the two capital providers do not come together.

The risk of discoverability can be significantly reduced by taking out insurance. The insurance comes into effect when the risk of discovery has occurred. The state funds are interest-bearing, whereby the conditions will not cover the risk. The amount of the own contribution and the further transfer of risk is important for the question of whether an insurance solution is acceptable for banks and investors. The bank tends to be the limiting factor, as it has to act more cautiously than an investor. In the following (see Figure 12.4), we present an insurance solution for the discovery risk:

			Equity	Debt	Total	Government
1	Total Cost 1st drilling		10.000.000	40.000.000	50.000.000	
2	Loss Probability		25%	25%	25%	
3	Loss Amount (before coverage)		2.500.000	10.000.000	12.500.000	
4	Interest Rate Bank			4,50%		
5	thereof: Margin of Bank			1,50%		
6	Interest Payments p.a.		1.800.000			
7	Interest Yield p.a.			1.800.000		
Insurance						
8	Share of Risk:	**20%**				
9	Insurance Premium:	1,50%	750.000			
10	Loss Amount (insurance included)		500.000	2.000.000	2.500.000	10.000.000
11	Loss Amount (insurance and costs included)		3.050.000			
12	Loss Amount (insurance and income included)			1.400.000		

Figure 12.4: Geothermal Project: Consequences of Exploration Risk (insurance as risk mitigation).

Again, we consider 10 wells, each costing € 5 million. Each individual well is financed proportionately by the investor and the bank (1). Assuming a 25% discovery risk, € 12.5m is lost across all 10 projects (2). The investor has to pay interest on the borrowed capital (€ 40 million) at 4.5% p.a., which amounts to an annual interest expense of € 1.8 million (6). An insurance company now assumes a large part of the risk, leaving only an own share of 20% (8): For this, the investor must pay an insurance premium (8) per year. Since 80% of the risk is assumed, the investor is left with a loss of 20%, which in the example is € 500 thousand (10); the bank incurs a loss of € 2 million. If the interest expense, insurance premium and loss are added together, the investor's loss amounts to € 3.05 million (11). From the bank's perspective, the loss is reduced by the margin income from the loan (1 and 5), resulting in a total loss of € 1.4 million (12).

The solution is probably acceptable for the investor, provided he can implement several projects and has sufficient capital. The situation is different for the bank: Although 80% of the risk is assumed, a bank will shy away from financing because it has to reckon with a loss-making transaction.

We now change the insurance scenario by having the bank pass on its remaining risk to the investor. The payout matrix is then as shown in Figure 12.5 below:

			Equity	Debt	Total	Government
1	Total Cost 1st drilling		10.000.000	40.000.000	50.000.000	
2	Loss Probability		25%	25%	25%	
3	Loss Amount (before coverage)		2.500.000	10.000.000	12.500.000	
4	Interest Rate Bank			4,50%		
5	thereof: Margin of Bank			1,50%		
6	Interest Payments p.a.		1.800.000			
7	Interest Income p.a.			1.800.000		
Insurance						
8	Deductible:	**20%**				
9	Insurance Premium:	1,50%	750.000	0	0	0
10	Loss Amount (insurance included)		500.000	2.000.000	2.500.000	10.000.000
11	Additional Coverage of Bank Risk by Investor:		2.500.000	0	2.500.000	10.000.000
12	Loss (after insurance and costs)		5.050.000			
13	Gain (after insurance and costs)			600.000		

Figure 12.5: Geothermal Project: Consequences of Exploration Risk (insurance as risk mitigation and full coverage of the risk of the bank).

There is a state insurance policy with an own contribution of 20% and the agreement that the investor also assumes the remaining risk of the bank (its deductible): The investor must pay an insurance premium (9) per year. As 80% of the risk is assumed, the investor retains a loss of 20%, which in the example is € 500 thousand (10); the bank incurs a loss of € 2 million (10). As the investor now also assumes the bank's loss (11), the investor incurs a loss of € 5.05 million from interest expense, insurance premium and loss assumption (12). From the bank's perspective, there is no loss, but a margin income of € 0.6 million (13).

This solution assumes that the bank is allowed to be reimbursed for its own deductible under the terms of the insurance. The solution may be acceptable for the investor, provided that he can implement several projects and has sufficient capital. This solution is likely to be acceptable for the bank.[17]

We now present an alternative safeguard, namely the use of a fund model (see Figure 12.6). The idea of a geothermal fund is that state funds can be accessed in advance to finance the exploratory drilling.

If the drilling is successful, the funds must be repaid by the project over time. If the drilling is unsuccessful, the funds do not have to be repaid and are converted into a grant.

17 A bank is normally not a big fan of insurance solutions. For them, it is not completely certain that the insurance will pay out (have all conditions been met for the insurance cover to be triggered?). In addition, an insurance company will only pay out if the insured loss can be quantified precisely. While the assessment of the loss can take a correspondingly long time, the exposure must continue to be managed accordingly and will burden the bank's balance sheet with a default rating. In this respect, a clear, rapid and reliable settlement of the claim will be central to the design of an insurance policy.

			Equity	Debt	Government
1	Total Cost 1st drilling		50.000.000	0	0
2	Loss Probability		25%	25%	0%
3	Loss Amount (before coverage)		12.500.000	0	0
Funds					
4	Deductible:	**20%**			
5	Funds Premium p.a.	1,50%	750.000	0	0
6	Total Loss		2.500.000		10.000.000
7	Total Loss (including Funds Premium)		3.250.000	0	0

Figure 12.6: Geothermal Project: Consequences of Exploration Risk (funds model); figure created by author.

Here, too, we are looking at 10 wells (Figure 12.7), each costing € 5 million. Each well is financed by the fund (1). Assuming a discovery risk of 25%, € 12.5 million is lost across all 10 projects (3). The investor must pay an insurance premium of 1.5% p.a., i.e. € 0.75 million p.a. (5). The fund finances the drilling and the deductible is only due in the event of failure (6). The loss of the deductible (6) and the fund premium (5) add up to a total loss of € 3.25 million. The state assumes a loss share of € 10 million.

The possible payout matrices are summarized below:

Total Drilling Costs 50.000.000

		Investor	Bank	Government
1	No Coverage	4.300.000	8.200.000	0
2	Insurance (with 20 % deductible)	3.050.000	1.400.000	10.000.000
3	Insurance (20 % deductible + Loss Coverage of Bank)	5.050.000	600.000	10.000.000
4	Risk Funds (20 % deductible)	3.250.000	0	10.000.000

Figure 12.7: Geothermal Project: Overview of different Coverage Concepts; figure created by author.

A situation without hedging the well risk will not be acceptable, especially for the bank (1). An insurance solution (2) will be acceptable for a bank if the investor also bears the banking risk (3). This risk is probably acceptable for the investor. The geothermal fund model allows realization without a bank in the discovery phase. The costs of hedging are identical for the state in both solutions.

The discussion can certainly be continued. In the following, we look at the aspects of target adequacy and political acceptance:

Target adequacy: We have already described this aspect in detail above by carrying out a risk assessment from the perspective of investors and banks. We see a clear advantage for a fund solution here.

Political acceptance: a) Establishment: An insurance solution does not immediately lead to a commitment of budget funds, whereas start-up financing requires a budget item. b) Maintaining the insurance model: With the insurance solution, success/failure is easily recognizable, so that a well-founded decision can be made about changing the insurance offer. With the assumed default rates, the fund model leads to a gradual depletion of funds and thus to continued political coordination processes.

The following applies to projects in Germany: If you want to access the BEW subsidies, the fund solution described above is not permitted for reasons of state aid. However, it is possible to eliminate a market failure by means of suitable insurance solutions.[18]

Following the presentation of the discovery risk and possible hedging instruments, in the following section we provide some basic considerations on functional risk.

12.5.2 The Functional Risk – Proven Technology

Geothermal Energy can be used for baseload electricity generation as well as sustainable heating and cooling. The mode of utilization of geothermal energy depends largely on the resource temperature. Electricity production is more favorable from geothermal resources of medium to high temperatures, which is in the range of 90 to 150 C.

Three primary power plant technologies are used to convert the energy in **geothermal resources to electricity**: dry steam, flash steam and the binary cycle (Figure 12.8).

Most geothermal plants in operation for electricity generation are dry steam or flash plants that harness geothermal resources at temperatures of more than 150°C. However, lower temperature resources are increasingly being developed for electricity generation or combined heat and electricity using binary cycle technology. Dry steam technology is applicable when dry steam is produced directly from the geothermal reservoir. With this technology, saturated or superheated geothermal steam at high pressure is obtained directly from the geothermal well and directed to a steam turbine coupled with a generator to produce electricity. The steam exhaust from the turbine is discharged into a condenser at low pressure or partial vacuum to optimise the efficiency of electricity generation.[19]

Direct utilization of **geothermal heat** covers a wider range of resource types. Geothermal systems can be divided into near-surface and deep geothermal systems in terms of the depth of heat extraction. This distinction is important for our purposes because different geoscientific parameters are required in addition to different energy generation techniques. In this publication, as already mentioned, we are concerned with deep geothermal energy.

18 In Germany, KfW/MunichRe is currently developing a national settlement concept, which is expected to come into force in 2025.

19 R. DiPippo: Geothermal Power Plants 2013.

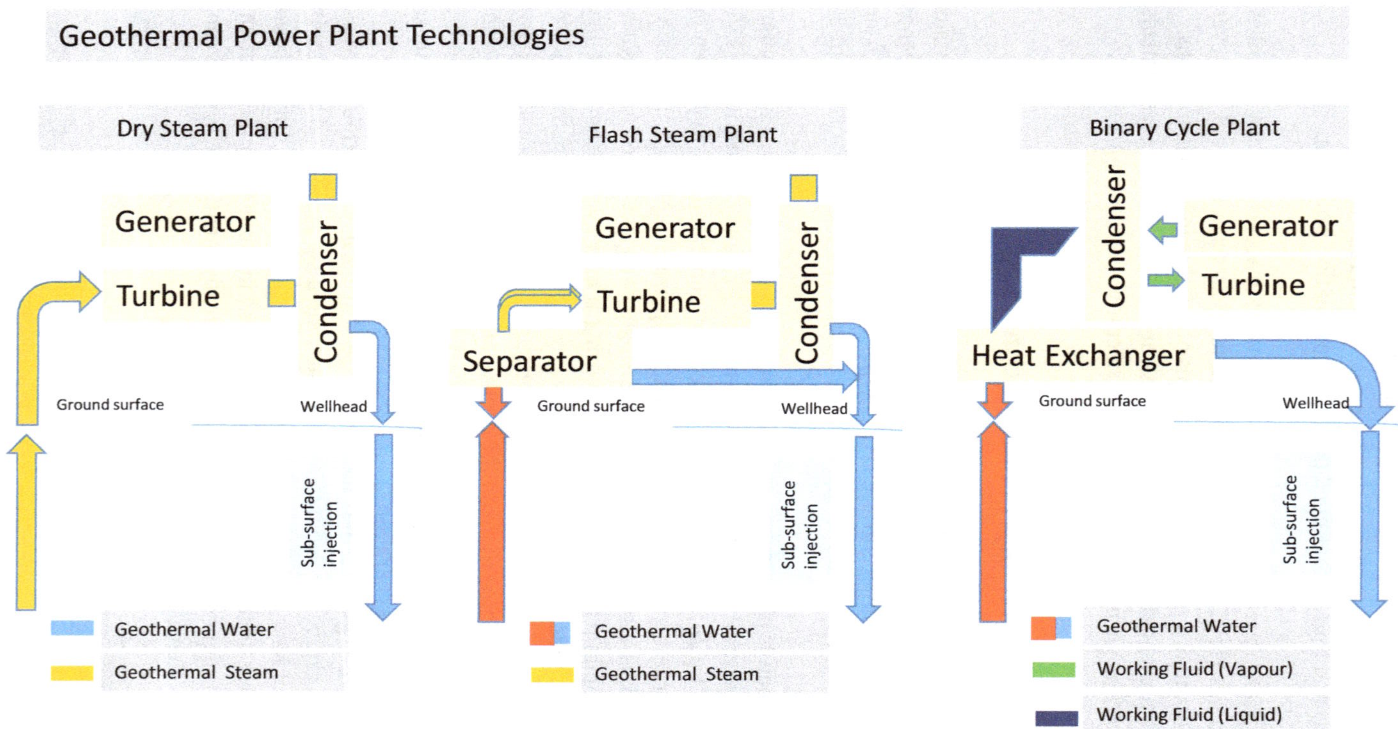

Figure 12.8: Principles of Geothermal Power Plant Technologies; figure created by author.

It is generally accepted to speak of deep geothermal energy at depths of more than 400 m and at temperatures of more than 60°C. Geothermal energy includes the following systems, as shown in Figure 12.9.[20]

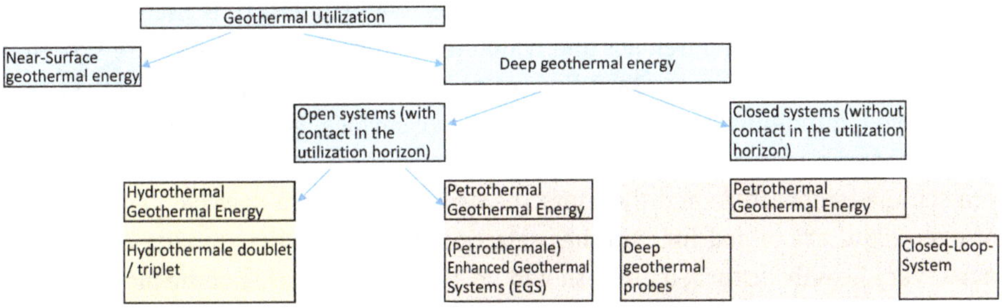

Figure 12.9: Geothermal Project: Overview of different Coverage Concepts; figure created by author.

In hydrothermal utilization, water is pumped from deep water-bearing rock layers, from which heat is extracted via a heat exchanger. The cooled water is usually returned to the same aquifer at a certain distance. Such a system consists of a production well and an injection well.

As deep waters often have a high mineralization and high gas content, reinjection is also necessary for disposal reasons. From a hydrogeological point of view, it is problematic if the injection does not take place in the same aquifer from which production takes place. The aim is to maintain the hydraulic balance and not to pump the thermal water reservoir dry.[21] As a rule, the production and injection wells are drilled from one well site, with the useful horizon being developed underground by deviated wells. The hydraulic connection to the aquifer is more favorable than with vertical wells.

The technology of hydrothermal utilization using doublets is tried and tested. The water, which is extracted and re-injected after cooling, circulates above ground in a closed circuit, which usually has to be kept under pressure to prevent the precipitation of minerals from the water. The thermal water pumped to the surface with the aid of a submersible pump is passed through a heat exchanger and the heat obtained is fed into a secondary circuit. At temperatures above 100 C, electricity can be produced using additional technologies such as an ORC system (**Organic Rankine Cycle**) or a Kalina system (water-ammonia mixture as a working medium). Both processes use low-boiling fluids. However, only ORC systems have longer-term practical experience.[22]

With the HDR process, the deeper underground can be used as a heat source for generating electricity. The extraction of geothermal energy is independent of water-

20 PK Deep Geothermal Energy 2007, p. 4 f.

21 GEFGA, p. 6.

22 PK Deep Geothermal Energy 2007, p. 11.

bearing useful horizons. A temperature range of 150 to 250 C and depths of around 5,000 m are usually targeted. According to current knowledge, the crystalline bedrock of the Earth's upper crust is fissured. These fissures are partially open and water circulates through them, albeit at very low flow rates. The crystalline bedrock therefore behaves like an aquifer with low permeabilities. Water pressure is used to open up flow paths between the boreholes and widen the naturally existing fracture system.[23] In order to achieve the necessary flow rates and temperatures, the fissure system must have a minimum size for the heat exchange surface. Water is sent through this heat exchanger via production wells in order to absorb the rock heat. For an HDR project, the temperature and therefore the drilling depth are decisive; temperatures of over 200 C are aimed for. Another selection criterion is the stability of the rock mass. Very heavily disturbed areas should be avoided in the area of the planned stimulation sections and circulation areas. Furthermore, water losses should be as low as possible or controllable and be less than 10%.

Provided that only the existing fracture network can be utilized, the natural fracture density of the rock should be medium to high. The fracture system should be relatively evenly distributed in order to achieve an optimum size for the heat exchange surfaces during stimulation under the specified stress field. Experience with HDR projects has shown that stimulation results in the formation of a steep, ellipsoid-shaped reservoir in accordance with the prevailing stress field. A sufficiently high permeability should be generated during the stimulation measures. If the permeability is too high, there is a risk of hydraulic short circuits and therefore insufficient heat transfer.[24]

12.5.3 The Completion Risk – Involvement of a General Contractor

The completion risk includes all risks and the resulting losses that are realized if the project plant is not completed with the contractual performance, delayed, at higher costs or not completed at all.[25] The completion risk is of particular significance for geothermal projects.[26]

In geothermal projects, the completion risk is closely related to the technical risk and the exploration risk. This can be seen, for example, in the distance between wells: in hydrothermal applications, there must be no hydraulic or thermal short-circuiting between the production and injection wells.[27] The distance between the injection and production wells must be large enough to ensure that no detrimental temperature re-

23 GEFGA, p. 7.
24 PK Deep Geothermal Energy 2007, p. 16.
25 J. Böttcher 2009, pp. 73–79.
26 See also section 16.
27 PK Deep Geothermal Energy 2007, p. 12 f.

ductions can occur in the production well within the planned project duration as a result of the cooled water being discharged into the productive horizon via the injection well. Therefore, certain minimum distances between the two boreholes in the aquifer must be maintained. However, the distance must not be too great to ensure a hydraulic connection between the two boreholes and thus a permanent yield of the production borehole. Numerical models are used to optimize the distance between the production and injection wells.[28]

The drilling risk is a key risk during the completion phase of a geothermal project. It manifests itself in increased drilling costs or a drilling time overrun, which can lead to the abandonment of the borehole and the termination of the project. The drilling risk includes damage due to technical causes – e.g. rod breakage – well as geological causes – borehole stability. As the damage often occurs deep below the earth's surface, it is often difficult to determine the cause of the damage. However, this is important as damage caused by geological factors is attributed to the client, i.e. the project company. Technical damage may have been caused by the drilling company's materials, for example. However, both causes of damage often work together. In view of this mixed situation, the only advice that can be given for project financing is to take out insurance with an insurer that covers both causes of damage.

A special geothermal risk is the subsidence risk, which describes the sinking of the ground due to tectonic and thermal processes. Here, the balance in the earth can be disturbed by the extraction of deep water. It must be ensured that the extracted water is pumped back into the subsoil so that there is no erosion of the soil and long-term subsidence. If the soil settles, this can lead to damage to the power plant and the surrounding buildings. However, if the extracted water is returned to the ground, damage can regularly be avoided.[29]

There are several ways to reduce the risks posed by induced seismicity: Firstly, reinjection conditions should be carried out during the operational test of the plant. In order to keep the imbalance as low as possible, there must be clarity about the fluid paths in the subsurface by means of hydraulic-thermal mass transport modeling. This is an investor risk that can be minimized through proper planning and implementation.[30] Furthermore, insurance policies can be taken out that take into account the probability of an earthquake and the resulting damage.[31] The risk can be monitored using metrological controls and seismological reports.[32]

The completion phase can be divided into the following phases (Figure 12.10):

28 PK Deep Geothermal Energy 2007, p. 12 f.

29 Schierenbeck/Trillig, p. 103.

30 Sass/Homuth, p. 19.

31 Schierenbeck/Trillig 2011, p. 103.

32 E. Huenges 2011, p. 39.

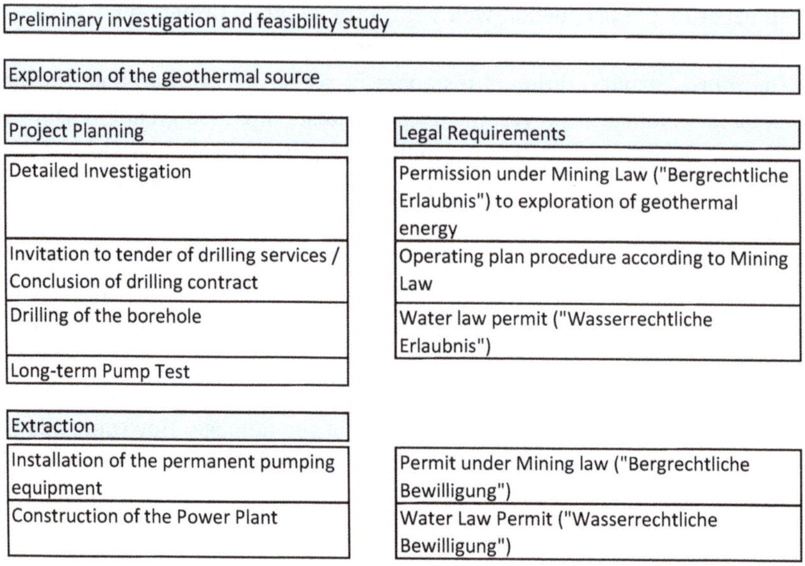

Figure 12.10: Flow chart completion phase of a geothermal project.[33]

The completion risk can have a considerable impact on the project and, in the worst case, make commercial operation impossible and thus lead to the project being abandoned. As banks will only grant project financing if the project cash flow is sufficiently high and stable, they generally demand extensive liability from one of the project participants if such a price risk is identified.

To counteract the completion risk, a number of contracts have been developed that allocate this risk – to varying degrees – to sponsors, borrowers and equipment suppliers. It is not uncommon for a penalty payment to be agreed to compensate for the shortfall in revenue if a deadline entitling a certain tariff is missed. The penalty can be selected in such a way that the resilience of the project remains constant from the bank's perspective.

In principle, the usual financial options for limiting the consequences of a completion risk, as shown in Table 12.2 can be classified as shown in Table 12.2.

Due to the very far-reaching scope of a completion guarantee on the one hand and the risks that are often difficult to manage during project creation on the other, rules are often agreed that limit the guarantor's obligations.

In the normal case of limited recourse financing, the risk assumption changes with the completion of the plant: while the sponsors or the plant constructor were previously responsible for the completion and at least partially liable to the lenders, it is now only

33 Figure based on A. von Dobschütz 2011, p. 793.

Table 12.2: Distribution of completion risks among investors.

	Completion guarantees	Obligation to make additional contributions
Object:	The sponsors are responsible for repaying the loans until the project is completed.	If the planned costs are exceeded, sponsors or lenders undertake to provide the project with additional equity or borrowed capital.
Scope or form:	The scope of the completion guarantee can relate to the total amount of the project loans or only to a certain percentage.	1. Completion Undertaking: The sponsors must provide additional capital until completion is achieved. If this obligation is unlimited, this corresponds economically to a completion guarantee 2. Pool-of-funds agreement: In economic terms, this is a post-financing obligation of the sponsors for a limited amount.

the project that is transformed into non-recourse project financing.[34] This temporal limitation of the sponsors' liability is the main economic reason for them to choose project financing instead of corporate financing. As this change in liability is crucial for the allocation of risk, great care is regularly taken to define when "completion" has been reached.[35] As a rule, completion is determined by an independent expert who, in addition to determining construction, also carries out certain performance tests.

Covering the completion risk through insurance is an option that is particularly important for geothermal projects. Both the exploration risk and the drilling risk can be insured.

When insuring the borehole risk, the insured items and damage are described in a positive list. The insurance also covers unforeseen events during the drilling phase, such as the loss of the drill pipe. The prerequisite for exploration insurance is that the probability of success of a geothermal project can be calculated, which is regularly done by means of a geothermal survey. From the point of view of project financing, it is ultimately crucial that the economically targeted thermal output is achieved. Ideally, the thermal output of the thermal water should therefore be insured in an explo-

34 This means for the project appraisal: The lenders must not only consider the viability of the project based on its expected cash flow stream in the operating phase, but also pay attention to the creditworthiness of the sponsors or the general contractor until the completion phase. It is important to bear in mind that the liability of the sponsors or the general contractor is not unlimited, but is usually limited in terms of amount for economic reasons. This consideration is therefore more about the right incentives.

35 The earliest possible point in time is the installation of the system, i.e. the end of construction and assembly work (physical completion). However, the value of a system depends on its functionality – completion in this context means the test run, during which certain performance parameters must be demonstrated. In addition, a certain operating time may be required, during which certain performance parameters must be proven in stages. The most far-reaching requirement is that certain economic criteria of the plant operation must also be demonstrated (**economic test**). If parameters are used that are not related to the plant itself (e.g. realized demand), the character of non-recourse project financing shifts back in the direction of corporate financing.

ration insurance policy. In addition, the influence of mineralization on the density of the thermal water must be taken into account: The higher the mineralization, the higher the density and the lower the necessary extraction rate to achieve the required thermal output. This means that in the Upper Rhine Graben, for example, a lower flow rate or lower temperature is required for a given thermal output due to the highly saline fluids than in the Molasse Basin with less mineralized thermal waters.[36]

Having outlined the topic of completion, we now turn our attention to operational risk.

12.5.4 Operating and Management Risk

Operational and management risk refers to all hazards that can lead to interruptions or even a shutdown of the system. The causes of an operational and management risk are usually errors in the planning, organization, implementation and control of operational processes (e.g. logistical weaknesses or miscalculations) or incorrect operation as well as inadequate maintenance and servicing by the system personnel. The operational risk is closely linked to the technical risk. Ultimately, how the components are loaded and the long-term performance of the project depends on the operation of the system.

Often the operational and management risk can also be attributed to the inexperience of the management itself.[37] Even hiring experienced staff does not guarantee good operational management. In the case of complex projects, it is important that the team works well together and is properly trained in the project, in addition to their qualifications.

Depending on their extent, the resulting restrictions on production operations have a negative impact on production volumes and therefore on sales and revenue. Furthermore, the operating risk can manifest itself in increased production costs, for example due to technical problems with the project system during the production process. If the earnings situation remains constant, these increased costs in turn reduce the cash flow.[38] As this represents the most important security after project completion due to the usual loss of the completion guarantee and is the primary source of repayment, lenders react sensitively to operational disruptions, so that banks prefer a management that has sufficient technical and commercial experience in the operation of a similar plant. If the sponsors do not have the necessary operational management experience, a professional operating and management company is required, which undertakes to ensure the continuous operation of the project and the functionality of the project plant. The selection of the operator should be based on the following criteria and is often based on the company's reference projects:

---

36 Enerchange 2012, p. 19 f.

37 M. Schulte-Althoff 1992, p. 118.

38 W. Schmitt 1989, p. 146; H. Uekermann 1993, p. 75.

- Reputation of the company,
- Ability to manage operations,
- Experience in the operation of comparable systems and
- Ability to provide suitable personnel.

The legal structure is provided by the management contract, which precisely defines the operator's rights and obligations. In order to provide an appropriate incentive for the operator, his remuneration should be at least partly variable: Profit shares and penalties act as an incentive for better management and efficient operation of the project facility and form the counterpart to the sponsors' return on equity.[39] When choosing an operator, a term corresponding to the project and the project loans should therefore be agreed. In the event that the suitability of the operator is misjudged or poor performance requires a change, the operating and management agreement should allow for a right of termination.

In many projects, a long-term maintenance contract is concluded with the manufacturer in order to avoid problems arising from the interface between the operator and manufacturer as far as possible. This means that important maintenance work is carried out outside of the operator's responsibility and signs of faulty operation can be detected at an early stage.

The use of the subsurface is particularly dependent on the efficient management of the reservoir. This requires a qualitative and quantitative understanding of the overall system of borehole and reservoir in the subsurface. With this understanding, the processes in the borehole, in the area close to the borehole and in the reservoir can be controlled.

After this brief introduction to the topic of operations, we now turn to one of the most important topics in project realization, the assessment of the legal and regulatory system.

12.5.5 The Legal and Regulatory Risk in Germany

As already described in the introduction, the stability and reliability of the regulatory environment is of paramount importance for the chances of realizing a geothermal project.

In this section, we limit ourselves to the heating market and ignore the use of heat for electricity generation.[40] In Germany, heat is largely marketed freely on the market and has to assert itself against alternative generation concepts.

39 For more detailed explanations of the structure of management contracts, see H. Uekermann 1993, p. 76 ff. and D. Tytko 1999, p. 84 et seq.

40 This is largely due to the fact that the electricity generation costs of alternative renewable energies have fallen significantly. However, this does not rule out the fact that there is also easily accessible potential worldwide that makes electricity generation the best solution from an economic perspective.

A remuneration system gives an initial impression of the attractiveness of a country for geothermal projects. In addition, however, it must be ensured that the project is provided with all rights in order to be constructed and operated as planned. The legal system must also allow the respective project contracts to be enforced. This means that the design of central project contracts is of paramount importance.

We will not go into the details of the legal regulations on mining law and water law here, but will only provide a short overview: There are two fundamental aspects of licensing law that must be observed when implementing deep geothermal energy in Germany. This is a two-stage process consisting of an exploration permit under mining law and a permit under mining law for the extraction of geothermal energy. If an area has been identified as promising in a feasibility study, a permit under mining law to explore for geothermal energy must be obtained for this area (**"exploration permit"**; "Aufsuchungserlaubnis").

Once the permit has been granted, the permit holder has the exclusive right to explore for geothermal energy in this area (**permit field**; "Erlaubnisfeld"). This makes it possible to carry out preliminary geophysical investigations, for example 3D seismics or exploratory drilling. The geographical extent and shape of a permit field varies depending on the location and population density, but is roughly in the region of 100 km^2. Permission under mining law is linked to conditions that must be observed in advance. The geothermal exploration permit is granted for three to five years depending on the federal state, although an extension is possible in justified cases.

A distinction must be made here between **mining permits** for the extraction of geothermal energy ("bergrechtliche Bewilligung"). The license gives the license holder the exclusive right to extract geothermal energy in the licensed field. The permit field ("Bewilligungsfeld") is generally part of the **exploration field** ("Aufsuchungsfeld"). As a rule, holders of the exploration permit also receive the permit.

If exploration permits or authorizations have already been granted elsewhere, it depends on the individual case. If the permit or authorization is not used for a longer period of time, the authority can reissue it. In practice, there is often also the possibility of poaching them from the holder. It is also possible for particularly large areas to be subsequently divided up by the authorities if this makes sense in order to allocate smaller areas to different holders.

12.6 Development of a Financing Structure from the Existing Risk Management

12.6.1 Fundamental Considerations

The risk identification and risk allocation process stages are followed by risk quantification, which also represents a review of profitability. For this purpose, the monetary consequences of the contractual and legal basis of a project are mapped using a cash flow model and examined with regard to possible changes to the schedule. As a rule, risk quantification does not end with a static cash flow assessment, but is supplemented by a rating tool that uses simulation calculations to map various environmental scenarios and arrive at a risk assessment of the project.

However, the cash flow model of a project is not only of paramount importance for the lenders, but also for the investors in a project. Both groups of investors are equally interested in the success of a project, although they have different levels and bases of entitlement. While the providers of debt capital have a fixed claim to servicing the debt service from the project that is independent of success, the providers of equity capital have a success-dependent and therefore variable claim to the remaining free cash flow. The methodological tool used by both groups to assess a project is a project-specific cash flow model.

However, the cash flow model does not mark the end point of the lenders' economic analysis. The next step is to carry out a simulation calculation of the cash flow progression, which provides information on how the project may develop under a variety of possible environmental scenarios. The result of these simulation calculations is a rating assessment that identifies a risk category and thus determines the interest costs via the risk premium and therefore also significantly influences the financing structure. In a second part, the aim is to work out which quantitative and qualitative factors can influence the rating.

In the following, a geothermal project will be examined in terms of its project financing capability by analyzing its risk potential. As the extent of the project risks depends to a large extent on the respective financing object, a case study from practice is examined and evaluated.

The assumptions of the project are as shown in Figure 12.11.

As a rule, the first step – starting from the base case – is to change various key cash flow-relevant parameters and examine their impact on cash flow. In this case, we have already assumed a 20% decline in revenue. This results in the following key economic parameters (Figure 12.12):[41]

The actual summary quantification of a project risk is carried out using a cash flow model which, in addition to assessing the project risks, also allows a certain optimization of the financing structure.

A somewhat more detailed analysis shows the following overview (see Figure 12.13).

Overall, this results in a very good debt service coverage ratio (min. DSCR in 2026 of 2.01) and an equally good return on equity of 14.26% p.a. Both investor groups are therefore very likely to be willing to participate in the project.

This raises the question of what happens if a discovery risk occurs and the technical performance is worse than expected. We will look at this question in the next section.

41 The return on equity calculated over 20 years is 14.54%.

Project Description	Project Name:	Geothermal Project		
Geothermal Project	1		0	1: yes, 0: no
Full Load Hours	6000,0	Insurance Percentage cost (in %)	10%	
Start of Operation	2025	Insurance Premium in €	1.800.000	
Heat Production in kWth	15.000	Partial Success starts at x %	80%	
Annual Amount of Heat (MWth)	90.000	Failure starts at y %:	60%	
Degradation p.a. (in %)	0	Deductible:	10%	
Heat Losses (in %):	20,00%	Insurance Payment:	0	
Production Level (in %)	100,00%	Insurance pays after year:	1	

Investment / Financing (Amounts in Euro)			
Drilling Expenses	18.000.000	Debt Service Reserve Target:	30%
Above Ground Facilities	7.500.000	Initial Amount DSRA:	0
Project Work	2.600.000		
Exploration	2.500.000	Depreciation in years:	20
Development Costs	1.000.000	Rate of Assessment in %:	330%
Other Costs	1.600.000	Basic federal rate:	5%
Preparation Works	500.000		
Insurance Premium	1.800.000		
Sum	35.500.000	Allowance	0

	Amount in Euro	Interest Rate in % p.a. (as of: 02.04.2024)	Start of Repayment	Maturity	ZB
Equity / Subsidy	14.200.000				
Senior Loan 1	0	4,67%	30.06.2026	10	10
Senior Loan 2	21.300.000	5,31%	30.06.2025	20	10
Sum	35.500.000				
Decommissioning Bonds	900.000	1,00%		20	
VAT Facility	6.745.000	5,31%		1	
Interest Rate for BWE-Interim		5,31%	final	1	

Prices in Cent/kWh			Inflation
Years 1 bis 5	7,00	1	
Years 6-10	7,00		
Years 11-15	7,00		
Years 16-30	7,00		2,0%

Operative Costs (in € p.a.)		Period	Inflation	Average Cost Quota	
Variable Costs	800.000		3,0%	36,73%	
Fixed Costs	390.000		3,0%		
Public Relation	50.000		3,0%	IRR 20J:	14,26%
Direct Marketing	0		3,0%	IRR 30J:	15,09%
Insurance	50.000	1	3,0%		
Personnel Costs	300.000		0,0%		
Sum	1.590.000				

Figure 12.11: Input Parameter for a Geothermal Project (own calculation).

12.6.2 Partial Discovery Risk – Consequences for the Financing Structure

In this section, we explain what happens if the project performs significantly worse than described in the planning. This can also be a manifestation of the risk of failure – often a situation may arise in which the risk of failure does not occur as a 0/1 risk, but as a gradual manifestation. This results in a certain degree of discoverability, but not the one that was planned.

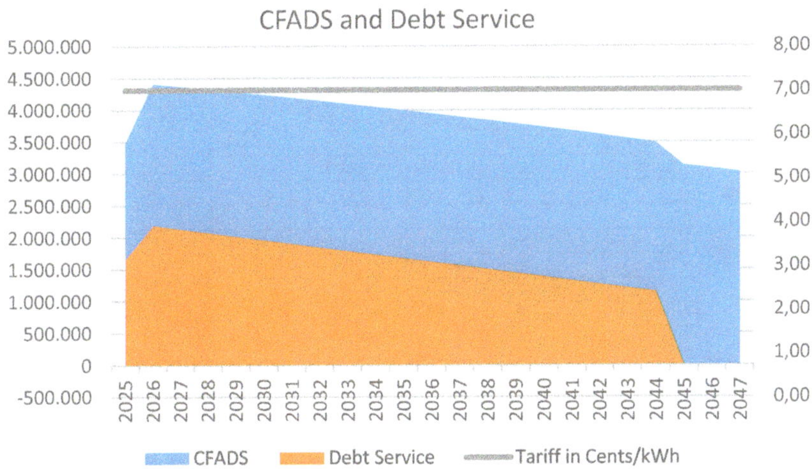

Figure 12.12: Geothermal Project (Cash Flow and Debt Service in a Base Case Scenario); figure created by author.

We are approaching this situation again using our model calculation (see Figure 12.14). However, we have now included a discount of 40%. At this point, the bank requires that its minimum DSCR target value of 1.40 is again adhered to. Without making any optimizations to the financing structure, the only remaining option is to reduce the amount of borrowed funds, in our example from € 21.3 million to € 15.2 million. This means, that the equity injection rises in parallel to 20.3 M€ (and now also includes real equity from the investor).

The financing bank thus achieves the same risk position as in the previous example with only a 20% discount on income. This is relatively easy for the bank to implement, for example by negotiating with the borrower that a deterioration in energy yield will result in the loan funds being adjusted so that the same minimum DSCR is achieved. The bank then only has to allocate the maximum possible loan amount in its internal credit decision, describe the adjustment mechanism and is then able to act in the possible scenarios.

However, it is now crucial for the implementation of a project that the investor, as well as the bank, is willing to continue to participate in the project. This is probably not the case in all scenarios. Let's take a look at the investor's situation in this scenario (see Figure 12.15).

The investor is now faced with the situation of having to replace the lost debt with equity. Due to the poorer technical performance and the leverage effect, the return on equity deteriorates considerably and now only amounts to 4.42% (over 20 years) or 6.21% over 30 years. These are values at which an investor may no longer be prepared to invest.

This means that, from the bank's point of view, a deterioration in the discovery rate can be planned for. Depending on the result of the heat production, the bank reduces its

Geothermal Project

	1	2	3	4	5	6	7
	2025	**2026**	**2027**	**2028**	**2029**	**2030**	**2031**
Energy Production in MWh	90.000	90.000	90.000	90.000	90.000	90.000	90.000
Price in Cent/kWh	7,00	7,00	7,00	7,00	7,00	7,00	7,00
Income:	6.300.000	6.300.000	6.300.000	6.300.000	6.300.000	6.300.000	6.300.000
Logik-Test Versicherung	1	0	0	0	0	0	0
Insurance Payment	0	0	0	0	0	0	0
Sum Income	6.300.000	6.300.000	6.300.000	6.300.000	6.300.000	6.300.000	6.300.000
Sum Incom 100 %	9.000.000	9.000.000	9.000.000	9.000.000	9.000.000	9.000.000	9.000.000
Variable Costs	800.000	824.000	848.720	874.182	900.407	927.419	955.242
Fixed Costs	390.000	401.700	413.751	426.164	438.948	452.117	465.680
Public Relation	50.000	51.500	53.045	54.636	56.275	57.964	59.703
Direct Marketing	0	0	0	0	0	0	0
Insurance	50.000	0	0	0	0	0	0
Decommissioning Bonds	9.000	9.000	9.000	9.000	9.000	9.000	9.000
Interest Payment VAT	358.160	0	0	0	0	0	0
Interest Payments BWE	754.020	0	0	0	0	0	0
Personnel Costs	300.000	300.000	300.000	300.000	300.000	300.000	300.000
Sum OPEX	2.711.180	1.586.200	1.624.516	1.663.981	1.704.631	1.746.500	1.789.625
EBITDA	3.588.821	4.713.800	4.675.484	4.636.019	4.595.369	4.553.500	4.510.375
Depreciation	1.775.000	1.775.000	1.775.000	1.775.000	1.775.000	1.775.000	1.775.000
Interest Payment	1.131.030	1.102.754	1.044.715	986.675	928.635	870.595	812.556
EBIT	682.791	1.836.046	1.855.769	1.874.344	1.891.734	1.907.905	1.922.819
./. Allowance	0	0	0	0	0	0	0
Business Income	682.791	1.836.046	1.855.769	1.874.344	1.891.734	1.907.905	1.922.819
Tax Measurement Amount	34.140	91.802	92.788	93.717	94.587	95.395	96.141
Trade Tax	112.660	302.948	306.202	309.267	312.136	314.804	317.265
CFADS (after tax)	3.476.160	4.410.852	4.369.282	4.326.752	4.283.233	4.238.696	4.193.110
Interest	1.131.030	1.102.754	1.044.715	986.675	928.635	870.595	812.556
Repayment	532.500	1.093.026	1.093.026	1.093.026	1.093.026	1.093.026	1.093.026
CFADS (after repayment)	1.812.630	2.215.072	2.231.541	2.247.051	2.261.572	2.275.074	2.287.528
Target Value DSRA	658.734	641.322	623.910	606.498	589.087	571.675	554.263
Allocation DSRA	658.734	0	0	0	0	0	0
Use of DSRA	0	0	0	0	0	0	0
Decrease of DSRA	0	17.412	17.412	17.412	17.412	17.412	17.412
Free CF	1.153.896	2.232.484	2.248.953	2.264.463	2.278.983	2.292.486	2.304.940

	2025	**2026**	**2027**	**2028**	**2029**	**2030**	**2031**
DSCR	2,09	2,01	2,04	2,08	2,12	2,16	2,20
DSCR (with DSRA)	2,09	2,31	2,34	2,38	2,42	2,46	2,50
DSRA	0	658.734	641.322	623.910	606.498	589.087	571.675

Figure 12.13: Geothermal Project – Cash flow overview; figure created by author.

loan amount and thus achieves the same debt servicing capacity as originally contemplated. However, this may mean that the investor is no longer prepared to implement the project as it is no longer sufficiently profitable. But if the investor is no longer prepared to participate in the project, the project will not come into existence. However, the stress case assumed here is also very pessimistic. And the risk of discovery can be significantly reduced by a good data basis and thorough preliminary investigation.

The example also shows that an insurance solution requires an economic assessment on the basis of a cash-flow-model. The model has to reflect all the current data

Project Description	Project Name:	Geothermal Project		
Geothermal Project	1		0	1: yes, 0: no
Full Load Hours	6000,0	Insurance Percentage cost (in %)	10%	
Start of Operation	2025	Insurance Premium in €	1.800.000	
Heat Production in kWth	15.000	Partial Success starts at x %	80%	
Annual Amount of Heat (MWth)	90.000	Failure starts at y %:	60%	
Degradation p.a. (in %)	0	Deductible:	10%	
Heat Losses (in %):	20,00%	Insurance Payment:	16.200.000	
Production Level (in %)	60,00%	Insurance pays after year:	1	

Investment / Financing (Amounts in Euro)			
Drilling Expenses	18.000.000	Debt Service Reserve Target:	30%
Above Ground Facilities	7.500.000	Initial Amount DSRA:	0
Project Work	2.600.000		
Exploration	2.500.000	Depreciation in years:	20
Development Costs	1.000.000	Rate of Assessment in %:	330%
Other Costs	1.600.000	Basic federal rate:	5%
Preparation Works	500.000		
Insurance Premium	1.800.000		
Sum	35.500.000	Allowance	0

	Amount in Euro	Interest Rate in % p.a.	Start of Repayment	Maturity	ZB
Equity / Subsidy	20.300.000	(as of: 02.04.2024)			
Senior Loan 1	0	4,67%	30.06.2026	10	10
Senior Loan 2	15.200.000	5,31%	30.06.2025	20	10
Sum	35.500.000				
Decommissioning Bonds	900.000	1,00%		20	
VAT Facility	6.745.000	5,31%		1	
Interest Rate for BWE-Interim		5,31%	final	1	

Prices in Cent/kWh			Inflation
Years 1 bis 5	7,00	1	
Years 6-10	7,00		
Years 11-15	7,00		
Years 16-30	7,00		2,0%

Operative Costs (in € p.a.)		Period	Inflation	Average Cost Quota	
Variable Costs	800.000		3,0%	61,51%	
Fixed Costs	390.000		3,0%		
Public Relation	50.000		3,0%	IRR 20J:	4,42%
Direct Marketing	0		3,0%	IRR 30J:	6,21%
Insurance	50.000	1	3,0%		
Personnel Costs	300.000		0,0%		
Sum	1.590.000				

Figure 12.14: Geothermal project (downside scenario); figure created by author.

and has to make assumptions regarding an acceptable internal rate of return as well as a minimum debt service cover ratio.

Overall, the insurance concept described above would in principle be suitable for overcoming the market failure caused by a discovery risk. And the geothermal project is proving to be extremely economical in continuous operation.[42]

42 This results in the interesting constellation that a hedging strategy for the discovery risk is absolutely necessary, but conversely, investment support in the form of subsidies is not necessary and not

Geothermal Project

	1	2	3	4	5	6	7
	2025	**2026**	**2027**	**2028**	**2029**	**2030**	**2031**
Energy Production in MWh	54.000	54.000	54.000	54.000	54.000	54.000	54.000
Price in Cent/kWh	7,00	7,00	7,00	7,00	7,00	7,00	7,00
Income:	3.780.000	3.780.000	3.780.000	3.780.000	3.780.000	3.780.000	3.780.000
Logik-Test Versicherung	1	0	0	0	0	0	0
Insurance Payment	16.200.000	0	0	0	0	0	0
Sum Income	19.980.000	3.780.000	3.780.000	3.780.000	3.780.000	3.780.000	3.780.000
Sum Incom 100 %	9.000.000	9.000.000	9.000.000	9.000.000	9.000.000	9.000.000	9.000.000
Variable Costs	800.000	824.000	848.720	874.182	900.407	927.419	955.242
Fixed Costs	390.000	401.700	413.751	426.164	438.948	452.117	465.680
Public Relation	50.000	51.500	53.045	54.636	56.275	57.964	59.703
Direct Marketing	0	0	0	0	0	0	0
Insurance	50.000	0	0	0	0	0	0
Decommissioning Bonds	9.000	9.000	9.000	9.000	9.000	9.000	9.000
Interest Payment VAT	358.160	0	0	0	0	0	0
Interest Payments BWE	1.077.930	0	0	0	0	0	0
Personnel Costs	300.000	300.000	300.000	300.000	300.000	300.000	300.000
Sum OPEX	3.035.090	1.586.200	1.624.516	1.663.981	1.704.631	1.746.500	1.789.625
EBITDA	16.944.911	2.193.800	2.155.484	2.116.019	2.075.369	2.033.500	1.990.375
Depreciation	1.775.000	1.775.000	1.775.000	1.775.000	1.775.000	1.775.000	1.775.000
Interest Payment	807.120	786.942	745.524	704.106	662.688	621.270	579.852
EBIT	14.362.791	-368.142	-365.040	-363.087	-362.319	-362.770	-364.477
./. Allowance	0	0	0	0	0	0	0
Business Income	14.362.791	0	0	0	0	0	0
Tax Measurement Amount	718.140	0	0	0	0	0	0
Trade Tax	2.369.860	0	0	0	0	0	0
CFADS (after tax)	14.575.050	2.193.800	2.155.484	2.116.019	2.075.369	2.033.500	1.990.375
Interest	807.120	786.942	745.524	704.106	662.688	621.270	579.852
Repayment	380.000	780.000	780.000	780.000	780.000	780.000	780.000
CFADS (after repayment)	13.387.930	626.858	629.960	631.913	632.681	632.230	630.523
Target Value DSRA	470.083	457.657	445.232	432.806	420.381	407.956	395.530
Allocation DSRA	470.083	0	0	0	0	0	0
Use of DSRA	0	0	0	0	0	0	0
Decrease of DSRA	0	12.425	12.425	12.425	12.425	12.425	12.425
Free CF	12.917.847	639.283	642.385	644.338	645.106	644.656	642.949

	2025	**2026**	**2027**	**2028**	**2029**	**2030**	**2031**
DSCR	12,28	1,40	1,41	1,43	1,44	1,45	1,46
DSCR (with DSRA)	12,28	1,70	1,71	1,73	1,74	1,75	1,76
DSRA	0	470.083	457.657	445.232	432.806	420.381	407.956

Figure 12.15: Geothermal Project – Cashflow in a Stress Scenario; figure created by author.

---

advisable from a regulatory perspective. I show an alternative to a mere injection of susidies in section 4.2.5.

Attakan Janpidok

13 Indonesia's Deep Geothermal Drilling Program: Background and Experiences

13.1 Introduction

At the end of 2021, 13% of the energy produced in Indonesia was generated from renewable energy. This is significantly below the target of 23% pledged by the government by the end of 2025 and 31% by the end of 2050 in the 2007 Energy Law (Law No. 20 of 2007).

Indonesia is blessed with many renewable potentials, including geothermal energy, which is believed to be as large as 25,386 MWe or equal to 21% of the world's geothermal potential (Directorate General of New, Renewable Energy and Energy Conservation, 2020).

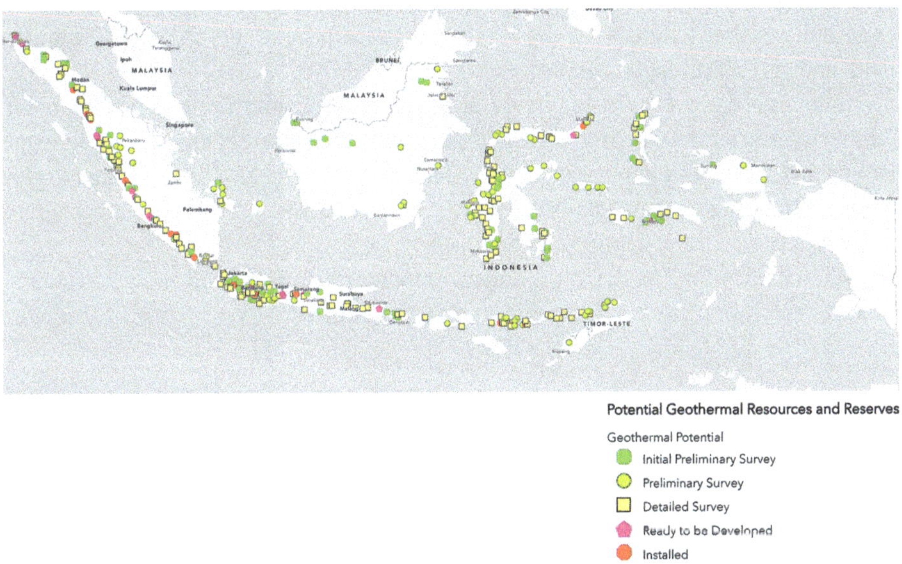

Figure 13.1: Indonesia map of geothermal resource potential. Source: geoportal.esdm.go.id.

Despite the geothermal potential and government efforts to support the industry, only 2.1 GW of geothermal capacity has been developed as of 2021. Approximately 8% of the total resources native to Indonesia are currently developed to produce power (see Figure 13.1).

According to the World Bank's report No. PAD1595, low levels of private sector participation have contributed to slower-than-desired geothermal development. This reflects high resource risk, a critical barrier to geothermal development which remains unaddressed in Indonesia.

https://doi.org/10.1515/9783110784299-013

Resource risk is intensified by high exploration drilling costs plus supporting infrastructure such as access roads and bridges. Considering a typical requirement of a minimum of three exploration wells for proper resource estimation, the total cost can be prohibitive to investors, especially when success cannot be assured. Exploration drilling is also considered the most significant barrier to obtaining financing as its high associated risks increase investors' equity return requirements.

Besides the resource risk, Indonesia also faces a structural issue where conflicting objectives exist among main government stakeholders, namely the Minister of Finance (MoF), Ministry of State-Owned Enterprises (MSOE) and state-own enterprises (SOEs), and the Ministry of Energy and Mineral Resources (MEMR). The MoF is concerned about the Public Service Obligation's size, while the subsidy must increase to achieve the geothermal target. MSOE and SOEs have their objective to deliver commercial performance of the enterprises, which includes other energy sources, e.g., oil and gas, that yield much higher returns. SOEs often face battles for resources from their parent company. Finally, MEMR is responsible for promoting geothermal energy, but whether its goals can be met is determined by others (Asia Development Bank and The World Bank, 2015).

Indonesia has a very long history of geothermal exploration and development. This chapter briefly summarizes the key activities and important changes in the regulations that impacted the market. The project economics will also be examined to give an understanding of the geothermal project cost breakdown, as well as the geothermal well's success rate, depth, and costs.

We will also look at the government's monetary support, and the government-funded exploration campaign recently commenced in 2020.

13.2 History of Deep Geothermal Drilling in Indonesia

This section splits Geothermal exploration and development activities in Indonesia into two periods, i.e., before the year 2000 and after the year 2000.

13.2.1 Before the year 2000

Hochstein and Sudarman (2008) summarized the history of geothermal exploration in Indonesia from 1970-2000, that geothermal drilling activities in Indonesia began in 1926 when three shallow wells (66, 123, and 128 meters deep) were drilled inside a large fumarole field in Kamojang Crater, West Java. The exploration was then followed by another shallow well drilled in the Dieng field (Sikidang Crater) in 1928 to a depth of 80 meters. After those wells, no drilling activities were made until 1972.

The Indonesian State Oil Company (PERTAMINA) entered geothermal exploration from 1974 onwards and became responsible for all geothermal exploration in Java and Bali, in line with Presidential Decree 16/1974. The exploration notably increased in various fields, including Kamojang, Dieng, and Darajat (Sudarman 2008).

In 1981, a Presidential Decree (No. 20/1981) allowed PERTAMINA to enter joint ventures with local and international partners.

In 1982 Unocal Geothermal Indonesia (UGI) signed a Joint Operating Contract (JOC) with PERTAMINA to develop the high-temperature reservoir of the Gunung Salak prospect. At the same time, UGI also agreed to an energy sales contract (ESC) involving UGI in supplying steam to PERTAMINA, which, in turn, would sell it to PT Perusahaan Listrik Negara (Persero) or PLN, a state-owned utility company, to operate power stations after successful field development.

The first geothermal power plant in Indonesia was developed in 1983. To encourage the development of other geothermal fields, the Government of Indonesia (GoI) allowed private companies to participate by issuing Presidential Decree No. 45 of 1991. During the 1990s, PLN offered high geothermal tariffs (around 7–10 US cents/kWh).

Geothermal drilling increased from 1990-1995, which may be caused by the assignment by the Government to PERTAMINA on almost all of Indonesia's geothermal working area (GWA) and the implementation of Joint Operating Contract (JOC) following Presidential Decree 22 of 1981 and 49 of 1991 that allow foreign and private investors to develop and to run so-called independent power projects (IPPs). IPPs were to sell geothermal power under Energy Sales Contracts (ESC) to the state electricity company PLN. At least seven Energy Sales Contracts (ESC) were signed between JOC and PT PLN.

Unfortunately, the Asian Financial Crisis of 1997 caused the Indonesian Rupiah to fall dramatically with respect to the US Dollar and compromised PLN's financial position. The Government of Indonesia was consequently forced to decrease the geothermal tariff to below 5 US cents/kWh. Thus, most geothermal projects became unfeasible, and many private geothermal companies withdrew from Indonesia.

During 1995-2000, six prospects were tested by deep drilling, which discovered productive high-temperature reservoirs at Patuha, Karaha, Namora-I-Langit, and Bedugul on Bali. All discovery wells were drilled by the private sector. A deep exploration well was drilled by Pertamina at Ulebulu; the Mataloko prospect on Flores was also explored by deep drilling, supported by government and aid funding (Sudarman, 2008).

13.2.2 After the year 2000

In 2000, the government issued Presidential Decree No. 76 (PR 76) concerning the Exploitation of Geothermal Resources for Power Generation. Following PR 76, PERTAMINA was no longer the only business entity managing geothermal energy in Indonesia. PERTAMINA returned 18 of the 33 Geothermal Concession Areas (WKP) that it had previously managed to the Government (PGE Annual report, 2021).

In 2003, the Geothermal Law (Law 27/2003) was declared, making geothermal the only renewable energy governed by its own law (ADB, 2015).

In 2005, the Directorate of Geothermal Enterprise Supervision and Groundwater Management was established by MEMR to strengthen sector management and support. This became the Directorate of Geothermal Energy in November 2010 (ADB, 2015).

In 2006, MEMR initiated the Master Plan Study for Geothermal Power Development in Indonesia, funded by the Japan International Cooperation Agency (JICA), further solidifying knowledge and understanding about developing Indonesia's geothermal resources. (ADB, 2015). This Fast-Track Program was designed to expedite the development of 10,000 MW capacity, equivalent to about one-third of PLN's then-total existing system capacity (ESMAP, 2014).

In 2006, PERTAMINA established PT Pertamina Geothermal Energy based on Deed No. 10 dated December 12th, 2006, which was approved by the Minister of Law and Human Rights of the Republic of Indonesia by Decree Number W7-00089HT.01.01-TH.2007 dated January 3rd, 2007 (PGE Annual report, 2021).

In 2012, the MEMR issued a feed-in tariff (FIT) policy for geothermal electricity based on the analytic work supported by the World Bank and/or Global Environment Facility Geothermal Power Generation Development Project (ADB, 2015).

In 2012, the Ministry of Finance (MoF) established a geothermal fund with more than $ 200 million of initial capitalization to mitigate resource risks related to geothermal development.

In June 2014, GoI revised the FIT, providing some relief to developers willing to take on exploration and development risks (ADB, 2015). Later in the year, a new Geothermal Law was issued, which allows centralizing geothermal concession tenders while securing the interest of the local government in geothermal development through a production bonus – a benefit-sharing mechanism – levied on top of any applicable taxes (ADB, 2015).

In 2015, GoI took the first step to transfer funds (about IDR 3.1 trillion or US$ 225 million) from what was previously known as the Geothermal Fund Facility (GFF) to the new Infrastructure Fund for Geothermal Sector (IFGS) in PT Sarana Multi Infrastruktur (PT SMI) for mitigation of geothermal exploration drilling risks, particularly in areas where development prospects are not attractive for purely private sector players. The original design of GFF was based on collateral-backed loans and failed to adequately address the high exploration risk issues since the GFF loans were to be paid back in full, even in the case of unsuccessful drilling. The design of IFGS will enable, among other things, government-sponsored drilling, which hinges on a more balanced approach to risk allocation in the overall geothermal development process. PT SMI has been given a government mandate to finance and facilitate exploration drilling with a specific focus on the eastern islands. However, it lacks the geothermal expertise needed to implement a pre-license drilling window and can only use a limited share of the IFGS funds for this purpose (ADB, 2015).

In 2017 MEMR Regulation 12/2017 set the ceiling price based on the Annual Average Costs of Generation (BPP). This policy was designed to reduce the overall electricity subsidies but keep the electricity end-user costs in Indonesia affordable to consumers. The resulting lower maximum geothermal tariffs make geothermal development less attractive in several (most) regions with low BPPs, especially the main islands (Java, Bali, Sumatra), where generation is dominated by cheap coal (Campen et al., 2017)

Until 2019, there were at least 766 wells drilled to confirm the existence of geothermal reservoirs and recover sufficient brine and steam to support geothermal power generation in Indonesia. Geothermal drilling activities in Indonesia up until 2019 are shown in Figure 13.2 below.

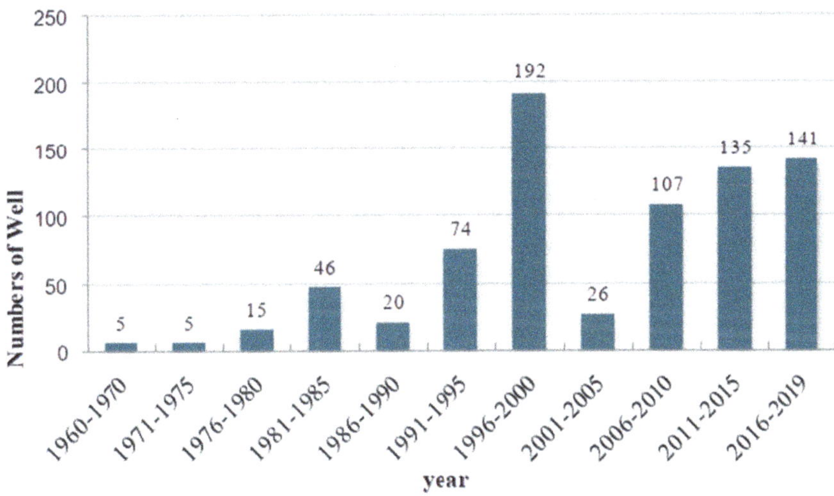

Figure 13.2: Geothermal Drilling Activity in Indonesia from the 1970s to 2019. Source: Purwanto, 2021.

In September 2022, a long-awaited Presidential Decree 112/2022 (PR112) was released. Assegaf Hamzah & Partners (2022) has summarized it as follows (see Table 13.1):

- PR112 exhibits the government's commitment to stop new development of coal-fired power plants as well as early retirement of the existing ones
- The ceiling tariff system remains. However, unlike MEMR Regulation 12/2017, the ceiling tariff, shown in Table 13.1, is no longer benchmarked against the Annual Average Costs of Generation (BPP). The new basis for the tariff is yet to be clarified
- The tariff depends on the location of the plant. A location factor ("F" factor) shall be applied to the base tariff. The "F" factor is listed in Table 13.2
- Base tariff is not subject to escalation, except for geothermal power plant
- The ceiling tariffs are subject to annual review by the Ministry

- For geothermal, PLN is allowed to procure the electricity or steam using the direct appointment method
- Appointment of the public agency or state-owned enterprise (SOE) to acquire additional geothermal data
- Appointment of a developer to conduct a preliminary survey and exploration in exchange for the Right of First Refusal in the tender
- Addition Incentives for all renewable energies mandated by PR 112 include fiscal and non-fiscal forms. Fiscal incentives encompass income tax facilities, import duty, land and building tax facilities, geothermal development support, and financing or project guarantees by Indonesian state-owned companies. Non-fiscal incentives are from central and regional governments. However, no clarity has been provided at this stage.

Table 13.1: Purchase prices of electricity and steam power. Note that the ceiling price is after the "F" factor is applied (sources: Presidential Decree 112/2022).

Purchase price of electricity from geothermal – 2022			
No.	**Installed Capacity**	**Ceiling Price (Cent USD/kWh)**	
		Years 1 to 10	**Year 11 to 30**
1	10 MW or below	(9.76 x F)*	8.30
2	>10 to 50 MW	(9.41 x F)*	8.00
3	>50 to 100 MW	(8.64 x F)*	7.35
4	100 MW or more	(7.65 x F)*	6.50

Purchase price of steam power (electricity equivalent) –2022			
No.	**Installed Capacity**	**Ceiling Price (Cent USD/kWh)**	
		Years 1 to 10	**Year 11 to 30**
1	10 MW or below	(6.60 x F)*	5.60
2	>10 to 50 MW	(6.25 x F)*	5.31
3	>50 to 100 MW	(5.48 x F)*	4.65
4	100 MW or more	(4.48 x F)*	3.81

Table 13.2: Location Factor or "F" factor (source: Presidential Decree 112/2022).

No	Location	"F" Factor
1	Java, Madura, Bali	1.00
	- Small island	1.10
2	Sumatra	1.10
	- Kep. Riau	1.20
	- Mentawai	1.20
	- Bangka Belitung	1.10
	- Small island	1.15
3	Kalimantan	1.10
	- Small island	1.15
4	Sulawesi	1.10
	- Small island	1.15
5	Nusa Tenggara	1.20
	- Small island	1.25
	North Maluku	1.25
6	- Small island	1.30
7	Maluku	1.25
	- Small island	1.30
8	West Papua	1.50
9	Papua	1.50

13.3 Recent Project Economics

Indonesia has had a long history of drilling deep geothermal wells since 1974. Although project financial data are not always made publicly available due to their nature of confidentiality, there are multiple studies conducted by several individuals and organizations. This section focuses on the economic aspect of recent geothermal projects in Indonesia.

13.3.1 Geothermal Project Investment Cost Breakdown

Directorate General NREEC published an analysis of investment costs for geothermal projects in Indonesia in 2016-2019. The analysis comprises four different project sizes, i.e., 10 MW, 30 MW, 90 MW, and 120 MW. From Figure 13.3, the three biggest cost components are EPCC power plant cost at an average of 40%, exploration dril-

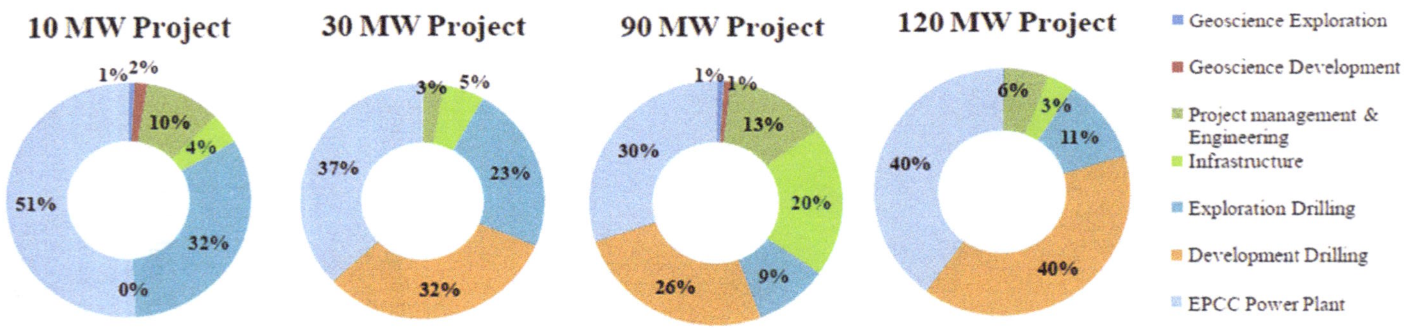

Figure 13.3: Investment cost distributions of four geothermal projects in Indonesia (modified from Directorate General of NREEC, 2020).

ling cost at an average of 19%, and development drilling cost at 25% of the overall project cost.

Combined exploration and development drilling cost ranges from 32% to 55% of investment costs, with an average of 43%. The drilling cost represents a significant portion of the investment. Unlike EPCC power plant cost, it also carries a large uncertainty due to complex subsurface risks, especially exploratory drilling.

The probability of success in drilling improves as the cumulative lesson learned is built over time (learning curve effect). Sanyal (2011) has studied the learning curve effect in Indonesia from all wells drilled in all stages (exploration, confirmation, and development) and reported that the success rate (successful wells being greater than 2 MW in capacity) stabilized at approximately 62% after 90 wells (see Figure 13.4 for more details). However, since the success rate naturally improves through different stages, it is better to analyze it on a project-to-project basis. Figure 13.5 shows the average drilling success rate at the Kamojang field. The drilling success rate increases with the number of wells, until a plateau of approximately 73% is reached after about 40 wells.

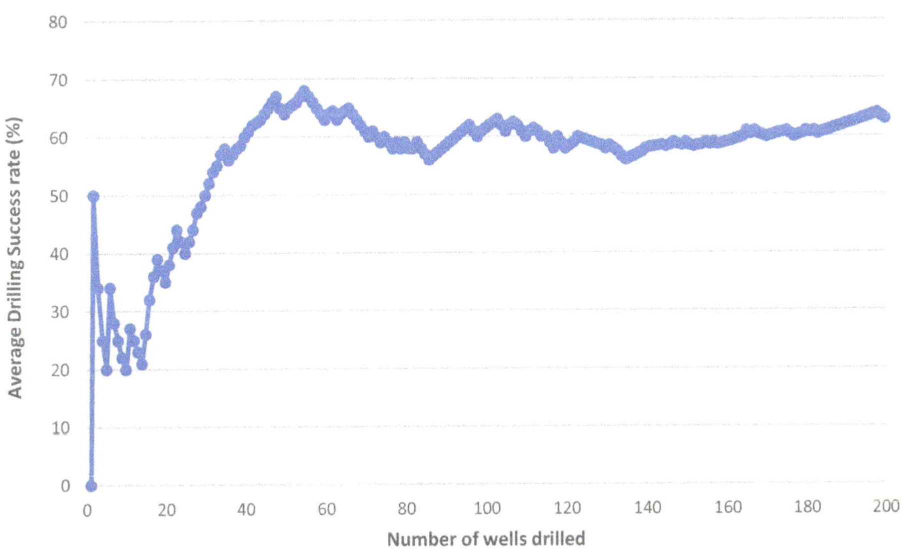

Figure 13.4: Average drilling success rate versus the number of wells drilled across Indonesia (Sanyal et al., 2011).

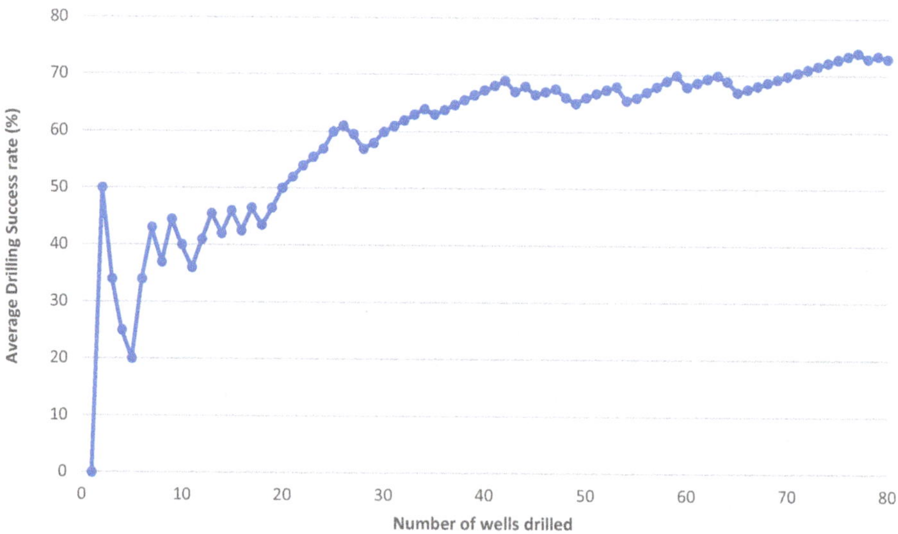

Figure 13.5: Average drilling success rate versus the number of wells drilled in the Kamojang field in Indonesia (Sanyal et al., 2011).

13.4 Deep drilling Cost Analysis

According to Sanyal (2011), depth is the main contributing factor to the drilling cost of a geothermal well, and drilling cost increases exponentially with depth. Figure 13.6 below shows a histogram of well depth from 609 wells in Indonesia. The chart indicates the range of 800-3500 mMD with a mean value of drilling depth of 1960 mMD.

Purwanto (2018) also examined 121 geothermal wells from 213 wells drilled from 2011 to 2018. Drilling costs were normalized using the US Department of Labor Bureau's Producer Price Index (PPI) to escalate the well cost to a 2018 US dollar value. The result is shown in Figure 13.7. Note that the drilling cost in the same field increases with depth. A large variety of well costs from wells within the same field can also be observed: For example, well costs in Hululias vary from $8 M to $18 M at a similar depth of approximately 3000 mMD. The reason for such variation could be due to various drilling challenges such as lost circulation, wellbore stability, stuck pipe, fishing, sidetracking, rig, or drilling equipment failure.

Zuhro & Arif (2015) suggested measuring drilling performance using geothermal drilling unit cost or GDUC (US$/m). As per Purwanto et al. (2021), statistical analysis suggests that the mean geothermal drilling unit cost (GDUC) of 203 geothermal wells in Indonesia from 2011 to 2019 is US$ 3,499/meter (see Figure 13.8). Note that the drilling cost for exploration holds the highest unit cost at US$ 3,947/meter, while the operational or make-up wells in a well-understood field can be drilled at about 19% lower cost or US$ 3,210/meter.

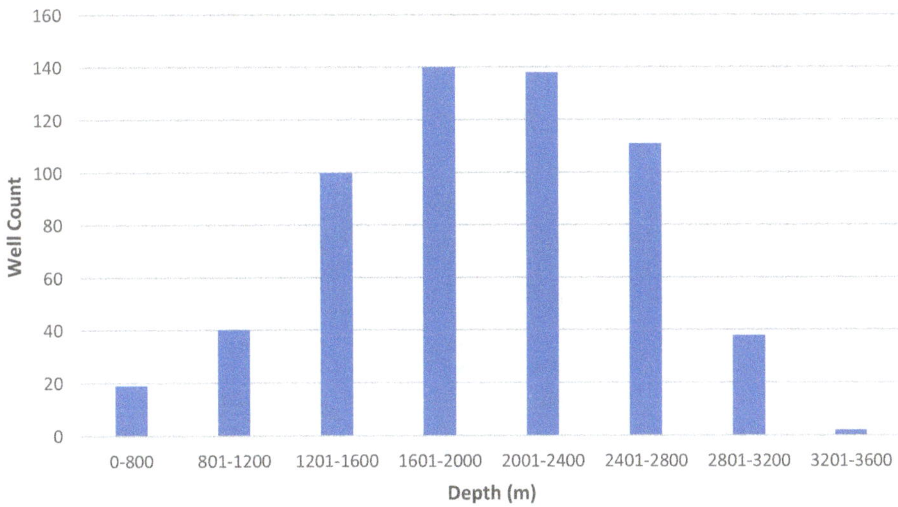

Figure 13.6: Histogram of geothermal wells drilled in Indonesia (Purwanto 2018).

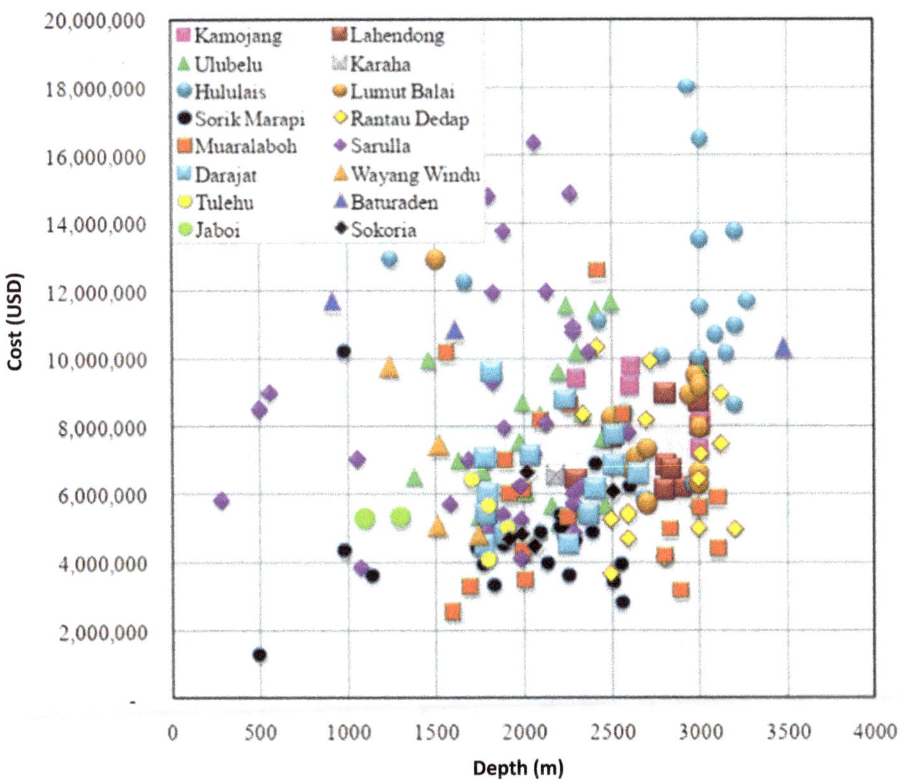

Figure 13.7: Geothermal Well Cost v.s. Depth. Source: Purwanto 2021.

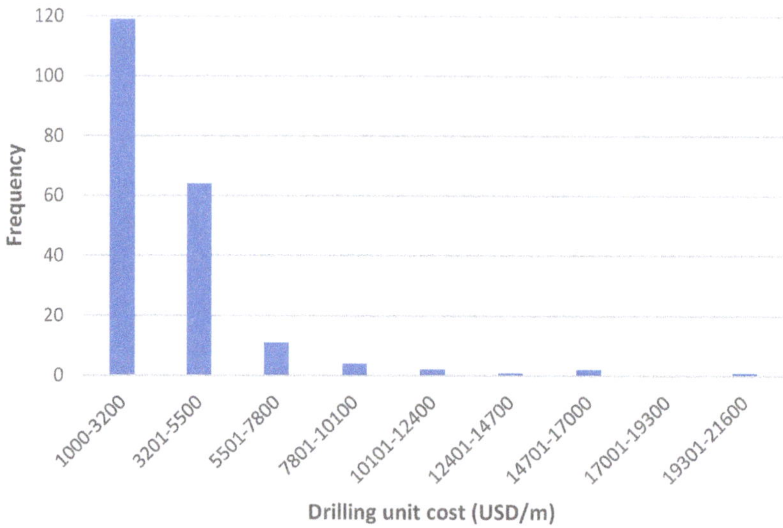

Figure 13.8: Geothermal Drilling Unit Cost (US$/m). Source: Purwanto 2021.

The end goal of drilling is to produce the geofluid to extract the heat; hence drilling cost per MW should also be evaluated. According to the study of geothermal drilling cost per MW well capacity conducted by Purwanto (2021), the cost per MW ranges from 0.4 to 1.5 MUSD per MW (see Table 13.3 for details).

Table 13.3: Geothermal drilling cost per MW in Indonesia (Purwanto 2021).

Field	Depth(m)	Cost (MUSD)	Output (MW)	Avg. Cost (MUSD/MW)	Cost (USD/m)	Wells
Kamojang	2308-3009	8.1–9.7	4–16.8	1.05	3516	4
Lahendong-Tompaso	2300-3003	6.1–8.8	2–14.8	1.33	2646	5
Ulubelu	1979-3000	5.7–9.7	5–30	0.72	3388	8
Lumut Balai	2495-2632	6.6–8.2	4.3–7.5	1.24	2851	4
Hululais	1668-3280	8.4–17.9	4.8–15.5	1.35	4067	5
Sorik Marapi	2089-2541	3.9–9.9	5–15	.74	3409	6
Rantau Dedap	2248-3123	3.6–10.2	2–20	1.48	2782	10
Muaralaboh	1600-3103	2.5–10.1	5–40	.43	2362	11
Sarulla	1799-2128	4.9–16.2	14–62	.35	5068	10
Darajat	1775-2652	4.5–6.8	4–37	.64	2610	8
Wayang Windu	1244-1747	4.7–9.7	6.7–18	.67	4691	4

In summary, drilling costs represent a significant portion of the overall project cost, i.e., 32–55%, per NREEC (2020). According to Sanyal et al. (2011), Indonesia's overall drilling success rate is 62%. Well depth ranges from 800-3500 m with an average drilling cost of US$ 3,499/m and US$ 0.4-$1.5M/MW.

Sudarman (2012) studied data from 16 geothermal tenders in Indonesia to generate a levelized cost of electricity. Arwin (2014) summarizes the findings and assumptions in Table 13.4 below. Note that Sudarman used a commercial loan from the international money market with a discount rate of 9% in his study. The outcome was extrapolated from various project sizes and plotted in Figure 13.9 below.

Table 13.4: Summary of parameters derived from the study of 16 tenders in Indonesia, source Sudarman (2012) and Arwin (2014).

Development Parameters	Value
POWER PLANT	
Plant Size (MWe)	2 x 55
Projected well output (MWe)	10
Drilling Success Rate:	
Exploration wells	50%
Appraisal wells	75%
Production wells	80%
Re-injection wells	
Ratio to production wells	1/3
Unsuccessful wells used for reinjection	30%
Ratio of wellhead main plant	0%
GENERATION PARAMETERS	
Excess steam at startup	10%
Steam decline rate	3%
Plant capacity factor	90%
Economic life (year)	30
Service life (year)	30
Operational hours per day	24
Operational days per year	365
Parasitic Load	5%
EQUITY RATE OF RETURN	
Expected rate of return on equity	16%
DEBT	
Interest	8.5 – 10 %
Grace period	7
Term	10–15
Arrangement fee	0.50%
Commitment fee	0.50%
Pay-back period	5–8
Debt ratio/equity	70%

Table 13.4 (continued)

Development Parameters	Value
TAX	
Corporate tax rate	25%
EXCHANGE RATE (Rp/US$)	10,000
SALES UNIT PRICE	0.094
O&M COST	0.016
PLANT SIZE	2 x 55 net
Ratio of production/injection wells	1/3
% steam wells used for reinjection	30%
Estimated well output (MWe)	10
Drilling cost (US$)	6,000,000
Drill pad cost(US$)	2,000,000
Exchange rate (Rp/US$)	10,000
PIPELINE COST (US$/MW)	400,000
PLANT COST (US$/MW)	1,500,000
Well testing cost to drilling cost	15%

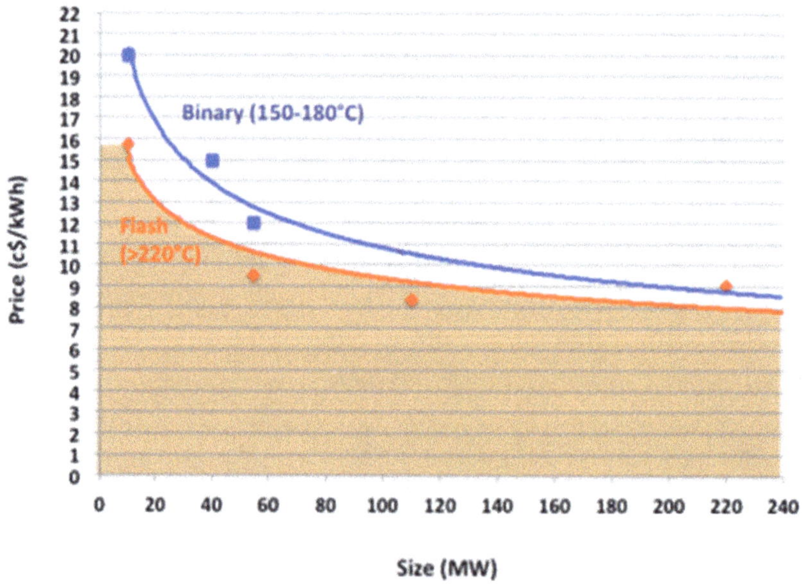

Figure 13.9: Levelized Cost of Geothermal Power. Source: Sudarman (2012).

From this study, the 2012 Levelized Cost of Electricity for geothermal is higher than those purchasing prices set in the 2022 Presidential Decree 122 (PR112), i.e., 7.65 USD cents/kWh for 2 x 55MW size. The result emphasizes the importance of using concession loans to make geothermal projects feasible.

13.5 Project Financing: Example of Ulubelu and Lahendong (Tompaso)

Let us examine a project from Pertamina Geothermal Energy (PGE) that appeared to be financially non-viable. The project was to develop two geothermal fields: Ulubelu, located in the Lampung district in the southern part of the island of Sumatra, and Lahendong (Tompaso), located in the northern part of the island of Sulawesi. PGE planned expansions of approximately 110 MW in Ulubelu (Units 3 & 4) and 40 MW in Lahendong (Tompaso) (Units 5 & 6).

According to World Bank, the economic analysis of the investment and operational costs of the project is not competitive with an equivalent scale coal-based project using the present values of the investment and operations only. The analysis suggested that capital investment cost per MW of capacity is significantly higher for Ulubelu (US$ 3.3 million) and Lahendong (Tompaso; US$ 4.6 million) compared to a medium-sized coal-fired power plant (US$ 1.4 million) (ESMAP, 2014).

The GoI took several actions to close the financial viability gap of this project. First, the government convinced PGE to accept a lower rate of return on its equity of 14% from the expected 20%. PGE agreed even though investments by its parent state-owned company, Pertamina, typically generate substantially higher returns in Oil & Gas business.

Second, the government agreed to facilitate power purchase agreement (PPA) negotiations between PLN and PGE, which, by existing law, would increase the PSO subsidy without the higher costs being passed through to electricity consumers. PGE was able to secure electricity purchase tariffs of US$ 7.53 cents/kWh for Ulubelu and US$ 8.25 cents/kWh for Lahendong (Tompaso).

Third, the GoI made low-cost public financing options available for implementing the project. To facilitate the financing, the GoI sought assistance from the World Bank. The World Bank was able to offer a concessional loan package of US$ 300 million, as detailed in Table 13.5.

Combining all the efforts, PGE could develop Ulubelu and Lanhendong geothermal fields with a Net Present Value (PNV) of 51.4 M USD, as shown in Figure 13.10 and Table 13.6 below.

Table 13.5: Project Financing by components and source of funding. Source: ESMAP, 2014.

Project Component	Total	PGE Internal Sources	World Bank Financing	
			IBRD	CTF
	(US$ Million)			
Investments in Geothermal Field Development and Power Generation				
a. Ulubelu	326,2	140,2	108,5	77,5
b. Lahendong (Tarnpaso)	191,8	105,8	50,2	35,8
Total Baseline Costs 518.0	**518**	**246**	**158,7**	**113,3**
Physical & Price Contingencies (10%)	51,8	23,8	16,3	11,7
Total Project Cost	**569,8**	**269,8**	**175**	**125**
Interest During Construction	4,1	4,1		
Front End/MOB Fee (0.25%)	0,8	0,8		
Total Cost and Financing Required	**574,7**	**274,7**	**175**	**125**

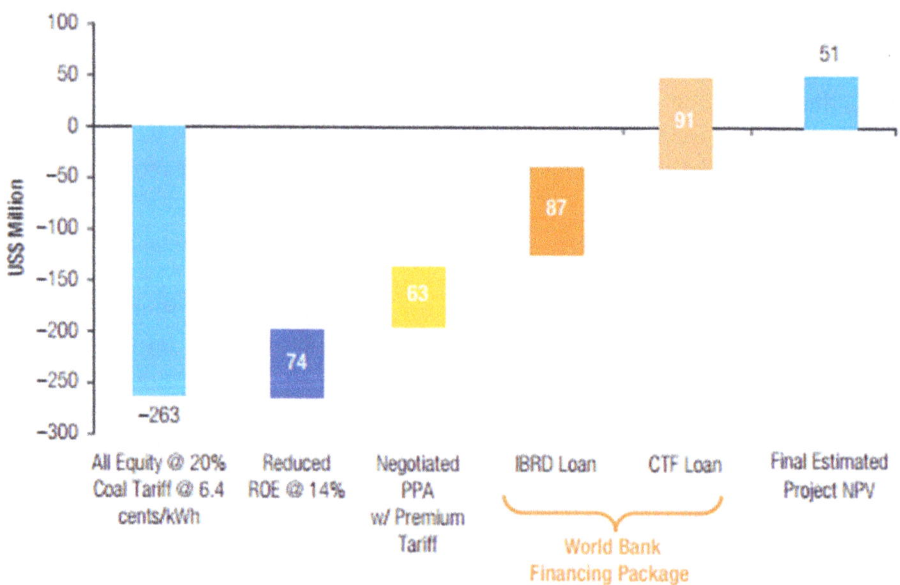

Figure 13.10: Factors of increasing NPV for a geothermal field.

Table 13.6: Bridging the financial viability gap (NPV from PGE's Equity point of view). Source: ESMAP, 2014.

	Financial Metrics	Ulubelu	Lahendong (Tompaso)	Combined Project
PGE all equity financing @14% ROE	Nominal FIRR	9,00%	6,80%	**8,30%**
Tariff @ US¢6.4 /kWh	NPV (US $ million)	−109,9	−79,0	**−188,8**
PGE all equity financing @14% ROE	Nominal FIRR	11,00%	9,40%	**10,40%**
Tariff @ PPA	NPV (US $ million)	−71,1	−55,8	**−126,2**
Project Scenario IBRD + CTF financing	Nominal FIRR	17,40%	14,60%	**16,50%**
Tariff @ PPA	NPV (US $ million)	46,8	4,0	**51,4**

13.6 Geothermal Resource Risk Mitigation Facility (GREM)

In October 2018, Green Climate Fund (GCF) approved a geothermal exploration financing facility for Indonesia's public and private sectors called Geothermal Resource Risk Mitigation (GREM). GREM aims to mitigate early-stage development risks through a de-risking or risk-sharing scheme.

The GREM Developer Manual (PT SMI, 2022) explains that the facility is envisioned to include three windows:

1) Government drilling
2) Public developer window
3) Private developer window

The first window capitalized $104.25M, while the latter two are planned to be capitalized with a total of US$651.25 million in the aggregate towards investments in geothermal exploration drilling.

Government Drilling

The first window (Government drilling) has been operationalized since 2017 under the existing World Bank supported Geothermal Energy Upstream Development Project (GEUDP), where PT Sarana Multi Infrastruktur (Persero) (PT SMI) acts as the financial implementing agent, and PT Geo Dipa Energi (PT GDE) acts as the implementing agency.

GEUDP is capitalized from the Government's Infrastructure Financing for Geothermal Sector (IFGS) and Clean Technology Fund under the P155047 project code for a total of US$98 million for geothermal exploration drilling. GEUDP also includes a $6.25M grant from the Global Environment Facility (GEF) under the P161644 project code for technical assistance and capacity building for PT SMI, PT GDE, and relevant stakeholders (see Table 13.7).

Table 13.7: Project components and source of funds for GEUDP.

Project Component	Project Cost	IBRD or IDA Financing	Trust Funds	Counterpart Funding
Component 1 – Risk Mitigation for Geothermal Exploration Drilling	98,00	0,00	49,00	49,00
Component 2 – Capacity Building on Geothermal Exploration and Environmental and Social Safeguards Management	6,25	0,00	6,25	0,00
Total Financing Required	**104,25**	**0,00**	**55,25**	**49,00**

Source: World Bank's Geothermal Energy Upstream Development Project (P155047).

Once the resources have been proven, the sites shall be tendered by the Ministry of Energy and Mineral Resources (MEMR). The successful bidder is required to re-pay the exploration drilling costs with a 25% premium to the revolving fund. The premium is designed to cover the drilling costs of the unsuccessful wells. Note that the fund assumes an 80% success rate as a basis for the 25% premium (see Figure 13.11).

13.6.1 Public Developer Window

Based on MOF Regulation No. 80/2022, public developers that are eligible to access exploration financing facility include (see Figure 13.12 and 13.13):

- Geothermal SOE (**"BUMN Panas Bumi"**) is an SOE established to conduct business in geothermal for indirect use.
- SOE in the energy sector

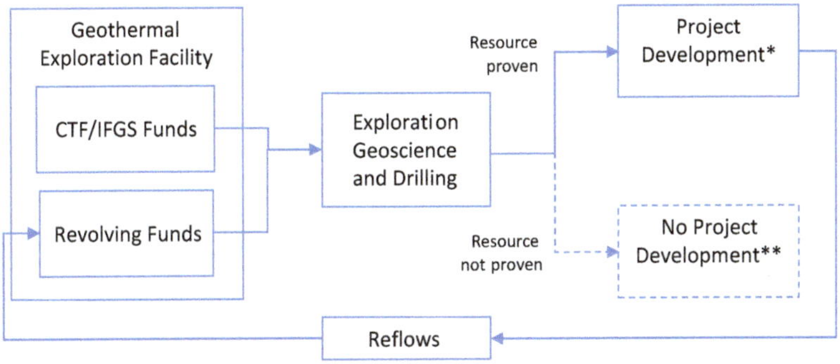

Figure 13.11: A Schematic of GEUDP. Source: World Bank's Geothermal Energy Upstream Development Project (P155047).
**Successful bidder pays for the Geothermal Data Package*
*** Geothermal Data Package transferred to EBTKE*

Geothermal Company ("**Badan Usaha Panas Bumi**") which is wholly or majority owned by a Geothermal SOE or SOE in the energy sector.

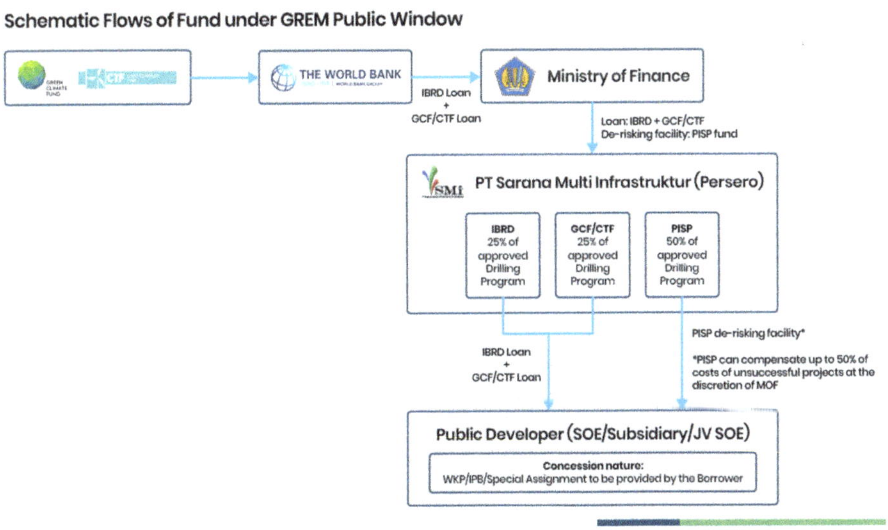

Figure 13.12: Financing Arrangement for GREM Public Sector Window. Source: World Bank's Geothermal Energy Upstream Development Project (P155047).

13.6.2 Private Developer Window

For the Private Developer Window, the sub-borrower can be any company that does not fall under the definition of public developers as per MOF Regulation No. 80/2022.

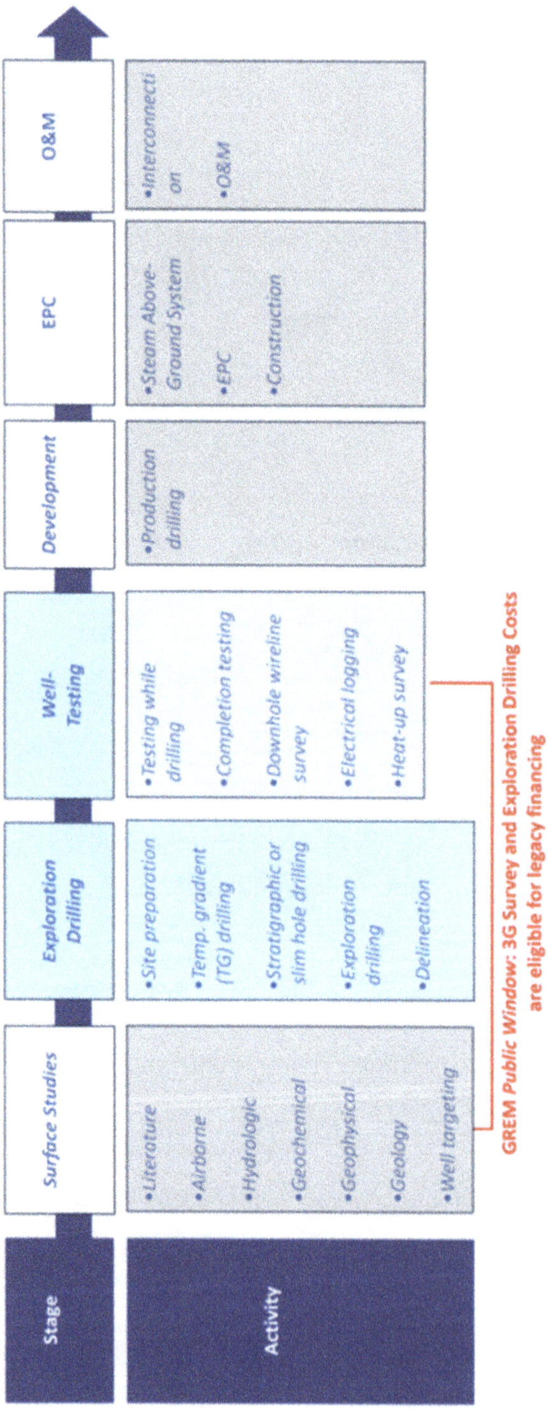

Figure 13.13: Eligible Activities for GREM Public Sector Window. Source: World Bank's Geothermal Energy Upstream Development Project (P155047).

However, they must meet the following criteria and requirements (see also Figure 13.14 and 13. 15):

1. The sub-borrower has already obtained geothermal data and information, both from the results of the Preliminary Survey or other activities, which cannot be used to conclude the feasibility of the Working Area to proceed to the development stage; and
2. All or part of the land for exploration purposes are in the Working Area:
   1. has been legally controlled or obtained a permit from the authority;
   2. not being the object of a case/dispute in the court that can hinder exploration activities
3. the Sub-Borrower must have a sound financial condition as evidenced by audited-financial statement for the past 2 (two) years
4. the controlling shareholder(s) and management of the Sub-Borrower have experience in the exploitation of geothermal for indirect use; and/or
5. the controlling shareholder(s) and/or other shareholders of the Sub-Borrower are willing to provide a guarantee for the payment obligations of the Sub-Borrower

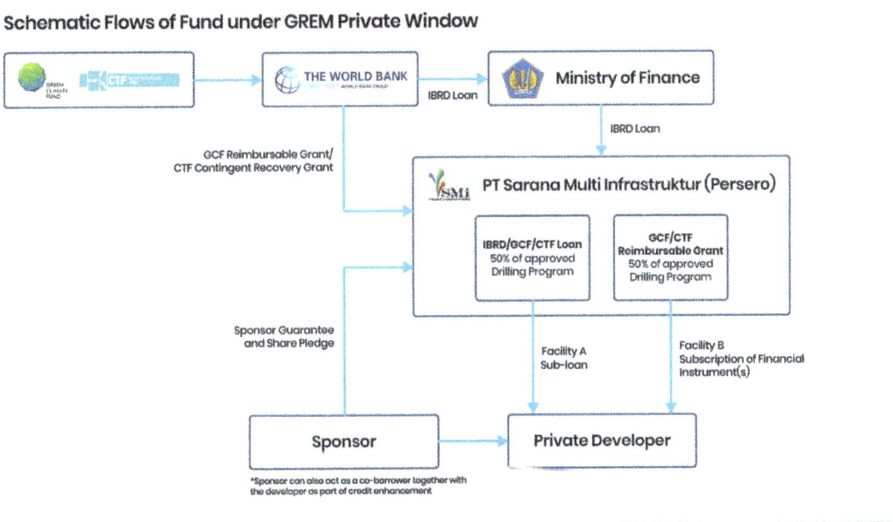

Figure 13.14: Financing Arrangement for GREM Private Developer Window. Source: World Bank's Geothermal Energy Upstream Development Project (P155047).

According to the GREM Developer Manual (PT SMI, 2022), PT SMI will provide support in a combination of a 50/50 basis of a conventional loan and subscription to a financial instrument(s) (FI) to be issued by the Developer with a maximum total combined limit of USD 30 million. The conventional loan will be funded from the IBRD loan and will be for four years to finance the exploration phase. The second part of the support – subscription of Developer's FI(s) – is financed by PT SMI from the proceeds of the GCF and CTF reimbursable grants. The private developers will be required to com-

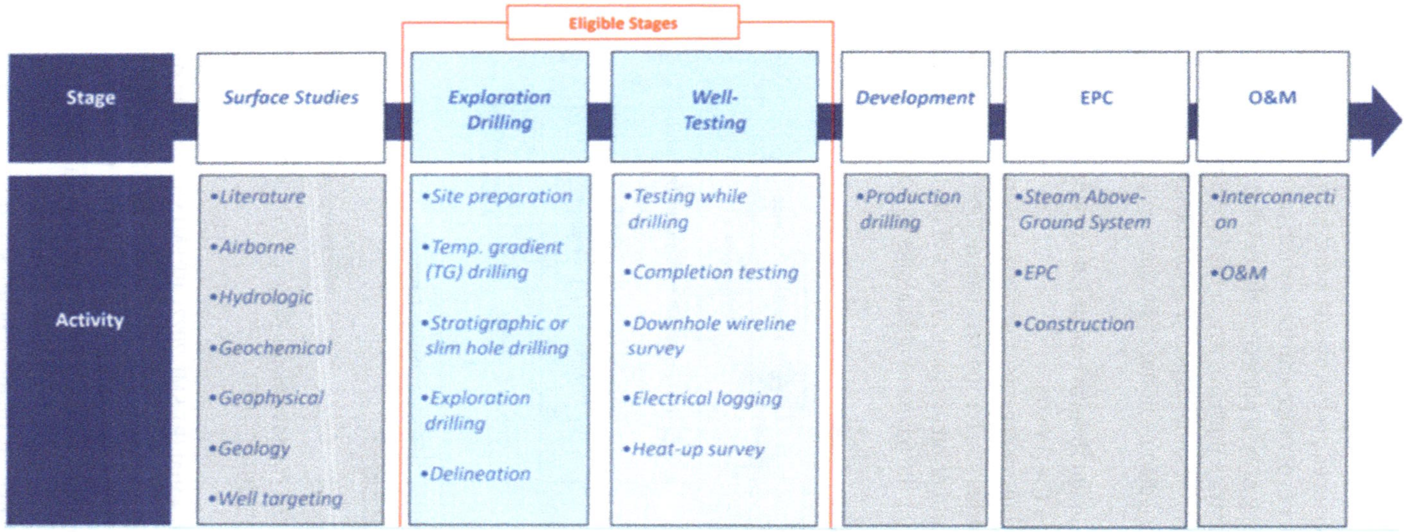

Figure 13.15: Eligible Activities for GREM Private Sector Window. Source: World Bank's Geothermal Energy Upstream Development Project (P155047).

mit Sponsor funds (i.e., equity) equivalent to at least 25 percent of the total cost of the Total Drilling Program.

The FI is an innovative way to track the additional value created from positive results of the exploration drilling by the private sector and allow capture of the upside from successful exploration by charging a 7.0% p.a. coupon (with a minimum of 30%). This facility allows, at the overall facility level, covering the losses from unsuccessful explorations where the Developer decides to stop the drilling program, and the value of the FI is less than the amount initially advanced under the FI. FI also allows the GREM Facility to share the risk if the resource exploration does not result in an economic well by lowering the FI price to the Developer's market value, which is expected in such instances to be close to nil.

13.7 Government-led Exploration Drilling Program

The Government-led Exploration Drilling Program utilizes the Geothermal Energy Upstream Development Project (GEUDP) fund. The primary objective was to improve the quality of geothermal geoscience data to reduce development risk before geothermal areas are offered to business entities. The scope of the program includes the following:

- Data acquisition and processing. This scope comprises additional geoscience data acquisition, LIDAR surveys, integration of data acquisition results, conceptual model analysis and numerical simulation of reservoir models, well targeting, and peer review
- Exploration drilling of two slim hole wells and one standard hole well per prospect, including completion and production testing of those wells
- Land acquisition, preparation of road infrastructure and well pads, as well as supporting facilities.

The GoI expects multiple effects on the economy. On a local scale, the GoI expects job creation and labor absorption. The estimated workforce required for geoscience data acquisition activities and geothermal exploration drilling is 483 people per field or 9,660 people for 20 fields, out of which 4,780 people are expected to be supplied by the local communities around the drilling sites. Another local effect is the development of infrastructure such as access roads, bridges, and electricity access.

On the national scale, the GoI expects investment increases, including domestic construction and material consumption. The project is expected to increase the installed capacity of geothermal power, contributing to the energy mix achievement. At the same time, boosting the state and regional income through taxes and the production bonus.

Per the Ministry of Energy and Mineral Resources (ESDM), 20 prospects listed in Table 13.8 have met the technical and non-technical criteria and are prioritized to be explored by 2024. Technical criteria includes heat source, reservoir temperature, geological structure, fluid type, manifestation, available geoscience data, quality of infrastructure. Non-technical criteria includes land use, electricity demand, and social acceptance conditions.

Table 13.8: List of prioritized Prospects for Exploration by 2024 (compiled by authors).

Prospects	Resource (MW)	Development Plan (MW)	Est. Temp. (c)
1 Cisolok Cisukarame, Jawa Barat	45	20	200
2 Nage, NTT	39	20	230
3 Bittuang, Sulawesi Selatan	28	20	230
4 Tampomas, Jawa Barat	100	45	208
5 Ciremai, Jawa Barat	60	55	210
6 Marana, Sulawesi Tengah	70	20	154
7 Gunung Endut, Banten	180	40	180
8 Sembalun, NTB	100	20	165
9 Guci, Jawa Tengah	100	55	281
10 Sipoholon Ria-Ria, Sumatera Utara	60	20	180
11 Bora Poiu, Sulawesi Tengah	123	40	220
12 Lokop, Aceh	41	20	210
13 Limbong, Sulawesi Selatan	20	5	220
14 Maritaing, NTT	190	30	200
15 Gunung Batur-Kintamani, Bali	58	40	230
16 Gunung Galunggung, Jawa Barat	289	110	>225
17 Papandayan, Jawa Barat	195	40	290
18 Banda Baru, Maluku	54	40	190
19 Sajau, Kalimantan Utara	17	13	190
20 Sumani, Sumatera Barat	100	30	190

The first three prospects are WKP Cisolok Cisukarame in West Java, WKP Nage in East Nusa Tenggara, and Bituang in South Sulawesi. The three prospects are planned to commence in Q3-2021 and completed in Q1-2022, followed by completion of the next seven by the end of 2022.

On September 3rd, 2021, the Geological Agency of the Ministry of Energy and Mineral Resources (ESDM) announced the commencement of the first geothermal exploration drilling. A year later, in September 2022, the Indonesian Ministry of Energy and Mineral Resources (ESDM) announced that the exploration drilling had been completed. One of the two prospects, Nage, has shown promising data indicating a temperature of 287 degrees Celsius, hotter than the previously forecasted temperature of 210 degrees Celsius. According to ESDM, the exploration drilling also educated the local community that such activity can be done safely when carried out according to the standards procedure.

A few points to note about the program: First, only two slim holes were drilled in both fields, and no standard holes were attempted as originally planned. Second, each of these slim holes was planned to be completed in approximately 60 days; however, the actual well times are much longer. Third, one of the four wells could not reach the desired total depth due to technical issues.

In December 2022, New and Renewable Energy and Energy Conservation (EBTKE) announced that the Ministry of Energy and Mineral Resources (ESDM) planned to put Nage geothermal working areas (WKP) on an auction by the end of the year (2022). The auction would mark an important milestone as Indonesia's first auction for a geothermal field furnished with data acquired by government-funded exploration.

Participation of the private companies in the next auction will indicate the success of such a campaign. Though the hope for success is high, as the first auction is still on its way, it is yet to see the impact of this government-led exploration drilling campaign and the achievement it can bring.

Bibliography

Asian Development Bank & The World Bank (2015). Unlocking Indonesia's Geothermal Potential [online]. Available at: http://hdl.handle.net/11540/3311

Assegaf Hamzah & Partners (2022), Presidential Regulation 112: Indonesia's Commitment to Renewable Energy – Renewables – Indonesia. [online]. Available at: https://www.mondaq.com/renewables/1233326/presidential-regulation-112-indonesia39s-commitment-to-renewable-energy

Campen, V. B., Pratama, A.R., Lawless, J., Archer, R.: Analysis Of New Indonesian Geothermal Tariff System And Impact of Feasibility Of Geothermal Plants In Eastern Indonesia In The Next Decade. Proceedings 39th New Zealand Geothermal Workshop, (2017)

Energy Sector Management Assistance Program, The World Bank (2014). Scaling-Up Renewable Geothermal Energy in Indonesia [online]. Available at: https://documents1.worldbank.org/curated/en/955541468148774504/pdf/893820ESM0P1300in0Indonesia0KS15013.pdf.

Hochstein, M. P., and Sudarman, S.: History of Geothermal Exploration in Indonesia from 1970 to 2000. Geothermics, 37(3), (2008), 220–266.

Ministry of Energy and Mineral Resources (ESDM), ESDM One Map (2022): Energy and Mineral Overview. [online] Available at: https://geoportal.esdm.go.id/emo/#

Nazif, H. (2021): Government Drilling: A Strategic Program for Reducing Geothermal Development Risks in Indonesia. [online]. Available at: https://www.youtube.com/watch?v=uOcXOGzKkJY

PT Pertamina Geothermal Energy (PGE), 2021 Annual Report [Online]. Available at: https://www.pge.pertamina.com//media/Uploads/laporan-tahunan/laporan-Tahunan-TB-2021.pdf

PT Sarana Multi Infrastruktur (PT SMI) (2022). Indonesia Geothermal Resource Risk Mitigation Project Developer Manual [Online]. Available at: https://www.ptsmi.co.id/cfind/source/files/grem/dm-revision-grem-developer-manual-november-2022.pdf

Purwanto, E., H., Suwarno, E., Hakama,C., Pratama, A. R., and Herdiyanto, B.: An Updated Statistic Evaluation of Drilling Performance, Drilling Cost and Well Capacity of Geothermal Fields in Indonesia. Proceedings World Geothermal Congress 2020+1, (2021)

Purwanto, E., H., Suwarno, E., Lukman, R., F., and Herdiyanto, B.: Geothermal Drilling in Indonesia: A Review of Drilling Operation, Evaluation of Well Cost and Well Capacity. Proceedings The 6th Indonesia International Geothermal Convention and Exhibition (IIGCE), (2018)

Sanyal, S. K., Morrow, J. W., Jayawardena, M. S., Berrah, N., Li, S. F., and Suryadarma.: Geothermal Resource Risk in Indonesia – A Statistical Inquiry. Proceedings, 36th Workshop on Geothermal Reservoir Engineering Stanford University, (2011)

Sudarman S.: Ekstrapolasi data-data tender panasbumi Indonesia untuk mengambil levelized cost panasbumi. PROCEEDINGS, National Energy Council, (2012)

Sumari, A. D.W., Mariani, S., Dewi, R.G.: Promoting Geothermal for Energy Security (A Case of Indonesia), Proceedings 2nd International Conference on Technology, Informatics, Management, Engineering & Environment. (2014)

Sumari, Arwin D.W., et al. "Promoting Geothermal for Energy Security (A Case of Indonesia). Proceedings, 2nd International Conference on Technology, Informatics, Management, Engineering & Environment Bandung, Indonesia, (2014)

The World Bank (2017), Project Appraisal Document for the Geothermal Energy Upstream Development Project (Report No: PAD1595) [online]. Available at: https://documents1.worldbank.org/curated/en/762021486868465688/pdf/PAD-Indonesia-Geothermal-Project-P155047-01232017.pdf

Zuhro, A.A., Arif, G.R., (2015). Pertamina Geothermal Energy Drilling Campaign. World Geothermal Congress 2015

Jens Kühne

14 Solar Thermal Systems Within a District Heating Concept

14.1 Introduction

Solarthermal heat generation is due to its low operation costs (low electricity demand, no fuels needed) one of the most cost stable renewable heat generation technologies. Hence there is almost no dependence of the costs of heat production on energy markets, taxes or other surtaxes.

The following chapter will give an overview on solarthermal heat generation while focusing on a description of the technological basics and explaining the economic fundamentals with characteristic values.

14.2 Fundamentals

14.2.1 Functionality

Solar thermal systems use the radiant energy of the sun and convert it into thermal energy, which can then be used for heating applications. The global irradiation on the earth's surface is made up of direct and diffuse radiation. Direct radiation is defined as the part of the global radiation that hits the earth's surface in a directed manner. Radiation that is undirected due to reflection and scattering processes is referred to as diffuse radiation. With solar thermal heat generation both radiation components and thus the global radiation is used in total.

The radiant energy hits the absorber surface of the collector as light, where it is converted into thermal energy. The selective coating of the absorber ensures the highest possible absorption rate of sunlight. At the same time, a heat conducting medium flows through the absorber to transfer the thermal energy from the collector. Subsequently, the heat conducting medium can be led through another collector to further heat the medium or the heat can be fed into a heat network by means of a heat exchanger. By connecting several modules, the available absorber area increases analogously, which can influence the performance and the solar yield (the amount of heat available) of the solar system.

Due to reflection, convection and thermal radiation processes, a large part of the theoretically available radiation energy is not usable, so that depending on the temperature level of the heat conducting medium, approx. 50% [Geiger] [Wesselak] of the radiation energy is available as useful heat.

https://doi.org/10.1515/9783110784299-014

14.2.2 Geometry of Collectors

Due to the relevance of reference areas for solar thermal systems, the most frequently mentioned areas according to DIN EN ISO 9488:2001 are differentiated in the following:

The gross collector area often just referred to as collector area, describes the largest projected area of a complete solar collector. This does not take into account the devices for mounting and pipe connections. In simplified terms, the gross collector area indicates the external dimensions or the dimensions of the collector. The aperture area describes the largest projected area through which unconcentrated solar radiation enters the collector. In simplified terms, the aperture area is the light entry area through which the solar radiation (direct and diffuse) enters the collector. For flat-plate collectors, the aperture area is the visible glass pane; for evacuated tube collectors, the aperture area is the sum of the longitudinal sections of all glass tubes. For evacuated tube collectors with mirrors on the back, the mirror surface counts. The absorber area describes the area of a collector through which the conversion of radiant energy into thermal energy takes place. Only in the case of round absorbers can this surface be larger than the gross collector area due to the cylindrical shape.

The decisive factor for the space utilization of a solar thermal system is the required installation area. The installation angle of the collectors is the most important factor, with the aim of maximizing the yield of the system. It must be considered that shading of the collectors from each other reduces the yield, but more collectors can be accommodated on a given area. Starting from this basis, an optimization problem arises that must be solved and for which, among other things, the orientation, the ground angle, etc. of the collectors must be taken into account.

Furthermore, when installing solar thermal systems, applicable building laws and constraints must be observed, including compliance with legal border distances and in addition a good accessibility of the collectors simplifies maintenance and servicing work.

Overall, it can be roughly estimated that there is a factor between the required installation area and the gross collector area of 2 to 3. This means that two to three times as much installation area is required as gross collector area is built up.

14.3 Solar Collector Types

14.3.1 Flat Plate Collectors

The collector types already used as district heating applications are flat plate collectors and vacuum tube collectors. Parabolic trough collectors are a less common but a suitable technology. The collector types described here use water or a water-glycol mixture as the heat transfer medium.

Flat-plate collectors consist of a thermally insulated housing, a highly transparent special glass – also called solar glass – and the absorber through which the heat conducting medium flows (see Figure 14.1). The heat conducting medium can be conducted either through channels directly in the absorber surface or through a pipe system connected to the absorber surface. The absorber of a flat-plate collector consists of a continuous surface.

The housing serves to protect the absorber, which at the same time offers sufficient options for the individual mounting types. The insulation of the housing reduces the heat losses to the environment, as the temperatures inside the collector are significantly higher than in the environment.

The requirements for the special glass are manifold. On the one hand, the glass must protect the absorber from external influences such as rain and hail. On the other hand, the glass must be highly transparent in order to allow as much light as possible onto the surface of the absorber and to reflect as little light as possible (e.g. further optimization by adjusting the surface properties). At the same time, a high level of thermal insulation must be achieved in order to keep thermal losses to the environment low here as well. Therefore, specially developed types of glass are used here that fulfil all requirements in the best possible way. The flat-plate collectors used in district heating are often also referred to as high-temperature flat-plate collectors (HT collectors). However, there is no uniform definition of the temperatures above which collectors belong to this group. It often only serves to distinguish the area of application: heating network or building.

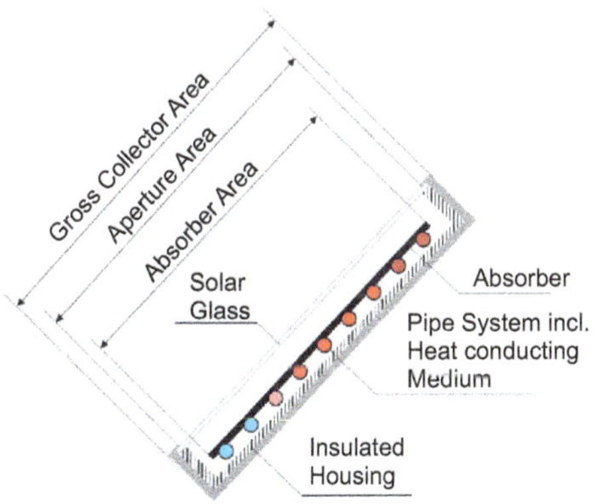

Figure 14.1: Schematic diagram of a flat-plate collector including main components as a sectional view (left) and photo of several installed flat-plate collectors (right).

14.3.2 Evacuated (Vacuum) Tube Collectors

Vacuum tube collectors essentially consist of several mostly hydraulically connected vacuum tubes. The housing of the vacuum tubes consists of a special glass, which must fulfil the requirements for flat-plate collectors in the same way: protection of the internal components, high light transmission, low reflection, and high thermal insulation. For the highest possible thermal insulation, there is also a vacuum inside the tubes. Each of these tubes has a defined absorber surface, whereby all the individual surfaces add up to the absorber surface of the collector. The heat transfer medium is conducted through a pipe system connected to the absorber surface. Based on the design and the resulting absorber surface, two main types of vacuum tubes can be distinguished: Strip absorbers and cylindrical absorbers (see Figure 14.2).

The spaces between the individual vacuum tubes of a collector can be provided with channel-shaped metal sheets to increase the radiation yield. This special design is called a CPC (compound parabolic concentrator) vacuum tube collector. The light hitting the reflectors is reflected onto the back of the tubes and can thus be used by the absorber, increasing the efficiency of the collector. Due to the low heat losses through the vacuum and the concentration of radiation, evacuated tube collectors tend to achieve higher temperatures and higher yields than flat-plate collectors.

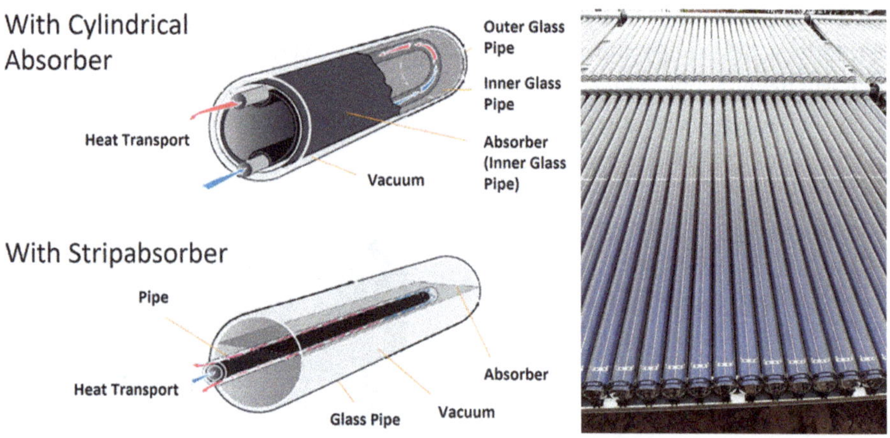

Figure 14.2: Schematic diagram of vacuum tubes with strip absorber and cylindrical absorber (left) (source: target GmbH [Geiger]), photo of a vacuum tube collector (right).

14.3.3 Innovative Collectors

Parabolic trough collectors have a parabolic trough in whose focal line the absorber pipe through which the heat transfer medium flows. The parabolic trough serves to concentrate the solar radiation in order to achieve higher temperatures of the heat-conducting medium, well above 150°C, by focusing it on a comparatively small area. In addition, the troughs of the collector can be equipped with tracking to adjust the alignment of the troughs according to the position of the sun and to achieve an ideal radiation concentration.

14.4 Dimensioning

The thermal output and the annual yield of a solar thermal system depend on factors that can be influenced as well as on factors that cannot be influenced. The most important aspects that must be considered when planning a solar thermal system are described below.

Solar thermal energy is a strongly weather-dependent generation technology. Output and yield depend directly on the weather conditions. The thermal output of a solar thermal system is highest in the summer months and lowest in the winter months. Because of this, the maximum output under ideal conditions is given as a parameter, which is referred to as peak output and specified in kWp (pronounced kW-peak). Even more frequently, however, the annual yield is used as a comparative value.

The achievable yield of a solar thermal plant depends on geographic location, since global radiation, i.e. the available radiation energy, depends on the location. The global radiation is given in kWh/(m^2 a). Regional differences in global radiation within Germany amounted to around 20% in 2023 (see Figure 14.3). Within a district heating supply area, the influence of regional fluctuations is negligible. More decisive here are aspects such as orientation and shading.

More relevant for individual projects, however, is the fluctuation of global radiation over several years. Even at the same location, global radiation is not constant over several years. For example, the average global radiation in Germany was 1,144 kWh/m^2 in 2023, 1,227 kWh/m^2 in 2022 and 1,096 kWh/m^2 in 2021. The actual heat yield will accordingly sometimes be above the design case and sometimes below the design case, depending on which assumptions the design is based on.

Other factors influencing the yield of a solar system are the grid temperatures of the heating network into which the solar system feeds. In general, low grid temperatures favor a high system yield: the lower the grid temperatures, the more often the collectors reach the required feed-in temperature.

Analogous to the specific heat yield, the efficiency of a solar collector also decreases with increasing grid temperatures. Accordingly, when planning solar thermal systems, consideration should be given to how the grid temperatures can be reduced in the short term (before commissioning until 2 years after commissioning of the system), in the medium term (2 to 5 years) and in the long term (5 to 10 years). A reduction of 1 °C can increase the yield by up to 2% [SDH Guidelines].

While the collectors usually have a non-variable orientation and a non-variable installation angle, the angle of incidence of the solar radiation changes due to the variable position of the sun during the course of the day and year. This means that both variables – orientation and installation angle – offer potential for optimization in the conception and design of solar thermal systems.

Solar collectors are generally aligned with the absorber surface facing south in order to be able to make effective use of the particularly radiation-intensive hours.

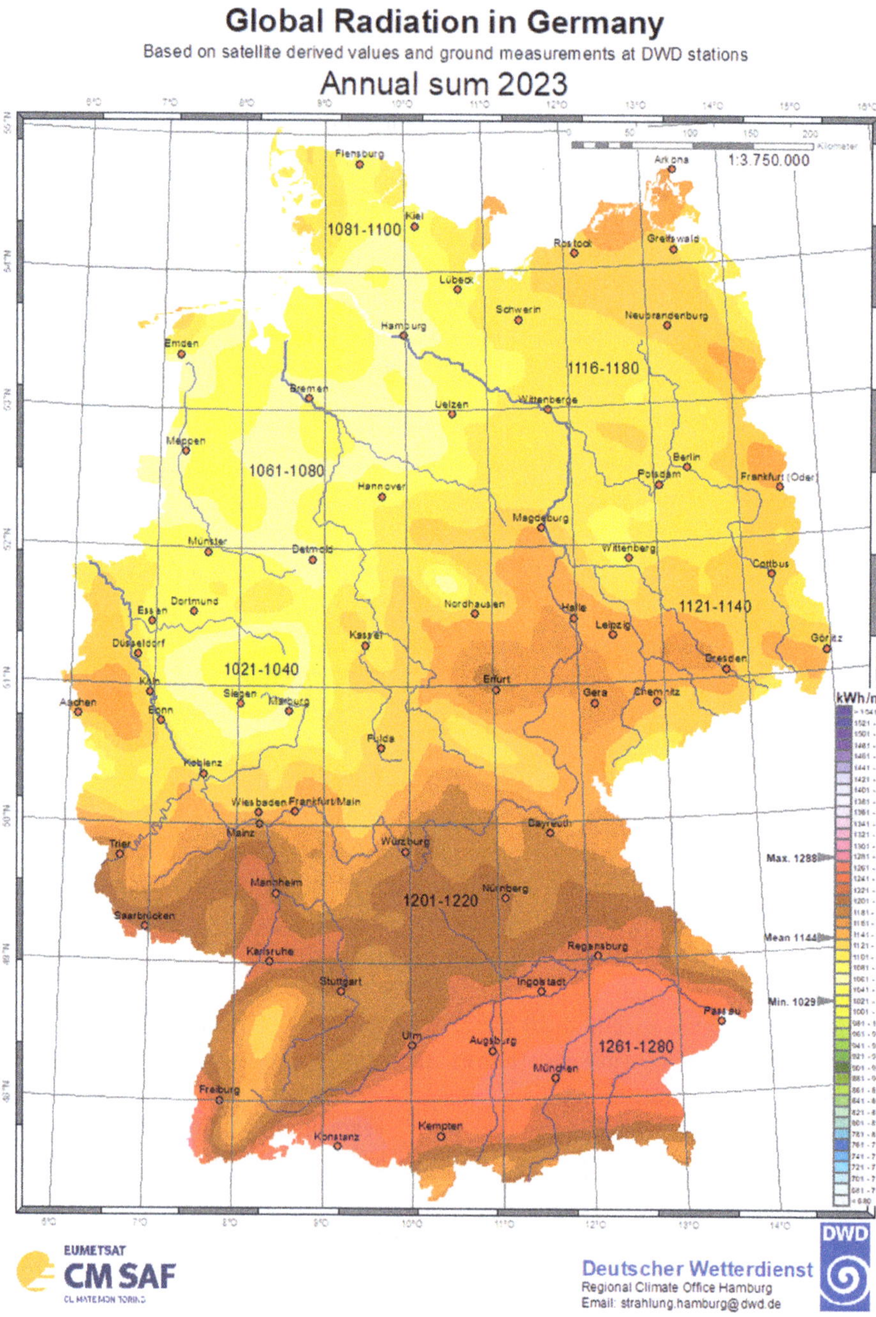

Figure 14.3: Annual Global Radiation in Germany in 2023 [DWD].

However, this is not always possible in practice, e.g. when installing solar collectors on noise barriers. With regard to the installation or tilt angle of a solar collector, the optimum depends on the individual application. The desired yield period (e.g. longer utilization period in the transitional periods spring and fall) is particularly relevant here. In the latitudes of Germany, a flat angle of inclination results in a higher yield in summer, while a steeper angle of inclination is also associated with higher solar yields in the transitional periods. Practice has shown that in Central European latitudes, the highest yield in relation to the gross collector area can be achieved with an installation angle of about 35°.

The (slightly) decreasing required gross collector area with an increasing installation angle shows that higher yields are achieved in the transitional periods. Over a year, this is accompanied by a higher specific yield per m^2_{gross}. However, a steeper installation angle also causes a greater shadow cast by the individual collectors when the sun is low. This shadowing, as shown in Figure 14.4, can then lead to shading of the other collectors, which reduces the specific yield. To counteract shading at a steeper installation angle β, the distance L between the collectors can be increased. These two degrees of freedom result in optimization potential with regard to the specific yield.

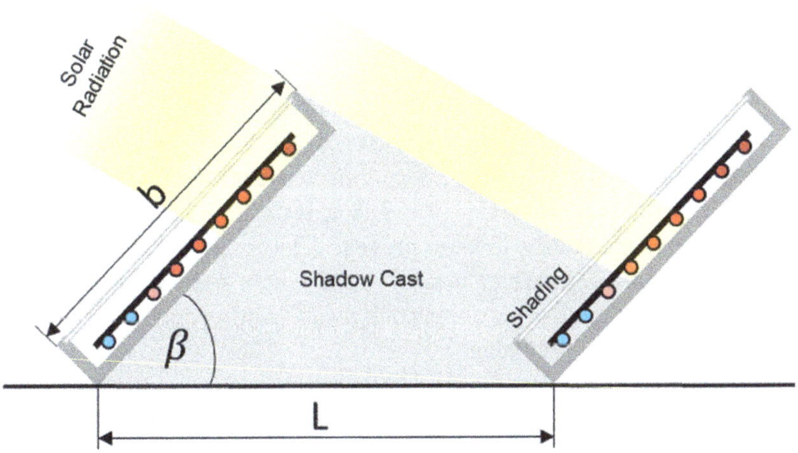

Figure 14.4: Effect of shading and qualitative representation of the relevant parameters for determining the optimum distance between solar collectors, based on [Witte-Humperdinck].

In practice, the limited, previously defined required installation area is often another degree of freedom that should be taken into account during optimization. This is because it is often not the specific yield of the collector, but the annual yield of a solar thermal field that is the variable to be optimized. With the optimization goal of maximum annual yield, it can make sense to allow temporary shading of the modules. In this way, the distance between the collectors can be reduced so that a larger number of modules/module rows fit onto the defined area. The intermittent shading reduces the specific heat yield per m^2_{gross}, but the overall higher yield due to a higher number of collectors outweighs this disadvantage. Based on this, an optimal installation angle and distance between the collectors can be determined for each area and each collector array. To calculate the optimal distances between rows of collectors, one can refer to [Witte-Humperdinck], for example. To further increase the solar yield, solar collectors can be adjusted to the position of the sun during the day. Tracking makes sense from a purely energetic point of view but is not currently implemented in the field of solar thermal energy due to various disadvantages (e.g. higher investment and maintenance costs, susceptibility to faults, possibly higher space requirements, etc.).

14.5 Operation Modes

The operation of solar thermal systems is limited by two phenomena. In the winter at low temperatures, frost limits operation, while in summer at high temperatures, the risk of overheating limits operation.

14.5.1 Antifreeze Measures

At operating temperatures below 0 °C, there is a risk that the heat-conducting medium will freeze, which can cause damage to the collectors and the periphery – such as pipes. Therefore, precautions are taken to prevent the heat-conducting medium from freezing. A distinction is made between passive and active measures to prevent frost. Both types of measures effectively serve to prevent damage but differ in terms of costs and implementation effort. Which measure makes more sense in each individual case therefore depends on the risk assessment, the costs incurred, the design case and the planned operation of the system.

Passive frost prevention includes the use of a water-propylene glycol mixture as a heat conducting medium (standard: 35% glycol content), whereby the added glycol acts as an antifreeze due to its lower freezing point. Due to the use of glycol, this type of frost prevention usually requires a water law permit. For this, certain technical requirements must be proven, in particular with regard to tightness, stability or leakage detection.

The active measures have the advantage that no antifreeze is required and thus district heating water or softened water can be used. To ensure the necessary reaction time as well as a certain safety margin, the corresponding measures should already be started at 4 °C for active frost prevention.

A frequently used measure of active frost protection is thermal frost protection. With thermal frost protection, the operating temperature is maintained above operating temperature of 0 °C if necessary, which prevents the heat-conducting medium from freezing. The external heat supply can be provided, for example, via the heating network to which the solar thermal system is connected. Since the danger of freezing only exists on a few days a year, the amount of heat required for thermal frost protection is small. Roughly speaking, the amount of heat needed for active frost protection can be assumed to be 1% of the amount of heat generated by the solar thermal system per year. However, this small "loss" is outweighed by a higher heat yield overall. By using water without glycol, the annual yield of a system can be increased by up to 5% due to the better thermal properties of water compared to water-glycol mixtures.

Alternatively, an electric heater can be used for thermal frost prevention, which acts exclusively on the circulating water in the collectors. For this purpose, an uninterruptible power supply (UPS) should also be integrated, which protects the entire system even in the event of a long-term power and/or system failure. Another measure of active frost protection is the planned idling of the collectors. If there is a risk of frost, the heat transfer medium is let out of the collectors early and stored frost-free. For this, a sufficiently dimensioned storage tank is required to be able to absorb the corresponding amount of heat-conducting medium. In addition, the system should be designed in such a way that the system can run empty without using compressed air.

14.5.2 Stagnation and Overheating

In the event of overheating, the collectors and the periphery may be damaged, e.g. by steam hammering of the heat conducting medium. When using water-glycol mixtures, glycol can also "crack". This decomposition of the glycol can lead to precipitation and thus cause blockages or other damage to the collectors. Overheating occurs if the heat that is introduced cannot be sufficiently dissipated to a sufficient degree. This is usually caused by stagnation in the system i.e. that the flow of the heat-conducting medium in the collectors comes to a standstill. Stagnation is usually triggered by a power failure.

Preventive measures to avoid overheating:
- Solar collectors with automatic cooling of the absorber (without the use of electricity)
- Active electrical cooling (rather for smaller system sizes)
- Battery-buffered emergency power solution

Measures to deal with overheating due to stagnation:
- Sufficiently dimensioned expansion vessels to hold the heat transfer medium.
- Safety valves with appropriate collecting vessels
- Installations that allow the system to run empty quickly

Irrespective of the measures taken, according to the requirements of the Drinking Water Ordinance, when using water-glycol mixtures, it must be ensured that escaping liquid and/or steam or condensate must be collected.

14.6 Installation

Solar thermal collectors that feed into the district heating system can be installed both on roof surfaces and on open spaces. When installing on roof surfaces, the surface loads and the acting wind loads must be considered in particular. Due to the mass of the required heat-conducting medium, the surface load is many times higher compared to photovoltaics. For ground-mounted installations there are various methods that can be used. Steel struts are often driven hydraulically into the ground, which serve as a frame for the collectors. If there are restrictions or requirements that prohibit pile-driving, the collectors can also be installed on non-anchored foundations. This method is mainly used on former landfill sites where soil penetration is not permitted. In terms of cost, roof installation is usually more expensive than ground-mounted installation. The exact cost depends on the specific boundary conditions on site and must be considered on a case-by-case basis.

The profitability analysis in section 14.9 is based on a ground-mounted installation, so that additional costs should be taken into account for rooftop installation.

14.7 Area Requirements

14.7.1 Energy-Economic Criteria

The economic viability of large-scale solar thermal systems can be positively influenced by:

- Distance to the district heating grid
- Geographical location, orientation (e.g. sloping surfaces)
- Possibilities of hydraulic integration into the district heating network
- Costs of land: high costs per unit area have a considerable effect on the economic efficiency of a plant.

14.7.2 Acceptance-Related Criteria

Large installations often lead to conflict situations. If the following aspects are taken into account at an early planning stage, acceptance can be increased, which can significantly reduce the potential for conflicts.

- Conflict potential glare: How should glare be assessed and how can it be avoided?
- Conflict potential for residents: What is the distance and orientation to the nearest residential buildings or recreational areas?
- Conflict potential for businesses: Is there direct competition for land with other commercial uses?
- Conflict potential nature conservation: What is the ecological value of the areas? Is there potential for ecological enhancement and compensation?
- Conflict potential agriculture: Can an existing agricultural use be continued? If necessary, on alternative land?

Acceptance-related conflict potential can often be resolved through direct contact with the relevant stakeholders. In direct contact, the ecological advantages of the use of solar thermal energy can be emphasized for those affected, possible joint solutions can be found (e.g. continuation of agricultural use, use of industrial roof areas) or sensible compromises can be worked out.

14.7.3 Legal Criteria

This section lists examples of questions that should support the legal classification of land use planning.

- Is there any existing planning law, e.g. unused provisions in land use and development plans for photovoltaic areas or unused commercial and industrial areas in the municipality?
- Where can planning law be created?
- Where are legal grounds for excluding individual areas?
- Where does the project promoter have land in its ownership?
- Is there a privilege in the external area?

14.7.4 Ecological Criteria

Based on the ecological criteria for finding areas for photovoltaic plants, it can generally be assumed that the following areas are suitable for the construction of solar thermal plants. However, the list does not represent an exhaustive list of all suitable areas:

- Previously polluted conversion areas from military, commercial or former residential use with a high degree of sealing.
- Areas along major traffic routes (e.g. motorways, railways)
- Intensively farmed arable land
- Rainwater infiltration areas or former sewage fields
- Disused landfills and dumps

When identifying areas for solar thermal plants, it should be borne in mind that areas close to district heating should be used for solar thermal energy instead of photovoltaics.

14.8 Characteristic Economic Values

The specific output of solar collectors is often stated in the literature as 0.7 kWp/m^2_{gross} collector area, irrespective of the collector and generally also irrespective of the location. However, the peak output alone is not a meaningful figure for an initial estimate, as full load hours and operating times cannot be planned in the same way as for conventional heat generators. More meaningful is the specific heat yield per square meter of gross collector area and year, which has proven itself in practice for planning purposes. For the specific heat yield per year, assumptions between 340–450 kWh/(m^2_{gross} a) can be found in the literature for flat-plate collectors. For vacuum tube collectors, assumptions between 400–540 kWh/(m^2_{gross} a) can be found.

14.9 Economic Efficiency

The economic efficiency of solar thermal systems in heating networks is influenced by various factors. On the cost side, the investment costs account for the largest share, while the operating costs are comparatively low. The yield is mainly determined by the choice of collector type and the heating network temperatures.

The following section provides general information on indicators for the operation of solar thermal systems in heating networks. Based on this data, an exemplary economic efficiency calculation is carried out and presented (see Table 14.2).

14.9.1 Total Investment

The total investment of a solar thermal system can be related to different sizes. Specific values are often found for the investment per square meter of gross collector area [EUR/m^2_{gross}]. Specific investments based on the yield in kWh are particularly suitable for comparing different collector types. The yield of a solar thermal system depends on the weather and therefore also on the location, so that the following key figures are to be understood as a rough estimation. The data was taken from scientific publications that take into account information from realized projects.

The specific total investment per square meter of gross collector area can be found in the following Table 14.1. Only the system-side costs are considered. Land costs are not considered.

Table 14.1: Guideline values for the specific total investment in solar thermal systems (compiled by authors).

Specific total investment in EUR/m² per gross collector area				
Total gross collector area		500 m²	10.000 m²	100.000 m²
Flat plate collectors	EUR/m^2_{gross}	389	318	295
Vacuum tube collectors	EUR/m^2_{gross}	566	424	365

The solar collectors themselves and the necessary system technology account for 40 to 45% of the total investment.

The connection costs to the district heating grid amount to around 10 to 15%.

The remaining costs for the electrical connection and I&C technology, construction costs and planning costs are roughly the same and each account for between 4 and 8% of the total investment.

The construction costs can vary depending on the location or nature of the available space.

14.9.2 Operating Costs

Operating costs for solar thermal systems include the costs for maintenance and servicing as well as costs for operating resources and electricity costs incurred for operation (pumps, measuring, control and regulation technology, etc.).

Maintenance and servicing costs can be recognized as annual costs at 0.7% of the total investment.

For the electricity costs incurred, the electricity requirement of the overall system can be assumed to be around 1 to 1.5% of the amount of heat generated. The specific electricity costs should be estimated based on the planned connection to the power supply (e.g. whether grid supply or self-supply is available).

14.9.3 Land and Development Costs (Land Costs)

Land and development costs are also part of a holistic profitability analysis and must be considered accordingly. However, these costs are project-specific and can therefore not be considered in this profitability analysis.

When analyzing a specific project, these costs can be taken into account either as an additional component of the overall investment or as additional costs in the operating costs. As part of the overall investment, the area costs are generally considered if the planned area is purchased for the construction of the solar energy system. As part of the operating costs, area costs are primarily taken into account when the planned area is leased.

Table 14.2: Exemplary Calculation of Heat Production Costs of Solar Thermal Systems.

Energy Quantities			
annually generated heat	MWh_{th}/a	10.000	
annualy consumed electricity	MWh_{el}/a	100	
Dimensioning Solarthermal System		**Flat plate collector**	**Vacuum tube collector**
specific annual yield	$MWh_{th}/(m^2_{gross}\cdot a)$	0,4	0,5
required gross collector area	m^2_{gross}	25.000	20.000
required installation area	m^2	75.000	60.000
Investment			
specific total investment	EUR/m^2_{gross}	318	424
- collectors	EUR/m^2_{gross}	127,2	190,8
- plant engineering	EUR/m^2_{gross}	95,4	106
- elec. connection and I&C technology	EUR/m^2_{gross}	25,44	21,2
- construction costs	EUR/m^2_{gross}	22,26	21,2
- planning costs	EUR/m^2_{gross}	15,9	21,2
- connection costs	EUR/m^2_{gross}	31,8	63,6
Total investment	EUR	7.950.000	8.480.000
Operating Costs			
maintenance (0,7 % of total investment)	EUR/a	55.650	59.360
electricity costs	EUR/a	15.000	15.000
Annual operating hours	EUR/a	70.650	74.360
Energy Costs			
specific electricity costs	EUR/MWh_{el}	150	150
Energy-economic Framework			
calculation interest rate	%	8%	8%
LCOH for 25 years	EUR/MWh_{th}	82	87

14.10 Funding Opportunities in Germany

There are some subsidy programs for the construction of solar thermal systems in Germany. The most important one currently is the Federal Funding Programme for Efficient Heating Networks (BEW). Other subsidy systems are aimed at heating systems in which the solar thermal system can be subsidized as part of a systemic subsidy. This applies in particular to subsidies via the Combined Heat and Power Act (KWKG).

Bibliography

[Geiger] Geiger, Solarthermie und Wärmenetze – eine Kombination mit Zukunft, Solites – Steinbeis Forschungsinstitut für solare und zukunftsfähige thermische Energiesysteme, Vortrag in: Schulung Erneuerbare und energieeffiziente Wärmenetze: Planung und Projektentwicklung. Stuttgart, 2019

[Wesselak] Wesselak et al., *Regenerative Energietechnik*, 2. Auflage, Springer Vieweg, ISBN 978–3-642-24165-9 (eBook), 2013

[DWD] https://www.dwd.de/EN/ourservices/solarenergy/maps_globalradiation_sum_new.html, accessed on February 22. 2024

[SDH Guidelines] Solar district heating guidelines, collection of fact sheets, within the frame of the SDH project, https://www.solar-district-heating.eu/en/knowledge-database/, 2012.

[Witte-Humperdinck] Witte-Humperdinck, Untersuchung der Auswirkungen von Einspeisung erneuerbarer Wärme auf KWK-Wärmenetze in urbanen Quartiersstrukturen, Dissertation, Universität Duisburg-Essen, http://dx.doi.org/10.17185/duepublico/70832, 2019

Jochen Weilepp

15 Levelized Cost of Electricity – Understanding the Concept and Setting the Right Priorities

15.1 Introduction

As a developer of a new technology, you are usually proud of the solution you and your team have developed. But instead of being interested in the exciting technical details, most people you talk to ask the inevitable question, “How much does it cost to generate a kWh of electricity with your technology?” or “When will it be economical?” Politicians in particular like to ask this question when you ask them for financial support for your technology development. As a technology developer at heart, I hate this question, but of course it has to be dealt with. Moreover, it is of utmost importance to understand which drivers are the most effective to reduce costs per kWh as fast and as cost-effective as possible.

Therefore, this chapter deals with the question: “How much does a kWh of electricity generated with a certain technology cost?” or, more precisely, it introduces the concept of “Levelized Costs of Electricity” (LCOEs). In the first part, the theoretical basis of the concept is discussed and the relevant components for the correct calculation of LCOEs are presented. The concept also explains which parameters have the greatest impact on LCOEs. The reader will understand the importance of cost of capital. The second part of this chapter introduces two important sources of learning – technical and financial learning. Both contribute to lowering LCOEs – but with different effectiveness and associated costs. The chapter closes with a set of recommendations for practitioners such as technology developers or policy makers.

15.2 The Concept of Levelized Costs of Electricity

The various power generation technologies have completely different cost structures. Fossil-fired power plants – especially gas-fired power plants – have relatively low investment costs (CapEx) but high operating costs (OpEx) due to the comparatively high fuel costs. Such a high OpEx ratio leads to these power plants being used for peak

Note: This article has first been published in: Jochen Weilepp: Einfluss des technischen Lernens und des finanziellen Lernens auf die durchschnittlichen Kosten für Strom, in: Ulf Lassen (Hrsg.): „Nachhaltiges Immobilien- und Energiemanagement“, Erich Schmidt Verlag, 1. Auflage 2024. Reprinted with the kind permission of Erich Schmidt Publishing House, Berlin, 2025.

https://doi.org/10.1515/9783110784299-015

power generation. Renewable energy power plants, on the other hand, are characterized by high CapEx and significantly low OpEx. Economically, these plants should be operated for as many hours as possible and are therefore ideally used as base-load power plants (e.g., hydropower). Since CapEx and/or OpEx of power plants are usually associated with high cash flow the timing of when those cashflows occur becomes relevant and, another important cost element must be included in the calculation, namely the **cost of capital**.

The typical cash flow pattern – e.g. for a renewable energy plant – looks like this (Figure 15.1):

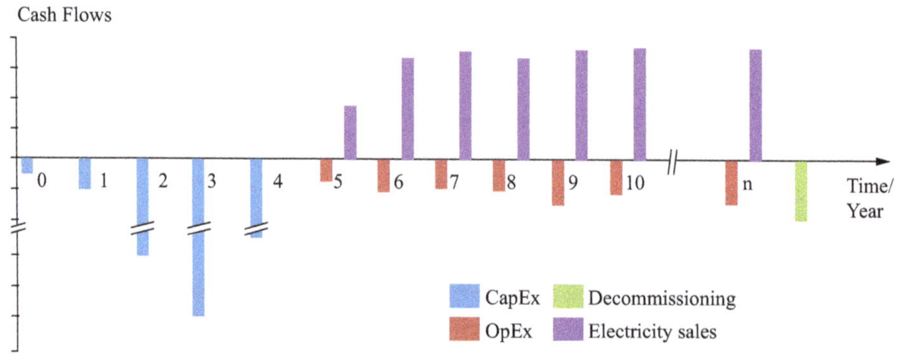

Figure 15.1: Schematic cash flow pattern of a power plant project figure created by author.

At the beginning, there is a project development phase in which site investigations, environmental impact analyses, discussions with stakeholders and permit applications have to be prepared. The associated cash outflows are small compared to the construction phase, which follows immediately after. Once the project is fully constructed, the commissioning phase follows, during which production is gradually ramped up to planned production and begins to generate cash inflows from (electricity) sales. During this phase, operating costs reduce the free cash flows available to project owners. Planned maintenance breaks may also occur, resulting in further cash outflows and lower cash inflows due to downtime. At the end of life, there is the decommissioning phase, which – depending on the type of power plant – can lead to significant cash outflows (e.g., for fossil, hydro, or nuclear power plants) or even slight cash inflows if the opportunity arises to sell old plants at scrap value (e.g., for wind power plants).

Therefore, the overall net cashflow pattern, as shown in Figure 15.1, can have a rather complex shape, even in the absence of unexpected conditions.

When investors know this expected net cashflow pattern, they must decide whether or not to invest in the project. To evaluate such investment opportunities, the technique of dynamic investment calculus was developed; cf. (Brealey, 2022). In simple terms, the Present Values of all future cash flows (future cash flows discounted with the investor's cost of capital *i*) associated with the project are calculated and summed up. If the resulting sum – called the Net Present Value (NPV) – of the investment's cash flows is zero, then the project has earned back all of its expenses (CapEx, OpEx) as well as its cost of capital, but it has not yet created value for the investor. Therefore, investors are – in this situation – indifferent of whether to invest or not. The Net Present Value of a Power Plant project with a lifetime of n years from the first project idea until decommissioning in its most general form reads:

$$NPV = -\sum_{t=0}^{n} \frac{CapEx_t}{(1+i)^t} - \sum_{t=0}^{n} \frac{OpEx_t}{(1+i)^t} - \sum_{t=0}^{n} \frac{Decom_t}{(1+i)^t} + \sum_{t=0}^{n} \frac{p_t \cdot W_{el,t}}{(1+i)^t}$$

Equation 3: Net Present Value of a Power Plant Project in the most general form

whereas $CapEx_t$, $OpEx_t$ and $Decom_t$ represent capital expenditures, operational expenditures and expenditures for decommissioning in year t, respectively. Usually *CapEx* occurs in earlier years of the project. In later years, there may be no further capital expenditures, hence $CapEx_t = 0$. The same applies reversely for $OpEx_t$ and $Decom_t$ accordingly. The electrical work/energy $W_{el,t}$ produced in year t is sold at a the per unit price p_t of the year t.

Equation 3 helps us in our efforts to determine the cost of electricity generation: If investors are compensated for selling the generated electricity at such a constant per unit price $p_t = const = p$ that the Net Present Value equals to zero, then all costs are recovered. This fictitious constant price is known as the Levelized Cost of Electricity (LCOE). It represents the life-cycle cost of one kWh of electricity. The expression for LCOE can be simplified as follows:

$$NPV = -\sum_{t=1}^{n} \frac{CapEx_t}{(1+i)^t} - \sum_{t=1}^{n} \frac{OpEx_t}{(1+i)^t} - \sum_{t=1}^{n} \frac{Decom_t}{(1+i)^t} + LCOE \cdot \sum_{t=1}^{n} \frac{W_{el,t}}{(1+i)^t} = 0$$

Now, this expression can be resolved for LCOE as follows

$$LCOE := \frac{\sum_{t=1}^{n} \frac{CapEx_t}{(1+i)^t} + \sum_{t=1}^{n} \frac{OpEx_t}{(1+i)^t} + \sum_{t=1}^{n} \frac{Decom_t}{(1+i)^t}}{\sum_{t=1}^{n} \frac{W_{el,t}}{(1+i)^t}}$$

Equation 4: Definition of Levelized Cost of Electricity

The interpretation of LCOE is now straight-forward: If the local feed-in tariff or the alternative electricity costs in a target region are known (e.g. production with diesel generators etc.), the economics of a new (renewable) power plant can now be easily evaluated: If the electricity price or the alternative electricity costs are higher than the LCOE, the new power plant can be operated economically and the investment is worthwhile – otherwise it is not.

At first glance, the denominator of the LCOE formula looks puzzling: A discounted W_{el} means that we are discounting the kWhs produced. What is the economic reason for this? This expression is a simple consequence of the time value of money concept, which states that money earned later in the life of an investment is worth less than money earned earlier. In the case of a power plant project, this means that electricity produced in later years is worth less than electricity produced in the early years, because it can only be sold later on the electricity market.

If we neglect interest rate effects in Equation 4, i.e. $i = 0$, then the expression becomes very intuitive:

$$LCOE = \frac{Total\ CapEx + Total\ OpEx + Total\ Decommissioning\ costs}{Sum\ of\ all\ Electricity\ produced\ during\ the\ project\ lifetime}$$

Equation 5: LCOE formula for $i = 0$

That is, LCOE is now the cumulative cash outflows associated with the power plant project over its complete lifetime divided by the cumulative electricity production over its lifetime.

Some technologies – such as photovoltaics – have a degradation effect, i.e. the expected average annual yield decreases each year by a certain percentage d. The degradation effect for PV technologies is currently estimated at $d = 0.2\%\ p.a.$ (cf. (Fraunhofer ISE, 2021). This degradation only affects the electricity production and hence only the electricity production term in the Net Present Value calculation for a given power plant project in Equation 4. Mathematically, this acts like a deflation of the amount of electricity produced annually. So, in the denominator of the LCOE formula, we need to replace the interest rate i with the deflated interest rate i^*, which is calculated as follows:

$$i^* = \frac{1+i}{1-d} - 1$$

Equation 6: Modified Interest Rate to account for Degradation Effects

Hence the LCOE formula for this case reads

$$LCOE = \frac{\sum_{t=1}^{n} \frac{CapEx_t}{(1+i)^t} + \sum_{t=1}^{n} \frac{OpEx_t}{(1+i)^t} + \sum_{t=1}^{n} \frac{Decom_t}{(1+i)^t}}{\sum_{t=1}^{n} \frac{W_{el,\ t}}{(1+i^*)^t}}$$

Equation 7: LCOE for a power plant experiencing degradation

But what is the cost of capital i used to discount the cash flows? Typically, a power plant project with a total capital requirement TC is financed partly with equity EQ from the project sponsors and partly with debt D from banks or bondholders. Because of the higher risk equity investors take, they usually expect a higher return k on their invested money than the fixed interest rate r charged by the banks.

In a real world, the government levies taxes (corporate tax rate t). Since interest on debt is usually tax deductible, the interest payment reduces the tax burden on a project company and further reduces the cost of capital for the project company.

Therefore, the cost of capital is the weighted average of the cost of equity k and the cost of debt r, adjusted for the tax saving effect:

$$WACC = ER \cdot k + DR \cdot (1 - t) \cdot r$$

Equation 8: Weighted Average Cost of Capital (WACC)

The weighting is given by the equity ratio $ER = EQ/TC$ or the debt ratio $DR = D/TC$.

For a reasonably mature technology such as offshore wind the debt ratio reaches $DR = 70\%$ at an interest rate of $r = 5\%$, while $ER = (1 - DR) = 30\%$ of the capital is contributed as equity with an expected return of $k = 10\%$; cf. (Fraunhofer ISE, 2021). That is, offshore wind turbines currently have a cost of capital of WACC = $30\% \cdot 10\% + 70\% \cdot 5\% \cdot (1 - 30\%) = 5{,}45\%$ *p.a.*

This cost of capital is nominal, i.e. it still includes inflation effects. If identical inflation rates $INFR$ for all cost and revenue elements in the calculation as well as a constant inflation over the lifetime of the project can be assumed than it is sufficient to set up the model at today's cost levels and consider inflation by calculating the real – deflated – $WACC^*$:

$$WACC^* = \frac{1 + WACC}{1 + INFR} - 1$$

Equation 9: Definition of the inflation adjusted WACC

This deflated expression does not necessarily need to be used. However, if the nominal $WACC$ is used, the appropriate inflation rates have to be applied to all cash flows in the LCOE formula. For simplicity, the deflated approach is used later in this chapter.

For some applications, such as the investment in a rooftop PV system, Equation 7 can be further simplified: If the cash flow associated with the project development and construction phases flows in period 0 to a good approximation (so-called "overnight costs" since the plant is notionally built overnight), if the annual operating costs $OpEx$ in all n periods of the plant life and the expected annual electricity production W_{el} over all periods can be considered constant to a good approximation and if decommissioning costs to a good approximation can be neglected,[1] the simplified cash flow pattern looks as in Figure 15.2:

Such a simplified pattern is often sufficient to model projects with fast execution times or when construction is about halfway complete. Under the above assumptions and in the case of negligible degradation ($d \approx 0$) Equation 7 can be simplified to an expression that is frequently used in engineering textbooks:

1 Since decommissioning usually occurs in the distant future and part of the decommissioning costs can be offset by the scrap value of the power plant, the present value of decommissioning usually does not have a large impact on the LCOE. Therefore, the approximation to neglect decommissioning costs is quite good in many cases.

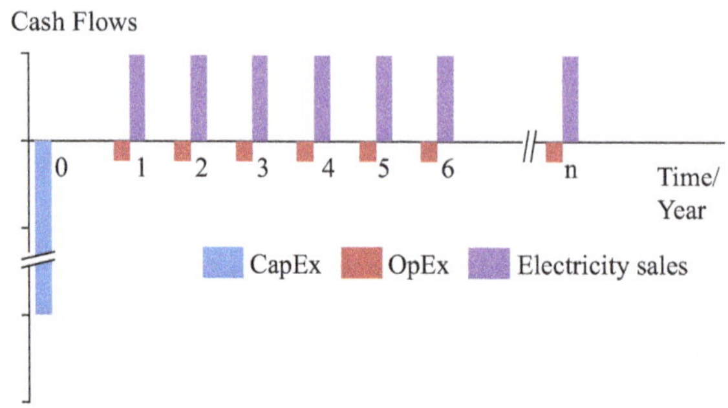

Figure 15.2: Simplified Cash Flow Pattern; figure created by author.

$$LCOE = \frac{Total\ Capex + OpEx \cdot PPVF(WACC,\ n)}{W_{el} \cdot PPVF(WACC,\ n)} = \frac{Total\ Capex \cdot AF(WACC,\ n) + OpEx}{W_{el}}$$

Equation 10: Annualized LCOE Definition

where the pension present value factor *PPVF* an the annuity factor *AF* are defined as follows:

$$PPVF(WACC, n) = \frac{1}{AF(WACC, n)} = \frac{(1 + WACC)^n - 1}{(1 + WACC)^n \cdot WACC}.$$

Equation 11: Definition of the pension present value factor and the annuity factor

Equation 10 and Equation 11 are derived by using the standard tools of financial mathematics. In the case of non-negligible degradation ($d \neq 0$) the corresponding expression transforms to

$$LCOE = \frac{Total\ Capex + OpEx \cdot PPVF(WACC,\ n)}{W_{el} \cdot PPVF(WACC*,\ n)}$$

Equation 12: Annualized LCOE Definition for Systems with Degradation

15.2.1 Understanding the Mechanics of the LCOE Formula

It is not surprising that a new (renewable) power plant technology starts with LCOEs far above the cost of an alternative existing technology: We humans tend to do the easy things first. For example, we started building water wheels or hydroelectric plants long before we looked at Offshore Wind or Tidal Power. Lower capital costs per MW, higher capacity factors, and much more technological experience have put large conventional hydro, for example, in a more economically favorable position than Offshore Wind. Some technologies (e.g. PV) will eventually be able to offset these disadvantages and become competitive at cost with other forms of power generation.

Others may operate under such difficult conditions that they will eventually reduce LCOEs but remain above competing technologies. Nevertheless, it may be politically desirable to develop such technologies further and therefore support them with subsidies. To prepare such a decision a deeper understanding of the constituents of the LCOEs of a new technology is extremely helpful. Depending on the nature of the technology under consideration the analyst may then be able to judge the target LCOEs in the mature state of the technology.

LCOE analyses for Offshore Wind and Large Rooftop Photovoltaics

To better understand the impact of different input parameters, the LCOE analyses of Offshore Wind and large-scale (> 30 kW installed capacity) rooftop photovoltaic plants are performed mainly for sites in Germany for the years 2010 and 2021.

Since detailed project cost data are difficult to obtain, we base our analysis on solid literature data from the Fraunhofer Gesellschaft; cf. (Fraunhofer ISE, 2010) and (Fraunhofer ISE, 2021) and – where necessary – plausible assumptions (see Table 15.1).

Table 15.1: Parameters used for Analyses (compiled by author).

Parameter	Unit	Offshore Wind		Large Rooftop PV (> 30 kW)	
		2010	2021	2010	2021
Capacity Factor CF	%	45	50	10.3	12.6
Annual Degradation d	% p.a.	0	0	0.3	0.3
CapEx	M€/MW	4	3.5	3	1
fixed OpEx	M€/MW	0	0.07	0.035	0.0215
variable OpEx	M€/GWh	0.03	0.008	0	0
Project Lifetime n	years	20	20	25	25
Equity Ratio ER	%	40	30	30	20
Debt Ratio DR	%	60	70	70	80
Corporate Taxes t	%	30	30	30	30
Cost of Equity k	% p.a.	14	10	10	6.5
Cost of Debt r	% p.a.	7	5	5	3
Inflation Rate INFR	% p.a.	2	1.2	2	1.2
Source		ISE 2010	ISE 2021	ISE 2010	ISE 2021

The calculations are performed using the overnight assumptions with the simplified model

The breakdown of the CapEx and OpEx contribution, as well as the contribution of the cost of capital to LCOE, can be determined if the discount rate *WACC* in Equation 10 is artificially set to zero. The resulting value is the contribution of CapEx and OpEx to LCOE under the assumption that the project would not incur any financing costs. If OpEx (CapEx) is then set to zero, the CapEx (OpEx) contribution to LCOE can be calculated. The financing costs can then be calculated by subtracting the CapEx and OpEx contribution from the derived LCOEs.

The result of this exercise is shown in Figure 15.3: The electricity production cost of an Offshore Wind Plant built in 2010 was 0.121 EUR/kWh and for a large rooftop PV plant 0.247 EUR/kWh. After 11 years, these costs have dropped to 0.076 EUR/kWh for Offshore wind and 0.066 EUR/kWh for PV, respectively. The figures derived here are consistent with those published in (Fraunhofer ISE, 2010) and (Fraunhofer ISE, 2021) although they generally appear somewhat low – especially for 2010.

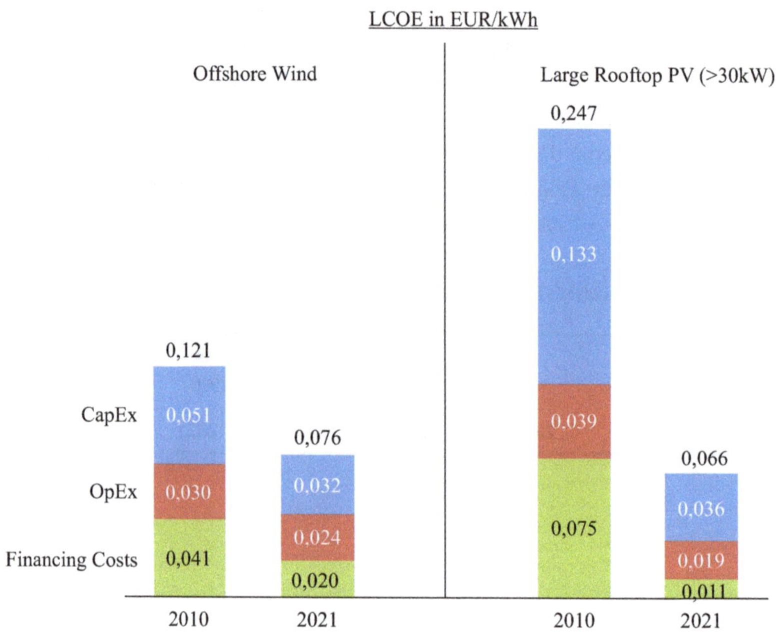

Figure 15.3: LCOE Breakdown for Offshore Wind and Large Rooftop PV (>30 kW) between 2010 and 2021; figure created by author.

It is worth noting the significant share of financing costs (Offshore Wind: 34% and large rooftop PV: 30%) in total costs in 2010 (note: for younger and riskier technologies such as Wave and Tidal Power, this share can be as high as 40% or even more). It is interesting to note, and a fairly common phenomenon, that the share of financing costs decreases with maturity, as seen in the case of 2021 (Offshore wind: 27% and large rooftop PV: 16%). The decrease in the absolute amount of the technical contribution (CapEx and OpEx) to the LCOE is due to **technical learning**, i.e., the unit cost decreases over time as technology and manufacturing processes etc. improve. The fact that the share of financial costs is decreasing is referred to as **financial learning**.

It suggests that the financial cost is decreasing over time not only because there is less material cost or OpEx to finance in more mature technologies, but also because financiers better understand the risk profile of a technology. As a result, they are now willing to provide more and cheaper capital.

Figure 15.4 shows the evolution of the LCOE for Offshore wind and large rooftop PV between 2010 and 2021 and assigns values to the different effects mentioned above: In a first step, we apply the 2021 financial parameters (debt-equity ratio and financing costs) to the 2010 technical parameters. This approach models the effect of **financial learning**. In a second step, we change the technical parameters to the values of 2021 and thus model the effect of **technical learning**.

We find that financial learning reduces financing costs by EUR 0.021/kWh, or more than 50%, in the case of Offshore Wind. This is only because financiers have more confidence in the technology. In the case of Large Rooftop Photovoltaics, this effect is even more pronounced at EUR 0.051/kWh or −67%.

The second step shows that the technical learning (reduction of CapEx and OpEx) in the case of Offshore Wind only contributes 0.025 EUR/kWh. The savings in financial costs due to the lower investment and operating costs per unit are eaten up by inflation effects, thus resulting in an overall saving of the financial contribution to the electricity production costs of 0.021 EUR/kWh. In the case of photovoltaics, the picture is different: While the technical contribution decreases by 0.117 EUR/kWh, the financial contribution decreases by only 0.064 EUR/kWh. Expressed as a percentage, however, the old picture emerges again: the technical costs fell by 68.0%, while the financial costs fell by 85.4%.

Two Sources of Learning and how they can be Influenced

We saw in the section above that there are two completely independent sources of learning that must be understood if LCOEs are to be managed most effectively.

Technical Learning and Experience Curves

The Wright Brothers discovered that their aircraft manufacturing time decreases by a constant factor when the cumulative number of aircraft of that type manufactured has doubled since the first one was built. This effect is called the **learning curve** effect; cf. (Wright, 1936).

Boston Consulting Group founder Bruce D. Henderson discovered that a similar phenomenon applies to the cost per unit produced of a given product. This effect is called the **experience curve effect**; cf. (Henderson, 1968). While the learning curve can be explained qualitatively by optimizing production processes, improving material selection, reducing waste, improving efficiency, etc., the origins of the experience curve are much broader. It can be due to learning, as just explained, but it can also be due to economies of scale, cost advantages due to procurement power, economies

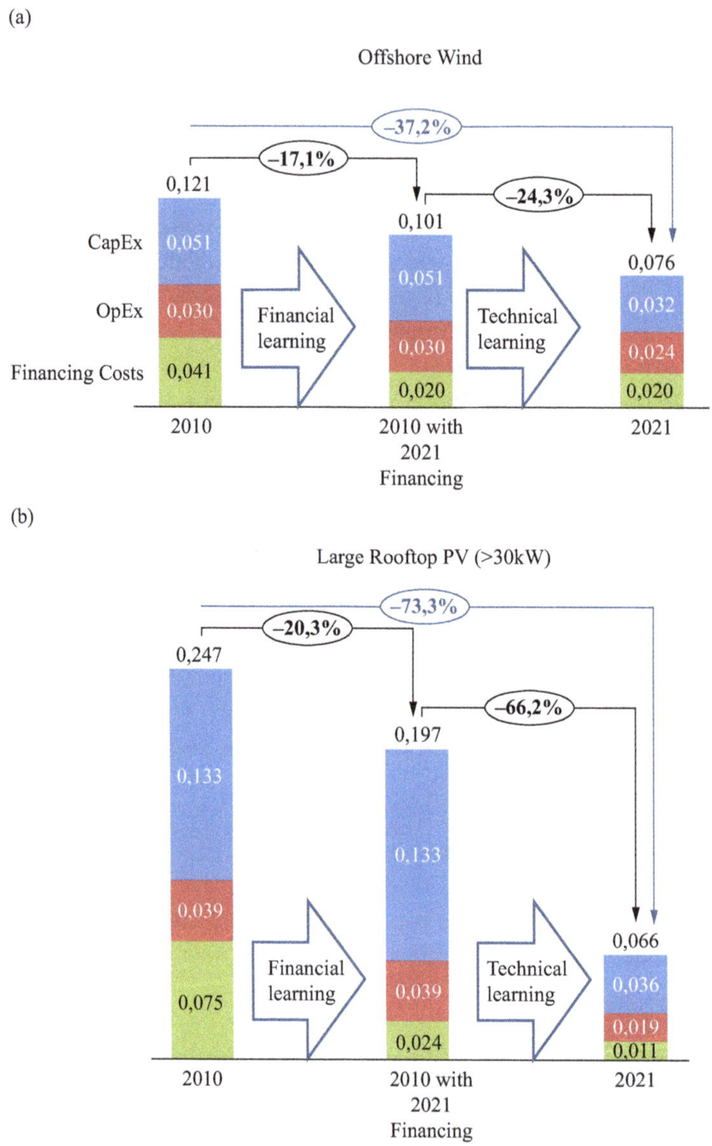

Figure 15.4: Decomposition of LCOE Development for Offshore Wind and Large Rooftop PV; figure created by author.

of scope (i.e., leveraging experience gained in adjacent technologies to which the company also has access), etc. Although the experience curve has a much wider range of ingredients, the phenomenological effect has been shown to be quite similar: Per doubling of cumulative output (since inception), unit costs decrease by a constant percentage.

Although, as explained above, the experience curve does not account only for learning effects, it has become customary to refer to this constant percentage as the learning rate *LR* and its complement $(1 - LR) = PR$ as the progress ratio.

Mathematically the development of the costs as a function of the cumulative output since inception $C(x_t)$ with x_0 as the first amount which has been produced, can be formulated as follows; cf. (Fraunhofer ISE, 2021)

$$C(x_t) = C(x_0)\left(\frac{x_t}{x_0}\right)^{-b}$$

where

$$b = -\frac{\ln PR}{\ln 2}$$

Equation 13: Mathematical Foundation of Experience Curve effects

It is quite common to plot experience curves on a double logarithmic scale, since the curves then become linear and the slope is a direct measure of LR or PR, respectively.

To forecast future LCOEs, experience curves are a very helpful tool and very often used. However, the difficulty is to forecast the correct values for LR. For this, it is helpful to examine neighboring industries or technologies. It is also very important to qualitatively understand the (potentially) underlying effects that eventually will lead to experience curve effects.

In retrospect, we are now conducting such a qualitative analysis of Offshore Wind and Photovoltaics:

The experience curve for solar modules is very well researched and the learning rate is about 22%; cf. (Samadi, 2018). This learning rate is generally at the upper end. The reason for this is a number of significant improvements over the last 20 years. The most important are:
- Manufacturing process optimization
- Optimization of ingot and wafer production
- Improvements of energy efficiency in the production processes
- Improvement of module efficiency
- Better use of semiconductor and non-semi conductor materials

Although semiconductors have already been used for microelectronics in the past, semiconductor manufacturing has had to expand significantly from a more laboratory-like production to high-volume production.

The situation is different for Offshore Wind power: Wind turbine manufacturing is largely an assembly business (if you ignore Enercon, which takes a completely different approach): Generators, gearboxes, couplings, bearings, shafts, and inverters etc. are mainly sourced from specialized manufacturers (unless those manufacturers have already been fully acquired by a wind OEM). However, these components are not new. Of course, they had to be adapted to the challenges of wind power (cf. series gearbox failures in the early days of wind power due to the very unusual load profile caused by wind gusts, etc.), but most of the experience curve effects for Offshore Wind have already been gained in other industries, such as Onshore Wind power, etc. Apart from completely new components – such as Offshore Wind foundations – and sophisticated operation and maintenance concepts that reduce the need for very expensive specialized installation and maintenance vessels, there are not too many learning effects that can be exploited. Moreover, the price of a lot of these components is largely determined by the cost of expensive input materials such as glass- or carbon-fiber reinforced polymers, copper or steel. The use of these materials can hardly be avoided, and the quantity used is already highly optimized. The strongest lever for experience curve effects in Offshore Wind energy can therefore almost only be scale. The larger the wind turbine, the better the fixed costs for foundations, grid connections, etc. are utilized. We also see this in Table 15.1. Between 2010 and 2021, investment costs have only decreased from 4 MEUR/MW to 3.5 MEUR/MW. A significant part of the technical progress is due to the increase of the operational lifetime from 20 to 25 years and due to higher capacity factors. The latter is mainly due to higher reliability (less downtime) and the ability to operate in higher energy areas. However, the overall effect is limited – especially since world market prices for raw materials are rising sharply. (Samadi, 2018) describes learning rates of 3% for offshore wind with a very wide error bandwidth (−5% < LR <10%).

Remark: The Special Role of Efficiency for Technical Learning
Many start-ups in the energy sector – in Tidal Current Power, for example – compare their technical solutions with those of other tidal companies. If their technology seems better or more cost effective, they believe they have a better strategic position. But what are these Tidal Power companies really competing for? It is not about generating the cheapest electricity from the tides. Rather, it is about producing cheap electricity, no matter what the technology. Electricity is a perfect commodity. Therefore, consumers do not care whether they use electricity from a photovoltaic system or a Tidal Power plant when they watch a soccer game on TV. In the end, only the lowest LCOE matters.

Looking at the LCOE formula in Equation 4, all discounted net cash flows are in the numerator and discounted kWh generated are in the denominator. If research leads to an improvement in system efficiency, this is usually associated with higher component costs. The tricky question now is: Does the improvement in efficiency justify the increase in cost?

Let us examine this using the example of an Offshore Wind electrical generator that costs about 8% of the total system's capital cost; cf. (Meißner, 2020). Due to the OpEx contribution this corresponds to about 5% of the total discounted cash outflows in the numerator of the LCOE formula. In other words, 95% of the system cost remains unchanged. If the generator efficiency increases by 1%, this means that the denominator of the LCOE formula increases by 1%. To keep costs constant, the total numerator can also increase by 1%. Since 95% of the discounted cash outflows are unaffected by this efficiency improvement of the generator, the cost of the generator can increase by 20% to keep the LCOEs constant.

This numerical example illustrates why efficiency improvements generally have an enormous impact on the cost of electricity. This effect is even more dramatic in the case of large hydropower plants, where around 80% of the investment costs are attributable to construction work (dams etc.) that has no direct influence on electricity generation. Efficiency gains are usually achieved either with an improved turbine runner or generator. These components usually only account for 20–30% and the overall mechanical and electrical CapEx for 4–6% of the overall CapEx. The higher the efficiency, the better use is made of the other 94–96% of "unproductive" components. It is therefore a clear recommendation to increase efficiency as much as possible when developing a power generation technology.

Financial Learning

Most investors follow a risk-averse investment behavior. This means that they want to be compensated with a higher return when they take a higher risk. There are three different types of investment money: Debt, mezzanine and equity. While debt capital – usually provided by banks or bond investors – is relatively low-risk, mezzanine (e.g. subordinated debt or profit participation certificates, etc.) or even equity entail much higher risk. As a result, expected returns rise sharply.

The perceived risk of an energy project determines the investors' expected returns, which are reflected in the *WACC*. For young technologies – such as Offshore Wind power in its early years or Wave and Tidal Power today – it is difficult to obtain loans from banks. Projects must be financed largely with expensive equity. Sometimes, a small mezzanine portion can be raised, but it has an almost negligible impact on the overall *WACC*. Consequently, the share of financing costs in the LCOE for such risky technologies is very high and can reach up to 50%.

However, mature technologies, such as utility-scale PV, are so low-risk that 80–90% of the required capital is typically provided by banks as low-interest project financing. The risk premiums charged by the debt and equity providers are very low.

A general rule of risk management in finance is that the person who understands a risk best should take it. The reason is obvious: Someone who knows exactly the risks and pitfalls of a technology will take on a smaller safety cushion and thus expect lower returns. At the beginning of a new technology, banks – even if some of them maintain technical due diligence departments – wait before taking larger, low interest stakes in a project and thus lowering financing costs. As the number of installations increases, a technology becomes more reliable and trustworthy, and eventually the financial sector's confidence in the new technology grows. Once the first banks have successfully provided debt capital for a new technology project, the others tend to quickly follow, as no one wants to miss out on a great new investment opportunity. This effect is called **financial learning**. It helps to dramatically reduce the cost of electricity when the risks of a technology are well under control and well understood. Unfortunately, there is not yet a formula to model financial learning.

15.2.2 How to Effectively Manage LCOE – Recommendations for Practitioners

The above findings result in a number of valuable recommendations for technology developers and policy makers, all with the same goal, namely, to reduce LCOE as effectively as possible.

Recommendations for Technology Developers

Technology developers operate in a space of uncertainty and are often influenced by stakeholders such as politicians, potential customers, funding agencies, as well as local residents of deployment sites, etc. On the one hand, it is important to consider the needs of these stakeholders, but on the other hand, it is equally important to deliver a carefully thought-out concept. From the point of view of reducing LCOE, but also from a technical perspective, it is very helpful to follow the recommendations below

– **Focus on reliability first – cost reductions shall come later**
The reliability of a new technology is the strongest and most important lever for reducing LCOE: A reliable concept has little downtime and therefore produces the maximum amount of electricity. In addition, a reliable concept is considered less risky by all types of investors – therefore, the cost of financing (*WACC*) and consequently LCOE decrease rapidly. The effect of cost savings due to cost reductions is much less pronounced. New technologies are often deployed in new and challenging environments where dynamic load conditions, etc. are not yet fully known. In such environments, it is better to provide for a certain level of safety and to reduce costs later.

– Do not build technology with an inherent efficiency disadvantage
The impact of efficiency on the cost of electricity has been discussed in detail above. To optimize a technology's cost of electricity, the last few percentage points of efficiency must eventually be squeezed out, and every player in the industry will do so. There are some technology solutions that have an inherent efficiency disadvantage compared to other solutions. Wind energy, for example, has experimented with different rotor designs and different numbers of blades. Research has shown that the three-bladed horizontal axis rotor has the highest peak efficiency. Vertical axis rotors, such as the Darrieus rotor, have a great advantage in terms of directionality because they can be operated from all directions without requiring additional yaw in the system. However, their power coefficient (fraction of kinetic energy that the rotor can extract from the flow) is limited to 40%, while fully optimized three-bladed horizontal axis rotors can achieve a power coefficient of up to 50%. Thus, a Darrieus rotor has an intrinsic efficiency disadvantage of 20%. This LCOE disadvantage is unbearable and will eventually lead to the need for a redesign with a better rotor configuration. This, however, will then be considered a higher risk new technology and hence experience higher capital costs by the financial markets. Back to square one . . .

Of course, this argument should be taken with a grain of salt: Even if gallium arsenide PV cells achieve efficiencies well above silicon solar cells, they have such a prohibitively large cost disadvantage that silicon PV cells will always be more economical (unless we need to use them in space).

– Understand your opportunities for technical and financial learning
Great potential for technical and financial learning is critical to significantly reducing LCOE. While financial learning will always depend on investment costs, operating costs, capital structure, and investors' risk-adjusted return expectations, the sources of technical learning can be very different. As mentioned earlier, at the beginning of its commercialization, photovoltaics offered significant potential for production process optimization that had never been similarly exploited in the past for semiconductor materials: microprocessors, memory chips, transistors, or other semiconductor-based applications required completely different production processes than PV panels. The proportion of semiconductor materials per Euro of sales in the PV industry is also much higher than it used to be in the "old semiconductor world". There was enormous need for learning that led to the large learning rates observed in photovoltaics. On the other hand, Offshore Wind is largely based on lessons learned from Onshore Wind or a variety of other mechanical equipment applications. As a result, initial CapEx C_0 per MW of installed capacity started much lower for Offshore Wind than for photovoltaics – with the consequence that learning rates for Off-Shore Wind have been relatively low.

If we apply these considerations to the emerging technology of Tidal Current Power as an example, we are dealing with a technology that at first glance looks like a hybrid of wind power and marine propulsion systems. In both cases, the applications are mechanical and electrical, largely using components that have already been optimized for other applications. However, foundation and subsea energy conversion technologies are largely tailored to Tidal Power applications. In addition, scaling will play a role similar to that of offshore wind. Here, water depths and the much higher thrust forces (per MW) in slow-moving fluids (wind speeds versus tidal speeds) may act as limiters. In summary, one should not expect too high learning rates for Tidal Power. They will most likely be higher than for Offshore Wind, but far below those of Photovoltaics.

– **Do not leave your development path unless your concept has a significant flaw or an inherent efficiency disadvantage**
Experience has shown that converting a particular energy source into electricity comes with a source-specific price tag. Interestingly, this price tag seems not vary too much across sensible technological solutions. A design may have slightly more capital cost but less operating cost, or slightly higher efficiency, but the resulting LCOE is relatively similar. Direct-drive and gear-driven wind turbines can serve as examples of this heuristic statement.

Technology developers are constantly questioning their concept and seeing other ideas from competitors that they may find attractive and would like to implement. We would encourage technology developers not to do this and rather stick to their design – unless it is not fundamentally flawed. There is another heuristic and unproven "law" that we would like to pass on from our development experience: It states that "the total amount of troubles is conserved": If you change your concept on one side and solve a problem in one part of the system, you will eventually get another problem on the other end. Usually you discover this problem much later in the development process and the result will be a botched design. Let us elucidate that with a practical example: Tidal currents are bidirectional, and at several the direction changes by about 180 degrees every six hours. Therefore, at first glance, it makes a lot of sense to eliminate the failure-prone yaw systems in the turbine design and instead use bi- or centrosymmetric rotor blades to operate the machine in both directions without yawing it. The consequence of this design change is a significant increase in equipment weight and therefore higher equipment and installation costs! Why is this? Well, bi- or centrosymmetric blades cannot be designed to be torque-free as engineers do i.e. for wind blades. To compensate for this torque-freeness blades, must be very stiff. However, stiff blades transmit vibrations caused by turbulence directly into the turbine, while soft blades dampen them. To ensure the longevity of the design, you now need to reinforce the components and therefore require a considerable amount of more material. The weight effect is dramatic.

– **Foster financial learning by banks and other financial intermediaries**
Banks and other financial intermediaries need to understand your technology as early as possible to assess the risk associated with the concept. Some banks, such as the European Investment Bank (EIB), have large technical departments that perform technical due diligence to assess whether a technology is bankable and therefore eligible to receive a cer-

tain amount of project financing. For example, under Horizon 2020, the EIB introduced several InnovFin programs, most notably InnovFin Energy Demo Projects, which provided debt financing for early-stage projects, or InnovFin SME Guarantee, which provided 50% loss protection to banks providing debt financing for early-stage projects. Another example are financial intermediaries such as the German reconcept, which provided a small mezzanine equity vehicle for a Canadian Tidal Power project by Scottish Tidal Power developer Sustainable Marine Energy Ltd. Organizing and structuring such financing schemes may be time consuming and may initially provide only a very limited amount of the needed capital at a still considerably high cost. However, they help the financial sector understand the risks associated with a new technology and thus promote financial learning. Once the first projects prove successful, more will usually follow – ideally under better financial conditions.

Recommendations for Policy Makers

Policy makers may intend to make a new energy converting technology happen. However, it is not always easy for them to understand what technology developers really need – especially if they themselves do not completely know. At the end policy makers are interested in low cost (low LCOE) and reliable power and in the generation of new jobs.

– Develop a clear technology strategy and provide sufficient financial resources for its implementation

Supporting technical learning is horrendously expensive. This is due to the nature of the experience curve. While the first doubling of cumulative power may mean going from 10 MW to 20 MW, the second step is already twice as big. Remember that only each doubling of cumulative output is associated with a constant percentage decrease in cost. Hence, the steps get larger and the decrease in costs shrinks. If we consider that the electricity production cost of PV was about 0.5 EUR/kWh at a wholesale price of about 0.03 EUR/kWh, it is obvious that the first kWs installed on rooftops had to be subsidized at 0.47 EUR/kWh over their complete lifetime. Otherwise nobody would have invested. Germany started subsidizing different types of renewable energy in 2000. The Renewable Energy Sources Act (EEG) established feed-in tariffs for the various sources of renewable energy. The difference between the feed-in tariff and the sales price was levied as a charge on the electricity price that consumers had to and still have to pay. The total amount charged to electricity customers for photovoltaics between 2006 and 2019 is 103 billion Euros. For onshore and offshore wind, this figure amounted to 71 billion. Euros; cf. (Bundesnetzagentur, 2020). This figure is significant and holds a lot of political explosive power. However, it was a clear strategy of the German government to make Photovoltaics and Wind Power economically competitive – with the goal of developing these technologies into core German industries. While the latter has failed, at least in the case of Photovoltaics, cost competitiveness has more or less been achieved for both technologies. If Germany – along with other countries such as Spain and Denmark etc. – had not decided to finance these developments, Photovoltaics and Wind Power would not be available as affordable green energies worldwide today. For the global climate, this was definitely a worthwhile investment.

It is very important for policy makers to understand the costs associated with technical learning due to the experience curve effect. If a country pursues a strategy of developing an energy conversion technology to market readiness, thereby (hopefully) creating an entire new industry and associated jobs, it must accept the costs associated with technical learning, which in many cases are on the order of 10–100 billion Euros. However, if the development of a technology is subsidized here and there with a "small" sum, say 10 or 20 million euros, this will not be enough for the technology developers to reduce the cost of electricity to a competitive level. If no follow-on project is subsidized at a comparable level, the company will give up after the first project is delivered. The money invested by this company's backers, as well as the funds provided by the government, are then wasted with no further benefit to anyone. Unfortunately, this fate has befallen many Wave and Tidal Power start-ups over the past 20 years even though they had promising technologies. A technology can only be developed to market readiness, if there is a clear political will to do so and if the subsidies are provided on a sufficient and reliable basis. To estimate the amounts needed in advance, it is critical to understand the mechanisms of technical and financial learning and their impact on LCOE. Unfortunately, there is no short-cut.

– **Provide funding for technology developers to capitalize on financial learning** Since financial learning can occur more quickly than technical learning, it is recommended that flanking measures be taken to train the banks. Activities such as those carried out by the EIB with project finance programs or guarantee funds can be emulated by national state investment banks. However, it is always important that commercial banks or other financial intermediaries are involved in the activities in parallel. However, this requires funds to be invested into demo-projects or equity participations. Banks need to be given the money to invest in maybe higher risk projects, if they are of strategic importance to the corresponding jurisdiction.

– **Pave the way for technology developers to achieve the required scale** New sources of renewable energy can have certain negative impacts on nature. With Wind Power, bird and bat strikes are frequently observed. With Hydro Power or Tidal Power, fish passing by turbines pose a major threat. In many cases, however, these impacts have not been fully studied, and local authorities are overwhelmed by designing the required permitting processes. Instead of building test turbines and objectively monitoring the potential impacts, complex and regional permitting processes are set up that vary from state to state or county to county and are often driven by a lack of understanding rather than by a tangible threat. It is nearly impossible and extremely expensive for a small start-up company to deal with all the restrictions and concerns of local stakeholders. We have seen startups driven to the edge of bankruptcy because they could not finance the cost of additional study requirements and delays by local officials who were reluctant to grant initial approval for a prototype. Pre-approved test centers – such as the European Marine Energy Center (EMEC) in Orkney – can be helpful for initial developments. However, for global deployments, it is highly recommended to agree on fast and simple, and ideally globally identical, approval procedures for initial projects. Pragmatic rather that bureaucratic approaches are much more helpful here.

Conclusion and Outlook

Levelized cost of electricity (LCOE) is a very important measure of an energy conversion technology's competitiveness. It is absolutely critical that technology developers understand the mechanics of their technology's LCOE. High initial costs for a new technology can be problematic, but can be overcome if there is significant technical and financial learning potential. If neither is present, it is most likely not worth investing much time in developing this new technology.

The concept of LCOEs can also be used for other energy-related technologies. For energy efficiency measures, for example, a similar figure – the Levelized Costs of Conserved Energy – can be defined. The formula Equation 4 or Equation 10 simply needs to be modified so that the denominator calculates the present value of the kilowatt-hours saved by using the energy-saving technology. The expression can also be modified to calculate the levelized cost of hydrogen production when substituting in the denominator the present value of the kilowatt-hours or kilowatt-hours of hydrogen produced with an electrolyzer. Most of the above discussion can be readily applied to these other applications.

We would like to conclude with a final remark: The years following the 2008 financial crisis have been characterized by extremely low interest rates. Due to the high inflation rates that can be expected, especially for the years following 2022, a significantly different interest rate level is expected. This increase in financing costs will undoubtedly drive up the LCOEs for all energy conversion technologies to be built. Hence, we will most likely see increasing rather than decreasing LCOEs for all energy conversion technologies, no matter how efficient technical learning will be.

Bibliography

Brealey, R. A. (2022). *Principles of Corporate Finance* (14 Ausg.). New York: McGraw-Hill Eduscation.
Bundesnetzagentur. (2020). *EEG in Zahlen 2019*.
Fraunhofer ISE. (2010). *Levelized Cost of Electricity Renewable Energies*. Freiburg.
Fraunhofer ISE. (2021). *Levelized Cost of Electricity*. Freiburg.
Henderson, B. D. (1968). The Experience Curve. (T. B. Group, Hrsg.) *BCG Perspectives*.
Meißner, M. (2020). Offshore wind turbine cost structure analysis. *High Technology Letters*, S. 24–38.
Samadi, S. (2018). The experience curve theory and its application in the field of electricity generation technologies – A literature review. *Renewable & Sustainable Energy Reviews*, S. 2346-2364.
Wright, T. (1936). Factors Affecting the Cost of Airplanes. *Journal of the Aeronautical Science*(3), S. 122–128.

Jörg Böttcher

16 Project Finance – The Finance Method for Established Technologies

16.1 Introduction

The imminent transformation of the energy markets and many of our living environments will lead to a considerable need for investment, which will require the involvement of banks as well as investors.

The transformation process has an impact on the banking world, but it is also the case that banks have to observe certain objectives and restrictions that must also be taken into account when shaping the transformation process. In this context, we are concerned with two things: firstly, due to their risk/reward structure, banks must ensure that they only invest in stable and predictable projects. Uncertainty regarding the borrower's performance can quickly lead to a withdrawal from the provision of debt capital. This statement applies to both corporate and project financing. However, we will refer to project financing, as it is the dominant form of financing when it comes to the realization of RE projects. On the other hand, several market failures must be taken into account in the context of transformation, which can lead to markets either not being able to establish themselves or not being able to establish themselves optimally. We want to point out and analyze these facts and show how they can be remedied.

But why is project financing so important for the energy transition? The main answer is that most renewable energy projects are planned by project developers, and their business models require a sale of the project to third parties. The financing alternative, a corporate financing for the developer, might fail due to an insufficient credit worthiness amid the long-payback time of the loan facilities. However, there are some developers which have amended their business model to a project-and-hold-strategy, but this strategy requires time. The second reason for the attrativeness of a project financing is the risik limitation of the project's owner, which we will describe in the following sections in more detail. And the third reason is that project financing is the financial platform which allows the wide-spread use of a technology at economical terms – we will come to this issue in the following sction.

We need to distinguish between two aspects: On the one hand, the requirements that banks place on the implementation of RE projects in the context of project financing. On the other hand, the market failures that can prevent the establishment of markets and justify government action. Both topics are interlinked. We want to start with the requirements that are placed on the design of RE projects in the context of project financing.

https://doi.org/10.1515/9783110784299-016

16.2 Project Finance – The Importance for the Roll-Out of the Transformation Process

In this section, we will look at the role of project financing in the implementation of the energy transition, a process that we will illustrate using Figure 16.1 below.

First of all, we have illustrated various stages of the technology development process in the left-hand section. The illustration is based on the concept of the development stages of technologies developed by NASA, which are referred to as **Technology Readiness Levels** (TRL).[1] The sequence shows a development from general basic research through increasing improvement of the technology until it has reached the stage of a tested and market-ready technology. However, technological development does not end at TRL 9; instead, it is crucial for the long-term success of the technology that efficiency – measured in terms of prime costs – is continuously improved.[2]

Financing must be seen in parallel to this technological development, which changes depending on the stage of development of the players and the rules of the game, and therefore also has repercussions on the necessary actions of the operators.

From an implementation perspective, the process between TRL 7 to TRL 9 is crucial: the aim here is to prove the technical and economic feasibility in real operation for the banking market. The transition from a functioning trial operation (TRL 7) to a market-ready project in real operation (TRL 9) is so difficult because it is not a gradual process but is characterized by several fundamental shifts:

1. **Scaling**: The step towards market maturity requires a much larger project, which triggers correspondingly large financing requirements. A tenfold increase in investment requirements is not unusual, and larger factors are also conceivable. This financing task can therefore neither be met by the existing funding providers nor by the existing instruments.
2. **New financial instruments**: The previous providers of funds will often be venture capital companies. Their business model is to bring such technology or projects to market maturity and then sell them to other investors, who then take over the implementation. They accept that individual projects may fail. The business model pays off for them if the profits from the sale of the successful projects exceed the write-downs on the aborted projects. However, their business model does not provide for permanent continuation in the market phase. In this respect, the project is faced with the task of finding new investors. This can and often will be project financing, but new financing requirements arise here.

1 The EU Commission has used this procedure, for example, to assess technologies as part of its Horizon 2020 program.

2 See J. Weilepp, especially section 2.4.4.

			Characterics	Financial Source
	TRL 9	System Test, Launch & Operation	Qualified System with proven technology and successful implementation (operational Use)	Equity, Project Finance
System / Subsytem Development	TRL 8		Qualified System with proven technology	Equity, Public Funds, Venture Capital
	TRL 7	Technology Demonstration	Prototype in use (1-5 years)	Equity, Public Funds, Venture Capital
	TRL 6		Prototype in use (start)	Equity, Public Funds
Technology Development	TRL 5		Experimental Set-Up (in real life environment)	Equity, Public Funds
	TRL 4	Research to Prove Feasibility	Experimental Set-Up (laboratory)	Equity, Public Funds
	TRL 3		Concepts demonstrated nalytically	Equity, Public Funds
Basic Technology Research	TRL 2		Technology Concept formulated	Equity, Public Funds
	TRL 1		Basic Principle Observed an reported	Equity, Public Funds

Figure 16.1: Technology Readiness Level and Financial Sources; figure created by author.

3. **New players**: Risk capital is typically provided by venture capital companies, and project financing is usually provided by commercial banks. This means we have new players with new rules of the game to which the operator must adapt.
4. **New rules of the game**: Project financing requires compliance with a set of rules that can be described as a future-oriented credit decision with a few strict rules. We will go into this in more detail below, but here (Table 16.1) is a brief comparison with the rules of the VC market.

Table 16.1: Comparison Venture Capital and Project Financing.

	Venture Capital	Project Finance
Focus of Success	Project can successfully be implemented in the Market, early sale	Debt Service has to be covered over the tenor of the loan (up to 20 years)
Chance of Capital Provider	1. Sale of the Project to third party with a gain 2. Comparatively high interest rate (or any other position which is closer to the requirement of an equity provider)	Timely repayment of the project loans (risk margin is the net yield component) Compared to Venture Capital, Project Finance Funds are considerably cheaper
Securities	A failure of the Project will almost always end in a full write-off. Securities will exist only as immateriell rights	All assets of the Project have to be secured in full. All of these assets have to be transferred to the financing as securities. This requires that a project company already exists and all the rights are located within the proejct company.

Project financing is only possible if – in addition to proven technology – other requirements are met. These include a functioning market and a suitable legal and regulatory environment.

The first decisive step is the technical maturity of the product, measured here by TRL as an example. Project financing is conceivable from TRL stage 9, but this requires preparatory work. First of all, a suitable regulatory environment is required: here it must be tested whether there are any **market failure circumstances** that justify state intervention.[3] With suitable technology and a suitable regulatory environment, a market should exist. The market then allows a viable business model to be created. If demand proves to be sufficiently stable, the prerequisite for the use of project financing is then also met (see Figure 16.2).

For a market to function, there must first be a need for the product, which must be flanked by a suitable legal environment. This means that the regulator must also

3 See for a discussion of different market failures section 4.1.

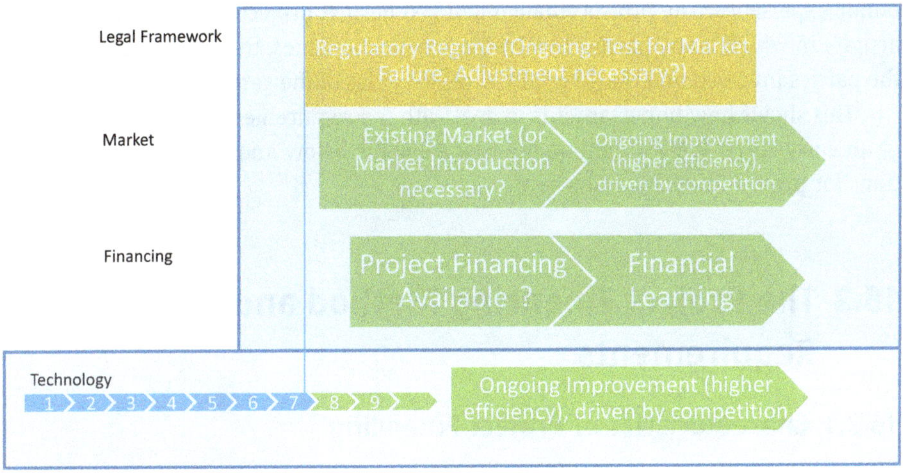

Figure 16.2: Side Conditions of Project Financing; figure created by author.

be aware of the objectives and ancillary conditions of the establishment and operation. This means that the operator must inform and lobby the regulator.

Further requirements arise from this constellation:

- A **new mindset** on the part of the operators: while the focus of action has so far been on achieving technical market maturity, completely different topics from the areas of law, economics and information provision are now coming to the fore. In addition, it is now a matter of successfully managing a production operation on a daily basis.
- **Integration of additional expertise**: due to the diverse new requirements, suitable personnel with complementary knowledge must be integrated into the project in good time.

The requirements for the transition from a research project to an operating project are diverse and require knowledge of the requirements of project financing and early consideration during the development phase.

The application of project financing depends very much on the aspect of experience: PV or wind energy projects are tried and tested at technical, legal and economic levels and are in a steady state. The steps just described have been taken here for years, if not decades. However, new fields of application still have to go through this learning process.

Geothermal energy is a subject area that is dealt with in very different degrees of detail in the various sub-areas: While the technical aspects have already been examined in great detail, there is still only a fairly small circle of specialized lawyers who have taken on the topic from a legal perspective. And since only a few geothermal projects have been realized in Germany, there is also a lack of a broad basis of economic experience. The project financing of geothermal projects is therefore a field of activity in which many parties involved are new: the issues, the cooperation between the parties involved and the generally accepted rules of the game.

This shows how important it is to deal with the requirements of project financing at an early stage. We will outline the requirements below and discuss market conditions for project financing elsewhere.

16.3 The Project Financing Method and its Requirements

16.3.1 Characteristics of Project Financing

There are different ways to finance projects. When a company raises funds via the financial markets to realize a project, this is not yet project financing, but classic corporate financing. The creditworthiness of the company and its ability to service the debt and provide adequate loan collateral are decisive for the approval of the financing. The loan provided and the fixed assets created in the course of realizing the project must be shown in the company's balance sheet and, in the case of major projects, have a negative impact on the company's key financial figures. The company's ability to take out further loans in the future is therefore limited in terms of corporate financing. Projects with a larger financing volume can therefore often no longer be realized via traditional corporate financing, especially as the risks associated with the realization of the project can no longer be borne by the financing company alone.

In contrast, project financing offers the possibility of realizing projects with a high financing volume that would not be possible with traditional corporate financing due to the restrictions described above. In the case of project financing, it is the project and its cash flow, but not a specific company that is responsible for the financing. The project must therefore be a closed, legally, technically and economically viable circle that offers investors a credible prospect of an appropriate return on equity and lenders sufficient security that the capital invested will be repaid: The project must be self-sustaining, self-financing. This is the basic idea behind project financing.

Various definitions of project financing can be found in the literature, with NEVITT/FABOZZI's definition being the most widely accepted:

> Project finance is the financing of a project where a lender initially focuses its credit assessment on the project's cash flows as the sole source of funds that will service the loans[4].

4 P. K. Nevitt; F.J. Fabozzi 2000, p. 1. Even though the definition places a clear emphasis on the role of the lenders, the method of project financing will be approached below from the perspective of the various project participants, as their efficient interaction is crucial to the success of project financing.

This definition incorporates the central principle of cash flow orientation (**cash flow related lending**): The project loans are made available in the confidence that the cash flows of the project are so stable that, in addition to the operating costs, the debt service can also be serviced safely. As the project forms a separate economic unit, the financing is structured in such a way that the projects are self-supporting and can pay the debt service for the loans provided from the cash flow generated (cash flow orientation). These are forecast in advance with the help of cash flow models.

As the stability of the cash flows requires that the project participants act in the interests of the project, the lenders closely examine the contractual and legal basis under which a project operates as part of their credit assessment. This is what is meant by "initially": in addition to the cash flow orientation, the assessment and structuring of the risk positions of the various project participants (**risk sharing** or **risk allocation**) takes place in the further course of the project appraisal. In addition to equity and debt capital providers, numerous project participants such as plant manufacturers, operating companies, suppliers and customers, consultants and government institutions are involved in the realization of the projects. The project risks are distributed among the individual project participants in the course of contract negotiations (risk distribution). Risk allocation is intended to counteract opportunistic behavior on the part of individual stakeholders and promote a common interest in making the project a success. Successful project financing requires appropriate contractual involvement of the project participants. The basic principle of risk sharing in project financing based on the incentives to act is to allocate the risk to the party that can best influence the occurrence of the risk. In the case of risk-averse project participants, however, the trade-off with the risk premium demanded by the respective contractual partner must be taken into account in this risk transfer: There are cases in which it is not worth incentivizing action because the premium for doing so would be too high. Ultimately, it is not a question of maximum risk transfer, but of optimal risk transfer that is just sufficient to create the desired efficient incentives to act.[5]

It is essential to provide the commissioned party with an incentive scheme that encourages it to act in the desired way in its own interest. As a rule, it must therefore share in the success and therefore also the risk of the respective project, regardless of its risk-bearing capacity. The agreements on risk allocation form a complex incentive scheme that is intended to harmonize the interests of the project participants and align them with the success of the project. Any remaining risks can then be distributed according to the criterion of risk-bearing capacity, e.g. outsourced to insurance companies or retained by the financiers. First of all, however, it is important to find a contract structure in which all parties are committed to the project. Which contracts

---

The clear emphasis on the role of the lenders is nevertheless sensible, as they should assume by far the largest share of the overall financing and their acceptance is therefore decisive in determining whether project financing comes about or not.

5 We have analysed the incentive structure in more detail in section 4.1.

are suitable for this depends on what can be established in court about the behavior of the individual parties.[6]

To ensure that the project can be largely shielded from influences outside the explicitly agreed contracts, an independent project company must be founded, which is the holder of all rights to the project and which books the project loans (**off-balance financing**).[7] Project financing takes place off the balance sheet of the project initiators (off-balance sheet financing). To this end, they establish a project company that acts as a borrower vis-à-vis the banks and is provided with equity by the project sponsors. The fixed assets created in the course of project realization and the borrowed funds raised for financing are then reported in the balance sheet of the project company. In the case of limited recourse financing, the project sponsors are liable not only for their equity contribution but also for other assurances, such as guarantees and additional funding obligations. If there is no further recourse to the sponsors beyond the equity contribution once the project has been commissioned, this is non-recourse financing. The collateralization of project financing is generally limited to the project assets and then no longer extends to the sponsors' assets.

As cash flows are the only source of loan servicing and return on equity, there are special requirements for their stability and reliability. In addition to risk identification, the aim is to allocate risks to individual project participants according to economic criteria. This is followed by a risk quantification in the form of a cash flow model and a rating procedure, which provides information on how much external funding can be made available to a project, what the repayment structure should look like and what other design elements should be included in the structure. It is important to be aware that the respective aspects of the risk management process – identification, allocation and quantification of risks – are not subject to a specific chronological sequence, but are interrelated. In order to adequately assess the statements on risk quantification, it is therefore necessary to consider the various aspects of risk management in equal measure. We will do this in this chapter where necessary and otherwise refer to the specific chapters.

To understand the methodological approach of a project financing, it is helpful to briefly outline the differences between **corporate financing** and **project financing** (**see** Figure 16.3**)**: If corporate financing is used, an investment project is viewed as part of the company. The assessment of the investment project from the perspective of the financing bank is based on the creditworthiness of the company as a whole and not on the expected cash flow of the project itself. If, on the other hand, project financing is implemented, the assessment of the lenders is essentially linked to the ability of the project to generate its own cash flow.

6 J. Böttcher 2012, pp. 77–92.

7 W. Schmitt 1989, p. 24.

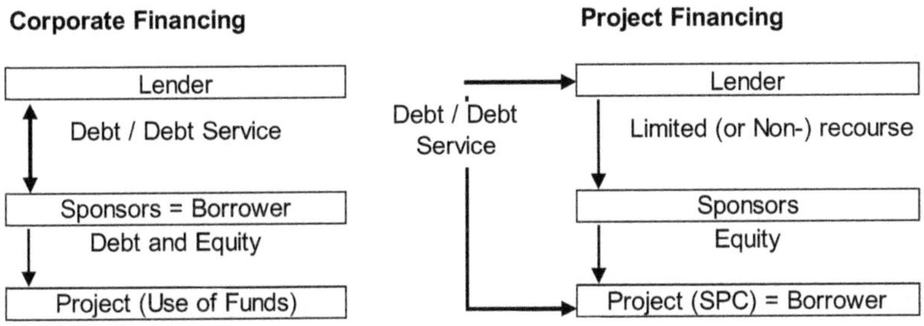

Figure 16.3: Comparison of Corporate Financing and Project Financing.[8]

The key point is that in the case of corporate financing, the borrower is also the equity provider and is responsible for servicing the loan for the entire term of the loan. This is in marked contrast to project financing: here, the equity providers are responsible for servicing the loan – usually for a limited amount – and only until the project has been successfully completed. This limitation of the sponsors' liability is – from their point of view – the main economic reason for the attractiveness of project financing.

As the sponsors do not assume unlimited liability for the borrowed capital in the case of project financing, the establishment of an independent project company by the sponsors as shareholders is regularly necessary for the realization of the projects. The sole business purpose of this project company is the realization, i.e. the construction and operation of the project. As a single-purpose company, it raises the external funds and has unlimited liability with its assets, so that from a formal point of view it is a corporate loan. In material terms, however, it is a loan for the specific project. The lenders expect the debt service to be serviced solely from the cash flow generated by the project. The assets and cash flow of the project are available to the creditors as collateral. However, this liability is typically difficult to realize, which does not need to be explained in detail in view of the high investment specifics – power plants, cell phone networks, transport systems, etc. – of the project. Therefore, in the event of a crisis in which the cash flow is not sufficient to service the debt, the focus will not be on realizing the collateral, but on continuing the project, if necessary at the financial sacrifice of all parties involved.[9]

An essential feature of any project financing is the focus on future cash flows and the involvement of the project participants, from which the following consequences can be derived:

1. **Factors influencing the cash flow**
   When assessing a project, particular attention must first be paid to the factors that influence cash flow. The key cash flow determinants for a project are the

8 Based on W. Schmitt 1989, p. 22.
9 J. Böttcher 2006, pp. 130–133.

procurement side, the sales markets, the operating costs, the financing conditions and, finally, influencing factors from the public sector.

2. **Principle of controllability**
   The allocation of risks to the project participants is normally based on the principle that the contracting party should assume the project risk that it is best able to assess and thus control on the basis of its business activities.
3. **Principle of risk-bearing capacity**
   The principle of controllability is supplemented by the principle of risk-bearing capacity: this involves the question of whether the contractually obligated project participants are also in a position to fulfill their obligations to the project due to their creditworthiness and economic capacity. In this respect, all project financing also includes elements of corporate financing, as the at least partial assumption of risk by the project participants is essential for project financing and in any case also requires a credit assessment of these risk carriers, as is typical for corporate financing. The more far-reaching the contractual commitment to the project, the more intensively the creditworthiness of the risk carrier must be examined. In this regard, reference is made to the relevant literature on borrower assessment.
4. **Incentive effects of the contract design**
   Finally, the incentive effects of the respective contract design must also be taken into account. From an ex-post perspective, the client may not care how a good project result was achieved. Ex ante, however, he would like to increase the probability of a good result, and he can only do this by influencing the behavior of the contracted party. If he could observe the contractor, he would force him to behave as desired by issuing appropriate instructions. As a rule, however, the client cannot check free of charge whether his instructions have been followed. It is therefore essential to provide the contractor with an incentive scheme that encourages him to behave as desired in his own interest. As a rule, the contractor must therefore share in the success and risk of the respective project, regardless of its risk-bearing capacity.

The special methodological features of project financing – focusing on the cash flow of the project, the early release of the sponsors from liability and the explicit contractual involvement of the various project participants – mean that the risk management of any renewable project is also of particular importance from a financing perspective. We outline the associated aspects in the following section 16.4.

At least two conditions must be met in order to realize project financing in an industry: The technology must deliver a stable and predictable energy yield in the long term and the state must provide a clear, predictable and reliable legal and regulatory environment that provides investors and lenders with sufficient planning security for economic operation. If these two fundamental requirements are met, the possibility for the commercial operation of a project opens up, typically in the form of project financing.

In order for project financing in the field of renewable energy to be realized, experts from the fields of technology, law and economics must come together and develop a tailor-made solution for a project.

16.4 Risk Management for Renewable Energy Projects

There are many different interpretations of the term risk in the business literature.[10] Here, risk should be understood as a negative deviation from the planned value of a target value, as this means a risk of loss for all parties involved.[11]

The aim of risk management is to achieve a systematic and success-oriented approach to dealing with risks. This applies in particular to project financing, as the novelty and uniqueness of each project are subject to unknown influencing factors that lead to risk positions.[12] Furthermore, the forward-looking cash flow orientation and the associated limitation of recourse to the sponsors result in special requirements for risk management, as this also regularly transfers entrepreneurial risks to the lenders.[13]

The importance of dealing with risks in connection with project financing arises directly from its nature: as it is the project alone that serves as the economic basis for the appropriate return on equity and servicing of the debt service, the value and robustness of the project is of crucial importance. However, as the project is only gradually being developed, the economic viability can only be determined by means of a forecast. As the outlook for the future is increasingly uncertain, the forecast must deal with the occurrence of all types of influences, assess their effect on the project and look for ways to determine whether and to what extent individual project participants are prepared to keep the project free of risks.

The success factors of a renewable energy project can be described as follows:

The first three aspects mentioned – stability of the legal and regulatory environment, use of proven technology and appropriate risk allocation – must be fully met for every project financing. The next step then involves a financial optimization task that must be solved depending on the volatility of the various influencing factors. The

10 For more details, see M. Hupe 1995, p. 43 ff.; D. Tytko 1999, p. 142 f.; H. Uekermann 1993, p. 23. On the concept of risk from a technical perspective, see P. Frohböse 2010, pp. 13–16.

11 Based on M. Hupe 1995, p. 46. In a broader understanding of the term, risk is understood as the danger that an actually realized resultdeviate positively ornegatively from the expected result Positive deviations are then referred to as "opportunity", negative deviations as "risk in the narrower sense". We will follow this latter interpretation of the concept of risk here.

12 M. Hupe 1995, p. 43 ff.

13 K.-U. Höpfner 1995, p. 166 ff.

Table 16.2: Success Factors for Project Financing (compiled by author).

1.		Reliability and Predictability of the Regulatory Regime / Enforcebility of Contracts	Project Assessment
2.		Adequate risk allocation among the project participants	
3.		Use of proven technology only	
4.		Profitability of the Project	
	4.1.	Volatiliy of the main risk drivers	
		4.1.1. Cash-Inflow and Cash-Outflow	CF-Model / Rating-Tool
		4.1.2. Volatility of the prices and the energy output	Rating-Tool
		4.1.3. Macroeconomic Factors (interest rates)	Rating-Tool
	4.2.	Unsecurity about the correct p(50)-level (Banking Case Uncertainty [BCU])	Rating-Tool
	4.3.	Correlation between the main risk drivers, especially between costs and income	CF-Model / Rating-Tool

first part of the project appraisal is therefore of a more fundamental nature, while the second part deals with risk quantification (see Table 16.2).

The use of project financing begins with the question of the fundamental suitability of the technology to be used, which must guarantee clear and stable long-term energy production. Individual geothermal technology systems have been in industrial use repeatedly and for many years, while the application of hot dry rock technology, for example, is still at an early stage.

The risks associated with project financing can vary greatly from project to project in terms of their content, cause, extent and probability of occurrence. Nevertheless, there are groups of risks that can lead to a threat to cash flow in the same or a similar way in most project financing and must therefore be the subject of risk management. It is often helpful to visualize the factors influencing the profitability of a project (as shown in Figure 16.4).

Risks can be appropriately subdivided in such a way that they largely do not overlap in terms of their content and causes and are based on the ways in which they can be influenced by the various project participants. Such a classification appears to make sense, as different measures have emerged in practice that usually handle the risks with as close a relationship as possible to their causes.[14] A distinction is there-

14 An economic analysis of the contractual relationships also suggests such a link between risk and risk-bearing capacity. From an efficiency point of view, it is better if the risk allocation is conditional on the occurrence of risk. See J. Böttcher 2009, pp. 67–69.

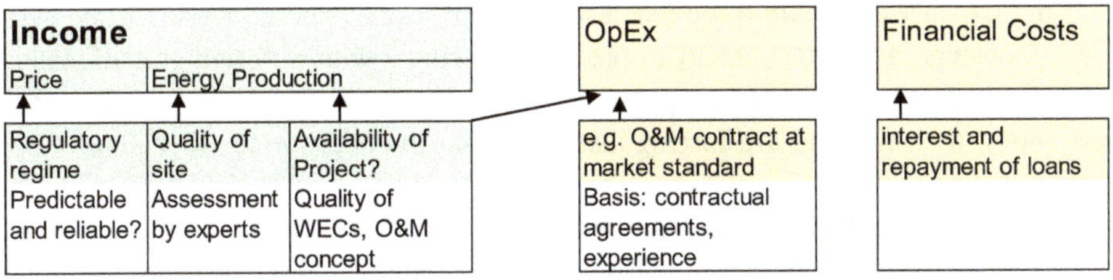

Figure 16.4: Factors influencing Economic Efficiency; figure created by author.

fore made below between risks that can be controlled by the project company or other project participants – project endogenous risks – and risks that affect the project outside of the project participants – project exogenous risks. A special feature of project-exogenous risks are risks that cannot be controlled by any of the parties involved in the project, so-called force majeure risks (see Table 16.3).

This subdivision is economically expedient, as the methodology of project financing essentially consists of structuring robust contracts between the project company and central project participants, which thus keep risks away from the project. This requires the contractual involvement of project participants in the project, or to put it another way: Endogenous risks are more manageable than exogenous risks from the project company's perspective.

Table 16.3: Overview of exogenous and endogenous risks (compiled by author).

Endogenous risks	Exogenous risks
Completion risk	Technical risk in the broad sense
Technical risk in the narrower sense	Resource risk
Management risk	Supply risk
Sales risk	Market risk
Operating risk	Contractual risk
Abandonment risk	Exchange rate risk
	Legal and regulatory environment
	Inflation risk
	Interest rate risk
Force majeure risk	

It is important to note that it is the contract structure that determines whether individual risk types are endogenous or exogenous risks: For example, it is only the contractual obligation of the buyer to purchase products from the project company at a certain price, quantity and quality that transforms an exogenous market risk into an endogenous sales risk. We have summarized the main project risks in Table 16.3 where we have also indicated in which section of this book these topics are dealt with.

In many areas, certain basic risk distribution rules have become established over time. However, as the technology of project financing – with different gradations for certain areas, e.g. geothermal projects – is relatively new, certain basic rules have not yet emerged clearly and are forcing discussions on the appropriate allocation of opportunities and risks.

The various individual risks can be addressed and their impact on the project at least mitigated by involving the various project participants. Nevertheless, residual risks remain that must be absorbed by higher-level security systems. These systems include the establishment of an efficient information structure and, above all, the development of a stable project and financing structure. The following Figure 16.5 illustrates the interrelationships:

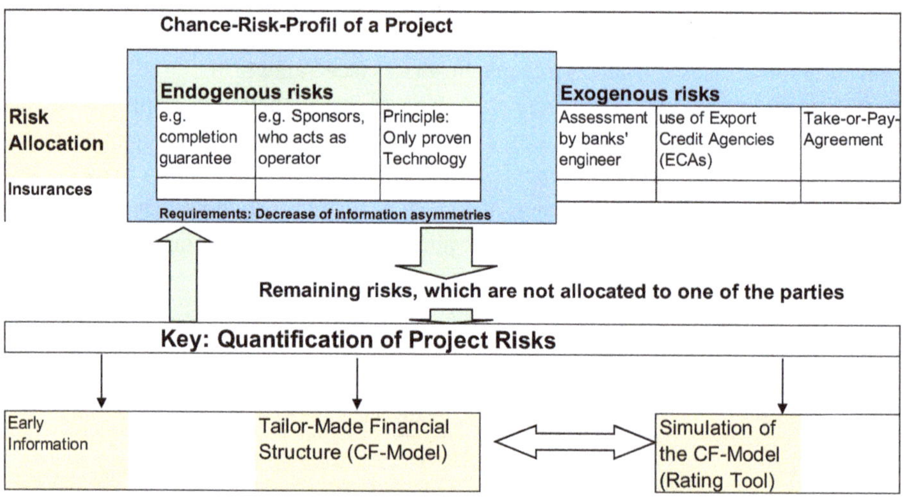

Figure 16.5: Risk Management Process for Project Financing; figure created by author.

For successful risk management, it is important to record the effects of identified project risks on the economic performance and resilience of the project. This provides insights into the selection and scope of risk policy measures and suitable contractual partners. Risk allocation is followed by risk quantification, which shows the impact of the individual project risks on the project's cash flow. On this basis, a financing structure is developed that gives investors the prospect of an appropriate interest rate and lenders the confidence that the debt service will be provided even under a stress scenario.

However, it is also clear that the topic of risk management requires a joint approach from legal, technical and economic specialists. The parties involved in a project are shown the partial aspects of their involvement in the above Figure 16.5 above, but it is only through their coordinated interaction that a viable project can be developed and realized.

Following this general description of the risk management process, we will discuss the risk management process in section 12.5 we will outline the various individual risks that are of particular importance in geothermal projects.

16.4.1 Risk Management for Project Financing

There are various definitions of the term risk in the literature, which will not be discussed in detail here. In the following, risk is understood as the danger that the result deviates from the expected target value due to incorrectly assessed or unconsidered factors. These deviations can be both positive and negative, whereby in project financing the negative deviations reflect the risk of loss and are therefore of real interest to the stakeholders of project financing. In order to control project risks, risk management must be implemented at the same time as project development. Risk management includes all activities, processes, structures and instruments that serve to manage risks. Figure 16.6 shows the risk management process for project financing.

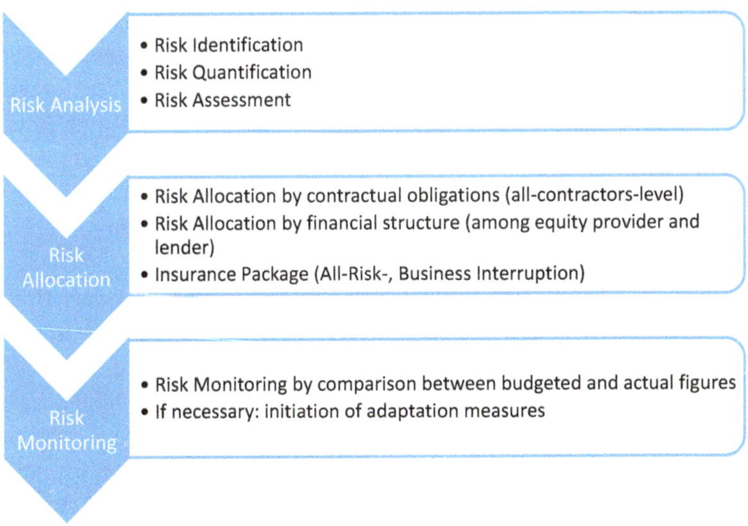

Figure 16.6: Risk Management Process for Project Financing; figure created by author.

The aim of risk management is to ensure that the forecast cash flows on the basis of which the project financing has been structured can also be reliably generated, thus guaranteeing the profitability of the project.

The individual process steps of risk management are dealt with in the following chapters.

16.4.2 Risk Analysis

The aim of the risk analysis is to first identify all factors that could have a negative impact on the generation of cash flow. The risk analysis is not only carried out by the sponsors, but in particular also by the creditors of a project financing. Together with the advisors, the banks must gain an overview of the risks that could negatively influence the success of the project in the individual phases of project implementation. This is particularly important in the project development phase, as only the preliminary project costs, which are usually borne by the sponsors, are incurred here, but no major loan-financed investment payments are made. It is therefore imperative for the investors to identify and assess all risks before financial close and to distribute the risks accordingly by structuring the contracts with the project's other stakeholders. During the construction phase, on the other hand, the financial resources for the project must be raised in line with the progress of construction, without cash flows already being generated in return. This phase is therefore associated with greater risks, particularly for lenders. It is only in the operating phase that the project generates cash surpluses that can be used to repay and pay interest on the debt capital. However, even in this phase, risks can mean that the payments generated from the sale of products are not high enough to cover the payments for ongoing project operations and debt servicing to creditors and to ensure an adequate payout to sponsors.

The following is a brief overview of the main sources of risk that need to be taken into account in project financing. It should be borne in mind that the risks vary from project to project and that a risk analysis must therefore always be carried out on a project-specific basis.

Technical risks can occur in all project phases and can lead to the planned cash flows being delayed, not being generated at all or only being generated to a reduced extent. In order to identify and assess risks in good time, technical consultants and experts are consulted as early as the project development phase. During the construction phase, there is a risk that the construction of the plant will be delayed or that the planned costs will be exceeded during the construction of the project and that completion will not be possible within the planned timeframe, taking into account the budgetary requirements. After commissioning, the technical equipment of the plant may fall short of expectations, meaning that the desired performance parameters, on which the calculation of future cash flows is also based, cannot be achieved. In order to reduce this risk, lenders generally prefer to finance technologies that are already established and for which empirical values are available. In the operating phase of the project, various factors can cause disruptions in operations, which can lead to increased cash outflows due to rising operating costs and falling cash inflows due to reduced output as a result of the downtime, thus significantly reducing cash flow. Management errors, such as inadequate maintenance or servicing, errors in operations management, the use of inferior or incompatible raw materials, but also technical problems with the plant itself can lead to business interruptions or defective products. In order to prevent business interruptions due to management errors, investors are more likely to select operating companies with a high level of expertise as contractual partners. For projects that rely on the use of natural resources, the availability of resources at the site must be ensured as early as the project development phase by means of appropriate expert opinions. This is important for projects involving the extraction of raw materials such as coal, natural gas or crude oil in order to be able to assess the extent of the raw material deposits and the geology of the site in advance. The successful operation of projects in the renewable energy sector, such as wind or solar parks, is highly dependent on the wind conditions and solar radiation at the selected locations, so that experts must also be commissioned to assess them. In the examples mentioned, low resource availability means that the expected output quantity cannot be extracted or produced, resulting in lower cash flows than planned.

In the operating phase of the project, the counterparty risks on the supplier and customer side must be examined.[15] For projects that depend on a continuous supply of raw materials, the supply risk on the input side must be taken into account. It must be ensured that the raw materials can be procured in the planned quantity, quality and at the planned price. Preliminary studies of the procurement market must therefore analyze which suppliers on the market can be considered as potential contractual partners and whether it may be possible to switch to other suppliers in the event of supply bottlenecks. The acceptance risk, on the other hand, is that the products manufactured with the system cannot be sold on the market to the planned extent and at the expected prices. In order to reduce the acceptance risk, the future sales market must be estimated on the output side so that the expected cash inflows from sales revenues can be better calculated. To this end, market studies must be carried out in order to forecast supply and demand in the relevant market. Polypolistic markets with many suppliers and buyers have the disadvantage that the opportunities to enforce one's own price expectations are limited and it is more difficult to agree long-term purchase agreements than in markets with less competition.

Financial risks are complex and must be taken into account in all project phases. Financial risks include interest rate and currency risks, inflation and default risks as well as liquidity risks. Interest rate risks arise if interest conditions deteriorate during the project term. Project financing is generally based on variable interest rates, i.e.

15 One example is described in section 2.2.5 and 2.2.6.

the loan interest rate is linked to a standard market interest rate. The project companies then receive loans from the banks at this reference interest rate plus a contractually negotiated margin. If the reference interest rate rises due to changes in market conditions, the loan interest rate for project financing also rises and with it the project company's financing costs. Exchange rate risks arise in the event of unfavorable changes in exchange rates between different currencies. The project is exposed to currency risks if project income or expenses have to be received or paid in other currencies. In addition, competitiveness can suffer if, for example, the domestic currency for an export-oriented company appreciates in value compared to that of the importing country, making the goods in the importing country more expensive. Inflation risks lead to an increase in the price level, which can have a particularly negative impact on cash flow if the prices for inputs are more strongly affected by inflation trends than the prices for the products manufactured with the plant. Default risks arise when the creditworthiness of stakeholders relevant to the success of a project deteriorates to such an extent that they are no longer involved in the further course of the project. If plant constructors or plant operators are at risk of default, there is a risk of delayed completion and commissioning of the plant or an interruption to operations. This would again result in a delay or reduction in the cash flows required to repay the capital service. The same consequences are to be expected if customers or suppliers default, meaning that the products cannot be purchased in accordance with the contract or the advance payments cannot be delivered. For its part, the project company must ensure its own solvency in order to prevent liquidity risks accordingly.

The country risks include government measures as well as the regulatory and legal environment, which can have a negative impact on the project. Government measures may, for example, result in the project company losing significant ownership rights to the project assets through expropriation, taxation may develop to the disadvantage of the project company, price controls may prevent the economic operation of the plant or a trade embargo may mean that necessary raw materials can no longer be imported and finished products cannot be exported. In addition, the extent to which the institutional and regulatory environment in the country is developed must be examined. If institutions are not adequately implemented or if existing institutions are not implemented, this can severely impair the success of the project. Examples of this include corruption or a delay in approvals that are urgently needed for the timely completion and operation of the project. An unreliable or complex legal environment would also impair the success of the project if, for example, the enforcement of claims in court can take years.

All force majeure risks are subsumed under force majeure risks. For example, fire, flooding, storms or other natural disasters such as volcanic eruptions or earthquakes, but also political unrest and (civil) war as well as terrorist attacks, theft or sabotage can destroy the project or at least severely impair the generation of the planned cash flow.

Major projects can have a significant impact on the environment. The construction of an airport, for example, is associated with considerable noise pollution for the surrounding population, while the construction of an offshore wind farm can have a negative impact on bird migration or fish stocks, to name just a few examples. For these reasons, protests and complaints from citizens and environmentalists are always to be expected with projects, which can lead to the project only being realized under stricter conditions, delayed or not at all.

16.4.3 Risk Quantification and Assessment

Once the individual sources of risk for a project have been identified, the risks are quantified and assessed. By deriving various scenarios in which individual risk parameters are varied, the lenders check whether the project cash flows are high enough, even under unfavorable conditions, to guarantee the debt service to the lenders and an appropriate return to the investors. The banks providing debt capital are generally more risk-averse than the sponsors of the project. The other project participants (such as plant constructors, operators, suppliers and customers) will also have to provide contractual assurances as the project progresses and must therefore also assess the opportunities and risks associated with their involvement during the project development phase.

The aim of risk quantification for all project participants is therefore to determine the economic viability and resilience of the project and to develop a project structure that ensures the long-term success of the project by properly allocating opportunities and risks. As the project progresses, the impact of risks on the cash flow generated is also monitored using key figures in order to be able to initiate adjustment measures in good time if certain predefined threshold values are exceeded or not reached.

16.4.4 Risk Allocation

Once the stakeholders, and in particular the project's investors, have assessed the risks, a decision is made during the contract negotiations on the appropriate distribution of opportunities and risks among the individual project participants. The risks of the project are distributed among the individual stakeholders to ensure that all participants have an interest in the successful creation and operation of the project. Risk allocation thus pursues the goal of ensuring that the project participants form a community of interests and do not strive to achieve their own advantage at the expense of the project's success by exploiting information asymmetries vis-à-vis other stakeholders or through opportunistic behavior.

As part of risk allocation, it should be ensured that a contractual party is not unilaterally burdened, but that all project participants are in a position to bear the risks assumed (principle of risk-bearing capacity) and have the ability to influence and manage the risks (principle of risk controllability). If the risks can be influenced by the stakeholders of a project, these are project endogenous risks; project exogenous risks exist if none of the stakeholders can actively control the project risks as they cannot be influenced by them.

As endogenous risks can be influenced much better than exogenous risks from the perspective of the project initiators, an attempt is made during the contract negotiations to distribute the individual risks among the individual project participants according to their risk-bearing capacity. By structuring the contract accordingly, originally exogenous risks can also be transformed into endogenous risks. A good balance must be struck between the assumption of risks by the project company and the other project participants. If too many risks remain with the project company, this makes it more difficult for it to take out loans to finance the project. If, on the other hand, almost all risks are transferred to the other stakeholders so that hardly any risks remain with the project company, this one-sided distribution of risk can also jeopardize the success of the project.

In the following section, the stakeholders of project financing are first introduced before the possibilities of risk allocation to the project participants through appropriate contract design are discussed.

16.4.5 Role of the Stakeholders

Project financing involves various stakeholders with whom relationships exist during the various phases of a project that must be actively maintained by the project team.

The initiators are responsible for the planning and implementation of the project, establish the project company and act as equity-providing sponsors in the further course of the project. The commitment and experience of the sponsors is a decisive success factor, as the optimal risk allocation among the other stakeholders must be driven by this group, especially during the negotiation phase. The partnership agreement between the sponsors of the project company defines not only the equity capitalization but also, in particular, the management of the project company.

In addition to financing projects, banks also play an advisory role in project financing. In the case of large-volume project financing, the arrangement is carried out by a consortium of banks. If a bank has been mandated as lead arranger, it is responsible for structuring the financing and administering the loan. Banks in the role of co-arranger are responsible for ensuring that the project company is provided with the required debt capital in line with the progress of construction. In addition, other banks in the role of partners can take on partial volumes of the syndicated loan. This assumption contributes to the further distribution of risk among the banks and ensures that one bank does not have to bear the entire financing risk alone. The loan agreement concluded between the project company and the banking syndicate sets out the terms and conditions of the loan (e.g. loan volume, tranches and number of drawings, loan costs, intended use, disbursement requirements, collateral and covenants).

In the plant installation contract, the plant constructor is obliged to meet technical, monetary and time-related framework conditions for the construction of the plant. It is essential that the technical equipment and performance parameters of the plant are specified and that the service package to be provided is set out precisely. In addition, the test runs to be carried out before final acceptance of the system must also be contractually stipulated. Contractual penalties and guarantees as well as the contribution of equity capital to the project company should create the necessary incentives for the plant manufacturer to successfully complete the project.

The operating and maintenance contract between the project company and the operating company stipulates which specific services are to be provided by the operator and which warranty guarantees and contractual penalties are provided in order to minimize the risk of interruption to operations. In addition to proven management qualifications, the operator is sometimes required to hold shares in the equity of the project company in order to be awarded the contract.

The quality, quantity and price of inputs and raw materials must be specified in supply contracts with suppliers. Banks make sure that the supply contracts are binding for as long as possible and thus provide a reliable planning basis for at least the term of the loan agreement. There is also the option of obliging suppliers to make equity investments in the project company.

The quality, quantity and price of the service to be provided by the buyer must also be determined as far as possible on a long-term basis within the framework of purchase agreements. Here too, the banks endeavor to align the term of the purchase agreements at least with the term of the loan agreement.

Government institutions are also among the stakeholders that can have a significant influence on the success of a project, as they influence key framework conditions for projects. They can help shape the financial framework conditions of the project by imposing conditions or providing corresponding investment incentives (such as guarantees or the prospect of tax relief or subsidies). For these reasons, letters of support are obtained from the authorities or institutions of a country during the project development phase. These are expressions of interest from the respective institutions in a country for the project to be created.

In order to be able to carry out as comprehensive a risk analysis as possible, experts are often called in to support the project company by preparing corresponding expert reports. Consultants are essential, especially for less established technologies and innovative project plans, in order to be able to shed more light on legal, technical and financial aspects in advance. Technical consultants are called in to assess the technical framework conditions (for example, when evaluating technical innovations, but also when selecting a location), financial consultants are called in to structure the financing and legal consultants are called in to draft contracts or provide expert opinions on international legal issues.

16.4.6 Contract Design for the Formation of an Alignment of Interest

Once the main sources of risk have been identified and the stakeholders of a project financing are known, the following section looks at how the individual risks can be distributed among the project participants by structuring the contracts. As the risks of a major project are too high to be borne by one stakeholder alone, a key objective is to distribute the risks among all project participants by structuring the contract in such a way that the risk-bearing capacity of a single party is not exceeded. Another key aspect of risk allocation is to ensure that the project stakeholders are incentivized to fully support the project by assuming risks, as the party may otherwise suffer financial disadvantages. As unresolved risk distribution can subsequently jeopardize the success of the project, the distribution of risk within the framework of the contract design contributes significantly to the success of the project. The different risk sources are therefore examined here against the background of which stakeholders are responsible for these risks and which contractual measures can be used to adequately allocate the risks among those responsible (see Table 16.4).

Table 16.4: Risk allocation for completion risks (compiled by author).

Risk	Responsibility	Measures
Completion	Sponsors Plant manufacturer	Completion guarantees Obligation to make additional contributions Turnkey clauses Equity participation Penalties Construction risk insurance

Sponsors and plant constructors are responsible to the lending banks for the completion of the plant. Through completion guarantees, contractual penalties and additional funding obligations, the plant constructors can indemnify the sponsors if delays or budget overruns jeopardize the success of the project. With the help of turnkey clauses, risks can be transferred from the project company to the plant constructor during contract negotiations, thereby transferring the risk to the party that can best influence it. The contribution of equity by the construction company and the insurance of construction risks contribute to the further distribution of risk (see Table 16.5).

Table 16.5: Risk allocation in the event of problems with technical equipment (compiled by author).

Risk	Responsibility	Measures
Technical equipment	Plant manufacturer	Involvement of experts Agreement on the use of proven technologies Performance guarantees

In order to ensure that the plant also achieves the planned output parameters during the operating phase, performance guarantees are agreed in the plant installation contract and, as a rule, proven technology is predominantly used in order to be able to draw on experience from reference projects (see Table 16.6). In addition, technical experts can be consulted if there is a lack of experience with innovative projects.

Table 16.6: Allocation of risks during the operating phase (compiled by author).

Risk	Responsibility	Measures
Operation of the system	Operating company	Use of experienced management Involvement of sponsors in management Incentivizing the management Warranty guarantees Equity participation Possibility of changing operator Business interruption and property damage insurance

Business interruptions can have a significant negative impact on cash flows. For this reason, particular attention must be paid to the management's experience in operating comparable plants and warranty guarantees must ensure that the operators are responsible for incidents and interruptions to operations. The possibility of a change of operator should also be provided for in the contract. Performance-related operator fees help to incentivize operators to a greater extent than fixed remuneration agreements. In addition, an equity participation of the operating company in the project company could be contractually agreed. Business interruption risk insurance helps to compensate for cash flow shortfalls in the event of disruptions (see Table 16.7).

The forecast cash flow can only be realized if raw materials and other inputs are available in the planned volume, quality and price and the products manufactured are also purchased by customers in the corresponding volume, quality and price. As has already become clear from the previous explanations of the stakeholders, these aspects must be defined in long-term contracts and enforcement must be ensured through appropriate clauses (see Table 16.8).

Table 16.8 provides an overview of the main financial risks. Interest rate and exchange rate risks can be hedged with the help of derivatives, while inflation and li-

Table 16.7: Risk allocation on the input and output side (compiled by author).

Risk	Responsibility	Measures
Counterparty risk input side	Suppliers	Long-term supply contracts Contractual definition of quality, quantity and, if applicable, price Deliver-or-pay clauses Equity participation
Counterparty risk output side	Customers/ purchasers	Long-term purchase agreements Contractual definition of quality, quantity and price Take-or-pay clauses Equity participation

Table 16.8: Treatment of financial risks (compiled by author).

Risk	Responsibility	Measures
Interest rate risks	Project company	Fixed interest agreement Use of derivatives
Exchange rate risks		Forward exchange transactions Use of derivatives
Inflation risks		Consideration of inflation risks when drafting contracts
Default risks		Customer rating Credit insurance
Liquidity risks		Financial planning and scenario analyses Structuring the financing Risk allocation to the project participants

quidity risks can be countered through careful calculation and risk distribution within the framework of the contractual arrangements. Default risks can be limited by appropriate evaluation of the customer and, if necessary, credit insurance. Liquidity risks arise in project financing if the project-related payments can no longer be made on time or in full by the sponsors or the project company. There are many reasons for this. For example, cost overruns during the construction phase can lead to a financing gap, or loans that have already been agreed cannot be paid out on time because certain disbursement requirements have not yet been met. Non-contractual behavior on the part of other project participants can also result in the cash flow not being generated on time or in the planned amount. In order to minimize the liquidity risk, extensive financial planning calculations and scenario analyses are carried out during the project development phase. These preliminary studies are intended to ensure that the solvency of the project company is guaranteed even under less favorable conditions. The risk allocation among the project participants explained in this section also plays a key role in ensuring that the project company remains solvent. The design and structure of the financing also have a considerable influence on whether the project company is able to make the payments due on time.

Table 16.9: Measures to reduce country risks (compiled by author).

Risk	Responsibility	Measures
Government measures Regulatory environment Legal environment	Project company	Letters of Support Legal Opinions Permits and licenses (import, export, foreign currency transfer) Export credit insurance Country ratings Regulation of other framework conditions (taxation, guarantees, subsidies)

Country risks (see Table 16.9) include all state influences that could negatively impact the success of a project. Restrictions on the free movement of foreign currency, restrictions on foreign trade, an unstable legal and regulatory environment, delays in approval procedures, stricter project requirements or the withdrawal of concessions are just a few examples that could jeopardize the success of a project.

The project initiators will therefore apply for the necessary approvals from the authorities in a timely manner and attempt to make the state institutions members of the community of interest through corresponding declarations of intent. Legal opinions and country ratings also help to better assess the framework conditions in the countries. In addition, further regulations are agreed with state institutions, which also shape the framework conditions of the project with regard to potential investment incentives (taxation, state guarantees, subsidies). Transfer and conversion risks can be countered with the help of insurance policies from export promotion institutions.

Table 16.10: Measures to manage force majeure risks (compiled by author).

Risk	Responsibility	Measures
Fire Flood Natural disasters Political unrest and wars	Project company	Taking out force majeure insurance (if possible)

Table 16.10 provides an overview of the main force majeure risks. Such force majeure risks, such as natural disasters, damage caused by fire and flooding or political unrest

Table 16.11: Measures to manage Environmental Risks (compiled by author).

Risk	Responsibility	Measures
Environmental risks	Project company	Expert opinion Early communication Taking out insurance (if possible)

and war, are characterized by the fact that the probability of these events occurring is very low, but the extent of the damage can be extremely high if they do occur. For this reason, no insurance is offered for some force majeure risks or they are only insured against payment of extremely high insurance premiums.

Planned turbines often affect the immediate surroundings. In addition, the plants can cause hazards or nuisances that can lead to corresponding citizen protests and lawsuits. This environmental risk is often underestimated or only taken into account once the first protest movements have formed, although this can severely delay and impair the completion and operation of a plant. Early environmental assessments and the information and involvement of citizens at the beginning of the planning process can help to reduce environmental risks (see Table 16.11).

16.4.7 Risk Monitoring

After analyzing and distributing the project risks among the stakeholders involved, the risks must also be constantly monitored during the construction and operating phases of the project. Before the individual loan tranches are disbursed, the bank checks whether the disbursement requirements agreed in the loan agreement have been met in full. During the operating phase, the bank checks compliance with the contractually fixed covenants in order to monitor the economic situation during the term of the loan and to be able to implement adjustment measures in good time in the event of a deterioration. Legal covenants include the assurances that the borrower must comply with during the term of the loan. Common legal covenants in the lending business are
- Obligation not to reduce equity
- Waiver to grant loans to third parties or to take out loans from other lenders
- Compliance with the intended use
- Guarantee of insurance cover
- Management of interest rate and currency risks
- Commitment to transparency vis-à-vis investors

The loan agreement also defines financial covenants that must be kept within contractually agreed limits during the term of the loan. As a rule, these ratios compare the cash flows generated from project operations with the payments to be made to the banks. If these cover ratios fall below a contractually agreed limit, there is a need for action, as the repayment of the loan funds may be at risk. In addition, the lending banks will monitor whether the debt service has been paid on time and in full by the due dates. The key financial ratios that are relevant for the evaluation of project financing are explained in more detail in the following section.

16.4.8 Cash flow Modeling

In the case of project financing, the source of the debt service and the satisfaction of the investors' return requirements are the returns generated by the operation of the project. The amount and distribution of these returns over time are subject to uncertainty and therefore cannot be calculated precisely in advance. Investors must therefore identify all factors that can influence the project and estimate the effects of variations in individual factor characteristics and their combination on the amount and distribution of future returns using cash flow models. The following section explains how project cash flows can be modeled and analyzed.

Determination of Cash Flow

The starting point for calculating the cash flows from the project is the cash flow available for servicing the debt to the lenders. This cash flow available for debt service (CFADS) is calculated as the difference between the forecast operating income and the expected operating expenses required to generate the income, including taxes but excluding financing costs. In principle, there are a number of approaches to calculating the cash flow available for debt service. The contracting parties must therefore first agree on the method for calculating the cash flow and set this out in the contract.

Structure of a Cash Flow Model

The aim of developing a cash flow model is to simplify complex relationships and illustrate assumptions and interdependencies that influence project performance and thus the expected cash flow. By identifying and varying the risk parameters and calculating key financial figures, the cash flow model can be used to assess the stability of the expected cash flow even under unfavorable conditions. The results of cash flow modeling thus form the basis for investors to decide whether a project should be financed under the given framework conditions and how potential financing should be structured.

Cash flow models are usually created with the help of spreadsheet programs. To ensure the clarity and comprehensibility of a cash flow model, it is advisable to use separate worksheets for entering the project data and assumptions, carrying out the calculations (e.g. income and operating expenses, financing with interest and repayment schedules, etc.), the output (cash flow, profit and loss account, etc.) and the summary and graphical presentation of the results and key figures. Figure 16.7 shows the model structure of a cash flow model with separation of input, calculation and result sheets.

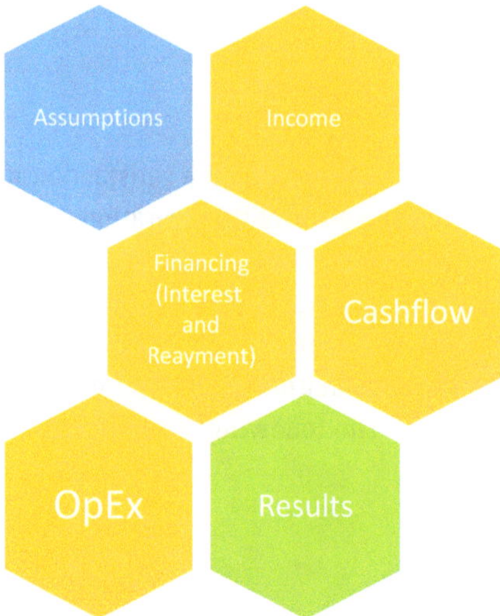

Figure 16.7: Structure of a Cash-Flow Model; figure created by author.

A financial modeler is always faced with the problem of depicting reality in the model in the best possible way, while at the same time ensuring that the model does not become too complex or too confusing for third parties.

The claims of lenders and sponsors are serviced from CFADS according to the waterfall principle (see Table 16.12). The debt service is first paid to the lenders from the CFADS. If it has been contractually agreed that a debt service reserve must be built up to cover possible future liquidity shortfalls, payments must be made to the debt service reserve account in accordance with the agreements. The residual amount would then represent the possible distribution potential to the equity providers.

In order to be able to finance a project, the cash flow generated from the project must be of a magnitude that guarantees the repayment of and interest on the loans granted by the lenders and ensures an appropriate dividend for the sponsors. To this end, the financial modeler determines key figures that can be used to estimate, on the basis of the cash flow model, whether the claims of all investors can be adequately satisfied.

Key figures relevant to cash flow

The key indicator for optimizing the financing structure is the Annual Debt Service Cover Ratio (ADSCR). From a banking perspective, this ratio can be used to determine

Table 16.12: Waterfall Principle for a Project Financing (compiled by author).

	1	2	3	4	5	6	7	8	9	10
Income	17.154.828	17.781.054	15.660.610	14.617.644	13.596.463	13.329.194	12.978.352	12.772.044	12.724.822	12.272.727
OpEx	2.353.000	2.422.990	2.495.080	2.569.332	2.645.812	2.724.586	2.805.724	2.889.296	2.975.375	3.064.036
Taxes	1.112.005	1.180.144	911.559	774.700	640.236	597.846	544.928	509.418	493.089	426.846
CFADS	13.689.822	14.177.919	12.253.971	11.273.612	10.310.415	10.006.762	9.627.700	9.373.330	9.256.359	8.781.845
Interest	1.900.000	2.590.000	2.302.222	2.014.444	1.726.667	1.438.889	1.151.111	863.333	575.556	287.778
Repayment	3.000.000	4.111.111	4.111.111	4.111.111	4.111.111	4.111.111	4.111.111	4.111.111	4.111.111	4.111.111
CF after Debt Service	8.789.822	7.476.808	5.840.638	5.148.057	4.472.637	4.456.762	4.365.478	4.398.886	4.569.692	4.382.957
Variation of DSRA	3.350.556	-143.889	-143.889	-143.889	-143.889	-143.889	-143.889	-143.889	-143.889	-2.199.444
Free Cash Flow	5.439.267	7.620.697	5.984.527	5.291.946	4.616.526	4.600.651	4.509.366	4.542.774	4.713.581	6.582.401

whether the project generates enough cash flow in each period to be able to service the debt to the banks. The ADSCR is also interesting from a sponsor's perspective, as it allows conclusions to be drawn as to whether there are sufficient funds left over from the expected cash flow to cover the investors' return requirements after the debt service has been paid. The ADSCR can also be translated as the debt service coverage ratio and shows whether and to what extent the cash flows generated by the project can cover the debt service to the lenders.

Equation 14: $$ADSCR_t = \frac{CFADS_t}{DS_t}$$

with

$ADSCR_t$ Annual Debt Service Cover Ratio in period t

CFADSt Cash flow available for debt service in period t

DS_t Debt service (interest and repayment) in period t

The ADSCR must always be greater than 1.00 so that at least the debt service can be paid from the current cash flow for a period. Banks require a certain excess cover as a risk buffer for this ratio, the size of which depends on the sector in which the project is to be realized. The ADSCR serves as an essential basis for structuring and optimizing project financing. The potential payout to the sponsors can also be derived from the cover ratio determined with it.

In addition to this short-term and period-based view of cash flows, key figures such as the loan life cover ratio (LLCR) or the project life cover ratio (PLCR) are used to assess whether the project will also pay off in the long term. The LLCR relates the present value of the cash flows generated over the remaining loan term before debt service to the loan amount still outstanding at the end of the loan term (Formula 2). The LLCR can be used to determine the extent to which the present value of the cash flows before debt service over the remaining term of the loan can cover the outstanding loan amount.

Equation 15: $$LLCR_t = \frac{\sum_t^T CFADS * (1+i)^{-t}}{L_t}$$

The PLCR shows a similar coverage ratio. With this ratio, the calculation period is not limited to the term of the loan, but is considered until the end of the project term (formula 3).

$$\text{Equation 16: } PLCR_t = \frac{\sum_t^N CFADS \ast (1+i)^{-t}}{L_t}$$

with

LLCRt	Loan Life Cover Ratio in the period t
PLCRt	Project Life Cover Ratio in the period t
CFADSt	Cash flow available for debt service in period t
i	Discount rate
Lt	Outstanding loan amount at the beginning of period t
T	Credit period
N	Project duration

The PLCR shows the extent to which repayment can be extended from the cash flows generated during the project term. The target values of the cover ratios, which are essential for structuring the financing, are industry-specific and range between 1.1 and 2.0 for the ADSCR and the LLCR. In addition to the cover ratios, the ratios between equity and debt are of interest to investors in order to ensure an adequate distribution of risk between banks and sponsors and to ensure that there is no risk of the project becoming overindebted.

Scenario and sensitivity analyses

Scenario analyses are used to develop different versions of the relevant project parameters and examine their impact on the decision-relevant key figures as part of cash flow modeling. Probabilities of occurrence are then determined for the various scenarios. The result of the scenario analyses gives the negotiating parties an idea of whether loan repayment is possible for the project even under less favorable conditions and whether an appropriate return can be generated for the sponsors.

First, the cash flow model is fed with the data and assumptions that are most likely to occur and are therefore also expected by the project participants in this form. This base case then serves as the starting point for further negotiations with all project participants. However, lenders in particular also want to check whether the debt service can still be provided under less favorable conditions before committing to financing. Downside scenarios are therefore used to check whether financing is still possible even if key drivers of the business model deviate from the base case, either alone or in combination. If this can also be ensured in worst-case scenarios, the investors are more likely to see themselves in a position to finance the project. Finally, the banking case would be the scenario on which the financing of the project is ultimately structured.

Sensitivity analyses illustrate how changes to individual project parameters affect the level of cash flow and the key figures presented above. Break-even values can also be calculated for the parameters. The aim of the modeling is to determine the extent to which a parameter or a combination of parameters may assume in order to keep the decision-relevant key figures at a level that is acceptable to the investors. The results of these analyses are then used as a decision-making basis for structuring the financing.

As part of their negotiations, the capital-providing sponsors and banks must agree on the framework conditions under which the project financing is to be structured based on the cash flow models generated. In addition to the loan interest rate, the amount of equity to be contributed, the number of redemption-free years, the loan term and the introduction of a debt service reserve determine the financing conditions for both parties. These parameters are varied in the course of the contract negotiations as part of scenario and sensitivity analyses in order to structure financing that is acceptable to both parties.

This chapter has outlined the fundamental aspects that need to be considered when planning and implementing project financing plans in order to successfully finance a project.

The stability and reliability of the cash flow is of paramount importance for the feasibility of project financing. Which aspects of project financing can fluctuate (see Table 16.13)?

Changes in these items have a direct impact on individual items within the cash flow waterfall and change various items such as CFADS and debt service, thereby changing the coverage ratios for debt service and free cash flow.

An increase in interest rates leads to a deterioration in the DSCR ratio while the borrowed capital remains the same. To ensure that this ratio remains the same, the bank will reduce the project financing loan. The investor now finds himself under pressure in two areas: firstly, the equity required to finance his project increases, which reduces his internal rate of return. And secondly, he has to spend more cash flow to meet the bank's interest rate requirements. Both aspects mean that his return on equity deteriorates very quickly.

Table 16.13: Profit Driver of a Project Finance Project (compiled by author).

	Description	Impact on Profitability	
		before Financial Close	**after Financial Close**
Amount	An increase of investment costs leads to a higher financing need, which has to be covered by equity and/or debt	High	not relevant
Price	Increase in price will lead to a linear increase of income and influences the CFADS as well as the Free Cashflows	High	depends: can be high, if no long-term contract is available
Volatility	The higher the volatility, the lower the debt capacity.	High	High
Energy Production (Amount)	The energy production can be assessed by experts	High	High
Energy Production (Volatility)	According to the asset classes the volatility differ	Typically a risk which is accepted by banks	High
Amount	Operative Costs are contractually fixed before Financial Close	Low	Low
Volatility of Operating Costs	For most project financings in renewable energy a volatility of operating costs is not a major item (apart from biomass)	mostly low	mostly low
Amount	Interest work in different directions: 1. Interest Increase will result in a lower debt capacity, 2. higher interest and lower debt will deteriorate the internal rate of return.	This issue has to be solved before Financial Close	not relevant (if interest risk has been covered)

Kathrin Langewald

17 The Strategic Role of Green Hydrogen in Germany's International Energy Policy: A South African Case Study

17.1 Introduction

> Germany wants to establish a strategic green hydrogen partnership with South Africa – and Africa at large – for mutual benefit. This partnership must boost sustainable economic development in Africa and supply Germany and the EU with additional clean energy. Another beneficiary of this partnership would be the world's climate as less CO_2 is emitted into the atmosphere.[1]

This quotation by the German ambassador to South Africa, Lesotho, and Eswatini comes from the third renewable green hydrogen webinar, held in 2021 in South Africa. It emphasizes the perceived contribution of green hydrogen (gH2) in German-South African energy relations, reflecting both its centrality to and the international dimension of Germany's energy transition, known as the Energiewende.

Germany's Energiewende is a significant element of its international policy and a frequent subject of academic discourse. Scholars have called for Germany to take an active role in global climate protection and have assessed its strategies and leadership on the global stage[2,3] Steinbacher (2019) explores Germany's interactions with various partner countries in the realm of renewable energy, focusing on the effectiveness of

1 This thesis is part of the research project Global-H2-Upscaling, carried out by IZES gGmbH, in the Department energy markets, on behalf of the Federal Ministry for Economic Affairs and Climate Protection (BMWK). Martin Schäfer, remarks at the third Renewable Green Hydrogen Webinar. 2021. in "Report Back: Third Renewable Green Hydrogen Webinar," Energize, last modified October 19, 2021, https://www.energize.co.za/article/report-back-third-renewable-green-hydrogen-webinar.

2 Messner, Dirk, and Jennifer Morgan. 2013. "Germany Needs an Energy Transformation Foreign Policy." The Current Column of 7 January 2013. Bonn: German Development Institute / Deutsches Institut für Entwicklungspolitik (DIE). Tänzler, Dennis, and Stephan Wolters. 2014. "Energiewende und Außenpolitik: Gestaltungsmacht auf dem Prüfstand." Zeitschrift für Außen- und Sicherheitspolitik 7 (2): 133–43. https://doi.org/10.1007/s12399-014-0408-x. Quitzow, Rainer. 2013. "Towards an Integrated Approach to Promoting Environmental Innovation and National Competitiveness." *Innovation and Development* 3 (2): 277–96. https://doi.org/10.1080/2157930X.2013.825070.

3 Röhrkasten, Sybille, and Kirsten Westphal. 2012. "Die IRENA: Schon Vergessen?" Stiftung Wissenschaft Und Politik (SWP), November. https://www.swp-berlin.org/en/publication/erneuerbare-energien-irena. Steinbacher, Karoline, and Sybille Röhrkasten. 2019. "An Outlook on Germany's International Energy Transition Policy in the Years to Come: Solid Foundations and New Challenges." *Energy Research & Social Science* 49 (March):204–8. https://doi.org/10.1016/j.erss.2018.10.013. Quitzow, Rainer, and Sonja Thielges. 2020. "The German Energy Transition as Soft Power." *Review of International Political Economy*, September, 1–26. https://doi.org/10.1080/09692290.2020.1813190.

https://doi.org/10.1515/9783110784299-017

its leadership in transferring renewable energy policies.[4] Similarly, Quitzow and Thielges (2020) discuss Germany's international energy partnerships, framing them as a form of "soft power" in contrast to traditional "hard power" concepts rooted in fossil energy policies.[5] More recently, scholarly attention has centered on the development of global gH2 value chains within Germany's international Energiewende and the geopolitical implications of its international policy strategies.[6] However, the implications of Germany's efforts to build a global value chain and ensure supply security with its partner countries remain underexplored.

Therefore, this thesis investigates how gH2 influences German energy cooperation with South Africa and its implications for Germany's Energiewende Narrative (EWN) as a soft power resource (cf. Thielges and Quitzow, 2020).[7] Through the lens of soft and hard power, I examine the strategies employed to promote future imports and market development, while South Africa's coal-dominated economy and its wider socio-technological environment serve as an entry point for assessing German engagement in South Africa. Using qualitative interviews, this case study explores the dynamics between national energy policies and global market strategies, emphasizing the role of gH2 within the larger framework of global energy transitions.

The thesis is structured as follows: the next chapter examines Germany's international Energiewende and highlights the role of gH2. This is followed by a discussion on South Africa's Minerals-Energy Complex (MEC) and its connection to German gH2 cooperation. Chapter four introduces my abductive case study approach, while chapter five explores the hydrogen factor in German-South African energy relations. The final chapter summarizes the findings and discusses their implications for German international energy policy.

---

4 Steinbacher, Karoline. 2019. "Case Study: South Africa." In Exporting the Energiewende: German Renewable Energy Leadership and Policy Transfer, edited by Karoline Steinbacher, 239–88. Wiesbaden: Springer Fachmedien. https://doi.org/10.1007/978-3-658-22496-7_7.

5 Quitzow and Thielges. 2020. "The German Energy Transition as Soft Power." *Review of International Political Economy*, September, 1–26. https://doi.org/10.1080/09692290.2020.1813190.

6 Nunez, Almudena, and Rainer Quitzow. 2023. "Germany's Hydrogen Strategy: Securing Industrial Leadership in a Carbon-Neutral Economy," April. https://doi.org/10.48481/rifs.2023.010. Quitzow, Rainer, Almudena Nunez, and Adela Marian. 2024. "Positioning Germany in an International Hydrogen Economy: A Policy Review." *Energy Strategy Reviews* 53 (May):101361. https://doi.org/10.1016/j.esr.2024.101361.

7 Cf. Quitzow and Thielges, "The German Energy Transition as Soft Power," 1–26.

17.2 Germany's Green Hydrogen Dimension in Promotion of the International Energiewende

Germany's Energiewende aims for a CO_2-neutral energy system by focusing on renewable energies, energy efficiency, and phasing out coal and nuclear energy, which should also ensure energy security and affordability.[8] To promote these national objectives, the Energiewende is also a cornerstone of its Foreign Energy Policy, targeting global climate goals and promoting renewable energy technologies through significant investments.[9] With gH2 as a newly emerging "key element" in energy transition,[10] this thesis assumes a geostrategic shift in its international Energiewende dimension, recognizing Germany's import dependence and the need for large scale market development.

17.2.1 Germany's Energiewende Leadership

Considered internationally as a role model – a concept known as the "lead market effect" – Germany uses its economic significance to encourage other nations to follow its example[11]. However, the global success of the energy transition and demand for renewable energy technologies hinges on the overall international movement towards sustainability. Only if successfully integrated into broader environmental and foreign policy, technical innovation can become a competitive advantage.[12]

Therefore, the Foreign Affairs Office (AA) highlights the role of renewables as a means to reduce dependency on fossil fuels, thus enhancing energy affordability and

---

8 BMWi (Bundesministerium für Wirtschaft und Energie). "Unsere Energiewende: sicher, sauber, bezahlbar." Accessed April 15, 2021. https://www.bmwi.de/Redaktion/DE/Dossier/energiewende.html.

9 AA (Auswärtiges Amt). 2019a. "Energieaußenpolitik." Auswärtiges Amt. August 20, 2019. https://www.auswaertiges-amt.de/de/aussenpolitik/klimaaussenpolitik/energie/energieaussenpolitik/205854.

10 BMWK (Bundesministerium für Wirtschaft und Klimaschutz). n.d.-a. "Wasserstoff: Schlüsselelement für die Energiewende." Accessed May 30, 2024. https://www.bmwk.de/Redaktion/DE/Dossier/wasserstoff.html.

11 Westphal, Kirsten. 2012. "Globalising the German Energy Transition." Stiftung Wissenschaft Und Politik (SWP) (blog). December 10, 2012. https://www.swp-berlin.org/en/publication/globalising-the-german-energy-transition. Quitzow, Rainer, Sybille Röhrkasten, and Martin Jänicke. 2016. "The German Energy Transition in International Perspective." https://doi.org/10.2312/iass.2016.009.

12 Quitzow, Rainer. 2013. "Towards an Integrated Approach to Promoting Environmental Innovation and National Competitiveness." *Innovation and Development* 3 (2): 277–96. https://doi.org/10.1080/2157930X.2013.825070.

their advantage for economic growth.[13] Quitzow and Thielges argue that the international dimension of the energy transition, including its emphasis on energy independence and affordability, acts as a solution to multiple interrelated challenges such as climate change, economic growth, and innovation.[14] This narrative enables Germany to leverage its energy transition as a soft power resource.

To promote the Energiewende internationally, Germany has developed bilateral and multilateral frameworks with countries in the Global South and fossil fuel-reliant economies. Since the 1970s oil crisis, renewable energies have been integral to German development cooperation, promoting energy independence and addressing political, economic, and social challenges in line with Sustainable Development Goal (SDG) 7[15] Energy partnerships and dialogues aim to advance the international energy transition by fostering cooperation with target countries on energy, climate change, and its market economy frameworks, including export opportunities and innovation.[16] Germany has also played a key role in establishing the International Renewable Energy Agency and the Renewable Energy Policy Network for the 21st Century, which disseminate knowledge and updates on renewable energy.[17] This strategy is not only intended to meet global climate targets but also helps Germany develop global markets for environmentally friendly technologies and to enhance its own transition goals.

---

13 AA (Auswertiges Amt) . 2019b. "Energiewende international." Auswärtiges Amt. August 20, 2019. https://www.auswaertiges-amt.de/de/aussenpolitik/klimaaussenpolitik/energie/energiewende/238782.

14 Quitzow and Thielges. 2020. "The German Energy Transition as Soft Power." *Review of International Political Economy*, September, 9–10. https://doi.org/10.1080/09692290.2020.1813190.

15 Steinbacher, Karoline. 2019. "Case Study: South Africa." In *Exporting the Energiewende: German Renewable Energy Leadership and Policy Transfe*r, edited by Karoline Steinbacher, 239–88. Wiesbaden: Springer Fachmedien. https://doi.org/10.1007/978-3-658-22496-7_7. BMZ (Bundesministerium für wirtschaftliche Zusammenarbeit und Entwicklung). n.d. "Energie und Klima." Bundesministerium für wirtschaftliche Zusammenarbeit und Entwicklung. Accessed May 30, 2024. https://www.bmz.de/de/themen/klimawandel-und-entwicklung/energie-und-klima.

16 Quitzow and Thielges. 2020. "The German Energy Transition as Soft Power." *Review of International Political Economy*, September, 1–26. https://doi.org/10.1080/09692290.2020.1813190. Auswärtiges Amt 2019c. "Energiepartnerschaften." Auswärtiges Amt. August 20, 2019. https://www.auswaertiges-amt.de/de/aussenpolitik/klimaaussenpolitik/energie/energiepartnerschaften/238784.

17 Quitzow, Rainer, Sybille Röhrkasten, and Martin Jänicke. 2016. "The German Energy Transition in International Perspective." https://doi.org/10.2312/iass.2016.009.

17.2.2 The Role of Green Hydrogen in the International Energiewende

The publication of the National Hydrogen Strategy (NHS) is the missing puzzle piece in Germany's energy transition, addressing challenges beyond the decarbonization of the electricity sector, see Figure 17.1 through the use of renewable energies.[18] In addition to establish the regulatory framework and continuing the energy transition, the NHS envisages 38 measures intending to ensure Germany's competitiveness and security of supply.[19]

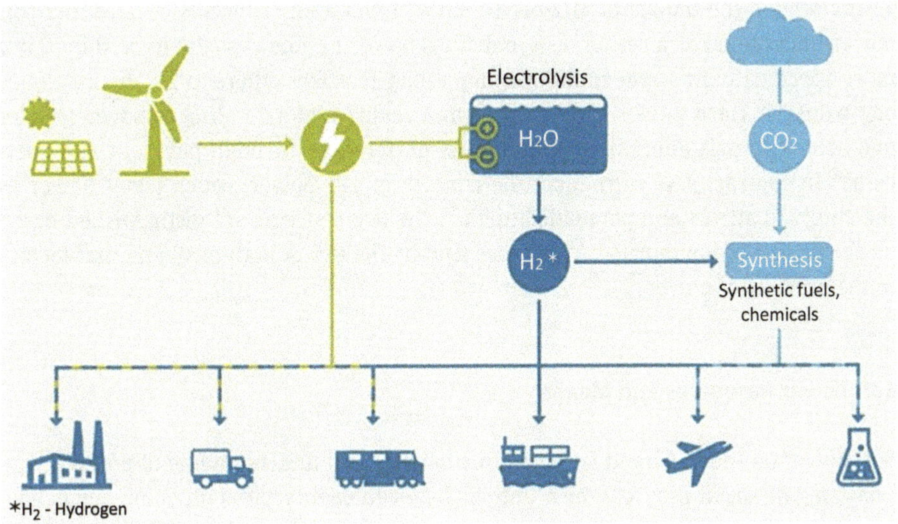

Figure 17.1: Production and Use of Green Hydrogen and PtX Products.[20]

Despite gH2's potential in industrial sectors, it is not yet competitive with fossil resources. To address this, Germany aims to accelerate the international market for gH2 and Power-to-X (PtX) products, leveraging economies of scale to reduce costs. This strategy sees Germany's technological leadership in gH2 and PtX as a chance to stimulate its own economy through technology exports. At the same time, Germany acknowledges not to meet its hydrogen needs domestically, thus emphasizing the import

18 BMWK, "Wasserstoff: Schlüsselelement." "Wasserstoff: Schlüsselelement für die Energiewende." Accessed May 30, 2024. https://www.bmwk.de/Redaktion/DE/Dossier/wasserstoff.html.
19 BMWi, "Die Nationale Wasserstoffstrategie." 2020. "Die Nationale Wasserstoffstrategie." https://www.bmwi.de/Redaktion/DE/Publikationen/Energie/die-nationale-wasserstoffstrategie.html.
20 KfW Development Bank. n.d.-a. "What Is Green Hydrogen?" Accessed May 30, 2024. https://www.kfw-entwicklungsbank.de/Our-topics/PtX/green-hydrogen/.

of hydrogen and PtX-products. Consequently, the NHS focuses on importing hydrogen from countries with abundant renewable energy resources.

This approach is assumed to promote climate protection in partner countries while fostering sustainable growth. It also aims to establish new supply chains and trade relationships, creating a win-win scenario aligning with the EWN.

17.2.3 Hypothesis 1: Soft and Hard Power Resources in the International Hydrogen Dimension

To investigate the impact of gH2 on the ENW, I will apply concepts of hard and soft power understood as a resource. Nye defines power as "the capacity to do things, but more specifically in social situations, the ability to affect others to get the outcomes one wants".[21] Hard power is derived from a realist understanding of power, where one actor compels another to abandon its preferences through payment and coercion.[22] In contrast, Nye introduces the concept of soft power, which views power as the ability to attract and persuade others.[23] The two concepts are distinguished based on the degree of voluntarism,[24] or more simply, the use of both push and pull factors in the exercise of power.[25]

Soft Power Resources and Means

Soft power on the one hand is rooted in a behavioural understanding of power,[26] emphasizing its social nature, which unfolds between counterparts on voluntary behaviour. In contrast to hard power, preferences are shaped through the attraction of culture, values and ideas.[27] In the context of Energy Transition the 'lead market effect'[28]

21 Nye, Joseph S. 2021. "Soft Power: The Evolution of a Concept." *Journal of Political Power* 14 (1): Page, 2. https://doi.org/10.1080/2158379X.2021.1879572.

22 Nye, Joseph S. 2011. The Future of Power. Public Affairs. 11.

23 Nye, Joseph S. 2005. Soft Power. First Edition. New York: Public Affairs. 5.

24 Nye, Joseph S. 2021. "Soft Power: The Evolution of a Concept." *Journal of Political Power* 14 (1): 8. https://doi.org/10.1080/2158379X.2021.1879572.

25 Nye, Joseph S. 2021. "Soft Power: The Evolution of a Concept." *Journal of Political Power* 14 (1): 8. https://doi.org/10.1080/2158379X.2021.1879572.

26 Nye, Joseph S. 2021. "Soft Power: The Evolution of a Concept." *Journal of Political Power* 14 (1): 6. https://doi.org/10.1080/2158379X.2021.1879572.

27 Nye, Joseph S. 2021. "Soft Power: The Evolution of a Concept." *Journal of Political Power* 14 (1): 8. https://doi.org/10.1080/2158379X.2021.1879572.

28 Quitzow, Röhrkasten, and Jänicke. 2016. "The German Energy Transition in International Perspective." 11. https://doi.org/10.2312/iass.2016.009. Cf. Quitzow and Thielges. 2020. "The German Energy

and the EWN constitute such ideas, promoting Germany's international leadership ambitions in hydrogen through win-win scenarios.

Germany's efforts to disseminate its ideas and values regarding electricity generation and energy efficiency involve development cooperation and partnerships as key strategies. Institutions such as the German Development Cooperation (GIZ) transfer expertise to partner country ministries, while Energy Partnerships leverage high-level dialogue and policy learning to bolster the energy transition.[29,30] Conceptually, soft power often manifests as public diplomacy, which, if effectively used, builds long-term relationships to strengthen policies and cultural exchange.[31] Supporting gH2 in partner countries like South Africa can build on Germany's legacy of promoting Energiewende based on mutual interests and voluntary measures.

Hard Power Resources and Means

Hard power on the other hand is fundamentally rooted in the idea that states possess various material resources that they can leverage against the preferences of others.[32] Germany has critical interests in energy security and market development. Hence, it has launched economic incentives like the H2Global auction model to secure hydrogen imports, providing investment certainty and ensuring the rapid deployment of hydrogen production facilities and related supply chains.[33] However, prioritizing export market creation may conflict with designing longer value chains, potentially hindering socio-economic development in partner countries,[34] thus its development goals. The H2UPPP instrument was launched to establish pilot projects in partner

Transition as Soft Power." *Review of International Political Economy*, September, 1–26. https://doi.org/10.1080/09692290.2020.1813190.

29 Steinbacher and Röhrkasten, 2019. "Case Study: South Africa." In *Exporting the Energiewende: German Renewable Energy Leadership and Policy Transfer*, edited by Karoline Steinbacher, 205. Wiesbaden: Springer Fachmedien. https://doi.org/10.1007/978-3-658-22496-7_7.

30 Quitzow and Thielges. 2020. "The German Energy Transition as Soft Power." *Review of International Political Economy*, September, 16. https://doi.org/10.1080/09692290.2020.1813190.

31 Nye, "Soft Power".

32 Nye 2021. "Soft Power: The Evolution of a Concept." *Journal of Political Power* 14 (1): 3. https://doi.org/10.1080/2158379X.2021.1879572.

33 H2Global Stiftung. n.d. "The H2Global Instrument." Accessed May 30, 2024. https://h2-global.de/project/h2g-mechanism.

34 Quitzow, Rainer, Clara Mewes, Sonja Thielges, Marina Tsoumpa, and Yana Zabanova. 2023. "Partnerschaften für eine internationale Wasserstoffwirtschaft – Ansatzpunkte für die europäische Politik.", 11. https://library.fes.de/pdf-files/a-p-b/20035.pdf. Quitzow et al., 2023. "Partnerschaftenfür eine internationale Wasserstoffwirtschaft – Ansatzpunkte für die europäische Politik." 11. https://library.fes.de/pdf-files/a-p-b/20035.pdf.

countries using German and European technologies.[35] Recognizing the immaturity and high cost of gH2 technologies, these support initiatives carry economic risks given uncertain demand for gH2. Hence, Germany could potentially use its economic clout to promote hydrogen and gain prioritized access to South Africa's gH2 market, possibly against South African interests.

In addition to the direct use of resources, there are more nuances to the notion of hard and soft power. Goldthau and Sitter differentiate between targeted and untargeted exercises of power.[36] Joining the European Union (EU) energy market is seen as an attractive prospect, constituting an untargeted form of soft power, whereas prescribing access terms, such as through standards, might verge on coercion when targeted at specific entities.[37] An example of this "soft power with a hard edge" is the Carbon Border Adjustment Mechanism (CBAM) under the "European Green Deal" and "Fit-for-55" package. It encourages green supply chains by restricting access to the EU market through CO_2 pricing.[38] While this regulation promotes climate protection, it may coerce countries and firms seeking to export to the EU to adhere to its sustainability standards. This could affect export-dependent countries concerning their financial stability. Such an understanding of CBAM aligns with Nye's definition of hard power, which involves involuntary changes in behavior through structural manipulation.[39]

From this perspective arises a threefold challenge to the promotion of gH2. Firstly, too strict standards could jeopardize Germany's climate ambitions if countries struggle to meet European criteria. Secondly, such measures could hinder the sustainable market development approach and import ambitions.[40] Thirdly they potentially

---

35 BMWK (Bundesministerium für Wirtschaft und Klimaschutz). n.d.-b. "International Hydrogen Ramp-up Programm – H2Uppp." Accessed May 30, 2024. https://www.bmwk.de/Redaktion/DE/Wasserstoff/Foerderung-International-Beispiele/10-h2uppp.html.

36 Goldthau, Andreas, and Nick Sitter. 2015. "Soft Power with a Hard Edge: EU Policy Tools and Energy Security." *Review of International Political Economy* 22 (5): 941–65. https://doi.org/10.1080/09692290.2015.1008547.

37 Goldthau and Sitter. 2015. "Soft Power with a Hard Edge: EU Policy Tools and Energy Security." *Review of International Political Economy* 22 (5): 16. https://doi.org/10.1080/09692290.2015.1008547.

38 Maat, Eva Pander. 2022. "Leading by Example, Ideas or Coercion? The Carbon Border Adjustment Mechanism as a Case of Hybrid EU Climate Leadership." *European Papers - A Journal on Law and Integration* 2022 7 (1): 55–67. https://doi.org/10.15166/2499-8249/546.

39 Nye. 2021. "Soft Power. The Evolution." *Journal of Political Power* 14 (1): 7–9. https://doi.org/10.1080/2158379X.2021.1879572.

40 Quitzow et al. 2023. "Partnerschaften für eine internationale Wasserstoffwirtschaft – Ansatzpunkte für die europäische Politik." 11 https://library.fes.de/pdf-files/a-p-b/20035.pdf. BMWi 2020. "Die Nationale Wasserstoffstrategie." 8. https://www.bmwi.de/Redaktion/DE/Publikationen/Energie/die-nationale-wasserstoffstrategie.html. Cf. Steinbacher. 2019. "Case Study: South Africa." 141. In *Exporting the Energiewende: German Renewable Energy Leadership and Policy Transfer*, edited by Karoline Steinbacher, 239–88. Wiesbaden: Springer Fachmedien. https://doi.org/10.1007/978-3-658-22496-7_7.

undermine the EWN as a soft power resource. To handle this trade off carefully will be a difficult task.

H1: Green hydrogen has an impact on the Energiewende Narrative as a soft power resource.

17.3 Locating Green Hydrogen in the Wider Economic Context of South Africa

As highlighted in the previous chapter, the NHS recognizes that Germany cannot meet its own needs wherefore a global market approach is deemed necessary. Under the geopolitical implications of a global market approach, the cooperation with today's fossil fuel exporting and dependent countries is highlighted as crucial concern to convert their supply chains for the use of renewables and the production of gH2. Furthermore, Germany's role in shaping the global energy transition is influenced by the preferences of powerful recipients and the disruptive impact of these efforts. Therefore, this chapter is concerned with the systemic role of gH2 promotion in a coal dominated economy such as South Africa.

17.3.1 The Minerals Energy Complex in the South African Economy

Largely shaped by its apartheid legacy, the MEC dominated South Africa's energy and economic landscape over a long period of time. Coined by Fine and Rustomjee the MEC describes the political-economic relations around coal and mining.[41] They identified rich coal mineral resources, cheap labor from the black population and therefore cheap electricity as the preconditions for the emergence of energy-intensive industries that formed the backbone of economic growth in South Africa. As shall be seen, the MEC continues to play a central role in the economy.

The MEC comprises core sectors, which, on the one hand, exhibit strong interconnections among themselves and weaker connections to other industries on the other. Fine and Rustomjee identified various core sectors of the MEC based on its input-output linkages compared to non-MEC sectors, see Table 17.1.[42]

41 Fine, Ben, and Zavareh Rustomjee. 1996. *The Political Economy of South Africa: From Minerals-Energy Complex to Industrialisation.* C. Hurst & Co. Publishers.

42 Fine and Rustomjee. 1996. *The Political Economy of South Africa: From Minerals-Energy Complex to Industrialisation.* C. Hurst & Co. Publishers. Ashman, Sam, Ben Fine, and Susan Newman. 2013. Systems of Accumulation and the Evolving MEC. 8. https://repub.eur.nl/pub/40424/.

Table 17.1: The Interdependence of the MEC input/output linkages 2010. Ashman, Fine, and Newman 2013, 8.

MEC subsector	Share of inputs from MEC sectors (% of total)	Share of output to MEC sectors (% of total)
Coal mining	26	90
Gold and uranium ore mining	55	5
Other mining	23	77
Coke and refined petroleum products	88	18
Basic chemicals	77	60
Other chemicals and man-made fibers	67	37
Plastic products	68	30
Non-metallic minerals	73	8
Basic iron and steel	82	59
Basic non-ferrous metals	91	59
Metal products excluding machinery	70	41
Machinery and equipment	63	53
Electricity gas and steam	53	47
Non-MEC manufacturing	23	6

The strong interconnection arising from these sectors indicates that "64.4% of productive inputs into the MEC sectors come from the MEC core itself and 53.0% of output from MEC sectors goes back into the MEC core as inputs", suggesting that MEC sectors operate relatively autonomously from other sectors.

The formation of the MEC has supported the development of state-owned enterprises, providing public infrastructure and the continuous supply and demand for coal. Transnet, a state-owned company, developed the country's rail, port, and pipeline systems,[43] primarily serving the mining industry's needs. Eskom and Sasol (a former state-owned enterprise) are the biggest coal consumers. Eskom produces and distributes electricity, while Sasol's coal-to-liquid synthetic fuel plants reduce dependence on international oil prices.[44]

43 Eberhard, Anton. 2011. "The Future of South African Coal: Market, Investment, and Policy Challenges." Program on Energy and Sustainable Development, January. 20. http://pesd.fsi.stanford.edu/publications/the_future_of_south_african_coal_market_investment_and_policy_challenges. Montmasson-Clair, Gaylor. 2015. "The Interplay between Mining and Green Economy in South Africa: An Energy Lens." SSRN Scholarly Paper. Rochester, NY. 21. https://doi.org/10.2139/ssrn.2748019.

44 Eberhard. 2011. "The Future of South African Coal: Market, Investment, and Policy Challenges." 6. Program on Energy and Sustainable Development, January. http://pesd.fsi.stanford.edu/publications/the_future_of_south_african_coal_market_investment_and_policy_challenges. Ashman, Fine, and Newman. 2013. Systems of Accumulation and the Evolving MEC. 16. https://repub.eur.nl/pub/40424/.

Historically, the MEC's contribution to the Gross Domestic Product has remained in the range of 20–30%, while the tertiary sector increasingly gained importance and non-MEC manufacturing declined.[45] The MEC accounts for 60% of foreign currency earnings. Despite its economic significance, employment in the MEC is relatively low compared to other sectors and has been declining with corresponding effects on the high unemployment and inequality in the country.

17.3.2 The Minerals Energy Complex in Transition

The concentration on these core sectors has led to a specific structure with relatively closed couplings and various social functions. However, in recent decades, the MEC has experienced internal and external transformative pressure:

- Firstly, the energy sector is in crisis with planned electricity shortfalls, known as loadshedding. These arise from aging infrastructure, mismanagement, and capacity constraints.[46] Eskom, the state-owned electricity producer, has struggled to meet increasing demand, resulting in significant economic damage to the mining sector, estimated at R2 billion (124 million euros) per day in 2008.[47]
- Secondly, the mining sector has declined, as indicative by dropping employment and production numbers.[48] Challenges relate to Eskom and Transnet include their mismanagement and decaying infrastructure.[49] Furthermore, platinum metals' (PGMs) role in the automotive industry is expected to decline, signifying the sector reliance on global commodity prices for its main export goods.

---

45 Montmasson-Clair. 2015. "The Interplay between Mining and Green Economy in South Africa, 22. An Energy Lens." SSRN Scholarly Paper. Rochester, NY. https://doi.org/10.2139/ssrn.2748019. Ashman, Fine, and Newman. 2013. Systems of Accumulation and the Evolving MEC. 11. https://repub.eur.nl/pub/40424/.

46 Baker, Lucy, Peter Newell, and Jon Phillips. 2014. "The Political Economy of Energy Transitions: The Case of South Africa." New Political Economy 19 (6): 791–818. https://doi.org/10.1080/13563467.2013.849674.

47 Lawrence, Andrew. 2020. "REIPPPP: Renewables' Rise, or REIPPPP RIP?" In South Africa's Energy Transition, edited by Andrew Lawrence, 72. Progressive Energy Policy. Cham: Springer International Publishing. https://doi.org/10.1007/978-3-030-18903-7_5.

48 Robinson, I., and R. Croll. 2018. "The Story of the Decline of the South African Mining Industry." *Journal of the Southern African Institute of Mining and Metallurgy* 118 (5): 0–0.

49 Eberhard. 2011. "The Future of South African Coal: Market, Investment, and Policy Challenges." 21. Program on Energy and Sustainable Development, January. http://pesd.fsi.stanford.edu/publications/the_future_of_south_african_coal_market_investment_and_policy_challenges. McClelland, Angela. 2014. South Africa and the Global Hydrogen Economy: The Strategic Role of Platinum Group Metals. Johannesburg: Real African Publishers Pty Ltd. 236.

– Thirdly, as one of the largest CO_2 emitters, the country faced increasing environmental pressures, creating a window of opportunity for renewable energy technologies. The Copenhagen Climate Accords and technological advancements in renewable energies led to a Renewable Energy Feed-in Tariff and its subsequent initiative the Renewable Energy Independent Power Producer Procurement Programme, which paved the way for renewable energy interests.[50] This split the energy industry into two segments: one featuring Independent Power Producers and another where Eskom remains the primary electricity provider.

Hypothesis 2: Regime Driven Sustainability Transition

Amidst internal and external pressures to the MEC and prevailing societal challenges, South Africa released its Hydrogen Society Roadmap (HSRM) to drive sustainable development. This leverages its abundant solar and wind resources, the world's largest PGM deposits crucial for gH2 technologies and expertise in the Fischer-Tropsch process. Research on sustainability transitions attempts to explain these processes of change in systems characterized by environmental problems.[51] The South African economy exemplifies such a socio-technological system where actors, technologies, and rules interact to provide societal functions particularly in energy-intensive industries spanning from electricity to transport with a focus on export, see "The Minerals Energy Complex in the South African Economy". However, transitioning to greener production methods is challenging due to entrenched structures and lock-in mechanisms favouring fossil energy.[52] Therefore, I want to assess the potential of Germany's international gH2 dimension for socio-technological change within the MEC.

---

50 Morris, M., and L. Martin. 2015. "Political Economy of Climate-Relevant Policies: The Case of Renewable Energy in South Africa." IDS Evidence Report IDS Evidence Report; 49–52. IDS/University of Cape Town. https://opendocs.ids.ac.uk/opendocs/handle/20.500.12413/5986. Ting, Marie Blanche, and Rob Byrne. 2020. "Eskom and the Rise of Renewables: Regime-Resistance, Crisis and the Strategy of Incumbency in South Africa's Electricity System." *Energy Research & Social Science* 60 (February):101333. 11. https://doi.org/10.1016/j.erss.2019.101333.

51 Geels, Frank W. 2004. "From Sectoral Systems of Innovation to Socio-Technical Systems: Insights about Dynamics and Change from Sociology and Institutional Theory." Research Policy 33 (6): 897–920. https://doi.org/10.1016/j.respol.2004.01.015. Verbong ans Geels. 2007. "The Ongoing Energy Transition: Lessons from a Socio-Technical, Multi-Level Analysis of the Dutch Electricity System (1960–2004)." Energy Policy 35 (2): 1025–37. https://doi.org/10.1016/j.enpol.2006.02.010.

52 Unruh, Gregory C. 2000. "Understanding Carbon Lock-In." *Energy Policy* 28 (12): 817–30. https://doi.org/10.1016/S0301-4215(00)00070-7.

The Multi-Level Perspective

Geels' Multi-Level Perspective (MLP) framework, with its three interacting dimensions, see Figure 17.2, offers a comprehensive view of transition processes.[53] The MLP posits that innovations, such as gH2 technologies, initially develop in niches, supported by a small number of actors like renewable entrepreneurs and research promoting the use of PGM in fuel cell development.[54,] These niches relate to socio-technical regimes characterized by established actor constellations who develop common belief systems, norms, rules, and structures.[55] Regime actors are part of the larger socio-technological system, which in the South African case refers to the MEC, with its established industries such as chemicals, steel, and mining and entrenched infrastructure networks, supply chains and export orientation. Socio-technical regimes are embedded in broader, less malleable landscape dynamics, including trends like globalization and climate change exerting transformative pressures.[56]

In the MLP, regimes and niches are similar in structure, comprising actor networks with shared rules but differ in size and stability.[57] Regimes typically dominate niches, dictating rules and stifling change to protect material interests and market control, leading to only incremental innovations.[58] Consequently, transitions are seen as niche-driven, eventually replacing existing regimes when landscape pressures are sufficient.[59]

53 Geels, Frank W. 2004. "From Sectoral Systems of Innovation to Socio-Technical Systems: Insights about Dynamics and Change from Sociology and Institutional Theory." *Research Policy* 33 (6): 897–920. https://doi.org/10.1016/j.respol.2004.01.015.

54 Cf. Verbong and Geels. 2007. "The Ongoing Energy Transition: Lessons from a Socio-Technical, Multi-Level Analysis of the Dutch Electricity System (1960–2004)." *Energy Policy* 35 (2): 1026. https://doi.org/10.1016/j.enpol.2006.02.010. cf. DSI 2021. "Hydrogen Society Roadmap for South Africa 2021." https://www.dst.gov.za/images/South_African_Hydrogen_Society_RoadmapV1.pdf.

55 Geels, Frank W., and Johan Schot. 2007. "Typology of Sociotechnical Transition Pathways." *Research Policy* 36 (3): 400. https://doi.org/10.1016/j.respol.2007.01.003.

56 Geels. 2004. "From Sectoral Systems of Innovation to Socio-Technical Systems: Insights about Dynamics and Change from Sociology and Institutional Theory." 913. *Research Policy* 33 (6): 897–920. https://doi.org/10.1016/j.respol.2004.01.015.

57 Geels and Schot. 2007. "Typology of Sociotechnical Transition Pathways." 402. *Research Policy* 36 (3): 399–417. https://doi.org/10.1016/j.respol.2007.01.003.

58 Markard, Jochen, and Bernhard Truffer. 2008. "Technological Innovation Systems and the Multi-Level Perspective: Towards an Integrated Framework." *Research Policy* 37 (4): 599. https://doi.org/10.1016/j.respol.2008.01.004.

59 Köhler, Jonathan, Frank W. Geels, Florian Kern, Jochen Markard, Elsie Onsongo, Anna Wieczorek, Floortje Alkemade, et al. 2019. "An Agenda for Sustainability Transitions Research: State of the Art and Future Directions." *Environmental Innovation and Societal Transitions* 31 (June): 5. https://doi.org/10.1016/j.eist.2019.01.004.

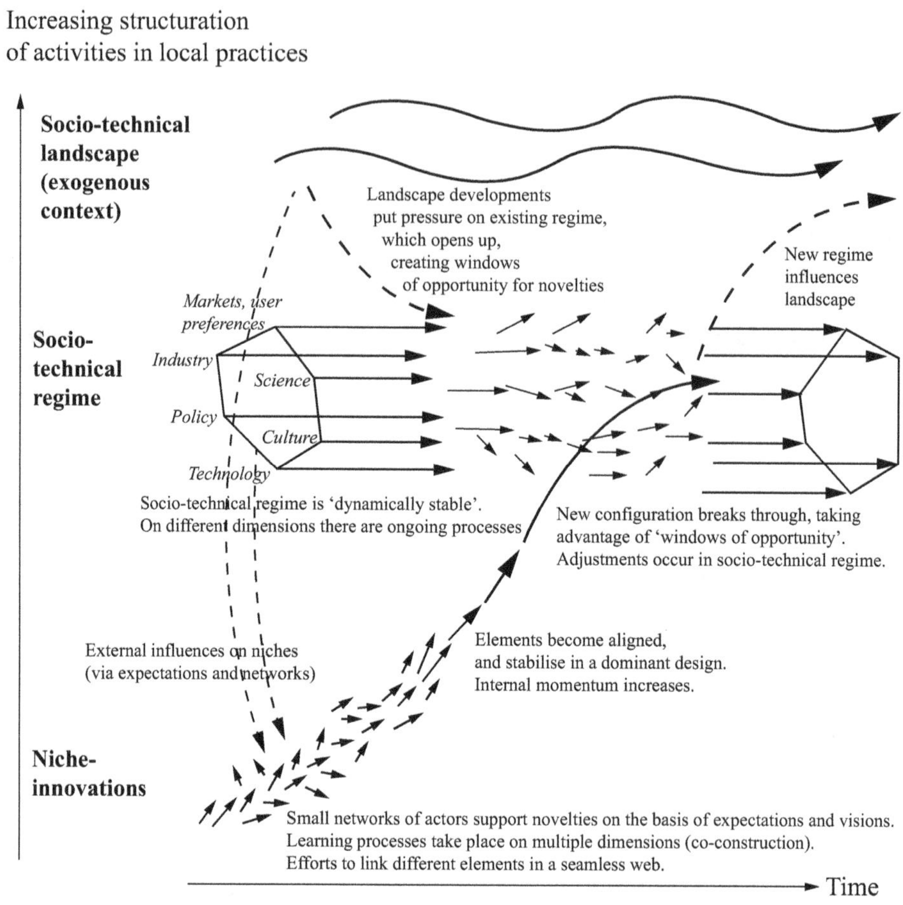

Figure 17.2: Multi-Level Perspective on Transitions.[60]

The Minerals Energy Complex in the GH2 Economy

Recent literature highlights regime actors' active role in socio-technical change. For example, Breggren et al. see the borders between niches and regimes blurring, as regimes may provide for well-integrated strategies to promote innovations.[61] This "re-

60 Geels and Schot, "Typology of Sociotechnical Transition," 401.

61 Berggren, Christian, Thomas Magnusson, and Dedy Sushandoyo. 2015. "Transition Pathways Revisited: Established Firms as Multi-Level Actors in the Heavy Vehicle Industry." *Research Policy* 44 (5): 1017–28. https://doi.org/10.1016/j.respol.2014.11.009.

configuration" pathway involves regimes adopting incremental changes across sectors.[62] The MEC, in comparison, is even contributing to the hydrogen value chain by supplying crucial inputs like carbon and mineral resources and has a legacy in in the provision of infrastructure, including electricity transmission lines, gas pipelines, and storage facilities. Thus, the HSRM makes clear that gH2 and derivatives align seamlessly with existing technology, infrastructure, and key actors within the MEC, positioning it as a transitional solution. Therefore, the MEC should be addressed in Germany's international hydrogen dimension as point for departure for transition.

H2: The MEC is crucial to the international dimension of German hydrogen goals in South Africa

To navigate this transformation, I leverage the socio-technological transitions framework and conceptualizations of niche and regime structures in the MLP as guiding tools, for a more nuanced understanding of the context and the targeted approach by which Germany promotes its international gH2 agenda, as shown in Table 17.2.

The MLP framework examines German institutions' support mechanisms in South Africa on various dimensions, including changing actor networks, technological transformations, and alterations in guiding rules. Based on that, this analysis identifies potential transformation pathways induced by the MEC.

17.4 Methodology

This methodological chapter lays out my exploratory case study approach to investigate the strategic role of gH2 within the EWN and its promotion in South Africa in relation to the MEC. To pursue this goal, the previous chapters have discussed gH2 in the light of its impact on the EWN as soft power resource as well as the usefulness of the MLP to familiarize myself with the policy implications within the MEC structures.

17.4.1 Research Design

With my research question, I want to understand the strategic role of gH2 in German international energy policy, but also how it is promoted. This question is explored as countries globally are trying to position themselves in this newly emerging market. With

62 Geels and Schot. 2007. "Typology of Sociotechnical Transition Pathways." 411. *Research Policy* 36 (3): 411. https://doi.org/10.1016/j.respol.2007.01.003. Cf. DSI, Hydrogen Society Roadmap, 2021. "Hydrogen Society Roadmap for South Africa 2021." https://www.dst.gov.za/images/South_African_Hydrogen_Society_RoadmapV1.pdf.

Table 17.2: Analytical Framework Multi-Level Perspective.[63]

Level	Changes in Actor Networks	Changes in Production through Technology	Changes in guiding Rules
Niche	Trying to get support from powerful actors	Building niche momentum through strategic learning	Changes in regulative rules cognitive routines and behavioral norms through new expectations
	The building of social networks		
Regime	Supporting Regime market entrants	Supporting incremental changes in the production pathway	Stabilization of existing regulative rules; cognitive routines and behavioural norms
	Supporting Institutional Entrepreneurs at the Regime level	Supporting wide scale niche deployments at the regime level; changes in the production lines	

my interest in this current phenomenon,[64] I followed Brinkmann's assessment of abductive reasoning in qualitative research,[65] interpreting a series of events within their context and incorporating new information and perspectives. Accordingly, I formulated initial expectations regarding hydrogen as puzzle piece within the respective country contexts. These expectations are "surprising" due to the normative aspects of energy transition and the emergence of vested interests in energy security. The German-South Afri-

63 a: Geels, F. W., and R. P. J. M. Raven. 2007. "Socio-Cognitive Evolution and Co-Evolution in Competing Technical Trajectories: Biogas Development in Denmark (1970–2002)." *International Journal of Sustainable Development & World Ecology* 14 (1): 63–77. https://doi.org/10.1080/13504500709469708.

b: Kern, Florian. 2012. "Using the Multi-Level Perspective on Socio-Technical Transitions to Assess Innovation Policy." *Technological Forecasting and Social Change*, Contains Special Section: Emerging Technologies and Inequalities, 79 (2): 298–310. https://doi.org/10.1016/j.techfore.2011.07.004.

c: Geels and Schot, "Typology of Sociotechnical Transition," 399–471.

d: Kern, "Using the Multi-Level Perspective," 298–310.

e: Markard and Truffer, "Technological Innovation Systems," 596–615.

f: Schwabe, Julian. 2024. "Regime-Driven Niches and Institutional Entrepreneurs: Adding Hydrogen to Regional Energy Systems in Germany." *Energy Research & Social Science* 108 (February):103357. https://doi.org/10.1016/j.erss.2023.103357.

g: Verbong and Geels, "The Ongoing Energy Transition," 1025–37.

64 Yin, Robert K. 2014. *Case Study Research*. SAGE Publications. 10. Cf. Rohlfing. 2012. "Types of Case Studies and Case Selection." In *Case Studies and Causal Inference: An Integrative Framework*, edited by Ingo Rohlfing, 61–96. Research Methods Series. London: Palgrave Macmillan UK. https://doi.org/10.1057/9781137271327_3.

65 Brinkmann, Svend. 2014. "Doing Without Data." *Qualitative Inquiry* 20 (6): 720–25. https://doi.org/10.1177/1077800414530254.

can energy relationship presents a unique case because of the longstanding cooperation in renewable energy. With the MEC central to energy-related decision-making in South Africa,[66] this relationship provides a fruitful ground for investigating the role of gH2 in the context of entrenched fossil fuel interests and German international energy policy.

17.4.2 Data Collection and Analysis

My case study builds on six expert interviews that were conducted between December 2023 and February 2024, supplemented where necessary by existing research and reports, see Table 17.3. This combination of sources is considered useful to triangulate information about the properties of energy transition on the ground. The interviewees were purposefully selected based on their occupation in German implementing institutions such as GIZ, German Development Bank (KfW) and German Chamber of Commerce (AHK) and their experience in the energy field and knowledge on gH2. One interview was conducted with a South African Institution as background material to assess my own assumptions regarding the role of German institutions in South Africa in the energy field and development of gH2 value chains. Each interview, lasting approximately one hour, was conducted online, transcribed, and coded via MaxQDA. Participants were informed about the confidentiality and anonymity of their responses.

Table 17.3: Interview Partners (own research).

Institution	Interview Style	Number of Interviews	Interview Code
German Chamber of Commerce (AHK)	Expert Interview	1	AHK
German Development Agency (GIZ)	Expert Interview	3	GIZ 0–2[67]
German Development Bank (KfW)	Expert Interview	1	KfW
Industrial Development Cooperation, National Bank South Africa (IDC)	Background Interview	1	/

The interview guide is built on two parts. First, questions were developed in order to capture possible shifts in German EWN implementation according to the perceived prevalence of national interests and its instruments.[68] Furthermore, a comparative

66 Baker, Newell, and Phillips. 2014. "The Political Economy of Energy Transitions: The Case of South Africa." *New Political Economy* 19 (6): 791–818. https://doi.org/10.1080/13563467.2013.849674.

67 GIZ 0: this interview was conducted in June 2022; therefore, the questions and answers deviate from the other interviews, but are nonetheless considered relevant to answer my research question and coded accordingly.

68 Ohnesorge, Hendrik W. 2020. "A Taxonomy of Soft Power: Introducing a New Conceptual Paradigm." In *Soft Power: The Forces of Attraction in International Relations*, edited by Hendrik W. Ohnesorge, 112, Cham: Springer International Publishing. https://doi.org/10.1007/978-3-030-29922-4_3.

approach for energy transition (renewable energy and energy efficiency only) and gH2 was applied in order to capture differences, with the publication of the NHS as a time marker. Second, with regard to its implementation in South Africa, I was particularly interested in actor network composition, technological production pathways, and rules and institutions due to the promotion of hydrogen, as identified by the MLP to steer transitions. After transcription, I developed a coding scheme according to criteria of qualitative content analysis,[69] integrating new codes when deemed necessary. Specifically, the dimension of power was initially identified and theoretically integrated only later. The content detailing German initiatives in promoting gH2, pertinent to the MLP, was evaluated for its niche and regime properties in conjunction with the literature in section "The Minerals Energy Complex in the GH2 Economy".[70]

My methodology's main limitations are that the selection of interview partners mostly reflects the assumptions of German implementation agencies within the context of South Africa and are subjective in nature. Therefore, assessments of shifts in the EWN as a soft power resource are limited due to lack of reciprocity such as through legitimacy, which is inherent to the relational assumption of soft power.[71] Besides, generalizability outside the South African context is restricted.

17.5 The Strategic Promotion of the Energiewende and the Green Hydrogen Factor

Germany has been actively involved in renewable energy and energy efficiency measures in South Africa for many years, primarily through various GIZ projects and KfW's financial assistance (GIZ 2; GIZ 1; KfW, AHK), see Table 17.4. Analogous, Germany now has become a key player in the development of South Africa's gH2 value chain (AHK).

Under the JET-P framework, Germany, and several international partners, support South Africa in its endeavors to decarbonize its economy and phase out coal. Central to Germany's involvement is also the provision of financial assistance aimed at fostering a

69 Mayring, Philipp. 2022. Qualitative Inhaltsanalyse: Grundlagen und Techniken. 13. Auflage. Weinheim Basel: Beltz.

70 Detailed information regarding the interview guideline and coding schemes were used as part of the assessment of a master's thesis. The original document can be accessed via Central European University Electronic Theses and Dissertations (ETD) services.

71 Ohnesorge. 2020. "A Taxonomy of Soft Power: Introducing a New Conceptual Paradigm." In *Soft Power: The Forces of Attraction in International Relations*, edited by Hendrik W. Ohnesorge, 112. Cham: Springer International Publishing. https://doi.org/10.1007/978-3-030-29922-4_3.

Table 17.4: German Implementing Institutions.[72]

Implementing Institution	Projects/Programmes	Ministry	Role
German Chamber of Commerce (AHK)	H2UPPP Export Initiatives	BMWK BMU [b]	Individual business consulting services; Support for the utilization of various German funding programs
German Development Cooperation (GIZ)	International PtX Hub H2.SA German-South African Energy Partnership	BMWK BMZ BMWK	Technical, market-related, and political expertise; Capacity building and learning; Dialog and Networking
German Development Bank (KfW)	JET-P PtX Growth Fund [a]	BMZ	Concessional financing in the form of loans or credits

gH2 economy.[73,74] Additionally, Germany is the primary provider of technical assistance to South Africa regarding hydrogen. This includes projects implemented by GIZ, such as the International PtX Hub, commissioned by the German Federal Ministry for Economic Affairs and Climate Action (BMWK) with works to identify sustainable PtX value chains for climate change mitigation;[75] the H2.SA project, financed by the German Federal Ministry for Economic Cooperation and Development (BMZ), which aims to improve the investment conditions for gH2 in South Africa;[76] as well as the Energy Partnership, financed by the BMWK, with the aim to ensure political dialogue between Germany and

72 Own research: a: KfW. n.d.-b. "Funktionsweise PTX-Plattform | KfW Entwicklungsbank." Accessed June 2, 2024. https://www.kfw-entwicklungsbank.de/Unsere-Themen/PtX/PtX-Plattform/. b: NOW GmbH. n.d. "Brennstoffzelle statt Dieselgenerator – BMU-Exportinitiative Umwelttechnologien auf den Punkt gebracht." NOW GmbH (blog). Accessed June 2, 2024. https://www.now-gmbh.de/aktuelles/pressemitteilungen/brennstoffzelle-statt-dieselgenerator-bmu-exportinitiative-umwelttechnologien-auf-den-punkt-gebracht/.

73 Cassidy, Christopher, and Rainer Quitzow. 2023. "Green Hydrogen Development in South Africa and Namibia: Opportunities and Challenges for International Cooperation." Forschungsinstitut für Nachhaltigkeit. 20. https://www.rifs-potsdam.de/de/ergebnisse/publikationen/2023/green-hydrogen-development-south-africa-and-namibia-opportunities-and.

74 The Presidency Republic of South Africa 2023. "JET Implementation Plan 2023–2027." 178. https://www.stateofthenation.gov.za/assets/downloads/JET%20Implementation%20Plan%202023-2027.pdf. Cassidy and Quitzow. 2023. "Green Hydrogen Development in South Africa and Namibia: Opportunities and Challenges for International Cooperation." 21. Forschungsinstitut für Nachhaltigkeit. https://www.rifs-potsdam.de/de/ergebnisse/publikationen/2023/green-hydrogen-development-south-africa-and-namibia-opportunities-and.

75 International PtX Hub. n.d. "South Africa." PtX Hub (blog). Accessed May 30, 2024. https://ptx-hub.org/south-africa/.

76 GIZ (Deutsche Gesellschaft für Internationale Zusammenarbeit (GIZ) GmbH). n.d.-a "Promoting a South African Green Hydrogen Economy (H2.SA)." Accessed May 31, 2024.

South Africa on energy issues.[77] The AHK implements the investment program H2UPPP led by GIZ to identify local value chains and is responsible for promoting foreign trade in the field of hydrogen.[78] These institutions work synergistically, with KfW providing substantial financial support and GIZ contributing to the strategic framework. While KfW and GIZ operate at macro and meso levels, focusing on the public sector, AHK operates at the micro level, advising companies directly (AHK). Together, these three institutions are pivotal in Germany's international hydrogen dimension.

17.5.1 Prevalence of National Interests

The EWN is not only promoted as a response to global crises such as climate change, but also presented as a motor for economic growth. The conducted interviews reinforce the notion that Germany's support for renewable energies in South Africa addresses critical social challenges such as energy security and poverty alleviation (GIZ 1). GIZ emphasizes environmental, social, and economic factors geared towards benefiting the broader population. KfW prioritizes infrastructural financing, focusing on decentralizing the energy system. Meanwhile AHK highlights foreign trade, technological solutions, and export initiatives related to energy security and economic decarbonization. Collectively, these efforts aim to develop local value chains from a socioeconomic perspective, underscoring Germany's engagement in South Africa.

Given its limited domestic production capabilities and subsequent reliance on imports, a vital aspect of the NHS lies in Germany's hydrogen ambitions on the international stage. One interviewee delineates the evolution of German interests in gH2 promotion across three key dimensions:

> At the beginning, German efforts were clearly focused on climate protection. The question was how to reduce CO_2 emissions by using hydrogen products, PtX products, etc. The COVID-19 pandemic changed this, and Germany now looked at green hydrogen as one of the growth paths that could lead to some kind of reindustrialization. [. . .] In addition, since Russia's war of aggression against Ukraine, there has also been a new component on the German side, namely energy security. Prior to this event, energy security was not an issue. (GIZ 1)

Although hydrogen aligns with the broader international energy transition, it also underscores Germany's increasing focus on energy security. Several interviewees stress

77 GIZ (Deutsche Gesellschaft für Internationale Zusammenarbeit (GIZ) GmbH). n.d.-b. "German-South African Energy Partnership." Accessed March 16, 2022. https://www.giz.de/en/worldwide/65749.html. Cassidy and Quitzow. 2023. "Green Hydrogen Development in South Africa and Namibia: Opportunities and Challenges for International Cooperation." 21. Forschungsinstitut für Nachhaltigkeit. https://www.rifs-potsdam.de/de/ergebnisse/publikationen/2023/green-hydrogen-development-south-africa-and-namibia-opportunities-and

78 GIZ. n.d.-c. "Promoting Hydrogen Projects in Developing Countries and Emerging Economies: H2Uppp." Accessed May 30, 2024.https://www.giz.de/en/worldwide/107551.html.

that particularly considering recent geopolitical events Germany's commitment to address the global climate crisis and security concerns (GIZ 1) has led to increased financing for hydrogen projects within bilateral energy frameworks (GIZ 1, AHK, KfW).

The promotion of gH2 value chains reflects coordinated cooperation between Germany and South Africa (AHK). There is a growing awareness of the necessity of energy transition, driven by the prospects of gH2 (GIZ 2). Mutual interests include importing and exporting gH2 and potential green reindustrialization (GIZ 1).

Germany also prioritizes social and environmental criteria to mitigate risks associated with developing the hydrogen value chain in South Africa (GIZ 0; KfW). For example, GIZ focuses on local economic value creation and capacity building (GIZ 0), while KfW mobilizes financial resources for decarbonization, job creation, and coal phase-out projects (KfW). A common goal is empowering the private sector to enhance its competitiveness in the global hydrogen economy.

National interests play a significant role in assessing hydrogen's impact on the energy transition as a soft power resource.[79] The interviews suggest promoting gH2 aims to create a "win-win situation" focused on decarbonization and economic recovery. Efforts involve close coordination with South African partners to ensure alignment with ongoing energy crisis mitigation and socio-economic value generation. Despite the energy security dimension as a hard power motive, gH2 primarily supports the EWN of sustainable development as a soft power resource.

17.5.2 Channels of Influence

On the one hand, the interviewees emphasize that gH2 continues previous renewable energy and energy efficiency efforts (GIZ 2, AHK), supporting the EWN. Through the provision of technical, market, and policy expertise (GIZ 1), GIZ addresses diverse needs identified in previous dialogue formats (GIZ 2). Therefore, capacity building measures, such as those introduced within the South African-German Energy Program, promote institutions and reform processes for the uptake of renewable energy technologies (GIZ 2). Expertise is also disseminated through commissioned studies and workshops on renewable energy additionality, ensuring hydrogen production does not hinder electricity sector decarbonization (GIZ 0). From a soft power perspective, GIZ leverages its institutional roots for agenda-setting, as it influences decision-making by framing topics in workshops.[80] Furthermore, GIZ emphasizes dialogue and consensus-building activities to share knowledge among stakeholders (GIZ 2). The En-

79 Cf. Ohnesorge. 2020. "A Taxonomy of Soft Power: Introducing a New Conceptual Paradigm." 112. In Soft Power: The Forces of Attraction in International Relations, edited by Hendrik W. Ohnesorge, 85–225. Cham: Springer International Publishing. https://doi.org/10.1007/978-3-030-29922-4_3.

80 Nye. 2021. "Soft Power: The Evolution of a Concept." 7. *Journal of Political Power* 14 (1): 7. https://doi.org/10.1080/2158379X.2021.1879572.

ergy Partnership facilitates technical exchanges and cooperation between political representatives through various formats, such as study trips and workshops.[81] This strengthens the energy transition narrative of the hydrogen economy in South Africa,[82] as one interviewee noted regarding the inclusion of gH2 in Jet-P conditions under German suggestions (KfW). The AHK supports businesses by offering consulting services, facilitating funding access, and providing networking opportunities with decision-makers (AHK). Ultimately these persuasive efforts build on the long-term relationship between South Africa and Germany in the field of energy with the goal to promote the market development of gH2 within the broader values of climate change and sustainable development.

On the other hand, South Africa is under pressure from the EU's legal framework for climate and energy to decarbonize its economy (AHK). CBAM incentivizes South Africa to produce gH2 to reduce export products' carbon content and to remain competitive through creating new value chains.[83] Under these structural conditions, the EU shapes South Africa's regulatory and economic environment, pushing production process changes.[84] However, the nascent market for gH2 requires substantial upfront investments (KfW). Considering this, the NHS and the provision of Covid-19 stimulus funds create demand and investment certainty for the South African gH2 economy (KfW, GIZ 0). In addition, the H2Global initiative intends to provide funding for large-scale industrial investments to bolster the market and to secure energy imports (AHK). With JET-P and the funds provided by KfW, the South African government can improve its borrowing capacity and investment climate for gH2,[85] as South Africa grapples with budget deficits and public debt since the Covid-19 recession. KfW typically provides concessional financing, with funds allocated to South African institutions under KfW oversight (KfW). KfW's provision of financial resources increasingly involves policy-related development loans tied to energy market reforms (KfW, GIZ 2).

81 Quitzow and Thielges. 2020. "The German Energy Transition as Soft Power." *Review of International Political Economy*, September, 13–14. https://doi.org/10.1080/09692290.2020.1813190.

82 Cf. Quitzow and Thielges, 2020. "The German Energy Transition as Soft Power." 17. *Review of International Political Economy*, September, 1–26. https://doi.org/10.1080/09692290.2020.1813190. Cf. Nye, Joseph S. 2005. Soft Power. First Edition. New York: PublicAffairs. 8.

83 Maimele, Seutame. 2023. "Responding to the European Union's Carbon Border Adjustment Mechanism (CBAM): South Africa's Vulnerability and Responses." TIPS. https://www.tips.org.za/research-archive/sustainable-growth/green-economy-2/item/4590-responding-to-the-european-union-s-carbon-border-adjustment-mechanism-cbam-south-africa-s-vulnerability-and-responses. Cf. Nye. 2021. "Soft Power: The Evolution of a Concept." 8. Journal of Political Power 14 (1): 196–208. https://doi.org/10.1080/2158379X.2021.1879572.

84 National Treasury Republic of South Africa. 2022. "Media Statement-Bilateral Loan Agreements AFD and KFW." https://www.treasury.gov.za/comm_media/press/2022/2022111001%20Media%20statement-Bilateral%20Loan%20Agreements%20AFD%20and%20KFW.pdf.

85 The Presidency Republic of South Africa. 2022. "South Africa's Just Energy Transition Investment Plan (JET IP) for the Initial Period 2023–2027." 97. https://pccommissionflo.imgix.net/uploads/images/South-Africas-Just-Energy-Transition-Investment-Plan-JET-IP-2023-2027-FINAL.pdf.

Infrastructure projects focus on export and port development. Both measures aim to attract foreign investors (KfW). Germany's financial support aligns with the EU's broader regulatory ambitions to promote sustainable value chains and South Africa's need for capital market access. Shaping South Africa's economic transformation and export orientation reflects an element of structural manipulation. However, Germany's influence on the institutional and economic context in which it cooperates with South Africa is limited. Following that, Germany's actions in promotion of the hydrogen factor in the EWN are based on persuasion and attraction, rather than on payment and coercion.[86]

Overall, the soft power approach in developing a gH2 economy in South Africa gears toward Germany's EWN. The outlined combination of strategic communication of gH2 with the infrastructural and capacity development programs aims to boost the market ramp-up. Tied to this are the provision of sustainability criteria and the outlook of importing hydrogen and PtX products in future. However, some interviewees have expressed their doubts considering the effectiveness of this approach. For example, Sasol got support from the German government under the H2Global funding mechanism to produce sustainable aviation fuels with the aim to export its products to Germany.[87] Therefore, considering the EU Renewable Energy Directive 2, sustainable carbon sources will be required. But from a technological point of view, the company will not be able to transition its whole production processes accordingly (AHK). Consequently, the export focus in South Africa is currently shifting away to other destinations such as South Korea and Japan (KfW). This example illustrates how the promotion of gH2 is currently interfering with the advancement of the EWN. Germanys involvement in South Africa is supporting its goal of the global development of gH2 value chains and respective export initiatives. But conflict arises with EU regulations fostering climate mitigation efforts and competitiveness of European industries. Measures like CBAM coupled with unrealistic certification requirements could deter South African industries from export to Europe and Germany. Thus, the German soft power approach potentially clashes with European hard power measures. There is a risk that a successful yet under-targeted promotion of gH2 in South Africa could undermine the ENW by failing as much to support climate ambitions as the goal of import security.

86 Nye. 2021. "Soft Power: The Evolution of a Concept." *Journal of Political Power* 14 (1): 8. https://doi.org/10.1080/2158379X.2021.1879572.

87 SASOL. 2021. "Sasol to Explore Potential of Cleaner Aviation Fuels with World Class Partners." 2021. https://www.sasol.com/media-centre/media-releases/sasol-explore-potential-cleaner-aviation-fuels-world-class-partners.

17.5.3 Changes in the Composition of Actors, Networks and Social Groups

By examining the strategic orientation of the international hydrogen dimension, this first part looks at German efforts on the promotion of different stakeholder networks in South Africa's gH2 value chain. The MLP assumes that technological niches and socio-technical regimes have similar structures but differ in size and stability and are clearly differentiated from each other. Measures to back the gH2 niche are targeted at existing MEC structures, through the support of institutional entrepreneurs, the stimulation of supply and demand, as well as by gaining support from powerful actors and the promotion of niche networks.

Support for Institutional Entrepreneurs

In South Africa, institutional entrepreneurs are mainly political actors advocating for gH2 to drive economic growth and reduce coal dependence. These actors, including the presidency and several ministries, play a central role in transforming existing regimes, allowing deviations from previous energy paths. At the same time, these political partners are responsible for German implementing institutions. Support services provided by German institutions include, for example, capacity building in the form of workshops on regulations and standards, as well as market-based aspects that bring together both policymakers and private sector actors (GIZ 0). Most importantly, the Presidency, taking patronage for strategic infrastructure and investment decisions, coordinates various ministries (GIZ 0). With the emergence of gH2, South Africa made the strategic choice to anchor hydrogen development under the Presidency (AHK), with a central role taken by the Minister of Electricity who worked previously in the presidential acceleration unit for Infrastructure South Africa and continues to oversee the issue in his current role (GIZ 2).

Support of New Market Entrants

South Africa's nascent gH2 market require stimulation of the demand side (GIZ 1, GIZ 2, GIZ 0) and supply side (GIZ 1), allowing established players and new market entrants to form a common network. Support includes feasibility studies, training, workshops, and funding advice (GIZ, n.d.). New market participants include MEC actors like the steel company ArcelorMittal (GIZ 0) and Anglo American with hydrogen-powered mining vehicles,[88] and Transnet, which has to adapt to the new conditions (KfW). However, established players, like the petrochemical companies SASOL and PetroSA also support the niche development (GIZ 2), aiming to transition from grey to gH2 (KfW, GIZ 0). New players on the supply side are renewable energy companies with significant interest in putting their electricity molecules into value (AHK, GIZ 0).

Niche Support from Powerful Actors

Niche technologies can also gain traction through support from powerful actors.[89] In South Africa, many have had bad experiences with international companies failing to keep their development promises (KfW). Key measures for their involvement include training courses and information events (GIZ 2) for communities, NGOs, and trade unions, strengthening discourse and building societal support (GIZ 1). Another group consists mainly of JET-P partners as most of them are largely reluctant to support gH2, except for the Netherlands, Denmark and Japan (GIZ 1). Collaboration with development banks, such as the European Investment Bank and World Bank, mobilizes funding and ensures financial security for international investors, as they strengthen South Africa as a business location (KfW). Overall, the support of these actors is crucial to increase the credibility of gH2 technologies in South Africa.

Building Networks

The MLP suggests that niche technologies break through primarily by networks facilitating exchanges between companies, decision-makers, and researchers.[90] German implementing institutions play a key role in connecting these groups and coordinating efforts (AHK) that is crucial due to the lack of a dedicated hydrogen interest group (GIZ 0, GIZ 2). Activities include matchmaking events such as the South African GH2 Summit, largely financed and organized by GIZ (KfW), which allows niche developers in the field of hydrogen production to build networks with the demand side. Develop-

88 Anglo American. 2022. "Driving the Hydrogen Economy in South Africa." 2022. https://www.angloamerican.com/our-stories/innovation-and-technology/driving-the-hydrogen-economy-in-south-africa.

89 Kern. 2012. "Using the Multi-Level Perspective on Socio-Technical Transitions to Assess Innovation Policy." *Technological Forecasting and Social Change, Contains Special Section: Emerging Technologies and Inequalities*, 79 (2): 303. https://doi.org/10.1016/j.techfore.2011.07.004. Cassidy and Quitzow. 2023. "Green Hydrogen Development in South Africa and Namibia: Opportunities and Challenges for International Cooperation." 20. Forschungsinstitut für Nachhaltigkeit. https://www.rifs-potsdam.de/de/ergebnisse/publikationen/2023/green-hydrogen-development-south-africa-and-namibia-opportunities-and.

90 Geels and Raven, 2007. "Socio-Cognitive Evolution and Co-Evolution in Competing Technical Trajectories: Biogas Development in Denmark (1970–2002)." *International Journal of Sustainable Development & World Ecology* 14 (1): 67. https://doi.org/10.1080/13504500709469708.

ers range from smaller local start-ups that are already producing gH2 to various international technology providers active in consortia, such as Enertrag together with Sasol, Bambili or Energie (GIZ 0). These efforts indicate that the market is still nascent, with the gH2 technologies network being internationally and systematically expanded across different industrial sectors.

17.5.4 Changes in Technological Pathways

In addition to stakeholder networks, the MLP framework highlights the technological dimension, embedded in niches and regime structures, driving the dissemination and innovation of technologies.[91,92] This section assesses Germany's contribution to the emergence of new production systems using gH2 technologies. Production processes, such as in the MEC sectors, aim to utilize gH2 technologies within their established applications, while niche-level innovations lead to new uses through learning processes.

Changes in Production Process

Hard-to-abate sectors are increasingly recognizing the role of green molecules in their decarbonization efforts, beyond the energy transition in the electricity sector. For example, the mining sector is investing in mini-grid installations to avoid the costs of loadshedding and to respond to international market incentives, such as those from the EU (GIZ 2). By focusing, e.g. on sustainable water usage in the development of a gH2 value chain, the AHK conceives strategies to link the hydrogen and mining sectors, converting former burdens into sustainable value (AHK). Furthermore, companies like Transnet are not only needed to connect important hubs such as mining areas with corresponding ports, but explore the use of gH2 for decarbonization themselves.[93] Notwithstanding, financial support is challenging due to profitability con-

91 Verbong and Geels. 2007. "The Ongoing Energy Transition: Lessons from a Socio-Technical, Multi-Level Analysis of the Dutch Electricity System (1960–2004)." 1025–37. *Energy Policy* 35 (2): 1025–37. https://doi.org/10.1016/j.enpol.2006.02.010.

92 Geels and Raven. 2007. "Socio-Cognitive Evolution and Co-Evolution in Competing Technical Trajectories: Biogas Development in Denmark (1970–2002)." *International Journal of Sustainable Development & World Ecology* 14 (1): 63–77. https://doi.org/10.1080/13504500709469708. Cf. Markard and Truffer. 2008. "Technological Innovation Systems and the Multi-Level Perspective: Towards an Integrated Framework." *Research Policy* 37 (4): 599. https://doi.org/10.1016/j.respol.2008.01.004.

93 InfraCo Africa. 2023. "Memorandum of Cooperation between Transnet and GreenCo to Conduct Research and Development on the Viability of Green Hydrogen to Power Freight Trains in South Africa." 2023. https://infracoafrica.com/africa-greenco-memorandum-of-cooperation-between-transnet-and-greenco-to-conduct-research-and-development-on-the-viability-of-green-hydrogen-to-power-freight-trains-in-south-africa/.

cerns (KfW). Thus, activities in sectors like mining focus on incremental innovations supporting decarbonization which does not result in changes in the final product.

Changes in Production Line

Changes in production lines also emphasize decarbonization but lead to different end products, where hydrogen technologies are integrated into production processes. For instance, many renewable energy project developers are looking into mini-grid applications, using electrolyzers to make use of their additional molecules (AHK). GH2 is also gaining value in sectors already using grey hydrogen, such as the petrochemical industry, including companies like Sasol and PetroSA, which produce synthetic fuels. As a result, the German government's H2Global initiative has provided Sasol with purchase guarantees for sustainable aviation fuel exports.[94] These initiatives demonstrate gH2's potential to generate green products within existing regimes, aiming to mainstream hydrogen technologies and develop large-scale applications.

Strategic Learning: Commercialization and Capacity Building

At the niche level, strategic learning is crucial to promote new production processes and build capacity for the emerging gH2 market. GIZ collaborates with scientific networks to develop skills and conduct studies on niche technologies, such as desalination (GIZ 2, GIZ 0). Accompanying, KfW offers non-repayable loans (KfW). The South African government has designated nine catalytic hydrogen projects, whose pilot characteristics are aimed at pooling resources and facilitating learning processes. These are indirectly financed by KfW through IDC development loans (KfW). Altogether, this demonstrates how German implementation organizations support decarbonizing existing production processes, developing new product lines, and promoting niche developments.

17.5.5 Changes in Guiding Rules

Finally, rules form the behavioral frameworks in which technologies and actors operate. These include not only formal regulations such as legislation, but also a variety of

94 SASOL. 2021. "Sasol to Explore Potential of Cleaner Aviation Fuels with World Class Partners." 2021. https://www.sasol.com/media-centre/media-releases/sasol-explore-potential-cleaner-aviation-fuels-world-class-partners.

cognitive expectations about market developments or normative rules of conduct.[95,96] With Germany's support for "renewable energy additionality", it seeks to transform these existing frameworks.

Regulative Rules

GH2 is currently addressed through national strategies like the HSRM and the GH2 Commercialization Strategy, which set targets for hydrogen use, including exports and local consumption (GIZ 1). The strategies were developed in close consultation with German stakeholders and formulated according to common goals (AHK). Germany supports the principle of additionality for renewable energies, ensuring that resources designated in current energy legislation are not diverted to gH2 production (GIZ 0). This is in line with recent regulatory changes, allowing for increased private renewable energy installations (GIZ 2). Such topics are discussed within training courses and workshops about regulations, codes and standards tailored for ministries, officials, and the private sector (GIZ 0). Additionally, KfW is using policy-based loans to support the reform of the energy market from a centralized to a decentralized system, enabling large-scale hydrogen production (KfW). Despite the lack of integration of gH2 into South Africa's existing regulatory framework (GIZ 1), new regulatory frameworks are emerging. Thus, support mechanisms reinforce ongoing changes in the overall energy system.

Normative Rules

Normative discussions focus on the transition from coal to renewable energies, shaped by concerns about energy security and the impact on local communities and existing coal sector jobs (KfW, AHK, GIZ 1, 2, 0). The JET-P agreement has prompted warnings against a neo-colonial agenda from the Global North, amplifying fears about the responsibility of international actors and skepticism about public money being used for export investments (AHK; GIZ 2). The Department of Mineral Resources and Energy (DMRE) and its current minister, a former trade union leader of the coal industry, reinforce these sentiments by remaining focused on coal, and opposed to renewable energy efforts (AHK, KfW, GIZ 0).

---

95 Verbong and Geels. 2007. "The Ongoing Energy Transition: Lessons from a Socio-Technical, Multi-Level Analysis of the Dutch Electricity System (1960–2004)." *Energy Policy* 35 (2): 1025–37. https://doi.org/10.1016/j.enpol.2006.02.010.

96 Geels and Raven. 2007. "Socio-Cognitive Evolution and Co-Evolution in Competing Technical Trajectories: Biogas Development in Denmark (1970–2002)." *International Journal of Sustainable Development & World Ecology* 14 (1): 63–77. https://doi.org/10.1080/13504500709469708.

Despite this, South Africa faces high unemployment of over 30%, especially among young people (AHK), with gH2 offering a partial solution. Developing a gH2 economy under the principle of additionality could create projected 20,000 additional jobs by 2030 without threatening existing coal jobs (GIZ 0).[97] To establish a suitable Technical and Vocational Education and Training system, GIZ collaborates with universities, aiming to provide skilled workers for maintenance and promote human capital for innovation (GIZ 1, GIZ 0).

Furthermore, implementing international environmental and social standards is crucial to minimize local community disruption. Information dissemination by German implementing institutions intend to prevent resource conflicts (GIZ 1, KfW) and inform about job opportunities (GIZ 0). Thus, gH2 is expected to drive sustainable development independent of exports.

Cognitive Rules

Since 2008, South Africa is experiencing recurring power outages, which have stunted economic growth (AHK) and shaped the cognitive perception of energy-related issues (GIZ 1, GIZ 2, GIZ 0, AHK; KfW). The energy crisis has heightened demand for electricity from all sources, prompting companies like Anglo American to invest in mini-grid systems powered by hydrogen (GIZ 2). These market-driven solutions are more prevalent in regions severely affected by loadshedding (GIZ 2). Although the current regulatory framework lacks sufficient market incentives, the development of a hydrogen economy serves as a possible catalyst to developing a decentralized energy system more immune to power outages.

Finally, the expansion of the gH2 economy is also based on the conviction that in a CO_2-free world the market for coal will decline (GIZ 1). The cooperation in the Energy Partnership, initially represented by the disinterested DMRE, and now headed by the presidency (AHK, KfW), displays this changing mindset. GIZ has made a significant contribution to this shift during a delegation trip to Brussels. The Presidency, the Department of Science and Innovation (DSI), and the Department of Trade, Industry and Competition engaged with the topic, while the DMRE remained in the background during the visit (KfW). This has fostered increased cooperation between German institutions and the Presidency, which has provided necessary attention to the country and indicates high expectations of the hydrogen market. However, many uncertainties remain, due to the lack of projects in the final development phase and because gH2 is not yet competitive with fossil fuel alternatives, challenging investor confidence (GIZ 1, KfW).

97 Cf. DSI, "Hydrogen Society Roadmap. 2021. "Hydrogen Society Roadmap for South Africa 2021." https://www.dst.gov.za/images/South_African_Hydrogen_Society_RoadmapV1.pdf.

17.6 Conclusion and Discussion

GH2 emerged as a cornerstone of Germany's energy transition with the goals to mitigate climate change and to decarbonize its industry and mobility sector. Germany's recognition of its import dependence and the need for large-scale market development emphasize the international orientation of the NHS. With this thesis, I investigated how the perceived necessity for hydrogen imports and global market development in the NHS influences Germany's efforts in South Africa, and its implications for international EWN as a soft power resource.

The promotion of gH2 in South Africa is a continuation of the EWN as a soft power resource. Efforts to stimulate gH2 are driven by its environmental benefits and as a strategic tool for economic growth and reindustrialization – therefore creating a "win-win" narrative for both South Africa's and Germany's own energy security. The narrative is supported under soft power tactics such as the long-term cooperation measures of capacity building, provision of expertise, and dialogue to foster policies and to increase gH2s legitimacy. But given the European Green Deal and its implementation of CBAM, economic incentives also show coercive elements with regard to the development of an export market in South Africa.

These measures are leveraged against South Africa's coal dominated economy and its reliance on MEC sectors for public services and export earnings. As this regime structure is under significant economic, environmental, and social pressure for greener production methods, German agencies support the architectural reconfigurations of the MEC through the promotion of gH2 technologies. In close cooperation with the Presidency, Germany forms a private sector alliance linking MEC sectors with green technology providers. Technologically, these initiatives aid in decarbonizing hard-to-abate sectors, introducing new product lines, and fostering incremental innovations within MEC sectors. The associated changes along technological trajectories and actor-network compositions are stimulated under the principle of renewable additionality, emphasizing the potential for a decentralized energy system, job creation, as well as energy security.

However, the results also indicate that the EWN is insufficient in terms of climate targets and securing hydrogen imports. German support measures do promote the MEC's business model, especially through the establishment of export structures for the global hydrogen value chain. Nevertheless, the technological prerequisites and the necessary standards for exporting to European markets have not yet matured sufficiently. Given the structural conditions established by the EU under CBAM and instruments like H2Global, the goal of future imports is not adequately assured. Consequently, South Africa is targeting other sales markets in Asia in the medium term, thereby thwarting the establishment of a hydrogen market aligned with European climate goals.

In conclusion, the success of the international hydrogen dimension of German energy policy faces a dilemma, between sustainability and import security. To address this, future research should focus on enhancing sustainability standards and effectively designing international energy policy to address this challenge.

Bibliography

Anglo American. 2022. "Driving the Hydrogen Economy in South Africa." 2022. https://www.angloamerican.com/our-stories/innovation-and-technology/driving-the-hydrogen-economy-in-south-africa.

Ashman, Sam, Ben Fine, and Susan Newman. 2013. Systems of Accumulation and the Evolving MEC. https://repub.eur.nl/pub/40424/.

AA (Auswärtiges Amt). 2019a. "Energieaußenpolitik." Auswärtiges Amt. August 20, 2019. https://www.auswaertiges-amt.de/de/aussenpolitik/klimaaussenpolitik/energie/energieaussenpolitik/205854.

———. 2019b. "Energiewende international." Auswärtiges Amt. August 20, 2019. https://www.auswaertiges-amt.de/de/aussenpolitik/klimaaussenpolitik/energie/energiewende/238782.

———. 2019c. "Energiepartnerschaften." Auswärtiges Amt. August 20, 2019. https://www.auswaertiges-amt.de/de/aussenpolitik/klimaaussenpolitik/energie/energiepartnerschaften/238784.

Baker, Lucy, Peter Newell, and Jon Phillips. 2014. "The Political Economy of Energy Transitions: The Case of South Africa." New Political Economy 19 (6): 791–818. https://doi.org/10.1080/13563467.2013.849674.

Berggren, Christian, Thomas Magnusson, and Dedy Sushandoyo. 2015. "Transition Pathways Revisited: Established Firms as Multi-Level Actors in the Heavy Vehicle Industry." *Research Policy* 44 (5): 1017–28. https://doi.org/10.1016/j.respol.2014.11.009.

BMWi (Bundesministerium für Wirtschaft und Energie). 2020. "Die Nationale Wasserstoffstrategie." https://www.bmwi.de/Redaktion/DE/Publikationen/Energie/die-nationale-wasserstoffstrategie.html.

———. n.d . "Unsere Energiewende: sicher, sauber, bezahlbar." Accessed April 15, 2021. https://www.bmwi.de/Redaktion/DE/Dossier/energiewende.html.

BMWK (Bundesministerium für Wirtschaft und Klimaschutz). n.d.-a. "Wasserstoff: Schlüsselelement für die Energiewende." Accessed May 30, 2024. https://www.bmwk.de/Redaktion/DE/Dossier/wasserstoff.html.

———. n.d.-b. "International Hydrogen Ramp-up Programm – H2Uppp." Accessed May 30, 2024. https://www.bmwk.de/Redaktion/DE/Wasserstoff/Foerderung-International-Beispiele/10-h2uppp.html.

BMZ (Bundesministerium für wirtschaftliche Zusammenarbeit und Entwicklung). n.d. "Energie und Klima." Bundesministerium für wirtschaftliche Zusammenarbeit und Entwicklung. Accessed May 30, 2024. https://www.bmz.de/de/themen/klimawandel-und-entwicklung/energie-und-klima.

Brinkmann, Svend. 2014. "Doing Without Data." *Qualitative Inquiry* 20 (6): 720–25. https://doi.org/10.1177/1077800414530254.

Cassidy, Christopher, and Rainer Quitzow. 2023. "Green Hydrogen Development in South Africa and Namibia: Opportunities and Challenges for International Cooperation." Forschungsinstitut für Nachhaltigkeit. https://www.rifs-potsdam.de/de/ergebnisse/publikationen/2023/green-hydrogen-development-south-africa-and-namibia-opportunities-and.

DSI (Department of Science and Innovation). 2021. "Hydrogen Society Roadmap for South Africa 2021." https://www.dst.gov.za/images/South_African_Hydrogen_Society_RoadmapV1.pdf.

Eberhard, Anton. 2011. "The Future of South African Coal: Market, Investment, and Policy Challenges." Program on Energy and Sustainable Development, January. http://pesd.fsi.stanford.edu/publications/the_future_of_south_african_coal_market_investment_and_policy_challenges.

Fine, Ben, and Zavareh Rustomjee. 1996. *The Political Economy of South Africa: From Minerals-Energy Complex to Industrialisation*. C. Hurst & Co. Publishers.

Geels, F. W., and R. P. J. M. Raven. 2007. "Socio-Cognitive Evolution and Co-Evolution in Competing Technical Trajectories: Biogas Development in Denmark (1970–2002)." International Journal of Sustainable Development & World Ecology 14 (1): 63–77. https://doi.org/10.1080/13504500709469708.

Geels, Frank W. 2004. "From Sectoral Systems of Innovation to Socio-Technical Systems: Insights about Dynamics and Change from Sociology and Institutional Theory." *Research Policy* 33 (6): 897–920. https://doi.org/10.1016/j.respol.2004.01.015.

Geels, Frank W., and Johan Schot. 2007. "Typology of Sociotechnical Transition Pathways." *Research Policy* 36 (3): 399–417. https://doi.org/10.1016/j.respol.2007.01.003.

GIZ (Deutsche Gesellschaft für Internationale Zusammenarbeit (GIZ) GmbH). n.d.-a "Promoting a South African Green Hydrogen Economy (H2.SA)." Accessed May 31, 2024. https://www.giz.de/en/downloads_els/Annex%201.%20H2.SA%20Factsheet.pdf.

———. n.d.-b. "German-South African Energy Partnership." Accessed March 16, 2022. https://www.giz.de/en/worldwide/65749.html.

———. n.d.-c. "Promoting Hydrogen Projects in Developing Countries and Emerging Economies: H2Uppp." Accessed May 30, 2024. https://www.giz.de/en/worldwide/107551.html.

Goldthau, Andreas, and Nick Sitter. 2015. "Soft Power with a Hard Edge: EU Policy Tools and Energy Security." *Review of International Political Economy* 22 (5): 941–65. https://doi.org/10.1080/09692290.2015.1008547.

H2Global Stiftung. n.d. "The H2Global Instrument." Accessed May 30, 2024. https://h2-global.de/project/h2g-mechanism.

InfraCo Africa. 2023. "Memorandum of Cooperation between Transnet and GreenCo to Conduct Research and Development on the Viability of Green Hydrogen to Power Freight Trains in South Africa." 2023. https://infracoafrica.com/africa-greenco-memorandum-of-cooperation-between-transnet-and-greenco-to-conduct-research-and-development-on-the-viability-of-green-hydrogen-to-power-freight-trains-in-south-africa/.

International PtX Hub. n.d. "South Africa." PtX Hub (blog). Accessed May 30, 2024. https://ptx-hub.org/south-africa/.

Kern, Florian. 2012. "Using the Multi-Level Perspective on Socio-Technical Transitions to Assess Innovation Policy." *Technological Forecasting and Social Change, Contains Special Section: Emerging Technologies and Inequalities*, 79 (2): 298–310. https://doi.org/10.1016/j.techfore.2011.07.004.

KfW Development Bank. n.d.-a. "What Is Green Hydrogen?" Accessed May 30, 2024. https://www.kfw-entwicklungsbank.de/Our-topics/PtX/green-hydrogen/.

———. n.d.-b. "Funktionsweise PTX-Plattform | KfW Entwicklungsbank." Accessed June 2, 2024. https://www.kfw-entwicklungsbank.de/Unsere-Themen/PtX/PtX-Plattform/.

Köhler, Jonathan, Frank W. Geels, Florian Kern, Jochen Markard, Elsie Onsongo, Anna Wieczorek, Floortje Alkemade, et al. 2019. "An Agenda for Sustainability Transitions Research: State of the Art and Future Directions." *Environmental Innovation and Societal Transitions* 31 (June):1–32. https://doi.org/10.1016/j.eist.2019.01.004.

Lawrence, Andrew. 2020. "REIPPPP: Renewables' Rise, or REIPPPP RIP?" In *South Africa's Energy Transition*, edited by Andrew Lawrence, 99–114. Progressive Energy Policy. Cham: Springer International Publishing. https://doi.org/10.1007/978-3-030-18903-7_5.

Maat, Eva Pander. 2022. "Leading by Example, Ideas or Coercion? The Carbon Border Adjustment Mechanism as a Case of Hybrid EU Climate Leadership." European Papers – A Journal on Law and Integration 2022 7 (1): 55–67. https://doi.org/10.15166/2499-8249/546.

Maimele, Seutame. 2023. "Responding to the European Union's Carbon Border Adjustment Mechanism (CBAM): South Africa's Vulnerability and Responses." TIPS. https://www.tips.org.za/research-archive/sustainable-growth/green-economy-2/item/4590-responding-to-the-european-union-s-carbon-border-adjustment-mechanism-cbam-south-africa-s-vulnerability-and-responses.

Markard, Jochen, and Bernhard Truffer. 2008. "Technological Innovation Systems and the Multi-Level Perspective: Towards an Integrated Framework." Research Policy 37 (4): 596–615. https://doi.org/10.1016/j.respol.2008.01.004.

Mayring, Philipp. 2022. Qualitative Inhaltsanalyse: Grundlagen und Techniken. 13. Auflage. Weinheim Basel: Beltz.

McClelland, Angela. 2014. *South Africa and the Global Hydrogen Economy: The Strategic Role of Platinum Group Metals*. Johannesburg: Real African Publishers Pty Ltd.

Messner, Dirk, and Jennifer Morgan. 2013. "Germany Needs an Energy Transformation Foreign Policy." The Current Column of 7 January 2013. Bonn: German Development Institute / Deutsches Institut für Entwicklungspolitik (DIE).

Montmasson-Clair, Gaylor. 2015. "The Interplay between Mining and Green Economy in South Africa: An Energy Lens." SSRN Scholarly Paper. Rochester, NY. https://doi.org/10.2139/ssrn.2748019.

Morris, M., and L. Martin. 2015. "Political Economy of Climate-Relevant Policies: The Case of Renewable Energy in South Africa." IDS Evidence Report IDS Evidence Report;128. IDS/University of Cape Town. https://opendocs.ids.ac.uk/opendocs/handle/20.500.12413/5986.

National Treasury Republic of South Africa. 2022. "Media Statement-Bilateral Loan Agreements AFD and KFW." https://www.treasury.gov.za/comm_media/press/2022/2022111001%20Media%20statement-Bilateral%20Loan%20Agreements%20AFD%20and%20KFW.pdf.

NOW GmbH. n.d. "Brennstoffzelle statt Dieselgenerator – BMU-Exportinitiative Umwelttechnologien auf den Punkt gebracht." NOW GmbH (blog). Accessed June 2, 2024. https://www.now-gmbh.de/aktuelles/pressemitteilungen/brennstoffzelle-statt-dieselgenerator-bmu-exportinitiative-umwelttechnologien-auf-den-punkt-gebracht/.

Nunez, Almudena, and Rainer Quitzow. 2023. "Germany's Hydrogen Strategy: Securing Industrial Leadership in a Carbon-Neutral Economy," April. https://doi.org/10.48481/rifs.2023.010.

Nye, Joseph S. 2005. *Soft Power*. First Edition. New York: PublicAffairs.

———. 2011. *The Future of Power*. PublicAffairs.

———. 2021. "Soft Power: The Evolution of a Concept." *Journal of Political Power* 14 (1): 196–208. https://doi.org/10.1080/2158379X.2021.1879572.

Ohnesorge, Hendrik W. 2020. "A Taxonomy of Soft Power: Introducing a New Conceptual Paradigm." In *Soft Power: The Forces of Attraction in International Relations*, edited by Hendrik W. Ohnesorge, 85–225. Cham: Springer International Publishing. https://doi.org/10.1007/978-3-030-29922-4_3.

Quitzow, Rainer. 2013. "Towards an Integrated Approach to Promoting Environmental Innovation and National Competitiveness." *Innovation and Development* 3 (2): 277–96. https://doi.org/10.1080/2157930X.2013.825070.

Quitzow, Rainer, Clara Mewes, Sonja Thielges, Marina Tsoumpa, and Yana Zabanova. 2023. "Partnerschaften für eine internationale Wasserstoffwirtschaft – Ansatzpunkte für die europäische Politik." https://library.fes.de/pdf-files/a-p-b/20035.pdf.

Quitzow, Rainer, Almudena Nunez, and Adela Marian. 2024. "Positioning Germany in an International Hydrogen Economy: A Policy Review." *Energy Strategy Reviews* 53 (May):101361. https://doi.org/10.1016/j.esr.2024.101361.

Quitzow, Rainer, Sybille Röhrkasten, and Martin Jänicke. 2016. "The German Energy Transition in International Perspective." https://doi.org/10.2312/iass.2016.009.

Quitzow, Rainer, and Sonja Thielges. 2020. "The German Energy Transition as Soft Power." *Review of International Political Economy*, September, 1–26. https://doi.org/10.1080/09692290.2020.1813190.

Robinson, I., and R. Croll. 2018. "The Story of the Decline of the South African Mining Industry." *Journal of the Southern African Institute of Mining and Metallurgy* 118 (5): 0–0.

Rohlfing, Ingo. 2012. "Types of Case Studies and Case Selection." In *Case Studies and Causal Inference: An Integrative Framework*, edited by Ingo Rohlfing, 61–96. Research Methods Series. London: Palgrave Macmillan UK. https://doi.org/10.1057/9781137271327_3.

Röhrkasten, Sybille, and Kirsten Westphal. 2012. "Die IRENA: Schon Vergessen?" Stiftung Wissenschaft Und Politik (SWP), November. https://www.swp-berlin.org/en/publication/erneuerbare-energien-irena.

SASOL. 2021. "Sasol to Explore Potential of Cleaner Aviation Fuels with World Class Partners." 2021. https://www.sasol.com/media-centre/media-releases/sasol-explore-potential-cleaner-aviation-fuels-world-class-partners.

Schäfer, Dr. Martin. 2021. "Dr Martin Schafers Opening Remarks." South Africa, June 8. https://www.energize.co.za/article/report-back-third-renewable-green-hydrogen-webinar.

Schwabe, Julian. 2024. "Regime-Driven Niches and Institutional Entrepreneurs: Adding Hydrogen to Regional Energy Systems in Germany." *Energy Research & Social Science* 108 (February):103357. https://doi.org/10.1016/j.erss.2023.103357.

SIPS (South Africa Green Hydrogen Summit. n.d. "Overview of Strategic Infrastructure Projects." Green Hydrogen Summit (blog). Accessed June 2, 2024. https://greenhydrogensummit.org.za/sips/.

Steinbacher, Karoline. 2019. "Case Study: South Africa." In *Exporting the Energiewende: German Renewable Energy Leadership and Policy Transfer*, edited by Karoline Steinbacher, 239–88. Wiesbaden: Springer Fachmedien. https://doi.org/10.1007/978-3-658-22496-7_7.

Steinbacher, Karoline, and Sybille Röhrkasten. 2019. "An Outlook on Germany's International Energy Transition Policy in the Years to Come: Solid Foundations and New Challenges." *Energy Research & Social Science* 49 (March):204–8. https://doi.org/10.1016/j.erss.2018.10.013.

Tänzler, Dennis, and Stephan Wolters. 2014. "Energiewende und Außenpolitik: Gestaltungsmacht auf dem Prüfstand." Zeitschrift für Außen- und Sicherheitspolitik 7 (2): 133–43. https://doi.org/10.1007/s12399-014-0408-x.

The Presidency Republic of South Africa. 2022. "South Africa's Just Energy Transition Investment Plan (JET IP) for the Initial Period 2023–2027." https://pccommissionflo.imgix.net/uploads/images/South-Africas-Just-Energy-Transition-Investment-Plan-JET-IP-2023-2027-FINAL.pdf.

———. 2023. "JET Implementation Plan 2023-2027." https://www.stateofthenation.gov.za/assets/downloads/JET%20Implementation%20Plan%202023-2027.pdf.

Ting, Marie Blanche, and Rob Byrne. 2020. "Eskom and the Rise of Renewables: Regime-Resistance, Crisis and the Strategy of Incumbency in South Africa's Electricity System." *Energy Research & Social Science* 60 (February):101333. https://doi.org/10.1016/j.erss.2019.101333.

Unruh, Gregory C. 2000. "Understanding Carbon Lock-In." *Energy Policy* 28 (12): 817–30. https://doi.org/10.1016/S0301-4215(00)00070-7.

Verbong, Geert, and Frank Geels. 2007. "The Ongoing Energy Transition: Lessons from a Socio-Technical, Multi-Level Analysis of the Dutch Electricity System (1960–2004)." *Energy Policy* 35 (2): 1025–37. https://doi.org/10.1016/j.enpol.2006.02.010.

Westphal, Kirsten. 2012. "Globalising the German Energy Transition." Stiftung Wissenschaft Und Politik (SWP) (blog). December 10, 2012. https://www.swp-berlin.org/en/publication/globalising-the-german-energy-transition.

Yin, Robert K. 2014. *Case Study Research*. SAGE Publications.

Thies Goldner, Christoph Torwegge, Felix Jaeger

18 Supportive Systems in a Decarbonized World With Fluctuating Renewable Resources: What Has to be Done?

18.1 Introduction

In 2022 approximately 56%[1] of gross electricity production (**Bruttostromerzeugung**) in Germany was made from nuclear power, firing coal, gas and oil and other energy sources while approximately 44%[2] of gross electricity production was made from renewable energy sources. Worldwide, only approximately 28%[3] of electricity production is made from renewable energy sources. Since it has been scientifically shown[4] that the world climate is rapidly changing and that this change will have an adverse effect on the flora and fauna of our planet and on mankind in particular, the majority of the industrialised states and many other states have begun the path into a future in which mankind meets its electricity needs entirely from non-fossil energy sources. The decarbonisation of electricity production will go hand in hand with a complete decarbonisation of industry and traffic. While some states rely on nuclear power in this transition process,[5] other states, particularly Germany, have decided to shut down their existing nuclear power plants[6] and to rely entirely on renewable energy sources for their future electricity production.[7] Now, that the route to decarbonised energy production seems to have been set and renewable energies appear to be on the verge of a global breakthrough, the question arises as to whether there is still a need for substantial subsidies and support schemes, especially for electricity production from renewable energy sources. In an industrialised world and especially in market-driven economies, significant processes of change are very rarely initiated and carried forward by pure scientific knowledge. Accompanying policy measures, that are translated into legislative action, are needed to oblige business and industry actors to adapt their behaviour to more long-term sustainability goals and less to short-term profit expectations which can be achieved more easily by conventional means in the existing economic and industrial order. Sustainable long-term processes of change

1 Statistisches Bundesamt, Bruttostromerzeugung in Deutschland 7 March 2024.

2 Statistisches Bundesamt, Bruttostromerzeugung in Deutschland 7 March 2024.

3 *Pawlik*/Statista, Study: Distribution of global electricity generation by energy source in 2023, 17 October 2024.

4 IPCC, Climate Change 2022 Mitigation of Climate Change.

5 Like e.g. France: IEA, France 2021 Executive summary.

6 BASE, Der Atomausstieg in Deutschland.

7 BMWK, Wie kann das Energiesystem der Zukunft aussehen? p. 26.

https://doi.org/10.1515/9783110784299-018

in democratic societies require balanced legislative action that is characterised by a constant political will to change while the economic strength of the national economy should remain unbroken. National and, above all, international or supranational legislation therefore plays a central role in the transformation. In addition, the security of electricity supply for industry and private households must be continuously ensured. Against this background, the problem of the fluctuation of renewable electricity generation must be tackled primarily in technical terms as soon as fossil and nuclear energy sources are no longer available in sufficient quantity to cover the base load. State support measures for the expansion of renewable energy generation capacities must therefore be designed in such a way that there is room for the development and deployment of technology that has not existed to a sufficient extent to date, in order to mitigate the fluctuating nature of energy generation. In addition, the legal framework must lead to energy production from renewable sources becoming more profitable than energy production from fossil sources in the foreseeable future. It therefore follows that the decarbonisation of electricity generation will continue to require state support measures.

18.2 Overview Over Existing Support Schemes

According to the definition in Article 2 para 5 RED III directive,[8]

> 'support scheme' means any instrument, scheme or mechanism applied by a Member State, or a group of Member States, that promotes the use of energy from renewable sources by reducing the cost of that energy, increasing the price at which it can be sold, or increasing, by means of a renewable energy obligation or otherwise, the volume of such energy purchased, including but not restricted to, investment aid, tax exemptions or reductions, tax refunds, renewable energy obligation support schemes including those using green certificates, and direct price support schemes including feed-in tariffs and sliding or fixed premium payments;

This shows various options for promoting renewable energy without defining the path to follow.[9] The choice between the promotional instruments lies within the discretion of each Member State. In line with Article 4 RED III directive, Member States are obliged to allocate support for electricity from renewable sources in an open, transparent, competitive, non-discriminatory and cost-effective manner, with exemptions for small-scale installations and demonstration projects as well as peripheral regions and small islands.[10]

8 Directive (EU) 2018/2001 of the European Parliament and of the Council of 11 December 2018 on the promotion of the use of energy from renewable sources.

9 For the following, see *Goldner/Torwegge*, The Legal Framework of Promoting Renewable Energies – a Cross-National Study, in: Boettcher, Green Banking, 2019, section 5.3.2 / pages 114 et. seq.

10 European Commission, Support schemes for renewable energy.

Support schemes can be grouped into (1) price-driven mechanisms, such as feed-in tariffs and feed-in premiums, (2) volume-driven mechanisms, such as quota obligations and auctions/tenders, (3) direct financing, such as investment grants/subsidies and "soft loans", and (4) tax incentives.

18.2.1 Price-Driven Mechanisms (Feed-in Tariffs and Premiums)

In a **feed-in tariff** regime, the electricity generated from renewable energies is rewarded with statutory fixed revenues which are guaranteed for a certain period. The costs arising due to the difference between the level of the tariff and the market price for electricity are borne either by the taxpayer or the electricity consumer. The feed-in tariff decouples the revenue of renewable energy from the market price and is often geared towards the chosen form of production of renewable energies. Most of the designs have a degressive approach, which means that funds for new renewable energy plants decrease over time. Due to the stable and predictable revenue, investment security is ensured for plant operators and investors. As all renewable energy technologies are subject to promotion, the plant operators do not face competition and do not have to offer their energy on competitive terms. Therefore, a cross-technology expansion is encouraged. In this way, all renewable energy technologies are supported in the same way. Hence, it is possible to expand technologies that would not be economically viable without promotion. The aim is to promote technologies until they can operate independently of any government support. This will ensure green electricity production by reaping the benefits from diversification of technologies as well as decentralisation.

However, a disadvantage of feed-in tariffs arises from the investment in all technology forms (not only the most economical) of plant operators under the promotion system. This causes a higher cost structure to develop compared to other promotion systems, for example, under a quota system. Through the stable and predictable promotion of renewable energy plants the operators act unbundled from the market and market laws. This can result in higher electricity prices for consumers and a market for renewable energies that is unbundled and unadjusted to the remaining electricity market.

Another variation of a price-driven mechanism is a **feed-in premium**. The producer of renewable energies sells the energy on the spot market or directly to a third party. In addition to the received market price a premium is paid. This premium is usually designed as a fixed or a sliding premium. Under a fixed premium scheme the plant operator is granted a premium which stays the same over a period of time, i. e. the premium is independent of the revenues generated at the market. The sliding premium is also granted on top of the market price, however in contrast to the fixed premium, the sliding premium depends on the achieved market price. The premium is calculated as the difference between the average wholesale level and a predetermined reference tariff level. How the reference tariff level is set depends.

Feed-in premiums bear advantages in comparison to feed-in tariffs. Feed-in premiums follow a more market-based approach: They incentivize to produce energy when demand is high or production from other sources is low. As a result, renewable energy sources are better integrated into the existing electricity market and energy supply and demand matches more efficiently. This can also help grid management by easing pre-existing grid bottlenecks. At last, those schemes establish competition among electricity producers from which the consumers can benefit. Nonetheless, feed-in premiums also bear disadvantages. The biggest disadvantage is that they are not equally suited as a promotion mechanism for all forms of renewable energy production. Under a feed-in premium, the control of energy supply and the control of production gain importance when attempting to maximise revenue by supplying and producing in times of high market prices, for example, when demand is high or production from other sources is low. However, for example for wind and solar energy, this control of supply is not possible as these energy sources are dependent on environmental factors. So additional costs for the procurement of balancing services will arise. Consequently, it is rather unlikely that wind and solar energy are able to adapt the market price signals cost-effectively.

Price-driven mechanisms are the most prevalent support systems in the EU. In 2020/2021, only three EU Member States did not use feed-in tariffs or premiums as a support system.[11]

18.2.2 Volume-Driven Mechanisms (Quota Obligations and Auction/Tender)

In the **quota model**, the national government determines a fixed share of renewable energies that must be generated and sold by producers and grid operators over a specified period of time. The adherence to the quota is ensured by green electricity certificates. For each produced unit of green energy, a tradable green energy certificate is issued. By purchasing those certificates, the quota obligated consumers are able to prove the adherence to the determined quota. Non-compliance with the quota will be punished by a fine or other sanctions. A volume model opens the possibility of achieving the expansion goals of renewable energies precisely. In this way, the society is protected from unnecessarily high costs caused by an excessive expansion of renewable energies and the statutory climate targets can be achieved more easily. Due to the absence of a statutory feed-in tariff, plant operators are forced to sell the elec-

11 CEER – Status Review of Renewable Support Schemes in Europe for 2020 and 2021, CEER report 2023 p. 12 et seq.

tricity on the market at competitive prices and conditions. In addition to the revenue generated from the traded electricity, they also receive the revenue from the certificates, which are also traded on competitive terms. The market for renewable energy, therefore, follows market economy rules. For individual plant operators to be able to survive in this market, they must realise cost reduction potentials, whereby only the most cost-effective plants and the best locations are used which leads to cheaper electricity from which consumers benefit.

Volume-driven support mechanisms also have disadvantages. The absence of statutory premiums and tariffs reduces the planning and investment security for plant operators. It is therefore reasonable to assume that price-driven mechanisms encourage a greater expansion of renewable energies. Also, the need of plant operators to choose the most favourable form of renewable energy production has an impact on the diversification of renewable energy. In Germany, a price-driven approach has led to a diverse range of renewable energy plants (such as wind, solar, biomass, water and offshore wind). Under a volume-driven approach onshore wind, the cheapest production form concentrated on certain regions, would most likely be the only expanded production form. As a result, advantages arising from a decentralised and diversified renewable energy production would remain unused.

After developing nearly equally for a long time, in the recent past, a worldwide increase of price-driven schemes and a decrease of volume driven schemes has been observed.[12] An upcoming development in EU and non-EU countries are auction/tender schemes. The outcome of the tender is the feed-in tariff, the reference value for the feed-in premium (e. g. in Germany), or, alternatively, the basis for a capacity payment per installed kW.[13] The majority of the tender schemes in Europe and also worldwide follow a **static sealed bid** approach.[14] In a static sealed bidding procedure each participant makes one bid, while no information regarding the price or the auctioned product are exchanged between the participants. The best bid wins. This might lead to the scenario that the bid winner undervalues the true project's value, therefore underbids but wins the bid and remains with an unprofitable project. However, due to the low transaction costs and simple execution they remain rather attractive.[15] The Netherlands for example use a different tender procedure, containing a dynamic (ascending) auction procedure. This form of bidding process goes through several rounds. In each round of the bidding process, the auctioneer suggests a lower price than in the previous round, while the bidders make their offer which states the quantity they are willing to produce at the suggested price. The bidding process ends when requested quantities and supplied quantities match. The main benefit of this bidding process is its transparency. Better pricing is achieved as the bidders can adjust their

12 Institut der deutschen Wirtschaft Köln – Einspeisetarife vs. Quotensysteme 2011.

13 CEER – Tendering procedures for RES in Europe: State of play and first lesson learnt 2018 p. 9.

14 Auctions for Renewable Support: Lessons Learnt from International Experiences p. 11.

15 EWEA – Design options for wind energy tenders 2015 p. 7.

bids. However, practical experience shows that energy suppliers usually do not adjust their bids during a bidding process.[16]

By the end of 2017 a tender scheme was already implemented by 13 European states, among them Belgium, Denmark and Greece.[17] This development is due to the European Commission State Aid Guidelines for Environmental Protection and Energy which state that Member States searching for state-aid clearance shall set up a competitive bidding process for the support of all new plants.[18]

Considering the rapid spread of tendering procedures and the demands of the European Commission to set up competitive bidding process to achieve state-aid clearance in their guidelines, tendering procedures may be put forward as the future promotion scheme, at least in Europe. As they allow the control of support costs and volume distribution, this is a welcome development.

18.2.3 Direct Financing (Investment Grants/Subsidies and Soft Loans)

Some countries give **grants** or subsidise renewable energy projects under certain conditions in addition to promotion schemes. In Finland, for example, under the so-called energy aid, grants are given for renewable energy production facilities and related research projects. The granted projects must promote the use or production of renewable energies, advance energy efficiency and the environmental effects caused by energy production and use. The basic condition for funding is that at least 25% of the project funding must come from non-governmental sources. Furthermore, investment aid for renewable energy and new renewable technologies exists, which can be granted against a fixed assets investment.

Another approach to promoting the diffusion of renewable energies is finance assistance through **soft loans**. An example for this is, the joint project facility of the International Renewable Energy Agency (IRENA) and the Abu Dhabi Fund for Development (ADFD). The project aims to support replicable, scalable, potentially transformable renewable energy projects in developing countries by providing soft loans. Further prerequisites for funding are a positive development impact, the improvement of energy access, addressing energy security and governmental support of the project.[19]

16 IRENA – Renewable Energy Auctions: A Guide to Design 2015 p. 12.

17 CEER – Tendering procedures for RES in Europe: State of play and first lesson learnt 2018 p. 10.

18 Auctions for Renewable Support: Lessons Learnt from International Experiences p. 6.

19 IRENA – Accessible Finance for Renewable Energy Projects in Developing Countries.

18.2.4 Tax Incentives

The promotion of renewable energies can also take the form of **tax incentives**, as found in the United States. Tax credits are granted, beside others, through a so-called Production Tax Credit (PTC). An inflation-adjusted per kWh tax credit for electricity generated by qualified energy resources and sold to a third party is granted to the taxpayer for ten years after the facility commenced production. Besides tax credits, certain tangible property might also be recovered for tax purposes through annual deductions in the Modified Accelerated Recovery System Depreciation Schedule (MARCS). Under MARCS a life time is set for various types of property, during which they may be depreciated. In this way, the property owner has the possibility to deduct his tax basis during the depreciation period. Further tax incentives can be found in Sweden. Land owners pay a reduced real estate tax in case a wind energy plant is constructed on the property. Furthermore, a reduced energy consumption tax applies to electricity produced in generators with a capacity lower than 50 kW. Electricity production from wind, wave and solar is rewarded with a higher capacity. In addition, micro producers of electricity from renewable energy are rewarded with a tax reduction on the basis of the kWh fed into the grid at the grid connection point during the calendar year.

18.3 The Need for Re Support Schemes in the Future

18.3.1 Commercial Background

The production of electricity from renewable energy sources is cheaper than from any fossil fuel or nuclear based source. It is even expected that levelized costs of electricity from renewable energy sources will decline even further.[20] Also, while grid parity has already been achieved in Germany in the fields of electricity from solar power and onshore wind, this is still pending for other sources.

However, when looking at both the costs of generating electricity from renewable energy sources and the consumer price, an apparent contradict emerges. On the one hand, the levelized costs of electricity from renewable energy sources are the cheapest in comparison with any source of electricity from fossil fuels or nuclear power. Nevertheless, the price to be paid by the consumers is much higher due to the merit order effect, where the consumer price is set by the most expensive source of electricity to meet the anticipated demand, which is usually based on fossil fuels and therefore more expensive.

20 Fraunhofer Institute for Solar Energy Systems – Levelized Cost of Electricity – Renewable Energy Technologies, June 2021, page 28 et seq. See also 2.1.

The consumer price is also significantly affected by network charges, taxes and other levies. In the year 2023, consumers in Germany paid about 45.19 (Euro-)Ct/kWh, of which 9.35 ct/kWh was network charge and 12.25 ct/kWh was for taxes and levies – totalling about 47.8% of the overall consumer price on average.[21]

Moreover, during the last years, the German electricity market experienced a time of crisis. After electricity prices had already risen on the German market from September 2021, the trend increased with the Russian invasion in Ukraine in February 2022 before reaching a peak around September 2022, when Russia terminated the supply of gas to the German and European markets entirely.

This correlation between rising gas prices due to a shortage of supply and rising prices for electricity is no coincidence on German markets. This is due to the fact that gas-fired power plants are often price-setting in the merit order of electricity prices. Therefore, a strong increase of gas prices usually also leads to rising prices for electricity.[22]

During the months that followed the end of the year 2022, prices for electricity declined due to various reasons. On the one hand, consumer demands dropped significantly year-on-year. On the other hand, increased producing capacities also enhanced the supply of electricity.[23] Furthermore, by the end of the year 2022, the German parliament passed the "gas and electricity price brake bill", which supported private households to bear the rising costs of electricity and gas supply.

18.3.2 Expansion of Reduction Targets Until 2050

A central benchmark for national legislators regarding decarbonisation are binding targets for the reduction of CO_2 emissions. In effect, these targets practically form the external framework for further concrete measures and, in particular, for introducing national policies and structuring the content of support schemes for renewable energies. The existence of binding obligations for the reduction of greenhouse gas emissions on an international and supra-national level is a key element for national legislators because the domestic and foreign policy hurdles to unilaterally terminating such commitments are high, so that the inter- and supranational commitments have a remarkable incentive effect. Such inter- and supranational commitments can be found in the following sources:

21 See: German Federal Network Agency – Monitoring Report 2023, Average Household Customer Price 2023.

22 See also the annual report of the German Federal Network Agency on the German electricity market 2022, 2 January 2023, SMARD | Der Strommarkt im Jahr 2022.

23 See also the annual report of the German Federal Network Agency on the German electricity market 2022, 2 January 2023, SMARD | Der Strommarkt im Jahr 2022.

18.3.3 International Emission Reduction Targets

The UN Convention on Climate Change

The United Nations first explicitly formulated the limitation of greenhouse gases in the atmosphere as a goal in the **Framework Convention on Climate Change** (United Nations Framework Convention on Climate Change –"**UNFCCC**") of 9 May 1992, which entered into force on 21 March 1994 and is today binding for 197 parties.[24] The objective of the UNFCCC was the stabilisation of greenhouse gas concentrations in the atmosphere at a level that would prevent dangerous anthropogenic interference with the climate system.[25] Such a level should be achieved within a time frame sufficient to allow ecosystems to adapt naturally to climate change, to ensure that food production is not threatened and to enable economic development to proceed in a sustainable manner.[26] The UNFCCC established the Conference of the Parties ("**COP**"), which convenes annually since 1995 and keeps under regular review the implementation of the UNFCCC and any related legal instruments that the Conference of the Parties may adopt. Furthermore, the COP makes, within its mandate, the decisions necessary to promote the effective implementation of the UNFCCC.[27] Within the UNFCCC, it was recognised that the parties had common but differentiated responsibilities and their own specific national and regional development priorities, objectives and circumstances. Within this context, the parties committed themselves, among other things, to take climate change considerations into account, to the extent feasible, in their relevant social, economic and environmental policies and actions, and employ appropriate methods, for example impact assessments, formulated and determined nationally, with a view to minimising adverse effects on the economy, on public health and on the quality of the environment, of projects or measures undertaken by them to mitigate or adapt to climate change.[28] Each party shall adopt national policies and take corresponding measures on the mitigation of climate change, by limiting its anthropogenic emissions of greenhouse gases and protecting and enhancing its greenhouse gas sinks and reservoirs.[29] Although the UNFCCC did not define certain percentages of emission reduction goals for all its parties, it must be considered as the basis for far-reaching future developments at an international level. It formed the basis for binding long-term obligations of the parties to implement and adopt policies on a national level.

24 United Nations Framework Convention on Climate Change (UNFCCC), BGBl. 1993, p. 1784.
25 Article 2 UNFCCC.
26 Article 2 UNFCCC.
27 Article 7 para 2 UNFCCC.
28 Article 4 para 1 lit. f) UNFCCC.
29 Article 4 para 2 lit. a) UNFCCC.

The Kyoto Protocol

On 11 December 1997, the third session of the Conference of the Parties resolved on the **Kyoto Protocol** to the UNFCCC which entered into force on 16 February 2005.[30] 192 states are currently parties to the Kyoto Protocol, including all Member States of the European Union, the European Union itself and important developing countries, such as Brazil, China and India.[31] The United States of America did not ratify the Kyoto Protocol. Canada terminated the Kyoto Protocol with effect from 15 December 2012.[32] Within the Kyoto Protocol, both developed countries and developing economies committed themselves to limit and reduce greenhouse gases emissions.[33] Individual reduction targets were agreed upon for single states.[34] The Kyoto Protocol primarily addresses developed countries as these countries are responsible for the majority of greenhouse gas emissions worldwide. The lead principle is the principle of "common but differentiated responsibility and respective capabilities". For 37 developed countries and developing economies and the European Union, the Kyoto Protocol provides for binding emission reduction targets.[35] The reduction target was in total an average of 5.2%[36] emission reduction compared to the emission level of the base year 1990 within the first quantified emission limitation and reduction commitment period between 2008 to 2012.[37] Within this period, the developed countries committed themselves to a reduction of greenhouse gas emissions by 5.2% in total.[38] Under the protocol, the EU and its Member States are considered as an "emissions community".[39] Within the community, the Member States have set new reduction quotas. These deviate considerably from the reduction targets set in the Kyoto Protocol and even provide for an increase in emissions for some states. For example, Portugal was allowed to increase its emissions by 27%,[40]

30 UN, 7. a Kyoto Protocol to the United Nations Framework Convention on Climate Change.

31 UNFCCC, What is the Kyoto Protocol.

32 See *Ludwig* in Säcker/Ludwigs, Berliner Kommentar zum Energierecht, Volume 3 5th edition 2022, Chapter A, no 17.

33 Article 3 Kyoto Protocol.

34 *Kay*/PoA, Terms and Impacts of the Kyoto Protocol.

35 See Kyoto Protocol, Annex B; See Article 4 Kyoto Protocol.

36 EU and all Member States: reduction target of 8%; Japan 6%.

37 Article 3 para 1 and 7 of the Kyoto Protocol.

38 *Kay*/PoA, Terms and Impacts of the Kyoto Protocol.

39 See Annex B Kyoto Protocol.

40 *Borrego/Martins/Lopes*, Portuguese industry and the EU trade emissions directive: development and analysis of CO_2 emission scenarios, p. 75.

while Denmark and Germany have committed themselves to a reduction of 21%[41] each and Luxembourg to a reduction of 28%.[42]

In addition to the definition of binding emission reduction targets, the Kyoto Protocol established several instruments to meet emission reduction targets by way of market-based mechanisms:

Article 4 of the Kyoto Protocol provides for the states' possibility of a regional economic integration organisation joining together to form an emissions community (so-called "**Joint Fulfilment**"). As a result of such an association, when calculating the reduction of trace gases (CO_2, CH_4 and N_2O), it was no longer the individual contracting state that is decisive, but only the emissions community as a whole, so that if one state fell short of its reduction obligations, this can be compensated by the surplus results of other states.[43]

Article 6 of the Kyoto Protocol implements a trade mechanism for emission reduction units between developed countries. Any of these can use the emission reduction units that it has acquired from other parties to the protocol for meeting its own commitments under the Kyoto Protocol. Pursuant to Article 6, the emission reduction units must result from projects aimed at reducing anthropogenic emissions by sources or enhancing anthropogenic removals by sinks of greenhouse gases in any sector of the economy, provided that: (a) any such project has the approval of the Parties involved; (b) any such project provides a reduction in emissions by sources, or an enhancement of removals by sinks, that is additional to any that would otherwise occur; (c) it does not acquire any emission reduction units if it is not in compliance with its obligations under Articles 5 and 7; and (d) the acquisition of emission reduction units shall be supplemental to domestic actions for the purposes of meeting commitments under Article 3. Because Article 6 provides for a mechanism by which one country can make use of emission reduction units that were created based on projects in another country, this mechanism is called "**Joint Implementation**".

Pursuant to Article 12 of the protocol, it is also possible for developing countries to trade emission reduction units that were created by projects within these developing countries. This so called "**Clean Development Mechanism**" – **CDM** – aims to boost the necessary technology and financial transfer from developed countries to developing countries.

The most famous instrument that has been acknowledged by the Kyoto Protocol is the instrument of international emissions trading pursuant to Article 17 of the Kyoto Protocol. The trading good under this mechanism are "**Assigned Amount**

---

41 Denmark e.g.: Danish Ministry of the Environment, The Kingdom of Denmark's Report on Assigned Amount – under the Kyoto Protocol, p. 3.

42 Luxembourgian Ministry of Sustainable Development and Infrastructure – Department of the Environment, Seventh National Communication of Luxembourg under the United Nations Framework Convention on Climate Change.

43 *Kloepfer*, in: Umweltrecht, 4th edition 2016, Section 17 no. 83.

Units"– AAUs – that can be traded between developed countries in particular. The total sum of the AAUs represents the allowed emissions divided into quotas and assigned to the relevant states. If AAUs are not used by the relevant country, it may sell the respective excess capacity to countries that have exceeded their emission reduction targets. The United Nations point out that by implementing the instrument of international emissions trading, emission reductions or removals were effectively transformed into a new commodity. The so-called "carbon market" was born.[44]

After several unsuccessful attempts to agree on a successor agreement to the Kyoto Protocol, the COP 18 conference in Doha on 8 December 2012 agreed on the so-called "**Doha Amendment**" or "**Kyoto-II**".[45] This (partial) prolongation of the Kyoto protocol entered into force not before 31 December 2020 after 148 parties accepted and ratified its terms.[46] It provided for a second commitment period from 2013 and until 2020.[47]

The Paris Agreement

The **Paris Agreement** can be considered as the successor of the Kyoto Protocol. It was adopted by 195 parties at COP 21 in Paris on 12 December 2015 and entered into force on 4 November 2016.[48] Even China and Canada participate, and US-President Biden has reversed the United States termination under US-President Trump in 2019.[49] Meanwhile, 192 states have ratified the Paris Agreement.[50] The core objectives of the Paris Agreement are described in Article 2 as follows: (a) Holding the increase in the global average temperature to well below 2°C above pre-industrial levels and pursuing efforts to limit the temperature increase to 1.5°C above pre-industrial levels, recognising that this would significantly reduce the risks and impacts of climate change; (b) Increasing the ability to adapt to the adverse impacts of climate change and foster climate resilience and low greenhouse gas emissions development, in a manner that does not threaten food production; and (c) Making finance flows consistent with a pathway towards low greenhouse gas emissions and climate-resilient development.

The reference year for the pre-industrial levels is the year 1750.[51] It should be noted, however, that the 1.5°C target of the Paris Agreement is probably more sym-

44 UNFCCC, Emissions Trading.

45 UNFCCC, The Doha Amendment.

46 UNFCCC, The Doha Amendment.

47 See Article 1 lit. F. Doha amendment to the Kyoto Protocol.

48 UNFCCC, What is the Paris Agreement?; *Böhringer*, ZaöRV 2016, 753 et seq.

49 *Denchak*, Paris Climate Agreement: Everything you Need to Know, February 19th, 2021, Natural Resources Defense Council.

50 *Ludwigs*, in: Säcker/Ludwigs, Berliner Kommentar zum Energierecht, Volume 3 5th edition 2022, Chapter A, no. 18.

51 *Hakenberg*, in: Weber, Rechtswörterbuch, 28. edition 2022, reference: Pariser Klimaabkommen.

bolic, as global CO_2 emissions would have to be reduced to zero by around 2050 in order to meet the 1.5°C target.[52]

In contrast to the Kyoto Protocol, the Paris Agreement addressed developed countries as well as developing countries. The Paris Agreement does not prescribe binding instruments or measures to be taken by the countries and does not set forth binding individual emission reduction targets for each country. To achieve its targets, the Paris Agreement pursues a so-called **bottom-up approach**. In contrast to the Kyoto Protocol, where concrete individual emission reduction commitments for each country were defined, so-called Nationally Determined Contributions ("**NDCs**") have been implemented under the Paris Agreement.[53] Each country successively develops the NDCs it wishes to achieve itself. For this purpose, the countries are to pursue domestic mitigation measures with the aim of achieving the objectives of such contributions, Article 4 para 2 of the Paris Agreement. All NDCs submitted shall be recorded in a public register, Article 4. As of today, 194 NDCs are listed in this publicly viewable register.[54]

The European Commission (the "**Commission**") updated its NDC on 17 December 2020 and stated that with a combination of measures the EU will deliver by 2030 an at least 40% reduction in greenhouse gas emissions as compared to 1990 levels.[55] The United States is setting an economy-wide target of reducing its net greenhouse gas emissions by 50–52% below 2005 levels in 2030.[56] Canada's updated NDC is to reduce emissions by 40–45% below 2005 levels by 2030.[57]

Under the Paris Agreement, the instruments implemented by the **Kyoto Protocol** (Joint Implementation, Joint Fulfilment and Clean Development Mechanism) were not continued with the same content. The idea of international cooperation and the goal of a carbon market were only continued in a general form and with a different focus. The Paris Agreement, for example, implemented a general principle according to which the use of internationally transferred mitigation outcomes to achieve nationally determined contributions shall be voluntary and authorised by the participating parties, Article 6.1 of the Paris Agreement. Furthermore, a mechanism was implemented which aims to contribute to the mitigation of greenhouse gas emissions and support sustainable development under the authority and guidance of the Conference

---

52 Rogelj, J., D. Shindell et al., 2018: Mitigation Pathways Compatible with 1.5°C in the Context of Sustainable Development. In: Global Warming of 1.5°C. An IPCC Special Report on the impacts of global warming of 1.5°C above pre-industrial levels and related global greenhouse gas emission pathways, in the context of strengthening the global response to the threat of climate change, sustainable development, and efforts to eradicate poverty.

53 See Article 4 Paris Agreement.

54 UNFCCC, NDC Registry, https://unfccc.int/NDCREG; dated 22 June 2022.

55 UNFCCC, Submission by Germany and the European Commission on behalf of the European Union and its Member States, p. 5.

56 UNFCCC, The United States of America National Determined Contribution, p.1.

57 UNFCCC, Canada's Nationally Determined Contribution Under The Paris Agreement, p.1.

of the Parties, Article 6.4 of the Paris Agreement. The instrument aims (a) To promote the mitigation of greenhouse gas emissions while fostering sustainable development; (b) To incentivise and facilitate participation in the mitigation of greenhouse gas emissions by public and private entities authorised by a party; (c) To contribute to the reduction of emission levels in the host party, which will benefit from mitigation activities resulting in emission reductions that can also be used by another party to fulfil its nationally determined contribution; and (d) To deliver an overall mitigation in global emissions. Emission reductions resulting from this mechanism shall not be used to demonstrate achievement of the host party's nationally determined contribution if used by another party to demonstrate achievement of its NDC. The parties to the Paris Agreement recognised in Article 6.8 the importance of integrated, holistic and balanced non-market approaches being available to parties to assist in the implementation of their nationally determined contributions, in the context of sustainable development and poverty eradication, in a coordinated and effective manner, including through, among other things, mitigation, adaptation, finance, technology transfer and capacity building, as appropriate.

It is evident that the exact guidance, rules, modalities, and procedures for engagement through Article 6 remain the subject of ongoing negotiations of the Conference of the Parties.[58] Finally, after six years of negotiations, the UN climate talks at COP 26, which took place in late 2021 in Glasgow, produced a rulebook as part of the Glasgow Climate Pact for international cooperation through carbon markets and non-market approaches enabling stronger cooperation between countries on mitigation and adaptation.[59]

EU-Wide Emission Reduction or Expansion Targets for Renewable Energies

Against the background of its international commitments to reduce greenhouse gas emissions, the EU has adopted several directives setting out binding obligations for its Member States.

The RED I Directive

With its Directive 2009/28/EC of the European Parliament and of the Council of 23 April 2009 on the promotion of the use of energy from renewable sources (**"RED I Directive"**), the European Union (**"EU"**) implemented binding obligations on its Member States to promote the generation of energy from renewable energy sour-

58 The Bonn Climate conference shows further efforts to shape Article 6: Al Amer, Bonn Climate Conference Outcomes, a Series on the role of CCS at COP 27 part 1, p.6 Global CCS Institute; UNFCCC, Rules, modalities and procedures for the mechanism established by Art 6, paragraph 4, of the Paris Agreement and referred to in decision 3/CMA.3.

59 UNFCCC, COP26 Outcomes: Market mechanisms and non-market approaches (Article 6).

ces.[60] A directive is generally the most suitable legislative instrument of the EU for considering the existing differences between the legal systems of the Member States.[61] Pursuant to Article 288 para 3 of the Treaty on the Functioning of the EU ("**TFEU**"), a directive shall be binding, as to the result to be achieved, upon each Member State to which it is addressed, but shall leave to the national authorities the choice of form and methods. Directives must be transformed into national law by the national legislators within the transformation period which is set out within the directive. However, the European Court of Justice ("**ECJ**") ruled that individuals may, in the absence of implementing measures adopted in due time, also directly rely on provisions of a directive that appear to be unconditional and sufficiently precise as to their content against all national provisions which do not comply with the directive.[62] Pursuant to Article 3 para 1 of the RED I Directive, each Member State shall ensure that the share of energy from renewable sources in gross final consumption of energy in 2020 is at least its national overall target for the share of energy from renewable sources in that year.[63] Such mandatory national overall targets are consistent with a target of at least a 20% share of energy from renewable sources in the Community's gross final consumption of energy in 2020. These targets for the share of energy from renewable sources in gross final consumption of energy in 2020 were for example: Germany 18%, Spain 20%, France 23%, Latvia 40%, Austria 34%, Finland 38%, Sweden 49% and the United Kingdom 15%. The Member States had until 5 December 2010 to transform the RED I Directive into national law. These targets constitute indirect emission reduction targets as they are related to the share of energy from renewable sources in gross final consumption. The higher the share of energy from renewable sources, the lower the share of energy from fossil sources and, thus, the lower the overall emission of greenhouse gases resulting from energy generation from fossil sources.

The RED II Directive

By the Directive (EU) 2018/2001 of the European Parliament and of the Council of 11 December 2018 ("**RED II Directive**") on the promotion of the use of energy from renewable sources, the Member States were collectively called to adjust their national legal framework on the promotion of energy from renewable sources in a manner that their share in the European Union's gross final consumption of energy in 2030 is at

60 The complete name is DIRECTIVE 2009/28/EC of the European Parliament and of the Council of 23 April 2009 on the promotion of the use of energy from renewable sources and amending and subsequently repealing Directives 2001/77/EC and 2003/30/EC.

61 *Scherer/Heselhaus*, in: Dauses/Ludwigs, Handbuch des EU-Wirtschaftsrechts, 55. EL, January 2022, Chapter O, no. 178.

62 ECJ, C-33/70, 17 December 1970 ("SACE").

63 The national overall target for the national share of energy from renewable sources is set out in the third column of the table in part A of Annex I of the RED I Directive.

least 32%.[64] This share could even be increased because Article 3 para 1 sentence 2 RED II Directive provided the following: The Commission should assess that target with a view to submitting a legislative proposal by 2023 to increase the share where there are further substantial cost reductions in the production of renewable energy, where needed to meet the European Union's international commitments for decarbonisation, or where a significant decrease in energy consumption in the European Union justifies such an increase. The Member States were to bring into force the laws, regulations, and administrative provisions necessary to comply with these goals by 30 June 2021.

The EU Green Deal and the EU Climate Law 2021

In the communication from the Commission to the European Parliament, the Council, the European Economic Committee and the Committee of the Regions (COM(2019) 640 final) dated 11 December 2019 the "The European Green Deal" was made public as a new growth strategy that aims to transform the EU into a fair and prosperous society, with a modern, resource-efficient and competitive economy where there are no net emissions of greenhouse gases in 2050 and where economic growth is decoupled from resource use. It was announced that by summer 2020 the Commission wanted to present an impact assessed plan to increase the EU's greenhouse gas emission reductions target for 2030 to at least 50% and towards 55% compared with 1990 levels. To deliver these additional greenhouse gas emissions reductions, the Commission announced to review and propose to revise, if necessary, all relevant climate-related policy by June 2021. The Green Deal contains a comprehensive timetable of legislative initiatives and measures for many sectors, among others, industry, mobility, energy and finance. The Green Deal was implemented through the European Climate Law,[65] which was passed by the European Parliament and the Council as a regulation. According to Article 288 para 2 TFEU, a regulation shall have general application. It shall be binding in its entirety and directly applicable in all Member States. Thus, the European Climate Law made the European Union's political climate change targets and commitments legally binding, particularly the climate-neutrality objective to be achieved latest by the year 2050, Article 2 para 1 EU Climate Law. Article 4 para 1 of the EU Climate Law sets an interim goal: binding EU-wide 2030 climate target shall be a domestic reduction of net greenhouse gas emissions (emissions after deduction of removals) by at least 55% compared to 1990 levels by 2030.

64 Article 3 para 1 RED II.

65 Regulation (EU) 2021/1119 of the European Parliament and the Council of 30 June 2021 establishing the framework for achieving climate neutrality and amending Regulations (EC) No 401/2009 and (EU) 2018/1999 ("European Climate Law").

The RED III Directive

On 14 July 2021 the Commission issued a communication regarding a proposal amending, among others, the RED II Directive.[66] In its proposal, the Commission stated that a significantly higher share of renewable energy sources in an integrated energy system is required to achieve the objectives of the EU Climate Law. The current EU target of at least 32% renewable energy by 2030, set in the RED II Directive, was not sufficient and needed to be increased from 38% to 40%, according to the Climate Target Plan ("**CTP**") which was already published as a communication on 17 September 2020.[67]

18.3.4 The Increasing Need for Electricity from Re/"Transformation"

In a decarbonised world, energy supply will rely heavily on electricity instead of fossil fuels. However, renewable energy sources not only have to replace energy from fossil fuels, but most likely must also meet an even higher demand for electricity. The expected increase in demand of electricity is mainly driven by a few certain factors, which must be considered when calculating future consumption of electricity.

One of the most important factors lays within the transportation sector. It is expected that the production and demand for electrical vehicles as well as hybrids will increase drastically. This will also lead to an increasing demand for electricity due to new battery factories. This shift from a fossil-based transportation to traffic driven by electricity will not only include cars, but also busses and motorcycles. Furthermore, due to a necessary and planned support of rail transport, also the demand in electricity of the rail sector will most likely increase.[68]

One other important aspect is the production of hydrogen through the electrolysis of water. It is estimated that up to 80% of the produced so called "green hydrogen" will be

66 Proposal for a Directive of the European Parliament and of the Council amending Directive (EU) 2018/2001 of the European Parliament and of the Council, Regulation (EU) 2018/1999 of the European Parliament and of the Council and Directive 98/70/EC of the European Parliament and of the Council as regards the promotion of energy from renewable sources and repealing Council Directive (EU) 2015/652.

67 Communication from the Commission to the European Parliament, the Council, the European Economic and Social Committee and the Committee of the Regions "Stepping up Europe's 2030 climate ambition", COM (2020 562 final).

68 See *Kemmler/Wünsch/Burret*, Short paper on the development of the total electricity consumption until 2030, commissioned by the German Federal Ministry for Economic Affairs and Energy, 21 October 2021.

consumed by the industry, transportation and energy sectors.[69] The production of hydrogen is a central keystone of the German strategy to obtaining a carbon free economy.[70]

Moreover, especially in private households, electrical heat pumps are replacing common fossil-based heaters in growing numbers, which also leads to an increasing demand in electricity.[71]

Outlook to Electricity Price Development in the Future

Nowadays, prices for electricity are stabilizing, but nevertheless remain high today in comparison to the time before 2021. It is also not expected that prices will decline much further. Instead, a study commissioned by the German government suggests that prices will remain merely static only for a short period of time before rising again.[72] Another study commissioned by the Association of the Bavarian Economy from the year 2023 also suggests that the prices for electricity are likely to remain stable, but nevertheless at a high level.[73]

Several aspects are essential for the assumption of higher future electricity prices.

Firstly, it is expected that gas prices will remain on a high level or even increase due to the shortage of supply caused by the Russian war in Ukraine. Therefore, also electricity prices are likely to remain high due to the already mentioned correlation of both prices.[74]

Secondly, the demand for electricity is likely to increase significantly during the following years due to an anticipated exit on fossil fuels, an increasing electrical mobility and the increasing production of hydrogen by hydrolysis of water.[75]

---

69 See *Kemmler/Wünsch/Burret*, Short paper on the development of the total electricity consumption until 2030, commissioned by the German Federal Ministry for Economic Affairs and Energy, 21 October 2021.

70 See the National Strategy on Hydrogen ("Nationale Wasserstoffstrategie"), German Federal Government.

71 See *Kemmler/Wünsch/Burret*, Short paper on the development of the total electricity consumption until 2030, commissioned by the German Federal Ministry for Economic Affairs and Energy, 21 October 2021.

72 See: Study of the ifeu Institute for Energy and Environmental Research Heidelberg, "Heating with 65% renewable energies – Accompanying analyses on the design of the regulation from the 2021 coalition agreement", 3 April 2023.

73 See: Forecast on electricity prices by the Association of the Bavarian Economy, "vbw / Prognos Strompreisprognose 2023", July 2023.

74 See: Forecast on electricity prices by the Association of the Bavarian Economy, "vbw / Prognos Strompreisprognose 2023", July 2023.

75 See: Forecast on electricity prices by the Association of the Bavarian Economy, "vbw / Prognos Strompreisprognose 2023", July 2023.

Thirdly, the burden of taxes and other statutory levies on the price of electricity also plays a significant role in the development of future prices: Up to approximately 48% of the average household price are usually set up by taxation and network charges.[76] Moreover, the German government also revoked a planned subsidy on network charges due to a household shortage.[77]

Lastly, not only are electricity prices expected to remain high, but also be affected by an increasing level of volatility due to an increasing share of solar power energy, which will lead to a high supply of energy and therefore lower prices during summer and less supply as well as higher prices during winter.[78]

Is There A Need for Capacity Markets?

In addition to the question of to what extent and how the generation of energy (especially electricity) from renewable energy sources should be promoted so that their share in the mix of energy sources is shifted to the detriment of (nuclear and) fossil energy sources, there is also the question of how the electricity market needs to be adjusted. In contrast to (nuclear,) coal and gas power plants, which can produce electricity continuously and, above all, controllably, the generation of energy from the fluctuating energy sources wind and sunlight is volatile. Generation cannot be adjusted to demand. In times of weak wind with little sunlight, the peak load cannot be met with electricity from renewable energy sources alone. During the discussion on promoting the expansion of renewable energy generation capacities, the question of security of supply must also be answered.

The UK Capacity Market

For example, in 2014, the United Kingdom (**"UK"**) estimated that the electricity market in Great Britain would reach critical levels of generation adequacy already around 2017/2018.[79] The UK therefore designed a capacity market where the system operator would organise centrally managed auctions to procure the level of capacity required to ensure generation adequacy.[80] Bidders who are successful in the auctions are awarded capacity agreements under which they will receive a steady payment for the duration of the capacity agreement in return for a commitment to deliver electricity

76 German Federal Network Agency, Monitoring Report 2023, page 31.

77 German Federal Government, Press Release 280/23, 19 December 2023.

78 See: Forecast on electricity prices by the Association of the Bavarian Economy, "vbw / Prognos Strompreisprognose 2023", July 2023.

79 See: Commission Decision (EU) 2020/348 of 24 October 2019 on the aid scheme SA.35980 – 2019/C United Kingdom – Electricity Market Reform: Capacity Mechanism, No. 2.

80 See: Commission Decision (EU) 2020/348 of 24 October 2019 on the aid scheme SA.35980 – 2019/C United Kingdom – Electricity Market Reform: Capacity Mechanism, No. 2.

at times of system stress if called upon to do so by the system operator.[81] Financial penalties apply if the capacity provider does not deliver the amount of energy required in accordance with its capacity commitment.[82] The UK capacity market is financed through a levy on electricity suppliers.[83]

An electricity market design without elements of a capacity market in which the generation capacity provided is traded, is the so-called energy only market in which only actually delivered and used electricity quantities are traded and remunerated.[84]

The discussion about capacity markets / Electricity Market 2.0 in Germany

In Germany, there was a broad discussion about the need of implementing a capacity market. In 2014, the German Federal Ministry for Economic Affairs and Climate Action (“**BMWK**”) published a discussion paper in which the current electricity market and the challenges for a future electricity market were discussed (“**Green Paper**”).[85] In Chapter III, in which possible solutions for sufficient, cost-efficient and environmentally compatible capacity provisions were discussed, there was an explicit comparison between an optimised energy only market (so-called “**Electricity Market 2.0**”) and the capacity market. The Electricity Market 2.0 is characterised as an energy only market with a capacity reserve. The Green Paper described the different models as follows:

> The introduction of a capacity market changes the existing electricity market design: An additional market is created alongside the existing electricity market. In capacity markets, only the provision of capacity (output) is traded and explicitly remunerated. For the remuneration costs are incurred in addition to the costs of procuring electricity on the electricity only market. The electricity suppliers bear these costs and pass them on to the consumers. On the existing electricity market, power is remunerated only implicitly on futures markets, spot markets and in electricity purchase contracts through unconditional obligations to deliver. Power is explicitly traded and remunerated, for example, on the balancing market, in option contracts or in supply contracts.[86]

---

81 See: Commission Decision (EU) 2020/348 of 24 October 2019 on the aid scheme SA.35980 – 2019/C United Kingdom – Electricity Market Reform: Capacity Mechanism, No. 2.

82 See: Commission Decision (EU) 2020/348 of 24 October 2019 on the aid scheme SA.35980 – 2019/C United Kingdom – Electricity Market Reform: Capacity Mechanism, No. 2.

83 See: Commission Decision (EU) 2020/348 of 24 October 2019 on the aid scheme SA.35980 – 2019/C United Kingdom – Electricity Market Reform: Capacity Mechanism, No. 2.

84 European Parliamentary Research Service, Capacity mechanisms for electricity, https://www.europarl.europa.eu/RegData/etudes/BRIE/2017/603949/EPRS_BRI(2017)603949_EN.pdf.

85 “An electricity market for the energy transition”, Discussion paper of the Federal Ministry for Economic Affairs and Energy (Green Paper), Federal Ministry for Economic Affairs and Energy, October 2014.

86 “An electricity market for the energy transition”, Discussion paper of the Federal Ministry for Economic Affairs and Energy (Green Paper), Federal Ministry for Economic Affairs and Energy, October 2014, p. 37.

In 2015, the BMWK published a results paper (**"White Paper"**),[87] summarising the results of the consultations, in which associations, trade unions, companies, research institutions, public authorities and citizens participated and submitted around 700 comments. The fundamental decision was that the existing electricity market should be further developed into an Electricity Market 2.0. The reasons behind that decision were the following:

> The Electricity Market 2.0 guarantees security of supply. In the Electricity Market 2.0, the required capacities can be remunerated through the market mechanisms. The remuneration works under two conditions: Firstly, electricity pricing must remain free; secondly, electricity suppliers must have strong incentives to fulfil their supply commitments. The Electricity Market 2.0 is cheaper than an electricity supply system with an additional capacity market, and it enables innovation and sustainability. Capacity markets are susceptible to regulatory failure and make it more difficult to transform the energy system. An Electricity Market 2.0 does not require any intervention in the market mechanism and is thus less susceptible to regulatory failure. A competitive system will bring out the cheapest solutions for the integration of renewable energy sources. As a result, the Electricity Market 2.0 creates incentives for new fields of business and sustainable solutions.

18.3.5 Will We See an Evolution or A Revolution for Support Schemes?

Requirements for Support Schemes Set Forth by the RED I Directive

The central goal of the RED I Directive is that at least 20% of the community's gross final energy consumption is covered by energy from renewable sources by 2020. In order to achieve this goal, the Member States must take measures, namely introduce support schemes or take measures of cooperation between different Member States and with third countries, Article 3 para 3 RED I Directive.

The term **"support scheme"** is defined in the RED I Directive for the first time on an EU-wide basis.[88] "Support scheme" means, according to the legal definition in Article 2 lit. k) of the RED I Directive, any instrument, scheme or mechanism applied by a Member State or a group of Member States that promotes the use of energy from re-

87 "An electricity market for the energy transition", Federal Ministry for Economic Affairs and Energy, July 2015.

88 Recital (7) of the RED I Directive clarifies that Directive 2001/77/EC of the European Parliament and of the Council of 27 September 2001 on the promotion of electricity produced from renewable energy sources in the internal electricity market and Directive 2003/30/EC of the European Parliament and of the Council of 8 May 2003 on the promotion of the use of biofuels or other renewable fuels for transport and Directive 2003/54/EC of the European Parliament and of the Council of 26 June 2003 concerning common rules for the internal market in electricity lay down definitions that are used in an identical or similar way in the RED I Directive.

newable sources by reducing the cost of that energy, increasing the price at which it can be sold, or increasing, by means of a renewable energy obligation or otherwise, the volume of such energy purchased. This includes, but is not restricted to, investment aid, tax exemptions or reductions, tax refunds, renewable energy obligation support schemes including those using green certificates, and direct price support schemes including feed-in tariffs and premium payments. A "renewable energy obligation" is legally defined in Article 2 lit. (l) as a national support scheme requiring energy producers to include a given proportion of energy from renewable sources in their production, requiring energy suppliers to include a given proportion of energy from renewable sources in their supply, or requiring energy consumers to include a given proportion of energy from renewable sources in their consumption. This includes schemes under which such requirements may be fulfilled by using green certificates. These legal definitions have given national legislators a wide room for manoeuvre. A commitment to a specific price- or volume-based support scheme has been avoided.[89] According to recital 25 of the RED I Directive, the reason for granting the greatest possible freedom of design and the combination of support schemes and cooperation measures is that the Member States have different renewable energy potentials and operate different schemes of support for energy from renewable sources at the national level. In addition, most Member States apply support schemes that grant benefits solely to energy from renewable sources that is produced on their territory. The aim of the RED I Directive is therefore to complement the national focus of support schemes with an international, cross-border component.

In order to introduce the support schemes, the Member States should first establish overall targets for the shares of energy from renewable sources consumed in the transport, electricity and heating and cooling sectors in 2020 in **national renewable energy action plans** and define the measures to be taken to achieve them, Article 4 para 1.[90] By 30 June 2009, the Commission should have adopted a template for the national renewable energy action plans. The RED I Directive sets out a procedure and dates for the notification of the national renewable energy action plans to the Commission and their publication, Article 4 of the RED I Directive.[91]

Particularly significant for the effective use of renewable energies in electricity supply are the rules for grid access and grid expansion in Article 16 RED I Directive. According to this, Member States are obliged to require transmission and distribution system operators to transmit and distribute electricity from renewable energies based on **priority or guaranteed grid access.**[92] This has the direct effect of putting renew-

89 *Mohr*, in: Säcker, Berliner Kommentar zum Energierecht, edition 2018, § 1 EEG no. 26.

90 *Gundel*, in: Dauses/Ludwigs, Handbuch des EU-Wirtschaftsrecht, 55th edition, EL January 2022, no. 258.

91 *Lehnert/Vollprecht*, ZUR 2009, issue 6, 307 (308).

92 *Gundel*, in: Dauses/Ludwigs, Handbuch des EU-Wirtschaftsrecht, 55th edition, EL January 2022, no. 260.

ables in a better position than other energy sources.[93] In addition, Member States should take the appropriate steps to develop transmission and distribution grid infrastructure, intelligent networks, storage facilities and the electricity system, to allow the secure operation of the electricity system as it accommodates the further development of electricity production from renewable energy sources, including interconnection between Member States and third countries.

Another valuable achievement of the RED I Directive is found in Article 15 RED I Directive. It provides for the mandatory issuance of **guarantees of origin** for electricity, heating and cooling generated from renewable energy sources.[94] The guarantee of origin is to be issued electronically by a suitable body, that must be reliable and fraud-proof. In addition, it is to be mutually recognised by the Member States. A guarantee of origin shall specify at least (a) The energy source from which the energy was produced and the start and end dates of production; (b) Whether it relates to: (i) electricity; or (ii) heating or cooling; (c) The identity, location, type and capacity of the installation where the energy was produced; (d) Whether and to what extent the installation has benefited from investment support, whether and to what extent the unit of energy has benefited in any other way from a national support scheme, and the type of support scheme; (e) The date on which the installation became operational; and (f) The date and country of issue and a unique identification number.

The end customer is thereby sufficiently informed about the share of energy from renewable sources in the energy mix of an energy supplier or in what quantity it is contained therein. In this way, guarantees of origin indirectly aim to change the market behaviour of consumers. In particular, the guarantees are intended to show the advantages of energy from renewable sources over grey electricity and thus motivate consumers to switch electricity suppliers.

In the area of **cooperation measures** for joint projects between Member States for the production of electricity, heating or cooling from renewable sources, Member States have notification obligations towards the Commission, Article 7 RED I Directive. The notification must describe the proposed installation or identify the refurbished installation, specify the proportion or amount of electricity or heating or cooling produced from the installation which is to be regarded as counting towards the national overall target of another Member State; identify the Member States in whose favour the notification is being made; and specify the period, in whole calendar years, during which the electricity or heating or cooling produced by the installation from renewable energy sources is to be regarded as counting towards the national overall target of the other Member State, Article 6 para 3 RED I Directive.

In the case of cooperation measures between Member States and third countries, Article 9 of the RED I Directive stipulates that electricity generated in a third country

93 *Mohr*, in: Säcker, Berliner Kommentar zum Energierecht, edition 2018, §1 EEG no. 31.
94 *Ringel/Bitsch*, NVwZ 2009, 807 (809).

must be consumed within the community of Member States. Opportunities for international cooperation are also identified in the area of support schemes. For example, Article 11 of the RED I Directive stipulates that two or more Member States may decide on a voluntary basis to merge or partially coordinate their national support schemes. In such cases, under certain conditions, a certain amount of energy from renewable sources produced in the territory of a participating Member State can be counted towards the national overall target of another participating Member State.

In addition to the preceding regulations, Article 3 para 4 RED I Directive provides for special provisions for the **transport sector**. The directive sets the expansion target that obliges Member States to increase their share of energy from renewable sources in the transport sector to at least 10% by 2020. This shall be made possible by the more intensive use of biogenic fuels and the increased use of electromobility. A prescribed charging mode for the use of road vehicles with electric operation makes it clear that the use of electric vehicles is preferred: the share of renewable energies in the electricity sector used for transport is taken into account and counted towards the achievement of the sector-specific expansion target by a factor of 2.5. Thus, the increased use of electrically powered vehicles leads to an accelerated growth of the share of renewable energies in the transport sector.[95]

In addition, Article 17 et seq. RED I Directive regulate the use of **biofuels and other bioliquids**. In order to be considered for inclusion in the national target values at all, biofuels and bioliquids must fulfil defined sustainability criteria. For biofuels and bioliquids produced from wastes and residues, it is required that the greenhouse gas emission savings achieved through their use must be at least 35%. In addition, biofuels and bioliquids must not be produced from land with high biodiversity value or from land with high carbon stocks. Furthermore, biofuels and bioliquids may not be produced from land that was peat bog in January 2008, unless it is demonstrated that the cultivation and harvesting of the feedstock in question does not require drainage of previously undrained land.[96]

In order to implement the measures to be taken, the Member States are to enact proportionate, necessary and non-discriminatory **procedural regulations.**[97] The focus is on making the use of renewable energies faster, simpler and more cost-effective. In particular, existing barriers to use for consumers and suppliers are to be counteracted and the use of renewable energies is to be made particularly attractive and consumer friendly. In concrete terms, for example, sufficient information on the processing of applications and support services is to be provided.

As we have just seen, there are promotion measures at numerous levels. For these to be promising, the topic of "renewable energies" must meet with acceptance and understanding among the population. To this end, Article 14 of the RED I Directive

95 *Ringel/Bitsch*, NVwZ 2009, 807 (808).

96 *Gundel*, in: Dauses/Ludwigs, Handbuch des EU-Wirtschaftsrecht, 55th edition, EL January 2022, no. 260.

97 *Ringel/Bitsch*, NVwZ 2009, 807 (810).

obliges the Member States to ensure both the dissemination and the provision of information. In particular, all important factors such as consumers, building contractors, installers, architects and suppliers of equipment and systems for the generation of heat, cold and electricity and of vehicles are to be provided with sufficient information on support measures. It also aims to create jobs in this sector and strengthen a "renewable economy".

Requirements For Support Schemes Set Out in the RED II Directive

The RED II Directive contains a new target for the European Union with regard to the share of energy from renewable sources in gross final energy consumption. By 2030, the share of renewable energies is to be at least 32%. To achieve this target, however, the RED II Directive does not create a pan-European support system.[98] Rather, it again sets minimum standards that must be implemented by the Member States when designing the support systems.[99]

Article 23 RED II Directive addresses the **heating and cooling sector**. In this sector, the share of renewable energies is to be increased by an indicative target of 1.3% compared to the share of renewable energy in the heating and cooling sector in 2020.[100] Depending on the already existing shares of renewable energy, different amounts of waste heat can be counted towards this target. What measures the Member States take to implement this is up to them. For example, this increase can be achieved through physical blending of renewable energy or waste heat, through direct reduction measures such as the installation of high-efficiency heating and cooling systems, or through indirect reduction measures such as certificates. Political measures that provide sufficient incentives for consumers to use energy are also possible.

The use of **district heating** is regulated by Article 24 RED II Directive. In this area, the increase in the share of renewable energies and the share of waste heat is to be 1% per year. The European Union obliges the Member States to implement this target. This gives the states the necessary leeway for flexible implementation. Individual actors, such as district heating suppliers, on the other hand, are not directly obligated.[101] In the case of district heating, the eligibility of waste heat is made even easier compared to Article 23. The reason for this is to avoid disproportionately high overall costs of this policy.[102] Special importance in this area is also attached to the exchange of information and communication with consumers. Member States shall ensure that information on the energy per-

98 *Vollprecht/Lehnert/Kather*, ZUR 2020, 204 (204); *Gundel* in: Dauses/Ludwigs, Handbuch des EU-Wirtschaftsrecht, 55th edition, EL January 2022, no 262.

99 *Pause*, ZUR 2019, 387 (388).

100 *Topp*, in: Theobald/Kühling, Energierecht, 115th edition, EL January 2022, no. 90.

101 *Topp*, in: Theobald/Kühling, Energierecht, 115th edition, EL January 2022, no. 103.

102 *Topp*, in: Theobald/Kühling, Energierecht, 115th edition, EL January 2022, no. 90.

formance and the share of renewable energy of their district heating and cooling systems is made available to final consumers in an easily accessible form.[103] With these measures, it is possible to increase citizens' understanding of the untapped potential of the heating and cooling sector, which is crucial for the decarbonisation of the energy system.

According to Article 19 para 2 RED II Directive, the Member States must ensure that **guarantees of origin** for energy from renewable sources are issued upon request. This is to fulfil obligations of proof that exist vis-à-vis the end customer. Article 15 para 2 RED I Directive already stipulated that "a guarantee of origin is issued in response to a request from a producer of electricity from renewable energy sources". Article 19 para 2 RED II Directive extends the personal scope of the regulation and requires that a guarantee of origin be issued in response to a request from a producer of energy from renewable sources. The amended wording in Article 19 para 2 RED II Directive means that guarantees of origin must now also be issued for gas and heating and cooling from renewable energy sources.[104] However, it is evident that the guarantees of origin are designed to support the sale of energy from renewable energy sources on the market which conflicts with the idea of a support scheme. Article 19 para RED II Directive provides that the guarantees of origin shall only be issued unless the Member States decide, for the purposes of accounting for the market value of the guarantee of origin, not to issue such a guarantee of origin to a producer that receives financial support from a support scheme.

For the first time, the directive contains provisions to enable and promote self-supply of electricity from renewable energy sources. Article 2 No. 14 RED II defines **"renewable self-consumer"** as a final customer operating within its premises located within confined boundaries or, where permitted by a Member State, within other premises, which generates renewable electricity for its own consumption, and which may store or sell self-generated renewable electricity, provided that, for a non-household renewables self-consumer, those activities do not constitute its primary commercial or professional activity. The special significance of the "self-consumer" is to be seen in the fact that the Member States are obliged in Article 21 para 1 RED II Directive to ensure that consumers are entitled to become self-consumers.[105] To make self-consumption of electricity from renewable energy sources as attractive as possible, self-consumers are to be exempted from levies, charges and fees for the electricity consumed on site. In addition, the directive requires that neither the electricity consumed, the electricity drawn from the grid, nor the electricity fed into the grid by producers may be subject to discriminatory surcharges and charges and grid fees that are not cost-reflective. However, Member States are entitled to impose proportionate surcharges, levies or charges on certain quantities of electricity used for self-supply in

103 *Schulz/Losch*, EnWZ 2017, 107 (113).

104 *Vollprecht/Lehnert/Kather*, ZUR 2020, 204 (206, 207).

105 *Pause*, ZUR 2019, 387 (389).

three cases. This is the case if self-generated renewable electricity is effectively promoted under support schemes, Article 21 para 3 lit. (a). Furthermore, in the case where, from 1 December 2026, the overall share of self-consumption installations exceeds 8% of the total installed electricity capacity of a Member State and it is demonstrated, by means of a cost-benefit analysis performed by the national regulatory authority of that Member State, which is conducted by way of an open, transparent and participatory process that the above exemption results in a significant disproportionate burden on the long-term financial sustainability of the electricity system, Article 21 para 3 lit. (b). Furthermore, in the case where the self-generated renewable electricity is produced in installations with a total installed electrical capacity of more than 30 kW, Article 21 para 3 lit. (c).

Another achievement of the directive is the promotion of "**renewable energy communities**", Article 22 RED II Directive. According to the legal definition in Article 2 No. 16, a "renewable energy community" is a legal entity which, in accordance with the applicable national law, is based on open and voluntary participation, is autonomous, and is effectively controlled by shareholders or members that are located in the proximity of the renewable energy projects that are owned and developed by that legal entity; (a) The shareholders or members of which are natural persons, small or medium sized entities or local authorities, including municipalities; (b) The primary purpose of which is to provide environmental, economic or social community benefits for its shareholders or members or for the local areas where it operates, rather than financial profits. Member States shall ensure that renewable energy communities are entitled to produce, consume, store and sell renewable energy, including through renewables power purchase agreements. In addition, it is to be ensured that energy produced with production units owned by the community can be shared within the community and that the communities have access to all energy markets directly or via aggregators. In doing so, Member States shall establish a regulatory framework to support and promote the development of such communities. Within this framework, Member States shall ensure that distribution system operators cooperate with these communities to facilitate energy transfers within them.

While the regulations for **grid access and grid expansion** in RED I were still highlighted as a significant achievement, there are no longer any regulations in Article 17 para 1 of the RED II Directive that grant priority or guaranteed grid access. However, a procedure of simple notification is to be introduced for grid access for installations of renewables self-consumers and demonstration projects with an electrical capacity of up to 10.8 kW. Upon notification, such installations shall be connected to the grid, unless justified grounds of safety concerns or technical incompatibility of the system components exist. Member States may allow a simple-notification procedure for installations or aggregated production units with an electrical capacity of above 10.8 kW and up to 50 kW, provided that grid stability, grid reliability and grid safety are maintained.

In addition, the directive also provides for the possibility of **cooperation between Member States** in Article 9 and between Member States and third countries, Article 11 of the RED II Directive. This is a continuation of the possibilities of the so-called flexible mechanisms that essentially already existed under the RED I Directive. This is an exception to the principle that only the amount of energy generated from renewable sources in the respective Member State is relevant for determining the renewable energy target.[106]

Requirements for the use of renewable energies in the **transport sector** are contained in Articles 25 to 28 of the RED II Directive. According to Article 25 para 1 RED II Directive, Member States must require fuel suppliers to comply with the sector-specific target that the share of renewable energy in final energy consumption in the transport sector is at least 14% by 2030. As in the predecessor directive, Article 15 of the RED II Directive contains provisions on administrative regulations. These largely correspond to the respective predecessor regulations, so that reference can be made to these explanations.[107]

Requirements for Support Schemes Set Out in the RED III Directive

The RED III Directive sets a new target to become climate neutral by 2050 in a way that contributes to the European economy and to growth and job creation in Europe.[108] The set EU target in the RED II Directive to achieve a share of at least 32% renewable energy by 2030 is not sufficient to enable the necessary energy transition. Therefore, in Article 3 para 1 as amended by the RED III Directive, the share of energy from renewable sources in the European Union's gross final consumption of energy in 2030 is raised to at least 40%. To achieve this objective, Member States should establish a framework that creates the conditions for achieving the targets set out in Union law.

A remarkable upgrade regarding the regulations of the RED II Directive is that so-called **power purchase agreements** are explicitly listed in Article 3 para 4a RED III Directive alongside support schemes as a further means of achieving the Member States renewable energy contributions.

The already existing regulations regarding **guarantees of origin** are to be adapted in order to promote consumer awareness regarding renewable energies, to advance the dissemination of power purchase agreements, as well as to inform consumers more comprehensively. For example, pursuant to Article 19 para 2 as amended by the RED III Directive, the Member States shall ensure that a guarantee of origin is issued in response to a request from a producer of energy from renewable sources. Member States may arrange for guarantees of origin to be issued for energy from non-renewable sources. Issuance of guarantees of origin may be made subject to a minimum capacity limit.

---

106 *Vollprecht/Lehnert/Kather*, ZUR 2020, 204 (205).
107 *Vollprecht/Lehnert/Kather*, ZUR 2020, 204 (208).
108 Recital 1 RED III Directive.

A guarantee of origin shall be of the standard size of 1 MWh. No more than one guarantee of origin shall be issued in respect of each unit of energy produced.

The newly inserted Article 22a RED III Directive is intended to regulate the inclusion of renewable energies in **industry**. According to this provision, the Member States are to strive to ensure that the share of renewable sources in the amount of energy sources used for final energy and non-energy purposes in the industry sector increases by an average of 1.1 percentage points by 2030. In addition, the target set for the contribution of renewable fuels of non-biological origin used for final energy and non-energy purposes shall be 50% of the hydrogen used for final energy and non-energy purposes in industry by 2030.

The **building sector** is also included with the new inclusion of Article 15a as amended by the RED III Directive. Member States shall set an indicative target for the share of renewables in final energy consumption in their buildings sector in 2030 that is consistent with an indicative target of at least a 49% share of energy from renewable sources in the buildings sector in the European Union's final consumption of energy in 2030. To achieve this target, Member States shall introduce measures in their building regulations and codes and, where applicable, in their support schemes, to increase the share of electricity and heating and cooling from renewable sources in the building stock, including national measures relating to substantial increases in renewables self-consumption, renewable energy communities and local energy storage, in combination with energy efficiency improvements relating to cogeneration and passive, nearly zero-energy and zero-energy buildings, Article 15a para 2 RED III Directive.

According to Article 23 para 1 RED III Directive, the share of renewable energies in the **heating and cooling** sector is to increase by an average of 1.1 percentage points as an annual average calculated for the periods 2021 to 2025 and 2026 to 2030, starting from the share of renewable energy in the heating and cooling sector in 2020. Measures that Member States can take to achieve this target include the physical blending of renewable energy or waste heat and cooling with energy sources and fuels for heating and cooling, as well as the possibility of installing highly efficient heating and cooling systems.

The development of infrastructure for district heating and cooling networks is to be accelerated and geared towards efficiently and flexibly using a wider range of sources for heating and cooling supply with renewable energies. Thus, Article 24 para 1 RED III Directive shall ensure that information on the overall energy efficiency and the share of renewable energy of their district heating and cooling systems is provided to the end user in an easily accessible form. In the district heating and cooling sector, the share of renewable energy in total consumption is to increase by at least 2.1 percentage points as an annual average calculated for the period 2021 to 2025 and for the period 2026 to 2030, starting from the share of energy from renewable sources and from waste heat and cold in district heating and cooling in 2020, instead of the current 1%. In addition, district heating and cooling operators are to grant third parties' access to district heating and cooling systems under certain conditions.

In the **transport sector**, greenhouse gas intensity is to be reduced by 13% by 2030, Article 25 para 1 lit. a) as amended by the RED III Directive. By using the greenhouse gas intensity as a new connecting factor, greenhouse gas emissions are to be saved more effectively and, in addition, innovative pressure for low-CO_2 fuels is to be built up. In Article 25 para 1 lit. b) RED III Directive, a minimum share of 2.2% for advanced biofuels and biogas by 2030 is prescribed with a partial target for 2022 and 2025. The share of renewable fuels of non-biological origin should be at least 2.6% in 2030. All modes of transport in the European Union should be increasingly supplied with renewable energies. In order to promote the development of electric mobility, a credit mechanism should be introduced. Operators of public e-charging stations receive credits that they can sell to fuel suppliers. They can use the credits to meet their renewable energy power quotas.

Measures under Article 20a as amended by the RED III Directive are intended to help with the system integration of renewable electricity. This results in the obligation of transmission and distribution system operators to provide digitally usable information on the share of renewable electricity and greenhouse gas emissions in the respective bidding zone in at least hourly intervals. In addition to these requirements, battery manufacturers are to be obliged (Article 20a II as amended by RED III Directive) to provide owners and users of batteries with access to information on capacity, condition and charging status to owners or third parties commissioned by the owner.

There is enormous potential in cooperation between Member States, which has not been sufficiently exploited so far. **Cooperation** enables the cost-efficient use of renewable energies throughout Europe and at the same time contributes to market integration. In order to make use of this opportunity, the Member States are encouraged, in Article 9 para 1 a as amended by the RED III Directive, to promote intra-European cooperation in the form of pilot projects. By 31 December 2025, each Member State shall agree to establish at least one joint project with one or more other Member States to produce renewable energy.

To counteract the shortage of skilled workers, the Member States are to take measures in accordance with Article 18 para 3 as amended by the RED III Directive to attract more people from groups currently underrepresented in the relevant occupational fields to these activities. Sufficient training programmes are to be offered in the field of renewable energy-based technologies for heating and cooling supply and the latest innovative solutions. In order to particularly strengthen the attractiveness of these professions, Member States shall introduce measures to encourage participation in such programmes. The focus is particularly on small and medium-sized enterprises and the self-employed.

PPAs

Overview of Types of PPAs on the Market

The types of PPAs on the market can be differentiated by certain underlying characteristics. Depending on the type of the PPA, different contractual provisions are of key importance. Characteristics that play a role in the contract design are the following:

Type of Buyer – Corporate or Utility

The purchaser of the electricity and the guarantees of origin can be either a utility or a corporate purchaser. A utility is an energy supply company or a direct marketeer that redistributes the electricity that it purchases from the plant operator via the PPA to third parties, e. g. end consumers, other utilities or the European Energy Exchange. A corporate buyer is a company that purchases the electricity for end-consumption for the operation of its production facilities, offices, branches or similar.

Type of Delivery – On Site or Off Site ("Sleeved")

An "on site-PPA" is one in which the supply takes place within a "factory grid" (a customer system, German: "Kundenanlage"; a closed distribution network, German: "geschlossenes Verteilernetz" or a direct line, German: "Direktleitung"). In this case, the public general supply network is not used. In the case of delivery within a customer system, the customer avoids the grid charges and grid-related levies/charges for the quantity of electricity supplied under the PPA. Both parties also benefit from the fact that the system is generally less likely to be curtailed by the grid operator, as there is no risk of grid overloads from other market participants and curtailment only occurs if electricity is fed back into the general supply grid.

An offsite PPA is characterised by a delivery via the public general supply network ("sleeved"). In this case, the delivery takes place via a balancing group designated by the customer, to which the market location ID of the power plant must be assigned. The balancing group must be "managed", i.e. it must be ensured that feed-in and feed-out off the balancing group are always in balance. If the customer is an end-consumer, it will generally commission a service provider responsible for the balancing group, such as the energy supply company that supplies it. For this purpose, the consumer and the energy supply company will conclude a so-called balancing agreement.

Type of Performance Fulfilment – Physical or Financial

The purpose of physical PPAs is to supply electricity and guarantees of origin in return for payment of a remuneration in EUR/MWh. The physical supply or procurement of electricity takes place via the general supply network or via a "factory network" (customer system, closed distribution network or a direct line).

Financial PPAs (also: "virtual", "financial" or "synthetic" PPAs, hereinafter "vPPA") do not involve the physical delivery of electricity, but merely a purely financial settlement mechanism in relation to a certain amount of electricity from a real generation plant and the delivery of guarantees of origin in return for payment of a fee. A financial PPA therefore only involves the exchange of purely monetary payments; no physical delivery of electricity is owed. Physical electricity trading takes place via the electricity markets (usually EPEX Spot).

PPA-drivers – Regulatory framework and market price development

The main driver for the development of the PPA market in Germany is the decline in state subsidies under the Renewable Energies Act ("EEG") – from a fixed feed-in tariff to competitive tenders in which the lowest bid is awarded a contract. Hence, a further decline in state subsidies would promote an expansion of the PPA-market.

However, how the PPA market in Germany develops depends not only on the regulatory framework but also to a large extent on the development of electricity market prices. Since 2019, the PPA market in Germany has been significantly driven by sharply rising and, especially in 2022, strongly fluctuating exchange electricity prices. This is because both customers and plant operators were hedging against rising and highly volatile exchange electricity prices with long-term (fixed-price) PPAs.[109]

18.2.6 International Cooperation in Order to Support Renewable Energies

The German Energy Partnership Programme

In addition to its commitment to cooperation measures set forth in the Kyoto Protocol and the Paris Agreement, Germany has implemented another instrument in striving towards energy transition and the goal of being carbon free. This instrument is the energy partnership. The energy partnership programme is a project initiated and maintained by the BMWK.[110] It is viewed as one of the key instruments of the BMWK energy foreign affairs. The key to establishing an energy partnership is a joint declaration. A joint declaration is the foundation of an energy partnership.[111] The joint declaration in form of a Memorandum of Understanding ("MoU") or Letter of Intent

109 For further information regarding PPAs see https://marktoffensive-ee.de/startseite/.

110 GIZ, Bilaterale Energiepartnerschaften und -dialoge.

111 IASS, Deutschlands Energiepartnerschaften in der internationalen Energiewendepolitik, p. 8; BMWK, Internationale Energiepolitik.

("LoI") serve as defining foundation for the objectives of the individual partnership.[112] The instrument without the joint declaration in form of a MoU or a LoI classifies as an Energy Dialogue ("ED"). The ED tends exclude a joint declaration of intent such as a memorandum of understanding or letter of intent.[113] The BMWK utilizes this closely related instrument flexible instrument sometimes with subnational stakeholders.[114] The energy partnership and the ED are a key function of energy foreign affairs due to the potential supply of energy for Germany's domestic energy demand.

Germany is dependent on energy imports to fulfil its domestic needs for electricity and other usages of energy.[115] Germany imports close to 70% of its energy resources.[116] The international goals require an adjustment to energy system and energy usage on a global scale. These challenges need to be addressed by Germany by forging new partnerships internationally and engage in dialogue.[117]

Through energy partnerships the BMWK intends to foster relations with countries that export energy resources and act as transit countries for Germany's and Europe's energy supply. This makes it possible to create favourable and stable economic conditions for energy projects and investors, and to diversify the sources of energy and the transport routes for energy.[118] There is also a need to intensify cooperation with energy consumers and producer countries. The BMWK pursues cooperation with major energy supplier countries in order to achieve progress in the field of low-emission climate technologies, renewable energies and energy efficiency. This is a contribution to defusing the global competition for dwindling resources for electricity generation and mitigating climate change. Therefore, the BMWK comprehends the energy partnership programme as a potential solution to the competition on resources and mitigation of climate change. Another role of the energy partnership is the energy partnerships contribution to the work of multilateral organisations and other initiatives. The BMWK views energy partnerships as a facilitator for work in multilateral organisations, forums and initiatives, because of the creation of competitive environmentally conscious global energy markets through energy partnerships.[119] These assumed benefits are a result of the topics discussed within energy partnerships.

---

112 IASS, Deutschlands Energiepartnerschaften in der internationalen Energiewendepolitik, p. 8; BMWK, Internationale Energiepolitik; e.g. Mexico Joint Declaration of Intent on the Energy Partnership between the Government of the United Mexican States and the Government of the Federal Republic of Germany.

113 IASS, Deutschlands Energiepartnerschaften in der internationalen Energiewendepolitik, p. 6.

114 IASS, Deutschlands Energiepartnerschaften in der internationalen Energiewendepolitik, p. 6.

115 *Engelkes/Schulz*, Out of Siberia, into the Desert; The Middle East and North Africa as Building Blocks of Europe's Energy Transition, p. 7.

116 *Heymann*, German energy supply at a historical turning point, p. 1.

117 BMWK, Internationale Energiepolitik.

118 BMWK, Internationale Energiepolitik.

119 BMWK, Internationale Energiepolitik.

The energy partnerships facilitate an exchange of information on an intergovernmental level and are a driving force for innovations in the energy industry and economic cooperation.[120] The topics reflect the drive for innovation. Energy partnerships include topics such as rules on the design of the energy market, market strategies for the implementation of hydrogen, the phase out of coal, energy efficiency and digitalisation in the energy sector serve as topics for energy partnerships.[121] Economic development and employment development is also a topic in energy partnerships.[122] A new topic are the innovative ways of storage solutions for electricity.[123] Carbon- free alternatives with regard to gaseous and liquid forms of energy become more important. Examples include a background study on the hydrogen debate in Australia.[124] The BMWK and its ministries in the partner countries employ a variety of activities in the energy partnership programme in order to engage with topics and to facilitate the pressing challenge of energy transition. Activities include regular meetings of working groups, workshops, bilateral conversations, and largescale events. Other working methods include Webinars and Hackathons, studies of the electricity market and pilot projects.[125]

The programme of energy partnerships is expanding. The energy partnership with Chile established in 2019 aims to capitalize on the country's great potential for energy from photovoltaics and other forms of renewable energies.[126] A new energy partnership has been formed with Ukraine.[127] Some of the previous existing energy dialogs with Japan, South Korea and Jordan have transitioned and expanded into energy partnerships[128] In general energy partnerships serve as an instrument to develop country specific solutions for the challenges of energy transition. Energy partnerships promote the expansion of renewable energy generation and the distribution of energy efficient technologies and provide a platform for continuous exchange on political and economic topics surrounding the energy transition.[129]

120 BMWK, Internationale Energiepolitik.

121 BMWK, Jahresbericht 2020 Energiepartnerschaften und Energiedialoge, p. 6.

122 IASS, Deutschlands Energiepartnerschaften in der internationalen Energiewendepolitik, p. 8; BMWK, Internationale Energiepolitik.

123 See footnote 89.

124 BMWK, Jahresbericht 2019 Energiepartnerschaften und Energiedialoge, p. 15.

125 BMWK, Internationale Energiepolitik.

126 BMWK, Internationale Energiepolitik.

127 BMWK, Jahresbericht 2020 Energiepartnerschaften und Energiedialoge, p. 7.

128 BMWK, Jahresbericht 2019 Energiepartnerschaften und Energiedialoge, p. 11.

129 BMWK, Internationale Energiepolitik.

What Limits Does EU State Aid Law Impose on the Design of Support Schemes?

At the level of primary Union law,[130] the Member States' national legislators' scope for shaping their national support schemes is limited above all by the general legal principle of proportionality[131] and by the rules protecting the free movement of goods between the Member States and protecting competition within the European Union from state intervention through state aid. In connection with the free movement of goods, the provision of Article 34 TFEU should be mentioned in particular, which provides for the prohibition of quantitative restrictions on imports and all measures having equivalent effect. Purely national support schemes are in principle likely to affect trade in electricity within the European Union between the Member States. In its judgment in the Ålands Vindkraft AB case,[132] the ECJ ruled that the domestic restriction of national support schemes for electricity from renewable energy sources was compatible with Article 34 TFEU in accordance with Article 3 para 3 of the RED I Directive. In connection with the protection of the internal market against state intervention that distorts competition, Article 107 TFEU provides that any aid granted by a Member State or through state resources in any form whatsoever which distorts or threatens to distort competition by favouring certain undertakings or the production of certain goods shall, insofar as it affects trade between Member States be incompatible with the internal market. Article 108 para 1 TFEU provides that the Commission shall, in cooperation with the Member States, keep under constant review all systems of aid existing in those states. It shall propose to the latter any appropriate measures required by the progressive development and functioning of the internal market. If, after giving notice to the parties concerned to submit their comments, the Commission finds that aid granted by a state TFEU or through state resources is incompatible with the internal market having regard to Article 107 TFEU or that such aid is being misused, it shall decide that the state concerned shall abolish or alter such aid within a period of time to be determined by the Commission.

The Commission made use of this provision, for example, in connection with the German Renewable Energies Act (**"EEG 2012"**). The Commission decided on 25 November 2014[133] that the feed-in tariffs and market premiums, which guaranteed producers of renewable electricity a higher price than the market price for the electricity they

130 These are, in particular, the founding treaties and related amending treaties of the European Union, the Treaty on the Functioning of the European Union (TFEU), protocols, agreements and conventions and ratified legal acts under international treaties, which form the EU Constitution. In addition, there is unwritten primary law in the form of customary law and general principles of law. Secondary law, on the other hand, contains the legal acts adopted by the institutions of the European Union based on the Treaties.

131 See ECJ, C-492/14 para 111 et seq., 29 September 2016 ("Essent Belgium II").

132 ECJ, C-573/12, 1 July 2014.

133 Commission Decision (EU) 15/1585 of 25 November 2014 on the aid scheme SA.33995 (2013/C) (ex 2013/NN), ABl. 2015 L 250/122.

generated, constituted state aid incompatible with the internal market. Furthermore, the reduction of the EEG levy for certain energy-intensive companies (so-called special equalisation scheme) constituted state aid which was only compatible with the internal market if it fell under certain categories.[134] As a consequence of this Commission decision, the aid granted had to be recovered. Corresponding recovery orders were issued for the years 2013 and 2014.[135]

The German Federal Government brought an action before the General Court against the Commission's decision on the EEG 2012. It took the view that neither the general support mechanism of the EEG nor the special equalisation scheme constituted state aid within the meaning of Article 107 TFEU. The compensation mechanism of the EEG does not provide for the direct or indirect use of state resources to favour the companies and is therefore to be understood as a pure price regulation. According to the provisions of the EEG 2012, the costs resulting from the promotion of renewable energies were passed on to the end consumers. The General Court dismissed the action of the German Federal Government and confirmed the decision of the Commission.[136] This is also of interest because the court had not classified the previous German regulation in the so-called Electricity Feed-in Act as state aid.[137] In the state aid proceedings on the EEG 2012, the ECJ finally ruled on 28 March 2019[138] that neither the support mechanism of the EEG nor the special equalisation scheme for electricity-cost-intensive companies under the EEG 2012 constitute state aid and followed the reasoning of the German Federal Government.

Under the influence of the state aid procedure for the EEG 2012 and since the legislative procedure for the German Renewable Energies Act 2014 ("**EEG 2014**"), the German legislator has been conducting a state aid notification procedure with the Commission for the EEG 2014. In order to avoid uncertainties associated with this, the German legislator aligned the special equalisation scheme in the EEG 2014 closely with the relevant criteria of the state aid guidelines applicable at the time. Accordingly, in its decisions on the EEG 2014 and its novelization in 2017, the Commission certified the compatibility of the EEG 2014 and the special equalisation scheme with European state aid law.[139]

The state aid guidelines[140] are explanations by the Commission on the conditions under which it considers aid to promote environmental objectives and the energy economy, in particular aid that serves an environmental objective, to be compatible

---

134 General Court, T-47/15, 10 May 2016.

135 Commission Decision (EU) 15/1585 of 25 November 2014 on the aid scheme SA.33995 (2013/C) (ex 2013/NN) p. 31; *Baumann/Todorovic*, in: Baumann/Gabler, Günther, EEG, § 63, no. 9.

136 General Court, T-47/15, 10 May 2016.

137 ECJ, C-379/98, 13 March 2001 ("PreussenElektra").

138 ECJ, C-405/16, 28 March 2019; EWeRK 2019, 92 with annotations from *Schwintowski*.

139 *Baumann/ Todorovic*, in: Baumann/Gabler/Günther, EEG, § 63, no. 13.

140 *Baumann/ Todorovic*, in: Baumann/Gabler/Günther, EEG, § 63, no. 9.

with the internal market based on Article 107 para 3 lit. (c) TFEU.[141] Legally, the guidelines have no direct binding effect vis-à-vis the Member States and the courts. However, they do have an indirect binding effect, as the Commission itself is bound by them when reviewing the compatibility of a measure with European state aid law and as they reflect a common administrative practice and thus have a certain binding effect in individual cases through the general principle of equality and for reasons of the protection of legitimate expectations, especially as deviations from the criteria laid down in the guidelines require more justification.[142]

Community Guidelines on State Aid for Environmental Protection ("EAG")

As early as 2008, i.e., shortly before the adoption of the RED I Directive, the Commission adopted the "Community guidelines on State aid for environmental protection",[143] which replaced the "Community guidelines on State aid for environmental protection" that entered into force in 2001.[144] The scope of these guidelines was limited to the field of environmental protection, the definition of which was, however, very broad. According to the definition of the "Community guidelines on State aid for environmental protection", the Commission takes environmental protection to mean any action designed to remedy or prevent damage to our physical surroundings or natural resources, or to encourage the efficient use of these resources. At that time, however, the Commission was already of the opinion that measures in favour of renewable sources of energy should also be classified as environmental protection measures. A key structural feature of the EAG was the distinction between investment aid and operating aid. In principle, investments are only eligible for aid if a company achieves a standard of environmental protection with them that goes beyond that of the community, be it that applicable community standards are exceeded, that binding community standards are lacking or that the investment is intended to adapt to applicable community law.[145] Operating aid is aid intended to relieve an undertaking of the expenses which it would normally have had to bear in its day-to-day management or its usual activities.[146]

141 See: EEAG: Guidelines on State Aid for Environmental Protection and energy 2014-2020 (2014/C 200/01).

142 *Baumann/Todorovic*, in: Baumann/Gabler, Günther, EEG, § 63, no. 9.

143 Community Guidelines on State Aid For Environmental Protection (2008/C 82/01).

144 Community Guidelines on State Aid For Environmental Protection, ABl Nr. C 37 v. 3.2.2001.

145 *Repplinger-Hach* in: Heidenhain, Handbuch des Europäischen Beihilfenrechts, § 17 No. 169.

146 *Repplinger-Hach*, in: Heidenhain, Handbuch des Europäischen Beihilfenrechts, § 17 No. 169; *Scheel*, DÖV, 2009,529 (532).

Guidelines on State Aid for Environmental Protection and Energy 2014-2020 ("EEAG")

EEAG as successor of the EAG, was drawn up by the Commission in parallel to the state aid procedure on the EEG 2012.[147] The EEAG are characterised on the one hand by the fact that they establish principles for the assessment of a measure under state aid law in order to clarify to the national states in general terms the relevant criteria to be examined by the Commission.[148] According to these principles, the Commission generally considers a national state aid measure to be compatible with the internal market only if it meets each of the following criteria:

- contribution to a well-defined objective of common interest: a state aid measure aims at an objective of common interest in accordance with Article 107 para 3 TFEU;[149]
- need for state intervention: the state aid measure is targeted towards a situation where aid can bring about a material improvement that the market alone cannot deliver, for example by remedying a well-defined market failure;
- appropriateness of the aid measure: the proposed aid measure is an appropriate policy instrument to address the objective of common interest;
- incentive effect: the aid changes the behaviour of the undertaking(s) concerned in such a way that it engages in additional activity which it would not carry out without the aid or which it would carry out in a restricted or different manner;
- proportionality of the aid (aid kept to the minimum): the aid amount is limited to the minimum needed to incentivise the additional investment or activity in the area concerned;
- avoidance of undue negative effects on competition and trade between Member States: the negative effects of aid are sufficiently limited, so that the overall balance of the measure is positive; and
- transparency of aid: Member States, the Commission, economic operators, and the public, have easy access to all relevant acts and to pertinent information about the aid awarded thereunder.

The EEAG also contain requirements that are specifically tailored to aid in certain areas, such as aid for the promotion of renewable energies. In this area, the EEAG call for a stronger market orientation of the support schemes for renewable energies to be approved. For example, EEAG state in Section 3.3.2.1, that aid recipients shall sell their electricity directly on the market and be subject to market obligations in order to create an incentive for the integration of electricity from renewable energy sources into the market. Since 1 January 2016 all new aid schemes and other aid measures

147 EEAG: Guidelines on State Aid for Environmental Protection and energy 2014-2020 (2014/C 200/01); *Scheel*, DÖV 2009, 529.

148 See Section 3.1 EEAG.

149 See Section 3.1 EEAG.

must fulfil all of the following conditions: a) the aid must be granted as a premium on top of the market price at which the electricity generators sell their electricity directly on the market b) aid recipients must be subject to a standard balancing responsibility, unless there are no liquid intraday markets, and c) measures must be taken to ensure that generators do not have an incentive to generate electricity at negative prices.

In addition to the mandatory introduction of the market premium instrument, the Commission has already heralded a system change towards tendering procedures in the EEAG by defining a transitional phase. This transitional phase covered the years 2015 and 2016. Within this period aid was granted for at least 5% of the planned new capacities for the generation of electricity from renewable energy sources within the framework of a tendering procedure based on clear, transparent and non-discriminatory criteria.[150]

Provided that all generators producing electricity from renewable energy sources can participate in these tenders on non-discriminatory terms, the Commission considers that the aid is proportionate and does not distort competition to an extent contrary to the internal market. Since 1 January 2017, aid must be granted in tenders on the basis of clear, transparent and non-discriminatory criteria.[151]

Furthermore, the EEAG stipulate that aid for renewable energies may also be granted as investment or operating aid. For investment aid schemes and individually notified investment aid, the general compatibility criteria of section 3.2 EEAG must be fulfilled.[152] The temporal scope of application of the EEAG was extended by one year until 2021.

Guidelines on State Aid for Climate, Environmental and Energy Protection 2022 ("CEEAG")

In the meantime, the Commission has published the CEEAG,[153] against which future support schemes of the Member States will have to be measured. The CEEAG entered into force with their adoption by the Commission on 27 January 2022.[154] Like the EEAG, Chapter 3 of the CEEAG sets out the general compatibility criteria for the categories of aid covered by the guidelines. Chapter 4 sets out the specific compatibility criteria for the aid measures covered by the different sections of the chapter. The compatibility criteria set out in Chapter 3 apply unless more precise provisions are contained in the specific sections of Chapter 4.[155]

150 See Section 3.3.2.1 EEAG.

151 Exemptions from the tendering requirement apply to installations with an installed electricity generation capacity of less than 1 MW and demonstration projects, except for wind turbines, for which the threshold is an installed electricity generation capacity of 6 MW or 6 generation units.

152 Section 3.3.1 no. 119 EEAG.

153 CEEAG: Guidelines on State aid for climate, environmental protection and energy 2022 (2022/C 80/01).

154 *Stöbener de Mora*, EuZW 2022, 195 (196).

155 Section 2.3 no. 17 CEEAG.

As regards aid, the Commission assesses compatibility with the internal market on the basis of the principle that, on the basis of Article 107 para 3 lit. (c) TFEU, aid may be considered compatible with the internal market if it facilitates the development of certain economic activities within the European Union (positive condition) where such aid does not adversely affect trading conditions to an extent contrary to the common interest (negative condition).[156] The CEEAG provide for a modified set of general compatibility criteria compared to the EEAG. With regard to the positive condition (the aid must facilitate the development of an economic activity), the following criteria must be fulfilled:[157] (a) identification of the economic activity, which is being facilitated by the measure, its positive effects for society at large and, where applicable, its relevance for specific policies of the European Union, (b) incentive effect, and (c) no breach of any relevant provision of Union law. The second, the negative condition (the aid measure must not unduly affect trading conditions to an extent contrary to the common interest) shall be assessed by the following criteria: (a) minimisation of distortions of competition and trade and (b) avoidance of undue negative effects on competition and trade. Ultimately, the positive effects of the aid must be compared with the negative effects on competition and trade.

Section 4.1 CEEAG contains the rules for the compatibility of sector-specific aid. While the EEAG regulates aid for the promotion of renewable energies (Section 3.3), the CEEAG generally deals with aid for the reduction and elimination of greenhouse gas emissions, among other things, through the promotion of renewable energies and energy efficiency. This significantly broadens the scope of this chapter compared with the EEAG. The section contains the compatibility rules for a total of thirteen aid measures to promote a wide range of other technologies primarily aimed at reducing greenhouse gas emissions. The EEAG covered only nine categories of aid, with the scope of the individual categories sometimes being much narrower.

Aid for the reduction of greenhouse gas emissions should normally be granted through competitive tendering in order to ensure that the objectives of the measure, such as the achievement of the Member State's decarbonisation targets can be adequately achieved while minimising distortions of competition and trade.[158] Exceptions to the obligation to grant aid on the basis of tenders and to determine its amount through tenders may be justified in the area of electricity generation or storage if evidence is provided that either there is insufficient potential supply or number of potential bidders to ensure competition or small electricity generation, or storage projects are affected.[159]

Strengthening the provisions of the EEAG, the CEEAG state that aid must be designed in such a way as to avoid undue distortions of the efficient functioning of the

156 Section 3 no. 21 CEEAG.

157 Section 3 no. 21 CEEAG.

158 Section 4.1.3.5 no. 103 CEEAG.

159 Section 4.1.3.5 no. 107 CEEAG.

market and, in particular, to maintain effective operating incentives and price signals.[160] For example, aid beneficiaries should continue to be exposed to price fluctuations and market risks unless this is contrary to the achievement of the objective of the aid. In particular, aid beneficiaries should not have an incentive to offer their production below their marginal costs and should not receive aid for it in periods when the market value of their production is negative. In contrast, the EEAG only required that electricity producers have no incentive to produce electricity at negative prices. According to the CEEAG, this is now to be prohibited.

As a specific support instrument in the context of decarbonisation, the CEEAG explicitly mention contracts for difference in para 121. The guidelines define these as contracts that entitle the beneficiary to a payment equal to the difference between a fixed 'strike' price and a reference price – such as a market price, per unit of output. It is explicitly stated that contracts for difference have been used in recent years for electricity generation measures but could also involve a reference price linked to the Emissions Trading System (ETS) – i.e., 'carbon' contracts for difference. Such carbon contracts for difference may be a useful tool for bringing to market breakthrough technologies that may be necessary to achieve industrial decarbonisation. Contracts for difference may also involve paybacks from beneficiaries to taxpayers or consumers for periods in which the reference price exceeds the strike price.

Thus, in addition to general and abstract guidelines for the design of subsidies, the guidelines also contain very specific indications of the desired instruments and designs of the subsidy schemes of the Member States.

18.4 Summary and Outlook

Most countries in the world today are already willing to reduce greenhouse gas emissions significantly. Numerous countries have already committed themselves to this aim on an international and supranational level and are already fulfilling these commitments, in some cases with considerable success. Nevertheless, these efforts are not yet sufficient to achieve a worldwide reduction in greenhouse gas emissions, so that even more ambitious and far-reaching requirements for the reduction of greenhouse gas emissions are to be expected in the future. This goes hand in hand with the need to further expand the use of renewable energy sources. Renewable energies can already be marketed economically today and there are already established markets, especially for electricity from renewable energies. The steadily increasing demand for renewable energy and the relatively narrow time frame in which an expansion of renewable energy is needed in order to achieve the reduction targets also call for the promotion of renewable energy in the future. This will no longer be a mere start-up

160 Section 4.1.4 no. 123 CEEAG.

financing for further development of a technology that is not yet ready for the market and for the development of a market as such. The current international regulations oblige a large part of the world's countries to reduce greenhouse gas emissions in many sectors. The pioneer in setting a regulatory framework for the promotion of renewable energies is currently the European Union with its far-reaching goals from the Green Deal. Other countries do not yet have such a differentiated and distinct set of rules as a framework for their future legislation. In the EU, obligations to reduce emissions already exist in the area of electricity supply and above all in the areas of transport, mobility, in the building sector, in the area of heating and cooling supply and in industry. It is to be expected that in the future further sectors and branches of the economy will become the subject of binding reduction targets and that measures for the promotion of renewable energies will be introduced there. It is becoming apparent that there will also be an expansion in the technologies promoted. While originally the generation of electricity from renewable energy sources was the main focus of national support systems, we will see an increase of supportive systems for storage technologies and for the generation and use of green hydrogen. The support mechanisms which will be seen in the EU will include a mix of market-based and non-market-based support mechanisms, with an increasing trend towards market-based support mechanisms such as tenders, CfDs and PPAs. Where capacity markets are not introduced, national electricity markets must provide for a capacity reserve to bridge grid congestion without jeopardising security of supply. These developments are already in place in many jurisdictions and will become even more important in the future as climate change continues steadily and unabated.

Bibliography

Al Amer, Global CCS Institute, Bonn Climate Conference Outcomes – A series on the role of CCS at COP 27, Part 1, July 2022, https://www.globalccsinstitute.com/wp-content/uploads/2022/07/Bonn-Climate-Conference-Outcomes-2022-Global-CCS-Institute-1.pdf, retrieved 15 May 2024

BASE, Der Atomausstieg in Deutschland, https://www.base.bund.de/DE/themen/kt/ausstieg-atomkraft/ausstieg_node.html, retrieved 15 May 2024

Baumann/Gabler/Günther, EEG-Handkommentar, 1st. Edition, Nomos, Düsseldorf 2019 [cited as *Author*, in *Baumann/Galber/Günther*, EEG, 1st edition 2019, §, no.]

Böhringer, Das neue Pariser Klimaübereinkommen Eine Kompromisslösung mit Symbolkraft und Verhaltenssteuerungspotential, ZaöRV 2016, 753

Borrego/Martins/Lopes, Environmental Science & Policy, Volume 8, Issue 1, Portuguese industry and the EU trade emissions directive: development and analysis of CO2 emission scenarios, February 2005

CEER – Council of European Energy Regulators, Status Review of Renewable Support Schemes in Europe for 2020 and 2021, CEER report 2023, p. 12 et seq., available at https://www.ceer.eu/wp-content/uploads/2024/04/RES_Status_Review_in_Europe_for_2020-2021.pdf#, retrieved 26 November 2024

CEER – Council of European Energy Regulators, Tendering procedures for RES in Europe: State of play and first lessons learnt 2018, https://www.ceer.eu/wp-content/uploads/2024/06/CEER-Report-on-Tendering-Procedures-for-RES-in-Europe.pdf, retrieved 15 May 2024

Danish Ministry of the Environment, The Kingdom of Denmark's Report on Assigned Amount – under the Kyoto Protocol, December 2006, https://unfccc.int/files/national_reports/initial_reports_under_the_kyoto_protocol/application/pdf/aareporttounfccc-20dec2006.pdf, retrieved 15 May 2024

Dauses/Ludwigs, Handbuch des EU-Wirtschaftsrechts, 55th Supplementary Delivery, January 2022 [cited as *Author*, in: *Daus/Ludwigs*, Handbuch des EU-Wirtschaftsrechts, 55. EL, January 2022, Chapter, no.]

Denchak, NRDC – Natural Resources Defense Council, Paris Climate Agreement: Everything you Need to Know, 19 February 2021, https://www.nrdc.org/stories/paris-climate-agreement-everything-you-need-know, retrieved 15 May 2024

Engelkes/Schulz, Konrad Adenauer Stiftung, Out of Siberia, into the Desert? – The Middle East and North Africa as Building Blocks of Europe's Energy Transition, 13 June 2022, https://www.kas.de/en/web/auslandsinformationen/artikel/detail/-/content/out-of-siberia-into-the-desert, retrieved 15 May 2024

European Commission, Support schemes for renewable energy, https://energy.ec.europa.eu/topics/renewable-energy/financing/support-schemes-renewable-energy_en, retrieved 26 November 2024

European Parliamentary Research Service, Briefing, Capacity mechanisms for electricity, May 2017, https://www.europarl.europa.eu/RegData/etudes/BRIE/2017/603949/EPRS_BRI(2017)603949_EN.pdf, retrieved 15 May 2024

EWEA – The European Wind Energy Association, Design options for wind energy tenders 2015, December 2015, https://www.ewea.org/fileadmin/files/library/publications/position-papers/EWEA-Design-options-for-wind-energy-tenders.pdf, retrieved 26 November 2024

Ferroukhi/Hawila/Vinci/Nagpal, IRENA – International Renewable Energy Agency, Renewable Energy Auctions: A Guide to Design, 2015, https://www.irena.org/-/media/Files/IRENA/Agency/Publication/2015/Jun/IRENA_Renewable_Energy_Auctions_A_Guide_to_Design_2015.pdf, 26 November 2024

Fritsche, GIZ – Deutsche Gesellschaft für Internationale Zusammenarbeit, Bilaterale Energiepartnerschaften und -dialoge, November 2022, https://www.giz.de/de/weltweit/53180.html, retrieved 15 May 2024

German Federal Government ("Bundesregierung"), Press Release 280/23, 19 December 2023, https://www.bundesregierung.de/resource/blob/975228/2250312/6fc279292e1cd7a71d62fa31d7aaf7bb/2023-12-19-haushalt-data.pdf?download=1, retrieved 26 November 2024

German Federal Ministry for Economic Affairs and Energy ("Bundesministerium für Wirtschaft und Klimaschutz"), Entwicklung der erneuerbaren Energien in Deutschland im Jahr 2021, https://www.erneuerbare-energien.de/EE/Redaktion/DE/Downloads/entwicklung-der-erneuerbaren-energien-in-deutschland-2020.pdf?__blob=publicationFile&v=33, retrieved 15 May 2024

German Federal Ministry for Economic Affairs and Energy ("Bundesministerium für Wirtschaft und Klimaschutz"), Wie kann das Energiesystem der Zukunft aussehen?https://www.bmwk.de/Redaktion/DE/Downloads/Monatsbericht/Monatsbericht-Themen/2021/2021-03-wie-kann-das-energiesystem-der-zukunft-aussehen.pdf?__blob=publicationFile&v=4, retrieved 15 May 2024.

German Federal Ministry for Economic Affairs and Energy ("Bundesministerium für Wirtschaft und Klimaschutz"), An electricity market for Germany's energy transition, Discussion paper (Green Paper), October 2014, https://www.bmwk.de/Redaktion/EN/Publikationen/gruenbuch.pdf?__blob=publicationFile&v=1, retrieved 15 May 2024

German Federal Ministry for Economic Affairs and Energy ("Bundesministerium für Wirtschaft und Klimaschutz"), Internationale Energiepolitik, https://www.bmwk.de/Redaktion/EN/Textsammlungen/Energy/international-energy-policy.html, retrieved 15 May 2024

German Federal Ministry for Economic Affairs and Energy ("Bundesministerium für Wirtschaft und Klimaschutz"), Energiepartnerschaften und Energiedialoge – Jahresbericht 2020, August 2021, https://www.bmwk.de/Redaktion/DE/Publikationen/Energie/jahresbericht-energiepartnerschaften-2020.pdf?__blob=publicationFile&v=4, retrieved 15 May 2024

German Federal Ministry for Economic Affairs and Energy ("Bundesministerium für Wirtschaft und Klimaschutz"), Energiepartnerschaften und Energiedialoge – Jahresbericht 2019, April 2020, https://www.bmwk.de/Redaktion/DE/Publikationen/Energie/jahresbericht-energiepartnerschaften

-2019.pdf?__blob=publicationFile&v=10#:~:text=Was%20sind%20Energiepartnerschaften%20bzw.,bei den%20Seiten%20unter%2D%20zeichnete%20Absichtserkl%C3%A4rung, retrieved 15 May 2024

German Federal Network Agency ("Bundesnetzagentur"), SMARD Energiemarkt aktuell, Der Strommarkt im Jahr 2022, 2 January 2023, https://www.smard.de/page/home/topic-article/444/209624#:~:text=Insgesamt%20ergab%20sich%20f%C3%BCr%20das,%2C13%20Euro%2FMWh, retrieved 26 November 2024

German Federal Network Agency ("Bundesnetzagentur"), Monitoring Report 2023, November 2023

Goldner/Torwegge, The Legal Framework of Promoting Renewable Energies – a Cross-National Study, in: *Boettcher*, Green Banking, 2019, section 5.3.2 / pages 114 et seq.

Heidenhain, Martin, Handbuch des Europäischen Beihilfenrechts, 1st edition, C.H. Beck 2003, München, cited as [*Author*, in Heidenhain, Handbuch des Europäischen Beihilfenrechts, 1st edition, 2003, §, No.]

Heymann, Deutsche Bank Research, Germany Monitor, German energy supply at a historical turning point, 25 March 2022, https://www.dbresearch.com/PROD/RPS_EN-PROD/PROD0000000000522499/German_energy_supply_at_a_historical_turning_point.pdf?undefined&realload=ZTf9VpsB5ODXQOI3xg8kd1cynIicy6PYRtZ2h7CAZX~OkeWu53dYhbVoVJoPrTva, retrieved 15 May 2024

Institut der deutschen Wirtschaft Köln, Umwelt-Service Nr. 4/2011, Einspeisetarife versus Quotensysteme, 13 November 2011

International Energy Agency, France 2021 Executive Summary, 2021, https://www.iea.org/reports/france-2021/executive-summary, retrieved 26 November 2024

IPCC, Climate Change 2022 – Mitigation of Climate Change, https://www.ipcc.ch/report/ar6/wg3/downloads/report/IPCC_AR6_WGIII_FullReport.pdf, 15 May 2024

Joint Declaration of Intent on the Energy Partnership between the Government of the United Mexican States and the Government of the Federal Republic of Germany, 12 April 2016, https://www.bmwk.de/Redaktion/DE/Downloads/E/gemeinsame-erklaerung-deu-mex-energiepartnerschaft-en.pdf?__blob=publicationFile&v=8, retrieved 15 May 2024

IRENA – International Renewable Energy Association, Accessible Finance for Renewable Energy Projects in Developing Countries, January 2018, https://www.irena.org/-/media/Files/IRENA/Agency/Publication/2017/Dec/IRENA-ADFD_Accessible_Finance_Brochure_2018.pdf, retrieved 26 November 2024

Kay, Terms and Impacts of the Kyoto Protocol, Current Issues Brief No. 10 1997 – 98, Parliament of Australia – Department of the Parliamentary Library, 9 March 1998

Kemmler/Wünsch/Burret, Short paper on the development of the total electricity consumption until 2030, commissioned by the German Federal Ministry for Economic Affairs and Energy, 21 October 2021, https://www.prognos.com/sites/default/files/2021-11/20211116_Kurzpaper_Bruttostromverbrauch2018-2030.pdf, retrieved 15 May 2024

Kloepfer, Umweltrecht, 4th Edition, C.H. Beck Munich 2016

Kost/Shammugam/Fluri/Peper/Memar/Schlegl, Fraunhofer Institute for Solar Energy Systems ISE, Levelized Cost of Electricity – Renewable Energy Technologies, June 2021, https://www.ise.fraunhofer.de/content/dam/ise/en/documents/publications/studies/EN2021_Fraunhofer-ISE_LCOE_Renewable_Energy_Technologies.pdf, retrieved 26 November 2024

Lehnert/Vollprecht, Neue Impulse von Europa: Die Erneuerbare-Energien-Richtlinie der EU, ZUR 2009, Issue 6, page 307

Luxembourgian Ministry of Sustainable Development and Infrastructure – Department of the Environment, Seventh National Communication of Luxembourg under the United Nations Framework Convention on Climate Change, 12 February 2018, https://unfccc.int/sites/default/files/resource/39752148_Luxembourg-NC7-1-LU_NC7_180212.pdf, retrieved 15 May 2024

Marktoffensive Erneuerbare Energien, https://marktoffensive-ee.de/startseite/, retrieved 26 November 2024

Pause, „Saubere Energie für alle Europäer" – Was bringt das Legislativpaket der EU?, ZUR 2019, 387

Pawlik/Statista, Study: Distribution of global electricity generation by energy source in 2023, 17 October 2024, https://de.statista.com/statistik/daten/studie/164193/umfrage/stromproduktion-weltweitnach-energietraeger-im-jahr-2009/#:~:text=Struktur%20der%20Stromerzeugung%20weltweit%20nach%20Energietr%C3%A4ger%202022&text=Im%20Jahr%202022%20wurden%20rund,Torf%20mit%20rund%2036%20Prozent, retrieved 26 November 2024

Pehnt, ifeu Institute for Energy and Environmental Research Heidelberg, Heating with 65 % renewable energies – Accompanying analyses on the design of the regulation from the 2021 coalition agreement, 3 April 2023, https://www.bmwk.de/Redaktion/DE/Downloads/Energie/heizen-mit-65-prozent-erneuerbaren-energien.pdf?__blob=publicationFile&v=8, retrieved 26 November 2024.

Prognos, commissioned by the Association of the Bavarian Economy(vbw), Forecast on electricity prices / Prognos Strompreisprognose 2023, July 2023, https://www.prognos.com/de/projekt/strompreisprognose-2023, retrieved 26 November 2024

Quitzow/Thielges/Helgenberger, IASS – Institute for Advanced Sustainability Studies Potsdam, Deutschlands Energiepartnerschaften in der internationalen Energiewendepolitik – Bestandsaufnahme und Optionen für die Zukunft, April 2019, https://publications.iass-potsdam.de/rest/items/item_4085890_11/component/file_4085893/content, retrieved 15 May 2024

Ringel/Bitsch, Die Neuordnung des Rechts der Erneuerbaren Energien in Europa, NVwZ 2009, 807

Rogelj/Shindell/Jiang/Fifita/Forster/Ginzburg/Handa/Kheshgi/Kobayashi/Kriegler/Mundaca/Séférian/Vilariño, Mitigation Pathways Compatible with 1.5°C in the Context of Sustainable Development, IPCC Special Report on the impacts of global warming of 1.5°C above pre-industrial levels and related global greenhouse gas emission pathways, in the context of strengthening the global response to the threat of climate change, sustainable development, and efforts to eradicate poverty, page 93–174, May 2022

Säcker, Berliner Kommentar zum Energierecht, 5th Edition 2022 [cited as *Author*, in: Säcker, Berliner Kommentar zum Energierecht]

Scheel, Benedikt, Klimaschutz durch Umweltschutz- und Energiebeihilfen: Neue Leitlinien der Europäischen Gemeinschaft, DÖV 2009, 529

Schulz/Losch, Die geplante Neufassung der Erneuerbare-Energien-Richtlinie, EnWZ 2017, 107

Schwintowski, Anmerkung zu EuGH 28.3.2019 – C-405/16, EWeRK 2019, 92

Statistisches Bundesamt, Bruttostromerzeugung in Deutschland, https://www.destatis.de/DE/Themen/Branchen-Unternehmen/Energie/Erzeugung/Tabellen/bruttostromerzeugung.html, retrieved 26 November 2024

Stöbener de Mora, Beihilferecht: Neue Klima-, Umwelt- und Energieleitlinien EuZW, 2022, 195

Theobald/Kühling, Energierecht, 123rd Edition, C.H. Beck Munich, November 2023 [cited as *Author*, in: *Theobald/Kühling*, Energierecht]

UNFCCC, Canada's 2021 nationally determined contribution under the Paris Agreement, https://unfccc.int/sites/default/files/NDC/2022-06/Canada%27s%20Enhanced%20NDC%20Submission1_FINAL%20EN.pdf, retrieved 15 May 2024

UNFCCC, COP26 Outcomes: Market mechanisms and non-market approaches (Article 6), https://unfccc.int/process-and-meetings/the-paris-agreement/the-glasgow-climate-pact/cop26-outcomes-market-mechanisms-and-non-market-approaches-article-6#eq-1, 15 May 2024

UNFCCC, Emissions Trading, https://unfccc.int/process/the-kyoto-protocol/mechanisms/emissions-trading, retrieved 15 May 2024

UNFCCC, NDC Registry, https://unfccc.int/NDCREG, retrieved 15 May 2024

UNFCCC, Submission by Germany and the European Commission on behalf of the European Union and its Member States, 17 December 2020, https://unfccc.int/sites/default/files/NDC/2022-06/EU_NDC_Submission_December%202020.pdf, retrieved 15 May 2024

UNFCCC, Subsidiary Body for Scientific and Technological Advice, Rules, modalities and procedures for the mechanism established by Article 6, Paragraph 4, of the Paris Agreement and referred to in decision

3/CMA.3, June 2022, https://unfccc.int/sites/default/files/resource/sbsta2022_L10E.pdf, retrieved 15 May 2024

UNFCCC, The Doha Amendment, https://unfccc.int/process/the-kyoto-protocol/the-doha-amendment, retrieved 15 May 2024

UNFCCC, The United States of America National Determined Contribution, Reducing Greenhouse Gases in the United States – A 2030 Emissions Target, https://unfccc.int/sites/default/files/NDC/2022-06/United%20States%20NDC%20April%2021%202021%20Final.pdf, retrieved 15 May 2024

UNFCCC, What is the Kyoto Protocol?, https://unfccc.int/kyoto_protocol, retrieved 26 November 2024

UNFCCC, What is the Paris Agreement?, https://unfccc.int/process-and-meetings/the-paris-agreement/the-paris-agreement, retrieved 15 May 2024

Vollprecht/Lehnert/Kather, Die neue Erneuerbare-Energien-Richtline (RED II): Steife Brise oder laues Lüftchen aus Europa?, ZUR 2020, 204

Weber, Rechtswörterbuch, 28th Edition, C.H. Beck Munich 2022 [cited as *Author*, in: Weber, Rechtswörterbuch 28th editition, reference: Pariser Klimaabkommen]

Wigand/Förster/Amazo/Tiedemann, Auctions for Renewable Support: Lessons Learnt from International Experiences, AURES Report D4.2, June 2016

Part 3: **General Toolbox**

Andreas Luczak

19 The Role of Hydrogen – Technology, Integration and Economics

19.1 Motivation

The EU aims to be climate-neutral by 2050, Germany by 2045. This means that all fossil based applications will have to be replaced by a climate neutral application by then. Green electricity from wind energy and photovoltaics are the only renewable resources that have enough potential to provide the required energy in a sustainable way and at reasonable costs.

In most of the countries the generation of electricity contributes less than the half of the current CO_2 emissions. The remaining emissions are mainly caused by burning fossil fuels to either generate heat for buildings and industry or for use in the mobillity sector. Many of these non-electric applications could be replaced by applications that use green electricity directly. However, this would mean significant and costly changes for the infrastructure (for example grid extensions) and for the end users (for example expensive heat pumps instead of the cheap traditional gas heating systems). Moreover, there are some applications as long distance mobility (aviation, ships, . . .) or special industrial processes (e.g. steel production) where a direct use of green electricity is not viable. And lastly and importantly green electricity has to be stored to cover days or even weeks of low wind energy and solar resources ("dark lulls") which is extremely expensive if it is done via batteries or other storage technologies.

Climate neutral hydrogen is a very elegant way to overcome the described challenges, as it can be used and stored in a similar way as the existing fossil fuels. Consequently, the climate neutral production of hydrogen and its derivates is seen as crucial part to reach climate neutrality.

For many fossil applications both electrification and use of hydrogen is possible. Moreover, many other emission reduction measurements (e. g. decarbonization of electricity generation, energetic modernization of buildings, . . .) are necessary to reach climate neutrality. Therefore the reasonable speed of the ramp up of production and use of climate neutral hydrogen has to be discussed. The key decision factor for the use of hydrogen should be economics in order to achieve the maximum emission reduction with the available financial resources. Since the future cost of climate neutral hydrogen is very hard to predict there is a significant uncertainty about the share of the energy supply hydrogen will have in a climate neutral energy system. Figure 19.1 shows the large bandwidth of the hydrogen share in Germany based on different studies. A common aspect of all studies is the fact that a significant ramp up of hydrogen is not expected before 2030. Neverthless, EU has defined as a strategic

https://doi.org/10.1515/9783110784299-019

objective an electricity based hydrogen production capacity of 40 GW by 2030 plus 40 GW built in the neighbourhood of the EU for import.[1]

In the following the different technologies to produce climate neutral hydrogen are described. Based on that the economic principles of the integration of hydrogen in the energy system will be shown with a focus on green hydrogen in the EU.

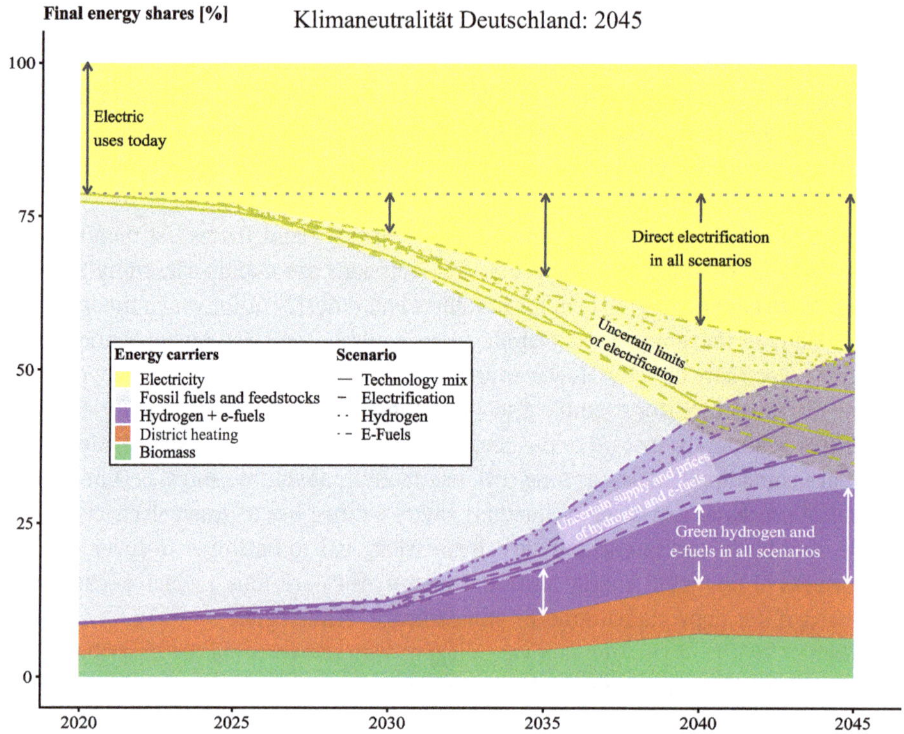

Figure 19.1: Share of hydrogen in different scenarios.[2]

19.2 Technology

In the public discussion hydrogen is sometimes assigned a color as a symbol of the way that particular hydrogen is produced:

Grey hydrogen: Nearly all hydrogen currently produced is grey hydrogen which is based on fossil resources. Natural gas is converted to hydrogen and CO_2 (steam re-

1 European Commission 2020.
2 Ariadne-Konsortium 2021 A.

forming) or is created as a by-product of refinery/chemical processes. As for now this is the cheapest way and hence globally the dominant way to produce hydrogen. In relation to its heating value it is causing about 50% higher CO_2 emissions than natural gas. Currently grey hydrogen is produced mainly for use in the chemical industry. Grey hydrogen is also the term used when hydrogen is produced by electrolysis of water using electricity from non-renewable energy sources.

Blue hydrogen: This is basically grey hydrogen produced by steam reforming where all CO_2 that is created during the production process is either stored (Carbon Capture and Storage – CCS) or used (Carbon Capture and Usage/Utilization – CCU).

- **CCS:** CO_2 can be stored in geologic rock formations. This has been successfully demonstrated in Norway. There are remaining risks, so it is unclear, if the technology will be generally accepted. For example in Germany CCS is strongly regulated and very restricted by law.
- **CCU**: Carbon utilization refers to using the separated CO_2 in an existing production process. As the CO_2 is ultimately emitted into the atmosphere this technology is not sufficient for reaching climate neutrality, but it has at least some potential for CO_2 reduction.

This technology has still some negative climate effect caused by potential leakages of the underground storage and during extraction and transportation of the required natural gas.

Turquiose hydrogen: Natural gas is split up into hydrogen and solid carbon at very high temperatures (methane pyrolysis). If the produced carbon remains permanently bound and is not combusted during further processing, this process has the potential of being climate-neutral. However, this hydrogen production process requires much natural gas (with the leakage problem mentioned above) and energy (which needs to be climate neutral). Moreover, this technology is far from mature.

Green hydrogen: Using electricity, an electrolyser splits up water into oxygen and hydrogen. There are two main types of electrolysis: Alkaline water electrolysis (AEL) and polymer electrolyte membrane (PEM). AEL electrolyers are currently cheaper than PEM electrolysers but it is expected that the costs of these two technologies will be quite similar in the future. As PEM electrolysis is better suited for operation with fluctuating power (which is typical for wind energy and solar power) it is expected that PEM-electroysers will prevail in the end. There are also other types of electrolysis, such as high-temperature electrolysis which are widely regarded as still being at an advanced R&D stage and are not yet commercially available.

In order to call the hydrogen "green" (also referred to as "renewable" or "clean") the electricity used by the electrolyser must not cause CO_2 emissions (otherwise it would be electrity based grey hydrogen). The definition when this is the case is not simple and is broadly discussed within the EU. If, for example, the generation of an existing wind energy park is used for the production of hydrogen, as a consequence this electrical energy is missing in the overall electrical system and therefore has to be replaced by additional fossil based generation causing additional CO_2 emissions. This means, that it has to be ensured that only *additional* renewable energy is used for the production of hydrogen. This condition is fulfilled, for example, if the electrolyser is directly connected to a wind energy farm and uses its output only if the wind energy park is not allowed to feed into the public grid due to a grid congestion situation. Another example would be if the electricity came from a new wind energy farm that was explicitely and only built to provide electricity for the production of green hydrogen. However, as long as the generation of electrictiy is not fully decarbonized in the overall electrical system, remains feeding the green electricity directly into the grid instead of producing hydrogen with costly electrolysers is the most efficient emission reduction (for further details see section 1.3.2).

Renewable hydrogen may also be produced through the reforming of biogas (instead of natural gas) or biochemical conversion of biomass, if in compliance with sustainability requirements. However, it is expected that its potential is very small compared to electricity based green hydrogen due to the very high land requirements compared to electricity generation from wind and solar. Therefore, the EU and especially Germany is focussing on green hydrogen using mainly wind energy and photovoltaics.[3] Given that, the following considerations will focus on electricity based green hydrogen.

Further processing of hydrogen: There are only two fossil applications where green hydrogen can be used directly: To replace grey hydrogen or to replace natural gas by mixing it to the natural gas network (currently allowed up to a fraction of 10%). For all other applications changes of the infrastructure and on the end user side are necessary. To avoid these changes hydrogen has to be converted via further processing steps either to synthetic natural gas or to liquid fuels (“e-fuels”) very similar to gasoline, kerosine, diesel or fuel oil.

19.3 Avoidance Costs of Green Hydrogen for Different Applications

In principle, green hydrogen can be used for all applications that require carbon based resources. The main applications are:

- Replacement of grey hydrogen in the chemical industry
- Replacement of fossil fuels in aviation, shipping and road traffic
- Replacement of natural gas in

3 European Commission 2020.

- the direct reduction process of steal production
- gas fired power plants (which will be still required to generate electricity during dark lulls even when huge installations of wind energy and photovoltaics exist)
- residential and industrial burners for heat generation (in cases where heat directly from electricity is not feasible)

Which of these applications will be decarbonized via hydrogen to which extent should depend on the economics of hydrogen compared to an alternative solution like the direct use of electricity. The key figures to compare the economics of different emissions reducing measures are the “avoidance costs”.

19.3.1 Principle of Avoidance Costs

Avoidance costs can be determined by estimating the additional costs of a certain emission reduction measure:

$$\text{Avoidance costs} = \frac{\text{Additional costs of climate measure}}{\text{Emission reduction of climate measure}}$$

The cost of the energy transition can be minimized if each fossil application is replaced by the technology with the lowest avoidance costs. In an ideal market based approach with a CO_2 price as sole instrument (based on CO_2 tax or a cross sector emission trading system) each market participant will always chose the technology with the lowest avoidance costs for his individual case.

If the government does not want to rely on a pure market-based approach and wants to support certain technologies (possible reasons for this see section 19.5) the avoidance costs have to be determined in order to prioritize the financial resources in the best way.

The optimum amount of green hydrogen in a climate neutral energy system depends on the avoidance costs of green hydrogen compared to alternatives for any given application. Since this prognosis is quite difficult, the predicted amount of required green hydrogen varies extremely in the different studies (see Table 19.1).

Table 19.1: Bandwidth of predicted green hydrogen demand (including synthetic derivates).[4]

2030	2040	2050
9 to 78 TWh	90 to 390 TWh	150 to 1000 TWh

4 M. Wietschel 2021.

As not all fossil applications can be decarbonized at the same time, there is also the question of prioritization. Since it is important to have the largest emission reductions as soon as possible, those fossil applications having climate neutral alternatives with low avoidance costs should be decarbonized first, and those, which can only be replaced with high avoidance costs, latest.

Based on this premise there is an optimum start time and end level of the ramp up curve of green hydrogen from an economic perspective.

In reality, a deviation from a strictly sequential implementation of climate measures according to avoidance costs might make sense. Reasons for this could be the fact that some climate measures might take a very long time (e.g. due to certain infrastructure needs) that they have to be implemented earlier. Another reason that could lead to a different prioritization could be that climate protection is not the only objective but also the support of certain industries.

Nevertheless, if politics decide to deliberately deviate from an optimum decarbonisation path the avoidance costs have to be known as accurately as possible in order to be able to weigh costs and benefits of such a deviation from the economic optimum.

As the timeframe for decarbonisation as well as the lifetime of systems and the required infrastructure are over two decades, not only the current avoidance costs but also a realistic prediction of its future development has to be taken into account. This depends on the cost development of required technologies and on external factors like market prices of fossil fuels or the amount of curtailed green electricity.

The avoidance costs specifically for the use of green hydrogen can be determined as follows:

$$\text{Avoidance costs} = \frac{\text{Costs of hydrogen application} - \text{Costs of fossil fuel application}}{\text{Emissions of fossil fuel} - \text{Emissions of hydrogen}}$$

This formula is only applicable if the emissions of hydrogen (= emissions of hydrogen production) are smaller than the emissions of the fossil fuel being replaced – otherwise the use of hydrogen wouldn´t be a climate measure.

The costs of a hydrogen application mainly consist of the sum of the hydrogen production costs and the costs of the systems and technology using the hydrogen (e.g. fuel cell, change of a steel production plant with hydrogen as reducing agent).

The costs of the fossil application being replaced are mainly the sum of the costs of the corresponding fuel and the costs of systems and technology using the fossil fuels (e.g. combustion engine, steel production plant with coke as reducing agent).

The costs of systems and technology using hydrogen are currently in most cases much higher than those using fossil fuel. The cost difference will probably diminish with increasing use of green hydrogen.

As an alternative hydrogen can be converted to synthetic natural gas or e-fuels which makes it unnecessary to change systems and technology. However, this leads to higher production costs compared to the direct use of green hydrogen.

Taking all this into account, the formula above can be approximated as follows:

$$\text{Avoidance costs} = \frac{\text{Costs of green hydrogen} - \text{Costs of fossil fuel}}{\text{Emissions of fossil fuel} - \text{Emissions of hydrogen}}$$

In the following, the four parts the avoidance costs of green hydrogen mainly depend on, are described in detail.

19.3.2 Emissions Using Green Hydrogen

Obviously, there is no direct CO_2 emission when hydrogen is used. However, depending on how hydrogen is being produced, more or less emissions may have to be taken into account as indirect emissions. The basic idea of green hydrogen is that the electrolysers uses green electricity exclusively. Since green electricity is produced without CO_2 emissions per definition, consequently also the use of green hydrogen is emission free.

If green electricity is produced by already existing eco-electricity plants it currently replaces electricity produced by fossil power plants and therefore it can not be used for the production of green hydrogen. The additional electricity demand of the electrolysers would lead to a higher production load of the fossil fuel plant and with that to significant CO_2 emissions.

In order to avoid this the EU is currently defining requirements regarding the electricity used to produce hydrogen so that it can be accounted as renewable or "green". The key idea behind these criteria is to make sure that the electricity used is truly additional green electricity to prevent that it has to be replaced by fossil based electricity. According to current legislative proposal[5] one of the following requirements has to be fulfilled:

- Direct connection between electrolyser and eco-electricity plant and commissioning of the eco-electricity plant not more than 36 months before the commissioning of the electrolyser
- The bidding zone where the electrolyser is located has an average renewable fraction of more than 90% (in the EU currently only bidding zones in Norway fulfil this requirement)
- The operator of the electrolyser has a power purchase agreement with an eco-electricity plant and is using its electricity only
- The day ahead electricity price is max. 20 €/MWh or lower than the 0.36 fold of the price of a THG emission certificate and the electrolyser is running at the same time as the eco-electricity plant is producing electricity
- Electricity is used that would have been curtailed otherwise during grid congestion periods

5 European Commission (A).

The less strict the final definition of these requirements is, the lower the cost of electricity required for the electrolysers but also the higher the risk that the electricity demand of the electrolysers is causing emissions. Consequently, organizations profiting from a fast and strong ramp up of green hydrogen are asking the EU for a less strict definition.

What is not taken into account with the criteria above (except the last one) are the emissions that could have been reduced if the additional green electricity had been fed into the public grid replacing fossil electricity instead of powering electrolysers. Currently one kilowatt hour wind energy power reduces 754 g CO_2 if it is fed into the public grid[6] whereas the green hydrogen produced by one kilowatt hour reduces only 247 g CO_2 if it replaces natural gas. Therefore, as long as the emission reduction by feeding the additional green electricity into the grid is higher than via the production of green hydrogen, the use of green hydrogen is actually increasing the CO_2 emissions compared to the cheap and simple grid feed in as alternative.

An actual emission reduction is only achieved if otherwise curtailed green electricity is used. However, since the amount and frequency of curtailed electricity is quite low (which is not expected to change much in the coming years) this leads to a very low usage factor of the electrolysers and hence to high hydrogen production cost and avoidance costs (see below).

19.3.3 Emissions of the Fossil Application Replaced with Green Hydrogen

The emission reduction achieved with hydrogen is the amount of CO_2 emissions related to production and use of the chemical that is replaced by hydrogen. If green hydrogen replaces natural gas it would be the CO_2 emission (including upstream chain) of that amount of natural gas with the corresponding heating value. Due to the production process, the CO_2 emission related to grey hydrogen is about 50% higher than natural gas with the same heating value, which makes the replacement of grey hydrogen by green hydrogen particularly reasonable.

19.3.4 Costs of Green Hydrogen

The specific costs of green hydrogen are determined by the sum of **CAPEX** (mainly the costs of the electrolyser) and **OPEX** (mainly the cost of the required electricity) divided by **the amount of hydrogen produced**.

6 T. Lauf et. al 2021.

CAPEX: About 45% of CAPEX are related to the electrolyser cell stacks where the chemical reaction takes place. The other 55% are related to system costs (balance of plant) like power electronics, gas handling and cooling system. A recent meta study shows the extreme large bandwidth of past, current and estimated future costs determined by different studies (Figure 19.2). The highest estimation of electrolyser cost is about a factor four higher than the lowest estimation. This meta study also shows that the proven alkaline electrolysis (AEL) is currently cheaper than the proton exchange membrane electrolysis (PEM), but this price advantage is expected to shrink in the future. Since the PEM-electrolysis is better suited for fluctuating electricity, it is expected that it will prevail. Most of current demonstration projects are realized with PEM-electrolysis, accordingly.

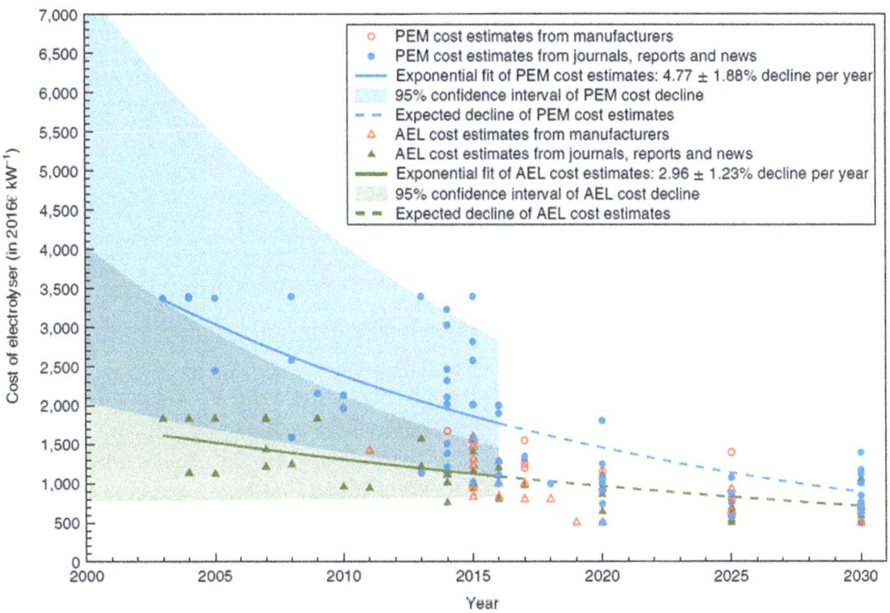

Figure 19.2: Cost development of electrolyers according to different studies.[7]

Opex: Annual maintenance and service costs are estimated with 4% of CAPEX. The cost of electricity strongly depends on the requirements that have to be fulfilled according to the definition of renewable hydrogen (see above). If electricity is used that would otherwise have been curtailed, it can be seen as free of charge. Apart from this, the cost of electricity corresponds to the LCOE of the electricity supplied by eco-electric plants.

7 G. Glenk et al. 2019. PEM and AEL are abbreviations for the two most common electrolysis technologies.

Amount of hydrogen being produced: The amount of hydrogen an electrolyser produces depends on its rated electrical power, efficiency, lifetime and utilization rate (full load hours). An efficiency of about 65% and a lifetime of about 60.000 operating hours or 20 years (whatever occurs first) can be expected. It has to be noted that there is some uncertainty with respect to these numbers as long term operational data of currently available commercial electrolysers is not yet available. If the electrolyser is operated with green electricity, which is otherwise curtailed, only full load hours of about 1000 can be expected. In this case calendrical lifetime is the determining limiting factor of lifetime.

Based on the assumptions on these three cost decisive elements the costs of green hydrogen may differ considerably (see Table 19.2).

Table 19.2: Bandwidth of predicted hydrogen costs in Europe.

	Year 2020	**Year 2030**
Prognos/BMWi[8]	6,74–10,09 €/kg or 224–336 €/MWh (LHV)	5,79–8,43 €/kg or 193–281 €/MWh (LHV)
Gutachten für das MELUND[9]	4,50–6,50 €/kg or 150–217 €/MWh (LHV)	3,90–5,70 €/kg or 130–190 €/MWh (LHV)
Agora Energiewende[10]	3,30–6,70 €/kg or 110–223 €/MWh (LHV)	1,90–5,40 €/kg or 63–180 €/MWh (LHV)

Especially countries like Germany with high energy demand in comparison to the availability of renewable resources will probably have to import significant amounts of green hydrogen from other countries. The reason for this is not necessarily the lack of sufficient potential for photovoltaics and wind energy, as studies showed that even densely populated countries like Germany have enough land potential for an exclusively domestic production.[11] But it seems very unlikely to realize this full potential due to public and political opposition to the building of large wind and solar farms. This opposition is already now a significant barrier for the expansion of renewable energy sources. Figure 19.3 shows the predicted worldwide hydrogen production volume potential together with hydrogen costs in 2050.

Lower production costs in certain regions (e.g. Spain, Norway or Chile) involve higher transportation costs. Transportation of hydrogen is quite challenging due to the very low volumetric density of hydrogen. If it is transported via dedicated hydrogen pipelines a pressure of 50–100 bar and an energy demand of about 5%/1000 km is

8 prognos 2020.

9 Umlaut Energy et al. 2021.

10 Agora Energiewende and Guidehouse 2021.

11 A. Luczak 2020.

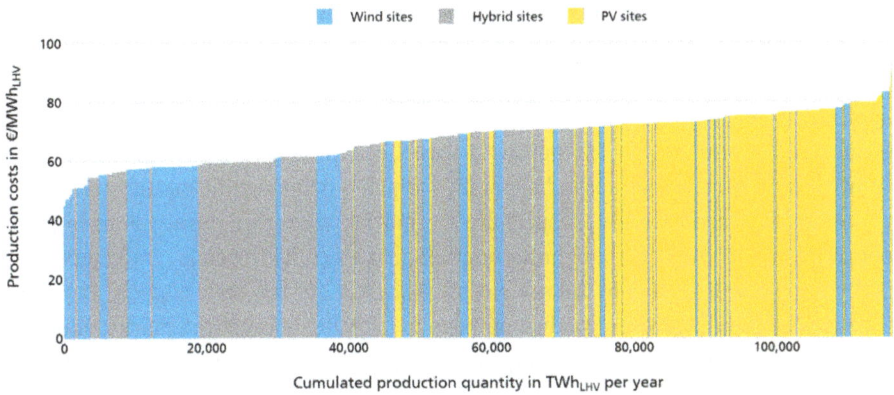

Figure 19.3: Hydrogen costs related to production potential at worldwide sites (hybrid: PV + wind).[12]

required. If it is transported via ship it has to be liquefied similar to LNG but at a much lower temperature leading to higher energy losses. An alternative could be to increase the volumetric density during transport with a liquid organic hydrogen carrier (LOHC). With this concept special organic molecules are used which can absorb and release hydrogen. In any case, shipping of hydrogen is more complex and costly than the transport of natural gas. This means that it is very unlikely that there will be a strong global market for hydrogen comparable to oil or gas. Taking all these considerations into account some studies don´t see a price advantage for imported hydrogen, some do. With distances >4000 km the costs are definitely higher compared to domestic production. Given that, the main reason for import of hydrogen is to reduce the required local additional installation of wind energy and photovoltaics power but not to reduce the costs of hydrogen.

It should be noted that from a global perspective, a country that is not yet fully decarbonized should use its green electricity and/or green hydrogen for its own decarbonisation instead of exporting green hydrogen to other countries. Also after a transformation to e-fuels like ammonia (which is required for the chemical industry anyway, especially for fertilizer production), the transport is easier and more economic than shipping hydrogen. In the long run, energy intensive parts of the industrial value chains might move to regions with high green hydrogen production to save transportation costs.

19.3.5 Cost of Fossil Fuel Replaced By Green Hydrogen

From an investor´s perspective, a specific profitability assessment has to take into account the charges and taxes applying to the fossil fuel being replaced. For a macro-

12 M. Pfennig 2022.

economic assessment of avoidance costs the comparability of different technologies is crucial. Therefore, taxes and charges (which are politically driven and quite different for different fuels) must not be included in the estimation of the fuel costs (this also applies for the calculation of hydrogen costs). The remaining production costs can be estimated from the commodity market prices. Since these prices are highly volatile, a future multi-year average has to be assumed which implies significant uncertainties.

With the war in Ukraine, especially the market costs of natural gas have risen enormously. This might lead to the assumption that green hydrogen is now competitive and investors will be able to finance electrolysers without any state support. But since investors in green hydrogen production have to take into account the market situation during the full lifetime of the electrolysers, temporary market price spikes don´t change the situation significantly.

19.3.6 Avoidance Costs – Results

Since the avoidance costs depend on cost and emission of the fossil fuel being replaced by green hydrogen the result is depending on the specific application hydrogen is used for emission reduction. Therefore there are three main applications resulting in different avoidance costs: Replacement of natural gas, replacement of liquid fossil fuels, replacement of grey hydrogen. Since grey hydrogen is more expensive and associated with significant higher CO_2 emissions than the other fossil fuels it can be expected that the replacement of grey hydrogen leads to the lowest avoidance costs. This is confirmed by the results of various studies (Table 19.3).

19.4 Project Examples

According to the project database available at the International Energy Agency[13] the worldwide electricity based hydrogen capacity is about 150 MW. In the following, there is a short description of some of the largest projects:

20 MW Puertollano (Spain)[14]

An alliance of the Spanish utility Iberdrola and Fertiberia, which is the largest manufacturer of fertilizers in Spain, have announced to build up a green hydrogen production capacity of 830 MW. As investment a sum of 1.8 billion Euros has been announced,

13 International Energy Agency 2021.

14 Iberdrola 2022.

Table 19.3: Comparison of the estimation of avoidance costs of green hydrogen (the years in brackets refer to the timeframe of the prognosis).

	Replacement of grey hydrogen	Replacement of natural gas	Replacement of liquid fossil fuel
Publication (timeframe)			
Luczak (2030)[15]	187–296 €/t	540–755 €/t	397–583 €/t
BCG / Prognos (2050)[16]		370 €/t	280 €/t
PIK (2020 – 2030)[17]		340–520 €/t	
Frontier[18]	190–320 €/t	300–700 €/t	
Agora Energiewende (2030)[19]	170–430 €/t		
TUHH (2020)[20]		3.964 €/t	1.149 €/t

which reflects quite high specific electrolyser costs taking the cost development curve shown above into account. The first project has just been commissioned and is one of the largest plants producing green hydrogen worldwide. As part of this project the hydrogen output pf a 20 MW electrolyzer system is used to produce ammonia which is required for manufacturing fertilizers. Moreover, a 100 MW photovoltaics plant and a 20 MWh lithium-ion storage system will be built at the same site.

20 MW Bécancour (Canada)[21]

Air Liquide has built a 20 MW electrolyzer using hydro power to generate green hydrogen for industrial use and mobility. It is located at Air Liquide´s hydrogen production complex of traditional grey hydrogen.

10 MW Cologne (Germany)[22]

At Shell´s Energy and Chemicals Park Rheinland near Cologne The REFHYNE consortium a 10 MW electrolyser is producing green hydrogen which is initially used for the production of e-fuels. Main objective of this project is technology testing and exploring of hydrogen applications in industry, power generation, heating and transport. The origin of the required electricity has not been mentioned, therfore it is unclear if

15 A. Luczak 2022.
16 P. Gerbert 2018.
17 Ariadne-Konsortium 2021 B, Figure 19.7.
18 C. Gatzen et al. 2021.
19 C. Schneider et al. 2019.
20 S. Drünert et al. 2019.
21 Air Liquide 2021.
22 REFHYNE 2021.

this project has a positive or negative effect on emissions. There are plans to expand the capacity up to 100 MW in the future.

10 MW Fukushima (Japan)[23]

A consortium including Toshiba has constructed a 10 MW electrolyser system using renewable energy including a hydrogen storage unit. The hydrogen volume production and storage is adjusted by an energy management system according to power grid balancing requirements without the use of storage batteries which are normally helping to balance the power grid. Hydrogen produced will be used to power stationary fuel cells, cars, busses and more.

19.5 Should the State Accelerate the Ramp up of Green Hydrogen?

As mentioned before there is a significant public disagreement about how much and how fast the production of green hydrogen should be ramped up. This disagreement is influenced by financial interests of private companies. The arguments in favour and against a fast ramp up of green hydrogen production are described in the following section.

19.5.1 Arguments not to Accelerate the Ramp up of Green Hydrogen Prodcution

From an economic perspective, production and use of green hydrogen should not start before its avoidance costs are lower than the existing marginal avoidance costs. An indication of the marginal avoidance costs in Europe is the CO_2 price of the European Emission Trading System which is currently in the range of 70 €/t. Certainly the marginal avoidance costs (and hence the CO_2 price in the ETS) will increase and the avoidance costs of green hydrogen decrease in the coming years. However, according to most projections it is highly unlikely that they will match before 2035 (Figure 19.4). As soon as this is the case, the ramp up of green hydrogen production and usage will automatically start without any other governmental intervention. The ramp up of the production of green hydrogen is strongly linked to the availability of surplus green electricity that cannot be used directly. If the annual installation of photovoltaics and wind energy capacity is not significantly increased, there won´t be much additional green electricity available for the production of green hydrogen even in 2035. With

23 Toshiba 2020.

large subsidized installations of electrolysers long before having sufficient green electricity there is a high risk that either the electrolysers can't be operated very often (converting them to stranded assets) or that the operation of these electrolysers prevents a faster decarbonisation of electricity generation.

Nevertheless, most studies show a relevant amount of use of green hydrogen already before 2030. The reason for this seems to be that due to the complexity of those simulations there is no economic optimization regarding start and ramp rate of green hydrogen taking possible alternative opportunities (e.g. faster/additional ramp up of wind energy and photovoltaics capacities) into account. In addition to that, the available studies do not include risk assessments about the consequences if the availability of green electricity is lower than expected compared to the consequences if the ramp up of green hydrogen production is postponed.

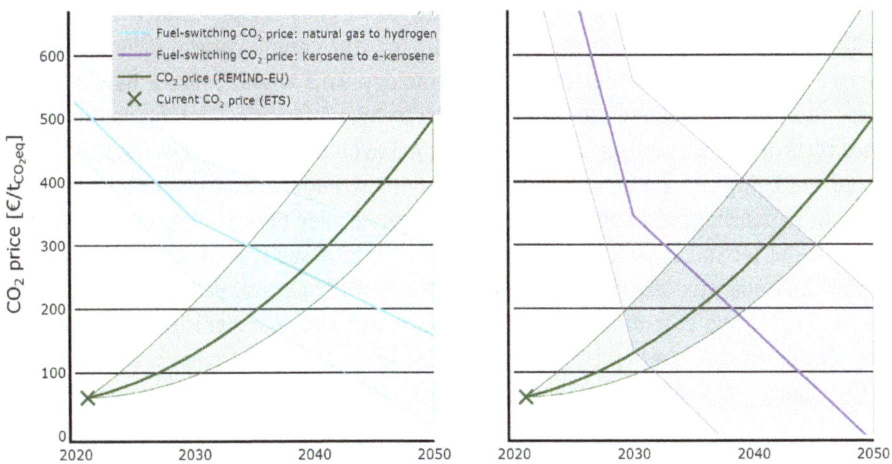

Figure 19.4: Prognosis of marginal avoidance costs (= CO2 price) and the avoidance costs of the replacement of natural gas or liquid fossil fuel (kerosene).[24]

19.5.2 Arguments to Ramp up Production of Hydrogen Already now Despite High Avoidance Costs

Sector Specific Climate Targets

In some EU countries, there are specific emission reduction targets for certain sectors like industry or mobility. Especially in the industry sectors a significant emission reduction without use of green hydrogen (e.g. in the steel production) is quite difficult.

24 Ariadne-Konsortium 2021 B Figure 19.7.

In order to achieve such sector specific targets, a significant amount of green hydrogen is required regardless of the costs.

However, sector specific targets contradict the principle of economic optimization mentioned at the beginning of this chapter. According to this principle emissions should be reduced across all sectors with lowest possible avoidance costs. Because of this principle, the EU is aiming to extend the ETS to the heating and mobility sectors. Following this principle it is unnecessary to enforce a certain emission reduction in the industry sector using technologies with very high avoidance costs as green hydrogen.

Avoidance of Fossil Lock-in

For applications with long term investment cycles it may happen that without availability of green hydrogen the renewal of infrastructure and technical facilities will be carried out based on the existing fossil technology. This would cause a barrier to replace these relatively new infrastructure and technical facilities as soon as the avoidance costs of green hydrogen reach the marginal avoidance costs. However, this risk of either not being climate neutral in 2045 or having stranded assets can also be avoided without expensive production of green hydrogen earlier than it makes economically sense. New technical facilities and production processes can be made “hydrogen-ready” and operated during a transitional period with natural gas or grey hydrogen which, e. g. in the case of steel production, already leads to a significant emission reduction. Later, a further emission reduction by using green hydrogen can be done exactly when the associated avoidance costs are lower than all other alternatives.

Limited Ramp up Rate of Green Hydrogen

Depending on how steep a possible ramp up rate of green hydrogen production is assumed the ramp up has to start sooner or later in order to reach the capacity required in 2045. Assuming a relatively small achievable ramp up rate would then lead to the necessity to start the ramp up as soon as possible. However, this logic does not take into account that most electrolysers that are built before 2030 will probably reach their end of lifetime before 2045 and cannot contribute to the required capacity for climate neutrality.

The ramp up of wind energy and photovoltaics energy (both are comparable to electrolysers regarding complexity and cost per kilowatt) has proven that the amount of annual installations mainly depends on achievable profits and can be increased by several gigawatts each year. With that, an electrolysis capacity of about 50 GW, which is seen as sufficient capacity required for a climate neutral energy system, can be reached within 10–15 years. A current prognosis shows a possible ramp up of annual installations of about 17 GW within four years.[25] That given, a significant ramp up starting immediately does not seem to be necessary.

Security of Energy Supply

The war in Ukraine has shown how important reliable energy resources are. So it sounds reasonable to replace the missing Russian natural gas as much and as fast as possible with green hydrogen. However, this only makes sense if the green hydrogen is produced with electricity otherwise curtailed, or if it is imported from countries that don´t have natural gas plants. The reason for this is the following: One kilowatt hour green electricity used for the production of hydrogen saves less than one kilowatt hour natural gas due to the losses of the electrolysis. On the other hand, one additional kilowatt hour green electricity fed into the grid saves more than one kilowatt hour fossil fuel due to the losses in the combustion plants. As most of the time gas plants are the marginal power plants reacting to marginal changes of electricity demand, more than one kilowatt hour natural gas is saved. Therefore, in order to save natural gas it is always better to feed green electricity directly into the grid (if this is technically possible) instead of producing green hydrogen.

Protection of Existing Gas Based Assets and Infrastructure

If green hydrogen is ramping up weaker due to economic optimization with a higher amount of direct electrification, the business model of owners and operators of gas based assets and infrastructure is jeopardized. Consequently, lobbyists representing the interests of gas grid owners and operators campaign for a fast and strong ramp up of hydrogen production.[26,27]

Support of Industries Benefitting from Green Hydrogen

One of the main targets of most hydrogen strategies is to support economic growth. By governmental funding of a fast ramp up of domestic hydrogen production and usage local industry should get a technology lead, which generates additional jobs, and increases tax income.

25 Zwei 2022.
26 Zwei 2021.
27 B. Balanyá et al. 2021.

A similar strategy has been applied regarding the support of Photovoltaic in Germany especially in the first decade of 2000. At that time, extremely attractive subsidies for photovoltaics had created the world´s biggest photovoltaics market in Germany. With such a large domestic market as competitive advantage, the German photovoltaics industry could gain worldwide technology and market leadership. However, competitors with lower production costs especially from Asia benefitted from innovations created at very high costs at that time ("technology spill over effect") and could finally squeeze out the German companies ("second/last mover advantages").

In addition to that, it should be noted that there are already significant investments in the range of 500 billion USD planned until 2030. Therefore, a crucial advantage of a strong domestic market generated by high governmental subsidies seems unlikely and has the risk of windfall profits for non-domestic companies.

19.6 How Could Green Hydrogen be Supported?

As mentioned above for at least the next ten years the expected price for CO_2 emission certificates will not be high enough that selling certificates based on CO_2 savings achieved with green hydrogen can finance the avoidance costs. If, for whatever reason, there is a political decision to ramp up green hydrogen faster than just based on market conditions, some additional kind of state subsidies and regulations are required. The following options are discussed or already used:

19.6.1 Investment Grants

There are several EU and national funding programs in place. Investors interested in such a funding have to apply with their project. The selection of projects is generally based on greenhouse gas (GHG) emission avoidance, innovation, maturity level, scalability and cost efficiency. The funding covers a certain percentage (typically in the range of 50%) of relevant project costs (CAPEX and for large-scale projects also OPEX).[28]

19.6.2 Carbon Contracts for Difference (CCfDs)

During an agreed time period, the state is paying the difference between the actual CO_2 price and a certain agreed CO_2 price required to finance green hydrogen. The amount of this subsidy fluctuates as it depends on the current CO_2 price. If the CO_2 price exceeds the agreed CO_2 price the producer of green hydrogen has to pay back

28 European Commission (B).

the difference. As a result the producer of green hydrogen gets a fixed amount of money per ton on top of the market price of green hydrogen. This concept is similar to the concept of a feed in tariff in the electric sector.

19.6.3 Hydrogen Supply Contracts

The state is buying a certain amount of green hydrogen at a fixed premium price and selling it at a lower price to the free market. In order to reduce funding costs this can be done via double auction: The hydrogen is bought from the cheapest suppliers (Hydrogen Purchase Agreement) and it is sold to end users willing to pay the highest prices (Hydrogen Service Agreement). The difference between the purchasing price and the selling price has to be covered by the funding entity. One example of this funding scheme is the "H2Global Stiftung".[29] Also this concept is similar to the concept of a feed in tariff in the electric sector.

19.6.4 Hydrogen Quota

Instead of financially supporting green hydrogen, a certain minimum use of green hydrogen could be legally enforced.[30]

As supporter of such an instrument a coalition of energy providers, shipping companies and NGO´s has called on the EU to introduce a minimum quota of 6% green hydrogen fuels by 2030.[31]

Germany has announced to establish binding minimum quota on aviation fuels sold in Germany combined with a purchase obligation. With this measure an annual production of at least 200.000 tons of kerosene for the German aviation industry by 2030 based on green hydrogen is expected.[32]

The EU is already using quota as a climate instrument to support a certain technology in the mobility sector. Although there is no direct quota for electric cars, the EU fleet target is set to such a low level which practically can only be achieved with a certain quota of sold electric cars.

29 H2Global Stiftung 2022.

30 Schlund, D. et al. 2021.

31 Transport & Environment 2022.

32 Federal Ministry for Digital and Transport.

19.7 Recommendations

Any strategic objectives of a certain hydrogen production capacity in a certain year (like the plan of the EU to install 40 GW of electrolysers by 2030) have to be transparently justified. With that, future energy system studies should include plausible estimations of the optimum ramp up rate of green hydrogen production. Risk assessments should compare additional costs and emissions of a slower ramp up compared to a ramp up that is too fast compared to the expansion of renewable energies.

If the state decides – for whatever reasons – to subsidize green hydrogen, the actual ramp up should be linked to external constraints. These are primarily the availability of surplus green electricity and the actual cost reduction achievements regarding the electrolyser costs.

Electroysers should only be operated with green electricity that otherwise would have been curtailed. Only then an increase of fossil generation leading to an increase of CO_2 emissions is avoided. An exception of this rule is only acceptable if there are other reasons for the operation of the electrolyser like getting operational experience in an R&D project.

Based on all state funded hydrogen projects a permanent, transparent and systematic cost analysis should be done. The knowledge of the actual development of the hydrogen costs is crucial for optimal prioritization of hydrogen compared to alternative emission reduction measures. This would also help to assess the cost efficiency of projects that apply for funding. There has to be a plausible reason to fund a project despite of higher avoidance costs, e.g. valuable scientific findings compared to existing hydrogen projects.

Green hydrogen should only be subsidized for use in applications where hydrogen is either the only viable climate neutral alternative or all other alternatives have higher long term avoidance costs (= "no-regret applications"). These are mainly: Production of ammonium for chemical use, direct reduction of steel and long distance road-, sea- and air transport. Otherwise the subsidized infrastructure investments and learning curves on the hydrogen consumption side might be practically worthless in the long run. If these "no-regret applications" are not feasible for a certain project, the green hydrogen should be simply mixed to the natural gas network. This is allowed up to a blending ratio of 10% without any change of infrastructure and therefore the easiest and cheapest way to use green hydrogen.

19.8 Conclusion

Hydrogen is one of the crucial technologies required to achieve climate neutrality. Unlike wind energy or photovoltaics it is not a renewable energy but an energy carrier to make surplus green electricity available for applications that are difficult to electrify. In addition to that, hydrogen will be required to store surplus green electricity for time periods when there is not enough green electricity during “dark lulls”.

There is a large uncertainty about the cost development of green hydrogen compared to alternative technologies. Therefore, there is also an uncertainty for which applications hydrogen should be used, and how fast its production should be ramped up. For an economic evaluation of this, CO_2 avoidance costs of hydrogen and its alternatives have to be determined. Since green electricity is converted with expensive electrolysers to a gas, which is comparable to relatively cheap natural gas, the CO_2 avoidance costs of green hydrogen are higher than most other possible emission reduction measures. If CO_2 emissions should be reduced as fast as possible, the most efficient emission reduction measures have to be implemented first. This would mean that a ramp up of green hydrogen production is not required before 2035. Mainly because of strong lobbying efforts of the industry related to gas infrastructure and the hope of several governments to gain a technology lead in the hydrogen industry there is a very ambitious target in the EU to produce and use significant amounts of green hydrogen already well before 2035.

Bibliography

Agora Energiewende and Guidehouse: Making renewable hydrogen cost-competitive: Policy instruments for supporting green H_2. Report written on behalf of Agora Energiewende (2021) https://static.agora-energiewende.de/fileadmin/Projekte/2020/2020_11_EU_H2-Instruments/A-EW_223_H2-Instruments_WEB.pdf, retrieved at 2022-08-22

Air Liquide (2021): Air Liquide inaugurates the world´s largest low-carbon hydrogen membrane-based production unit in Canada https://energies.airliquide.com/air-liquide-inaugurates-worlds-largest-low-carbon-hydrogen-membrane-based-production-unit-canada, retrieved at 2022-08–22

Ariadne-Konsortium (A): Durchstarten trotz Unsicherheiten: Eckpunkte einer anpassungsfähigen Wasserstoffstrategie. Potsdam-Institut für Klimafolgenforschung (PIK), 2021; https://ariadneprojekt.de/media/2021/11/Ariadne_Kurzdossier_Wasserstoff_November2021.pdf, retrieved at 2022-01–11

Ariadne-Konsortium (B): Durchstarten trotz Unsicherheiten: Eckpunkte einer anpassungsfähigen Wasserstoffstrategie. Potsdam-Institut für Klimafolgenforschung (PIK), 2021, https://ariadneprojekt.de/media/2021/11/Ariadne_Kurzdossier_Wasserstoff_November2021.pdf, retrieved at 2022-08–22

Balanyá, B. et al.: The hydrogen hype: Gas industry fairy tale or climate horror storry? Corporate Europe Observatory (CEO), Food and Water Action Europe (FWAE) and Re:Common, Brüssel 2020. https://corporateeurope.org/sites/default/files/2020-12/hydrogen-report-web-final_3.pdf, retrieved at 2022-08–22

Drünert, S. et al.: Power-to-X (PtX) aus „Überschussstrom“ in Deutschland – Ökonomische Analyse. Z Energiewirtsch (2019) 43:173–191. https://doi.org/10.1007/s12398-019-00256-7, retrieved at 2022-08–22

European Commission: A hydrogen strategy for a climate-neutral Europe, Brussels 2020, https://eur-lex.europa.eu/legal-content/EN/TXT/?uri=CELEX:52020DC0301, retrieved at 2022-08–22

European Commission (A): Commission delegated regulation (EU) . . . /. . . of supplementing Directive (EU) 2018/2001 of the European Parliament and of the Council by establishing a Union methodology setting out detailed rules for the production of renewable liquid and gaseous transport fuels of non-biological origin, Ref. Ares(2022)3836651 – 20/05/2022 Brussels 2022

European Commission (B): Innovation Fund, 2022 https://ec.europa.eu/growth/industry/strategy/hydrogen/funding-guide/eu-programmes-funds/innovation-fund_en, retrieved at 2022-08–22

Gatzen, C., Lenz, A.-K. (2021). Wasserstoff zur Dekarbonisierung des Wärmesektors. Frontier economics im Auftrag der DVGW. http://www.frontier-economics.com/media/4590/wasserstoff-im-waermemarkt.pdf, retrieved at 2022-08–22

Gerbert, P. et al. (2018). Klimapfade für Deutschland. Boston Consulting Group und Prognos im Auftrag des Bundesverbandes der Deutschen Industrie e. V. (BDI). https://web-assets.bcg.com/e3/06/1c25c60944a09983526ff173c969/klimapfade-fuer-deutschland.pdf, retrieved at 2022-08–22

Glenk, G., Reichelstein, S.: Economics of converting renewable power to hydrogen. *Nature Energy* **4**, 216–222 (2019). https://doi.org/10.1038/s41560-019-0326-1, retrieved at 2022-08–22

H2Global Stiftung: The H2Global Mechanism, 2022 https://www.h2global-stiftung.com/project/h2g-mechanism, retrieved at 2022-08–22

Federal Ministry for Digital and Transport: Federal government, states and industry agree on roadmap for market ramp-up of climate friendly electricity-based aviation fuel 2021, https://www.bmvi.de/SharedDocs/EN/PressRelease/2021/044-scheuer-roadmap.html, retrieved at 2022-08–22

Iberdrola (2022): Iberdrola commissions the largest green hydrogen plant for industrial use in Europe, https://www.iberdrola.com/about-us/lines-business/flagship-projects/puertollano-green-hydrogen-plant, retrieved at 2022-08–22

International Energy Agency (2021): Hydrogen Projects Database https://www.iea.org/data-and-statistics/data-product/hydrogen-projects-database, retrieved at 2022-08–22

Lauf, T., Memmler, M., Schneider, S.: Emissionsbilanz erneuerbarer Energieträger. Umweltbundesamt, 2021 https://www.umweltbundesamt.de/sites/default/files/medien/1410/publikationen/2021-12-13_climate-change_71-2021_emissionsbilanz_erneuerbarer_energien_2020_bf.pdf, retrieved at 2022-08–22

Luczak, Andreas: Deutschlands Energiewende – Fakten, Mythen und Irrsinn, Wiesbaden 2020

Luczak, Andreas: Zukünftige Erzeugung und Nutzung von grünem Wasserstoff in Schleswig-Holstein: Abschätzung der CO2-Vermeidungskosten, report written on behalf of EKSH, 2022, https://www.eksh.org/fileadmin/redakteure/user_upload/Kurzstudie_Prof._Dr._Andreas_Luzcak__2022_.pdf, retrieved at 2022-08–22

Maximilian Pfennig, Diana Böttger, Benedikt Häckner, David Geiger, Christoph Zink, André Bisevic, Lukas Jansen, Global GIS-based potential analysis and cost assessment of Power-to-X fuels in 2050, Applied Energy, Volume 347, 2023, 121289, ISSN 0306-2619, https://doi.org/10.1016/j.apenergy.2023.121289.

prognos:. Kosten und Transformationspfade für strombasierte Energieträger. Report written on behalf of the Federal Ministry for Economic Affairs and Innovation, 2020, https://www.bmwi.de/Redaktion/DE/Downloads/Studien/transformationspfade-fuer-strombasierte-energietraeger.pdf?__blob=publicationFile&v=4, retrieved at 2022-08–22

REFHYNE (2021): Shell starts up Europe´s largest PEM green hydrogen electrolyser https://refhyne.eu/shell-starts-up-europes-largest-pem-green-hydrogen-electrolyser/, retrieved at 2022-08–22

Schlund, David and Schönfisch, Max: Analysing the Impact of Renewable Hydrogen Quota on the European Electricity and Natural Gas Markets, EWI Working Paper, No 21/03 2021 www.ewi.uni-koeln.de/cms/wp-content/uploads/2021/05/EWI_WP_21-03_Analysing_the_Impact_of_a_Renewable_Hydrogen_Quota_Schlund_Schoenfisch.pdf, retrieved at 2022-08–22

Schneider, C. et al. (2019). Klimaneutrale Industrie: Ausführliche Darstellung der Schlüsseltechnologien für die Branchen Stahl, Chemie und Zement. Analyse im Auftrag von Agora Energiewende. https://static.agora-energiewende.de/fileadmin/Projekte/2018/Dekarbonisierung_Industrie/166_A-EW_Klimaneutrale_Industrie_Ausfuehrliche-Darstellung_WEB.pdf, retrieved at 2022-08–22

Toshiba Energy Systems & Solutions Corporation (2020): The world´s largest –class hydrogen production, Fukushima Hydrogen Energy Research Field (FH2R) now is completed at Namie town in Fukushima, https://www.global.toshiba/ww/news/energy/2020/03/news-20200307-01.html?utm_source=www&utm_medium=web&utm_campaign=since202202ess, retrieved at 2022-08–22

Transport & Environment: Broad industry-NGO coalition calls for EU hydrogen quota for shipping 2022, https://www.transportenvironment.org/discover/broad-industry-ngo-coalition-calls-for-eu-hydrogen-quota-for-shipping/, retrieved at 2022-08–22

umlaut energy, EMCEL, Becker Büttner Held, Forschungszentrum Jülich: H_2-Erzeugung und Märkte Schleswig-Holstein. Report written on behalf of the Ministry of Energy Transition, Climate Protection, Environment and Nature 2021. https://wasserstoffwirtschaft.sh/file/201217_wasserstofferzeugung-und–maerkte-schleswig-holstein_gutachten.pdf, retrieved at 2022-08–22

Wietschel, M. et al: Metastudie Wasserstoff – Auswertung von Energiesystemstudien. Report written on behalf of the National Hydrogen Council. Fraunhofer ISI, Fraunhofer ISE, Fraunhofer IEG, 2021. https://www.ise.fraunhofer.de/content/dam/ise/de/documents/publications/studies/Metastudie_Wasserstoff.pdf

Zwei – Das Magazin für Wasserstoff und Brennstoffzellen, 22. Jahrgang, Heft 1 Jan 22, S. 20 Hydrogeit Verlag 2022

Zwei – Das Magazin für Wasserstoff und Brennstoffzellen, 21. Jahrgang, Heft 4 Okt 21, S. 12–14 Hydrogeit Verlag 2021

Nicolai Herrmann

20 Pricing and Trading CO_2: Concepts, Assessment and Impacts on Competition

20.1 Introduction

This chapter starts with an introduction of the concept of external costs and describes the problem of market inefficiencies (market failure) and the aim to internalise external costs. Then the basic instruments for the internalisation of external cost are introduced: environmental standards, and price and volume control instruments. They are briefly compared along the main evaluation points in the field of environmental economics and the main arguments for a volume-based cap-and-trade system, as the standard market-based instrument for greenhouse gas emission control in practice, are presented. The practical implementation of an emission trading scheme regarding its set-up, operation, volume (and price) control are then highlighted, using the European Union's Emission Trading Scheme (EU ETS) as an example. The key regulatory and market-related points of the EU ETS are described. Finally, impacts of the introduction of a cap-and-trade system on competition and trade between nations and corporations are discussed.

20.2 Internalisation of External Costs

The neoclassical theory of markets and competition is the foundation of all market economies in the European Union and beyond. It is centred around the concept of a competitive balance between supply and demand that comes about in a market. A functioning market allows for effective price-formation and market clearing and is expected to lead to the most efficient allocation of resources and in parallel the maximisation of consumer and producer surpluses. However, markets do only deliver efficient results if certain preconditions such as the homogeneity of products, perfect competition and information, absence of specific preferences and a full internalisation of external effects are fulfilled. If this is not the case, markets regularly fail to deliver efficient results.[1] However, such market failures not only reduce the efficiency of a market, but they can also lead to largely unsustainable outcomes that become a concern for policymakers and society. Energy markets are particularly vulnerable for market failure because of their structure (e. g. natural or grown monopolies, high capital intensity paired with path-dependency) and the proximity of some of their products

1 See Chapter 4.2 for further information regarding market failure.

https://doi.org/10.1515/9783110784299-020

and services to the nature of public goods. But the single largest challenge that energy markets and policymakers as their 'market designers' must face, is undoubtedly the internalisation of external costs to mitigate anthropogenic climate change.

External effects are defined as costs (or benefits) that are a result of the behaviour of market actors but that are not reflected in the market price. Negative external effects are defined as external costs. External costs can be caused by producers but also consumers in a market. Because external costs are not reflected in the calculation of market prices, they lead to prices that are too low to account for the full costs associated with the specific product – the market price for that product is distorted. This again leads to a situation in which the demand and the resulting market-clearing volume for that product are too high, as the supply function only accounts for internal costs and ignores external effects. A missing internalisation of external costs hence distorts price, production, and allocation.

If producers are now forced by policy intervention to take into account the previously external cost into their marginal production function (i. e. to internalise them), then prices in the market will increase. This leads to higher prices, less demand and less product volume that is cleared in the market. By that mechanism internalisation (by what means will be discussed later in this chapter) reduces the occurrence of external costs – ideally to a level that matches the social marginal cost function.

While the (local) pollution of water and soil were the cases for which environmental economist initially developed and first applied the concept of external costs, greenhouse gas emissions and the problem of man-made climate change are nowadays the main field for application for this concept. The external cost theory and especially the policy instruments to enforce their internalisation are at the core of the battle against global warming. That is why it is necessary that we first discuss the theory of external costs and the instruments to internalise them and then turn to a discussion of how CO_2 can be priced and traded in markets like the EU ETS.

20.3 Instruments for an Internalisation of External Costs

In the following, three different instruments for an internalisation of external costs are presented and briefly compared in regards to their advantages and disadvantages as discussed in environmental economics literature. We will focus on policy instruments that are designed to control emissions that drive anthropogenic climate change.

20.3.1 Environmental Standards

The setting of environmental standards by an authority is one of the standard instruments for the reduction of negative external effects (environmental damages), such as harmful emissions. This can be realised for example by an explicit emission limit per unit of production and can apply for processes or products. There are manifold examples for the instrument of environmental standards such as the regulations for air pollution control (e. g. power plants that are obliged to not exceed an externally set limit on sulphur dioxide emissions per Megawatt hour electricity produced) as well as environmental standards for car manufacturers to reach externally set limits on the carbon emissions (CO_2/per kilometre) for the cars or the fleet they produce. A direct limitation of production volumes for a product with environmentally harmful impact is also understood as a type of environmental standard setting.

As the main advantages of this instrument, environmental economics literature lists the comparably good effectiveness (the mechanism produces fast and direct results), the practicability, and the verifiability of environmental standards. Environmental standards can also be particularly helpful to regulate new production units that are subject to regulatory approval. However, environmental standards do not aim for the explicit internalisation of external costs in the meaning of an internalisation into the target market's pricing mechanism (i. e. "putting a price on emissions"); so they are per se not market-based but have a purely regulatory nature.

Hence, the limited incentives for technical progress are mentioned as one of the main disadvantages of this instrument – however, if standards are set dynamically (e. g. using state-of-the-art technology as a benchmark), this downside can become less relevant. Summing up, setting environmental standards represents an important measure in reducing external effects, however, it is not a market-based mechanism and does not lead to direct price or volume effects in the market.

20.3.2 Price Control Mechanisms: Taxes and Levies

In a market economy, keeping or restoring the functionality of the price mechanism should be at the core of policy interventions. Therefore, the internalisation of external effects in (existing) markets is the guiding principle of market-based climate policy, as this is the only way to restore and strengthen the functioning of the price mechanism and use it as a lever to directly incentivise market actors to behave in a way that is efficient in economic terms and meets the environmental objective.

Therefore, most instruments for greenhouse gas emission control apply a market-oriented approach and focus either on price control (i. e. instruments that influence prices by environmental taxes or charges that are levied in proportion to the amount of pollution) or volume control (i. e. instruments where the state specifies a maximum permissible emission quantity and issues tradeable certificates to operationalise the emission cap in the market).

Economic theory argues that market-based instruments meet a desired level of emission reduction with higher efficiency than instruments that follow a purely regulatory "command-and-control" logic. The main reason for the higher efficiency of market-based mechanisms is that market players will operate within their individual cost structure to realise the most cost-effective emission reduction options individually and with minimal political and regulatory interference. However, also market-based instruments need a certain regulatory framework to enforce, manage and control the intended internalisation effect.

Price control mechanisms like environmental levies and taxes are one of the two standard instruments for a market-oriented internalisation of external effects. To the extent that such taxes or levies are used to internalise external costs, they are also referred to as Pigouvian Taxes. In comparison to environmental standards (see above), price control is an indirect instrument, as is does not directly limit the relevant production activity but puts a price on the utilisation of environmentally harmful production factors (input tax), or on emissions (output tax). The price is set externally, while emission volumes emerge from economic optimisation within the market.

The necessary tax or levy needs to be set by a regulatory authority – i. e. the state. By applying a price to the relevant input or output factors (production factors, products, emissions), rational producers will reduce their emissions as long as their abatement costs are below the cost set by the tax or levy. A price control mechanism can be applied to either price input factors (e. g. fuels, respectively their carbon content) that lead to emissions or the relevant output factor (i. e. emissions) directly.

The fundamental function of a price control mechanism is illustrated in the following Figure 20.1 for two sample firms A and B in a sample market:

Figure 20.1 illustrates the different marginal abatement cost (= "demand for pollution") of two firms A and B. Without any emission control, both firms operate at their maximum emission volumes of V_{A0} and V_{B0} and the total emission volume for the market amounts to V_0 which is the sum of V_{A0} and V_{B0}. Because there is no price or volume cap for those emissions, their quantity or cost is not considered.

The two firms have different marginal abatement cost. This is reflected by the different slopes of the two abatement cost functions in the graph. The abatement cost function generally describes the costs for the respective firm to reduce one unit of emissions. Due to the unequal abatement cost functions, it would be inefficient to demand equal emission volume reductions from both firms – for example require an emission reduction of 30% from each firm. Instead, it proves more efficient to distribute abatement volumes between firms along their individual marginal abatement cost. This is especially relevant when abatement costs are unknown, and the sector targeted by the emission control instrument consists of many players, which obviously happens quite often in practice.

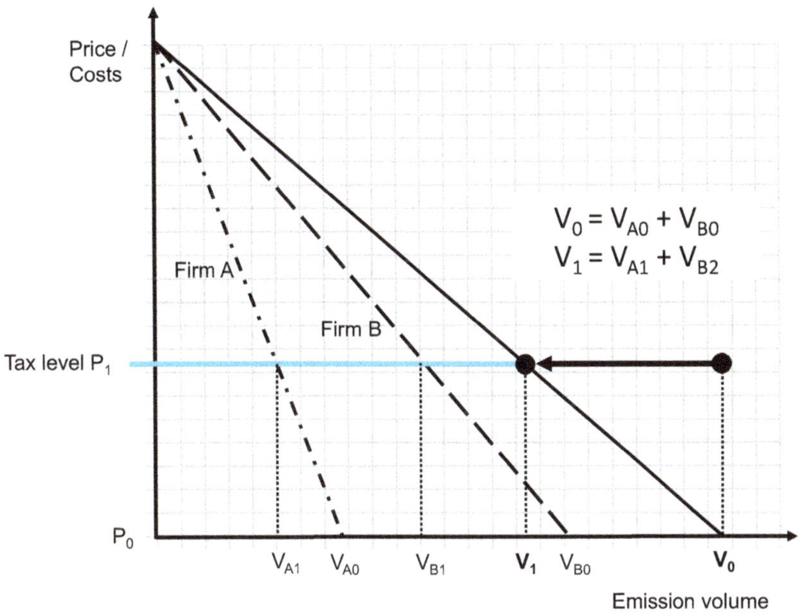

Figure 20.1: Price control mechanism; figure created by author.

To reduce emissions, the regulator sets a price for emissions (P_1) by introducing an emission tax that must be paid by every firm affected by the mechanism. Both firms A and B can either pay the emission price or reduce their emissions to avoid the payment for the reduced amount. This leaves each market player with the central question: what are the most cost-effective emission reduction options and what are their specific costs? This will then help to answer how much emission reduction is economically viable.

As rational market players, firms A and B will reduce their emissions along their individual abatement cost curve until marginal abatement costs are at the level of the tax where reduction of the next marginal unit of emissions is more expensive than paying the tax. Figure 20.1 illustrates that by setting an external price for emissions of P_1, overall emissions are reduced from the prior volume of V_0 to V_1 (which is $V_{A1} + V_{B1}$). If V_1 is the desired outcome for the policy maker (i. e. the tax level was chosen rightly), the price control mechanism in our example worked perfectly. We can therefore summarise that a price control mechanism is highly effective in defining the cost (the price) of the targeted pollution unit (e. g. carbon emissions), while the regulator has only limited control over the resulting emission volumes. Examples of typical price control mechanisms are energy taxes, levies on waste and CO_2 taxes that put a regulated price on emissions e. g. from fuels for transport or heating.

20.3.3 Volume Control Mechanisms: Cap-and-Trade

The second typus of market-oriented approaches for internalisation of external effects are volume control mechanisms like certificate-based cap-and-trade systems. Instead of setting an external price (tax, levy) to internalise the external costs, a maximum emission volume (emission cap) is defined by the regulator, and emission rights are issued. Efficient volume control mechanisms therefore require the institutionalisation of a market for pollution rights in the form of tradeable certificates (so-called emission allowances) and a system to monitor and exchange these certificates. Following the definition of an emission volume cap and the issuance of certificates, the certificate price emerges from economic optimisation within the market.

The fundamental function of a volume control mechanism with a cap-and-trade system is illustrated in the following Figure 20.2 for two sample firms A and B in a sample market:

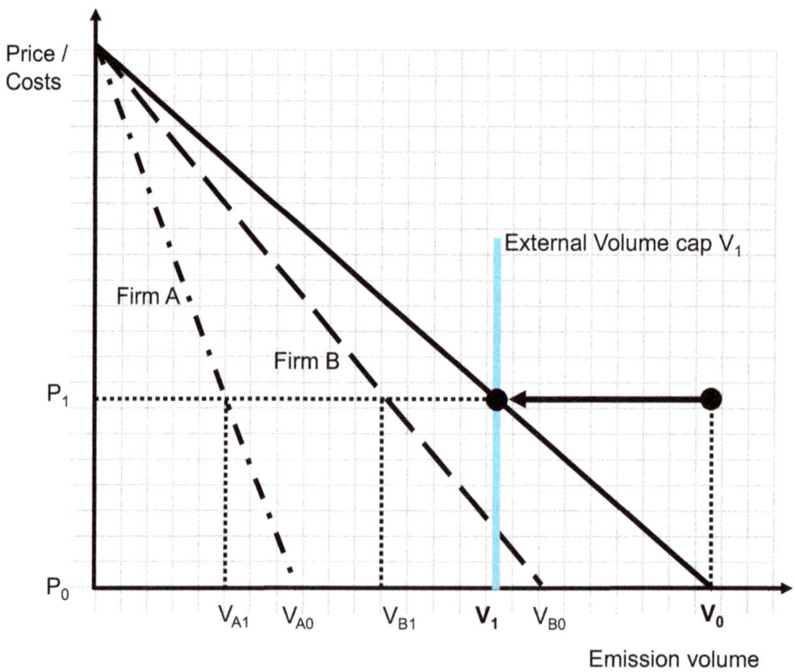

Figure 20.2: Volume control mechanism; figure created by author.

Figure 20.2 again uses the different marginal abatement cost functions (= "demand for pollution") of two exemplary firms A and B to illustrate the functioning of the mechanism. Like already outlined above, without any emission control, both firms operate at their maximum emission volumes of V_{A0} and V_{B0} and the total emission volume for the market amounts to V_0. The two firms have different marginal abatement cost, which is reflected by the individual slope of the two abatement cost functions.

Under a volume control approach the regulator introduces a maximum volume for emissions (V_1) by setting an emission cap and issuing emission rights in the form of tradable emission allowances (certificates) for each emission unit, e. g. one ton of CO_2. Firm A and firm B must acquire and surrender emission allowances to the regulator at the end of each trading period (normally a year) that exactly match their emitted volumes over that period. If they fail to do so, a substantial penalty applies. Like under a price control mechanism, the firms will examine their emission reduction options and the related costs along their abatement cost curve and check this against the price for allowances and the penalty. Then they decide for the least expensive cost option.

As rational market players, firms A and B will then reduce their emissions along their individual abatement cost curve until marginal abatement costs are at the level of the market's emission allowance price where the individual reduction of another unit of emissions would be more expensive than buying an emission allowance.

The price of emission allowances is determined by supply and demand, where supply is the quantity of emission allowances issued by the regulator and demand is the aggregated demand function of all emitters covered by the cap-and-trade system.

Hence, when the regulator sets an external volume cap for emissions of V_1, overall emissions are reduced from the prior volume of V_0 to V_1 ($V_{A1} + V_{B1}$) via the cap-and -trade system. If the system is designed right, the resulting emission level of all firms will not exceed the cap set by the regulator and the system has delivered the aspired results. Hence, a volume control mechanism is highly effective in reaching a maximum overall emission volume, while the regulator has only limited influence on the emission allowance price, which is defined within the certificate market based on abatement costs. An example of a comprehensive volume control mechanism to reduce carbon emissions is the EU's Emission Trading Scheme, which is described later in more detail in this chapter.

20.3.4 Design Choices for Market-Based Mechanisms

No matter if a volume or a price control mechanism is chosen, there are some fundamental design choices that the regulator needs to make. These choices concern the mechanisms scope, level of the set price or volume, point of regulation, monitoring, verification and enforcement, mitigation of unwanted behaviour of market participants, and the mechanism's competition and cross-border effects.

The following Table 20.1 (adapted from Handbook on Carbon Pricing, 2017, p. 23) discusses some fundamental design choices for emission control mechanisms:

Table 20.1: Central design choices for volume and price control mechanisms.

Design choices	Volume control (cap and trade)	Price control (emission tax)
Mechanism scope	Both types of mechanisms involve trade-offs between choices of greenhouse gases to regulate, which sectors to include, whether to allow relatively small emitters to remain unregulated or to include them, and whether the regulation occurs downstream or upstream. In both cases, these choices need to be made, considering political and economic factors.	
Volume cap or price rate	The emission quantity is a function of regulatory choice (cap) and needs to correspond to emission reduction targets (continuous reduction of the cap).	The emission price rate is a function of the pre-set tax rate. The tax rate should ideally be set at marginal abatement cost but in practice it is subject to political discussion and only a best guess.
Point of regulation	Can be upstream, downstream, or midstream (i. e. either addressing fuel importers, fuel converters, or end users)	
Verification and enforcement	Rather complex design: – Registry of emissions to manage issue, trade, and surrender of allowances – Organisation of certificate allocation (grand-fathering, auctioning, secondary markets), also considering risk for carbon-leakage – Banking of allowances for future use or borrowing for current use	Rather easy design: – Regulated entities must report emissions or a proxy for emissions such as fuel quantities and pay the pre-set tax based on those emissions.
Risk mitigation policies and enhancement of predictability for market actors	– Price ceilings – Price floors – Banking and borrowing across trading periods – Generally, less volume risk in comparison to price control (but higher price risk)	– Subsidies to mitigate economic hardship – Generally, less price risk in comparison to volume control (but higher volume risk)
Cross-border effects	Easier to link different emission control systems via the use of tradable or exchangeable emission allowances – but still complex; must account for political barriers and regulatory coherence in treatment of emissions allowances	More difficult to link.

20.3.5 Interim Conclusion: Comparison of Instruments

Environmental standards are said to be quite effective within certain application cases; however, they are not market-based, which is the main point of critique.

Price control mechanisms such as an emission tax are market-based and rather simple to implement and require less organisational effort than a volume control mechanism. It is therefore argued that they are especially suitable to be implemented in areas where there are many small emission sources – like in the household and commercial sector or for private transport. To monitor emissions from these diffuse sources would be highly complex. In addition, a pre-set price (tax rate) provides cost predictability.

However, the achievement of quantitative environmental goals like a specific emission level, which might also decrease over time, requires that policymakers choose the right tax rate and correctly estimate the abatement costs of market players. This is a main challenge of price control mechanisms, as the regulator has no detailed information about the specific abatement costs of individuals. Generally, the tax level must be defined by estimating or modelling the emission reduction cost of the relevant actors for a specific reduction target. Inaccuracies in the assumptions or the model result in actual emissions reductions that differ from the aspired target.

Meeting a desired level of emissions by setting an external price (tax) is indirect and might only be realised with a "trial and error" approach that entails multiple adaptions of the tax level – reducing predictability for the firms addressed by the mechanism. Hence, the main weakness of a price control mechanism is its lack of clarity on the resulting emission volumes, as meeting a certain emission volume would require the regulator to estimate abatement costs of all actors correctly and set the tax rate accordingly.

Volume control mechanisms, mostly organised as cap-and-trade systems require increased effort to monitor and manage the issuance, trade, and cancellation of emission allowances. This is, why volume control mechanisms are predominantly implemented in cases where there is a limited number of large and ideally stationary emission sources, like power plants and industrial production facilities. However, by setting a cap for emissions and reducing the cap over time allows the regulator to meet an emission reduction target (also in the long-term) with high accuracy. This is the pivotal strength of a volume control mechanism in relation to price control.

To sum up, there is not one emission control mechanism that is always superior to the other – it depends on the use case. However, based on the above considerations, it does not come as a surprise that many of the world-wide emission control schemes that cover especially the energy and industrial sector with a limited number of large emission sources are largely organised as volume control mechanisms in the form of a cap-and-trade system. The World Bank provides an overview of implemented and planned emission control mechanisms, price as well as volume control (World Bank, 2022).

20.3.6 Hybrid Instruments

Furthermore, not many market-based emission control mechanisms that are implemented in practice are designed with a pure price or a pure volume control logic – they rather combine elements of both. Such a mix of elements from a volume and a price instrument in one emission control mechanism is known as a hybrid instrument. Hybrid design is increasingly popular with decision makers as they are more flexible and promise to combine advantages of both underlying generic control instruments while addressing some of their major drawbacks.

Especially if the number of emission abatement options is large and their costs are not well known – which is the case in many real-world situations when design choices need to be made – hybrid instruments can be advantageous as they allow, for example, to combine quantity targets with price floors or caps or use quantity targets that are related to an emission tax or price. The possibilities for combinations are large and most of the existing emission control approaches – also the EU ETS – are in fact some sort of hybrid mechanism.

20.4 The EU ETS: A Cap-and-Trade System in Practice

The European Union's Emission Trading Scheme is designed as a classical cap-and-trade system. Hence, the EU ETS is a pure volume control mechanism by initial design; although some design elements like the market stability reserve were added over time that now make the EU ETS a hybrid system.

The cap is set by the regulator (the EU commission) and defines the total amount of a set of greenhouse gases that are allowed to be emitted by the emission sources covered by the system. The cap is reduced year after year so that overall emissions decrease and long-term targets in the EU can be monitored, operationalised, and finally realised.

After each year, an emission source (e. g. a power plant or an industrial production site) under the EU ETS is obliged to present the amount of emission allowances that corresponds to its emissions to the regulator. A failure to present sufficient emission allowances leads to a penalty payment for the excess emissions not covered by allowances. If an installation reduces emissions below its cap and therefore needs less allowances for itself, it generates excess allowances that can be kept for future emissions or can be sold to the market. These banking and trading options enable participants to plan and manage their emissions with high technical, temporal, and economic flexibility. In the end, the system ensures that emissions are reduced wherever the reduction costs are minimal.

20.4.1 Participating Countries, Sectors and Covered Greenhouse Gases

The EU ETS is operational in all EU member states plus Iceland, Liechtenstein and Norway that joined the system. In 2021, the United Kingdom withdraw from the system due to Brexit. Northern Ireland remains a member despite it being part of the UK due to the withdrawal agreement with the EU. The majority of countries joined the system already with its initiation in 2005. Four countries joined for the second trading phase in 2008 and one in 2013.

The following map (see Figure 20.3) gives an overview of the current members and the history of countries in the EU ETS:

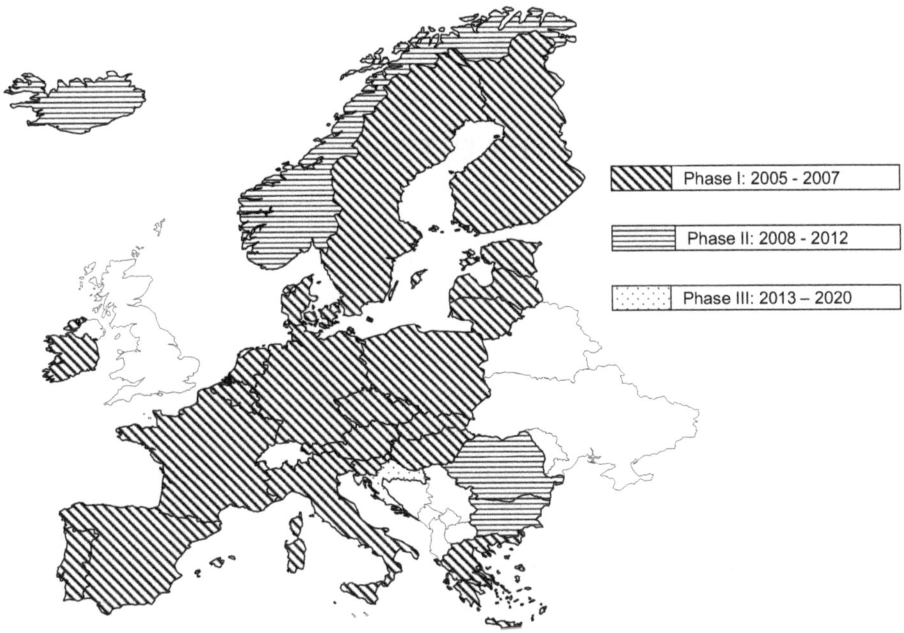

Figure 20.3: Countries in the EU ETS and years they joined the system; figure created by author.

Currently, the EU ETS covers the emissions of around 15,000 stationary installations in the power sector and manufacturing industry, as well as about 1,500 aircraft operators for flights between these countries. In total, the scheme covers around 40% of the EU's greenhouse gas emissions – with an increasing tendency.

The following greenhouse gases, emission sources and sectors are covered by the EU ETS:

- Carbon dioxide (CO_2) from large electricity and heat generation, energy-intensive industry sectors (such as oil refineries, production of steel, iron, aluminium, metals, cement, lime, glass, ceramics, pulp, paper, cardboard, acids and bulk organic

chemicals), commercial aviation (until end of 2023 only for flights within the European Economic Area).
- Nitrous oxide (N_2O) from production of nitric, adipic and glyoxylic acids and glyoxal.
- Perfluorocarbons (PFCs) from production of aluminium.

The maritime sector is planned to be included in the EU ETS in a stepwise phase-in from 2023 to 2026. Companies operating in the EU in the mentioned sectors are obliged to participate in the EU ETS, however for some sectors a minimum installation size is defined (de-minimis rule). For small installation, governments might use alternative (e.g. fiscal) measures to ensure the necessary emission reductions. Therefore, emissions from the household sector, commercial sector and individual transport are not (yet) directly covered by the EU ETS. This explains why the system currently covers only about 40% of the EU's total emissions.

20.4.2 Legal Framework of the EU ETS and Trading Phases

The European ETS Directive 2003/87/EC of 2003 represented the initial legislative basis for the EU ETS. Since its introduction in 2005 the scheme has undergone several revisions to adapt it to the increasingly ambitious EU climate policy objectives and to ensure the systems functionality despite external influences like the banking and financial crisis. Adaptations mainly addressed the faster reduction of the emission cap, the integration of additional emitters and the introduction of a volume-price-mechanism, the so-called market stability reserve (that in fact makes the EU ETS a hybrid mechanism).

The EU ETS is designed in continuous trading phases. It is currently in its fourth trading phase (2021–2030). Phase 1 run from 2005 to 2007 and marked the start of the EU ETS. It covered only emissions from power generators and energy-intensive industries and was designed as a pilot phase to implement the infrastructure needed to monitor, report, and verify emissions. ETS allowances were mainly grand-fathered (given away for free by the regulator) in the first phase. Phase 2 run from 2008 to 2012; it realized a gradual introduction of auctioning of ETS allowances (10%) and the penalty for non-compliance was increased to € 100 per ton of CO_2. The aviation sector was integrated into the EU ETS in 2012, but only for intra-EU flights due to concerns about competition disadvantages. Phase 3 run from 2013 to 2021; within this phase, the free allocation of allowances except for the carbon leakage list to energy-intensive industries was abolished. Emissions from the maritime sector will be phased-in stepwise from 2023 to 2026.

After the EU ETS started with decentralised planning and decentralised emission registers, the whole operation of the system was centralised in 2012 in one single EU registry that is now operated by the European Commission. It covers all countries that are members of the EU ETS. The registry has accounts for all stationary emission sources and for aircraft operators. It collects data about the national implementation measures, lists of all installations covered by the EU ETS, and the amounts of free allocations to these installations. Furthermore, the registry holds allowances accounts for all companies participating in the EU ETS; this allows the market actors to safely transact (buy and sell) and surrender allowances within the system.

20.4.3 Emission Reduction Targets and Caps

The EU has set ambitious targets to reduce its greenhouse gas emissions. With the "European Green Deal" and the "Fit for 55" package, proposed by the EU Commission in 2021, the reduction targets were further increased.[2] The current aim is to reduce greenhouse gas emissions by at least 55% (against 1990 emission levels) by 2030 across the EU and become climate neutral until 2050. These targets define the quantity framework for the EU ETS.

For a cap-and-trade system like the EU ETS it is pivotal to define the allowed emission quantity (the cap). For phase 1 and phase 2 of the EU ETS this was done via so-called national allocation plans (NAPs) in which each EU country decided individually on the allocation of emission allowances to its national sectors. NAPs were used to plan the EU-wide emission cap bottom-up from country level, based on historical emissions. This approach allowed the member states to decide upon individual allocation rules for their sectors or even single corporations. The national allocation plans then needed to be assessed and accepted by the EU commission, which often demanded changes to address competition issues and to enforce the overall cap. As the bottom-up planning process via NAPs proved to be time intensive and showed limited effectiveness, the EU decided to introduce an EU-wide cap and abandon the individual planning via NAPs.

During phase 3 of the EU ETS from 2013 to 2020 the EU-wide cap for stationary emissions decreased on a yearly basis by 1.74%. The initial cap in 2013 corresponded to the previous caps from phase 2 (2008 to 2012). Also, during the current phase 4 (2021 to 2030) the EU ETS emission cap is decreased year by year by a linear reduction factor of 2.2% (current legislation). The overall quantity of allowances for stationary emissions in 2021 was set to 1,571,583,007 allowances equaling 1 ton of CO_2. The annual reduction factor equals 43,003,515 tons of CO_2 (EU Commission 1, 2022). For reaching a 55% reduction of emissions in 2030 (Fit for 55 target), the reduction factor is proposed to increase to 4.2% per year.

Since 2013 (start of phase 3), emission allowances for stationary emission sources shall generally be auctioned by the EU and only allocated for free for predefined

2 See especially Chapter 1.13.

cases like carbon leakage prevention and other support instruments. For example, in 2013 80% of allowances for the manufacturing industry in the EU WTS was allocated for free. This share decreased to only 30% in 2020.

Auctioning allowances means that emitters need to pay the market price to cover their demand of allowances. In phase 3 and in the current phase 4 of the EU ETS, around 57% of emission allowances for the EU-wide cap are sold to the market by the EU (auctioned), 43% are provided for free to selected sectors and countries under specific grand-fathering rules. The amount of free allocation is decided based on the risk of carbon leakage of a certain sector, its historical emissions, and pre-defined benchmark values for the 10% installations with the lowest specific emissions for a certain product. Currently, there are 54 such benchmarks.

As a deviation from the general rule that no free allocation is allowed for the EU electricity sector, 10 EU member states that showed a comparably low GDP performance will be allowed to continue grand fathering in their energy sectors until 2030. From those 10 countries, only Bulgaria, Hungary and Romania have opted for this exemption. Finally, a small share of allowances is used to finance the EU's Innovation Fund that support technological development and the Modernisation Fund that provides support to 10 lower-income EU countries for modernization of their energy systems.

Between 2008 and end of 2023 the EU ETS member states generated a revenue of about 210 billion Euros from auctioning emissions allowances. The revenues are strongly depending on the allowance price, so due to the strong price increase in since the year 2021 a surge in revenues from emission allowance auctions could be observed (47 billion Euros in 2023 alone).

The ETS Directive requires member states to use at least half of the allowance revenues for climate and energy-related purposes. The EU Commission states that "around 78% of revenues in 2013–2019 were used for climate and energy related purposes" (EU Commission 2, 2022). The central data on auction amounts, revenues and revenue utilisation is reported in an annual Carbon Market Report.

20.4.4 Price Formation and Price Development

As described in the above section on volume control mechanisms, the price for EU ETS allowances is determined by the regulated supply of emission certificates in relation to the aggregated demand by emitters. Hence, prices for CO_2 are dependent on the provided number of certificates, which is reduced over time and the demand from industry that also varies and is driven, for example, by the economic development. The amount of free allocation, however, has no direct impact on the allowance price as based on the concept of opportunity costs, also allowances allocated for free are priced at the market.

The following Figure 20.4 gives an overview of historical prices in the EU ETS between January 2009 and November 2024:

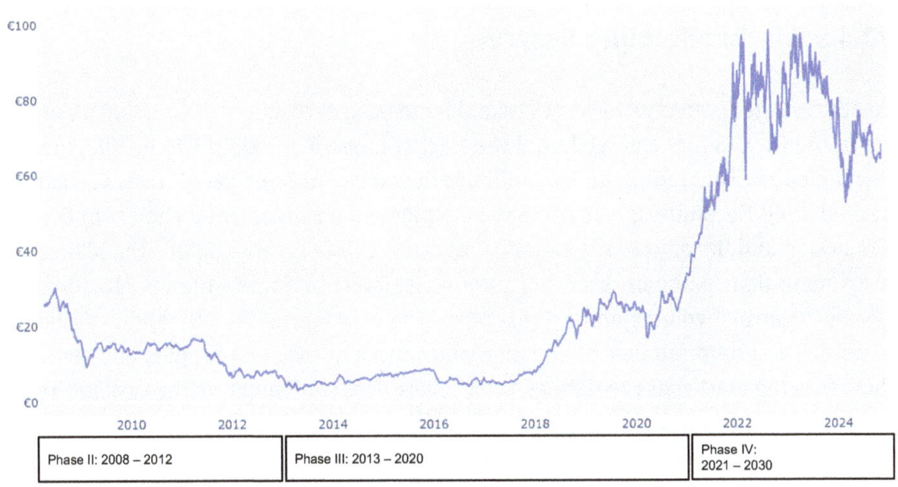

Figure 20.4: Allowance prices in € per ton CO_2 between January 2009 and November 2024 (price graph: https://sandbag.be/index.php/carbon-price-viewer).

In Phase 1 (2005 – 2007, not displayed here) allowance prices ranged at around 20 €/t CO_2 in 2005 and 2006 but then collapsed in 2007 at the end of the first trading period because there was no banking mechanism in place and all phase 1 allowances lost their validity and the end of 2007. Phase 2 started in 2008 with prices around 20 €/t CO_2 (see chart above) but then prices decreased due to structural allowance oversupply during the financial crisis and ranged at a level of about 15 €/t CO_2 with prices below 10 €/t CO_2 at the end of phase 2. Phase 3 showed the lowest allowance prices with the years 2013 to 2017 all trading below 10 €/t CO_2 on average; this was the time when the EU brought forward some design changes to the EU ETS, namely the instrument of backloading (non-issuance of 900 million allowances) and the introduction of the market stability reserve. From 2018 on allowance prices started to increase and rose above the threshold of 20 €/t CO_2 also despite the COVID pandemic and the related economic slowdown. Form mid-2021 on allowance prices increased strongly. Allowance prices in 2022 ranged at an average level of about 80 €/t CO_2. At the time of writing (July 2022) the future prices for EU ETS allowances trade at a level of about 120 €/t CO_2 for the direct front years and about 90 to 100 €/t CO_2 for the long-term up to ten years ahead. The very strong price increase in the spot and futures markets was presumably driven by several trends like a general price increase in the wake of the price hike on natural gas and power markets in the fourth quarter of 2021 and in 2022, the general expectation of stricter regulation and higher reduction targets in the

EU (e. g. the “Green Deal”) and the regulatory changes in the EU ETS, mainly the implementation of the market stability reserve (see next section).

20.4.5 Market Stability Reserve

To address the comparably low prices and increasing oversupply of emission allowances in the EU ETS that emerged until the start of phase 3 of the EU ETS in 2013, the EU postponed the auctioning of 900 million allowances to later years, this was called back-loading. Secondly, it was decided to implement a fundamental change to the EU ETS design and introduced the so-called market stability reserve (MSR). The MSR is an instrument that combines price and volume flexibility and made the EU ETS a hybrid system regarding volume and price control. The MSR started in 2019 and is designed to absorb a certain number of emission allowances in case of a surplus and re-inject them into the market at pre-defined price levels. This shall improve the system’s resilience to external shocks like an economic crisis by temporarily reducing the number of allowances available to the market and by that stabilising the price.

The 900 million allowances that were foreseen to be re-injected into the market from back-loading were initially transferred to the MSR in 2019 and by that kept out of the market. Further allowance quantities that are not yet auctioned could also be transferred to the reserve in future. To keep a maximum of predictability, the MSR is operated entirely based on pre-defined rules. These rules state clearly under which conditions allowances will be moved to the reserve or released into the market. The basis for this decision is a threshold number of allowances circulating in the market (MSR indicator). In addition, from 2023 onwards, allowances that are placed in the MSR will lose their validity if they remain in the reserve longer than a year. With the market stability reserve the EU has implemented a measure that aims to ensure the functionality of the EU ETS by giving some rule-based flexibility to the regulator regarding allowance volumes. Since its introduction in 2019 the MSR has helped to bring carbon prices in the EU ETS to a substantially higher level than in the years before and enforce the economic pressure for climate change mitigation within the system.

20.4.6 Link to Other Cap-and-Trade Schemes and International Carbon Credits

As defined in the Kyoto Protocol, there are two mechanisms through which international certificates for greenhouse gas reductions can be generated: First, the Clean Development Mechanism (CDM) that certifies greenhouse gas reductions from direct investments of industrialised countries (so-called Annex 1 countries) in projects in developing countries. And second, Joint Implementation (JI) that allow payments for emission reduction projects between two Annex 1 countries. Until 2020, the EU ETS allowed emitters to use international credits from CDM and JI projects to meet their obligations under the EU ETS. The utilisation of international credits was already limited in quantity and the international certificates had to meet certain quality requirements and needed to be approved by the EU regulator.

With the start of trading phase 4 in 2021, the EU has abolished the possibility to utilise any international credits – all emission reductions now need to be realised within the EU ETS itself. However, the Paris Agreement allows for a linkage of emission trading schemes across the world, but with reliable accounting rules. These rules are not yet defined and therefore the linkage of the EU ETS to other cap-and-trade systems around the world remains a future task.

20.4.7 Carbon Leakage

As the EU and other countries raise their climate ambitions and have put in place cap-and-trade or other emission control systems, less stringent environmental and climate policies prevail in other countries around the world. These differentials in climate policy across the different jurisdictions creates a risk of so-called 'carbon leakage'. Carbon leakage describes the situation when companies (emission sources) shift parts of their value chain or fully remove their production facilities and the related emissions from one country or region that has an emission control mechanism in place and source their production input or relocate their production facilities to a country or a region without such a mechanism. This means that the carbon emissions of these emitters 'leak' out of the emission control system due to cost pressure and competition effects. However, these emissions obviously are not reduced but just shifted to another jurisdiction. Therefore, carbon leakage can seriously undermine the success of any emission control instrument that has regional boundaries and also threatens global efforts to combat climate change.

The issue of carbon leakage is strongly linked to the impact of an emission control mechanism on competitiveness. The introduction of such a mechanism – no matter if price or volume control – introduces a new cost element for the emitters covered by the mechanism. With the introduction of the emission control mechansim, each entity addressed by the mechanism will evaluate the new cost and how it affects its competitive situation in its markets. In case the introduced emission costs affect all competitors in a market in a similar way, their competitive situation is not fundamentally changed as in general the newly introduced emission costs can and will be transferred to the market (price increase). Not all entities might be affected in exactly the same way, because they have different emission reduction cost – however, this selection is exactly what the emission control mechanism is designed for: to set economic incentives for efficient emission reductions within the targeted region. A good example for this situation is the power sector: As long as all power generators in a market or region are covered by the mechanism, they cannot avoid it because if they move their production site (power plant) they will lose access to the market, e. g. because there is no physical grid connection. However, the region covered by the emission control mechanism needs to be large enough to allow for such a result. This is for example the case for the EU ETS, as it covers almost all countries of the UCTE grid area, so there are no options for power generators to move away from the EU's power markets as they would not be able to deliver their generation to the offtakers anymore. In such a case the introduction of an emisson control mechanism does not entail the risk for carbon leakage and no measures are needed to avoid it.

Concerns about carbon leakage are much more relevant for industries that have a high exposure to global competition. These are especially energy-intensive and trade-exposed sectors, like the cement, chemical and steel sector. Without measures against carbon leakage, producers will not be able to pass-through the additional cost for emissions to their (global) markets and face the decision to either lose out on their competitors or move to other jurisdictions without the emission costs. To address the carbon leakage problem while introducing an emission control mechanism, the regulator has several options. The most direct measure is the allocation of free allowances that widely eliminates the cost pressure for energy-intensive industries from emission control.

This is why the EU will focus its free allocation of emission allowances now on sectors that have a high risk of leaving the EU with their production facilities.These sectors will get 100% of their allowances for free also in future. Less exposed sectors will receive free certificates only for a share of their emissions (maximum of 30%) with a pathway to reduce free allocation to zero between 2026 and 2030 (end of phase 4). Sectors that are at high risk of carbon leakage are continuously monitored by the EU and put on a specific carbon leakage list (c.f. EU Commission 3, 2022, p. 24 ff.). The decision, if a sector is deemed at high risk for carbon leakage mainly depends on the cost impact of emission allowances on the sector's products and the share of its of non-EU trade. For example, a high risk sector in phase 3 was defined by at least 5% production cost increase if allowances are not provided for free and a share of trade with non-EU countries of at least 30%. As a consequence, sectors being expected at high carbon leakage risk receive all or part of their allowances for free.

However, as emission cost are also present in the whole value chain (indirect emission costs), e. g. in power prices, financial compensation payments like the German "Strompreiskompensation" (compensation for carbon cost in the industrial power price) are allowed within the EU ETS. Each member state may decide about the use of such additional options individually, but must follow respective EU guidelines and report to the EU. More general options are that states provide investments tax credits, increased depreciation options, long-term securities or loans, funds research and development activities to reduce emissions in the affected sectors.

20.4.8 Outlook: Expansion of the EU ETS to Other Sectors and Carbon Contracts for Difference (CCfD)

The road ahead for the EU ETS will be marked by a tightening of reduction targets to meet the Fit for 55 ambitions that require a further cutback of the allowance quantities. Furthermore, an expansion of the scheme to other sectors like buildings and transport is discussed. Until these sectors can be included into the EU ETS, they will be targeted by national market-oriented sub-systems, like the German "Brennstoffemissionshandel" (national fuel emissions trading system) that adresses corporations like refineries that sell fuels like gasoline, diesel, heating oil, liquefied petroleum gas, and natural gas to end-users. Such national solutions are seen as necessary and efficient to bridge the time until the EU ETS is designed in such a way to include the additional sectors across all member states (Blum et al., 2019, p. 61).

A quite new instrument to reduce uncertainty about long-term development of carbon prices in a cap-and-trade system like the EU ETS, are the carbon contracts for difference (CCfD), as described for example by Richstein and Neuhoff (2019). For the industry, investing into emission reduction options requires a business case. As the costs of an emission reduction measure are quite well known and occur in the present, the revenues of this measure are rather uncertain and depend on many factors such as the future regulatory framework and especially the evolution of carbon prices. As there is no liquid long-term futures market for emission allowances, corporations are often reluctant to realise investments in emission reduction options that have a long time horizon. CCfDs are a way to minimize this price uncertainty.

A CCfD is a contract by which a regulator (the state) agrees with a private entity on a fixed carbon price (the strike price) for a pre-set period. For the contract period, the entity can now sell any emission allowances for that strike price – the uncertainty about price evolution is eliminated. As the contract is designed on a strike price to market difference, the CCfD is evaluated against the carbon market: if the market price for allowances is higher than the strike price, the entity is obliged to pay the difference to the regulator, and in case the market price for allowances is lower than the strike price, the entity is entitled to receive the difference to the agreed strike price from the regulator. This is illustrated in the following Figure 20.5:

In this way the CCfD ensures that exactly the agreed carbon price – not less, but also not more – is paid to the corporation as a revenue for the underlying emission reduction measure as a foreseeable revenue stream that enables long-term investments. Based on this understanding it is argued that "to date, carbon contracts are therefore the best tool to provide certain revenue streams to decarbonized technologies, which investors could then capitalize over the lifetime of the project" (Gerres and Linares, 2020, p. 2). CCfDs are compatible with all types of volume control mechanism in which a transparent carbon price is present.

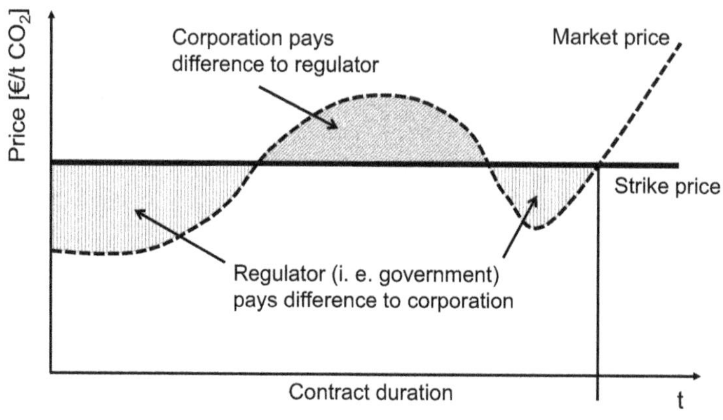

Figure 20.5: Functionality of a Carbon Contract for Difference; figure created by author.

Bibliography

Blum, J. et al.: Zur Bepreisung von CO2-Emissionen: Ergebnisse aus dem Ökonomenpanel, ifo Schnelldienst 16 / 2019 72, Jahrgang 22, August 2019

EU Commission 1: website on EU ETS caps and allowances, retrieved 27th July, 2022 from https://ec.europa.eu/clima/eu-action/eu-emissions-trading-system-eu-ets/emissions-cap-and-allowances_en

EU Commission 2: website on EU ETS auctioning and revenues, retrieved 27th July, 2022 from https://ec.europa.eu/clima/eu-action/eu-emissions-trading-system-eu-ets/auctioning_de

EU Commission 3: Official Journal of the European Union, Volume 62, 8 May 2019, retrieved 27th July, 2022 from https://eur-lex.europa.eu/legal-content/EN/TXT/PDF/?uri=OJ:L:2019:120:FULL&from=EN

Gerres, T., Linares, P.: Carbon Contracts for Differences: their role in European industrial decarbonization, Climate Friendly Materials Platform, 2020

Richstein, C., Neuhoff, K.: CO2-Differenzverträge für innovative Klimalösungen in der Industrie, DIW Aktuell, Nr. 23, 26th September 2019

The Climate Reality Project: 2017 Handbook on carbon pricing instruments, retreived on 5th April 2022 from https://www.climaterealityproject.org/sites/climaterealityproject.org/files/HandbookonCarbonFinancing_Final_May16.pdf

World Bank, website Carbon Pricing Dashboard, retrieved 27th July, 2022 from https://carbonpricingdashboard.worldbank.org

Jasmine Ramsebner

21 The Concept of Sector Coupling: Where is the Business Model?

21.1 Introduction

The start of the industrial revolution around 1760 brought about many pioneering innovations that pushed living standards and created much-needed jobs and opportunities. At the same time, however, this upswing mainly relied on fossil resources that replaced mechanic with thermal energy and reduced the physical effort while improving efficiency and economic performance for the striving economies. In the last decades, with global economic growth, the debate on the long-term effect of emissions from fossil fuel burning on the environment has received more and more attention, and the damage of ignoring these aspects in economical market frameworks turns out to be severe (Ramsebner, 2022).[1]

To achieve climate goals and the decarbonisation of all end-consumption sectors, specifically the efficient use of wind and solar power as large-scale, variable renewable energy sources (VRE), will be of great importance. However, the management of their intermittency imposes new challenges on the energy system (Ramsebner, Linares, et al., 2021).[2] The variability of these natural resources affects the energy system within a day and throughout the seasons. Excess electricity production can frequently occur, considering the known demand profiles. Especially in, for example, continental Europe during summer, high solar irradiance with low energy demand, or even in combination with high wind availability, may cause such a situation (Ramsebner, Haas, Ajanovic, et al., 2021).[3] In winter, however, this region may experience a scarcity of renewable electricity generation with low solar irradiance and higher demand. In northern countries, by contrast, the dominance of wind power generation may lead to excess electricity generation, especially during winter (Ramsebner, Haas, Ajanovic, et al., 2021),[4] (Ramsebner, 2022).[5]

1 Ramsebner, J. (2022). The role of sector coupling in the energy transition [Thesis, Technische Universität Wien]. https://repositum.tuwien.at/handle/20.500.12708/80280.

2 Ramsebner, J., Linares, P., & Haas, R. (2021). Estimating storage needs for renewables in Europe: The correlation between renewable energy sources and heating and cooling demand. Smart Energy, 3, 100038. https://doi.org/10.1016/j.segy.2021.100038.

3 Ramsebner, J., Haas, R., Ajanovic, A., & Wietschel, M. (2021). The sector coupling concept: A critical review. *WIREs Energy and Environment*, 10(4), e396. https://doi.org/10.1002/wene.396.

4 Ramsebner, J., Haas, R., Ajanovic, A., & Wietschel, M. (2021). The sector coupling concept: A critical review. *WIREs Energy and Environment*, 10(4), e396. https://doi.org/10.1002/wene.396.

5 Ramsebner, J. (2022). The role of sector coupling in the energy transition [Thesis, Technische Universität Wien]. https://repositum.tuwien.at/handle/20.500.12708/80280.

https://doi.org/10.1515/9783110784299-021

The global development of primary energy consumption and the recent and expectedly growing replacement of fossil fuels with renewables is shown in Figure 21.1.

Questions emerge on how to manage these non-controllable sources, handle excess electricity generation, and use it efficiently in terms of economic, ecologic, and social welfare aspects (Ramsebner, Haas, Ajanovic, et al., 2021).[6]

The solution can only lie in a move from former silo based sectors electricity, heat and gas to a seamless integration of those. This, however, represents a totally new approach in the operation of renewable energy systems and, apart from technical novelties also depends on an adjustment of the economic framework and business model to support this new situation.

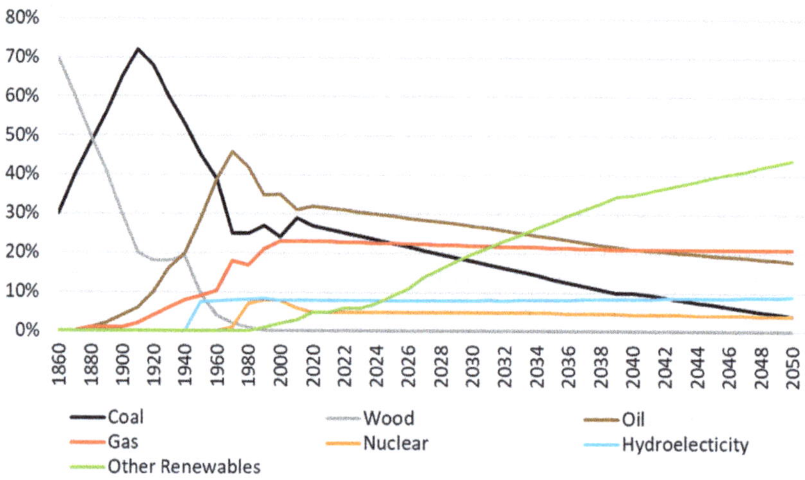

Figure 21.1: Current and expected global primary energy consumption (Sources: BP, 2019b, 2019a; Nakicenovic et al., 1998).

The following Chapters 21.2 and 21.3 outline the objectives and the role of sector coupling in renewable energy systems and are mostly based on (Ramsebner, 2022).[7] Chapter 21.4 provides examples of the economic perspectives of long-term financing based on (Haas et al., 2021),[8] the impact of variable renewable feed-in on the business case

6 Ramsebner, J., Haas, R., Ajanovic, A., & Wietschel, M. (2021). The sector coupling concept: A critical review. *WIREs Energy and Environment*, 10(4), e396. https://doi.org/10.1002/wene.396.

7 Ramsebner, J. (2022). The role of sector coupling in the energy transition [Thesis, Technische Universität Wien]. https://repositum.tuwien.at/handle/20.500.12708/80280.

8 Haas, R., Ajanovic, A., Ramsebner, J., Perger, T., Knápek, J., & Bleyl, J. W. (2021). Financing the future infrastructure of sustainable energy systems. *Green Finance*, 3(1), 90–118. https://doi.org/10.3934/GF.2021006.

of sector coupling considering (Haas et al., 2023)[9] and the techno-economic attractiveness of renewable hydrogen for the industry by (Ramsebner et al., Mimeo).[10] Finally also the impact of the recently observed multiple energy crises on the sector coupling business case is outlined considering (Ramsebner, 2022).[11]

21.2 Objectives of Sector Coupling

Apart from a supply side increase in the renewable energy share, the decarbonisation and ongoing electrification in all end-consumption sectors create new electricity demand sources. They consume the formerly identified electricity surplus. The electrification can be enabled either directly or through conversion technologies (e.g., power-to-gas) and requires a paradigm shift away from considering individual energy sectors (gas, electricity and heat), storage and demand towards a more integrated energy system (Iqbal, 2012;[12] Ramsebner, Linares, et al., 2021).[13] The concept of Sector Coupling (SC) or Sector Integration represents an approach to couple renewable electricity with the end-consumption sectors by electrifying transport, residential heating and cooling, and the industry. The concept was first mentioned during the energy transition in Germany. According to (IRENA, 2022),[14] the energy transition is "a pathway toward the transformation of the global energy sector from fossil-based to zero-carbon by the second half of this century" (Ramsebner, 2022).[15]

This includes the reduction of energy-related carbon emissions, a process called decarbonisation, addressed by specific goals for climate change mitigation. The main pillars of the energy transition are the use of renewable energy and energy efficiency measures (IRENA, 2022).[16] The EU has also attempted to promote investments into processes and activities that support the energy transition by providing clarity with a

9 Haas, R., Duic, N., Auer, H., Ajanovic, A., Ramsebner, J., Knapek, J., & Zwickl-Bernhard, S. (2023). The photovoltaic revolution is on: How it will change the electricity system in a lasting way. *Energy*, 265, 126351. https://doi.org/10.1016/j.energy.2022.126351

10 Ramsebner, J., Linares, P., Hiesl, A., & Haas, R. (2024). Techno-economic evaluation of renewable hydrogen generation strategies for the industrial sector. *Hydrogen Energy Journal*, Mimeo-under review. in International Journal of Hydrogen Energy. https://doi.org/10.1016/j.ijhydene.2024.02.167

11 Ramsebner, J. (2022). The role of sector coupling in the energy transition [Thesis, Technische Universität Wien]. https://repositum.tuwien.at/handle/20.500.12708/80280.

12 Iqbal, M. (2012). *An Introduction To Solar Radiation*. Elsevier.

13 Ramsebner, J., Linares, P., & Haas, R. (2021). Estimating storage needs for renewables in Europe: The correlation between renewable energy sources and heating and cooling demand. *Smart Energy*, 3, 100038. https://doi.org/10.1016/j.segy.2021.100038

14 IRENA. (2022). Energy Transition. Energy Transition. https://www.irena.org/energytransition

15 Ramsebner, J. (2022). The role of sector coupling in the energy transition [Thesis, Technische Universität Wien]. https://repositum.tuwien.at/handle/20.500.12708/80280.

16 IRENA. (2022). Energy Transition. Energy Transition. https://www.irena.org/energytransition.

classification of sustainable projects, the EU taxonomy (European Commission (EC), 2022).[17] Any actions that contribute to renewable energy generation, enable other activities to mitigate climate change, such as the storage of renewable energy or represent a significant improvement to the industry standard in terms of efficiency and CO_2 emissions. In this respect, SC is the tool that links renewable energy with the demand sector that requires a transition. The EU taxonomy is an essential signal that provides guidance for investment decisions into renewable technologies. SC supports establishing 100% renewable energy systems by substituting fossil fuels with renewable electricity. Furthermore, the concept may add flexibility through transformation options from power to heat or gas that can be stored and distributed easier than electricity (Ramsebner, Haas, Ajanovic, et al., 2021;[18] Schaber, 2013),[19] (Ramsebner, 2022).[20]

Despite the growing attention to SC, however, the scope and main objectives vary broadly throughout the literature, and the term is often used inaccurately (Robinius, Otto, Syranidis, et al., 2017).[21] Figure 21.2 provides an overview of the potential pathways of SC. While direct electrification is a possible SC approach, for example, in individual passenger transport, many processes rely on the conversion of electricity into gas or further into liquid fuels (power-to-gas/liquid [P2G/L]), into heat, known as power-to-heat (P2H), and storage in the respective form (Ramsebner, 2022).[22]

Definitions of SC range from limiting it to the one-way path from the power to another end consumption sector, to including cross-energy-carrier integration, even with, for example, excess heat use and biomass energy. A requirement for future renewable energy systems is the determination of new demand sources for renewable electricity by substituting fossil energy. The final goal is fulfilling the demand of end-consumption in industry, transport and residential homes. In contrast, the processes in these end-consumption sectors can be adapted to be powered by another form of energy carrier. A definition of end-consumption sectors, as in (Robinius,

17 European Commission (EC). (2022). EU taxonomy for sustainable activities. European Commission - European Commission. https://ec.europa.eu/info/business-economy-euro/banking-and-finance/sustainable-finance/eu-taxonomy-sustainable-activities_en

18 Ramsebner, J., Haas, R., Ajanovic, A., & Wietschel, M. (2021). The sector coupling concept: A critical review. *WIREs Energy and Environment*, 10(4), e396. https://doi.org/10.1002/wene.396.

19 Schaber, K. (2013). *Integration of Variable Renewable Energies in the European power system: A model-based analysis of transmission grid extensions and energy sector coupling*. Technische Universität.

20 Ramsebner, J. (2022). The role of sector coupling in the energy transition [Thesis, Technische Universität Wien]. https://repositum.tuwien.at/handle/20.500.12708/80280.

21 Robinius, M., Otto, A., Syranidis, K., Ryberg, D. S., Heuser, P., Welder, L., Grube, T., Markewitz, P., Tietze, V., & Stolten, D. (2017). Linking the power and transport sectors—part 2: Modelling a sector coupling scenario for Germany. *Energies*, 10(7), 957ff. https://doi.org/10.3390/en10070957

22 Ramsebner, J. (2022). The role of sector coupling in the energy transition [Thesis, Technische Universität Wien]. https://repositum.tuwien.at/handle/20.500.12708/80280.

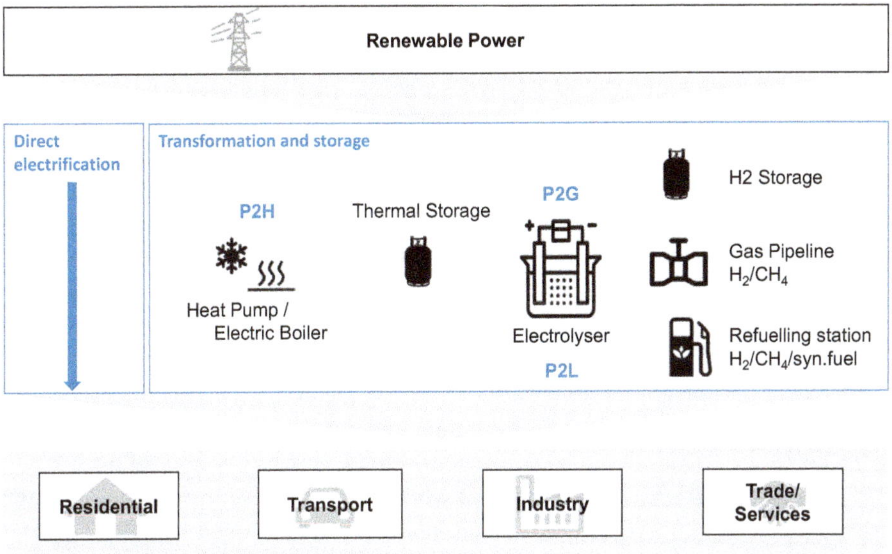

Figure 21.2: Overview of the main pathways of Sector Coupling in different end-consumption Sectors (Ramsebner, 2022).[23]

Otto, Syranidis, et al., 2017),[24] provides a technology-neutral approach and allows for a broader range of solutions, independently of the energy carrier applied (electricity, gas, or heat). To avoid thinking based on the existing infrastructure and trying to establish a new design of future energy systems, specifically supply, represents one crucial paradigm shift attempted through the energy hub by (Geidl et al., 2007).[25] We, therefore, suggest using the common sector definition in energy economics for techno-economic research on SC. More specifically, in technological research, it is evident that the perspective may focus on transforming power into a specific energy carrier. This would largely meet the approach by the (European Parliament (EP), 2018),[26] differentiating so-

23 Ramsebner, J. (2022). The role of sector coupling in the energy transition [Thesis, Technische Universität Wien]. https://repositum.tuwien.at/handle/20.500.12708/80280.

24 Robinius, M., Otto, A., Syranidis, K., Ryberg, D. S., Heuser, P., Welder, L., Grube, T., Markewitz, P., Tietze, V., & Stolten, D. (2017). Linking the power and transport sectors—part 2: Modelling a sector coupling scenario for Germany. *Energies*, 10(7), 957ff. https://doi.org/10.3390/en10070957.

25 Geidl, M., Koeppel, G., Favre-Perrod, P., Klöckl, B., Andersson, G., & Fröhlich, K. (2007). The energy hub: A powerful concept for future energy systems. 3rd Annual Carnegie Mellon Conference on the Electricity Industry: Ensuring that the Industry has the Physical and Human Resources Needed for the Next Thirty Years. https://www.research-collection.ethz.ch/handle/20.500.11850/3133.

26 European Parliament (EP). (2018). Sector coupling: How can it be enhanced in the EU to foster grid stability and decarbonise?

called end-use SC and integrating energy carriers as a cross-vector coupling (Ramsebner, 2022).[27]

The role of SC in the energy transition and the opportunities and challenges concerning successful VRE integration with new demand sources differ among the end-consumption sectors.

In this context, a new electricity demand source concerns the electrification of individual passenger transport, which is characterised by dynamic growth and requires early investigation of appropriate system integration. Efficient load-management solutions control the charging processes from supply side and avoid additional demand peaks. From demand side, end-users may be incentivized financially if they agree to provide flexibility and charge based on flexible electricity or grid tariffs. Together with the integration of local PV electricity generation, a growing number of actors will participate in the energy system and can support to match supply and demand.

As a result, SC may be implemented without an extensive distribution grid and electricity generation capacity expansion (Ramsebner, Haas, Ajanovic, et al., 2021).[28] As already mentioned, SC also includes converting renewable electricity into gas to provide a feedstock and fuel for the decarbonisation of hard-to-abate sectors, such as industrial steel production. Companies already aim to transform the former coal-based production process into a renewable-hydrogen or electricity-based one. On-site renewable hydrogen production offers a non-regret option for upscaling production volumes to decrease the overall cost independently of distribution infrastructure. It is essential to compare different operation strategies and their characteristics, such as conversion efficiency, investment cost and market participation affecting hydrogen production cost, to make informed decisions and promote the successful VRE integration for industrial purposes (Ramsebner, 2022).[29]

The energy transition depends on appropriate SC solutions to align VRE supply with demand in all end-consumption sectors. The efficient use of limited VRE capacities, load or demand side management, electricity grid expansion and the creation of flexibility through cross-sectoral integration are options to develop integrated renewable energy systems that may operate without fossil backup capacities. SC as a one-way path from power-to-X represents one part of this solution. However, the application of sector coupling remains challenging in historically built energy systems and market frameworks. The successful integration and broad application of renewable

27 Ramsebner, J. (2022). The role of sector coupling in the energy transition [Thesis, Technische Universität Wien]. https://repositum.tuwien.at/handle/20.500.12708/80280

28 Ramsebner, J., Haas, R., Ajanovic, A., & Wietschel, M. (2021). The sector coupling concept: A critical review. *WIREs Energy and Environment*, 10(4), e396. https://doi.org/10.1002/wene.396.

29 Ramsebner, J. (2022). The role of sector coupling in the energy transition [Thesis, Technische Universität Wien]. https://repositum.tuwien.at/handle/20.500.12708/80280.

energy in new applications require new market frameworks and precise regulatory guidelines to develop attractive business models.

21.3 The Role of Sector Coupling in Renewable Energy Systems

The principles of SC have already been known from the beginning of the 20th century. The first battery electric vehicle (BEV) produced in the US was demonstrated at the 1892 World's Fair in Chicago, Illinois (Warner, 2015).[30] The developments in Europe, not only in electric bicycles and automobiles but also in public transport, disappeared in the middle of the 20th century. In the US BEVs gained early market acceptance despite their short driving range and the speed limit of direct current (DC) motors to 32 km/h. The internal combustion engine of Henry Ford's low-priced Model T in 1908, the introduction of the automatic starter by Charles Kettering in 1911 and the US highway system expansion (Warner, 2015) defeated BEVs.[31] Driven by the desire for an all-electric American home at the beginning of the 20th century, the residential sector experienced significant electrification using electric stoves, hot plates, washing and ironing machines, dishwashers, electric doorbells and a vast number of lightening devices (Foy & Schlereth, 1994).[32] (Ramsebner 2022)[33]

(Robinius, Otto, Heuser, et al., 2017) found further early applications in energy generation, which applied the primary approach of SC:[34]

> Furthermore, while SC has always been practised, it has hitherto been in the context of fossil fuels such as kerosene, methane, oil or coal, rather than renewable energy sources (RES). For example, a combined heat and power plant that generates electricity would supply excess heat that would otherwise be wasted.

In the 1980s and 1990s the so-called integrated energy systems (IES) were investigated aiming to integrate individual fossil fuel streams through new ways of distribution

30 Warner, J. T. (2015). *The handbook of lithium-ion battery pack design: Chemistry, components, types and terminology*. Elsevier Science.

31 Warner, J. T. (2015). *The handbook of lithium-ion battery pack design: Chemistry, components, types and terminology*. Elsevier Science.

32 Foy, J. H., & Schlereth, T. J. (1994). Foy, J: *American Home Life 1880-1930* (1st ed.). University of Tennessee Press.

33 Ramsebner, J. (2022). The role of sector coupling in the energy transition [Thesis, Technische Universität Wien]. https://repositum.tuwien.at/handle/20.500.12708/80280.

34 Robinius, M., Otto, A., Syranidis, K., Ryberg, D. S., Heuser, P., Welder, L., Grube, T., Markewitz, P., Tietze, V., & Stolten, D. (2017). Linking the power and transport sectors—part 2: Modelling a sector coupling scenario for Germany. *Energies*, 10(7), 957ff. https://doi.org/10.3390/en10070957.

and conversion to reduce the environmental burden (Belyaev et al., 1987;[35] Häfele & Nebojsa, 1983).[36] The main idea included the decomposition of energy carriers into their elementary components to fulfil energy demand flexibly. The authors, however, point out that today the focus lies in using renewable electricity in an increasing amount of end-consumption applications to increase the share of renewable energy in these sectors (assuming that the electricity supply is or can primarily be renewable). There are two options for electrification: adapting fossil-fuel-based processes for direct electrification or converting electricity into a more suitable or flexible fuel type. In this context, SC has become popular during the German energy transition (Schaber, 2013),[37] (Ramsebner, 2022).[38]

By investigating the integration of large-scale, variable wind power in future renewable energy systems, the first mention of this approach under the term SC was found in studies by (Schaber, 2013),[39] (Schaber et al., 2013)[40] and later by (Richts et al., 2015).[41] In the course of the energy transition in 2017, several German ministries and international energy agencies developed thorough guidelines and information on SC (BMWi, 2016;[42] BMUB, 2016;[43] BDEW, 2017;[44] IRENA, 2018).[45] Nevertheless, as often as the term "sector coupling" is used in energy policy debates today, it is not used clearly

35 Belyaev, L. S., Kaganovich, B. M., Krutov, A. N., Filippov, S. P., Martinsen, D., Müller, M., Wagner, H. J., & Walbeck, M. (1987). *Ways of transition to clean energy use: Two methodological approaches*. IIASA.

36 Häfele, W., & Nebojsa, N. (1983). The contribution of oil and gas for the transition to long range novel energy systems. *World Demand for and Supply of Oil and Gas*, 9(3).

37 Schaber, K. (2013). *Integration of Variable Renewable Energies in the European power system: A model-based analysis of transmission grid extensions and energy sector coupling*. Technische Universität.

38 Ramsebner, J. (2022). The role of sector coupling in the energy transition [Thesis, Technische Universität Wien]. https://repositum.tuwien.at/handle/20.500.12708/80280.

39 Schaber, K. (2013). *Integration of Variable Renewable Energies in the European power system: A model-based analysis of transmission grid extensions and energy sector coupling*. Technische Universität.

40 Schaber, K., Steinke, F., & Hamacher, T. (2013). Managing temporary oversupply from renewables efficiently: Electricity storage versus energy sector coupling in Germany. 22.

41 Richts, C., Jansen, M., & Siefert, M. (2015). Determining the economic value of offshore wind power plants in the changing energy system. *Energy Procedia*, 80, 422–432. https://doi.org/10.1016/j.egypro.2015.11.446.

42 Bundesministerium für Wirtschaft und Energie (BMWi). (2016). Green paper on energy efficiency: Discussion paper of the Federal Ministry for Economic Affairs and Energy.

43 Bundesministeriums für Umwelt, Naturschutz, Bau und Reaktorsicherheit (BMUB). (2016). Climate Action Plan 2050 – Principles and goals of the German government's climate policy.

44 Bundesverband der Energie-und Wasserwirtschaft e.V. (BDEW). (2017). Positionspapier: 10 Thesen zur Sektorkopplung. http://www.bdew.de/service/stellungnahmen/10-thesen-sektorkopplung.

45 International Renewable Energy Agency (IRENA). (2018). Hydrogen from renewable power: Technology outlook for the energy transition.

and uniformly (Scorza et al., 2018).[46] One of the first scientific papers to specifically develop a more uniform interpretation of the concept was established by (Robinius, Otto, Heuser, et al., 2017).[47] Towards 2018, more peer-reviewed scientific papers were explicitly dedicated to SC, such as the work by (Buttler & Spliethoff, 2018)[48] and (Stadler & Sterner, 2018).[49] By 2019, the contributions to the concept were multiplied.

However, future research still needs to define SC uniformly and discuss critically which of its scopes are reasonable for future energy systems (Wietschel et al., 2018).[50] There is still discussion about whether SC only includes renewable electricity or if the coupling of conventional electricity generation may also be considered. Furthermore, the literature does not always agree on the question of whether biogas and biofuels, combined (cooling) heat and power (C(C)HP) plants and the use of excess heat are part of the concept. Additionally, the definition of the sectors referred to in SC widely differs throughout the literature depending on the research perspective, which may be technical or economical. The following sub-sections first provide an analysis of the interpretation of the sectors within SC and then continue with available definitions of SC. Once this basic understanding is established, a detailed literature review on SC and its common research focus is conducted (Ramsebner, 2022).[51]

21.3.1 Enabling Technologies

According to current research, P2X technologies will be inevitable to decarbonise various endues applications, such as industrial process heat, chemical production, international air, or maritime transport. Many of these processes cannot be powered by electricity directly but depend on specific chemical conditions that can only be provided by a material energy source, such as gas and liquid fuel. P2X technologies com-

46 Scorza, S. A., Pfeiffer, J., Schmitt, A., & Weissbart, C. (2018). Kurz zum Klima: »Sektorkopplung« – Ansätze und Implikationen der Dekarbonisierung des Energiesystems. 71(10), 49–53.
47 Robinius, M., Otto, A., Syranidis, K., Ryberg, D. S., Heuser, P., Welder, L., Grube, T., Markewitz, P., Tietze, V., & Stolten, D. (2017). Linking the power and transport sectors—part 2: Modelling a sector coupling scenario for Germany. *Energies*, 10(7), 957ff. https://doi.org/10.3390/en10070957.
48 Buttler, A., & Spliethoff, H. (2018). Current status of water electrolysis for energy storage, grid balancing and sector coupling via power-to-gas and power-to-liquids: A review. *Renewable and Sustainable Energy Reviews*, 82, 2440–2454. https://doi.org/10.1016/j.rser.2017.09.003
49 Stadler, I., & Sterner, M. (2018). Urban energy storage and sector coupling. In *Urban Energy Transition: Renewable Strategies for Cities and Regions* (2nd ed., pp. 225–244). Elsevier.
50 Wietschel, M., Plötz, P., Pfluger, B., Klobasa, M., Eßer, A., Haendel, M., Müller-Kirchenbauer, J., Kochems, J., Hermann, L., Grosse, B., Nacken, L., Küster, M., Pacem, J., Naumann, D., Kost, C., Kohrs, R., Fahl, U., Schäfer-Stradowsky, S., Timmermann, D., & Albert, D. (2018). Sektorkopplung – Definition, Chancen und Herausforderungen.
51 Ramsebner, J. (2022). The role of sector coupling in the energy transition [Thesis, Technische Universität Wien]. https://repositum.tuwien.at/handle/20.500.12708/80280.

pete with direct electrification in many energy-consuming processes. These, for example, include railway and long-distance truck transport or low-temperature heating. For private vehicles, local truck transport, and short-term flexibility within the electricity system, however, P2X transformation technologies are only competitive under certain conditions. Transformation processes from electricity into, for example, gas or fuel ideally are operated at times of excess electricity generation during which electricity prices are low (IRENA, 2018).[52] (Matthes, 2018) point out that if full decarbonisation and 100% supply from renewables shall be achieved, more powerful battery storage systems and P2G options for seasonal storage and sector integration will play a crucial role.[53] These applications require intensive research to become mature technologies in the next two decades (Ramsebner, 2022).[54]

21.3.2 Important is the Consideration of Efficiency

As an example of drive types in transport, Figure 21.3 shows that direct electrification through BEVs represents the most efficient overall resource use in mobility. Indirect electrification via liquid diesel fuels results in an overall efficiency Well-to-Wheel of only about 20%, and hydrogen fuel cell electric vehicles (FCEV) provide about 33% efficiency (Transport&Environment, 2020).[55] Fuel provision and the transformation into motion are much more efficient through direct electrification in a BEV, which achieves an overall efficiency of almost 80%. Any conversion toward renewable gas or liquid fuels will cause substantial efficiency losses. Still, on some occasions, they might be required to enable timely decarbonisation of hard-to-abate processes such as the industrial steel, cement or chemicals sector or heavy-duty transit transport (Ramsebner, 2022).[56]

52 International Renewable Energy Agency (IRENA). (2018). Hydrogen from renewable power: Technology outlook for the energy transition.

53 Matthes, F. Chr. (2018). Energy transformation in Germany: Progress, shortfalls and prospects.

54 Ramsebner, J. (2022). The role of sector coupling in the energy transition [Thesis, Technische Universität Wien]. https://repositum.tuwien.at/handle/20.500.12708/80280.

55 Transport&Environment. (2020). Decarbonising the EU's transport sector with renewable electricity and electrofuels. https://www.transportenvironment.org/wp-content/uploads/2020/12/2020_12_Briefing_feasibility_study_renewables_decarbonisation.pdf.

56 Ramsebner, J. (2022). The role of sector coupling in the energy transition [Thesis, Technische Universität Wien]. https://repositum.tuwien.at/handle/20.500.12708/80280.

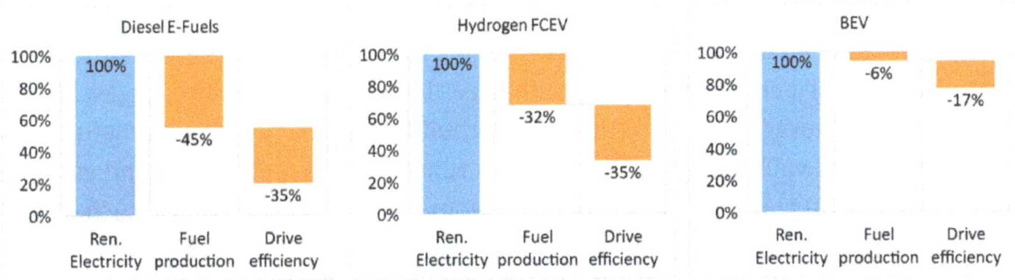

Figure 21.3: Efficiency of different electricity-based drive options (Transport&Environment, 2020);[57] Renewable (Ren.).

21.3.3 Direct Electrification

One potential application of SC refers to the direct electrification of processes. Formerly fossil-fuel-based applications and technologies may be adapted to be powered by electricity. Some currently directly electrified processes have already used electricity a long time ago. The mid-to-long-term goal nowadays, however, has to be using renewable electricity. Common examples include switching from traditional vehicles with combustion engines powered by fossil fuels to BEVs. Although usually direct electrification is not defined as a transformation process similar to P2G etc., (Wietschel et al., 2015) define e-mobility as power-to-move.[58] Moreover, industrial processes often have vast direct electrification potential. Steel production, for example, can be transformed using an electricity-based electric arc furnace instead of a coal-fired blast furnace and a basic oxygen furnace (BOF). Direct electrification is also an option for smaller-scale residential heating and cooling (Ramsebner, 2022).[59]

21.3.4 Power-to-Gas/Liquids

Not only is the importance of integrating large-scale renewable electricity generation mentioned frequently in the literature, but also the importance of integrating the gas sector. The existing gas infrastructure may already enable large-scale distribution and storage of renewable energy (BMUB, 2016; DVGW, 2020; DVGW & VDE, 2016). Water

57 Transport&Environment. (2020). Decarbonising the EU's transport sector with renewable electricity and electrofuels. https://www.transportenvironment.org/wp-content/uploads/2020/12/2020_12_Briefing_feasibility_study_renewables_decarbonisation.pdf

58 Wietschel, M., Haendel, M., Schubert, G., Köppel, W., & Degünther, C. (2015). Kurz- und mittelfristige Sektorkopplungspotentiale.

59 Ramsebner, J. (2022). The role of sector coupling in the energy transition [Thesis, Technische Universität Wien]. https://repositum.tuwien.at/handle/20.500.12708/80280

electrolysis converts power into gas – an electrochemical process that splits water into hydrogen (H_2) and oxygen (O_2) (IEA, 2019). H_2 is an energy carrier with a high energy density or energy-to-weight ratio, three times higher than that of gasoline or diesel, and low storage cost (Lewandowska-Bernat & Desideri, 2018;[60] Schiebahn et al., 2015).[61] This renewable gas is not only a promising alternative for diesel and gasoline in the transport sector but is also used in various industrial applications, such as ammonia synthesis, fertilizer production, glass, and fibre production, etc. (Lewandowska-Bernat and Desideri, 2018).[62] Today, three-quarters of H_2 demands are produced from natural gas, followed by coal (IEA, 2019).[63] Water electrolysis today accounts for less than 0.1% of dedicated H_2 production globally; however, it is expected to rise substantially with an increasing share of renewable electricity (Ramsebner, 2022).[64]

Electrolysers currently operate between 60% and 81% efficiency, depending on the technology type and load factor. H_2 may be processed to methane using renewable H_2 and a CO_2 source and even into liquid, synthetic fuels with similar characteristics to diesel or gasoline, such as methanol. CO_2 as a required input for methanation may either be produced from organic matter, for example, biomass, from fossil power plants, industrial processes, or captured from the air (Schiebahn et al., 2015).[65] While hydrogen may be fed into the natural gas grid only at a particular share – depending on local technical characteristics – methane can function as a direct substitute. The liquefaction process is usually defined as P2L (Danish Energy Agency (DEA), 2017).[66] However, some authors even specifically name it power-to-fuel (Robinius, Otto, Heuser, et al., 2017;[67] Schemme et al., 2017).[68] (Ridjan et al., 2016) reviewed the differ-

60 Lewandowska-Bernat, A., & Desideri, U. (2018). Opportunities of power-to-gas technology in different energy systems architectures. Applied Energy, 228, 57–67. https://doi.org/10.1016/j.apenergy.2018.06.001.

61 Schiebahn, S., Grube, T., Robinius, M., Tietze, V., Kumar, B., & Stolten, D. (2015). Power to gas: Technological overview, systems analysis and economic assessment for a case study in Germany. *International Journal of Hydrogen Energy*, 40(12), 4285–4294. https://doi.org/10.1016/j.ijhydene.2015.01.123.

62 Lewandowska-Bernat, A., & Desideri, U. (2018). Opportunities of power-to-gas technology in different energy systems architectures. Applied Energy, 228, 57–67. https://doi.org/10.1016/j.apenergy.2018.06.001.

63 International Energy Agency (IEA). (2019). The future of hydrogen [Report prepared by the IEA for the G20, Japan].

64 Ramsebner, J. (2022). The role of sector coupling in the energy transition [Thesis, Technische Universität Wien]. https://repositum.tuwien.at/handle/20.500.12708/80280.

65 Schiebahn, S., Grube, T., Robinius, M., Tietze, V., Kumar, B., & Stolten, D. (2015). Power to gas: Technological overview, systems analysis and economic assessment for a case study in Germany. *International Journal of Hydrogen Energy*, 40(12), 4285–4294. https://doi.org/10.1016/j.ijhydene.2015.01.123

66 Danish Energy Agency (DEA). (2017). Regulation and planning of district heating in Denmark.

67 Robinius, M., Otto, A. D.-I., Heuser, P., Welder, L., Syranidis, K., Ryberg, D. S., Grube, T. J., Markewitz, P., Peters, R., & Stolten, D. (2017). Linking the power and transport sectors—part 1: The principle of sector coupling. https://doi.org/10.3390/EN10070956

68 Schemme, S., Samsun, R. C., Peters, R., & Stolten, D. (2017). Power-to-fuel as a key to sustainable transport systems – An analysis of diesel fuels produced from CO 2 and renewable electricity. Fuel, 205, 198–221. https://doi.org/10.1016/j.fuel.2017.05.061

ence between synthetic fuels and electro-fuels, finding that the first does not directly indicate the type of resource or production process used. The authors suggest using electro-fuels if a large amount of electricity is included in the production process, as is the case for hydrogen or methanol production via electrolysis. We will refer to electro-fuels for gaseous or liquid fuels generated from P2G or P2L (Ramsebner, 2022).[69]

21.3.5 Power-to-Heat

The transformation of power into heat can be realized using heat pumps (HP) and electric boilers (EB) either on a central level for DH purposes or on a decentralized level directly at the consumption location. For small applications, even direct electric heating is possible. The flexible heating and cooling generation from renewable electricity is gaining importance in future renewable energy systems (Bloess et al., 2018). Currently, at least in Denmark, a decrease in taxes on electricity use for these technologies is improving economic performance (Nielsen et al., 2016). HPs are known to be more efficient compared with EBs. HPs transform heat from a low-temperature heat source (e.g., ammonia, air, seawater, industrial waste heat, etc.) to a higher temperature level output heat (Danish Energy Agency (DEA), 2016) and can also function the other way around for cooling.[70] There are so-called "compression HPs" – powered by electricity or combustion engines consuming fuel or biogas and "absorption HPs" – driven by steam, hot water, or flue gas. They can be used in industrial processes, individual space heating/cooling, or DH. Compression HPs achieve a seasonal performance factor or average coefficient of performance (COP) of 3–5 times the utilized electricity input (Danish Energy Agency (DEA), 2016),[71] (Ramsebner, 2022).[72]

(Nielsen et al., 2016) claim that EBs, in contrast, are easier to implement from a financial perspective because of lower investment costs and more flexible operation without any startup cost or constraints. They use electricity to generate hot water or steam, similarly to oil or gas boilers (Danish Energy Agency (DEA), 2016).[73] EBs operate at an average COP of 1 and are primarily implemented for ancillary services owing to

69 Ramsebner, J. (2022). The role of sector coupling in the energy transition [Thesis, Technische Universität Wien]. https://repositum.tuwien.at/handle/20.500.12708/80280.

70 Danish Energy Agency (DEA). (2016). Technology data for electricity, district heating, energy storage and energy conversion. https://ens.dk/sites/ens.dk/files/Statistik/technology_data_catalogue_for_el_and_dh_-_0009.pdf.

71 Danish Energy Agency (DEA). (2016). Technology data for electricity, district heating, energy storage and energy conversion. https://ens.dk/sites/ens.dk/files/Statistik/technology_data_catalogue_for_el_and_dh_-_0009.pdf

72 See for further details to heat pumps section 2.9 (O. Opel et al.).

73 Danish Energy Agency (DEA). (2016). Technology data for electricity, district heating, energy storage and energy conversion. https://ens.dk/sites/ens.dk/files/Statistik/technology_data_catalogue_for_el_and_dh_-_0009.pdf

their systemic benefits promoting wind energy utilisation and efficient heat source use. EBs are using electrical resistance for 1–2 MW; for example, a standard hot-water heater or electrode boilers for larger applications of more than a few MW directly connected to the high voltage grid (Danish Energy Agency (DEA), 2016).[74] The technologies described above enable various pathways for using renewable power to achieve the decarbonisation of the energy system. Figure 21.4 and Figure 21.5 illustrate the principles of SC from a technological perspective.

In Figure 21.4, the authors call direct electrification of transport using BEVs power-to-move, sometimes referred to as power-to-mobility or power-to-transport. Figure 21.4 outlines the integration of the energy carriers, namely, electricity, gas, and heat, as well as their cross-sectorial utilisation and energy storage options. Because the original figure by (Stadler & Sterner, 2018) focused on integrating energy carriers, the end-consumption sectors (transport, industry, and residential) were added as final demand sources.[75]

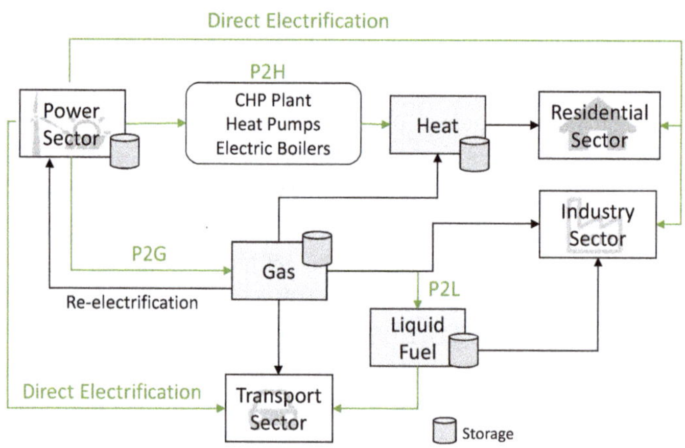

Figure 21.4: Power-to-X technologies and cross-sectorial energy storage for sector coupling (adapted from [Stadler & Sterner, 2018][76]).

Thus far, the SC concept has been discussed as a measure to integrate renewable electricity into a broad range of applications in different energy end-consumption sectors, enabled by transformation technologies and an adaption of traditional processes.

74 Danish Energy Agency (DEA). (2016). Technology data for electricity, district heating, energy storage and energy conversion. https://ens.dk/sites/ens.dk/files/Statistik/technology_data_catalogue_for_el_and_dh_-_0009.pdf.

75 Stadler, I., & Sterner, M. (2018). Urban energy storage and sector coupling. In *Urban Energy Transition: Renewable Strategies for Cities and Regions* (2nd ed., pp. 225–244). Elsevier.

76 Stadler, I., & Sterner, M. (2018). Urban energy storage and sector coupling. In *Urban Energy Transition: Renewable Strategies for Cities and Regions* (2nd ed., pp. 225–244). Elsevier.

Apart from a link between the power and other sectors, integrating different energy carriers and grids for electricity, gas, and heat is also gaining attention. The interactions between these energy carriers are essential to provide flexibility and handle renewable resources efficiently. These concepts can be found in the literature under several terms, such as multi-energy system, integrated energy system, or smart energy system (SES) (Connolly et al., 2016;[77] Lund et al., 2014;[78] Witkowski et al., 2020).[79] The common isolated view of energy grids is changing, and hybrid or integrated multi-energy systems are gaining importance with increasing VRE input (Ramsebner, Haas, Auer, et al., 2021).[80] (Geidl et al., 2007) claim that the traditionally grown energy grids are not ready for the new requirements.[81] Considering existing frameworks is a barrier when modelling this new view of optimal renewable energy systems. (Mancarella, 2014) understand MES as the optimal interaction of systems and energy carriers on different levels,[82] for example, geographically within a district, city or region (Ramsebner, 2022).[83]

(Lund et al., 2014) define an integrated energy system enhanced by new technologies and infrastructure as smart energy system.[84] However, the "smart" in SESs does not explicitly indicate many intelligent technologies but implies a radical shift in the understanding of energy system infrastructure. The authors also point out that the combination of smart electricity, thermal, and gas networks is a benefit for individuals and the whole energy system. SESs are considered an option to efficiently use a larger share of VRE sources in future energy systems by using new types of flexibility

77 Connolly, D., Lund, H., & Mathiesen, B. V. (2016). Smart energy Europe: The technical and economic impact of one potential 100% renewable energy scenario for the European Union. *Renewable and Sustainable Energy Reviews*, 60, 1634–1653. https://doi.org/10.1016/j.rser.2016.02.025.

78 Lund, H., Mathiesen, B. V., Connolly, D., & Østergaard, P. A. (2014). Renewable energy systems—A smart energy systems approach to the choice and modelling of 100 % renewable solutions. *Chemical Engineering Transactions*, 39, 1–6. https://doi.org/10.3303/CET1439001

79 Witkowski, K., Haering, P., Seidelt, S., & Pini, N. (2020). Role of thermal technologies for enhancing flexibility in multi-energy systems through sector coupling: Technical suitability and expected developments. *IET Energy Systems Integration*, 2(2), 69–79. https://doi.org/10.1049/iet-esi.2019.0061.

80 Ramsebner, J., Haas, R., Auer, H., Ajanovic, A., Gawlik, W., Maier, C., Nemec-Begluk, S., Nacht, T., & Puchegger, M. (2021). From single to multi-energy and hybrid grids: Historic growth and future vision. https://doi.org/10.1016/J.RSER.2021.111520.

81 Geidl, M., Koeppel, G., Favre-Perrod, P., Klöckl, B., Andersson, G., & Fröhlich, K. (2007). The energy hub: A powerful concept for future energy systems. 3rd Annual Carnegie Mellon Conference on the Electricity Industry: Ensuring that the Industry has the Physical and Human Resources Needed for the Next Thirty Years. https://www.research-collection.ethz.ch/handle/20.500.11850/3133.

82 Mancarella, P. (2014). MES (multi-energy systems): An overview of concepts and evaluation models. *Energy*, 65, 1–17. https://doi.org/10.1016/j.energy.2013.10.041.

83 Ramsebner, J. (2022). The role of sector coupling in the energy transition [Thesis, Technische Universität Wien]. https://repositum.tuwien.at/handle/20.500.12708/80280.

84 Lund, H., Mathiesen, B. V., Connolly, D., & Østergaard, P. A. (2014). Renewable energy systems—A smart energy systems approach to the choice and modelling of 100 % renewable solutions. *Chemical Engineering Transactions*, 39, 1–6. https://doi.org/10.3303/CET1439001.

and integrating the infrastructure of smart or bidirectional electricity, smart gas, and smart thermal grids (Lund et al., 2014).[85] The smart electricity grid connects the flexible electricity demand of HPs and EBs and the VRE sources. The smart thermal grid enables the integration of electricity and heating and cooling and provides flexibility through thermal energy storage. Finally, the smart gas grid connects electricity, heating and cooling, and transport by considering renewable gas generation through electrolysis (Lund et al., 2017),[86] (Ramsebner, 2022).[87]

Modelling such MESs for a time in the future is highly complex. According to (Arent et al., 2021),[88] the interconnectivity of energy supply and demand sources is the value and challenge of hybrid energy systems. Also, the successful realization of such system change via sector integration depends on technologies, interfaces, supportive regulatory frameworks and market design (Raux-Defossez et al., 2018).[89]

While (Geidl et al., 2007) suggest designing everything newly,[90] the risk of stranded investment favours the adjustment of existing infrastructure, e.g. for the distribution of renewable hydrogen or methane. The ongoing indirect or direct electrification of formerly fossil-fuel-based processes will result in a tighter connection between the relevant end-consumption sectors and the electricity sector (Mancarella, 2014).[91] The more prominent share of electricity in a renewable system may substantially change the role of traditional energy grids, such as gas and heat, and the share between grid and non-grid-related energy consumption. An increase in distributed solar, solar-thermal PV or heat pump systems, for example, increases non-grid-related energy consumption with a specific share of own electricity consumption on site. These renewable energy technologies also represent competition for the gas grid, which distributed 41% of the grid-related energy in 2015 (Ramsebner, Haas, Auer,

85 Lund, H., Mathiesen, B. V., Connolly, D., & Østergaard, P. A. (2014). Renewable energy systems—A smart energy systems approach to the choice and modelling of 100 % renewable solutions. *Chemical Engineering Transactions*, 39, 1–6. https://doi.org/10.3303/CET1439001.

86 Lund, H., Østergaard, P. A., Connolly, D., & Mathiesen, B. V. (2017). Smart energy and smart energy systems. *Energy*, 137, 556–565. https://doi.org/10.1016/j.energy.2017.05.123.

87 Ramsebner, J. (2022). The role of sector coupling in the energy transition [Thesis, Technische Universität Wien]. https://repositum.tuwien.at/handle/20.500.12708/80280.

88 Arent, D. J., Bragg-Sitton, S. M., Miller, D. C., Tarka, T. J., Engel-Cox, J. A., Boardman, R. D., Balash, P. C., Ruth, M. F., Cox, J., & Garfield, D. J. (2021). Multi-input, Multi-output Hybrid Energy Systems. *Joule*, 5(1), 47–58. https://doi.org/10.1016/j.joule.2020.11.004.

89 Raux-Defossez, P., Wegerer, N., Pétillon, D., Bialecki, A., Bailey, A., & Belhomme, R. (2018). Grid Services Provided By The Interactions Of Energy Sectors In Multi-Energy Systems: Three International Case Studies. *Energy Procedia*, 155, 209–227. https://doi.org/10.1016/j.egypro.2018.11.055.

90 Geidl, M., Koeppel, G., Favre-Perrod, P., Klöckl, B., Andersson, G., & Fröhlich, K. (2007). The energy hub: A powerful concept for future energy systems. 3rd Annual Carnegie Mellon Conference on the Electricity Industry: Ensuring that the Industry has the Physical and Human Resources Needed for the Next Thirty Years. https://www.research-collection.ethz.ch/handle/20.500.11850/3133.

91 Mancarella, P. (2014). MES (multi-energy systems): An overview of concepts and evaluation models. *Energy*, 65, 1–17. https://doi.org/10.1016/j.energy.2013.10.041.

et al., 2021).[92] Still, the storage characteristics of gas will be the key to the seasonal balancing of renewable availability. They will remain a significant energy carrier in the future in terms of flexibility and security of supply (Ramsebner, 2022).[93]

Hybrid energy systems can flexibly handle multiple inputs and supply energy as multiple outputs (Arent et al., 2021).[94] To optimised all energy streams, intelligent communication technology (ICT) is essential in the form of a hybrid demand-control system to achieve environmental and economic benefits. Real-time data is collected and evaluated through sensors and communication technology for optimal decision-making and forecasting (Masera et al., 2018),[95] (Ramsebner, 2022).[96]

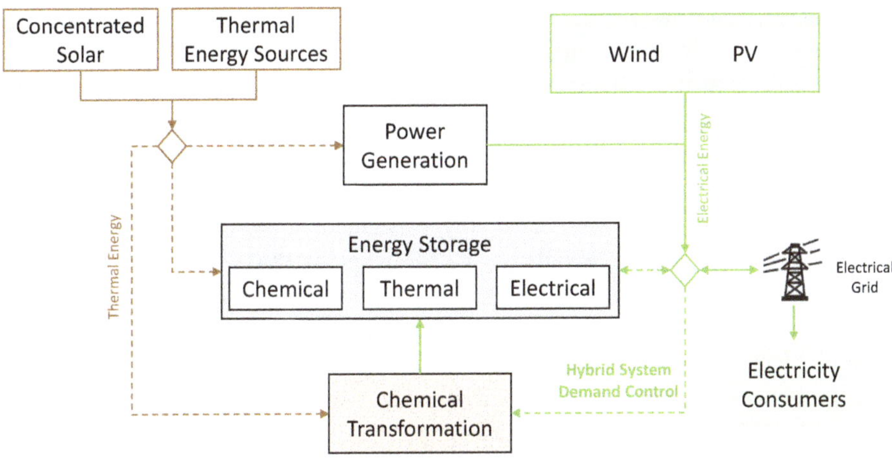

Figure 21.5: Integration of the electricity and thermal energy flows in a hybrid energy system (Ramsebner, Haas, Auer, et al., 2021), adapted from (Masera et al., 2018).[97]

92 Ramsebner, J., Haas, R., Auer, H., Ajanovic, A., Gawlik, W., Maicr, C., Nemec-Begluk, S., Nacht, T., & Puchegger, M. (2021). From single to multi-energy and hybrid grids: Historic growth and future vision. https://doi.org/10.1016/J.RSER.2021.111520.

93 Ramsebner, J. (2022). The role of sector coupling in the energy transition [Thesis, Technische Universität Wien]. https://repositum.tuwien.at/handle/20.500.12708/80280

94 Arent, D. J., Bragg-Sitton, S. M., Miller, D. C., Tarka, T. J., Engel-Cox, J. A., Boardman, R. D., Balash, P. C., Ruth, M. F., Cox, J., & Garfield, D. J. (2021). Multi-input, Multi-output Hybrid Energy Systems. Joule, 5(1), 47–58. https://doi.org/10.1016/j.joule.2020.11.004.

95 Masera, M., Bompard, E. F., Profumo, F., & Hadjsaid, N. (2018). Smart (Electricity) Grids for Smart Cities: Assessing Roles and Societal Impacts. Proceedings of the IEEE, 106(4), 613–625. https://doi.org/10.1109/JPROC.2018.2812212.

96 Ramsebner, J. (2022). The role of sector coupling in the energy transition [Thesis, Technische Universität Wien]. https://repositum.tuwien.at/handle/20.500.12708/80280.

97 Masera, M., Bompard, E. F., Profumo, F., & Hadjsaid, N. (2018). Smart (Electricity) Grids for Smart Cities: Assessing Roles and Societal Impacts. Proceedings of the IEEE, 106(4), 613–625. https://doi.org/10.1109/JPROC.2018.2812212.

Nevertheless, MESs and efficient VRE integration rely on the basic P2X transformation technologies for renewable electricity. Therefore, SC does not only represent an independent system but can also support fully integrated MESs. (Witkowski et al., 2020)[98] specifically analyze the ability of SC to provide flexibility in the form of renewable heat or gas. The transport, residential, and industry sectors demand mobility, electricity, and heating and cooling. Therefore, the SC concept is essential to the future 100% renewable energy systems. Hybrid energy systems require a successful link between formerly separated infrastructure through ICT and data-based action and prediction (Ramsebner, Haas, Auer, et al., 2021),[99] (Ramsebner, 2022).[100]

Looking at the business case, establishing such a new approach to the existing environment results in several challenges, such as complicated interdependencies between energy systems on technical, economic and market perspectives; a lack of commercial tools to operate them; isolated market structures and high operational and management complexity through the different energy grids managed together (Abeysekera, 2016).[101] Furthermore, substantial investments must be made in distribution grid expansion and installing new renewable power and transformation plants such as electrolysers. However, (Haas et al., 2021) conclude that the current framework for financing grid infrastructure and renewable technologies is not always setting the right incentives.[102] This is specifically true for applications where the technology has not yet been decided and imposes uncertainty on investors. As a result, pilot plants need to generate a specific volume to decrease the prices of new technologies, e.g. from grid operators to guarantee supply security in renewable energy systems (Ramsebner, Haas, Auer, et al., 2021),[103] (Ramsebner, 2022).[104]

98 Witkowski, K., Haering, P., Seidelt, S., & Pini, N. (2020). Role of thermal technologies for enhancing flexibility in multi-energy systems through sector coupling: Technical suitability and expected developments. *IET Energy Systems Integration*, 2(2), 69–79. https://doi.org/10.1049/iet-esi.2019.0061.

99 Ramsebner, J., Haas, R., Auer, H., Ajanovic, A., Gawlik, W., Maier, C., Nemec-Begluk, S., Nacht, T., & Puchegger, M. (2021). From single to multi-energy and hybrid grids: Historic growth and future vision. https://doi.org/10.1016/J.RSER.2021.111520.

100 Ramsebner, J. (2022). The role of sector coupling in the energy transition [Thesis, Technische Universität Wien]. https://repositum.tuwien.at/handle/20.500.12708/80280.

101 Abeysekera, M. (2016). Combined analysis of coupled energy networks [Phd, Cardiff University]. https://orca.cardiff.ac.uk/id/eprint/97879/.

102 Haas, R., Ajanovic, A., Ramsebner, J., Perger, T., Knápek, J., & Bleyl, J. W. (2021). Financing the future infrastructure of sustainable energy systems. *Green Finance*, 3(1), 90–118. https://doi.org/10.3934/GF.2021006.

103 Ramsebner, J., Haas, R., Auer, H., Ajanovic, A., Gawlik, W., Maier, C., Nemec-Begluk, S., Nacht, T., & Puchegger, M. (2021). From single to multi-energy and hybrid grids: Historic growth and future vision. https://doi.org/10.1016/J.RSER.2021.111520

104 Ramsebner, J. (2022). The role of sector coupling in the energy transition [Thesis, Technische Universität Wien]. https://repositum.tuwien.at/handle/20.500.12708/80280.

21.4 Sector Coupling: Where is the Business Model?

Heading towards sustainable economies means heading towards a sustainable energy system, hence providing energy services in an optimal way from society's point-of-view (Haas et al., 2021).[105]

Despite all the promising possibilities in setting up a renewable energy system with the help of renewable electricity, the business model of sector coupling or in general renewable energy sources remains challenging from the installation of power plants down to the optimization of energy supply and monetary incentives for growing decentral demand sources and their operators. With the right concept, grid stability and renewable power plant profitability can be achieved, while providing profits for all different actors such as services providers and operators as well as end users. In such a scenario with a market driven approach, every actor can benefit from managed demand to support the energy system efficiency and avoid distribution grid extension. On the supply side, profitability of storage technologies needs to be supported in a way to provide more flexibility and buffering short term and seasonal variability.

The approval process of wind power plants remains slow, with lengthy elaboration on social and environmental aspects, but finally attempts can be seen to speed up this process. At the same time, financing of renewable projects faces huge difficulties with long payback periods (see following section "Long-term financing situation of renewables").

Once the power plant is in place to supply green electricity, the management of its variable supply needs to find its respective demand to establish an equilibrium. As the elaboration on sector coupling applications make clear, the variability of renewable energy supply requires a flexible reaction of energy demand to achieve a stable and efficient renewable energy system or a type of storage. While large-scale storage options of electricity come at high cost or efficiency losses and therefore seasonal variability between high renewables availability in summer and scarcity in winter is very difficult to overcome, solving short-term discrepancies can be achieved more easily. However, demand and supply sources need to interact much closer than in a traditional fossil fuel based energy system.

Flexible demands, such as electric vehicle charging, green hydrogen production or heating can be managed by providing incentives to the consumers. So far, it seems that solutions to reduce the pressure on the distribution grid are often searched for on a technical level, with regulations demanding an automated reaction of flexible demand sources to grid instabilities. However, these end-consumer devices are often not meant to fulfil such requirements. An aggregator, such as mobility or other ser-

105 Haas, R., Ajanovic, A., Ramsebner, J., Perger, T., Knápek, J., & Bleyl, J. W. (2021). Financing the future infrastructure of sustainable energy systems. *Green Finance*, 3(1), 90–118. https://doi.org/10.3934/GF.2021006.

vice operators, could manage an amount of consumer devices much better on a more central level. The huge gap, that still needs to be filled and defined, is a market driven solution in which price signals represent the current demand and supply situation and trigger the right answer on supply or demand side in favour of the energy system. Firstly, the reaction based on variable electricity prices incentivises to shift electricity consumptions to a time at which prices are low. This usually coincides with a high availability of renewable electricity in the system and therefore supports the match of demand and supply in the distribution grid. An example of the impact of solar PV feed-in on electricity prices is provided in the section "The impact of variable energy feed-in on the business case: A solar PV example".

The downside of this situation is the fact that high feed-in and high consumption power increase the need for grid expansion. This can only be solved by direct consumption of distributed energy on site, such as in energy communities or a distribution of supply peaks through storage or centralized Power2X technologies relieving the distribution grid from power flow. Therefore, the variable electricity price should as well be accompanied by variable grid tariffs, representing the actual grid performance. During times of high pressure onto the grid e.g. through PV feed-in peaks, the grid tariff for feed in shall be higher for compensation of the cost. These two price indicators together, can steer demand for the benefit of the renewable power availability and the distribution grid stability, incentivized by cost savings for the end-user. For example, offers for spot price based charging of BEVs in which the consumed charging power varies with the hourly electricity spot price are requested more and more. This is of course only possible with the respective variable electricity tariff, which ideally only is valid for the flexible device, such as the BEV, and not necessarily covers the household electricity. Suitable tariffs are still hardly found but emerging in Nordic countries and slowly approaching central Europe. Examples are tibber and awattar.

One more challenge for renewable energy business cases, however, lies in the current market logic. Renewable energy feed-in always receives the current electricity market price based on the merit order. Low market prices for consumers also mean low profit for PV plant operators. As a result, the more solar PV is fed in and the more fossil fuel plants are pushed out of the market, the less profit renewable power plants can make. In the current market framework, this represents a cannibalization of solar PV feed-in, which at a tipping point doesn't make additional capacities of this technology profitable. At some point, this might also avoid excessive PV feed-in at noon times from harming the grid situation. Additional profitability could only be established if PV electricity can be fed into the grid outside of the usual noon peak, which can make storage capacities more attractive.

Only few countries have already started to open the market to smaller decentral energy consumers, such as EV charging stations, for example the Nordics and the UK. In these countries, for example charge point operators are starting to aggregate the power volume of their infrastructure and offer it to the reserve market as a grid service. Monitoring grid stability on an aggregated level and managing decentral devices via their common communication interfaces enables an attractive offer to the market, which is turned into profit and can be reimbursed partly to the end-user for their flexibility, while supporting the distribution grid stability.

In other areas of Europe, however, decentral energy supply devices (heat pumps, charging stations, air conditioning . . .) are confronted more and more with regulatory requirements to technically guarantee grid stability automatically. The downside in such an approach is the fact that it does not result in any monetary benefit for any one along the supply chain, from the distribution grid down to the end user but creates development cost for the manufacturer of the respective devices. Therefore, it can only be regarded as an intermediate and emergency solution.

21.4.1 Long-term Financing Situation of Renewables

In general, more environmentally benign energy services require higher inputs of capital to finance technologies and lower expenses for the commodity energy. In addition, energy services provided by non-commodity energy carriers, such as wind or solar require less (almost zero) flow energy inputs. The development of suitable financing models plays an important role in the implementation of long-term investments green energy infrastructures necessary to achieve climate goals through greenhouse gas (GHG) emission reduction (Haas et al., 2021).[106] Within the energy system there are three major areas in which long-term financing models play an important role:

- Investments in renewable energy (RE) technologies, e.g. large solar thermal, photovoltaic and wind power systems,
- Investments in network infrastructure for electricity, district heating, charging stations for electric vehicles (EVs) and hydrogen refuelling stations for fuel cell electric vehicles (FCEVs), storage, etc. (see e.g. (Ajanovic, 2015;[107] Ajanovic et al., 2020;[108] Ajanovic & Haas, 2018);[109]
- Investments in reducing energy demand and energy efficiency, e.g. by applying "Energy Performance Contracting", such as thermal insulation of buildings.

106 Haas, R., Ajanovic, A., Ramsebner, J., Perger, T., Knápek, J., & Bleyl, J. W. (2021). Financing the future infrastructure of sustainable energy systems. *Green Finance*, 3(1), 90–118. https://doi.org/10.3934/GF.2021006.

107 Ajanovic, A. (2015). The future of electric vehicles: Prospects and impediments. *WIREs Energy and Environment*, 4(6), 521–536. https://doi.org/10.1002/wene.160.

108 Ajanovic, A., Hiesl, A., & Haas, R. (2020). On the role of storage for electricity in smart energy systems. *Energy*, 200, 117473. https://doi.org/10.1016/j.energy.2020.117473.

109 Ajanovic, A., & Haas, R. (2018). Economic prospects and policy framework for hydrogen as fuel in the transport sector. *Energy Policy*, 123, 280–288. https://doi.org/10.1016/j.enpol.2018.08.063.

This means that there also are long-term market segments of a much larger financial scale, since investment is intensive and requires appropriate financing strategies. These investments are however essential as a public service to enable renewable energy supply. Short-term, operational energy markets therefore do not automatically provide long-term investment incentives, which harms renewable electricity installations (Haas et al., 2021).[110]

Traditional credit financing by banks for municipalities and public enterprises is becoming increasingly difficult. On the one hand, in the course of the global financial crisis, some credit institutions withdrew from the municipal finance business to a large extent. On the other hand, Basel III and currently already Basel IV have made lending conditions more stringent. Therefore, it is becoming increasingly interesting to involve citizens in the financing of projects, and thus also to increase the acceptance of energy (infrastructure) projects. In order to implement projects for which own funds or direct subsidies are not available in sufficient quantities, it is necessary to borrow from outside funds, e.g., from Green Bonds (see next chapter). The procurement of capital to cover the financial requirements for the realization of investment projects is referred to as financing. The choice of the right form of financing depends on the object of financing. Tax motives or products linked to services (e.g., leasing) can also be decisive. In principle, the forms of financing can be divided into classic and special forms of financing (Haas et al., 2021).[111]

While bottom-up financing is tailored to the individual project, top-down financing models such as "green bonds" are offered on the market and are characterized by predefined preconditions/targets that a project must meet in order to obtain financing. In recent years there has been remarkably growing sensitivity to the issue of financing energy technologies with a long life-time. In addition, the number and variety of possible financing models has increased significantly. This gives hope for an optimistic dissemination of efficient and sustainable financing solutions for the provision of energy services in the next years (Haas et al., 2021).[112]

In any financing case, the decision ultimately boils down to whether there is an appropriate balance between risk and return-on-investment. The risk can be reduced by an appropriate regulatory body. The lower the risk is estimated, the lower is the expected return. In the end the state or public authorities will always have to bear

---

110 Haas, R., Ajanovic, A., Ramsebner, J., Perger, T., Knápek, J., & Bleyl, J. W. (2021). Financing the future infrastructure of sustainable energy systems. *Green Finance*, 3(1), 90–118. https://doi.org/10.3934/GF.2021006.

111 Haas, R., Ajanovic, A., Ramsebner, J., Perger, T., Knápek, J., & Bleyl, J. W. (2021). Financing the future infrastructure of sustainable energy systems. *Green Finance*, 3(1), 90–118. https://doi.org/10.3934/GF.2021006.

112 Haas, R., Ajanovic, A., Ramsebner, J., Perger, T., Knápek, J., & Bleyl, J. W. (2021). Financing the future infrastructure of sustainable energy systems. *Green Finance*, 3(1), 90–118. https://doi.org/10.3934/GF.2021006.

the risk. Hence, the central question is: What amount of risk remains to be carried by the state? Of course, the risk should be as low as possible (Haas et al., 2021).[113]

Hence, at least three major issues play a crucial role in assessing the performance of financing a specific project:

- Are the costs of capital (interest rate or WACC) based on Eq. 18 justified (from societies' point-of-view)?
- What is the depreciation time accepted by the provider(s) of capital?
- What is the corresponding risk to get the money invested back?

The WACC interest rate (Equation 17) is a weighted average of the costs of equity *(E)* and debt *(D)* capital.

Equation 17: $$WACC_{afer\ taxes} = r_E \frac{E}{E+D} + r_D(1-T)\frac{D}{E+D} \tag{18}$$

r_E Cost of equity
r_D Cost of debt
T Tax rate
D Debt (market value)
E Equity capital (market value)

According to Figure 21.6, for interest rates from 0% to 10%, the capital recovery factor according to Eq. 20) varies between 0.13 and 0.07 for a depreciation period of 15 years (range of factor 2), and between 0.02 and 0.10 for a depreciation period of 50 years (range increases to factor 5).

Equation 19: $$CRF = \frac{i(1+i)^n}{(1+i)^n - 1} \tag{20}$$

Figure 21.7 shows how the sensitivity of an investment can be estimated as a function of interest rate (or WACC) and depreciation period using the example of an investment of EUR 100,000. The influence of the time factor at an interest rate of 5% is 1.8 comparing 15 years and 50 years depreciation periods, and the influence of the interest factor (WACC) at 50 years depreciation period is 3.2 (between 10% and 2% interest), (Haas et al., 2021).

The most important conclusion of this work is that there is practically no degree of freedom in the market regarding choice of financing parameters as interest rate and depreciation time. Another important conclusion is that the lower the investment risk the lower the expected returns-on-investment by the investors. Hence, practically all long-term investments, e.g. in the area of electricity or district heating networks,

113 Haas, R., Ajanovic, A., Ramsebner, J., Perger, T., Knápek, J., & Bleyl, J. W. (2021). Financing the future infrastructure of sustainable energy systems. *Green Finance*, 3(1), 90–118. https://doi.org/10.3934/GF.2021006.

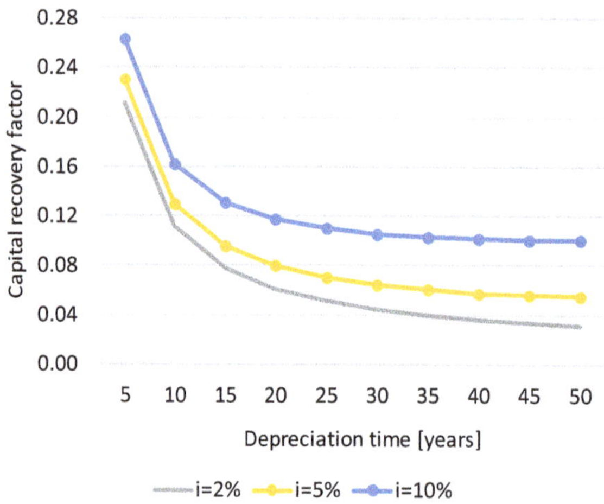

Figure 21.6: Sensitivity of the capital recovery factor depending on interest rate and depreciation time (Source: Haas et al., (2021).

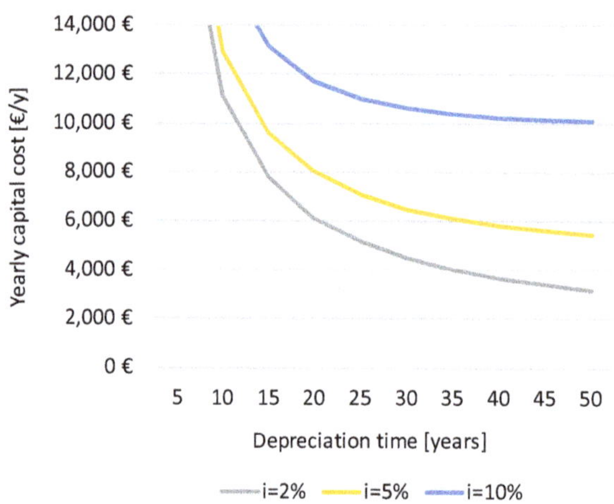

Figure 21.7: Example of the sensitivity of the repayment of an investment of EUR 100,000.

are located in a largely strictly regulated environment. The positive aspect is that this allows to apply lower interest rates, because it reduces risk for the investors (Haas et al., 2021).[114]

In the end, the three most important characteristics are:

114 Haas, R., Ajanovic, A., Ramsebner, J., Perger, T., Knápek, J., & Bleyl, J. W. (2021). Financing the future infrastructure of sustainable energy systems. *Green Finance*, 3(1), 90–118. https://doi.org/10.3934/GF.2021006.

1. The more durable the technology, the more important the influence of public regulation;
2. private initiatives can practically only play a relevant role in niches,
3. foundations are the only realistic private opportunity for funding.

The three most interesting issues for the future are:
- Who will cover the risk? It is very likely that also in future only public bodies will be available here.
- How will green bonds develop and how will these stimulate the investments? The great expectation on green bond is that cheaper, easier access to larger quantities of capital for green investments will be available. And finally:
- Which new model approaches will be provided by the "free" competitive capital market? With respect to this issue currently no reliable forecasts are possible. (Haas et al., 2021)[115]

One initiative that may support the payback time of renewable projects is the EU Regulation (EU) 2018/2001 known as the Renewable Energy Directive (RED III) to promote the use of energy from renewable sources. Among other things, RED III obliges EU member states to achieve a renewable share of at least 29% for fuels by 2030 or an average reduction in greenhouse gas intensity in transport of at least 14.5% by 2030. It also sets quotas for different alternative fuels. By setting targets for increasing the share of low-emission or emission-free fuels in total fuel consumption, this regulation also enables providers of fossil fuels to offset any amount exceeding a threshold by buying in on alternative renewable fuels. Similar to CO_2 certificates, after a certain budget of fossil fuels is used up, offsetting can be done by investing into renewable fuels of another company. As a result, providers of renewable fuels can generate income by offering amounts to a market and invest that into their infrastructure. This leads to a cash flow from fossil fuel providers towards renewable fuel providers – a positive sign in the economy.

On the other hand, there are inspiring pilot projects such as the approach chosen by the Island Samsø in Denmark. Eleven wind power plants mainly power the whole Island with some balancing transfers with the main land.[116] To increase the acceptance of the residents, they were offered shares to participate in the investment and benefit from the profit. This way, the residents' electricity bill is backed up by their shares. As a result, the residents feel like a part of the project and the wind power plants turn from an intrusion into the land scape into which the individuals have no interest, into an attractive investment opportunity that can reduce bills through green electricity.

115 Haas, R., Ajanovic, A., Ramsebner, J., Perger, T., Knápek, J., & Bleyl, J. W. (2021). Financing the future infrastructure of sustainable energy systems. *Green Finance*, 3(1), 90–118. https://doi.org/10.3934/GF.2021006.
116 https://www.visitsamsoe.dk/de/inspiration/energiakademiet/.

21.4.2 The Impact of Variable Energy Feed-In on The Business Case: A Solar PV Example

Variable energy feed-in can affect the profitability of sector coupling and related technologies in different ways. With the increasing use of variable renewable energy sources matching supply with demand is more and more dependent on storage and other supply and demand-side flexibility options. Nevertheless, it is obvious from economic and environmental point-of-view that not every peak in solar PV electricity production should be stored. From the demand side the change of the role of households from pure consumers to 'prosumagers' – consumers who also produce and store electricity – represents a great opportunity to relief the distribution grid to the financial benefit of households but also requires different approaches to ensure their most beneficial integration into existing systems. Properly designed new pricing and tariff systems could provide an incentive to shift consumption to match electricity supply and the grid situation. Hirschhausen (2017) argues that beside the potential of prosumage for decarbonisation,[117] this type of change to the energy system is also often regarded as a threat, especially by utilities (Haas et al., 2023).[118]

Optimal solutions regarding storage and grid interaction still need to be found. In 1994, when PV was not even a niche technology, Haas (1995) analyzed the value of PV electricity for utilities and already more than 20 years ago documented the challenges utilities face with PV today.[119] Schill et al. (2017) highlight the economic system inefficiency as a potential disadvantage of growing solar prosumage and self-consumption using decentral storage.[120] Obviously, this trend adds hardly controllable consumption and production units to an energy system. System cost can be reduced if decentralized storage is operated following requirements in the distribution grid and made available to further electricity market activities instead of mainly focusing on self-consumption. All these challenges therefore ask for a solution and an optimal strategy – from society's point of view – for integrating PV in the electricity system. In this context an effective regulatory framework and appropriate tariff systems are required for all electricity users to guide future consumer choices (Perez-Arriaga, 2017), (Haas et al., 2023).[121]

---

117 Hirschhausen, C. V. (2017). Prosumage and the future regulation of utilities: An introduction. Economics of Energy & Environmental Policy, Volume 6 (Number 1). https://ideas.repec.org/a/aen/eeepjl/eeep6-1-introduction.html.

118 Haas, R., Duic, N., Auer, H., Ajanovic, A., Ramsebner, J., Knapek, J., & Zwickl-Bernhard, S. (2023). The photovoltaic revolution is on: How it will change the electricity system in a lasting way. *Energy*, 265, 126351. https://doi.org/10.1016/j.energy.2022.126351.

119 Haas, R. (1995). The value of photovoltaic electricity for society. *Solar Energy*, 54(1), 25–31. https://doi.org/10.1016/0038-092X(94)00099-Y.

120 Schill, W.-P., Zerrahn, A., Kunz, F., & Kemfert, C. (2017). Decentralized solar prosumage with battery storage: System orientation required. *DIW Economic Bulletin*, 7(16), 141–151.

121 Haas, R., Duic, N., Auer, H., Ajanovic, A., Ramsebner, J., Knapek, J., & Zwickl-Bernhard, S. (2023). The photovoltaic revolution is on: How it will change the electricity system in a lasting way. *Energy*, 265, 126351. https://doi.org/10.1016/j.energy.2022.126351.

Effect of Solar PV Feed-In on Electricity Prices

The current market structure and rules of price determination first need to be understood to grasp the effect of PV feed-in on electricity market prices (Haas et al., 2013).[122] Prices in a functioning market are usually defined where supply and demand match and are balanced – at the intersection of the merit order supply curve and electricity demand at every point in time. As all power plants in a market are involved in determining the electricity price, also RESs have their specific impact, which may differ from experience with traditional and more flexible generation technologies. This phenomenon has already been known since volatile hydropower was first used for electricity generation (Haas et al., 2023).[123]

RESs influence electricity prices in formerly regulated markets at least at the marginal costs of electricity generation. Later, with the first experience of wind booms (about 2007–2009, in Denmark already earlier), temporarily strong winds in the systems even lead to negative electricity prices (Liu et al., 2011;[124] Moreno et al., 2012;[125] Nicolosi, 2010).[126] However, these effects mostly happened at off-peak times, sometimes because of wrong or careless wind forecasts. This fundamental approach has led to a quite different price development in several European electricity sub-markets between 2000 and 2017 (see Figure 21.8), (Haas et al., 2023).[127]

High volatility and considerable differences between electricity spot market prices have been observed in different submarkets within this period. Italy tended to experience higher prices and volatility throughout the horizon due to its over-reliance on imported electricity and congested cross-border transmission lines. In the case of

---

122 Haas, R., Lettner, G., Auer, H., & Duic, N. (2013). The looming revolution: How photovoltaics will change electricity markets in Europe fundamentally. *Energy*, 57, 38–43. https://doi.org/10.1016/j.energy.2013.04.034.

123 Haas, R., Duic, N., Auer, H., Ajanovic, A., Ramsebner, J., Knapek, J., & Zwickl-Bernhard, S. (2023). The photovoltaic revolution is on: How it will change the electricity system in a lasting way. *Energy*, 265, 126351. https://doi.org/10.1016/j.energy.2022.126351.

124 Liu, W., Lund, H., & Mathiesen, B. (2011). Large-scale integration of wind power into the existing Chinese energy system. *Fuel and Energy Abstracts*, 36, 4753–4760. https://doi.org/10.1016/j.energy.2011.05.007.

125 Moreno, B., López, A. J., & García-Álvarez, M. T. (2012). The electricity prices in the European Union. The role of renewable energies and regulatory electric market reforms. *Energy*, 48(1), 307–313. https://doi.org/10.1016/j.energy.2012.06.059.

126 Nicolosi, M. (2010). Wind power integration and power system flexibility–An empirical analysis of extreme events in Germany under the new negative price regime. *Energy Policy*, 38(11), 7257–7268. https://doi.org/10.1016/j.enpol.2010.08.002.

127 Haas, R., Duic, N., Auer, H., Ajanovic, A., Ramsebner, J., Knapek, J., & Zwickl-Bernhard, S. (2023). The photovoltaic revolution is on: How it will change the electricity system in a lasting way. *Energy*, 265, 126351. https://doi.org/10.1016/j.energy.2022.126351.

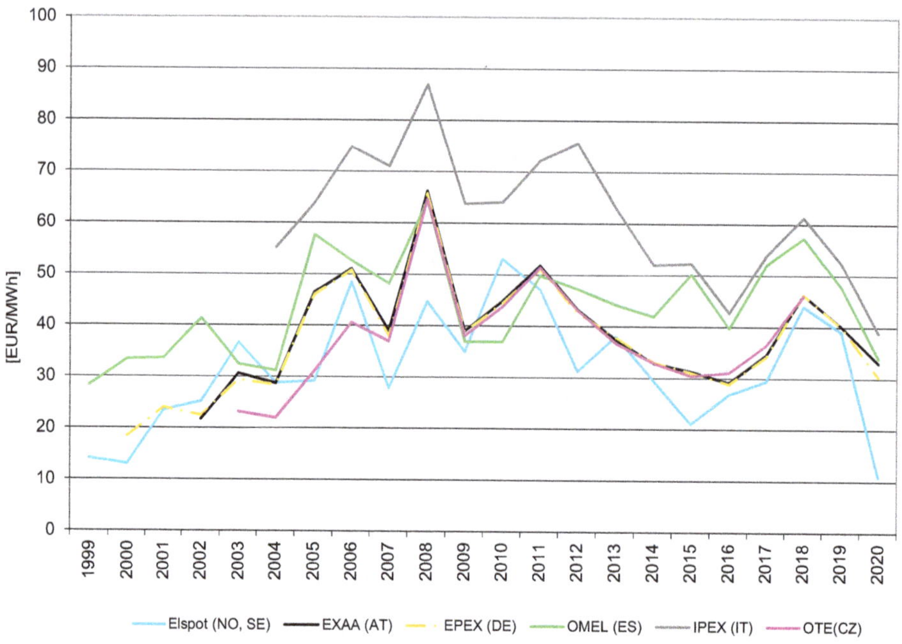

Figure 21.8: Price Development of day-ahead electricity prices in European electricity markets 1999–2020. Source: (Haas et al., 2023) .

the ELSPOT, which includes Sweden, Norway, Finland, and shares of Denmark, the pattern is different with high shares of hydroelectric power and a weak interconnection with continental Europe. Other markets – even the isolated Spanish market – show price convergence. The reason for high prices in 2008 in Continental Europe was the low hydropower availability, while the price decrease after 2008 may, at least to some extent, be associated with the economic crisis. On the one hand, the merit order effect played a role in increasing renewable electricity sources pushing traditional power plants out of the market. On the other hand, the low CO_2 prices do not lead to higher pricing of the remaining fossil-fuel-based plants. The price dip in 2020 is associated with the reduced demand caused by the COVID-19 pandemic. The massive integration of electricity from RES (especially wind and PV) was most pronounced between 2011 and 2016 (Haas et al., 2023).[128]

The increase in electricity generation from these sources led to the displacement of conventional power plants and this, together with low prices for emission allowances, led to a significant and relatively long-lasting decline in electricity prices. Similar

128 Haas, R., Duic, N., Auer, H., Ajanovic, A., Ramsebner, J., Knapek, J., & Zwickl-Bernhard, S. (2023). The photovoltaic revolution is on: How it will change the electricity system in a lasting way. *Energy*, 265, 126351. https://doi.org/10.1016/j.energy.2022.126351.

challenges may arise with the further feed-in of PV electricity, which differs substantially in timing and variability compared to wind availability. Figure 21.9 shows the merit order supply curve with and without PV capacities during the peak time of a summer day and the short-term marginal costs of conventional electricity capacities. On such a sunny day with strong solar irradiance, PV electricity generation shifts the supply curve to the right, which essentially pushes nuclear and fossil fuel generation "out of the market". Suppose the impact of PV electricity feed-in, e.g. on a sunny day in October, which is not a peak period for solar generation in Germany, can be as dramatic as is shown in Figure 21.9. In that case, one can expect much more dramatic impacts on market prices during summer months with even negative prices (Haas et al., 2023).[129]

Figure 21.10 shows the impact of increasing PV capacity on electricity prices on an exemplary day due to a change in the residual load by PV feed-in, especially during noontime. The result is a shape resembling a duck – the so-called "duck curve". With more electricity consumed onsite or storage technologies, the grid can be relieved from the noon peaks, and distributed producers can optimize their economic benefit. (Haas et al., 2023)[130]

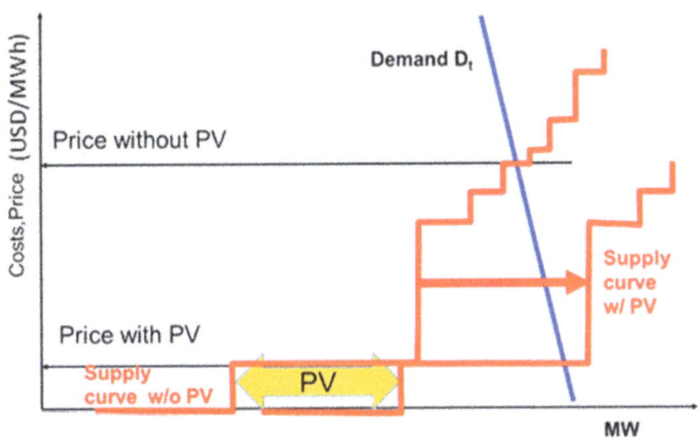

Figure 21.9: Merit order without and with PV feed in during a sunny summer day (Source: Haas et al., 2023).[131]

129 Haas, R., Duic, N., Auer, H., Ajanovic, A., Ramsebner, J., Knapek, J., & Zwickl-Bernhard, S. (2023). The photovoltaic revolution is on: How it will change the electricity system in a lasting way. *Energy*, 265, 126351. https://doi.org/10.1016/j.energy.2022.126351.

130 Haas, R., Duic, N., Auer, H., Ajanovic, A., Ramsebner, J., Knapek, J., & Zwickl-Bernhard, S. (2023). The photovoltaic revolution is on: How it will change the electricity system in a lasting way. *Energy*, 265, 126351. https://doi.org/10.1016/j.energy.2022.126351.

131 Haas, R., Duic, N., Auer, H., Ajanovic, A., Ramsebner, J., Knapek, J., & Zwickl-Bernhard, S. (2023). The photovoltaic revolution is on: How it will change the electricity system in a lasting way. *Energy*, 265, 126351. https://doi.org/10.1016/j.energy.2022.126351.

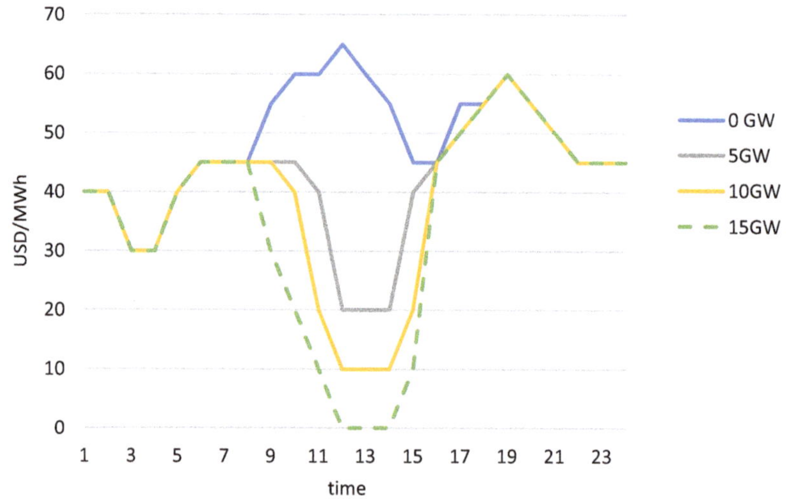

Figure 21.10: Impact of solar PV feed-in on electricity price (Source: Haas et al., 2023).

New Tariff Schemes for Efficient Renewable Integration

The increasing participation of distributed energy producers in the market add a substantial amount of complexity to a formerly rather centralized energy system and bring common energy management approaches towards the edge. An additional challenge within a renewable energy system is the vast variability of power that is fed-into the grid from a large amount of, e.g. local solar PV systems, or consumed from the grid at evening times when there is no local PV production.

Grid tariffs are usually imposed with an aim of an equal distribution of the network operation costs among all customers (Schittekatte et al., 2017).[132] Haas et al. (2023) claim that with more and more consumers installing distributed electricity generation systems, such as solar PV systems and battery storage to maximize their on-site consumption, the historical practice of volumetric, energy-based network tariffs do no more meet the newly arising requirements (Haas et al., 2023).[133]

To replicate the dependence of prosumers on infrastructure capable to handle high power capacity, they suggest an additional capacity tariff to add a price tag to the maximum electricity power consumed during the year. This would make the maximum

132 Schittekatte, T., Momber, I., & Meeus, L. (2017). Future-proof tariff design: Recovering sunk grid costs in a world where consumers are pushing back. EUI RSCAS Working Paper (FSR), 70. https://doi.org/10.1016/j.eneco.2018.01.028.

133 Haas, R., Duic, N., Auer, H., Ajanovic, A., Ramsebner, J., Knapek, J., & Zwickl-Bernhard, S. (2023). The photovoltaic revolution is on: How it will change the electricity system in a lasting way. *Energy*, 265, 126351. https://doi.org/10.1016/j.energy.2022.126351.

power that is either fed-in or consumed from the grid with prosumers of variable renewable electricity more expensive and would incentivize to avoid the transaction of high power amounts. This way, expanding the grid for the peak electricity power of very few hours of the year would be avoided while maintaining grid performance.

These prosumers would be incentivized to use even more of their locally produced peak electricity themselves, by demand response of their household devices, heat pump, charging their BEV at that time of the day in case the car is available or even offering it to the public, and local storage could become more economically feasible.

On the other hand, measures need to be taken to reduce the maximum power that is drawn from the grid – probably during the evening times when household demand is high and PV availability low. Flexible household devices should not operate at that time, storage should be consumed if available, BEV charging needs to be postponed as described by Ramsebner et al. (2020) in their ToU load management approach to keep the overall power capacity low.

In order to do so, however, the appropriate management tools and devices need to available in order to respond dynamically. Currently, easy to use energy management systems are gaining importance, with the recent energy price increase in European countries and the following boom of solar PV installations specifically with a focus on solar PV prosumage.

BEV charging based on PV excess electricity is addressed by many charging infrastructure manufacturers and energy management providers and also home and building energy management supply is steadily improving answering the growing requirements of individual consumers also participating in the market.

If thinking of storage, an evolution that is currently on its way will also be the use of BEV batteries as a variability buffer between e.g. solar PV supply and electricity demand. In a rather small scope, this can support solar PV peak consumption with a vehicle to home or building scenario. Depending on vehicle availability at the house, solar PV excess could be charged during noon and consumed in the evening to reduce grid-dependency.

For these approaches, however, the legislative framework and hence business models are still uncertain. The increase in battery re-and discharging cycles through bidirectional charging could lead to faster battery deterioration, which needs to be compensated monetarily and also requires warranty specifications. At least, the battery capacity range needs to be managed to avoid early battery defects. Furthermore, technical requirements on feeding into buildings and even more the grid remain undefined.

It becomes clear that all these alternatives to high PV power feed-in or grid consumption lead to additional investments for the individual prosumer. The cost of grid expansion is therefore avoided but shifted to the individual. Therefore, instead of "punishing" a prosumer, who already has added value to the energy transition by investing into a solar PV system, with an additional capacity tariff, it might be valuable to support individual flexibility measures even more. In a next step, building energy communities will help to balance variability of demand and supply on a micro level before depending on the mutual distribution grid to many challenges of the energy transition.

21.4.3 Techno-Economic Attractiveness of Renewable Hydrogen Generation in the Industrial Sector

Some relevant industrial processes already heavily rely on hydrogen as an input, such as oil refining or ammonia production, with an expected growing trend in the coming years (International Energy Agency (IEA), 2019).[134] However, so far, 96% of this hydrogen is produced from fossil fuels (natural gas 48%, fossil oil 30%, coal 18%), causing a substantial amount of CO_2 emissions (830 Mt CO_2/a in the industrial sector (Tang et al., 2021).[135] As described in Chapter 21.3, green hydrogen, produced through the electrolysis of water with renewable energy, is a promising alternative to decarbonise these processes (Ramsebner et al., 2024).[136]

Such hydrogen production can be considered either centrally in huge volumes, or decentrally close to the consumption site. Both have different significant challenges for an attractive business case. Above all, however, the efficiency losses inherent to the electrolysis are a major trigger of renewable hydrogen cost. While centralized production benefits from high production volumes and the pooling of many varying demand profiles, it also relies on an appropriate hydrogen transport and distribution infrastructure, which entails significant costs. For example, a Hydrogen Backbone could be built in Europe using existing and new dedicated H_2 pipelines (A. Wang et al., 2020),[137] (Ramsebner et al., Mimeo 2024).[138]

An alternative option, which would prove to be more robust in the early stages, would be to produce hydrogen locally in what is termed “hydrogen valleys” or industrial hubs (Tu et al., 2021).[139] This decentralized H_2 production might be beneficial due to the elimination of distribution infrastructure requirements and associated cost as well as the possibility to optimize electricity consumption according to specific demand situations. On the other hand, it poses several challenges, such as coupling H_2

134 International Energy Agency (IEA). (2019). The future of hydrogen [Report prepared by the IEA for the G20, Japan].

135 Tang, O., Rehme, J., Cerin, P., & Huisingh, D. (2021). Hydrogen production in the Swedish power sector: Considering operational volatilities and long-term uncertainties. *Energy Policy*, 148, 111990. https://doi.org/10.1016/j.enpol.2020.111990.

136 Ramsebner, J., Linares, P., Hiesl, A., & Haas, R. (Mimeo). Techno-economic evaluation of renewable hydrogen generation strategies for the industrial sector. *Hydrogen Energy Journal*, Mimeo-under review.

137 Wang, C., Yan, J., Marnay, C., Djilali, N., Dahlquist, E., Wu, J., & Jia, H. (2018). Distributed Energy and Microgrids (DEM). *Applied Energy*, 210, 685–689. https://doi.org/10.1016/j.apenergy.2017.11.059.

138 Ramsebner, J., Linares, P., Hiesl, A., & Haas, R. (Mimeo). Techno-economic evaluation of renewable hydrogen generation strategies for the industrial sector. *Hydrogen Energy Journal*, Mimeo-under review.

139 Tu, K., Deutsch, Matthias, & Flis, Gniewomir. (2021, May 6). Industrial clusters using green hydrogen can drive clean energy transition in Europe and China. World Economic Forum. https://www.weforum.org/agenda/2021/05/industrial-clusters-green-hydrogen-europe-china/.

production with industrial demand or optimizing the cost of electricity supply, which in turn will affect the cost of the hydrogen produced. Although Matute et al. (2019) observe that a large part of the cost of green hydrogen depends on the cost of electricity,[140] the potential need for hydrogen storage, the definition of the capacity of the electrolyser and the operation regime will also influence the final cost of the hydrogen delivered to industry to a large extent (Ramsebner et al., Mimeo).[141]

In (Ramsebner et al., 2024),[142] the techno-economic performance of different decentral hydrogen production strategies for industrial use are evaluated in detail. It assesses different technical configurations and strategies that may be used to provide renewable hydrogen for the industry by looking at different on-site operation strategies and storage and electrolyser configurations and the implications that these may have for the cost of producing hydrogen in industrial hubs. While the importance and potential of a deep decarbonisation of the industrial sector is evident, the knowledge among companies, analysts and policymakers lags behind the ambitions and specific goals for decarbonisation of other end-consumption sectors such as transport and heating (Bataille et al., 2018;[143] Fischedick et al., 2014;[144] Loftus et al., 2015),[145] (Ramsebner et al., 2024).[146]

For the same amount of yearly hydrogen produced, the following strategies are considered for a centralized production case with constant demand and a hydrogen valley characterized by variable hydrogen demand. Figure 21.11 describes the H_2 generation

140 Matute, G., Yusta, J. M., & Correas, L. C. (2019). Techno-economic modelling of water electrolysers in the range of several MW to provide grid services while generating hydrogen for different applications: A case study in Spain applied to mobility with FCEVs. *International Journal of Hydrogen Energy*, 44(33), 17431–17442. https://doi.org/10.1016/j.ijhydene.2019.05.092.

141 Ramsebner, J., Linares, P., Hiesl, A., & Haas, R. (Mimeo). Techno-economic evaluation of renewable hydrogen generation strategies for the industrial sector. *Hydrogen Energy Journal*, Mimeo-under review.

142 Ramsebner, J., Linares, P., Hiesl, A., & Haas, R. (Mimeo). Techno-economic evaluation of renewable hydrogen generation strategies for the industrial sector. *Hydrogen Energy Journal*, Mimeo-under review.

143 Bataille, C., Åhman, M., Neuhoff, K., Nilsson, L. J., Fischedick, M., Lechtenböhmer, S., Solano-Rodriquez, B., Denis-Ryan, A., Stiebert, S., Waisman, H., Sartor, O., & Rahbar, S. (2018). A review of technology and policy deep decarbonization pathway options for making energy-intensive industry production consistent with the Paris Agreement. *Journal of Cleaner Production*, 187, 960–973. https://doi.org/10.1016/j.jclepro.2018.03.107.

144 Fischedick, M., Marzinkowski, J., Winzer, P., & Weigel, M. (2014). Techno-economic evaluation of innovative steel production technologies. *Journal of Cleaner Production*, 84, 563–580. https://doi.org/10.1016/j.jclepro.2014.05.063.

145 Loftus, P. J., Cohen, A. M., Long, J. C. S., & Jenkins, J. D. (2015). A critical review of global decarbonization scenarios: What do they tell us about feasibility? *WIREs Climate Change*, 6(1), 93–112. https://doi.org/10.1002/wcc.324.

146 Ramsebner, J., Linares, P., Hiesl, A., & Haas, R. (Mimeo). Techno-economic evaluation of renewable hydrogen generation strategies for the industrial sector. *Hydrogen Energy Journal*, Mimeo-under review.

process assumed in our research. The electrolyser sources green electricity from the grid or a dedicated RES plant to split water (H_2O) into H_2 and oxygen (O_2). H_2 can either directly cover demand or be stored in tanks intermediately (Ramsebner et al., 2024).[147]

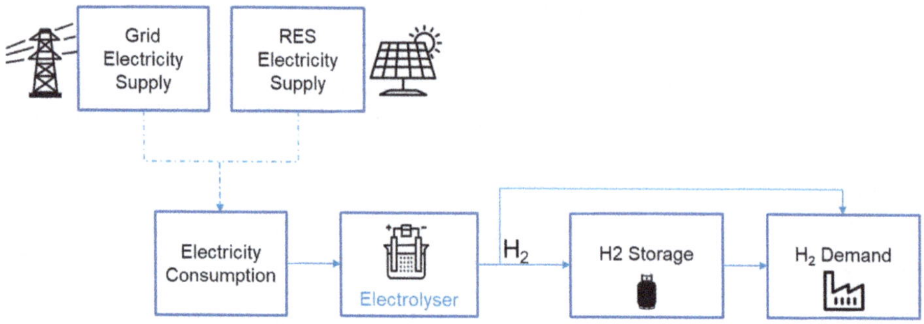

Figure 21.11: Hydrogen Production process; figure created by author.

Hydrogen Supply Strategies

Grid connected electrolyser

1. **Electricity price optimization**: The electrolyser consumes electricity at times of low electricity prices. Hydrogen is stored after transformation to balance against demand. An advantage of this approach is the possibility to exploit low and avoid high electricity prices through hydrogen storage. This approach may require greater electrolyser and storage capacity to increase electricity consumption at low prices but minimizes electricity cost. Nevertheless, the substantial variability in the production of the electrolyser could reduce actual H_2 conversion efficiency, and the storage cost must not offset the benefit of price optimization.
2. **Just in time sourcing:** H_2 is produced JIT and electricity is sourced according to demand. Just-in-time production does not consider any hydrogen storage facility. This scenario, which could be an outcome of the earlier price optimization (if the model chooses not to build storage), is explicitly included as a reference for a passive strategy, by exogenously setting a storage tank of zero. Depending on the hydrogen demand characteristics, production variability may be negligible. With JIT H_2 production, electricity consumption cost is subject to the electricity price variations (Ramsebner et al., 2024).[148]

---

147 Ramsebner, J., Linares, P., Hiesl, A., & Haas, R. (Mimeo). Techno-economic evaluation of renewable hydrogen generation strategies for the industrial sector. *Hydrogen Energy Journal*, Mimeo-under review.
148 Ramsebner, J., Linares, P., Hiesl, A., & Haas, R. (Mimeo). Techno-economic evaluation of renewable hydrogen generation strategies for the industrial sector. *Hydrogen Energy Journal*, Mimeo-under review.

Island solution – electrolyser connected to a dedicated RES power plant

This scenario presents the **direct connection of the electrolyser to a dedicated wind or PV power plant** without any grid support. Electricity supply and H_2 demand are only balanced through hydrogen storage. There is one scenario carried out for each Austrian and Spanish PV and wind availability using the generation profile from 2019.

The cost of electricity is evaluated based on an average LCOE. Fraunhofer ISE (2021) arrive at an LCOE for isolated PV of about 30 €/MWh and for wind at minimum 40 €/MWh.[149] According to IEA (2020) and IRENA (2021),[150,151] based on an exchange rate of 1/1.2 USD/EUR in 2021, large-scale PV power has an LCOE of about 15–30 €/MWh and wind of 25–37 €/MWh. In this scenario, the electrolyser optimization depends on the RES availability and storage is used to balance against demand. The RES capacity in the model is minimised to fulfil hydrogen demand (Ramsebner et al., Mimeo).[152]

While the exact methodology and parameters can be accessed in the mentioned full paper (Ramsebner et al., 2024)[153] the main conclusions concerning a business case are outlined herein: Renewable hydrogen is expected to play a major role in decarbonizing the industrial sector as feedstock and fuel to many processes and products. On-site production, or production limited to "hydrogen valleys", has been proposed as a non-regret, robust way of starting the deployment of this technology and achieving higher production scales to decrease the cost. However, this option may imply having to follow a variable hydrogen demand pattern, which in turn may affect the conversion efficiency of the electrolyser, or require investing in hydrogen storage. This situation may be particularly acute in the case of hydrogen production fed exclusively with off-grid renewable plants. In this paper they have looked at the economics of these configurations (Ramsebner et al., 2024).[154]

In a grid-connected scenario, constant demand representing centralized bulk production always allows for cheaper hydrogen production by optimizing electrolyser

149 Fraunhofer ISE. (2021). Studie: Stromgestehungskosten erneuerbare Energien. Fraunhofer-Institut für Solare Energiesysteme ISE. https://www.ise.fraunhofer.de/de/veroeffentlichungen/studien/studie-stromgestehungskosten-erneuerbare-energien.html.

150 IEA. (2020). Projected Costs of Generating Electricity 2020. IEA. https://www.iea.org/reports/projected-costs-of-generating-electricity-2020.

151 IRENA. (2021). Renewable Power Generation Costs in 2020. /Publications/2021/Jun/Renewable-Power-Costs-in-2020. https://www.irena.org/publications/2021/Jun/Renewable-Power-Costs-in-2020.

152 Ramsebner, J., Linares, P., Hiesl, A., & Haas, R. (Mimeo). Techno-economic evaluation of renewable hydrogen generation strategies for the industrial sector. *Hydrogen Energy Journal*, Mimeo-under review.

153 Ramsebner, J., Linares, P., Hiesl, A., & Haas, R. (Mimeo). Techno-economic evaluation of renewable hydrogen generation strategies for the industrial sector. *Hydrogen Energy Journal*, Mimeo-under review.

154 Ramsebner, J., Linares, P., Hiesl, A., & Haas, R. (Mimeo). Techno-economic evaluation of renewable hydrogen generation strategies for the industrial sector. *Hydrogen Energy Journal*, Mimeo-under review.

operation. In this case, electricity price optimization does not provide economic benefits unless the electrolyser investment cost decreases or price profiles change significantly. Grid-connected zero-storage (JIT) production achieves the lowest renewable hydrogen cost with constant hydrogen demand. In this case, maximizing asset utilisation with more than 8,600 full load hours leads to a cost-optimal solution. Power purchase agreements would also be helpful in reducing the cost of the electricity sourced from the grid while ensuring its renewable character (Ramsebner et al., 2024).[155]

In case hydrogen demand is variable, then electricity price optimization and hydrogen storage can be used to achieve lower costs. These results are valid under the assumption of no demand access fee applied for the maximum power sourced from the grid. Cost advantages may be intensified with higher electricity price spreads through higher renewable shares. Electricity price optimization would also, in this case, become a demand response tool to help manage the system (Ramsebner et al., 2024).[156]

Cost differences are significant when comparing grid-connected and island configurations. According to our results, wind power feed-in is 2.5 times, and PV power up to 5 times more expensive than a grid-connected scenario for hydrogen production based on the Austrian electricity market price for 2019. The intermittency of renewable resources leads to substantial electrolyser and storage capacity requirements and huge variability, harming actual operation efficiency. The RES profiles drive hydrogen production cost and capacity requirements more than the hydrogen demand profile. Wind power seems to be more favourable than PV in terms of availability (Ramsebner et al., 2024).[157]

In both island cases, the grid is missing as a means of storage to have flexible sourcing of the electrolyser. An electrolyser's isolated RES connection may only become competitive if electricity market prices skyrocket, as currently observed. With the current average Austrian electricity price reaching 207 €/MWh by the end of May 2022, even the Spanish PV power could almost compete in the presented base scenario. Additionally, a mid-to-long-term expected decrease in electrolyser investment cost would also support the economic performance of direct renewable electricity feed-in. These results are however subject to the existence or not of demand

155 Ramsebner, J., Linares, P., Hiesl, A., & Haas, R. (Mimeo). Techno-economic evaluation of renewable hydrogen generation strategies for the industrial sector. *Hydrogen Energy Journal*, Mimeo-under review.

156 Ramsebner, J., Linares, P., Hiesl, A., & Haas, R. (Mimeo). Techno-economic evaluation of renewable hydrogen generation strategies for the industrial sector. *Hydrogen Energy Journal*, Mimeo-under review.

157 Ramsebner, J., Linares, P., Hiesl, A., & Haas, R. (Mimeo). Techno-economic evaluation of renewable hydrogen generation strategies for the industrial sector. *Hydrogen Energy Journal*, Mimeo-under review.

charges imposed on electrolysers, something currently discussed in several countries (Ramsebner et al., 2024).[158]

Hence, our results show that, although the production of green hydrogen in grid-connected configurations is more competitive in normal circumstances, this is subject to several uncertainties, such as the evolution in the cost and efficiency of electrolysers, or the regulation of the power system. Including these uncertainties into the analysis to produce more robust investment decisions would be a natural next step in this research (Ramsebner et al., 2024).[159]

21.4.4 The Challenges and Value of Sector Coupling

An increasing share of electricity from VRE sources requires an energy system with more flexible consumption, distribution, and storage. In contrast, VRE sources represent an excellent chance to decarbonise the end-consumption sectors and are essential to reaching climate goals. (IRENA, 2018) considers P2G the most promising transformation technology that enables SC in various ways.[160] Thanks to the large capacity of gas pipelines in Europe, even low blending shares of H_2 would lead to the absorption of substantial quantities of VRE. Renewable gas may replace fossil fuels in many end-use applications while fully utilizing existing infrastructure. Olczak & Piebalgs (2018)[161] are convinced that SC will exploit the rising share of electricity by distributing and storing it after a transformation into H_2 and synthetic methane (Ramsebner, 2022).[162]

While P2G seems promising for various end-consumption sectors and as a large-scale storage technology, the importance of all potential transformation technologies together needs to be emphasised. The development of transformation technologies faces multiple uncertainties such as electricity prices and competition through transmission grid expansion. However, its economic feasibility remains challenging with a few full load hours if considered to operate only at times of excess electricity generation. Environmental effects of operation processes need to be internalised in new market frameworks to set the right incentives. To promote the economic feasibility of

---

158 Ramsebner, J., Linares, P., Hiesl, A., & Haas, R. (Mimeo). Techno-economic evaluation of renewable hydrogen generation strategies for the industrial sector. *Hydrogen Energy Journal*, Mimeo-under review.

159 Ramsebner, J., Linares, P., Hiesl, A., & Haas, R. (Mimeo). Techno-economic evaluation of renewable hydrogen generation strategies for the industrial sector. *Hydrogen Energy Journal*, Mimeo-under review.

160 International Renewable Energy Agency (IRENA). (2018). Hydrogen from renewable power: Technology outlook for the energy transition.

161 Olczak, M., & Piebalgs, A. (2018). Sector coupling: The new EU climate and energy paradigm?

162 Ramsebner, J. (2022). The role of sector coupling in the energy transition [Thesis, Technische Universität Wien]. https://repositum.tuwien.at/handle/20.500.12708/80280.

renewable transformation processes, transformation plants need to be regarded as producers/transformers and not as consumers when sourcing from the electricity grid and markets need to handle gas, electricity and heat flexibly. Low electricity prices may promote direct electrification within the existing market environment instead of P2G or P2H, whenever possible (Ramsebner, 2022).[163]

Brown et al., (2018) detect SC and transmission grid expansion as competing concepts concerning renewable energy integration.[164] In their study, many scenarios favour expanding transmission network capacity, specifically in the North and Baltic seas. They also find that direct electrification of individual transport is more efficient without the losses of P2G and P2L (Brown et al., 2018).[165] Furthermore, beyond sector coupling, the electricity grid competes with the traditional gas and heating grids, whose role may change. Nevertheless, P2G provides flexibility and long-term storage to renewable energy systems. We interpret these results as a suggestion to implement transmission network expansion and SC to integrate renewable energy efficiently (Ramsebner, 2022).[166]

Robinius et al. (2018) compare the cost efficiency of P2G and transmission grid expansion and conclude that investment in the latter is currently more cost-effective.[167] Achieving a certain amount of flexibility through transmission network expansion only costs 30% of the same capacity implementation of an electrolyser; however, they do not consider an income from selling the hydrogen at a competitive price. Hörsch & Brown, (2017),[168] in contrast, argue against a substantial cost benefit of transmission grid expansion compared with P2X technologies. They conclude that an almost fully renewable energy system without grid expansion is only about 20% more expensive. They assume the cost-benefit to be even smaller if investment in line vol-

163 Ramsebner, J. (2022). The role of sector coupling in the energy transition [Thesis, Technische Universität Wien]. https://repositum.tuwien.at/handle/20.500.12708/80280.

164 Brown, T., Schlachtberger, D., Kies, A., Schramm, S., & Greiner, M. (2018). Synergies of sector coupling and transmission reinforcement in a cost-optimised, highly renewable European energy system. *Energy*, 160, 720–739. https://doi.org/10.1016/j.energy.2018.06.222.

165 Brown, T., Schlachtberger, D., Kies, A., Schramm, S., & Greiner, M. (2018). Synergies of sector coupling and transmission reinforcement in a cost-optimised, highly renewable European energy system. *Energy*, 160, 720–739. https://doi.org/10.1016/j.energy.2018.06.222.

166 Ramsebner, J. (2022). The role of sector coupling in the energy transition [Thesis, Technische Universität Wien]. https://repositum.tuwien.at/handle/20.500.12708/80280.

167 Robinius, M., Raje, T., Nykamp, S., Rott, T., Müller, M., Grube, T., Katzenbach, B., Küppers, S., & Stolten, D. (2018). Power-to-Gas: Electrolyzers as an alternative to network expansion – An example from a distribution system operator. *Applied Energy*, 210, 182–197. https://doi.org/10.1016/j.apenergy.2017.10.117.

168 Hörsch, J., & Brown, T. (2017). The role of spatial scale in joint optimisations of generation and transmission for European highly renewable scenarios. 2017 14th International Conference on the European Energy Market (EEM), 1–7. https://doi.org/10.1109/EEM.2017.7982024.

ume was included to avoid the public rejecting a growing amount of new lines (Ramsebner, 2022).[169]

Robinius et al., (2018) later acknowledged that their results highly depended on the regional situation and scenario assumptions.[170] Decreasing capital investment costs and an expected decrease in wholesale electricity prices may eventually support the economic feasibility of P2G, at least for seasonal storage (Robinius et al., 2018).[171] This argument stands against the likely trend of using low-cost renewable electricity directly, reducing the need for P2G. Other transformation technologies also face uncertainties in the long term. For example, the importance of P2H may change along with the consequences of global warming, whereas, simultaneously, requirements for cooling may increase. This uncertainty asks for an appropriate design of DH using lower temperatures that allow efficient heating and cooling in the same network (Ramsebner, 2022).[172]

Additionally, further research is required in interface technologies for integrating new and more distributed demand sources, e.g. BEVs, heat pumps etc., in currently isolated energy grids. The future energy grid planning on the national and international level does not yet follow integrated system modelling but instead happens uncoordinatedly. This uncertainty also harms decision-making regarding the right technology for producers. Flexibility, furthermore, needs to be integrated into market structures and classified as a service instead of an energy consumer to avoid consumption-based energy fees. Eventually, the large amounts of data gathered and evaluated in such integrated energy systems impose a technical challenge. Still, they may also change demand behaviour and increase control and awareness by consumers. Pfeiffer et al., (2021) found that residential consumers do accept a substantial amount of demand-side flexibility or optimization of their demand and potential on-site energy production if rewarded appropriately (Ramsebner, 2022).[173,174]

169 Ramsebner, J. (2022). The role of sector coupling in the energy transition [Thesis, Technische Universität Wien]. https://repositum.tuwien.at/handle/20.500.12708/80280

170 Robinius, M., Raje, T., Nykamp, S., Rott, T., Müller, M., Grube, T., Katzenbach, B., Küppers, S., & Stolten, D. (2018). Power-to-Gas: Electrolyzers as an alternative to network expansion – An example from a distribution system operator. *Applied Energy*, 210, 182–197. https://doi.org/10.1016/j.apenergy.2017.10.117.

171 Robinius, M., Raje, T., Nykamp, S., Rott, T., Müller, M., Grube, T., Katzenbach, B., Küppers, S., & Stolten, D. (2018). Power-to-Gas: Electrolyzers as an alternative to network expansion – An example from a distribution system operator. *Applied Energy*, 210, 182–197. https://doi.org/10.1016/j.apenergy.2017.10.117.

172 Ramsebner, J. (2022). The role of sector coupling in the energy transition [Thesis, Technische Universität Wien]. https://repositum.tuwien.at/handle/20.500.12708/80280.

173 Pfeiffer, C., Puchegger, M., Maier, C., Tomaschitz, I. V., Kremsner, T. P., & Gnam, L. (2021). A Case Study of Socially-Accepted Potentials for the Use of End User Flexibility by Home Energy Management Systems. *Sustainability*, 13(1), Article 1. https://doi.org/10.3390/su13010132.

174 Ramsebner, J. (2022). The role of sector coupling in the energy transition [Thesis, Technische Universität Wien]. https://repositum.tuwien.at/handle/20.500.12708/80280.

Despite all these risks, some arguments justify the continuous investment into SC technologies and appropriate infrastructure. First, with the path towards CO_2 neutrality in Europe, direct or indirect use of large-scale renewable electricity remains a requirement in many cases. The flexible transformation between energy carriers provides a basis for efficiently managing renewable energy systems. This can be realized partly within the SC concept. Second, Novak, (2017) explains the value of energy carriers based on the amount of exergy,[175] defined as the net energy provided. Renewable energy from wind and solar power is pure exergy with zero variable fuel cost. Therefore, it needs to be used most efficiently without being wasted. Third, because of the increasing share of electricity generation from wind and PV, the demand for seasonal renewable storage, for example, in the form of H_2, is growing. These arguments only make sense if the input electricity for SC is produced from renewable sources and does not include grey electricity from coal, oil, or gas (Ramsebner, 2022).[176]

Nevertheless, hybrid solutions may be necessary to develop the required technologies in the long term before a sufficient VRE capacity is available. Another important aspect is the political risk related to the dependency on oil and gas extracting countries. Historically, severe crises have been caused by a lack of alternatives through national energy provision, such as the oil crisis in 1973. SC represents an option to fulfil former fossil-based fuel demand with renewable hydrogen, methane, and liquid electro-fuels. Because the electricity market does not currently provide the incentives to install P2G capacities, a possible solution is an initiation within the regulatory framework of national TSOs and DSOs. With the capital investment largely being refunded to these regulated institutions, they could install electrolyser capacity at a low cost and auction it to power plant operators for excess VRE feed-in. Renewable hydrogen may be stored or support the decarbonisation of the end-consumption sectors (Robinius et al., 2018),[177] (Ramsebner, 2022).[178]

Such projects have already been initiated to develop the technology towards maturity, establish economic feasibility, and guarantee the security of supply. One example is the hydrogen valley in the northern Netherlands, where emission-free hydrogen will become cost-competitive throughout the upcoming decade (Foresight Climate &

---

175 Novak, P. (2017). Exergy as measure of sustainability of energy system. *International Journal of Earth & Environmental Sciences*, 2(2). https://doi.org/10.15344/2456-351X/2017/139.

176 Ramsebner, J. (2022). The role of sector coupling in the energy transition [Thesis, Technische Universität Wien]. https://repositum.tuwien.at/handle/20.500.12708/80280.

177 Robinius, M., Raje, T., Nykamp, S., Rott, T., Müller, M., Grube, T., Katzenbach, B., Küppers, S., & Stolten, D. (2018). Power-to-Gas: Electrolyzers as an alternative to network expansion – An example from a distribution system operator. *Applied Energy*, 210, 182–197. https://doi.org/10.1016/j.apenergy.2017.10.117.

178 Ramsebner, J. (2022). The role of sector coupling in the energy transition [Thesis, Technische Universität Wien]. https://repositum.tuwien.at/handle/20.500.12708/80280.

Energy, 2019).[179] The generation scale is the measure to reduce the cost, thereby achieving energy expertise and using the available pipeline infrastructure. The project aims at generating 3–4 GW of wind energy for hydrogen production until 2030, which could be extended to 10 GW up to 2040. 800,000 t of renewable hydrogen could avoid 7 Mt of CO_2 emissions in 2040 (Foresight Climate and Energy, 2019),[180] (Ramsebner, 2022).[181]

21.5 Impact of Multiple Energy Crises on the Sector Coupling Business Case

The dependence of many countries on natural gas currently causes turbulences in energy markets, leaving natural gas and electricity prices skyrocketing since mid-2021 (see Figure 21.12 and Figure 21.13). A situation that has multiple impacts on the attractiveness of different renewable and conversion technologies and can significantly affect the development of the SC applications outlined in Chapter 21.3 (Ramsebner, 2022).[182]

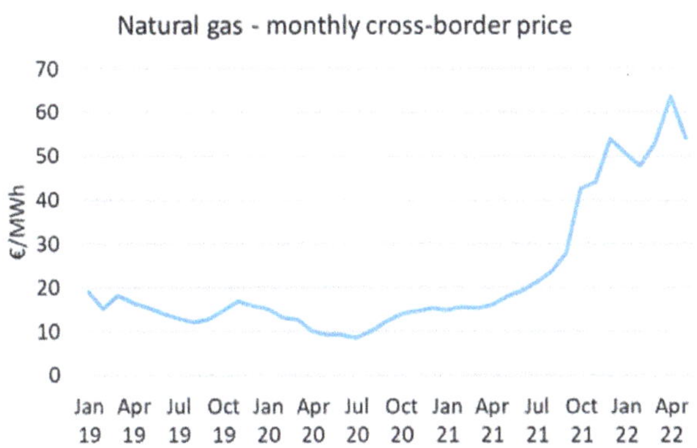

Figure 21.12: Monthly cross-border price for natural gas (BAFA, 2022).

179 Foresight Climate & Energy. (2019, May 3). Hydrogen: The Northern Netherlands is ready. https://foresightdk.com/hydrogen-northern-netherlands-is-ready/.

180 Foresight Climate & Energy. (2019, May 3). Hydrogen: The Northern Netherlands is ready. https://foresightdk.com/hydrogen-northern-netherlands-is-ready/.

181 Ramsebner, J. (2022). The role of sector coupling in the energy transition [Thesis, Technische Universität Wien]. https://repositum.tuwien.at/handle/20.500.12708/80280.

182 Ramsebner, J. (2022). The role of sector coupling in the energy transition [Thesis, Technische Universität Wien]. https://repositum.tuwien.at/handle/20.500.12708/80280.

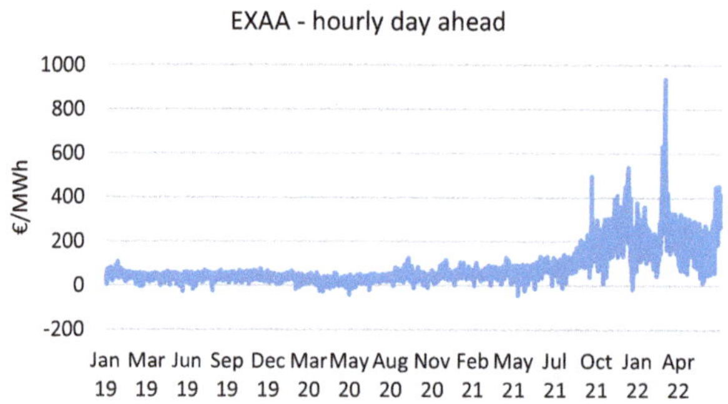

Figure 21.13: Hourly EXAA day ahead price for electricity (APG, 2022).

Rising electricity prices can severely affect the economic performance of on-site renewable hydrogen production evaluated above. Under the Austrian spot market price of 2019, with an average of 40 €/MWh, electrolysis with direct feed-in from variable renewable electricity (evaluated with an LCOE of 26 €/MWh for PV and 36 €/MWh for wind power) was not competitive due to highly variable operation and respective capacity needs and efficiency losses. However, the almost 5-fold market price increase could change this situation. While the cost for grid-connected hydrogen production accounted for about 109 €/MWh with price levels of 2019, an average price of 150 €/MWh would increase the cost to 267 €/MWh, almost making wind power feed-in competitive. With a 250 €/MWh market price, even highly variable Spanish solar PV feed-in could compete. Direct renewable feed-in at the LCOE would therefore be an economically attractive approach from the perspective of conversion plants such as electrolysers (Ramsebner, 2022).[183]

Distributed local PV, in combination with BEV charging, could continue to benefit from higher electricity market prices. On-site electricity production from rooftop PV for single-family buildings or in energy communities is a promising alternative to escape the market uncertainty and instead rely on locally produced renewable energy. Apart from the strategies described above, an appropriate load management approach could optimise BEV charging with solar PV generation. In that case, vehicle-to-grid can also be an interesting consideration to optimize locally controlled distributed energy systems or micro-grids (Marnay & Lai, 2012;[184] C. Wang et al., 2018),[185] (Ramsebner, 2022).[186]

---

183 Ramsebner, J. (2022). The role of sector coupling in the energy transition [Thesis, Technische Universität Wien]. https://repositum.tuwien.at/handle/20.500.12708/80280.

184 Marnay, C., & Lai, J. (2012). Serving electricity and heat requirements efficiently and with appropriate energy quality via microgrids. *The Electricity Journal*, 25(8), 7–15. https://doi.org/10.1016/j.tej.2012.09.017.

185 Wang, C., Yan, J., Marnay, C., Djilali, N., Dahlquist, E., Wu, J., & Jia, H. (2018). Distributed Energy and Microgrids (DEM). *Applied Energy*, 210, 685–689. https://doi.org/10.1016/j.apenergy.2017.11.059.

186 Ramsebner, J. (2022). The role of sector coupling in the energy transition [Thesis, Technische Universität Wien]. https://repositum.tuwien.at/handle/20.500.12708/80280.

However, rising electricity market prices also lead to a higher market value of renewable electricity. As a result, own consumption of cheaper on-site electricity competes with the remuneration for grid feed-in (Hartner, 2016). For distributed energy production, this could also be in favor of a relatively large PV system installation for more feed-in remuneration. Concerning the profit optimisation of the renewable power plant, grid feed-in at market prices will, therefore, also be more economical than the supply to conversion plants at a lower price (Ramsebner, 2022).[187]

The economic performance of renewable power generation plants is boosted, providing money for further investments (Hartner, 2016). Renewable energy providers will prosper under these conditions and can grow significantly. Electrolysers, however, will mainly rely on electricity from the grid and energy consumption costs will increase noticeably unless long-term price agreements have been made upfront. These dynamics could impose further challenges on the economic performance of enabling technologies for SC, such as P2H and P2G. Another trade-off in this upswing in renewable electricity from rising market prices is the social perspective and exposure of many households to extremely rising costs without any ability to change their energy source. These contrary situations may provide room for compensation between a prospering renewable energy sector and disadvantaged households (Ramsebner, 2022).[188]

The substantial dependence of many countries on natural gas causes severe uncertainty concerning the security of supply, mainly for residential heating and the industry, currently amplified by bottlenecks and consequently high prices. This situation triggers investments into decentral energy generation and district heating connections, especially to substitute individual gas boilers for residential heating. For example in Austria, the high share of natural gas connections needs to decrease and be substituted by local renewable sources to reduce dependency on energy imports, as well as the natural gas share in district heating. The latter comprised around 35% in the whole country in 2021 and even 40-70% in many Austrian capital cities, as illustrated in Figure 21.14 (BMK, 2021;[189] GLOBAL 2000, 2022),[190] (Ramsebner, 2022).[191]

---

187 Ramsebner, J. (2022). The role of sector coupling in the energy transition [Thesis, Technische Universität Wien]. https://repositum.tuwien.at/handle/20.500.12708/80280.

188 Ramsebner, J. (2022). The role of sector coupling in the energy transition [Thesis, Technische Universität Wien]. https://repositum.tuwien.at/handle/20.500.12708/80280.

189 BMK. (2021). Energie in Österreich 2021: Zahlen-Daten-Fakten. https://www.bmk.gv.at/themen/energie/publikationen/zahlen.html.

190 GLOBAL 2000. (2022). Klimareport: So heizen die Landeshauptstädte. GLOBAL 2000. https://www.global2000.at/publikationen/klimareport-so-heizen-die-landeshauptstaedte.

191 Ramsebner, J. (2022). The role of sector coupling in the energy transition [Thesis, Technische Universität Wien]. https://repositum.tuwien.at/handle/20.500.12708/80280.

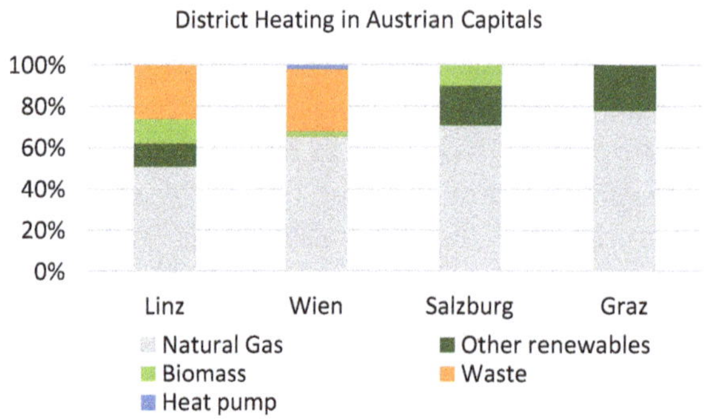

Figure 21.14: Energy carrier shares for district heating in Austrian capitals (BMK, 2021; GLOBAL 2000, 2022).

Therefore, the timely correlation between renewable energy sources and heating and cooling demand, as indicated in (Ramsebner, Linares, et al., 2021),[192] becomes even more valuable. Additionally, the political debate around short-term security of supply focuses on extended toleration or claimed amplified dependence of several countries on coal, natural gas or nuclear energy, which represents a considerable setback to the efforts around climate change mitigation and environmental protection. The European Commission has even suggested including nuclear energy and natural gas in the EU Taxonomy list of environmentally sustainable economic activities, which was accepted in a vote by the Members of the European Parliament in July 2022 (European Commission (EC), 2022;[193] European Parliament (EP), 2022),[194] (Ramsebner, 2022).[195]

The burden toward the installation and successful integration of renewable energy in the short term seems too high and the response too slow compared to reactivating or extending the use of known, fossil-fuel-based processes. The risk and long-term environmental impact associated with nuclear power and related waste handling seem to be assessed as justifiable. The required energy transition through the integration of much higher renewable shares still is subject to uncertainty. It requires immediate guidance to quickly and successfully implement research and develop-

192 Ramsebner, J., Linares, P., & Haas, R. (2021). Estimating storage needs for renewables in Europe: The correlation between renewable energy sources and heating and cooling demand. *Smart Energy*, 3, 100038. https://doi.org/10.1016/j.segy.2021.100038.

193 European Commission (EC). (2022). EU taxonomy for sustainable activities. European Commission - European Commission. https://ec.europa.eu/info/business-economy-euro/banking-and-finance/sustainable-finance/eu-taxonomy-sustainable-activities_en.

194 European Parliament (EP). (2022, July 6). Taxonomy: MEPs do not object to inclusion of gas and nuclear activities | News | European Parliament. https://www.europarl.europa.eu/news/en/press-room/20220701IPR34365/taxonomy-meps-do-not-object-to-inclusion-of-gas-and-nuclear-activities.

195 Ramsebner, J. (2022). The role of sector coupling in the energy transition [Thesis, Technische Universität Wien]. https://repositum.tuwien.at/handle/20.500.12708/80280.

ment findings. The mere installation of renewable capacities is not the solution yet. Answering current supply and market uncertainties with renewable systems requires comprehensive support by renewable storage options, digital interfaces for integrating all energy producing and consuming sectors, appropriate grid capacity, easy market entry of new actors in energy markets, and new business models (Ramsebner, 2022).[196]

21.6 Conclusion

In fact, sector coupling in practice is still in the beginnings. Renewable energy capacities are steadily increasing globally and with it, the number of use cases for sector coupling are increasing and being developed in more detail.

First steps are being made in many areas of former fossil fuel based energy sectors:

1. in transport the transformation towards battery electric vehicles and first steps using hydrogen in FCEVs is visible;
2. the heating sector is making an effort in replacing natural gas with heat pumps not only on decentral level but also in cities' district heating;
3. the energy extensive industrial sector is striving for a transformation towards renewable electricity and the use of green hydrogen;
4. and decentral renewable energy production is increasingly replacing grid electricity dependency.

However, in many cases these are isolated projects which have not yet reached state of the art. The market framework is still based on a situation of totally independent energy sectors and needs to adapt to the ongoing integration of gas, electricity and heating streams. The increasing possibilities for more and more actors, such as private homes, to participate in the market and to not only consume but also produce and store electricity still needs to be accounted for better in the known frameworks. Additionally, applications such as sharing of locally produced electricity in energy communities or joint charging infrastructure installation in multi-family homes are often still blocked by administrative and legal constraints.

Furthermore, the "competition" for renewable hydrogen among different sectors such as industry or transport but also the use for seasonal storage to achieve a renewable electricity or heating sector requires cross-sectoral methods to automate the deployment to consumption, transformation or storage. Finally, also the pressure on the distribution grid by a growing number of grid bound producers and consumers,

196 Ramsebner, J. (2022). The role of sector coupling in the energy transition [Thesis, Technische Universität Wien]. https://repositum.tuwien.at/handle/20.500.12708/80280.

needs a price tag for proper distribution. While currently most energy providers still mostly offer static tariffs, the development of variable energy prices needs to go hand in hand with variable grid fees to make the grid situation transparent and incentives different actors to act accordingly.

Apparently, a functioning market framework accounting for the increasing integration of formerly isolated energy sectors can provide great support to guaranteeing energy security. At the same time, the market based management of the energy system capacities and deployment as well as the CO_2 footprint of energy consumption can have a positive effect for people and planet. Integrating different actors from operators down to end users into the investment and management of renewable energy systems by financial incentives can make a big difference with respect to the system's profitability, sustainability and finally overall acceptance.

Eventually, the energy transition is, apart from technological progress, heavily relying on flexibility for "uncontrollable" renewable energy sources, digitalization to achieve cross-sectoral integration through informed decisions, and new, supportive business models accounting for demand, supply, CO_2 footprint and the grid situation.

Bibliography

Abeysekera, M. (2016). *Combined analysis of coupled energy networks* [Phd, Cardiff University]. https://orca.cardiff.ac.uk/id/eprint/97879/

Ajanovic, A. (2015). The future of electric vehicles: Prospects and impediments. *WIREs Energy and Environment*, *4*(6), 521–536. https://doi.org/10.1002/wene.160

Ajanovic, A., & Haas, R. (2018). Economic prospects and policy framework for hydrogen as fuel in the transport sector. *Energy Policy*, *123*, 280–288. https://doi.org/10.1016/j.enpol.2018.08.063

Ajanovic, A., Hiesl, A., & Haas, R. (2020). On the role of storage for electricity in smart energy systems. *Energy*, *200*, 117473. https://doi.org/10.1016/j.energy.2020.117473

Arent, D. J., Bragg-Sitton, S. M., Miller, D. C., Tarka, T. J., Engel-Cox, J. A., Boardman, R. D., Balash, P. C., Ruth, M. F., Cox, J., & Garfield, D. J. (2021). Multi-input, Multi-output Hybrid Energy Systems. *Joule*, *5*(1), 47–58. https://doi.org/10.1016/j.joule.2020.11.004

Bataille, C., Åhman, M., Neuhoff, K., Nilsson, L. J., Fischedick, M., Lechtenböhmer, S., Solano-Rodriquez, B., Denis-Ryan, A., Stiebert, S., Waisman, H., Sartor, O., & Rahbar, S. (2018). A review of technology and policy deep decarbonization pathway options for making energy-intensive industry production consistent with the Paris Agreement. *Journal of Cleaner Production*, *187*, 960–973. https://doi.org/10.1016/j.jclepro.2018.03.107

Belyaev, L. S., Kaganovich, B. M., Krutov, A. N., Filippov, S. P., Martinsen, D., Müller, M., Wagner, H. J., & Walbeck, M. (1987). *Ways of transition to clean energy use: Two methodological approaches*. IIASA.

Bloess, A., Schill, W.-P., & Zerrahn, A. (2018). Power-to-heat for renewable energy integration: A review of technologies, modeling approaches, and flexibility potentials. *Applied Energy*, *212*, 1611–1626. https://doi.org/10.1016/j.apenergy.2017.12.073

BMK. (2021). *Energie in Österreich 2021: Zahlen-Daten-Fakten*. https://www.bmk.gv.at/themen/energie/publikationen/zahlen.html

BP. (2019a). *Energy Outlook downloads and archive | Energy economics | Home*. Bp Global. https://www.bp.com/en/global/corporate/energy-economics/energy-outlook/energy-outlook-downloads.html

BP. (2019b). *Statistical Review of World Energy | Energy economics | Home*. Bp Global. https://www.bp.com/en/global/corporate/energy-economics/statistical-review-of-world-energy.html

Brown, T., Schlachtberger, D., Kies, A., Schramm, S., & Greiner, M. (2018). Synergies of sector coupling and transmission reinforcement in a cost-optimised, highly renewable European energy system. *Energy*, *160*, 720–739. https://doi.org/10.1016/j.energy.2018.06.222

Bundesministerium für Wirtschaft und Energie (BMWi). (2016). *Green paper on energy efficiency: Discussion paper of the Federal Ministry for Economic Affairs and Energy*.

Bundesministeriums für Umwelt, Naturschutz, Bau und Reaktorsicherheit (BMUB). (2016). *Climate Action Plan 2050 – Principles and goals of the German government's climate policy*.

Bundesverband der Energie-und Wasserwirtschaft e.V. (BDEW). (2017). *Positionspapier: 10 Thesen zur Sektorkopplung*. http://www.bdew.de/service/stellungnahmen/10-thesen-sektorkopplung/

Buttler, A., & Spliethoff, H. (2018). Current status of water electrolysis for energy storage, grid balancing and sector coupling via power-to-gas and power-to-liquids: A review. *Renewable and Sustainable Energy Reviews*, *82*, 2440–2454. https://doi.org/10.1016/j.rser.2017.09.003

Connolly, D., Lund, H., & Mathiesen, B. V. (2016). Smart energy Europe: The technical and economic impact of one potential 100% renewable energy scenario for the European Union. *Renewable and Sustainable Energy Reviews*, *60*, 1634–1653. https://doi.org/10.1016/j.rser.2016.02.025

Danish Energy Agency (DEA). (2016). *Technology data for electricity, district heating, energy storage and energy conversion*. https://ens.dk/sites/ens.dk/files/Statistik/technology_data_catalogue_for_el_and_dh_-_0009.pdf

Danish Energy Agency (DEA). (2017). *Regulation and planning of district heating in Denmark*.

DVGW. (2020, March 5). *Sektorenkopplung*. Sektorenkopplung – Synergien sinnvoll nutzen. https://www.dvgw.de/themen/energiewende/sektorenkopplung/

DVGW & VDE. (2016). *Eckpunkte zur Begriffsdefinition Sektorenkopplung*.

European Commission (EC). (2022). *EU taxonomy for sustainable activities*. European Commission – European Commission. https://ec.europa.eu/info/business-economy-euro/banking-and-finance/sustainable-finance/eu-taxonomy-sustainable-activities_en

European Parliament (EP). (2018). *Sector coupling: How can it be enhanced in the EU to foster grid stability and decarbonise?*

European Parliament (EP). (2022, July 6). *Taxonomy: MEPs do not object to inclusion of gas and nuclear activities | News | European Parliament*. https://www.europarl.europa.eu/news/en/press-room/20220701IPR34365/taxonomy-meps-do-not-object-to-inclusion-of-gas-and-nuclear-activities

Fischedick, M., Marzinkowski, J., Winzer, P., & Weigel, M. (2014). Techno-economic evaluation of innovative steel production technologies. *Journal of Cleaner Production*, *84*, 563–580. https://doi.org/10.1016/j.jclepro.2014.05.063

Foresight Climate & Energy. (2019, May 3). *Hydrogen: The Northern Netherlands is ready*. https://foresightdk.com/hydrogen-northern-netherlands-is-ready/

Foy, J. H., & Schlereth, T. J. (1994). *Foy, J: American Home Life 1880-1930* (1st ed.). University of Tennessee Press.

Fraunhofer ISE. (2021). *Studie: Stromgestehungskosten erneuerbare Energien*. Fraunhofer-Institut für Solare Energiesysteme ISE. https://www.ise.fraunhofer.de/de/veroeffentlichungen/studien/studie-stromgestehungskosten-erneuerbare-energien.html

Geidl, M., Koeppel, G., Favre-Perrod, P., Klöckl, B., Andersson, G., & Fröhlich, K. (2007). *The energy hub: A powerful concept for future energy systems*. 3rd Annual Carnegie Mellon Conference on the Electricity Industry: Ensuring that the Industry has the Physical and Human Resources Needed for the Next Thirty Years. https://www.research-collection.ethz.ch/handle/20.500.11850/3133

GLOBAL 2000. (2022). *Klimareport: So heizen die Landeshauptstädte*. GLOBAL 2000. https://www.global2000.at/publikationen/klimareport-so-heizen-die-landeshauptstaedte

Haas, R. (1995). The value of photovoltaic electricity for society. *Solar Energy*, *54*(1), 25–31. https://doi.org/10.1016/0038-092X(94)00099-Y

Haas, R., Ajanovic, A., Ramsebner, J., Perger, T., Knápek, J., & Bleyl, J. W. (2021). Financing the future infrastructure of sustainable energy systems. *Green Finance*, *3*(1), 90–118. https://doi.org/10.3934/GF.2021006

Haas, R., Duic, N., Auer, H., Ajanovic, A., Ramsebner, J., Knapek, J., & Zwickl-Bernhard, S. (2023). The photovoltaic revolution is on: How it will change the electricity system in a lasting way. *Energy*, *265*, 126351. https://doi.org/10.1016/j.energy.2022.126351

Haas, R., Lettner, G., Auer, H., & Duic, N. (2013). The looming revolution: How photovoltaics will change electricity markets in Europe fundamentally. *Energy*, *57*, 38–43. https://doi.org/10.1016/j.energy.2013.04.034

Häfele, W., & Nebojsa, N. (1983). The contribution of oil and gas for the transition to long range novel energy systems. *World Demand for and Supply of Oil and Gas*, *9*(3).

Hartner, M. (2016). *Economic aspects of photovoltaic from a strategic, end user and electricity system perspective* [Thesis, Wien]. https://repositum.tuwien.at/handle/20.500.12708/4762

Hirschhausen, C. V. (2017). Prosumage and the future regulation of utilities: An introduction. *Economics of Energy & Environmental Policy*, *Volume 6* (Number 1). https://ideas.repec.org/a/aen/eeepjl/eeep6-1-introduction.html

Hörsch, J., & Brown, T. (2017). The role of spatial scale in joint optimisations of generation and transmission for European highly renewable scenarios. *2017 14th International Conference on the European Energy Market (EEM)*, 1–7. https://doi.org/10.1109/EEM.2017.7982024

IEA. (2020). *Projected Costs of Generating Electricity 2020*. IEA. https://www.iea.org/reports/projected-costs-of-generating-electricity-2020

International Energy Agency (IEA). (2019). *The future of hydrogen* [Report prepared by the IEA for the G20, Japan].

International Renewable Energy Agency (IRENA). (2018). *Hydrogen from renewable power: Technology outlook for the energy transition*.

Iqbal, M. (2012). *An Introduction To Solar Radiation*. Elsevier.

IRENA. (2021). *Renewable Power Generation Costs in 2020*. /Publications/2021/Jun/Renewable-Power-Costs-in-2020. https://www.irena.org/publications/2021/Jun/Renewable-Power-Costs-in-2020

IRENA. (2022). *Energy Transition*. Energy Transition. https://www.irena.org/energytransition

Lewandowska-Bernat, A., & Desideri, U. (2018). Opportunities of power-to-gas technology in different energy systems architectures. *Applied Energy*, *228*, 57–67. https://doi.org/10.1016/j.apenergy.2018.06.001

Liu, W., Lund, H., & Mathiesen, B. (2011). Large-scale integration of wind power into the existing Chinese energy system. *Fuel and Energy Abstracts*, *36*, 4753–4760. https://doi.org/10.1016/j.energy.2011.05.007

Loftus, P. J., Cohen, A. M., Long, J. C. S., & Jenkins, J. D. (2015). A critical review of global decarbonization scenarios: What do they tell us about feasibility? *WIREs Climate Change*, *6*(1), 93–112. https://doi.org/10.1002/wcc.324

Lund, H., Mathiesen, B. V., Connolly, D., & Østergaard, P. A. (2014). Renewable energy systems – A smart energy systems approach to the choice and modelling of 100 % renewable solutions. *Chemical Engineering Transactions*, *39*, 1–6. https://doi.org/10.3303/CET1439001

Lund, H., Østergaard, P. A., Connolly, D., & Mathiesen, B. V. (2017). Smart energy and smart energy systems. *Energy*, *137*, 556–565. https://doi.org/10.1016/j.energy.2017.05.123

Mancarella, P. (2014). MES (multi-energy systems): An overview of concepts and evaluation models. *Energy*, *65*, 1–17. https://doi.org/10.1016/j.energy.2013.10.041

Marnay, C., & Lai, J. (2012). Serving electricity and heat requirements efficiently and with appropriate energy quality via microgrids. *The Electricity Journal*, *25*(8), 7–15. https://doi.org/10.1016/j.tej.2012.09.017

Masera, M., Bompard, E. F., Profumo, F., & Hadjsaid, N. (2018). Smart (Electricity) Grids for Smart Cities: Assessing Roles and Societal Impacts. *Proceedings of the IEEE*, *106*(4), 613–625. https://doi.org/10.1109/JPROC.2018.2812212

Matthes, F. Chr. (2018). *Energy transformation in Germany: Progress, shortfalls and prospects.*

Matute, G., Yusta, J. M., & Correas, L. C. (2019). Techno-economic modelling of water electrolysers in the range of several MW to provide grid services while generating hydrogen for different applications: A case study in Spain applied to mobility with FCEVs. *International Journal of Hydrogen Energy*, *44*(33), 17431–17442. https://doi.org/10.1016/j.ijhydene.2019.05.092

Moreno, B., López, A. J., & García-Álvarez, M. T. (2012). The electricity prices in the European Union. The role of renewable energies and regulatory electric market reforms. *Energy*, *48*(1), 307–313. https://doi.org/10.1016/j.energy.2012.06.059

Nakicenovic, N., Grubler, A., & McDonald, A. (1998). Global Energy Perspectives. *Global Energy Perspectives, Edited by Nebojsa Nakicenovic and Arnulf Grübler and Alan McDonald, Pp. 250. ISBN 0521642000. Cambridge, UK: Cambridge University Press, October 1998.*, *1*.

Nicolosi, M. (2010). Wind power integration and power system flexibility–An empirical analysis of extreme events in Germany under the new negative price regime. *Energy Policy*, *38*(11), 7257–7268. https://doi.org/10.1016/j.enpol.2010.08.002

Nielsen, M. G., Morales, J. M., Zugno, M., Pedersen, T. E., & Madsen, H. (2016). Economic valuation of heat pumps and electric boilers in the Danish energy system. *Applied Energy*, *167*, 189–200. https://doi.org/10.1016/j.apenergy.2015.08.115

Novak, P. (2017). Exergy as measure of sustainability of energy system. *International Journal of Earth & Environmental Sciences*, *2*(2). https://doi.org/10.15344/2456-351X/2017/139

Olczak, M., & Piebalgs, A. (2018). *Sector coupling: The new EU climate and energy paradigm?*

Perez-Arriaga, I. (2017). Improved Regulatory Approaches for the Remuneration of Electricity Distribution Utilities with High Penetrations of Distributed Energy Resources. *The Energy Journal*, *38*. https://doi.org/10.5547/01956574.38.3.jjen

Pfeiffer, C., Puchegger, M., Maier, C., Tomaschitz, I. V., Kremsner, T. P., & Gnam, L. (2021). A Case Study of Socially-Accepted Potentials for the Use of End User Flexibility by Home Energy Management Systems. *Sustainability*, *13*(1), Article 1. https://doi.org/10.3390/su13010132

Ramsebner, J. (2022). *The role of sector coupling in the energy transition* [Thesis, Technische Universität Wien]. https://repositum.tuwien.at/handle/20.500.12708/80280

Ramsebner, J., Haas, R., Ajanovic, A., & Wietschel, M. (2021). The sector coupling concept: A critical review. *WIREs Energy and Environment*, *10*(4), e396. https://doi.org/10.1002/wene.396

Ramsebner, J., Haas, R., Auer, H., Ajanovic, A., Gawlik, W., Maier, C., Nemec-Begluk, S., Nacht, T., & Puchegger, M. (2021). *From single to multi-energy and hybrid grids: Historic growth and future vision.* https://doi.org/10.1016/J.RSER.2021.111520

Ramsebner, J., Hiesl, A., & Haas, R. (2020). Efficient Load Management for BEV Charging Infrastructure in Multi-Apartment Buildings. *Energies*, *13*(22), 1–23.

Ramsebner, J., Linares, P., & Haas, R. (2021). Estimating storage needs for renewables in Europe: The correlation between renewable energy sources and heating and cooling demand. *Smart Energy*, *3*, 100038. https://doi.org/10.1016/j.segy.2021.100038

Ramsebner, J., Linares, P., Hiesl, A., & Haas, R. (2024). Techno-economic evaluation of renewable hydrogen generation strategies for the industrial sector. *Hydrogen Energy Journal, Mimeo-under review.* in International Journal of Hydrogen Energy. https://doi.org/10.1016/j.ijhydene.2024.02.167

Raux-Defossez, P., Wegerer, N., Pétillon, D., Bialecki, A., Bailey, A., & Belhomme, R. (2018). Grid Services Provided By The Interactions Of Energy Sectors In Multi-Energy Systems: Three International Case Studies. *Energy Procedia*, *155*, 209–227. https://doi.org/10.1016/j.egypro.2018.11.055

Richts, C., Jansen, M., & Siefert, M. (2015). Determining the economic value of offshore wind power plants in the changing energy system. *Energy Procedia*, *80*, 422–432. https://doi.org/10.1016/j.egypro.2015.11.446

Ridjan, I., Mathiesen, B. V., & Connolly, D. (2016). *Terminology used for renewable liquid and gaseous fuels based on the conversion of electricity: A review | Elsevier Enhanced Reader*. https://doi.org/10.1016/j.jclepro.2015.05.117

Robinius, M., Otto, A. D.-I., Heuser, P., Welder, L., Syranidis, K., Ryberg, D. S., Grube, T. J., Markewitz, P., Peters, R., & Stolten, D. (2017). *Linking the power and transport sectors – part 1: The principle of sector coupling*. https://doi.org/10.3390/EN10070956

Robinius, M., Otto, A., Syranidis, K., Ryberg, D. S., Heuser, P., Welder, L., Grube, T., Markewitz, P., Tietze, V., & Stolten, D. (2017). Linking the power and transport sectors – part 2: Modelling a sector coupling scenario for Germany. *Energies*, *10*(7), 957ff. https://doi.org/10.3390/en10070957

Robinius, M., Raje, T., Nykamp, S., Rott, T., Müller, M., Grube, T., Katzenbach, B., Küppers, S., & Stolten, D. (2018). Power-to-Gas: Electrolyzers as an alternative to network expansion – An example from a distribution system operator. *Applied Energy*, *210*, 182–197. https://doi.org/10.1016/j.apenergy.2017.10.117

Schaber, K. (2013). *Integration of Variable Renewable Energies in the European power system: A model-based analysis of transmission grid extensions and energy sector coupling*. Technische Universität.

Schaber, K., Steinke, F., & Hamacher, T. (2013). *Managing temporary oversupply from renewables efficiently: Electricity storage versus energy sector coupling in Germany*. 22.

Schemme, S., Samsun, R. C., Peters, R., & Stolten, D. (2017). Power-to-fuel as a key to sustainable transport systems – An analysis of diesel fuels produced from CO 2 and renewable electricity. *Fuel*, *205*, 198–221. https://doi.org/10.1016/j.fuel.2017.05.061

Schiebahn, S., Grube, T., Robinius, M., Tietze, V., Kumar, B., & Stolten, D. (2015). Power to gas: Technological overview, systems analysis and economic assessment for a case study in Germany. *International Journal of Hydrogen Energy*, *40*(12), 4285–4294. https://doi.org/10.1016/j.ijhydene.2015.01.123

Schill, W.-P., Zerrahn, A., Kunz, F., & Kemfert, C. (2017). Decentralized solar prosumage with battery storage: System orientation required. *DIW Economic Bulletin*, *7*(16), 141–151.

Schittekatte, T., Momber, I., & Meeus, L. (2017). Future-proof tariff design: Recovering sunk grid costs in a world where consumers are pushing back. *EUI RSCAS Working Paper (FSR)*, *70*. https://doi.org/10.1016/j.eneco.2018.01.028

Scorza, S. A., Pfeiffer, J., Schmitt, A., & Weissbart, C. (2018). *Kurz zum Klima: »Sektorkopplung« – Ansätze und Implikationen der Dekarbonisierung des Energiesystems*. *71*(10), 49–53.

Stadler, I., & Sterner, M. (2018). Urban energy storage and sector coupling. In *Urban Energy Transition: Renewable Strategies for Cities and Regions* (2nd ed., pp. 225–244). Elsevier.

Tang, O., Rehme, J., Cerin, P., & Huisingh, D. (2021). Hydrogen production in the Swedish power sector: Considering operational volatilities and long-term uncertainties. *Energy Policy*, *148*, 111990. https://doi.org/10.1016/j.enpol.2020.111990

Transport&Environment. (2020). *Decarbonising the EU's transport sector with renewable electricity and electrofuels*. https://www.transportenvironment.org/wp-content/uploads/2020/12/2020_12_Briefing_feasibility_study_renewables_decarbonisation.pdf

Tu, K., Deutsch, Matthias, & Flis, Gniewomir. (2021, May 6). *Industrial clusters using green hydrogen can drive clean energy transition in Europe and China*. World Economic Forum. https://www.weforum.org/agenda/2021/05/industrial-clusters-green-hydrogen-europe-china/

Wang, A., van der Leun, K., Peters, D., & Buseman, M. (2020). *European Hydrogen Backbone: How a dedicated hydrogen infrastructure can be created*. https://gasforclimate2050.eu/sdm_downloads/european-hydrogen-backbone/

Wang, C., Yan, J., Marnay, C., Djilali, N., Dahlquist, E., Wu, J., & Jia, H. (2018). Distributed Energy and Microgrids (DEM). *Applied Energy*, *210*, 685–689. https://doi.org/10.1016/j.apenergy.2017.11.059

Warner, J. T. (2015). *The handbook of lithium-ion battery pack design: Chemistry, components, types and terminology*. Elsevier Science.

Wietschel, M., Haendel, M., Schubert, G., Köppel, W., & Degünther, C. (2015). *Kurz- und mittelfristige Sektorkopplungspotentiale*.

Wietschel, M., Plötz, P., Pfluger, B., Klobasa, M., Eßer, A., Haendel, M., Müller-Kirchenbauer, J., Kochems, J., Hermann, L., Grosse, B., Nacken, L., Küster, M., Pacem, J., Naumann, D., Kost, C., Kohrs, R., Fahl, U., Schäfer-Stradowsky, S., Timmermann, D., & Albert, D. (2018). *Sektorkopplung – Definition, Chancen und Herausforderungen*.

Witkowski, K., Haering, P., Seidelt, S., & Pini, N. (2020). Role of thermal technologies for enhancing flexibility in multi-energy systems through sector coupling: Technical suitability and expected developments. *IET Energy Systems Integration*, *2*(2), 69–79. https://doi.org/10.1049/iet-esi.2019.0061

Tomas Haug, Philipp Hiemann, Lorenz Wieshammer

22 Improvement of Grid Infrastructure

22.1 Executive Summary

Energy networks connect energy supply and demand, for example for electricity. They are usually considered natural monopolies due to the combination of high fixed cost and low variable cost of operating them, which gives rise to the need for regulation. Decarbonisation and the energy transition have a particularly large impact on electricity networks. Renewable generation is more volatile than fossil fuel-based generation and the increasing electrification of sectors such as transport and heating will likely increase electricity demand further, creating a need for investments into the existing networks and a changing role and tasks of network operators. While the expansion of physical network capacity will be necessary in parts of the network, other "smart" solutions can also help facilitate the energy transition in a cost-efficient way. For example, smart grid operation can smoothen consumption patterns and reduce peak demand, whereas increased interconnection capacity throughout Europe can help distribute renewable energy from regions where it is abundant to regions where it is scarce. The changing role of network operators likely requires changes to the regulatory framework, while questions arise around the financing of the necessary investments.

22.2 Introduction

Energy networks are an integral part of the energy sector and are necessary to supply energy to households, industrial customers and any other type of end user. Electricity networks are probably the type of energy networks that first comes to mind. However, there are other types of energy networks, too, such as natural gas networks, district heating networks and, increasingly, hydrogen networks. While each of these network types has individual characteristics, they share some common features. As natural gas networks and electricity networks are the most common types of energy networks, the following introduction into energy networks focusses on these two.

The term energy network describes the infrastructure that connects energy supply and energy demand, i.e., mainly electric lines and natural gas pipelines, but also any other assets necessary to operate the network, like electricity transformers or gas compressors. For both electricity and natural gas, a differentiation is usually made between the larger, long-distance parts of the networks and the smaller, more local parts of the network. The large parts with high capacity are called transmission net-

https://doi.org/10.1515/9783110784299-022

work. This term is used for high-voltage electricity networks or high-pressure natural gas pipelines that are used for transit of large energy volumes. Usually, large electricity generators and natural gas suppliers are connected to this type of infrastructure. The smaller parts of the networks are called distribution network. The distribution network connects to the transmission network and is used to distribute energy to end users, such as households and industrial customers. The distribution network is usually much more ramified, and each distribution line or pipeline has a lower capacity than a typical transmission line or pipeline. In most countries, different entities are responsible for operating transmission and distribution networks.

From an economic point of view, energy networks have characteristics that set the energy network sector apart from most other sectors. Most importantly, energy networks are characterised by high fixed costs. Before being able to operate the network, a company needs to install a significant number of cables or pipelines. This often requires groundwork to lay part of the network underground. Hence, the investments necessary to start an energy network company are usually worth several millions, if not billions, of euros. However, once the network is in operation, the variable cost are very small and typically close to zero.[1] For this reason, energy networks are usually considered so-called "natural monopolies".[2] The term "natural monopoly" characterises a situation in which it is cheaper for one single company to supply a market than for multiple different companies.[3] For energy networks, this is intuitive: having two or more parallel networks that connect the same households to the same generation facilities would be a waste of resources. However, while a natural monopoly might seem beneficial for consumers at first sight, the opposite is usually the case. The reason is that a natural monopoly would normally exploit its position as the only company in the market and charge excessive prices to its customers. The customers would have no choice other than to accept these prices, which the natural monopolist would hold at levels just low enough to deter market entry from potential competitors. Market entry is also deterred by the fact that investments into energy networks are often irreversible: Once a cable or pipeline is installed at one location, it is usually not viable to demolish it and rebuild it somewhere else. This is in contrast to investments in other types of assets, such as machinery or IT equipment. The characterisation of energy networks as natural monopolies often necessitates public control and

1 It is important to note that the variable costs of an energy network do not comprise the cost of the energy that is transported via the network. Energy network companies are usually not responsible for, and in fact in Europe usually not permitted to, buying and selling the energy that they transport. Hence, the variable cost of energy networks mainly comprises energy to keep the network running, for example energy to operate natural gas compressors. This cost varies when the amount of energy transported via the network increases, but it does not vary significantly.

2 Other network industries are often considered natural monopolies, too, such as railway infrastructure or communications infrastructure.

3 See https://stats.oecd.org/glossary/detail.asp?ID=3267 [11 July 2022].

regulation, which tries to reach a balance of fair prices for consumers and appropriate returns for energy network operators. We will discuss the issue of energy network regulation in more detail in section 22.5.

Electricity networks exhibit some further special characteristics. For example, electricity networks always need to maintain a balance between supply and demand. If one exceeds the other by more than a small margin, the frequency of the electricity in the network deviates from its standard (in Europe: 50 hertz), which can lead to blackouts. For instance, this can happen when large power plants suddenly turn off, without an equivalent drop in electricity demand, or when electricity generation exceeds demand. Hence, electricity network operators are responsible for balancing supply and demand, which implies anticipating demand patterns and accommodating sufficient supply at all times. Energy networks therefore share responsibility for security of supply in the energy system.

The remainder of this section will focus on electricity networks. Electricity networks have a central role for the energy transition and the move towards a zero-carbon economy.[4] The shift towards renewable electricity and further increases in electrification and, consequently, electricity demand constitute significant challenges and imply changing responsibilities and an expanding set of tasks for electricity network operators. Mastering these challenges will necessitate large investments, innovative solutions and regulatory changes. In the following, we aim to present an overview of the role of electricity networks for the energy transition and the energy transition's implications for energy networks (section 22.3), discuss potential solutions to the challenges arising for electricity networks (section 22.4), and describe the implications of the changing role of electricity networks for issues such as regulation, financing of the infrastructure, and network tariffs (section 22.5). We will discuss the above with a focus on Europe and, in particular, the experience in Germany. Germany poses a suitable example as it was one of the early movers in the energy transition, thus having some degree of experience with problems and solutions for accommodating this transition.

4 The energy transition poses a significant challenge for natural gas networks, too. To reach the climate goals, natural gas will largely have to be replaced by renewable gases such as hydrogen. This raises questions such as whether the existing networks can be fitted for hydrogen transport or to what extent gas networks will be needed in the future. If the gas network were to be significantly reduced in size, questions around asset stranding, i.e., the non-recovery of infrastructure investments, may arise as well. For a brief description of challenges and key issues to be considered for the future of natural gas networks, see Oxford Institute for Energy Studies 2019.

22.3 Problem

Before describing the impact of the energy transition on energy networks and explaining the importance of energy networks for a successful transition, it is necessary to explain what changes the energy transition brings about.

The energy sector is one of several sectors that need to decarbonise in order to reach carbon neutrality in the future. Examples of other sectors are transport and heating. The term “energy transition” is commonly used to describe the shift towards renewable, zero carbon energy generation and consumption. While energy does not only comprise electric energy, discussions of the energy transition often centre around the decarbonisation of the electricity sector.

Electricity can be generated by different means, i.e., by using different so-called “fuels”. Historically, mainly fossil fuels such as coal, lignite, oil and natural gas have been used to generate electricity. Moreover, nuclear energy is a widely used way to generate electricity, albeit concerns about the long-term security and disposal of nuclear waste have led to some countries phasing out nuclear energy lately, including Germany.[5] Besides fossil and nuclear fuels, electricity can also be generated using renewable fuels. The main renewable fuels are wind (both onshore and offshore), solar photovoltaic and water (e.g. run-of-river or pump storage). While the latter has already been used for more than a century, e.g. in the Alps, wind and solar have only seen a notable application in recent decades.[6] Within the electricity sector, the term “energy transition” mainly concerns the shift from using fossil fuels for electricity generation to using renewable sources.

The energy transition also entails the electrification of sectors other than the energy sector. Most prominently, the transport sector needs to move away from fossil fuels. Besides alternative, renewable fuels such as hydrogen or synthetic fuels, electric vehicles seem to play the central role in the future of (at least) individual road transport.[7] The heating sector, which in 2018 accounted for around 50% of global final energy consumption, is another important sector in need of decarbonisation.[8] Here, too, it seems that electricity will account for a larger market share in the future, e.g. via the increased use of heat pumps. Therefore, the energy transition does not only bring about a change in the way that electricity is generated but will also likely lead to further increases in electricity demand through electrification of other sectors.

5 Germany announced its nuclear phase-out by the end of 2022 after the events in Fukushima in 2011.

6 First hydroelectric power plants have been operated in the Alps as early as in the 1890s, see e.g. https://tirolatlas.uibk.ac.at/maps/thema/query.py/text?lang=de;id=1502#:~:text=Zahlreiche%20M%C3%BChlen%20und%20mit%20Wasserkraft,die%20Sch%C3%A4chte%20trocken%20zu%20halten. [11 July 2022].

7 For large freight traffic, be it on land, sea or in the air, some experts consider renewable fuels (e.g. hydrogen, synthetic fuels) to be more viable than electricity.

8 See https://www.iea.org/reports/renewables-2019/heat [11 July 2022].

There are some notable differences between the conventional generation from fossil fuels and the generation using renewable sources which have consequences for energy networks. Firstly, probably the most prominent difference is that generation from renewable sources depends on the availability of a renewable energy source, e.g. wind or sunshine. The power plant operator cannot control this availability. The dependence on availability results in rather volatile patterns of generation from renewable sources. To a certain degree, these patterns are predictable, e.g. solar energy will only be generated during the day, not during the night. For single renewable energy plants, however, it is challenging to reliably forecast a generation pattern more than a few hours into the future. Generation from conventional power plants, in contrast, depends on the availability of the respective fossil fuel used, which the power plant operator usually controls, often via long-term contracts with fuel suppliers. As long as fuel is available, the plants can generate electricity continuously.[9] Hence, conventional power plants are suited to generate so-called baseload electricity because their generation pattern can be scheduled and forecast reliably.[10] Decreasing generation from conventional power plants and increasing amounts of renewable energy, thus, constitute a challenge for network operators, which are responsible for the stability of the network. If generation is less predictable and more volatile, it becomes more difficult and more costly to balance supply and demand in the network.

The second difference between conventional and renewable electricity derives directly from the first: renewable energy plants are best located where sufficient renewable energy (i.e., wind, sun, water) is available. It is only economically viable to operate renewable energy plants at certain locations that offer enough full load hours.[11] In contrast, the location of conventional power plants is flexible.[12] Typically, they are

9 Note that, due to the specifics of the market for electricity generation, the availability of fuel does not necessarily imply that the plant is also producing. Whether or not a plant is generating electricity at a given time depends on the market price for electricity at that time. For further information on the functioning of the wholesale electricity market, see e.g. Joskow 2006, pp. 8–15.

10 Baseload electricity means the amount of electricity that needs to be supplied to the electricity network at any time to meet the minimum electricity demand in the network. One can think of appliances such as fridges and servers or certain industrial processes that are operated continuously.

11 Full load hours are the hours of a year in which a power plant is operated. For conventional power plants that supply baseload electricity, such as nuclear power plants, this number is typically close to 8,760 hours (the number of hours in a year with 365 days). For renewable power plants, the value is currently around 4,500 hours for very good offshore locations in Germany (offshore wind), around 2,000 hours for German inland wind locations (onshore wind), or around 1,000 hours for German solar modules. See https://de.statista.com/statistik/daten/studie/224720/umfrage/wind-volllaststunden-nach-standorten-fuer-wea/ [11 July 2022] and Fraunhofer ISE 2022a, p. 43.

12 Locations for conventional power plants are not entirely flexible. For natural gas plants, a connection to the natural gas grid is required. However, such a connection is often purpose-built if a new plant is constructed. Lignite plants are usually located close to lignite fields because lignite is not as easily transportable as other energy fuels. Nonetheless, the location of conventional power plants depends far less on the local conditions than renewable power plants.

built close to demand centres. Therefore, in Germany most conventional power plants are located in the West and South, where most of German industry is centred and population density is comparably high. In contrast, wind plants are typically located at offshore locations or in rather flat landscapes. Hydropower plants can only be operated close to rivers or lakes or at the coast. And solar power plants are better located at sunny places, e.g. in the South of Europe. Hence, the location of renewable plants is usually not determined based on where demand is largest but based on where availability is best. Therefore, the energy transition means that in Germany, most renewable generation is centred in the North and East, where the landscape is rather flat and the North Sea and Baltic Sea offer (offshore) wind potential. However, this also means that the distance between the demand centres and generation locations grows, necessitating higher transmission capacities to transport the electricity to the demand centres.

Thirdly, power plants using fossil fuels usually have a significantly higher capacity than renewable plants. In conventional power plants, the capacity sometimes exceeds 1 GW, e.g. in the case of lignite or coal. Germany's biggest power plant has a capacity of more than 4 GW.[13] Renewable energy, on the other hand, is usually generated on a much smaller scale. Today, the newest and largest single offshore wind turbines have capacities of up to 15 MW, but the average installed turbines on land are considerably smaller.[14] While there are larger wind farms that combine multiple turbines, in particular in suitable offshore locations, often only a few turbines form one combined wind farm. For solar energy, the case is similar. There are some large solar parks that cover several square kilometres of land and have a capacity of around 2 GW.[15] However, many solar units are operated in small groups or on their own, with many units being installed on private household roofs. This difference between conventional and renewable energy generation means that the total number of generating units that need to be connected to the network increases as the energy transition continues.[16] Network operators need to monitor more generating units and potentially intervene if too much generation from renewable energy at a time would endanger system stability. Another consequence of the smaller size of renewable energy plants is that they are increasingly often connected to the lower voltage distribution network instead of the transmission network, which conventional power plants

---

13 Neurath Power Station in Grevenbroich in the West of Germany. See Neuß-Grevenbroicher Zeitung 2020.

14 See https://www.siemensgamesa.com/products-and-services/offshore/wind-turbine-sg-14-222-dd [11 July 2022].

15 See https://www.nsenergybusiness.com/features/largest-solar-power-plants/ [11 July 2022].

16 Note that not all renewable generation units, especially those at private household roofs, are connected to the network. An advantage of such off-grid solutions is that they do not cause costs for the network. The downside is that they cannot be used to stabilise the network and the electricity generated can only be utilised at the location where it is generated.

have traditionally been connected to. A good example are private, small-scale solar PV units that are located at household roofs. The distribution network needs to work bi-directionally now, whereas several years ago, it mainly accommodated flows from the transmission networks directly to the end-users, but not the other way around. As distribution networks were originally not intended to transmit electricity into the transmission network, costs for technical solutions are incurred.

Most academic and practitioners' studies expect electricity demand to increase further in the next years and decades, partly attributable to the ongoing electrification of sectors such as transport and heating as a means of decarbonisation. For Germany, some studies even expect an increase in electricity demand by close to 100% until 2045/2050 in scenarios with a high degree of electrification.[17] However, existing electricity networks have a maximum capacity that cannot be exceeded. Thus, rising electricity demand necessitates investments into the network to increase capacity. When it comes to electricity demand, an important issue needs to be noted: what determines the necessary capacity of an electricity network is not the average amount that flows through the network over, say, one day. Instead, it is the peak flow of electricity. If a significant share of the daily electricity shall flow through a network within a few hours, the network needs to have a higher capacity than when the same amount of daily electricity flows through it steadily over the day. Hence, peak electricity demand is the main factor to consider when determining the necessary amount of investment into the electricity network. However, rising electricity demand is not the only factor that drives investment. There is also a continuous need for investments for replacement and maintenance of existing lines. Overall, a significant need for investments into electricity network infrastructure arises. In Germany, the electricity TSOs estimate an investment need (for the transmission grid including offshore connections, but without the distribution grid) of 74.5 to 80 billion euros until 2040.[18] For Austria, the electricity TSO Austrian Power Grid estimates investment volumes of 3.5 billion euros until 2031.[19]

Investments are needed for both the distribution and the transmission network. This fact is often overlooked in public debate, which centres around the investment needs for the large transmission lines. However, without distribution networks that are fit for rising electricity demand and bi-directional flows, investments into trans-

17 See dena 2018, p. 250. The scenario "EL95" shows an increase in electricity demand (including distribution loss and demand from electrolyzers, without power plants' own consumption) between 2015 and 2050 of 104%. The reference scenario "RF" shows an increase of only 8%, but this scenario does not include a climate target. Another study by the German transmission system operators estimates an increase in gross electricity consumption of 36% to 97% between 2018 and 2045. See 50Hertz et al. 022, p. 32. A more recent dena study estimates an increase in gross electricity demand of c. 53% between 2018 and 2045. See dena 2021, p. 21.

18 50Hertz et al. 2021, p. 151.

19 Austrian Power Grid, Vorarlberger Übertragungsnetz GmbH 2021, p. 15.

mission lines offer little benefit. The large investment needs create a set of questions that network operators, but also politicians and regulators, need to contemplate and answer. Amongst others, these are questions of how to finance these investments, of how these investments should affect network tariffs, and of how regulation can support the viability of these investments. Another issue to consider is how to decide what investments should be pursued by network operators and what investments should not. We will have a closer look at these questions surrounding regulation, financing, and network tariffs in section 22.5.

All of the above are not merely issues for the future, but the energy transition already affects network operators today. Certain countries, like Germany, already manage to generate a significant proportion of their electricity from renewable sources. This increasing share of renewable electricity leads to more load intermittency over the seasons and within days, mainly depending on the weather. In Germany, in recent years, the electricity network has struggled to increase capacity in line with increasing demand and increasingly volatile generation. To manage congestion occurring on their respective network, the German transmission system operators increasingly have to shut down ("curtail") power plants in certain hours when generation is too high for the network to handle. In other cases, when imbalances in the system arise, network operators can instruct power plants in certain parts of the network to curtail their generation, while others need to increase their generation, creating flows that counteract the congestion. These so-called "redispatch" and other "congestion management" measures come at a cost as network operators need to compensate power plant operators for the hours in which they are curtailed.[20] In Germany, the annual cost for these measures regularly lay above 1 billion euros in recent years. In 2020, the cost for congestion management amounted to roughly 1.4 billion euros. The curtailment of renewable energy plants, called "Einspeisemanagement" in Germany, was responsible for more than 50% of this, or c. 760 million euros.[21] More than 6 TWh of electricity from renewable sources had to be curtailed. Moreover, the redispatch of conventional power plants cost c. 440 million euros, which includes so-called countertrading measures, while the cost for network reserves were responsible for the re-

20 Note that there are different types of congestion management measures besides what is called "redispatch", such as "feed-in energy" or "countertrading". Congestion management measures form a large part of the so-called "system services" or "ancillary services". For a more detailed overview of the types of system services, see e.g. https://www.next-kraftwerke.com/knowledge/ancillary-services [11 July 2022].

21 See a report by the German Federal Network Agency (Bundesnetzagentur) and Federal Cartel Office: Bundesnetzagentur, Bundeskartellamt 2022, p. 28.

maining circa 200 million euros. The above costs form part of the network tariffs and, hence, form part of the electricity bills of German households and industry.[22]

Besides the redispatch measures, network operators, in particular transmission system operators, increasingly take over further tasks that effectively make them participants in the electricity wholesale market. For example, network operators often have to procure energy to cover grid losses, making them large actors on the electricity market. This involves both the procurement several days in advance as well as daily on the spot market. Network operators are also often responsible for procuring balancing energy and, sometimes, even for the marketing of energy.[23] All of the above measures are basically service or trading activities, which have traditionally only played a minor role for electricity network operators. Originally, network operators were mainly (and nearly exclusively) responsible for the construction and operation of the physical network. Nowadays, network operators are more than the historical "classic network operators" as they have become market participants on the wholesale market, too.

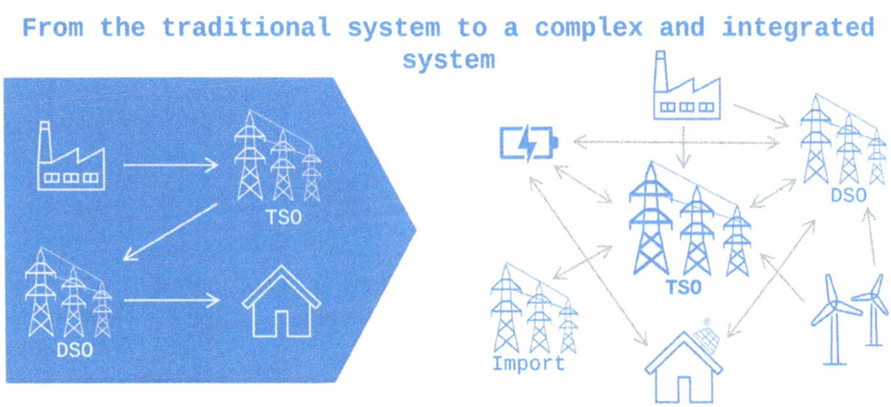

Figure 22.1: Evolution of the Electricity System (Source: Terna S.p.A.).[24]

It is apparent that the energy transition constitutes a challenge for network operators for several reasons. The structure of the electricity system is changing, and its complexity is growing, as depicted in Figure 22. 1. These changes affect all European network operators, although the impact and speed of the evolution might be differing. The structure of the network changes as generation increasingly relocates away from

22 Albeit, as the Bundesnetzagentur notes, part of the 760 million euros is offset by the fact that less money needs to be paid for the renewable energy levy when less renewable electricity is being generated. This levy used to be paid by German electricity consumers, too.

23 In Austria, the electricity transmission system operator Austrian Power Grid is responsible to market excess wind energy since 2015.

24 Own illustration based on Terna S.p.A.: Corporate Strategy 2020–2024.

demand centres towards areas where renewable energy sources are abundant. Moreover, generation evolves towards a system with a growing number of smaller generating units that need to be connected to the transmission network, but increasingly also to the distribution network, for instance in the case of households. Hence, the distribution network increasingly needs to function as a bi-directional network, for which it is currently not entirely fit. Decarbonisation drives the electrification of further sectors such as heating and transport, implying future rises in electricity demand, which the network needs to accommodate. Moreover, flexible components on the demand side, such as storage solutions, gain importance and can also help to stabilise supply. Overall, the changes described increase the complexity of the system and make guaranteeing security of supply more challenging.[25] If the network struggles to keep up with these developments, this will be costly because it will ultimately slow down the energy transition, which is associated with high external costs. Making the electricity network fit for the challenges described comes at a cost, too. From an economic point of view, it is necessary to find cost-efficient solutions, potentially by combining established measures such as physical capacity expansion with new technologies and an optimisation of network usage. Potential solutions to accommodate the energy transition in electricity networks are discussed in the following subsection 22.4.

22.4 Solutions

As described in subsection 22.3 above, the shift towards a carbon-free economy creates a set of challenges for electricity network operators because the electricity system becomes more complex. It is likely that there is not one perfect solution for all these challenges. Short-term redispatch measures need to stabilise the network, but physical network capacity expansion will also be inevitable. A smart combination of these two and further measures could be the most promising and cost-efficient solution from an economic point of view. Below, we depict different measures to accommodate more volatile generation and higher demand. Some of these measures are already widely applied, while others are more innovative and might require further technological progress or regulatory and legislative changes. The possible solutions we describe in this subsection are:

- Network expansion: This is the most established solution. However, costs for network investments are eventually borne by network users. From an economic point of view, it is inefficient to focus solely on physical network capacity expansion. We discuss this solution in subsection "**Network Expansion**".

---

25 The increasing share of renewable energy in the system, as well as the growing risk of extreme weather conditions, make it harder to guarantee security of supply, compared to a system with conventional power plants that are not weather-dependent.

- Design of network tariffs: As physical network expansion is costly, incentivising reductions in peak demand and a smoothing of demand over time could reduce the need for network expansion and, thereby, reduce the costs for network users. A different, dynamic design of network tariffs might be a suitable measure. We discuss dynamic electricity tariffs and potential distributional effects in subsection "**Design of Network Tariffs**".
- Increase of energy efficiency: Besides changing demand patterns, reducing demand or, more realistically, limiting demand growth would also reduce the need for network expansion. We discuss this briefly in subsection "**Increase of Energy Efficiency**".
- Increase of storage capacity: As electricity supply becomes more intermittent and, therefore, more volatile, increasing storage capacity and enhancing sector coupling could be part of the solution to reduce the need for curtailment and accommodate volatile generation patterns. We provide more insights in subsection "**Increase of Storage Capacity**".
- Flexibilisation of demand: To account for increasing volatility of electricity generation, demand can be made more flexible, too. Network tariffs and other market-based instruments, for instance, can incentivise a flexibilisation of demand, while storage and sector coupling can contribute to the flexibilisation as well. We discuss this briefly in subsection "**Flexibilisation of Demand**".
- Electricity market design: different approaches to pricing, such as nodal pricing, could help to guide investments into the network to optimal locations and avoid excessive investments. We discuss this in subsection "**Electricity Market Design**".
- Further technological solutions to expand the existing network capacity are discussed briefly in subsection "**Other Solutions**".

22.4.1 Network Expansion

Increasing the physical capacity of the existing network by adding new lines might be seen as the "natural" solution for the challenges posed by the energy transition. However, physical network expansion is always associated with costs, both direct and indirect. Direct costs are the money that needs to be invested by network operators to expand network capacity, e.g. for the technical components and the construction work. Network users will eventually have to pay for these investments via their tariffs. Besides these costs that will be reflected on consumers' electricity bills, there are also indirect costs caused by physical network expansion because building new lines has an impact on the environment. Therefore, particularly larger transmission lines frequently face resistance from local communities and environmentalist groups. If they are built underground, this may mitigate some resistance as it will not visibly affect the environment; however, the costs for underground lines can exceed the cost for overground lines by a significant margin. In some cases, indirect environmental costs can translate into actual, direct costs. This could be the case, for instance, when complaints by local residents and environmentalist groups lead to delays in the permitting process and delay projects. This, in turn, would likely imply costs for the network operator, but also costs for the wider energy consumers if, for example, a much-needed new connection is delayed and further congestion costs are incurred. However, there are also external costs associated with not expanding the network if the lack of expansion causes the non-achievement of the decarbonisation targets and, hence, higher global warming. The external cost of not achieving the decarbonisation targets is likely to be very high as global warming affects virtually every aspect of the economy and everyday life.

A significant degree of physical capacity expansion will be necessary in most countries to accommodate rising electricity demand and increasingly volatile generation. In Germany, the largest investments relate to large transmission lines connecting the demand centres in the West and South with the centres of renewable energy generation in the North and East. As generation had originally been centred close to demand, existing transmission lines to the new generation centres are insufficient. The German experience with recent large investments into the transmission network showcases the potential conflict between pleasing environmentalist concerns and rising network tariffs: most of Germany's new North-South transmission lines are built as underground lines, which was a political decision to please environmentalists.[26] This has increased costs significantly. For example, the flagship project SuedLink will cost TSO TransnetBW approximately EUR 10 bn, which is more than 10 times the current value of TransnetBW's regulatory asset base.

In the distribution network, investments will be inevitable, too, to accommodate rising levels of household (peak) demand, e.g. resulting from electrical vehicles or electric heating. Existing distribution networks are usually not suited to private individuals charging one or several electric vehicles, in addition to potentially heating their house and water with electricity. Moreover, investment may be necessary to enable a higher quantity of bi-directional flows if more and more households generate electricity themselves and feed it into the network, for which distribution networks had originally not been designed.

Interconnection between European countries is also part of the solution. The national electricity networks in the EU member states are (some more, some less) connected to the networks of their neighbouring countries. The physical cross-border connections are called interconnectors. The EU has a set of rules for its internal electricity market that facilitate the cross-border trade and transmission of electricity. This, in turn, enables the transmission of electricity from regions where it is abundant

26 Gesetz zur Änderung von Bestimmungen des Rechts des Energieleitungsbaus. According to the legislator, underground lines are more expensive, but increase acceptance from environmentalists because underground lines interfere less with the environment.

to regions where it is scarce. One example is France, which has exported a notable proportion of its electricity generation to neighbouring countries in the last years, making use of its large fleet of nuclear power plants. Scarcity of electricity might become a more important issue for some regions in the years to come, especially in winter days with little to no wind, when little renewable electricity is available, while conventional generation capacities are being phased out. Increasing the interconnector capacity mitigates the risk of scarcity problems because it would become easier to transport abundant electricity across borders. For example, if there was a lot of sun on the Iberian Peninsula, excess solar electricity could be transported to France and, from there, further towards Central, Northern and Eastern Europe. Or if countries in the Nordics had abundant electricity from their hydropower plants, it could be transported south towards Central Europe. Of course, this already happens nowadays, but a higher interconnection capacity could help to remove bottlenecks and, thereby, increase security of supply. One such bottleneck, for example, currently exists on the border of Germany and Austria.[27] Increasing interconnection capacity also comes with the benefit of increasing competition at the generation level.

One innovative idea in recent years have been so-called “hybrid offshore projects”. These combine offshore wind generation and interconnectors between several countries. By connecting a new offshore wind farm directly to a new interconnector between two or more countries, an unnecessary doubling of infrastructure (i.e., an onshore connection and a separate interconnector to another country) can be avoided. Thereby, costs for new infrastructure can be reduced, increasing the attractiveness to potential investors. However, as such hybrid offshore projects combine generation and transmission, several legal and regulatory issues can arise. To our knowledge, no such project is yet operational; however, Denmark plans to build large “energy islands” in the North Sea and Baltic Sea, which shall feature large offshore wind generation, and assesses whether these islands should be connected to other countries via interconnectors.[28]

In summary, the energy transition will not be successful without physical network expansion. This will entail an expansion of both the transmission and the distribution network as well as increased interconnection capacities with neighbouring countries. In theory, this could be the one single solution to apply. In the absence of budget and space restrictions and resistance from residents and environmentalists, each country could just build as much network infrastructure as it needs to accommodate changing demand patterns and increasing volumes of electricity flows. However, budget restrictions do exist, for network operators and network users (who eventually pay for the investments) alike. Moreover, building new infrastructure takes time.

27 The so-called “Deutschland-Leitung”, which is in the planning stage at the time of writing, will reduce congestion at this bottleneck.

28 For more information on the current plans for the energy islands, see https://ens.dk/en/our-responsibilities/wind-power/energy-islands/denmarks-energy-islands [11 July 2022].

In fact, approval processes for large infrastructure projects can take several years, such that new infrastructure does not necessarily constitute the best solution for the imminent, short-term challenges.

Hence, it is worthwhile from an economic perspective to consider other solutions for the issues that the energy transition is causing for electricity networks. These solutions might help to limit physical network expansion to the extent that is cost-efficient. In the following subsections, we will explain alternative concepts and discuss their advantages and disadvantages as solutions to support the energy transition in energy networks.

22.4.2 Design of Network Tariffs

Physical grid expansion, as discussed in the previous subsection, is a solution clearly directed at accommodating increasing (peak) demand in the network. Implicitly, this solution takes rising demand as given. An alternative approach is to try to influence demand such that physical network expansion can be limited to lower levels. As discussed further above, what really defines the need for physical network expansion is peak demand, i.e., the maximum flow of electricity at a given time. Hence, reducing peak demand reduces the need for physical capacity expansion of the network. The aim should be to try to find an economically efficient balance between investments into the physical capacity of the network and measures to avoid further investments.

This is easier in theory than it is in practise. Demand in the network consists of the sum of demand of every single household and industrial customer. Network operators, however, do not usually have the authority to determine how much each network user is allowed to demand or at what times they are allowed to consume.[29] Thus, there is a coordination problem. It would be optimal for all network users if every network user would optimise their demand, i.e., shift their demand over time to reduce peak demand. This would reduce network operators' costs and, thereby, reduce network tariffs. But many network users will think that their portion of total demand is very small, such that it will barely make any difference. This situation is similar to that of carbon emissions: One person by itself cannot change much about global carbon emissions by cutting their own emissions. But if everyone thought that way, nothing would change, whereas if every person did its part, the total reduction would be significant.

Economic theory provides a solution to this problem: Incentives can induce behavioural changes. If politics or network operators would incentivise optimal behav-

29 Of course, in times of extreme scarcity, the legislator may enact measures that limit the offtake by certain end users. Luckily, this is not a common situation, albeit the wider public has become aware of it in spring and summer 2022, when natural gas supply in several European countries was considerably limited as a consequence of the Russian invasion of Ukraine and subsequent sanctions.

iour by network users, network users might actually change their behaviour in a way that is beneficial for the network. Optimal behaviour in this situation means optimal behaviour from an overall peak demand and network usage perspective. A good example is that not every network user should charge their electric vehicle directly at 6pm when they get back home from work. If people would charge their cars more sequentially overnight, peak demand would be reduced, reducing the need for capacity expansion. The classic example of incentives are monetary incentives. For instance, high taxes on tobacco are meant to reduce demand for tobacco, while rising carbon prices in the EU shall reduce the amount of carbon emissions and induce a switch to less carbon-intensive means of production. Hence, if network tariffs would reflect the actual cost that network use imposes at a given time, for example by higher prices in times of high demand, this may induce beneficial behaviour by network users.[30]

Nowadays in most European countries, tariffs for the distribution and transmission network are charged based on the amount of energy consumed. That means that end users usually pay per kilowatt-hour, even though the variable cost of electricity flows is very close to zero, as discussed further above. In contrast, network operators' costs are mainly determined by the fixed investments into the capacity, not by the volumes that flow through the network. Hence, normally the users' decision to consume more electricity does not influence network costs considerably. But there is one important exception: in times of very high (peak) demand and congestion on the network, every extra unit of electricity flowing through the network has a comparably high cost. The so-called marginal cost of an extra unit of electricity is high at times of congestion. But if prices are fixed and based on total consumption, not on the timing of consumption, they do not offer a signal to network users to reduce their demand when congestion occurs. The idea of marginal cost pricing, i.e., setting electricity network prices that reflect the additional cost that one additional unit of electricity flowing through the grid causes, might constitute a solution. However, due to its complexity and (to our understanding) low importance in the European context, we do not discuss marginal cost pricing in more detail.[31]

A more innovative solution to react to the fact that both generation (via increasing renewable energy) and demand (via electric vehicles, smart household appliances, etc.) are becoming more volatile, intermittent and, therefore, unpredictable is so-called dynamic pricing for transmission and distribution. New technologies like so-called smart meters can enable consumers to track their patterns of electricity consumption (nearly) in real time and adjust them accordingly. Network operators can track demand, too, such that more dynamic pricing is suitable. Dynamic pricing is not

30 Much of the following discussion about dynamic pricing and cost-reflective tariffs is partly based on a white paper by some of our colleagues at NERA Economic Consulting. See Druce et al. 2020.
31 For more information on marginal cost pricing, see Druce et al. 2020, p. 3–4.

an entirely new concept. For example, a differentiation between the cost of electricity during the day and the cost at night is relatively common in many countries. Prices overnight are often cheaper because demand is lower at night, but certain types of power plants generate electricity (base load) 24 hours a day, such as nuclear power plants. The lower price overnight might induce people to shift some electricity consumption that would otherwise occur during the day overnight. Today's idea of dynamic pricing takes this concept further and makes it more granular. Prices could be adjusted multiple times a day in reaction to the current situation on the network. In times of congestion, prices could be increased to induce consumers to reduce their consumption and, thereby, reduce congestion. Several designs are possible. Prices could be adjusted at fixed times or at flexible times, to fixed levels or to levels determined on the spot, based on the amount of congestion. In contrast to Germany, dynamic pricing is already more established for instance in France, where distribution tariffs can vary based on the consumption profile and time of use. Another example is the Australian state of Victoria, where an early roll-out of smart meters allowed for the introduction of time-of-use tariffs as early as 2013.[32]

Regardless of the potential benefits of dynamic pricing, it may also cause new issues that need to be taken into account. Dynamic, more cost-reflective pricing can imply that different consumers pay different amounts of money for the same quantity of energy flowing through the network, based on consuming their energy at different times. At first sight, this may be seen as fair and economically desirable because it means that those users whose consumption causes more costs in the network (i.e., those who consume in times of high or peak demand) pay a higher proportion of the costs of the network.

But dynamic pricing could also have distributional effects on society. It depends on the degree of users' engagement with dynamic pricing. Some users may be very engaged and happy to follow prices in real time and adjust their consumption accordingly. However, there will also be disengaged users that may not do so and, thereby, may end up paying even more money than before. While for some of them, this might be a conscious decision, others may simply not have the opportunity to monitor prices throughout the day and adjust their consumption. The issue would be particularly severe if the disengaged users were, on average, low-income consumers, while the engaged users would have a high income. In such circumstances, the benefits from dynamic pricing would be claimed by users for whom electricity tariffs are less of an issue regardless. This may rightfully be seen as unfair. And indeed, there have been studies that found out that a significant proportion of end users are disengaged in the energy retail market.[33] Hence, it is not a priori clear that dynamic pricing really pro-

32 For more information and other examples, see Druce et al. 2020, p. 5–7.

33 For example, an investigation by the British Competition and Markets Authority (CMA) in 2016 found that there was a "material percentage of customers who are disengaged in the domestic retail energy markets". See CMA 2016, para 9.60. This example is based on Druce et al. 2020.

vides a sufficient incentive for many users to change their demand profile for electricity.[34] Whether or not disengaged consumers are in danger of actually paying more under dynamic prices than before depends on the specifics of the dynamic tariff design. It may well be that engaged consumers decrease overall network costs by so much that the tariffs for disengaged consumers decrease, too.

Apart from distributional effects, certain other issues should be considered as well when introducing dynamic pricing. Price differentiation means an increase in tariff complexity for end users. Moreover, dynamic pricing can only be successful when end users have the technology to make use of it, i.e., if they have smart meters or similar hardware to follow their consumption in real time. Most importantly, there are also legal or regulatory aspects. In many European countries network operators are legally obliged to ensure sufficient capacities with a certain degree of certainty for reasons of security of supply.[35] Traditionally, network operators have expanded their physical network capacities if they expected future increases in demand, which was generally accepted by regulatory authorities. However, regulatory authorities would need to be convinced that the introduction of dynamic pricing constitutes a suitable alternative to physical network expansion to guarantee security of supply.

22.4.3 Increase of Energy Efficiency

The ideas in the subsection above are directed at changing the pattern of consumption and, thereby, reducing peak demand. Besides, there is another, more classical behavioural solution: a general decrease in demand. More precisely, as electrification makes rises in electricity demand seem inevitable, reductions in demand growth. Energy saving, e.g. through more energy-efficient appliances or behaviour (which could be as simple as switching lights off when you leave a room), is a proven method of demand reduction and can, thus, limit the need for physical network capacity expansion, too. In light of electricity becoming more important for heating in the future, housing insulation could be another promising area to limit demand growth. Again, financial incentives may be necessary to induce energy-saving behaviour, like state subsidies on insulation material or on high energy efficiency household appliances.

Overall, physical network expansion can be successfully limited if the consumption behaviour of end users can be influenced. This holds true for established methods like saving energy to reduce demand (or, more suitably today, to limit demand

34 Of course, this abstracts from extreme cases. If prices would vary by thousands of euros per kilowatt-hour, one would expect nearly every market participant to react to price changes.

35 In Germany, for example, network operators regularly need to create network development plans in which they set out their assumptions about future demand and the measures (i.e., investments into the grid) that they are planning to implement to accommodate changing demand.

growth) as well as for methods to smoothen a given level of demand over the hours of the day or over periods of the year in order to reduce peak demand.

22.4.4 Increase of Storage Capacity

The previous subsections focussed on ways to expand the network or reduce (peak) demand to avoid situations when the network's capacity is insufficient. In this subsection, we focus instead on storage as a solution for situations when the network has sufficient capacity, but supply and demand do not match. In times of increasingly volatile supply from renewable energy, this is likely to happen more frequently. As noted above, however, mismatches of supply and demand can have serious, and expensive, consequences. If demand exceeds supply, some consumers may have to be taken off the network, which can be expensive if industrial processes were interrupted or appliances damaged. If supply exceeds demand, network operators have to curtail certain power plants and compensate them for lost revenues. The costs for compensation are, eventually, paid by network users via their tariffs. Storage capacity can provide redress for such situations, in particular for situations when supply exceeds demand. Hence, storage can help to limit network costs.

Storage itself is not an invention. For more than one-hundred years, reservoirs have been used to generate electricity using the power of water. A specific type of these hydroelectric power plants are so called pump-storage plants. They make use of two reservoirs at different elevations. Using electricity, water is pumped upwards from the lower reservoir to the upper reservoir. Electricity can then be generated by releasing water from the upper reservoir and passing it through one or several turbines back into the lower reservoir. When electricity is abundant, for example when much renewable electricity is generated at a time, this excess electricity can be stored as water in the upper reservoir. When there is excess demand, on the other hand, the water can be released and the plant can generate additional electricity to support network stability and cover demand. Pump-storage plants are an established method of storing electricity and have been used in Switzerland and Italy since the 1890s.[36] However, pump-storage plants, as other types of hydroelectric power plants, face public criticism as their construction and operation have a severe impact on the environment. Moreover, the construction of pump-storage plants is limited to favourable locations, which makes them a suitable solution only at certain locations.[37]

Of course, other ways of storing electricity or making use of excess electricity exist, too. On a small scale, the storage of electricity in batteries is very common. Such

36 See https://www.energy.gov/eere/water/pumped-storage-hydropower [13 July 2022].

37 A certain level of elevation is needed to operate pump-storage plants. They are usually built at locations with natural elevation, such as in the Alps. However, space for such power plants is naturally limited, and so is the acceptance of local residents.

batteries can also be built on a larger scale (i.e., with a larger capacity) and be used to store excess electricity from renewable sources. In times of high demand, these large battery storage plants can supply electricity to the network on very short notice. However, the market for large-scale battery storage is still developing; in Germany, for instance, only a few projects on a scale of multiple megawatts of capacity are already realised.[38] Therefore, large-scale battery storage plants may only become a relevant factor to support the energy transition in the medium- to long-term. But batteries can also support the energy transition on a smaller scale. Small-scale batteries to store excess electricity from solar panels see increasing popularity among households. More and more solar panels on private roofs, combined with small-scale private battery storage, can reduce household electricity demand and, thereby, reduce total and peak demand in the network. This reduces the need for physical network expansion and the cost of electricity provision, too.

Although puzzling at first sight, electric vehicles, too, could help reduce the need for network expansion. In recent years, electric vehicles have seen a significant uptake in industrialised nations such as Germany, which is partly attributable to government subsidies. Now that more and more electric vehicles are privately owned, more and more households want to charge their cars at home, which yields increases in electricity demand. But as each of these cars comes with a battery, these batteries could collectively be used as short-term storage for excess electricity, most useful if they were also able to supply their electricity back to the distribution network, if necessary. Vehicle batteries might offer a particularly good temporary storage as cars are stationary most of the time and only move for a fraction of time each day. The use of vehicle batteries, however, poses several questions of technological, economic and regulatory nature. Vehicle owners would need to be remunerated for offering their car batteries to the wider network, the car batteries would need to be able to quickly discharge their energy into the network, and batteries would tend to wear out faster through the repeated charging and discharging.[39]

One last, and probably the most prominent example for storage in recent years is hydrogen. Hydrogen can be produced from water using electrolysers. If the electricity used to power the electrolysers comes from renewable sources, the resulting hydrogen is called "green hydrogen". This is seen by many as a solution to decarbonise sectors like heavy duty transport or certain industrial processes like steel production, for which full electrification currently seems non-viable. But hydrogen also constitutes a storage solution for electricity. When excess electricity is being generated, hydrogen could be produced and stored. In times when demand is high or supply is low, the hydrogen could then be used as a fuel in power plants to generate electricity. How-

38 Fraunhofer ISE 2022b, p. 16.

39 For a short discussion of using electric vehicles for temporary storage, see for example information by the German Association of Electrical Engineering: VDE 2021.

ever, while this process allows the storage of renewable energy and reduces the need for curtailment, it is criticised for its low efficiency. During the electrolysis as well as the generation in a power plant, part of the energy is lost. Nonetheless, many people see hydrogen as an important factor for the success of the energy transition, and governments around the world have prepared extensive hydrogen strategies and hydrogen support policies.

Hydrogen is a good example for what is summarised under the keywords "sector coupling" or "energy system integration": the interconnection of energy consuming sectors with the electricity generating sector.[40] While the energy consuming sectors, such as transport, heating and industry, can partly be electrified, a full electrification of all processes does not seem to be viable. However, by using excess electricity to produce e.g. synthetic gaseous fuels, other sectors can still be decarbonised. This widens the array of possibilities for the use of excess electricity. Solutions of using excess electricity to generate gases or fuel batteries are also referred to as "power-to-X".[41]

22.4.5 Flexibilisation of Demand

The aforementioned storage solutions and the idea of increasing sector coupling contribute to the flexibilisation of demand. The flexibilisation of demand is an important idea to better accommodate the rising intermittency in electricity supply. Batteries can flexibly store electricity, electric vehicles could flexibly be charged, and renewable gases could flexibly be produced in times of excess electricity generation. Sector coupling creates possibilities for a further flexibilisation of demand. Electricity supply and demand in the network still have to be in balance, but the actual use of the electricity that is being generated can increasingly be decoupled from the timing of electricity generation when sufficient flexibility, e.g. through batteries and storage, is available on the demand side.

Of course, this idea is also interlinked with the design of network tariffs. After all, dynamic pricing or other means of reducing peak demand are precisely meant to influence demand and make use of flexibility potential, not only of households, but also of industrial customers. However, tariff design need not be the only way to activate the industry's existing potential for demand flexibilisation. The creation of new market-based instruments, such as new flexibility and balancing markets, could accompany changes in tariff design and constitute an incentive for further demand flexibilisation. Such instruments would need to entail appropriate compensations for marketing the

40 For a detailed discussion of sector coupling, see for example van Nuffel et al. 2018. See also 3.3.1.

41 The "X" in this case summarises the array of possibilities for which excess electricity can be used. Power-to-gas, for example, means using electricity to generate (renewable) gases like hydrogen.

company's flexibility; otherwise, existing flexibility potential would not be activated. The flexibilisation of demand via market-based instruments can be seen as the evolution of classic markets for flexibility that were primarily focussed on the supply side of the electricity market.[42] Compared to dynamic pricing, new flexibility or balancing markets would likely only be attractive for industrial customers of a certain size and for certain industrial processes that can be flexibly shifted over time.

Technological advancements, such as smart network infrastructure, could be the enabler of a further flexibilisation of demand. Through real-time monitoring of prices, supply and demand, companies and storage suppliers are in a better position to provide their flexibility optimally to the network. Moreover, smart technology also offers the opportunity to automate flexible and network-beneficial behaviour, in particular for specific industrial processes.

22.4.6 Electricity Market Design

While all the above are solutions that are relatively widely known, there are a few more ideas that may not be as popular yet, but could prove valuable in the future, too. These ideas relate to changes to electricity market design. One possibility is to incentivise the strategic positioning of power plants through the market design. As discussed previously, at least in Germany, there is an increasing distance between the centres of generation and the centres of demand. This has plausible reasons. However, it also requires the construction of large new transmission lines, which is very costly. The idea behind incentivising strategic positioning of power plants is that it might be economically efficient to make use of existing transmission infrastructure instead of building at locations where transmission capacity is already insufficient. The idea of so-called nodal pricing could make this happen. A node in the electricity system can be thought of as an intersection in the network, e. g. a point where the transmission or distribution lines intersect with a generating unit, an off-taker such as a household or industrial user, or another line. European electricity markets usually work based on price zones, where multiple of these nodes are summarised in geographical zones. Under the assumption that the electricity generated within each zone can be consumed at every node within the zone, without congestion occurring, electricity prices in the zone are uniform. The idea of nodal pricing is that market clearing prices are defined for every node in the system.[43] The prices at each node would reflect the value of electricity at that particular location, which includes the cost of the

42 A detailed discussion of the potential benefits of the flexibilisation of industry demand and examples of industrial processes that are particularly suited for a flexibilisation can be found in Heffron et al. 2020.

43 For a more detailed description of nodal pricing, see for instance Dietrich et al. 2005, pp. 13–15, or Antonopoulos et al. 2020, pp. 7–13.

electricity as well as the cost of delivering it (losses and congestion). Prices are calculated as the additional cost of providing one additional megawatt of electricity at each location, subject to the system constraints at the respective location, e.g. line capacity limits and maximum generation capacity. Therefore, prices differ between locations where congestion occurs, as compared to locations where capacity is sufficient. In simplified terms, the higher prices at locations where congestion occurs attract further investments into expanding transmission or distribution capacity. The capacity expansion, in turn, reduces congestion and, thereby, reduces prices. This might come at the cost of lower levels of generation (i.e., if wind is less abundant at a certain location compared to other locations) and, therefore, lower revenue from the generation and sale of electricity, but the reduction in necessary network investments may overcompensate for this. However, power plant operators would only have an incentive to optimise their locations from this network cost perspective if they would not be worse off financially. Hence, they need to be either incentivised for choosing the "optimal" location or would need to be penalised for choosing sub-optimal locations, e.g. by not being paid 100% compensation in case of curtailment by the network operator.

22.4.7 Other Solutions

Furthermore, there are some technical solutions to expand the capacity of existing lines. Instead of trying to provide an exhaustive list of technical solutions being discussed, we focus on two examples. One first solution are so called "phase-shifting transformers" (PST). A PST is a type of transformer that allows for a more efficient use of an existing system by mitigating problems with congestion. While not increasing the capacity of lines directly, PSTs can optimise power flows if certain lines are congested, while other, parallel lines are not. By doing so, investments into increasing networks' physical capacities can be postponed or, potentially, avoided.[44] A second, innovative solution is the so-called real-time thermal rating of overhead lines. This procedure makes use of the fact that an overhead line's maximum transmission volume increases as the surrounding temperature falls or the wind speed increases. According to German and Dutch TSO TenneT, overground transmission lines normally assume an outside temperature of 35°C and a wind speed of 0.6 m/s. However, at least in Germany, this temperature is not reached on the majority of days. Hence, if temperatures are lower than 35 C or wind speed is above 0.6 m/s, overground transmission lines could transmit more than their standard capacity. According to TenneT, at −5 °C and a wind speed of 2 m/s, a line could transmit up to 180% of its standard capacity. As of the end of 2018, TenneT uses a dynamic real-time thermal rating method for 40% of its German overhead lines

44 For more information on phase-shifting transformers, including technical details, see https://www.entsoe.eu/Technopedia/techsheets/phase-shifting-transformers [13 July 2022].

by monitoring weather conditions around the lines and adjusting the maximum flows according to this information.[45] Hence, depending on the weather, investments in parts of the network might be avoided or delayed by allowing a higher volume to flow through existing lines.

22.5 Regulation, Financing, Network Tariffs

The need for regulation arises from the fact that energy networks are considered natural monopolies.[46] Without public interference in the market, the network operators would charge excessive prices, detrimental to consumer welfare. Regulation is intended to mitigate this potential market failure. Very broadly, regulation comprises any rules that are meant to change the behaviour of firms and individuals in the private sector.[47] Regulation is frequently applied in the energy and infrastructure sector, but also plays an important role in sectors such as banking or agriculture. It comes in different fashions, too. Market entry can be regulated, as exemplified by banking licenses that businesses need to obtain in many countries to perform banking activities. Prices can be regulated as well, which is sometimes the case for end users' energy prices. Regulation can also prescribe the way that certain telecommunication spectrums are to be auctioned off.

Regulation is usually performed by an independent regulatory agency, the so-called regulator, which is often a dedicated public entity.[48] In Germany, the regulator for the energy sector is the Bundesnetzagentur (Federal Network Agency), which is a federal agency that also regulates the German telecommunications and post sector. Usually, the national governments appoint key personnel and the legislators lay out key regulatory principles in law and ordinances. Politics more generally also predefines overarching principles and targets of energy policy, such as the move towards carbon neutrality, and thereby defines the broad environment that regulators operate in. However, when it comes to more specific rules, such as the approval of costs, regulators usually assume some degree of independence, which is also mandated by European law.

45 For more information, see TenneT 2018.

46 This general statement comes with a few caveats. While electricity and natural gas networks are almost exclusively considered natural monopolies, district heating networks, for instance, are not always considered natural monopolies as it can be argued that district heating is competing with other heating solutions. For example, if prices for district heating would increase too much, some households might switch to heating with wood, coal or electricity. In contrast, there is no real alternative to electricity, which is why electricity network operators would have strong incentives to charge excessive prices in the absence of regulation.

47 For a detailed definition, see https://stats.oecd.org/glossary/detail.asp?ID=3295#:~:text=Regulation%20is%20broadly%20defined%20as,firms%20in%20the%20private%20sector. [18 July 2022].

48 However, this does not need to be the case. In certain smaller countries, for instance, regulation is overseen by ministries, e.g. the ministry for the economy.

This subsection focusses on regulation in the energy sector. In Europe, broad regulatory principles are defined on a European level. More detailed rules are subsequently set on a national level, such that the details of energy sector regulation differ across European countries. An important part of European energy regulation is ownership regulation and so-called unbundling. Many activities along the energy sector value chain, such as generation, transmission, distribution and retail, were historically combined in large vertically integrated utilities. This hindered effective competition, which, in turn, harmed consumers. Nowadays, unbundling requirements mandate that a single entity can only perform one of the activities along the value chain, e.g. only be active in electricity generation or in electricity distribution, but not in both.[49] Activities such as generation or retail are, in theory, competitive markets.[50] On the other hand, distribution and transmission, respectively, are best performed by one respective single company because this is economically efficient, as noted above. Hence, more detailed regulation is necessary to make sure that consumers are not negatively affected by the monopolist's behaviour. For example, many countries regulate access to the energy networks to avoid discrimination of producers that want to connect their plants to the network. Moreover, the costs and tariffs of energy networks are usually regulated. This is one of the most important, if not the most important, aspect of energy network regulation.

Tariff regulation is a key driver for the profitability of network operators. Regulators try to achieve a balance between the needs of consumers and network operators. As consumers prefer tariffs that reflect efficient costs and network operators prefer high tariffs, this implies finding a level of tariffs that is (financially) viable for network operators, but not excessive for consumers. As a guiding principle, many regulators scrutinise network operators' costs and, if they deem them efficient and justified, they allow operators to charge these costs to consumers. In turn, if certain costs are deemed unjustified, network operators cannot include these costs in their tariff calculations. Regulators often set an annual revenue cap that determines the amount of justified costs, which network operators are then allowed to charge their consumers via network tariffs.[51] Abstracting from

49 However, it is permitted to perform multiple of these activities within one group, as long as the activities are performed in separate group entities.

50 Admittedly, energy generation in particular is very capital intensive, which in practice means that a few large companies still dominate most domestic markets. However, the energy transition and its switch towards smaller generating units have enabled smaller companies to stay in the market for electricity generation, too.

51 The literature differentiates between three different approaches to regulation of electricity networks. Regulators can limit the allowed costs, limit the allowed revenues, or limit the allowed prices. If regulators were able to perfectly control network operators, these approaches would in theory all yield the same result: the network operators would only recover those costs that are deemed efficient. However, information asymmetries exist between regulator and network operators. These information asymmetries are one reason for the development of "incentive regulation", which aims to influ-

any more complex elements,[52] the revenue cap is often determined as the sum of justified operating costs and justified capital costs. Operating costs comprise costs such as salaries, material costs and office supplies. Capital costs, on the other hand, comprise the cost of the assets, i.e., mainly the cost of the network. These investments are primarily recovered via depreciation allowances.[53] On top of the depreciation allowance comes a return on the capital invested. This regulatory rate of return, also called regulatory cost of capital, is usually determined by the regulator and is a key driver of the profitability of network operators.[54]

There are a number of principles that are generally accepted in today's European energy regulation:[55]

- Regulation needs to enable network operators to earn their efficient costs, including a fair remuneration for their invested capital. Without a fair remuneration, network operators would be unable to attract the capital needed for their future investments.
- Regulation needs to incentivise efficiency increases. Regulation should incentivise network operators to lower their costs over time, providing a better service to consumers at lower prices and, in the long-term, reducing the financial burden for network users.
- Regulation needs to be transparent, stable and predictable. As energy networks are very capital-intensive and investments are expensive, investors need to be able to rely on the environment they are investing in. Unstable regulation would increase the risk premium that investors demand for investments into network infrastructure, increasing the capital costs and overall costs of energy networks.
- Regulation needs to promote security of supply. Energy networks are essential for guaranteeing security of supply, which regulation should not endanger.

The last of these four principles illustrates an important asymmetric risk: if the regulatory cost of capital is set too low, security of supply and the energy transition more widely might be put at risk because network operators might be unable to cover part of their costs via network tariffs, which lowers the willingness of network operators

ence the network operators' behaviour. For an introduction into network regulation, see for instance Zweifel et al. 2017, chapter 13, pp. 297–314.

52 Such complex elements comprise, for example, so-called efficiency benchmarking and allowances from regulatory accounts. For a discussion of further elements of regulation in Germany, in particular efficiency benchmarking, see e.g. Waidelich et al. 2022.

53 Regulatory depreciation and asset lives can differ from statutory depreciation and the asset lives used by companies for tax and accounting purposes. Regulatory asset lives are usually determined by the regulator and prescribe the depreciation periods for different asset types.

54 For this reason, the rate of return is very relevant for investors and also heavily debated. Depending on the country, legal proceedings are common when it comes to the determination of the allowed regulatory rate of return.

55 For a more detailed description of the criteria for the assessment of regulatory regimes, see for example Bonbright et al. 1988, pp. 382–387 or Phillips 1993, pp. 172–173.

and investors to invest additional capital for long-term investments. This holds true for equity and debt financing alike. If the regulatory cost of capital is too low, necessary investments might not be performed and, thereby, security of supply might be endangered. In contrast, the risks created by a regulatory cost of capital that is set too high are less fundamental. The demand for electricity is relatively inelastic, i.e., it does not change significantly in response to moderate price increases. Moreover, the assessment and confirmation of network development plans by regulators, which is common across European countries, avoids inefficiently high investments by network operators. While the resulting higher end-user prices in case of a high regulatory cost of capital are undesirable, the system stability and energy transition are not put at risk.[56]

Based on recent political developments regarding the energy transition and the move towards a zero-carbon economy, two further regulatory principles could be added to the above list:

- Support of the energy transition and sustainability;
- Support of innovation in the energy network sector. Innovations enable cost reductions and can contribute to reaching dynamic efficiency.

The classical focus of current network regulation is not suited for today's market environment anymore. European energy network regulation was designed for a relatively static network, with a focus on efficiency increases and cost reductions through means such as efficiency benchmarking and the budget principle.[57] Current regulation assumes a relatively stable asset base and a relatively low level of operating costs, as network operators have historically mainly been active in the very capital-intensive construction and operation of the electricity network. Of course, this is still their main business today. However, the structure of the electricity networks is significantly changing, both in terms of new investments to increase capacity as well as through a shift of generation towards more granularity and larger distances between centres of generation and electricity demand. In addition, the tasks of network operators have changed and will change further in the future. For example, new tasks related to system stability, e.g. redispatch measures, and the trade of electricity, e.g. for balancing energy or, in some countries, the marketing of electricity, have been added to the responsibility of network operators. These tasks require less physical capital (fixed assets) than the "classic" network business but require a relatively high level of liquidity (current assets), which have traditionally not played an important role for network operators. Hence, network operators increasingly combine the characteris-

56 For a detailed discussion of the asymmetric risk described, see e.g. Romeijnders and Mulder 2022.

57 The budget principle applies to the portion of network operators' costs that the regulator deems influenceable. Over a regulatory period of several years, the allowed revenues of the network operator are decoupled from its actual costs. This implies that if the regulated company is able to decrease its costs compared to the costs that the regulator allows, it earns additional profits.

tics of “classic” network operators (high capital intensity) with the characteristics of trading or service companies (high degree of operational costs).

Consequentially, the focus of network regulation should now be placed on the facilitation and acceleration of the energy transition. Regulatory instruments from the past should be subject to scrutiny and it needs to be analysed whether and to what extent they should still be applied. For example, the regulatory cost of capital, earned on investments into fixed assets, is still a key driver of the profitability of network operators. Network operators can earn a return on the capital they invest in their classic network operations, while operating costs are often only passed through without a margin. Without the prospect of earning profits on these operations, network operators lack incentives for innovation in the “new” areas of operation.

The structural changes in the energy network sector and technological progress cause changing requirements for regulation. Dynamic efficiency gains importance relative to technical and allocative efficiency, which have been very important when the electricity network sector used to be more static. Dynamic efficiency in this case means that a product or service is always provided using the optimal production technology to use resources optimally over time. Available technology depends on innovation activity and spending on research and development. Static cost minimisation, although still important, should be less of a focus compared to the support of innovative solutions. Below, we summarise a few principles and ideas that could play a larger role in the years to come to better account for the changes that the energy transition brings about.[58]

22.5.1 Output-oriented regulation and new incentives

Firstly, current regulatory frameworks focus on short-term operating cost reductions. This focus has been sensible in the relatively static environment where the structure of the network and the operators’ tasks were relatively stable. However, necessary long-term investments into new technologies might be discouraged when the short-term perspective is most important. The development of new technologies requires spending for research and development. Once applied, new technologies are often less capital-intensive than the conventional physical network expansion and instead come with a higher share of operating costs. Under most current regulatory frameworks, investments into new technologies might be discouraged because operators mainly generate profit from fixed asset investments. If network operators are allowed to earn a return on their invested capital but are only allowed to recover their operating costs without a margin, it would be more attractive for network operators to con-

58 For a more detailed discussion of potential regulatory changes, see a NERA study for the Austrian transmission system operator. NERA 2020.

tinue to focus on capital-intensive physical network expansion (i.e., increase the asset base on which they earn a return) and disregard expenses for research and development.[59] Innovative solutions would yield efficiency gains for the electricity system as a whole but developing them delivers no (financial) benefit for network operators under most current regulatory frameworks. To tackle this issue, regulators need to create (financial) incentives for network operators to develop and apply new, efficient technologies, without predefining what technologies to choose. This would constitute a switch towards focussing on long-term, dynamic efficiency and a more output-oriented, rather than cost-oriented regulation. Regulatory success should be measured more in terms of reaching certain milestones, e.g. delivering better services, rather than in terms of reaching certain cost reductions.

The British regulator Ofgem and the Spanish regulator CNMC have already implemented certain incentives for the services business of network operators. They regulate the classical network business and the system services separately, focussing the incentives for the latter area more on long-term system benefits instead of short-term cost reductions. In Spain, CNMC sets incentive payments and penalties. Incentive payments are paid out for three categories of targets: cost reductions in congestion management, an improvement of demand forecasts (which reduces the need for redispatch measures), and improvements of supply forecasts for wind and solar energy. The incentive payments will amount to up to 5% of the basic revenue cap for system services in the medium-term.[60]

Moreover, allowing network operators to earn a margin on their service operations is another potential solution to encourage network operators to invest in novel, less capital-intensive solutions. Such service margins are already applied in regulatory frameworks for sectors with a high share of operating costs, such as electricity retail. In contrast, service margins are rare in energy network regulation. One of the few examples where network operators can earn a margin on their system service business is Spain, where the regulator CNMC allows a mark-up of 5% on the operating costs for "system control" since 2020. CNMC explicitly argues that the regular regulated cost of capital does not reflect the network operators' risks in low capital-intensity operations.[61]

59 In the academic literature, the problem of a bias towards capital-intensive solutions is called "gold plating". It is particularly present if the regulated rate of return is higher than the company's opportunity costs (the so-called Averch-Johnson effect). See Averch, Johnson 1962.

60 For a description of the regulatory framework, see Comisión Nacional de los Mercados y la Competencia 2019, p. 132182.

61 Comisión Nacional de los Mercados y la Competencia 2019, p. 132182.

22.5.2 Importance of the regulatory Cost of Capital

A second consideration is related to the cost of capital. It can be used to actively incentivise investments. In Italy, for example, investments that the regulator considers strategically important receive a higher regulatory cost of capital than other projects, i.e., they are more attractive for network operators to pursue.[62] By increasing the return to be earned on important projects, regulators can try to incentivise the prioritised completion of projects with the highest social impact. In future years, when (operational or financial) bottlenecks are likely to persist, simply because many projects are developed simultaneously, prioritisation might be inevitable. In Belgium, Italy and the UK, early completion of projects or cost reductions compared to initial plans are financially rewarded as well, incentivising early and cost-efficient completion.

22.5.3 Remuneration of current assets

Thirdly, regulation does currently often not allow network operators to earn a return on all their invested capital. While for fixed assets, there is agreement that a return needs to be earned, the case is different with current assets. In Austria, for instance, current assets do not form part of the regulated asset base and are, therefore, not remunerated. However, the new system tasks of network operators necessitate a higher liquidity, e.g. for buying and selling balancing energy on the market. Regulation does not reflect this. Regulators should consider enabling network operators to earn a return on current assets. Otherwise, network operators might be discouraged from performing their service operations in the most efficient way due to essentially being penalised for holding sufficient cash.

More generally, regulation will need to keep up with the developments in the market that is being regulated. The evolving role of network operators needs to be reflected in network regulation. If network operators need to take on new tasks, regulation needs to make sure that it is financially viable for operators to perform these tasks; otherwise, the tasks will not be performed in an efficient or sufficient manner. Even better, if regulators actively incentivise beneficial improvements in the way that network operators perform their new tasks, this will likely improve the electricity system overall and yield better outcomes for network users. Changing the regulatory framework in line with developments in the market does not contradict the requirement of regulatory stability. Stability should not be interpreted as stability of regulatory rules, but rather as stability and predictability of regulatory outcomes. An important outcome for regulated companies is a stable and reliable return, which can only be maintained if new requirements and the expanding set of network operators' tasks

62 ARERA 2019, Art. 5.3.

are reflected in regulation, such that there is a fair chance that network operators can generate a fair remuneration on their new operations, too.

In light of high investment needs to make the networks fit for the challenges of the energy transition, financing becomes an increasingly important issue. The revenues from the network business will likely not suffice to finance the expansion and reinforcement of the network as well as the necessary investments into new technologies in due time. Therefore, network operators are looking at alternative, additional sources of financing. Increasingly, they finance investments via the capital market, for instance by placing bonds on the market. On the bond market, network operators compete with companies from other sectors for investors' money. If the network operators' financial situation is challenging, this will worsen the conditions for debt financing, e.g. via an increased interest rate that investors will demand to reflect the perceived higher risk. The same holds true for equity financing; investors will only invest in network operators if the expected return on their investment is attractive. Otherwise, they will invest their money in other sectors that promise higher returns. However, the returns of network operators are regulated in most countries, and in recent years regulatory cost of capital has exhibited a negative tendency in most European countries. The decreasing rates of return are partly attributable to a decline in interest rates but also to a focus of regulators on short-term cost reductions. Network operators have long raised their concerns around the low levels of the regulatory cost of capital, indicating that these would endanger their ability to attract funding for the necessary investments and, thereby, discourage investments, particularly large, long-term investments. As described above, the risk in setting the regulatory cost of capital is asymmetric: a return level that is too low has more severe consequences than a return level that is set too high. Since 2022, market interest rates have started to increase again, a factor that should lead to rising regulatory cost of capital in the future. This certainly is in the interest of network operators and will facilitate that network operators can attract sufficient levels of capital to finance their investments.[63]

Lastly, the level of network tariffs might become a more important issue in the future. Usually, network operators can pass their costs on to network users via network tariffs, provided that the regulator deems these costs efficient. This implies that network tariffs will rise if investments increase, which they are projected to do. Record investments will, subsequently, be followed by increasing network tariffs, which have to be borne by network users. The issue is that higher network tariffs might not be viable for all groups of users. Low-income households might struggle to handle increasing costs for electricity transmission and distribution, which would add further to the drastic increases in energy retail prices that have taken place throughout 2021 and 2022 due to increasing wholesale prices. If network tariffs would increase signifi-

63 Discussions around the level of the regulated rate of return are not the focus of this section. For a more detailed discussion, see e.g. the relevant chapters in NERA 2020.

cantly, too, energy prices might become even more of a political issue. Therefore, it is possible that direct transfers from the government to network operators could be discussed in the future to limit the imminent rise in network tariffs to an acceptable level.

22.6 Summary

Energy networks, in particular electricity networks, have a central role for a successful energy transition and the move towards a zero-carbon economy. However, the changing energy environment also entails several challenges for electricity network operators. Operating the network becomes more and more complex in light of increasing volumes of renewable electricity, with network operators taking over new tasks.

The shift towards more renewable generation brings about multiple changes. Electricity generation becomes more granular and more volatile, decreasing the predictability of generation patterns and increasing the challenge of guaranteeing security of supply. Generating units are more often connected to the distribution network, increasing bi-directional flows to and from households and to and from the transmission network, for which the distribution network was originally not intended. Moreover, the shift in generation from demand centres towards regions with abundant renewable energy sources requires the construction of large new transmission lines, resulting in record investment needs. On top of that, the decarbonisation brings about further increases in electricity demand through the electrification of sectors such as transport and heating. All this adds to the future investment needs.

All of the above has already expanded the set of tasks that network operators need to perform. Historically, they were mainly responsible for the construction and operation of the physical network. Nowadays, they are increasingly taking over service tasks, such as performing redispatch measures to stabilise the system. They have also become active participants in the wholesale market, e.g. through having to secure balancing energy on short-term and long-term markets and, in some jurisdictions, even through marketing renewable energy themselves.

The rising complexity of the energy system as well as growing demand for electricity will likely require a combination of different solutions. Short-term measures to stabilise the network, such as redispatch and curtailment, will be inevitable to some degree. Moreover, a certain level of investments to expand the physical network capacity will be necessary, too. Part of these investments will be investments into further interconnection capacities to neighbouring countries. However, investments into the physical network should be limited to economically efficient levels. They can be supplemented by investments into "smart networks" and the development of dynamic pricing approaches. By influencing the demand patterns of end users and limiting peak demand, these measures could help to limit the physical capacity expansion. Storage solutions and flexible capacities, such as electric vehicles or power-to-X, could help to reduce the intermittency of supply and reduce curtailment by utilising renewable electricity in times of excess generation. Technological progress might bring about further solutions that expand the set of possibilities.

Because of their characterisation as natural monopolies, energy networks are subject to regulation. Historically, European energy regulation has been relatively stable, as it was designed for a relatively static network. Today, regulators still focus on cost reductions and efficiency increases. However, the larger variety in network operation tasks and the changing structure of the overall energy system necessitate that regulation evolves, in particular for electricity networks. The new service tasks that electricity network operators perform are less capital-intensive than the classical tasks but require more current assets and liquidity. Unless regulation reflects this properly, network operators may be disincentivised to perform sufficient investments. Moreover, their regulated returns may be insufficient to attract enough capital from investors to meet record-level investment demands. A general shift towards a more output-oriented, rather than cost-oriented, regulation might be a promising approach. A few European countries, such as the United Kingdom, are already developing their regulation in this regard by providing incentives for network operators to fulfil tasks that are deemed important and beneficial for network users.

In light of record investment needs in many European countries, the issue of financing the investments and innovation activities gains importance for network operators. To be able to secure funding, both from debt and equity investors, their business needs to offer attractive returns because they compete for funding with other sectors. Hence, the regulatory cost of capital, which largely determines the profitability of the network business, needs to be set at appropriate levels that reflect the return expectations of investors, but are also acceptable for network users. Under current regulation, network users eventually end up paying for network investments because network operators pass on their costs to them via network tariffs. High investments, therefore, will eventually translate into rising network tariffs. If network tariffs increase to levels that are non-viable for certain groups of network users, energy prices might become more of a political issue. In that case, other solutions might be discussed, such as transfers from the government to network operators to limit the rise in network tariffs to acceptable levels.

Bibliography

50Hertz, Amprion, TenneT, TransnetBW: Netzentwicklungsplan Strom 2035, Version 2021. Zweiter Entwurf der Übertragungsnetzbetreiber, Berlin / Dortmund / Bayreuth / Stuttgart 2021

50Hertz, Amprion, TenneT, TransnetBW: Szenariorahmen zum Netzentwicklungsplan Strom 2037 mit Ausblick 2045, Version 2023. Entwurf der Übertragungsnetzbetreiber, Berlin / Dortmund /Bayreuth / Stuttgart 2022

Antonopoulos, Georgios; Vitiello, Silvia; Fulli, Gianluca; and Masera, Marcelo: Nodal pricing in the European Internal Electricity Market. JRC Technical Reports, Luxembourg 2020

ARERA: Decision 118/2019/R/gas, Attachment A: Tariff methodology for regulatory period 2020–2023, Rome 2019

Austrian Power Grid, Vorarlberger Übertragungsnetz GmbH: Netzentwicklungsplan 2021. Einreichversion. Planungsstand August 2021, Vienna 2021

Averch, Harvey; and Johnson, Leland L: Behavior of the Firm under Regulatory Constraint, American Economic Review, Vol. 52, No. 5, 1962, pp. 1053-1062

Bonbright, James C; Danielsen, Albert L; and Kamerschen, David R: Principles of Public Utility Rates, Arlington, Virginia 1988

Bundesnetzagentur, Bundeskartellamt: Monitoringbericht 2021. Monitoringbericht gemäß § 63 Abs. 3 i.V.m. § 35 EnWG und § 48 Abs. 3 i.V.m. § 53 Abs. 3 GWB, Bonn 2022

Comisión Nacional de los Mercados y la Competencia: Circular 4/2019, de 27 noviembre, de la Comisión Nacional de los Mercados y la Competencia, por la que se establece la metodologia de retribución del operador del sistema eléctrico, Boletin Oficial del Estado, No. 290, Sec. 1, pp. 132179-132188, Madrid 2019.

Competition and Markets Authority: Energy Market Investigation. Final Report, London 2016

Dena: dena-Leitstudie integrierte Energiewende. Impulse für die Gestaltung des Energiesystems bis 2050, Berlin 2018

Dena: dena-Leitstudie Aufbruch Klimaneutralität. Eine gesamtgesellschaftliche Aufgabe, Berlin 2021

Dietrich, Kristin; Hennemeier, Uwe; Hetzel, Sebastian; Jeske, Till; Leuthold, Florian; Rumiantseva, Ina; Rummel, Holger; Sommer, Swen; Sternberg, Christer; and Vith, Christian: Nodal Pricing in the German Electricity Sector – A Welfare Economics Analysis, with Particular Reference to Implementing Offshore Wind Capacities, Trends in German and European Electricity Working Papers WP-GE-08, Dresden 2005

Druce, Richard; Anstey, George; Olive, Laura; Taylor, Will; and Dickx, Leen: The Role of Cost-Reflective Distribution Tariffs and Flexibility Contracts in Supporting the Energy Transition, London 2020

Fraunhofer ISE: Aktuelle Fakten zur Photovoltaik in Deutschland, Freiburg 2022a

Fraunhofer ISE: Batteriespeicher an ehemaligen Kraftwerksstandorten. BAT4CPP, Freiburg 2022b

Heffron, Raphael; Körner, Marc-Fabian; Wagner, Jonathan; Weibelzahl, Martin; and Fridgen, Gilbert: Industrial demand-side flexibility: A key element of a just energy transition and industrial development, Applied Energy, Vol. 269, 2020, 115026Joskow, Paul L.: Competitive Electricity Markets and Investment in New Generating Capacity, MIT CEEPR Working Paper 06–009 WP, Cambridge, Massachusetts 2006

NERA: Effektive Anreize für neue Herausforderungen. Ansätze zur Regulierung der Austrian Power Grid AG, Berlin 2020

Neuß-Grevenbroicher Zeitung: Seit 50 Jahren steht Neurath unter Strom, Neurath 2020

Oxford Institute for Energy Studies: The Future of Gas Networks. Key Issues for Debate, Oxford 2019

Phillips, Charles F: *The Regulation of Public Utilities: Theory and Practice*, Arlington, Virginia 1993

Romeijnders, Ward; and Mulder, Machiel: Optimal WACC in tariff regulation under uncertainty, *Journal of Regulatory Economics*, Vol. 61, No. 2, 2022, pp. 89–107

TenneT: Freileitungsmonitoring. Witterungsabhängiger Freileitungsbetrieb, Bayreuth 2018 van Nuffel, Luc; Gorenstein Dedecca, Joao; Smit, Tycho; and Rademaekers, Koen: Sector coupling: how can it be enhanced in the EU to foster grid stability and decarbonise?, Brussels 2018

VDE: VDE Fact Check. Bidirectional Energy Flows, Frankfurt 2021

Waidelich, Paul; Haug, Tomas; and Wieshammer, Lorenz: German efficiency gone wrong: Unintended incentives arising from the gas TSOs' benchmarking, *Energy Policy*, Vol. 160, 2022, 112595

Zweifel, Peter; Praktiknjo, Aaron; and Erdmann, Georg: *Energy Economics. Theory and Applications*, Berlin 2017

Torsten Cziesla, Alexander Stuckenholz, Jens Thorn

23 Digitalization and Climate Change – Business Models and Services

23.1 Introduction

The climate protection is one of the essential tasks of the society. The transformation of the energy industry towards climate-friendly solutions supports climate protection. An essential element is the use of renewable energies such as wind power or photovoltaics in electricity generation. However, electricity generation through the use of renewable energies is subject to fluctuations, as solar radiation and wind are uneven. This results particularly in the challenge how to deal with the fluctuations in electricity generation, as the storage capacity of electricity is low. In addition, the increased use of renewable energies leads to a more decentralized energy supply compared to centralized conventional large-scale power plants. Therefore, without the widespread use of information technology in the energy sectors, it is not possible to reconcile fluctuating electricity generation with the corresponding electricity demand. In other words, no energy transformation to climate-friendly solutions without digitalization.

This chapter starts with a brief explanation of the individual components of the energy sectors and the traditional value chain. The following is an elaboration of the transformation of the energy system required, particularly to integrate renewable energies to a greater extent into the energy sectors. The energy transition can be broken down into three dimensions of change: Defossilization, decentralization, and digitalization. Concepts are required that can be used to compensate for the increasingly frequent phases of temporary electricity surpluses and shortages. These are the flexibility options. The widespread use of information technology supports the perception of flexibility options and leads to the concept of an intelligent Smart Grid.

However, the transformation and the implementation of digitalization in the energy sectors is a rather slow process. The main reasons are the long-term nature of investments in the energy infrastructure such as power plants or networks and the comprehensive regulation of the sectors. The transformation process is expected to support by the Smart Grid Architecture Model (SGAM) as a reference model. In this reference model the technologies being currently used are presented based on the levels of the SGAM. The final step of a transformation process is the formulating of appropriate business models. The individual technical ideas for the implementation of climate-friendly solutions are only possible if economically viable business models for companies can be derived from them. Therefore, possible business models to support climate-friendly solutions are presented with a focus on business models derived by the digitalization. In total, the implementation of different business models is the deci-

https://doi.org/10.1515/9783110784299-023

sive piece of the puzzle so that digitalization can ultimately contribute to the success of the energy transition through the increased use of renewable energies.

23.2 The Energy Sectors and the Value-added Chain

From the physical point of view, energy can be described as the ability of a system to produce an effect. As an example, we can observe a force moving a technical item from one point to another. The here corresponding form of energy is called technical work (W_{t12}), which can be expressed by the physical context of the product "force (F) times distance (s)":[1]

Equation 21: $$W_{t12} = \int_1^2 \vec{F}\, d\vec{s}$$

Furthermore, each energy conversion process is governed by the principles of thermodynamics like the first law, describing the so-called energy conversation, and the second law, treating the imperfection of energy usage.[2] Thus, the total thermodynamic energy of a physical system can neither be created nor destroyed. It is only possible to convert forms of energy into each other. According to the second law of thermodynamics technical processes are defined by so called energy loss. That means, that the amount of the energy form, which is gained by the conversion process, is always smaller than the amount of the required energy form. The ratio of produced and converted energy amount can be interpreted as the technical efficiency, what is an important parameter for evaluating the quality of the conversation process.

From the economic point of view, the meaning of energy can be described by differentiation between primary, final (produced) and useful energies. Primary energies represent the forms of energy, which are originally occur in nature.[3] They exist without any technical conversation process. An overview of the primary energies is shown by Table 23.1.

The final energies are produced by converting primary energies via defined technical processes (e.g. power plants, refineries) for satisfying the needs of energy in the economical consumer areas such as households, commerce/ trade/services, industry and transportation.[4] The different applications of energy usage on the consumption level are based on the useful energies, which are converted by final energies.[5] That means, electricity, gasoline, prepared (e.g. dried) natural gas or steam (in a pipeline for process

1 Herwig, Heinz; Kautz, Christian; Moschallski, Andreas 2016, p. 44.
2 Herwig, Heinz; Kautz, Christian; Moschallski, Andreas 2016, p. 9.
3 Umweltbundesamt 2023a.
4 Umweltbundesamt 2023b.
5 Umweltbundesamt 2023c.

Table 23.1: Overview of Primary Energies (compiled by authors).

Renewable	Non-renewable
Solar Power	Coal
Hydro Power	Oil
Wind Power	Natural Gas
Biogenic Energy Sources	
Environmental Heat	Uranium (fission)
Geothermal Heat	Hydrogen (fusion)
Tidal Power	

heat supply) are examples of final energies, which can be transformed into useful energies e.g. light (through lamps), kinetic energy (through engines), room heat (through gas heating) or process heat (through heat exchangers). So final energies are forms of energy, which are requested for in energy markets to supply (useful) energy demand.

Thus, final energies, which are at first produced by converting primary energies, are necessary in order to ensure life, work, nutrition etc. This sequence of energy conversion processes can be described by the energy flow chain. The latter one represents the basis for the development of the value-added chain, which characterizes the present energy market structure (see Figure 23.1).

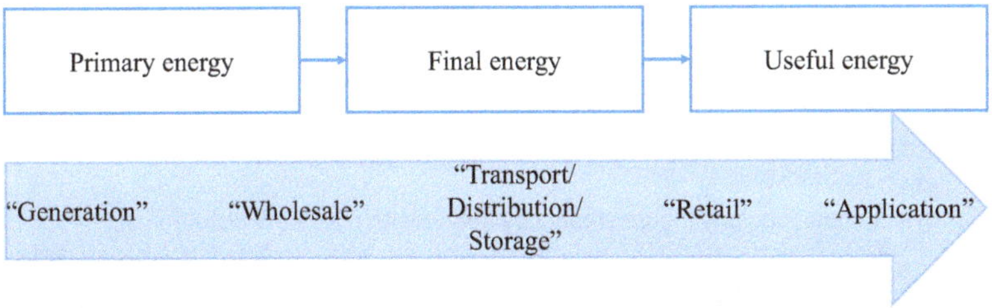

Figure 23.1: Value-added stages and energy flow chain; figure created by author.

Due to this the following value-added stages have been established in the energy sector:

- Generation
  - Refinement of primary energy into final energy
- Wholesale
  - Procurement (for sales divisions)
  - Marketing (for generating assets)
- Transport/distribution/storage and metrology
  - Spatial balancing of imbalances between energy supply and demand based on grid-bound energy transport (in the natural gas industry also balancing of temporal imbalances through storage usage)
  - Organization of the recording of energy consumption data

- Retail
  - Sales of commodities and service for end customers (conclusion of contract and securing energy supply)
  - Offering energy services (application advice, implementation energy efficiency measures)

The refinement of primary into final energies takes place on the generation stage. To that technical processes are operated, which are adjusted to the properties of the used primary energies. So solar energy is converted into electricity by photovoltaic modules, whereas crude oil is converted into fuels like gasoline or diesel by refineries and raw natural gas is converted into marketable natural gas by drying and cleaning processes.

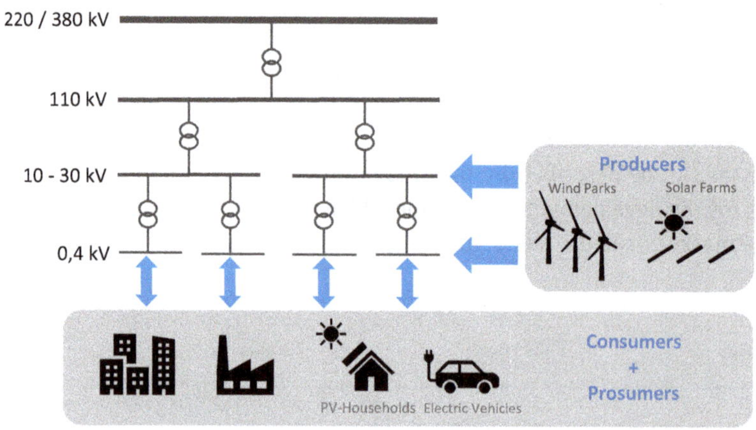

Figure 23.2: Power Grid System Topology (figure created by author with use of MS Pictograms).

In the grid-bound energy market the balance of supply and demand for the final energies such as electricity and natural gas is ensured by the trading activities on the wholesale market level. Energy quantities, which can be physically produced by the generation divisions, are offered by the trading divisions, which are responsible for the marketing of generation assets (e.g. power plants, wind parks). Vice versa there are traders interested in such offers in order to fulfill their energy demand. This corresponds with the energy quantities, which sales divisions have to guarantee for their (end) customers on retail level in case of successful contract conclusions. The trading deals on whole sale market level can be operated by energy exchanges or over-the-counter. For ensuring network stability it is important, that the trading divisions ensure the balance between their purchased and sold energy quantities at each point of time. Thus, traders are responsible to make sure an even balancing group concerning their trading activities. This requirement helps the grid operators to make sure, that there is a stable state for the whole transmission and distribution network, what means to keep the feed-ins an off-takes of electrical energy continuously in balance. The infrastructure, which is used for bridging long distances, is called transmission level. Because of energy efficiency here current flow takes place at higher voltages. In contrast final supply of the end customers is organized via low voltage levels, named distribution network. In context with the energy transition more and more quantities of electricity produced by renewables like solar and wind power have to be integrated into the supply system, i.e. electrical grid. Such a development of power generation is primary going to be fed into the lower voltage levels (see Figure 23.2). At the same time the number of exit points for supply of new end customers increase as well. This is due to the ramp-ups of heat pumps in the heat market on the one hand of charging stations for electric vehicles in the mobility sector on the other hand. In addition, new behaviour of former only customers appears, what is called prosumption and means temporary consumption and production of electricity (e.g. households with photovoltaics).

The energy consumers complete the discussed value-added chain in the energy sector. Regarding this the following economic categories for energy consumption can be differentiated:

- households
- commerce/trade/services
- industry
- transport

For example, the final energy consumption in Germany in 2022 was 2,368 TWh.[6] Here the transport sector had the highest consumption (698 TWh), followed by households (678 TWh), whose demand is dominated by heating, and just behind industry (667 TWh). The final energy consumption of companies in the sector for commerce/trade/services was significantly lower (325 TWh).

It is remarkable that energy demand in the individual consumption sectors can be very often satisfied via using different conversion processes. That means, that for the same application there is sometimes the implementation of more than one technology, which converts final in useful energy, possible. For instance, motor drive in the transport sector can be provided by combustion of fossil fuels like gasoline or diesel, causing greenhouse gas emissions, but also by using ecological electricity in combination with electric vehicles. Another example is given by the heat market, where electricity driven heat pumps represent promising alternatives for the usual application of gas- and oil-fired boilers. From the economic point of view, that opens up the perspective to establish a sustainable system of supply in the energy industry, when fossil energies are substituted through renewable energy forms.

Such a development corresponds with the goals of the energy transition, for which it is very important to enable the transfer of ecological electricity into other sectors like transportation and heat market, called sector coupling. As Figure 23.3

6 Umweltbundesamt 2024d.

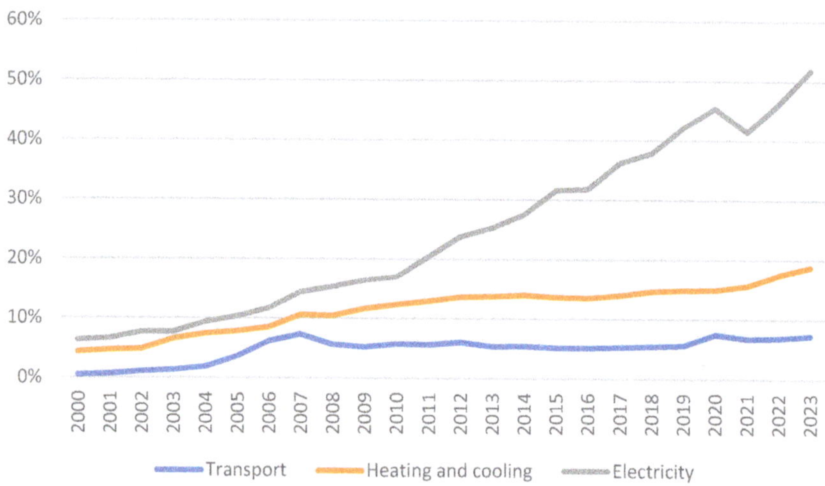

Figure 23.3: Shares of Renewable Energy in the Primary Energy Demand in different Sectors in Germany; figure created by author.[7]

shows for Germany in 2023, the ratio of renewable energy in the electricity sector (51.8%) was significantly higher than in the heating and cooling (18.8%) and transportation (7.3%) sectors.[8] The low ratios of renewables for heating (including cooling) and transportation are caused by the mentioned fossil fuels driven technologies, which are still dominating the energy applications in these sectors (natural gas for heating, gasoline and diesel for transportation). This comparison also demonstrates, that the transition of the electricity sector is relatively well-advanced. In contrast, it seems very ambitious to achieve the climate protection goals in the heating sector and, especially, in the transportation sector to the same extent. So cross-sectoral usage of electricity, e. g. produced by solar and wind power, is an important factor regarding the defossilization of technical processes in heating and transportation sectors. Today such an endeavor is reflected by the market developments concerning electric vehicles and heat pumps. For example, the latter are meanwhile the preferred heating technology for new buildings in Germany.[9] Simultaneously this means an enhancement of the demand for electricity requiring enough solar and wind power generation capacities.

7 Data base see Umweltbundesamt 2024e.

8 Umweltbundesamt 2024e.

9 Bundesverband Wärmepumpe e.V. 2020.

23.2.1 Energy Transition and System Transformation

Energy transition is generally interpreted as a complex change process targeting a sustainable energy supply system based on the usage of renewables. In this context the regulatory policy specifies the following areas of activities:

- Reduction of greenhouse gas emissions on 0% by 2050[10]
- Achieving 32% for renewable energy sources in the EU's energy mix by 2030[11]
- Additional reduction of energy consumption of 11.7% by 2030 compared to the 2020 reference scenario projections[12]

Against this background energy transition represents a game changer for the energy industry, similar to the meaning of digitalization as a further very important factor for change processes concerning all areas of today´s work and life. Because the generation of sufficient quantities of ecological electricity means a success factor, German policy addresses the enhancement of capacities for solar and wind power plants. Figure 23.4 exemplary shows the present (from 2023) structure of power generation by renewables in Germany.

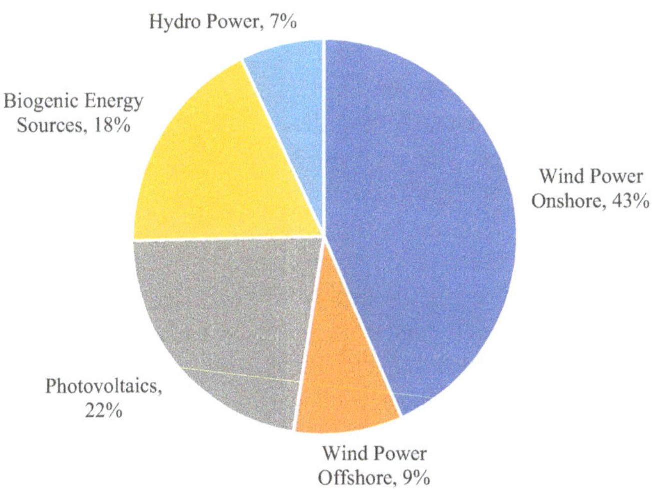

Figure 23.4: Shares (values rounded) of Renewables for Power Generation in Germany, 2023; figure created by author.[13]

10 Umweltbundesamt 2023.
11 European Commission 2018.
12 European Commission 2023.
13 Data base see Umweltbundesamt 2024e.

Due to substitution of non-renewable energies through mainly fluctuating sources like solar and wind power the energy transition leads to increasing amounts of volatile electricity generation. This corresponds with more and more fluctuating electricity feeds into the network. Furthermore, in comparison with fossil energy sources the renewables contain lower "energy densities", i.e. lower specific amounts of energy. Against this background suitable measures for the grid development like adaption and expansion become necessary, expressed by decentralization of the whole supply system.

Such an infrastructural transformation causes additional expense, which the grid operators have to handle for ensuring stability of the network. One the one hand the number of also small power generators increases rapidly. This especially affects the level of distribution grids because of the appearance of electricity producers such as private households with photovoltaic systems.[14] On the other hand, the number of exit points for new consumers like electric vehicles and heat pumps grow as well. Furthermore, the emerging prosumers will increase complexity. For example, to this category belong households generating electricity at high noon by photovoltaics and consuming it during the night. Charging stations can also represent prosumers, when not only used for charging electric vehicles but also for discharging their batteries (as practiced by the vehicle -to-grid-concept). So, the whole system of assets feeding electricity into the network as power generators or rather feeding electricity out of it as prosumers become more and more difficult to control. So far the transmission grid operators have been responsible for balancing the incoming and outcoming electricity flows, what is necessary to ensure grid stability. Nevertheless, an increasing effort for network control can be expected especially what the lower voltage levels is concerned. Because system decentralization mainly affects the distribution level, as mentioned above. And according to the developments of increasing volatile production and consumption, technical solutions based on sector coupling, which enhance system flexibility for consuming electricity in case of its surplus and reducing the demand in case of its shortage, become more and more important.

As a consequence of the decentralization of the electrical network the digitalization of the whole energy supply system become necessary. Because only the sufficient digital connection of almost all infrastructure components, which are used for production, transportation, distribution, storage and consumption of electricity (including the energy forms, which are converted through electricity, e. g. warm water by heat pumps), enables the needed transparency of grid data for creating suitable network control decisions. Consequently, the energy transition can be described by the sequence of the following transformation paths:

- Defossilization
- Decentralization
- Digitalization

14 PricewaterhouseCoopers, 2017, p. 365.

Digitalization of the energy supply system contributes the establishment of so-called smart grids.[15] That means the digital connection of different components of the energy supply system including their smart control like smart generation by virtual power plants, smart prosumption by smart homes or charging stations as well as smart metering, see Figure 23.5. At the end of this system transformation it could be possible, that smart grids melt with the appearance of the so-called internet of things (IoT).

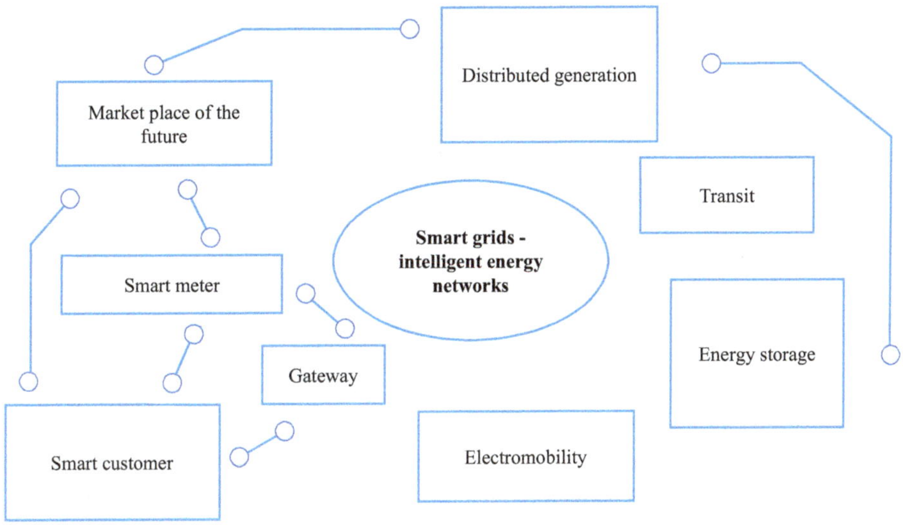

Figure 23.5: Smart Grid as an intelligent Energy Network (figure created by author following BDEW 2017 and also IEA 2017, p. 85).[16]

However, significant progress in digital connection of technical assets for production and consumption of electricity is evident for keeping energy transition on track. Because smart control of infrastructure components of the energy supply system is not only necessary for balancing the electricity flows as fundament for ensuring security of supply in future. But it also enables the efficient and cross-sectoral integration of energy forms based on renewables even in energy applications alongside the electricity sector. Thus, digitalization promotes sector coupling, what, in turn, offers promising solutions for the Defossilization of the heat and transportation sectors. Equally further technical development and market integration of cross-sectoral technologies support the provision of flexibilities for the electricity market.[17] For sector coupling the following concepts can be classified:

15 International Energy Agency 2017, p. 86.

16 Bundesverband der Energie- und Wasserwirtschaft e.V. (BDEW) 2017, also International Energy Agency (IEA) 2017, p. 85.

17 See also Ramsebner, section 3.3.

- Power to Heat (P2H)
  - heat pumps
  - electrode boilers
  - night storage heaters
- Power to Mobility (P2M)
  - electric vehicle charging
  - discharging via vehicle-to-grid technology
- Power to Gas/Fuels/Chemicals (P2G/P2F/P2C)
  - electrolytic generation of hydrogen or rather from hydrogen refined hydrocarbons (in gas or liquid phases and for energetic or material usage)

The assessment of the conversion losses for individual sector coupling categories demonstrates, that concepts for power-to-heat give the highest efficiencies. For example, heat pumps are able to supply more than three times more heat than they consume electricity to drive the compressor.[18] And even resistance heatings (e.g. electrode boilers) allow almost lossless conversion of electricity into heat.[19] In contrast the transfer from electricity into the transportation sector (via power-to-mobility for charging electric vehicles) gives a conversion loss of about 20%, i.e. an efficiency of still 80%.[20] The lowest efficiencies are given by conversion of electricity into synthetic gases or fuels via electrolysis (like hydrogen or hydrocarbons refined by hydrogen), e.g. 74%.[21] In case of further conversions, e.g. for generating so-called efuels by power-to-fuels or for applications concerning power-to chemicals, the energy losses are still higher.[22]

Nevertheless, the conversion of ecological electricity, e.g. generated by solar and wind power, into synthetic hydrogen or rather hydrocarbon allows the supply of climate-friendly fuels. That can help to achieve the sustainability goals in fields of applications, where the usage of electricity, i.e. concepts based on power-to-heat or power-to-mobility, are limited from the technological point of view. Examples for this are industrial processes, where not the energetic but material usage of hydrogen or hydrocarbon is significant, as well as long distance transport in the transportation sector.

Against this background sector coupling represents the key factor for the success of the energy transition. So, the expansion of capacities for generating electricity through solar and wind power is in focus of action. But only if it is possible to supply not only the electrical, but also the neighbouring sectors with enough ecological energy, the addressed climate protection goals can be achieved. Therefore, it will become more important to compensate for the fluctuations, which are induced in the supply system by solar and wind power. For damping such instabilities, corresponding with the tempo-

18 Umweltbundesamt 2024 f.
19 Umweltbundesamt 2024 f.
20 Umweltbundesamt 2024 f.
21 Umweltbundesamt 2024 f.
22 Umweltbundesamt 2024 f.

rary appearance of surplus or rather shortage, all the mentioned above concepts, finally based on sector coupling, have to be used in a suitable manner. All these flexibility options can be categorized by four clusters, see Figure 23.6.[23]

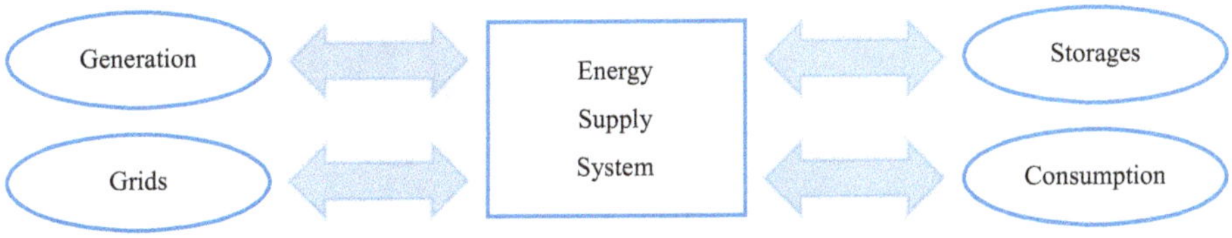

Figure 23.6: Four Categories for Energy Flexibility Options (figure created by author following BWK 2015, p. 13 f.).[24]

One category is the usage of flexible power plants, as they are very common in the energy market so far. This type of flexibility is based on the ability of generators to change its produced electrical power, which they feed into the network, very fast (e.g. gas turbines). Smart grids represent another category for flexibility, based on the ability to adapt the electricity flows very fast and in a suitable manner, e. g. via operations like switching on or off of consumption areas. Furthermore storages (e.g. batteries or pumped-storage power plants) for (electrical) energy are an own category for flexibility. Demand side management, which allows the network or rather market orientated control of electrical demand of consumers, concludes the list of clusters for flexibility.[25]

Digitalization is the key factor for the suitable application of all the concepts, which belongs to the mentioned flexibility categories. Only the digital connection of the technical components, on which the flexibility concepts are based, allows the establishment of a market information level, which ensures suitable decisions by the grid operators (including the here needed other market players) to guarantee security of supply. In addition, the impact of (new) data could induce the development of innovative business models, what allows market players to create new kinds of value-added. For example, there are possibilities for energy supply companies to include energy customers into arrangements for demand side management, so that the customers flexibility to reduce and enhance their energy consumption could be sold by the energy company at the flexibility markets of energy sector.[26] Forthcoming market developments like introduction of time-depending tariffs will enhance the potential for new forms of collaboration.

23 Bundesministerium für Wirtschaft und Klimaschutz (BMWK) 2015, p. 13 f.

24 Information base see Bundesministerium für Wirtschaft und Klimaschutz (BMWK) 2015, p. 13 f.

25 Deutsche Energieagentur GmbH 2012, p. 8 f.

26 Deutsche Energieagentur GmbH 2012, p. 23.

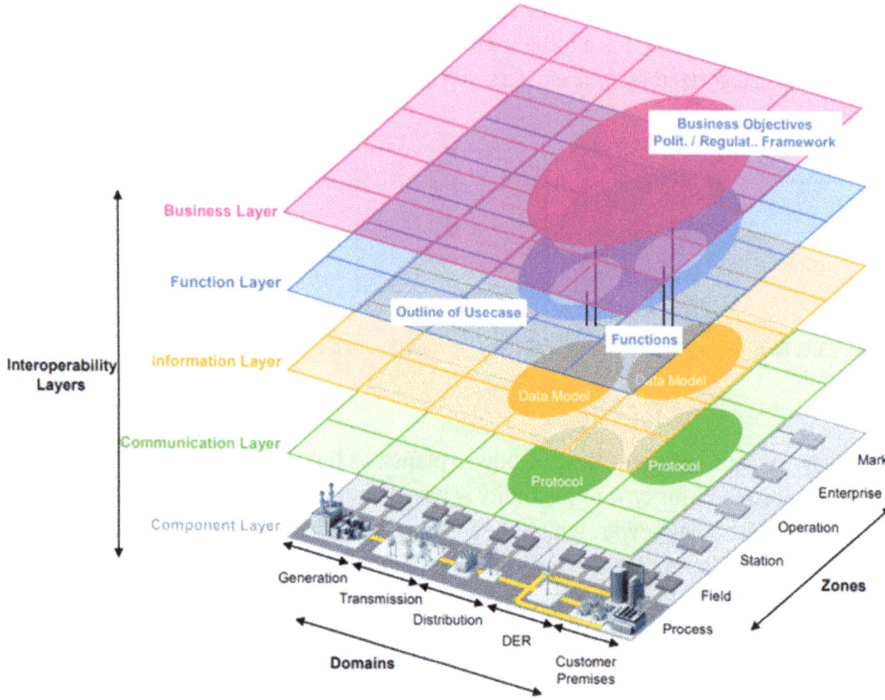

Figure 23.7: Smart Grid Architecture Model (SGAM) Framework.[27]

Against this background it is described in the next chapters, how the here mentioned market developments can take place considering, in particular, the different dimensions of the digital framework, illustrated in Figure 23.7.

23.3 Market Players and Business Processes

As already explained in the previous section, the energy value chain follows the physical energy flow chain in terms of the conversion of primary via final into useful energy. Based on this, there are business processes that are to be handled in the market by the individual market players according to its position in the value-added structure.

For the players at the generation level, this means organizing the provision of final energies by refining primary energies using suitable technical conversion technologies, such as the conversion of natural gas into electricity by gas turbines or the

27 CEN-CENELEC-ETSI Smart Grid Coordination Group 2012, p. 30.

conversion of biomass into district heating (hot water) by heating plants. In terms of the market economy, the generation of added value is reflected in the higher price level that market participants have to pay for the final energies in contrast to the primary energies. This price difference between "low-quality" primary energy and "high-quality" final energy forms the basis of the added value for the business model of e.g. power or heating plants.

The final energies that can be generated through generation plants represent the supply side. In this context, it is up to the wholesale energy traders to suitably adjust these offers with the demand side, so that demand in the market can always be met with an adequate supply. This is an essential prerequisite for the existence and functioning of a market. On the one hand, trading units buy quantities from the generation companies ("power plant operators"), which companies compare to the demand items to be covered as part of its portfolio management. The demand items here result in turn from the energy requirements of the sales companies serving the energy customers via corresponding supply contracts (e.g. the municipal utilities). In this respect, a process chain that is oriented in the opposite direction to the energy flow applies here.

The sales departments use load forecasts to determine the customer requirements they have contracted (e.g. for the customer segments private households, commercial or industrial companies), which are addressed to the energy trading companies as open procurement positions. In this role, the energy trading companies take on the "market access" service function for the sales companies – just like in this context this role for the generation companies to market generation volumes. As part of the portfolio management for which trading is responsible, the open procurement positions of the sales companies can then be compared with the open market positions of the generation companies and closed by corresponding purchase or sale transactions on the market. For this purpose, transactions in electricity and natural gas trading are carried out on the energy exchange or bilaterally with third parties in the sense of over-the-counter trading.

For the purpose of portfolio optimization and in terms of risk management, trading transactions are carried out in three time periods. There is the possibility of trading in the so-called futures market for several years into the future, in particular to hedge against falling prices from a generation perspective or against rising market prices from a sales perspective. Buy and sell transactions can also be placed in the short-term spot market in order to optimize the portfolio, e. g. by the more accurate load forecasts or generation forecasts are induced in the temporal area close to the fulfillment date. The same applies analogously to the use of intra-day trading, in which on the day of fulfillment (for previously contracted energy procurements or sales) there is still a reaction possible by depending on trading balance. Because such differences at the grid level contribute to disturbances in grid stability, which in turn have to be compensated for by costly technical arrangements such as the provision of balancing power plants – a process that is located in the so-called balancing power market and that is coordinated by the transmission system operators. In this context the trading companies are responsible to keep their trading balances balanced at all times. This is done using the mechanism of penalties, according to which trading companies have to pay the more to the transmission system operator, the more their trade balance deviates from zero. Specifically, the trade balances are recorded in an account that shows the deviations as balancing energy. The incentive for trading companies is now to keep the balancing energy costs passed on to them from the control reserve costs as low as possible. This is done as part of their balancing zone management, in which they map all generation and demand quantities in their trading portfolio and compare them per trading time interval – and try to conclude counter-transactions if there are differences that are not equal to zero.

The market roles and business processes outlined here make it clear that – regardless of which value-added stage is being looked at – there is always a large amount of information that needs to be exchanged in the structure of the business players involved. And so, since the beginning of the liberalization of the grid-bound energy markets for electricity and natural gas, in the course of which the market processes mentioned above have developed up to the present day, the importance of market communication processes has increased massively. The relevance of suitable and high-performance information and communication technologies has increased to the same extent. And so, it becomes obvious here that digital instruments and concepts are of great importance, even against the background of the orchestration of processes in the liberalized energy market. This applies all the more if the transformational developments of the energy transition, how the decentralization of the infrastructure and the development of regional and local markets, should be organized suitably in the future.

23.4 Digitalization of the Energy Industry

The previous sections gave an overview how the grid-bound energy supply is organized in the energy system and what challenges the energy transition will bring to the structure of the value chain. Overall, the challenge of reconciling consumption and generation in real time is becoming more and more complex due to the increasingly fluctuating energy production from renewable generators and its distributed nature. Without information technology, this task cannot be tackled.

In context of the energy value chain the term digitalization refers to the use of digital technologies in order to transform business processes, to gain new knowledge and evaluate and optimize options for action by collection, processing, connecting, transmitting and analyzing digital data.[28] The digitalization of the energy industry serves as an enabler on the way to a Smart Grid, i.e. an intelligent, self-regulating network of energy producers and consumers.

28 Schütz, Johann; Uslar, Mathias; Clausen, Marie 2022, p. 35.

In order to get an overview of the IT resources used in the energy domain the Smart Grid Architecture Grid (SGAM) proves to be helpful. The SGAM was originally created to identify and define gaps in the standardization process on the way to a Smart Grid. However, the SGAM is also helpful in understanding which tasks fundamentally occur in a grid-bound energy supply system and how information technology can support these tasks at different levels of abstraction. The European market model is located at the top level, the business layer.

Within the European energy system, the delivery of grid-bound energy is processed via virtual energy quantity accounts, the balancing groups. This is a core idea of the European market model and is part of European law. All participants in the European energy markets must align themselves with this market model. Hence, more than 40 roles with their corresponding tasks arise from the existence of those balancing groups. The exact definition of which tasks are associated with the roles is made by the three relevant organizations, ENTSO-E, the association of European Transmission System Operators, EFET, the European association of energy traders and EBIX, the European forum for energy business information eXchange.[29]

In the concrete implementation, information technology is then required to map this market model to operational processes. In the energy companies, this then manifests itself in an application landscape, in information and data models, communication protocols and transmission paths. These artifacts of information technology can be found in the middle three levels of the SGAM, namely the function layer, the information layer, and the communication layer. In the following the technologies being currently used are presented based on the levels of the SGAM.

23.4.1 Functional Layer

The function layer of the SGAM is an abstract collector for all the different use cases that ultimately make up the operational business of the energy value chain. It heavily depends on the specific market role what use cases a company must fulfill. A balancing group coordinator has different tasks in the value chain as an energy trader or a distribution network operator.

It also makes a big difference whether a use case requires close coordination with other market participants or if a company can handle the use case on its own. Compatibility and interoperability play particularly an important role in so-called market communication, i.e., those business processes that run across several companies. Accordingly, standards are required to which all market participants must adhere. Companies can only interact with each other in an automated and error-free

29 ENTSO-E - EFET - EBIX 2022.

manner through standardization of whatever kind, in which processes, responsibilities, message formats and transmission paths are defined in detail.

At the European level, however, most use cases of market communication are not specified to such an extent that a technical implementation in the form of processes would already be possible. Important details on processes, deadlines, errors, data formats and communication paths are missing. The design of specific details regarding to those details is the responsibility of the national energy bodies.

One of the few applications where pan-European standardization has taken place is cross-control-area network nomination (1:1 nomination). This involves an energy trader notifying the balancing group coordinators about an energy flow from one control area to another. To do that, certain messages must be exchanged. Each of the messages contain at least one schedule, a time series that shows what amount of energy is to be transported in which period (quarter hourly values). Technically, this message can be sent in different ways. The somewhat older entso-e Scheduling System (ESS) is still used, usually. Such a message is encoded in a specific XML format and transmitted via an email attachment.[30]

In contrast to cross-control-area network nomination, most of the applications that require coordination between market players are specified in national regulations. In Germany, for example, there are the market rules for the implementation of balancing group accounting for electricity and the business processes for customer supply with electricity.[31] In these documents, the use cases are described in such detail that technical processing is possible in the first place. A customer changing from energy supplier to another is an example of such use cases. It is certainly understandable that there is a high need for coordination between the distribution system operator and the new and the old electricity provider to execute that use case. Accordingly, it must be determined who has to send messages to whom in which order, how these messages have to be encoded, how they are to be transmitted and how to proceed in the various error situations. The premise here is that each market participant must be able to communicate with each other (n:m communication).

In Germany, the technical specifications for data formats and transmission paths are defined by the EDI@Energy working group of the BDEW. The core of that specification is given by the definition of message formats based on the Electronic Data Interchange for Administration, Commerce and Transport (EDIFACT). These messages are usually sent between the market actors as an attachment of emails. Soon, the exchange of those messages is to be handled based on web services (AS4) instead of emails.

As the technical design of the market processes is the responsibility of the respective countries, the details differ greatly from country to country. While EDIFACT and

30 ENTSO-E 2012.

31 Bundesnetzagentur 2019; Bundesnetzagentur 2020.

emails are used for market communication in Germany, completely different procedures are used in other countries. In Austria, for example, there is a central communication platform (energy industry data exchange, EDA) which is used by the market players to exchange messages based on Electronic Business using XML (exXML). So, it must be said that there are no European standards regarding market communication. The application systems that ultimately map these processes in the energy companies are therefore mostly implemented by national software firms or self-implemented by the major market players. A European market for IT solutions can hardly establish itself because of the fragmented process world in this area. One can speak of an immense heterogeneity of the IT system landscape.

23.4.2 Information Layer

Every data processing system needs an implicit or explicit idea of what the data to be processed means and how individual statements are related to one another. Otherwise, data cannot be processed meaningfully. For example, an average value of several prices may make sense, but an average value of postal codes probably does not. An information model gives the data a meaning, so it serves as an abstract description of objects, their properties, and relationships with each other. The application systems that implement the use cases of the function layer therefore always implement at least one such information model. The information layer of the SGAM serves as a collector for such information models being used in the energy industry.

For special applications, such an information model can be rather lightweight. For example, the ESS format mentioned above describes a manageable number of different messages for a single use case. Each message transports at least one time series. The information model must also specify what is meant by the term time series and which values are permissible. The term data model is to be distinguished from an information model. A data model specifies the mapping of an instance of an information model to a specific character sequence, e.g., from ESS to an XML document. There can be an arbitrary number of data models for each information model. Since an information model without a data model is of little use, both models are usually defined together (see Figure 23.8).

Many different information and data models are used in the energy industry to model, represent and transmit data. Some of these models have been introduced in the context of communication protocols, e.g., in the context of building automation, others are more general.

The Common Information Model (CIM) is a standardized information model specifically for the energy industry, which was introduced without a specific reference to a single use case or application (e.g., IEC 61970, IEC 61968, etc.). The CIM is based on the object-oriented programming paradigm and introduces a large number (more than 1,600) classes, their attributes, and different types of relationships in between.

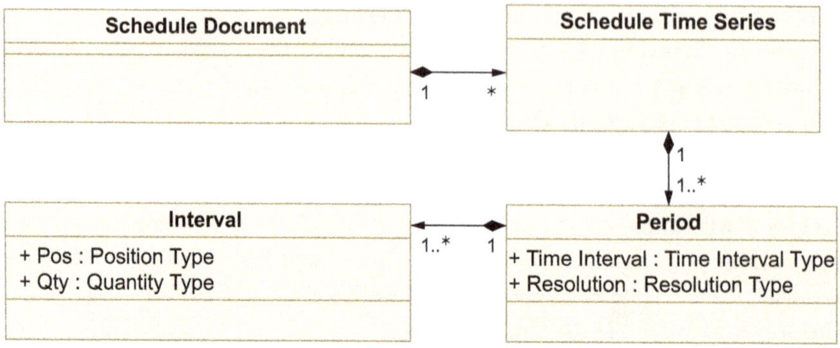

Figure 23.8: The Information Model of ESS as a Class Diagram of the UML; figure created by author.

One can think of CIM as a language to express all kinds of master and transactional data of the energy industry. With the help of CIM, physical assets (e.g., lines, transformers, and generators) can be represented, network models mapped, but also more virtual concepts such as invoices, auctions and information on operation and maintenance can be described (see Figure 23.9).[32]

The CIM has at least one mapping to an XML-based data model. It is particularly meant to be used to exchange energy management data at a high level of abstraction across company boundaries. As an example, CIM is currently used by the European transmission system operators to create a common network model on a daily basis in order to carry out a load flow analysis. As far as the type of data to be expressed, the CIM has a large degree of freedom.

In the context of network automation or the control and regulation of distributed regenerative energy systems, other aspects come to play, e.g., standardization and thus the interchangeability of physical devices in the systems (preferably plug-and-play). Devices should be standardized in their functions and be self-descriptive. In IEC 61850, a separate information model is defined, which is e.g., able to map the usual artefacts of switchgear (e.g., circuit breakers) and the data that such devices may have to offer. In this information model the so-called logical devices represent a specific function of a physical asset. They provide data and functions as servers in a communication network. Initially, the manufacturers can freely decide which data and functions such devices may provide. However, in a further step the standard specifies which data and functions a specific device must offer exactly to be allowed to be regarded as e.g., a circuit breaker. Manufacturers must adhere to these specifications so that their devices can be certified according to IEC 61850 (see Figure 23.10).

To transmit the data of a logical device in a communication network, several protocol mappings between the information model and concrete communication protocols are

32 Uslar, Mathias et al. 2012.

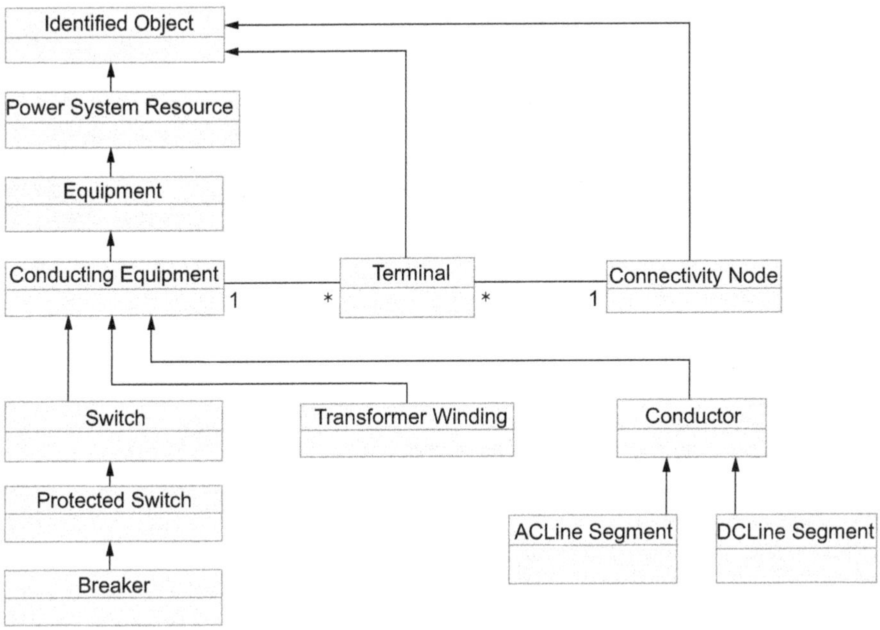

Figure 23.9: Extract from the CIM – Relationships between Conducting Elements; figure created by author.

described in IEC 61850 for different scenarios. Here, the possibilities range from real-time communication on a process bus within a powerplant or a control unit (SV and GOOSE) to communication with a wind turbine via a public internet connection (MMS).[33]

IEC 61850 solves problems in the field of network automation and the control and regulation of distributed energy assets. This is, of course, an important aspect of the energy value chain. In addition to that, the timely and digital recording and transmission of energy consumption data, the creation of a price-based incentive through time-based tariffing and the possibility of granting network operators limited access to distributed generators (e.g., PV systems) or consumers (e.g. charging electric cars) are other important prerequisites on the way to the Smart Grid. In principle, all these tasks can be fulfilled by digital measuring systems.

In Germany, an intelligent measuring system consists of a digital meter connected to a communication device, a Smart Meter Gateway (SMG), the actual core of the measuring system. In a Local Meteorological Network (LMN), the SMG retrieves data from potentially multiple meters, aggregates it, applies different profiles to the measured values and forwards the data to the meter operator via a Wide Area Network (WAN). The SMG can also serve as a proxy server between external market participants and

33 Stuckenholz, Alexander 2020, pp. 62–65.

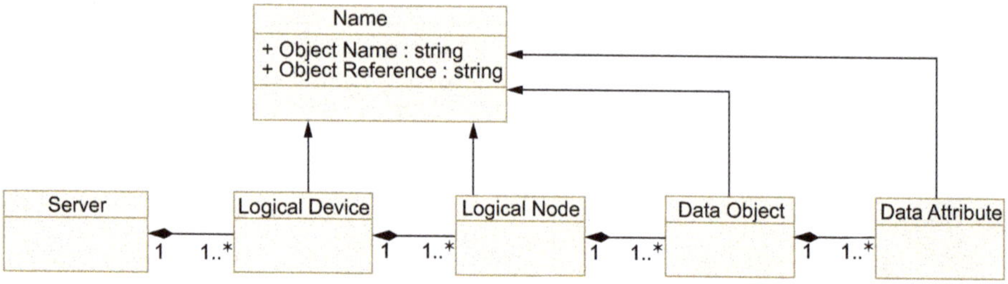

Figure 23.10: The Information Model of the ACSI of IEC 61850; figure created by author.

local actors (PV systems, storage, . . .) via the controllable local system interface (CLS).

In Germany, the Federal Office for Information Security (BSI) was responsible for the design and standardization of the Smart Meter Gateway. The result is an amazingly complex standard that is partly influenced by the Open Metering System Specification (OMSS). The Smart Meter Gateway represents the data based on another object-oriented information model called Companion Specification for Energy Metering (COSEM, see Figure 23.11). A physical device consists of several logical devices providing objects with attributes that hold the actual measured values. A device provides at least one logical device, the management logical device. As this logical device provides information about which objects and data the device offers, the physical device is self-descriptive. The meaning of the data is represented by OBIS codes on each attribute.[34]

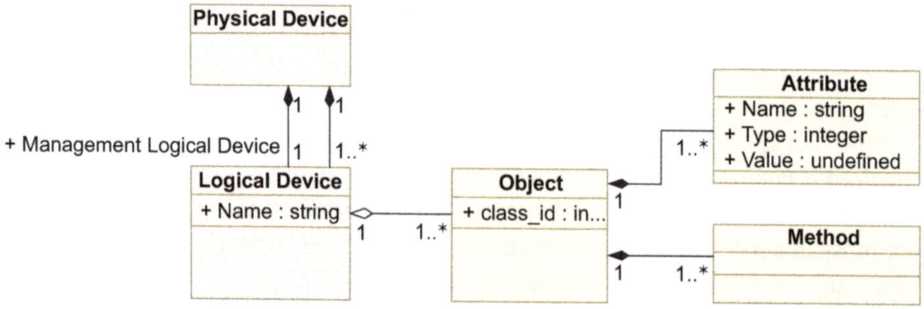

Figure 23.11: COSEM as a Class Diagram of the UML; figure created by author.

34 IEC.

Communication Layer

On the one hand, the use cases of the function layer force market players to coordinate processes across companies. On the other hand, data must also be collected, aggregated, and transmitted via communication networks internally. For both scenarios, the meaning of the data is defined by appropriate information models at the level of the information layer. The data is then transmitted via communication networks using technologies that can be assigned to the communication layer of the SGAM. A secure and sometimes real-time transmission of data is one of the basic requirements for being able to implement an integrated energy system. A veritable cornucopia of technologies is currently being used in the energy industry. To some extent, each use case uses its own method. A single, integrated, and centralized data hub does not exist, and the odds are currently not very good for the origination of such a platform.

To be able to technically transmit data, all the challenges described in the seven layers of the OSI reference model for data communication must be solved. The information from the application layer is passed on to the lower layer in each case to enable robust and error-free transmission over a potentially disrupted channel (e.g., a cable or a radio link). In the best case, the protocols and technologies on the layers can be exchanged for other variants without the higher levels noticing.

It heavily depends on the specific application which protocols and transmission technologies are used. If small sensors or actuators are to communicate with each other in context of process control or building automation, so-called fieldbus systems are often used. For example, smart meters often implement standards such as Modbus or M-Bus to transmit data. In room automation with KNX or LonWorks, proprietary transmission methods are used. KNX is based on RS-485 and encodes data in the form of voltage differences on a two-wired cable. In new buildings, such transmission lines are therefore laid separately from the power supply in a tree-like topology. If, on the other hand, room automation is to be retrofitted in existing buildings, radio transmission is often used. Zigbee is usually used for this.

In cases where there are no real-time requirements, the protocol stack of the Internet (TCP or UDP and IP) is mostly used for data transmission for many reasons. Software support is widespread and the hardware (routers, etc.) is inexpensive and can be purchased independent of single manufacturers. A variety of methods can then be used as the underlying transmission technology, e.g., wired Ethernet, WLAN or LTE. Some metering operators also use PowerLine to communicate with a Smart Meter Gateway. Then, the power grid is not only used for energy transmission, but also for data transmission. The data is modulated at high frequency onto the 50 Hz of the AC voltage. The advantage is that the end customer does not need to have their own telecommunications network. However, data transmission can only be guaranteed over a short distance and must be decoupled at the next transformer station. Due to the release of the frequencies of the former radio telephone network C, a separate radio network is currently being created in Germany, which is operated based on LTE and is to be used exclusively for the energy industry. The system promises a transmission rate of up to 100 Mbit/s, comparatively long ranges of up to 20 km, good building penetration and high availability. Another alternative to monitor or control highly distributed generation plants like offshore wind farms is to use satellite communication.

23.5 Digital Transformation of the Energy Domain and beyond

The digitalization of business processes creates the conditions for completely new business models, not only in the energy domain. The process that follows digitalization meaning an ongoing and fundamental change in the economy and society as a whole is referred to digital transformation.

The digital transformation of the energy system is certainly linked to the concept of the Smart Grid. A Smart Grid is understood to be an energy system that can reconcile increasingly fluctuating and distributed generation with consumption in real time and to cover dynamic peak loads using flexibility and storage. This goal has certainly not been reached, yet. The recently published report on the results of the SINTEG research program on the topic of digitalization lists four core requirements for the further transformation of the energy system on the way to the Smart Grid: infrastructure, interoperability/standardization, platforms, and IT security.[35]

All those technologies that are intended to carry out the safe, secure, and partially real-time transmission of data for the energy industry certainly belong to that infrastructure. Furthermore, the measurement and collection of data through intelligent measuring systems, can also be included into the infrastructure. As mentioned before, there are hardly any technical gaps in this area. In principle, IT systems are more or less able to cover the requirements of a Smart Grid as of today. One of the few gaps in the communication infrastructure arise through the need to interconnect devices like electric cars, wall boxes, storage systems, PV systems, white goods, etc. in private households in order to optimize energy consumption. With the EEBus system that gap is currently closed. However, this promising system is not yet available in the form of products for end customers.

Another prerequisite on the way to the Smart Grid is the availability of data-driven platforms. Such platforms can, for example, implement digital marketplaces where supply and demand meet with a low entry hurdle for all market participants. On such platforms both classic energy products as well as flexibility can be traded. Although digital platforms in the energy industry have often been the subject of research and development, they have not been arrived in reality so far. On the one

35 Schütz, Johann; Uslar, Mathias; Clausen, Marie 2022.

hand, the high degree of IT heterogeneity in the energy industry established a technical hurdle for the development of market platforms. On the other hand, the interest in a platform economy among the energy companies in (at least in Germany) does not seem to be too high. The business models that could develop in this area are discussed in a subsequent section.

To allow all actors in the Smart Grid to interact seamlessly with each other via the communication infrastructure, the interoperability of the systems must be guaranteed. Hence, standardization plays a key role on the future path to a Smart Grid. From an IT-perspective that means to define business processes, data formats and transmission paths on a detailed level. However, as mentioned before, the standardization process mostly ends at the borders of the national states. Furthermore, the energy industry is rather conservative in the usage of state-of-the-art technologies. The market processes (MABIS and GPKE) are technically implemented at a lowest common denominator. The software architecture of many energy companies is dominated by fat-client systems making it very difficult to adapt to new business processes. It can only be recommended to develop a little more innovative spirit in the (German) energy industry.

With the increasing use of information and communication systems in the energy industry, the operational risk due to system failures increases as well. The availability of the IT systems is threatened by very different dangers. In addition to force majeure, negligence and technical failure, cyber-attacks pose an increasing threat today. In terms of the number of cyber-attacks registered in Germany, energy supply is already in second place. In the context of the war in Ukraine, there were recently failures in communication with wind turbines, which were apparently also due to cyber-attacks.

In Germany, as part of the KRITIS Ordinance, energy supply companies are being forced to deal much more closely with the topic of IT security. In particular, the facility operators are obliged to implement an Information Security Management System (ISMS), which includes the formulation of security goals, the determination of the affected assets, the risk assessment and risk treatment and continuous improvement. The topic poses a particular challenge for smaller market participants (e.g., municipal utilities). In reality, the risks and threads increase with the complexity and heterogeneity of the IT systems used. Another argument for the widest possible standardization.

Not only the energy system itself is the subject to digital transformation. Many other economic sectors are currently undergoing major changes due to digital technologies. Some of these transformation processes also have feedbacks on the generation and especially on the use of energy.

It is well known that the production and transport of physical products is particularly energy intensive. If products or services can be completely digitalized, it has considerable potential to save energy and thus also reduce CO_2 emissions. An example of such a shift is the way many people consume music today. Instead of having to produce and transport sound carriers such as CDs or records, music is consumed almost exclusively digitally today. This has a demonstrable effect on energy consumption.[36] A corresponding effect can therefore also be assumed for the digitalization of printed matter, films and so on. One can assume that sooner or later all areas in which information forms the core of value creation will be digitalized completely. Administrative and communication processes are particularly predestined, e.g., banks, insurance companies, media, etc. New technologies, such as video conference systems, social media and virtual reality will make their contribution.

Other areas can also benefit from energy savings through digitalization. For example, online trading reduces the energy required for logistics processes. It uses considerably less energy, if delivery vans transport goods, as if many people would transport their purchases home with their individual vehicles. Surprisingly, this reduction is compensated by the fact that the free time gained is mostly spent on energy-intensive activities.

A major disruption is sure to come as all types of vehicles become autonomous in the foreseeable future. Not only in logistics, but also in private transport, this will bring a drastic change. The need to own a car is eliminated as mobility will be available as-a-service by multiple vendors, most likely. What means a radical change for vehicle manufacturers can have a positive effect on energy consumption. In principle, autonomous vehicles can always be in motion. While significantly fewer vehicles must be produced and operated, vehicle utilization can therefore increase drastically. Many parking spaces, especially in the inner cities, can then be used better. As Waymo, a subsidiary of Google, already provides such a service in Phoenix (AZ), it is only a matter of time before this will also be offered everywhere else.

23.6 Digitalization in the Business Models

As explained in the previous sections, the described defossilization as well as the decentralization of electricity generation can only be achieved through an increased use of digitalization due to the fluctuating generation and the low storage capacity of electricity. In a market-oriented economy, digitalization can contribute to climate-friendly energy sectors if appropriate business models can be created by companies. In general, a business model is a simplified representation of how a company generates an offer by compiling resources, possibly with partners.[37] This offer must stand out from the competition in order to achieve added value for the customer. This should make it possible to generate revenues for the offering company in order to be able to survive on the market. A single company can also include different business models.

36 Weber, Christopher L.; Koomey, Jonathan G.; Matthews, H. Scott 2010.
37 Schallmo, Daniel R.A. 2018, p. 18.

The traditional business models in the energy sectors follow the logic of a predominantly central power generation in large power plants and a subsequent distribution of the generated electricity to the consuming units (private households, companies from trade, commerce and services, industry). These traditional business models in the energy sectors are designed to ensure the highest possible economic efficiency through central large-scale power plants. Security of supply and grid stability is secured by sufficient capacities in the central energy supply. The basis of this efficient and stable energy supply is formed by capital-intensive generation plants and grid infrastructures with its long-term investment periods of usually several decades. Such infrastructures are stable systems that require linear coordination processes of a few large players. In recent years, these structures have adapted through smaller generation plants primarily for electricity generation from renewable energies as well as stronger electricity trading between production and distribution or consumption.[38]

Companies in the energy sectors have to adapt their business models due to the explained changes in the market environment described. Digitalization plays a decisive role in these changes. The adapted or new business models in the energy sectors have to meet the objectives of security of supply, grid stability and cost-effectiveness for affordable electricity as far as possible. Flexibility plays an important role in achieving security of supply and cost-effectiveness. In terms of flexibility, there are several possibilities to consider, as shown in the previous sections. This includes flexibility in power generation, storage, power consumption or in the grid. To achieve flexibility, intelligent measurement in the Smart Grid is required, enabling intelligent and real-time control systems. From the point of view of information technology, this is also associated with changed information flows.

Developing a business model in the context of a Smart Grid, three key topics are to consider.[39] What exactly does a company sell and what are the customers or the customer segments? What exactly is the added value for customers and the revenue model? What exactly are the resources required to develop and successfully execute such a business model to generate a value proposition? This raises the question: Which company can offer such business models in principle? On the one hand, these are the established companies that are changing and have to offer new or adapted business models. This requires a high degree of adaptability on the part of established companies. On the other hand, the new business models can also be offered by companies that have so far been active in other industries, such as information technology. In addition, new companies, i.e. start-up companies, can offer new business models in the field of a Smart Grid. Implementing new or adapted business models can lead to additional companies in the entire value chain of the energy sectors. Thus, it can also happen that companies push themselves as an additional value creation step,

38 Löbbe, Sabine; Hackbarth, André 2017, p. 14.

39 Giehl, Johannes et al. 2020, p. 7; PricewaterhouseCoopers 2017, p. 369.

for example between customers and established companies in the energy industry.[40] As a result, there is a challenge for established companies to lose direct customer contact. This in turn means that partnerships play an increasingly important role in the value chain. Potential customers of these business models in the new environment can be diverse. In principle, this includes all institutions in the energy sectors such as industrial and commercial companies, energy suppliers, network companies, housing associations, municipal customers or private individuals.[41]

Business models require a sufficient willingness to pay on the part of potential customers. That reflects the revenue model. A business model can only be successful in the long term if the activities generate sufficient revenues that can cover the costs incurred and lead to profits. Otherwise, a business model cannot survive in the long term. The question is, what triggers a sufficient willingness to pay on the part of customers? What and in what amount are customers willing to pay? This, in turn, depends on the benefits generated for the customers. It may be that customers pay money to be provided with, for example, produced useful energy. This may include customers being willing to pay money to ensure that customers consume less energy through the value proposition of the business model. Perhaps this is also due to the fact that customers buy certain product bundles such as energy as well as service and maintenance. In particular, the discussed integration of fluctuating renewable energies into the energy sectors can lead to a willingness to pay by dealing with flexibility. Flexibility offers that arise from sector coupling, for example, can be reflected in new business models in the future.[42] In this way, supply and demand can be balanced by market mechanisms.

Another important element for the development of business models are the key resources. These resources are required to generate a value proposition to customers. Depending on the business model, other resources are required. These resources can be created alone or with partner companies and can be divided into financial, physical, human and technological resources. Technological resources play an important role in the implementation of adapted or new business models in the energy sectors. The increased digitalization of activities can lead to the need for new key resources for a successful business model. This includes resources such as knowledge of IT security or of digital customer interfaces.

A reference system for different business models in the energy sectors offers the possibility of a structured recording of conceivable business models that can result from digitalization and thus contribute to climate friendliness. The following Figure 23.12 shows a description of business model classes in the energy transition. It is a comprehensive presentation of the business models that result from the energy transition.[43] The business

40 PricewaterhouseCoopers 2017, p. 319.

41 Burger, Scott P.; Luke, Max 2016, pp. 7–8.

42 Liggesmeyer, Peter; Rombach, Dieter, Bomarius, Frank 2018, p. 353.

43 Giehl, Johannes et al. 2019, p. 10.

models that are assigned to the area of modern value creation result from the aforementioned trends of defossilization, decentralization and digitalization.

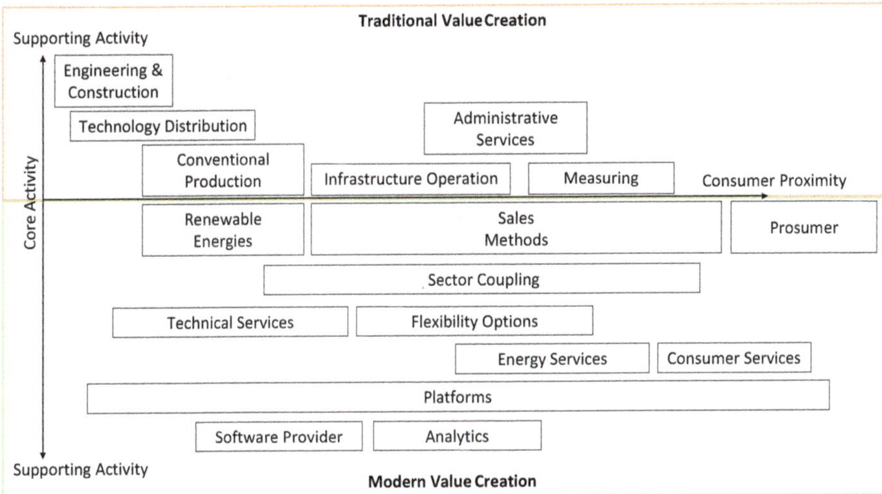

Figure 23.12: Overview of Classification of Business Models in the Energy Sectors.[44]

The upper part of Figure 23.12 covers the traditional value creation of the energy sectors.[45] As described in the previous remarks, the individual activities can be assigned to the traditional value chain of the energy industry with the engineering and construction of large-scale plants (power plants), conventional production, trade, transport and sales.

The lower part of Figure 23.12 shows business model classes resulting from the energy tranition with the associated decentralization and digitalization.[46] This area includes under the class renewable energies business models that deal with the generation of electricity from renewable energies. In modern value creation, business models in the class of sector coupling play an important role as already explained. The stronger interaction of the previously separate sectors of electricity, heat and cooling as well as mobility lead to new business models. Sector coupling is also important for the activities described above to increase flexibility in the energy sectors. With sector coupling, the importance of electricity production increases, which in turn must be associated with grid construction. The Smart Grid plays a decisive role here, so that the use of digitalization plays an important role for the business models of the sector coupling. However, while from a technical point of view the necessity of sector cou-

44 Giehl, Johannes et al. 2019, p. 10.
45 Giehl, Johannes et al. 2019.
46 Giehl, Johannes et al. 2019.

pling for the energy transition is not called into question, the economic implementation of the corresponding business models represents a certain challenge at this juncture.

Business models in the class of flexibility options include business models that deal for example with storage technologies and demand side management. Demand side management summarizes the flexibilization of electricity demand, i.e. demand response measures on the consumer side.[47] This is also an area in which digitalization plays an essential role. Business models in the area of modern value creation in the energy sector also include solutions that can be originally assigned to information and communication technologies. This includes the class of analytics, which involves data analytics, for example from generation, grid and consumption. Especially data analysis for consuming units is an important element of the Smart Grid. The increasing use of smart meters will further increase the amount and quality of available data. From this, individualized tariffs can also be established as a business model.[48] At the same time, suitable data analysis makes it possible to carry out activities to improve energy efficiency. Higher energy efficiency can be achieved by intelligently designing customers' appliances. It is conceivable that technologies can optimize heating and air conditioning depending on the time of day and occupancy level.[49] The class of platform includes, for example, business models such as virtual power plants or crowd storage. The platforms can be used for all stages of value creation in the energy sectors. The class technical services includes, for example, business models such as energy management systems or maintenance and retrofitting. Here, too, digitalization plays a central role, for example in the field of predictive maintenance. In the different stages of the value chain, activities from the class of sales methods can take place. This includes traditional sales to end customers. However, digitalization opens up new opportunities in sales. These contain, for example, the implementation of a digital customer interface or the aforementioned possibilities for new tariff systems that are made feasible by smart meters. The class of sales methods also includes storage trading activities or system services.

Basically, the presentation shows that the illustrated effects of digitalization create new or adapted business models, which in turn can contribute to the energy transition and thus to climate protection through a market implementation. However, the challenge for a rapid implementation of the individual business models is partly due to strong regulation, which can also change in the short term. This strong regulation can inhibit new or adapted business models, as a necessary adaptation of the regulation usually lags behind technical innovations. In addition, the long-term nature of investment decisions in the energy sectors hampers innovation.[50] On the one hand,

47 PricewaterhouseCoopers 2017, p. 266.

48 PricewaterhouseCoopers 2016, p. 15.

49 Lang, Thorsten; Ewald, Johannes 2020, p. 12.

50 Liggesmeyer, Peter; Rombach, Dieter; Bomarius, Frank 2018, p. 349.

companies will try to use the existing infrastructures for as long as possible, as significant capital is tied up in these investments. On the other hand, new business models will require highest possible predictability in order to be able to justify possible significant investments in new infrastructures and business models. In addition, for some business models, the revenue model is difficult to represent. This currently applies, for example, to business models in the field of making the energy system more flexible including activities in the class of sector coupling.

23.7 Conclusion

This chapter gave an overview of the means that digitalization offers to cover the ever-increasing requirements to use information technologies in the energy sectors. On the one hand, the need for coordination among different market players is increased due to the rising specialization of energy companies. On the other hand, fluctuating energy production by decentralized regenerative assets increases complexity to monitor and control energy assets to balance production and consumption.

The information technology artifacts that are used can be assigned to the three levels of the SGAM: function layer (use cases), information layer (information and data models) and communication layer (communication protocols and processes). Real IT solutions in the energy industry are perpendicular to the interoperability levels of the SGAM and use concepts from all the three levels mentioned. In recent years, various business models have been developed and implemented in the context of the energy transition, but they were more regionally oriented.

The degree of pan-European standardization is currently relatively low at all levels. This makes cross-border integration of market players and business models more difficult and thus prevents a pan-European platform economy for energy. The details for handling many use cases are only specified at national level. Accordingly, hardly any standard solutions can be established here. On the information layer, there are some internationally standardized information models. The CIM, in particular, needs to be mentioned here, which aims to express data from the widest possible range of physical assets (cables, transformers, switches, generators, . . .) or virtual concepts (invoices, contracts, . . .). However, the CIM has a relatively low degree of dissemination. It also currently plays no role in the handling of market processes (at least in Germany).

A wide range of communication standards is used for data transmission in the energy industry. Roughly each use case uses its own communication system. A corresponding number of systems can be assigned to the communication layer of the SGAM. Some of these are protocols that have been specially defined for the energy industry (or related areas) (e.g. IEC 61850, KNX, LonWorks, CDMA 450, Smart Meter Gateway, . . .). On the other hand, general communication standards are also used (Internet protocols, fieldbus systems, . . .). A single, central, and integrated communication platform does not exist. Communications is not a technical problem in the energy domain, standardization and interoperability is.

While the energy industry is struggling with a digital transformation on the way to the Smart Grid, facts are being created in many other areas. The way of consuming music or films has already changed dramatically during the last years. The same degree of change will happen to mobility with autonomous cars. However, such leaps of innovation are currently difficult to implement in the energy sectors. The energy value chain is far too closely aligned with national regulations so that digital solutions cannot scale to a European level.

Bibliography

Bundesministerium für Wirtschaft und Klimaschutz (BMWK) (formerly Bundesministerium für Wirtschaft und Energie), Weißbuch – Ein Strommarkt für die Energiewende (03.07.2015), https://www.bmwk.de/Redaktion/DE/Publikationen/Energie/weissbuch.html, 26.03.24, Berlin, 2015

Bundesnetzagentur (BNetzA): Geschäftsprozesse zur Kundenbelieferung mit Elektrizität (GPKE), Bonn 2019.

Bundesnetzagentur (BNetzA): Marktregeln für die Durchführung der Bilanzkreisabrechnung Strom (MaBiS), Bonn 2020.

Bundesverband der Energie- und Wasserwirtschaft e.V. (BDEW): Smart Grids (10.08.2017), https://www.bdew.de/energie/smart-grids/, 26.03.2024, Berlin, 2017.

Bundesverband Wärmepumpe e.V.: Baugenehmigungen 2019: Wärmepumpe verteidigt Spitzenposition, Press Release May 26th 2020, https://www.waermepumpe.de/presse/pressemitteilungen/details/baugenehmigungen-2019-waermepumpe-verteidigt-spitzenposition/, 26.03.2024, Berlin, 2020.

Burger, Scott P.; Luke, Max: Business Models for Distributed Energy Resources: A Review and Empirical Analysis, An MIT Energy Working Paper, Cambridge 2016.

CEN-CENELEC-ETSI Smart Grid Coordination Group: Smart Grid Reference Architecture, November 2012. https://www.cencenelec.eu/media/CEN-CENELEC/AreasOfWork/CEN-CENELEC_Topics/Smart%20Grids%20and%20Meters/Smart%20Grids/reference_architecture_smartgrids.pdf, 26.03. 2024, 2012

Deutsche Energieagentur GmbH (dena): Handbuch Lastmanagement (Stand 12/2012), https://www.dena.de/fileadmin/dena/Dokumente/Pdf/1408_Lastmanagement_Handbuch.pdf, 26.03.2024, Berlin, 2012.

ENTSO-E European Network of Transmission System Operators for Electricity: The ENTSO-E Scheduling System (ESS) Implementation Guide, 2012.

ENTSO-E-EFET–EBIX: The Harmonized Electrical Role Model, 2022.

European Commission: Clean energy for all European package, Renewable Energy Directive 2018/2001/EU, https://energy.ec.europa.eu/topics/energy-strategy/clean-energy-all-europeans-package_en, 26.03.2024, Brussels, 2018.

European Commission: Energy Efficiency Directive, DIRECTIVE (EU) 2023/1791 OF THE EUROPEAN PARLIAMENT AND OF THE COUNCIL of 13 September 2023 on energy efficiency and amending Regulation (EU) 2023/955 (recast), https://energy.ec.europa.eu/topics/energy-efficiency/energy-efficiency-targets-directive-and-rules/energy-efficiency-directive_en, 26.03.2024, Brussels, 2023.

Giehl, Johannes; Göcke, Hayri; Grosse, Benjamin; Kochems, Johannes; Müller-Kirchenbauer, Joachim: Vollaufnahme und Klassifikation von Geschäftsmodellen der Energiewende, Working Paper Energie und Ressourcen, 2019.

Giehl, Johannes; Göcke, Hayri; Grosse, Benjamin; Kochems, Johannes; Müller-Kirchenbauer, Joachim: Survey and classification of business models for the energy transformation, Energies, 2020, Volume 13, Issue 11, 2981.

Herwig, Heinz; Kautz, Christian; Moschallski, Andreas: Technische Thermodynamik – Grundlagen und Anleitung zum Lösen von Aufgaben, Wiesbaden 2016.

IEC [620]56–61:2002-02 – Electricity metering – Data exchange for meter reading, tariff and load control – Part 61: Object identification system (OBIS).

International Energy Agency (IEA): Digitalization & Energy (10/2017), https://iea.blob.core.windows.net/assets/b1e6600c-4e40-4d9c-809d-1d1724c763d5/DigitalizationandEnergy3.pdf, 26 03.2024, Paris, 2017.

Lang, Thorsten; Ewald, Johannes: Digitalisierung und Energieeffizienz, IW Consult, Köln 2020.

Liggesmeyer, Peter; Rombach, Dieter; Bomarius, Frank: Smart Energy. In: Neugebauer, Reimund (eds) Digitalisierung, Berlin, Heidelberg 2018, pp. 347–363.

Löbbe, Sabine; Hackbarth, André: Geschäftsmodelle in der Energiewirtschaft: Ein Kompendium von der Methodik bis zur Anwendung, Reutlingen Working Papers on Marketing & Management, Reutlingen 2017.

PricewaterhouseCoopers: Deutschlands Energieversorger werden digital 2016.

PricewaterhouseCoopers: Regulierung in der deutschen Energiewirtschaft, Band II Strommarkt, Freiburg, München, Stuttgart 2017.

Schallmo, Daniel R.A.: Geschäftsmodelle erfolgreich entwickeln und implementieren – mit Aufgaben, Kontrollfragen und Templates, 2nd edition, Wiesbaden 2018.

Schütz, Johann; Uslar, Mathias; Clausen, Marie: Digitalisierung, Synthesebericht 3 des SINTEG Förderprogramms, Studie im Auftrag des Bundesministeriums für Wirtschaft und Klimaschutz, Berlin 2022.

Stuckenholz, Alexander: Basiswissen Energieinformatik: Ein Lehr- und Arbeitsbuch für Studierende und Anwender, Wiesbaden 2020.

Umweltbundesamt: Glossar beginnend mit P, https://www.umweltbundesamt.de/service/glossary/p, 24.03.2024, Dessau, 2024a.

Umweltbundesamt: Glossar beginnend mit E, https://www.umweltbundesamt.de/service/glossary/e, 24.03.2024, Dessau, 2024b.

Umweltbundesamt: Glossar beginnend mit N, https://www.umweltbundesamt.de/service/glossary/n, 24.03.2024, Dessau, 2024c.

Umweltbundesamt (UBA): Endenergieverbrauch nach Energieträgern und Sektoren (19.01.2024), https://www.umweltbundesamt.de/daten/energie/energieverbrauch-nach-energietraegern-sektoren#allgemeine-entwicklung-und-einflussfaktoren, 25.03.2024, Dessau, 2024d.

Umweltbundesamt (UBA): Erneuerbare Energien in Zahlen (08.03.2024), Download Excel-File: Zeitreihen zur Entwicklung der erneuerbaren Energien in Deutschland unter Verwendung von Daten der Arbeitsgruppe Erneuerbare Energien-Statistik (AGEE-Stat) (Stand Februar 2024), https://www.umweltbundesamt.de/themen/klima-energie/erneuerbare-energien/erneuerbare-energien-in-zahlen#uberblick, 26.03.2024, Dessau, 2024e.

Umweltbundesamt (UBA): Wasserstoff – Schlüssel im künftigen Energiesystem (15.04.2023), https://www.umweltbundesamt.de/themen/klima-energie/klimaschutz-energiepolitik-in-deutschland/wasserstoff-schluessel-im-kuenftigen-energiesystem#Rolle, 26.03.2025, Dessau, 2024f.

Umweltbundesamt: Treibhausgasminderungsziele Deutschlands (02.05.2023), https://www.umweltbundesamt.de/daten/klima/treibhausgasminderungsziele-deutschlands#nationale-treibhausgasminderungsziele-und-deren-umsetzung, 26.03.2024, Dessau2023.

Uslar, Mathias; Specht, Michael; Rohjans, Sebastian; Trefke, Jörn; González José M.: The Common Information Model CIM: IEC 61968/61970 and 62325 – A Practical Introduction to the CIM, Wiesbaden 2012.

Weber, Christopher L.; Koomey, Jonathan G.; Matthews, H. Scott: The Energy and Climate Change Implications of Different Music Delivery Methods, in: *Journal of Industrial Ecology* Vol. 14, Issue 5, 2010, pp. 754–769.

Jan Lüken

24 Towards a Circular Economy – Business Rationale and Steps Forward

24.1 Introduction

The circular economy is a topic of increasing awareness – among the public, in politics, in business and in science. However, scientific publications on this topic come more often from technical disciplines than from business administration and even less often from economics. Yet policymakers and business leaders are already developing strategies and roadmaps for the transition to a circular economy. So what does "circular economy" mean, where does it lead, and how can companies prepare for it? This paper attempts to answer these questions by combining a political, an economic and a business perspective.

24.2 What is a Circular Economy?

24.2.1 From Cowboys to Spaceships – an Introduction

The discussion of the circular economy as an alternative method of production starts with spaceships and cowboys. With the publication of an article in the 1960s by Kenneth E. Boulding (Boulding 2013),[1] the concept of a circular economy first emerged. He described the future economy as a spaceship. Resources in a spaceship must be used repeatedly in a self-renewing system. It is this way, because in space there is no possibility to gain newly harvested so called "virgin" materials (Grafström and Aasma 2021;[2] Kirchherr et al. 2018;[3] Ayres 1998).[4] The system of a spaceship is closed, and in a material sense so is planet earth with its atmosphere. Neither substantial materials are leaving the planet earth nor entering it from space. But people have been residing on earth for thousands of years as if resources were filled up from a hidden source.

1 Boulding, Kenneth E. 2013. "The Economics of the Coming Spaceship Earth." In *Interdisciplinary Economics*, edited by Kenneth E. Boulding, Wilfred Dolfsma, and Stefan Kesting, 3–14. Routledge Studies in the History of Economics. Abingdon, Oxon: Routledge.

2 Grafström, Jonas, and Siri Aasma. 2021. "Breaking Circular Economy Barriers." *Journal of Cleaner Production* 292: 126002. https://doi.org/10.1016/j.jclepro.2021.126002.

3 Kirchherr, Julian, Laura Piscicelli, Ruben Bour, Erica Kostense-Smit, Jennifer Muller, Anne Huibrechtse- Truijens, and Marko Hekkert. 2018. "Barriers to the Circular Economy: Evidence from the European Union (EU)." *Ecological Economics* 150: 264–72. https://doi.org/10.1016/j.ecolecon.2018.04.028.

4 Ayres, Robert U. 1998. "Eco-Thermodynamics: Economics and the Second Law." *Ecological Economics* 26.

https://doi.org/10.1515/9783110784299-024

Figure 24.1: A cowboy constructing a spacehip, figure created by author with DALL.E).

Boulding described this way of living as a "cowboy economy", who stated that "the cowboy being symbolic of the illimitable plains and also associated with reckless, exploitative, romantic, and violent behavior [. . .]" (Boulding 2013, 339).[5] The cowboy economy operates as if there were no material or environmental restrictions (see Figure 24.1).

We now understand that there are such limits. But our spaceship is still under construction. In this sense the circular economy is a vision of a more ecologically sustainable economic system. Many companies embrace this idea. There are entrepreneurs and businesses with business models based on circular economy concepts such as cradle-to-cradle (Drabe 2022).[6] Such businesses are certified by the C2C-Centre. Currently, nearly 200 firms from various industries are listed, including apparel, building materials, ceramics, concrete, consumer goods, electrical and electronic manufacturing, fashion, food, furniture, textiles, and so on. As we will see, these branches do not

5 Boulding, Kenneth E. 2013. "The Economics of the Coming Spaceship Earth." In *Interdisciplinary Economics*, edited by Kenneth E. Boulding, Wilfred Dolfsma, and Stefan Kesting, 3–14. Routledge Studies in the History of Economics. Abingdon, Oxon: Routledge.

6 Drabe, Viktoria. 2022. Innovating in a Circular Economy. Nomos Verlagsgesellschaft mbH & Co. KG. https://doi.org/10.5771/9783828878426.

produce waste in an environmentally friendly manner.[7] There are also investment funds who buy assets from companies that are considered to have chances for a successful circular transformation. We might be surprised to find well-established companies in the portfolio. The ECPI Circular Economy Leaders Equity Index promises to "offer investors exposure to listed companies in Global developed markets, characterized by a positive ESG profile and that are the ones best placed to grasp the benefits deriving from the adoption of circular economy models and companies that have been able to translate circular economy principles into business practices." (ECPI Group 2019)[8]

But the composition of the index may not be what one would expect when reading this description. Top of the list comes General Mills, a company that manufactures convenience foods and thus likely generates a lot of packaging waste. The index also includes Coca-Cola, IBM, Colgate, and Linde. On the other hand, companies that we would expect to see are listed, such as AutoZone, an automotive parts retailer, and Waste Management, the largest recycling company in the United States. Companies such as BMW have implemented circular economy strategies in Germany. The chemical industry, including major players such as BASF, Covestro, and Wacker, sees circular economies as a potentially lucrative field of business.

Meanwhile, activists, non-governmental organizations, and environmental movements applaud circular economy approaches and demand political or regulatory support for their implementation. Many national governments and the European Union include support for the circular economy in their economic and environmental policies. How is it that politicians, managers, business owners, and environmentalists all support the same idea? In the following sections, I will attempt to provide an answer, with a particular emphasis on the business rationales that could set incentives for companies to adopt a circular transformation.

First, I will argue that the circular economy is appealing to many stakeholders and political actors because it promises to soften the previously existing trade-off between a thriving economy and an intact environment. Second, I will provide an overview of the circular economy in Europe. Third, based on the findings of the second section, I will discuss business rationales for the transition to a circular economy.

24.2.2 Circular Economy as a Narrative

> Scaling up the circular economy from front-runners to the mainstream economic players will make a decisive contribution to achieving climate neutrality by 2050 and decoupling economic

7 See http://www.c2c-centre.com/ for products and projects certified.

8 ECPI Group. 2019. "ECPI Circular Economy Leaders Quantity Index." https://www.ecpigroup.com/wp-content/uploads/rules/ECPI_Circular_Economy_Leaders_Equity_INDEX_RULES.pdf.

> growth from resource use, while ensuring the long-term competitiveness of the EU and leaving no one behind. (European Commission 2020a)[9]

The preceding quote is an example of circular economy political communication. It promises both economic prosperity and environmental sustainability. It's no surprise that the circular economy concept is gaining traction in the public sphere, as evidenced by Figure 24.2. The first graph shows how frequently the term “circular economy” appears in published books that are part of the GoogleBooks database. This database can be searched for keywords and used to create diagrams like these. Everyone can use the “GoogleBooks Ngram Viewer” (https://books.google.com/ngrams/) to check it out for themselves. A ngram is a text fragment such as the term “circular economy.” The graph was created with the ngram-package in R (Drew Schmidt and Christian Heckendorf 2013)[10] and with the ggplot2-package (Wickham 2016),[11] that is used for most figures in this text.

Economic narratives, according to Shiller (2017),[12] are constructs that appear to spread like a disease in the public sphere. Such as a virus spreads from body to body narratives spread from mind to mind – growing on an exponential path and fading away when they have become less effective. The circular economy narrative's isolated development has an impressive exponential shape. However, its true potential becomes clearer in the second chart when the time series is compared to another economic narrative, bitcoin. Bitcoin is obviously mentioned in more books, but the circular economy is within sight. The time series of scientific publications contained in the database Webofscience about the circular economy in the chart on the right is also growing at an exponential rate. This implies a greater academic investment in research in this area. More and more scientific resources are being directed toward studying the circular economy.

However, there has been relatively little economic research on the circular economy. Only about 10% of the articles on topics related to the circular economy are classified as management, business, or economics literature – as one can see in Figure 24.3. Today, the circular economy is primarily a technical concept aimed at improving production methods. Although the concept is gaining popularity in management sciences, there is still relatively few “economics” in the circular economy.

---

9 European Commission. 2020a. Circular Economy Action Plan: For a Cleaner and More Competitive Europe. Luxembourg: Publications Office of the European Union.

10 Drew Schmidt, and Christian Heckendorf. 2013. “Ngram: Fast n-Gram 'Tokenization'.” https://cran.r-project.org/web/packages/ngram/ngram.pdf.

11 Wickham, Hadley. 2016. Ggplot2: Elegrant Graphics for Data Analysis. Second edition. Use r! Switzerland: Springer.

12 Shiller, Robert. 2017. “Narrative Economics.” *American Economic Review*, no. 107 (4): 967–1004. https://www.aeaweb.org/articles?id=10.1257/aer.107.4.967.

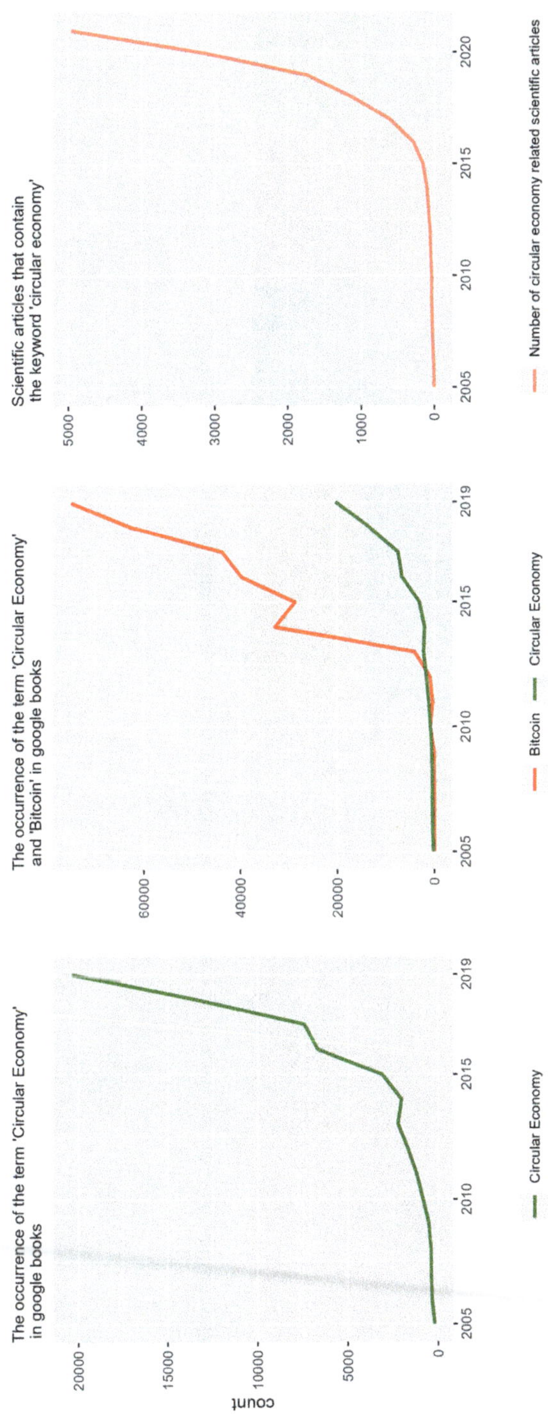

Figure 24.2: The term "circular economy" in books and articles, sources: GoogleNgram and webofscience.

Figure 24.3: Distribution of circular economy articles across scientific disciplines; Source: webofscience as of August 2022.

24.2.3 The Concept of a Circular Economy

But what are all these publications about, what is the circular economy? A popular definition of the circular economy is made by Geissdoerfer et al. (2017).[13] Circular economy is defined as a "regenerative system in which resource input and waste, emission, and energy leakage are minimised by slowing, closing, and narrowing material and energy loops. This can be achieved through long-lasting design, maintenance, repair, reuse, remanufacturing, refurbishing, and recycling."

Consider the inverse to grasp the essential aspects of the circular economy. A linear economy Sørensen (2018)[14] is nothing more than Bouldings' "cowboy economy." In general, it is a simplistic production model that implies natural resources are employed as inputs in a manufacturing process that produces goods. These items are utilized by economic subjects. On the one hand, using the products provides utility or value; on the other side, it generates waste. This garbage is released into the environment. Dumping can take many different shapes and occur through many different mechanisms. Plastics, for example, end up in the ocean, in the atmosphere as carbon dioxide, or as residual garbage in the earth.

This model is commonly referred to as "Take, Make, Waste" (Drabe 2022).[15] This strategy of dealing with industrial wastes is increasingly at odds with the environment's ability to recover. According to ecological economists such as Dasgupta (2021),[16] this production model results in an overuse of natural capital. Natural capital is comprised of more than only natural resources such as fossil fuels, minerals, metals, and fertile soils. Natural capital also includes ecosystems that provide free services and products to civilization, such as breathable air, a stable atmosphere, and oceans that govern climate. Waste generated by manufacturing and consumption endangers these ecosystems. As Pearce and Turner (1990) pointed out in a textbook more than three decades ago,[17] overuse is driven by the fact that natural commodities and services are not priced. They are free to use for both businesses and consumers. The ecosystem has become ecologically unsustainable as a result of this overuse of nature.

It should be noted that the "linear model" is oversimplified. Sectors of economic systems manage waste, recycle materials, and coordinate product reuse and sharing.

---

13 Geissdoerfer, Martin, Paulo Savaget, Nancy M. P. Bocken, and Erik Jan Hultink. 2017. "The Circular Economy – a New Sustainability Paradigm?" *Journal of Cleaner Production* 143: 757–68. https://doi.org/10.1016/j.jclepro.2016.12.048.

14 Sørensen, Peter Birch. 2018. "From the Linear Economy to the Circular Economy: A Basic Model."

15 Drabe, Viktoria. 2022. Innovating in a Circular Economy. Nomos Verlagsgesellschaft mbH & Co. KG. https://doi.org/10.5771/9783828878426.

16 Dasgupta, Partha, ed. 2021. *The Economics of Biodiversity: The Dasgupta Review*: Abridged Version. Updated: 2 February 2021. London: HM Treasury.

17 Pearce, David W., and R. Kerry Turner. 1990. *Economics of Natural Resources and the Environment.* Baltimore: Johns Hopkins University Press.

Product or material use cycles already exist in these activities. What distinguishes the circular economy as a techno-economic model is its emphasis on these activities. They are no longer at the end of a chain, but rather at the heart of circular economy manufacturing models. As a result, the "circular economy" as shown in Figure 24.4 is the polar opposite of the linear model. A circular economy creates a material flow that generates as little waste as possible by keeping materials within a production-use -reuse cycle. Instead of extracting materials from nature and returning them to it after use as waste, the materials are kept within the production system and used as frequently as possible. The manufacturing process and the products themselves should be designed to minimize the flow of waste into nature from the use or production of goods. This flow is actively designed, for example through the implementation of remanufacturing and recycling possibilities for products. The circular economy as pictured by the European Environment Agency in Figure 24.4 has four interlocking circles.

The Big "Rs"

To obtain a sense of the various components of the circular economy strategy, read the image from the inside out. The five "Rs" – reuse, repair, redistribute, refurbish, and remanufacture – make up the center of the circle. Recycling is another very important "R" which is directly related to trash treatment. As such, recycling is seen as a crucial element of a circular economy in the literature (Geissdoerfer et al. 2017).[18] The Rs are used to create material cycles that have five stages: design, production and distribution, consumption and stock, waste and materials. The manufacturing and usage of things should be organized in such a way that one or more of the Rs are utilized.

Two things should be noted. First, that the "Rs" are used to skip phases of the cycle. A refurbished product is not redesigned, but it must be given to the new consumer. Ideally, reuse and refurbishment are considered from the beginning of the design process for new items, or the product itself can bypass the material and design steps. The product is then directly prepared for usage by other households or firms in the production and distribution stages. Second, that companies can produce in a partially circular economy. They use both recycled and newly extracted materials, as Schlosser, Chenavaz, and Dimitrov (2021) points out.[19]

---

18 Geissdoerfer, Martin, Paulo Savaget, Nancy M. P. Bocken, and Erik Jan Hultink. 2017. "The Circular Economy – a New Sustainability Paradigm?" *Journal of Cleaner Production* 143: 757–68. https://doi.org/10.1016/j.jclepro.2016.12.048.

19 Schlosser, Rainer, Régis Y. Chenavaz, and Stanko Dimitrov. 2021. "Circular Economy: Joint Dynamic Pricing and Recycling Investments." *International Journal of Production Economics* 236: 108117. https://doi.org/10.1016/j.ijpe.2021.108117.

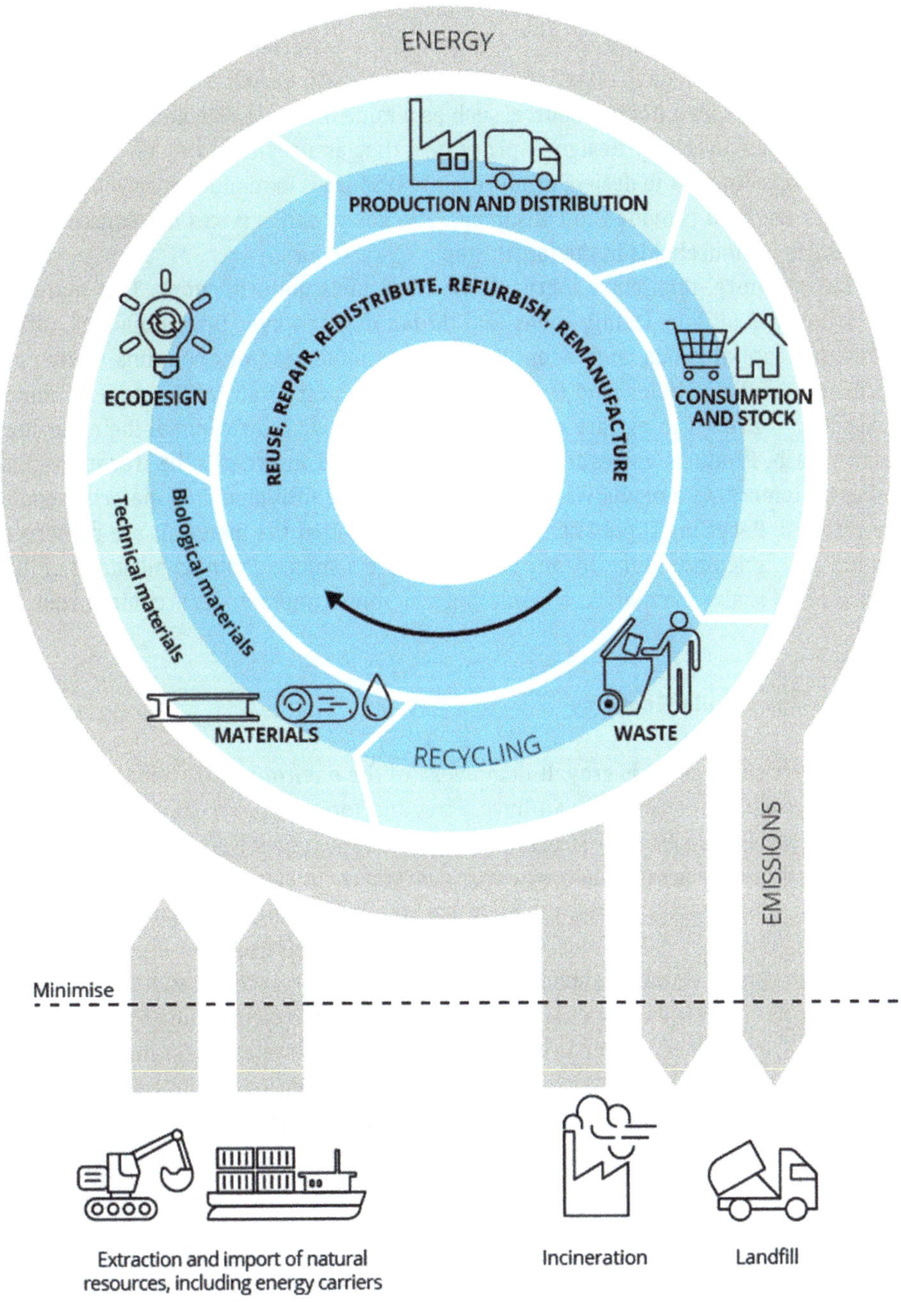

Figure 24.4: The Circular Economy as a System, source: European Environment Agency.

The employment of the "Rs" results in a biological or technical material cycle, depending on which materials are used. The stages of the material cycles are the same, no matter if wood (biological material) or iron (technical material) is used. Materials are first extracted from a natural source, such as a mine or forest. The materials are then employed in a precisely designed procedure that assures they can be reused in a cycle.[20] According to the design, producers use resources to manufacture goods. These products are then provided to consumers. The goods and services are consumed by consumers or households in the fourth step.

Furthermore, this phase increases an economy's material stock. This material stock includes all long-lasting goods that do not degrade over time: buildings, infrastructure, capital goods, and so on. Because this material stock contains many raw elements, Schiller, Müller, and Ortlepp (2017) refers to it as an "anthropogenic material stock."[21] Even if things are reconditioned, reused, or redistributed, their use generates trash. Products, for example, must be packaged in order to be transported to other customers. As a result, waste is generated. In the fifth phase, all material should be recycled. Recycling should be done in such a way that the materials are preserved within the circle once more. Ideally companies use a supply chain on all stages of this process that is also circular. As a result material inputs and outputs are minimized.

Limits of the Circular Economy

The figure's outer circle is grey. It demonstrates the environmental impact that even circular production strategies can have. First and foremost, even in a circular economy raw materials must be extracted for industrial purposes from natural resources. Recycling is hardly possible in a way that guarantees an endless material flow inside the circle. Ayres (1998) specifies the limit that comes from thermodynamic fundamental rules as follows:[22] "All resources extracted from the environment must eventually become unwanted wastes and pollutants. [. . .] Materials recycling can help (indeed, it must) but recycling is energy intensive and imperfect, so it cannot fully compensate."

Second, going through the processes demands a lot of energy. Let me give an example. An automobile supplier who offers recycled components to manufacturers must ship these parts from numerous locations to the manufacturing side. As a result, the supplier must consume energy. As long as the energy system is not totally CO_2-

20 The model is not consistent at this point because the materials used or the products generated from the materials are not redesigned each time they are utilized in a production cycle.

21 Schiller, Georg, Felix Müller, and Regine Ortlepp. 2017. "Mapping the Anthropogenic Stock in Germany: Metabolic Evidence for a Circular Economy." *Resources, Conservation and Recycling* 123: 93–107. https://doi.org/10.1016/j.resconrec.2016.08.007.

22 Ayres, Robert U. 1998. "Eco-Thermodynamics: Economics and the Second Law." *Ecological Economics* 26.

neutral, the use of energy produces carbon emissions. Third, after materials can no longer be used because their qualities have degraded in numerous usage circles, they must be kept somewhere. They are either land filled or burned, releasing energy and increasing carbon emissions.

Even in a circular economy there are waste and emissions. Many economists have identified such facts as a barrier to circularity in general (Korhonen, Honkasalo, and Seppälä 2018;[23] Georgescu-Roegen 2013).[24] The reason is the second law of thermodynamics which applied to the circular economy, predicts that endless amounts of energy are needed to close the circular material loop. This has led to the skepticism towards the circular economy. However, "solar energy is not a finite resource," as Ayres (1998) noted out.[25] That means, in principle solar energy is not endless but practically unlimited. The discussion emphasizes the importance of abundant energy in the move to a circular economy. This explains why renewable energy sources are an essential component of most circular economy approaches. Because a zero-emission circular economy necessitates emission-free energy inputs. Otherwise, weird incentives could exist: in order to develop a circular economy, a country could subsidize energy demanding recycling, which would result in greater rather than less CO_2 emissions.

Other potential constraints of a circular economy have been recognized by scientists like social, political or cultural barriers (Korhonen, Honkasalo, and Seppälä 2018).[26] The circular rebound is another quite mind blowing one. According to Zink and Geyer (2017) a circular rebound means that the overall production in an economy increases because it has become more circular.[27] This occurs when circularly created commodities are manufactured alongside traditionally produced goods. People just buy more goods in general. And since recycling enables additional material flows the overall material use increases which results in even higher emissions. In short, a circular rebound occurs when increases in eco-efficiency have no positive environmental impact since production and consumption grow stronger (Castro et al. 2022).[28] How can we imagine this rebound effect? When high end smartphones are refurbished, people might buy more often one. Or another example, when plastics are recy-

---

23 Korhonen, Jouni, Antero Honkasalo, and Jyri Seppälä. 2018. "Circular Economy: The Concept and Its Limitations." *Ecological Economics* 143: 37–46. https://doi.org/10.1016/j.ecolecon.2017.06.041.

24 Korhonen, Jouni, Antero Honkasalo, and Jyri Seppälä. 2018. "Circular Economy: The Concept and Its Limitations." *Ecological Economics* 143: 37–46. https://doi.org/10.1016/j.ecolecon.2017.06.041.

25 Ayres, Robert U. 1998. "Eco-Thermodynamics: Economics and the Second Law." *Ecological Economics* 26.

26 Korhonen, Jouni, Antero Honkasalo, and Jyri Seppälä. 2018. "Circular Economy: The Concept and Its Limitations." *Ecological Economics* 143: 37–46. https://doi.org/10.1016/j.ecolecon.2017.06.041.

27 Zink, Trevor, and Roland Geyer. 2017. "Circular Economy Rebound." *Journal of Industrial Ecology* 21 (3): 593–602. https://doi.org/10.1111/jiec.12545.

28 Castro, Camila Gonçalves, Adriana Hofmann Trevisan, Daniela C. A. Pigosso, and Janaina Mascarenhas. 2022. "The Rebound Effect of Circular Economy: Definitions, Mechanisms and a Research Agenda." *Journal of Cleaner Production* 345: 131136. https://doi.org/10.1016/j.jclepro.2022.131136.

clable, but burned after the last cycle, the overall carbon emissions might rise, because there is recycling. The alternative to depose plastics in the ground is clearly not circular, but emits no CO_2.

Summarizing Zink and Geyer (2017)[29] and Castro et al. (2022),[30] a circular economy causes rebound effects when there is insufficient substitution of regular by circular output. The circular economy has the potential to exacerbate environmental damage. Despite the fact that it is intended to be more environmentally friendly. Figure 24.5 from Castro et al. (2022) summarizes the evolution, classifications, and mechanisms that can result in a circular economy rebound effect.[31]

Circular Economy as a Policy Goal

The circular economy concept – although nowhere fully implemented so far – is a fundamental concept of national, international and supranational economic or environmental policies. The circular economy concept is one of the 17 Sustainable Development Goals (SDGs) set by the United Nations in 2017. It is also a component of the European Green Deal. The EU fit for 55 program makes explicit use of circular economy ideas. The European Commission describes their motivation:

> The enormous appetite for resources (energy, food and raw materials) is putting extreme pressure on the planet, accounting for half of greenhouse gas emissions and more than 90% of biodiversity loss and water stress. Scaling up the circular economy will be vital to achieve climate neutrality by 2050, while decoupling economic growth from resource use and keeping resource use within planetary boundaries (European Commission 2020b)[32]

---

29 Zink, Trevor, and Roland Geyer. 2017. "Circular Economy Rebound." *Journal of Industrial Ecology* 21 (3): 593–602. https://doi.org/10.1111/jiec.12545.

30 Castro, Camila Gonçalves, Adriana Hofmann Trevisan, Daniela C. A. Pigosso, and Janaina Mascarenhas. 2022. "The Rebound Effect of Circular Economy: Definitions, Mechanisms and a Research Agenda." *Journal of Cleaner Production* 345: 131136. https://doi.org/10.1016/j.jclepro.2022.131136.

31 Castro, Camila Gonçalves, Adriana Hofmann Trevisan, Daniela C. A. Pigosso, and Janaina Mascarenhas. 2022. "The Rebound Effect of Circular Economy: Definitions, Mechanisms and a Research Agenda." *Journal of Cleaner Production* 345: 131136. https://doi.org/10.1016/j.jclepro.2022.131136.

32 European Commission 2020b. "Critical Raw Materials Resilience: Charting a Path Towards Greater Security and Sus- tainability: COMMUNICATION FROM THE COMMISSION TO THE EUROPEAN PARLIAMENT, THE COUNCIL, THE EUROPEAN ECONOMIC AND SOCIAL COMMITTEE AND THE COMMITTEE OF THE REGIONS." Edited by European Commission. https://eur-lex.europa.eu/legal-content/EN/TXT/?uri=CELEX:52020DC0474.

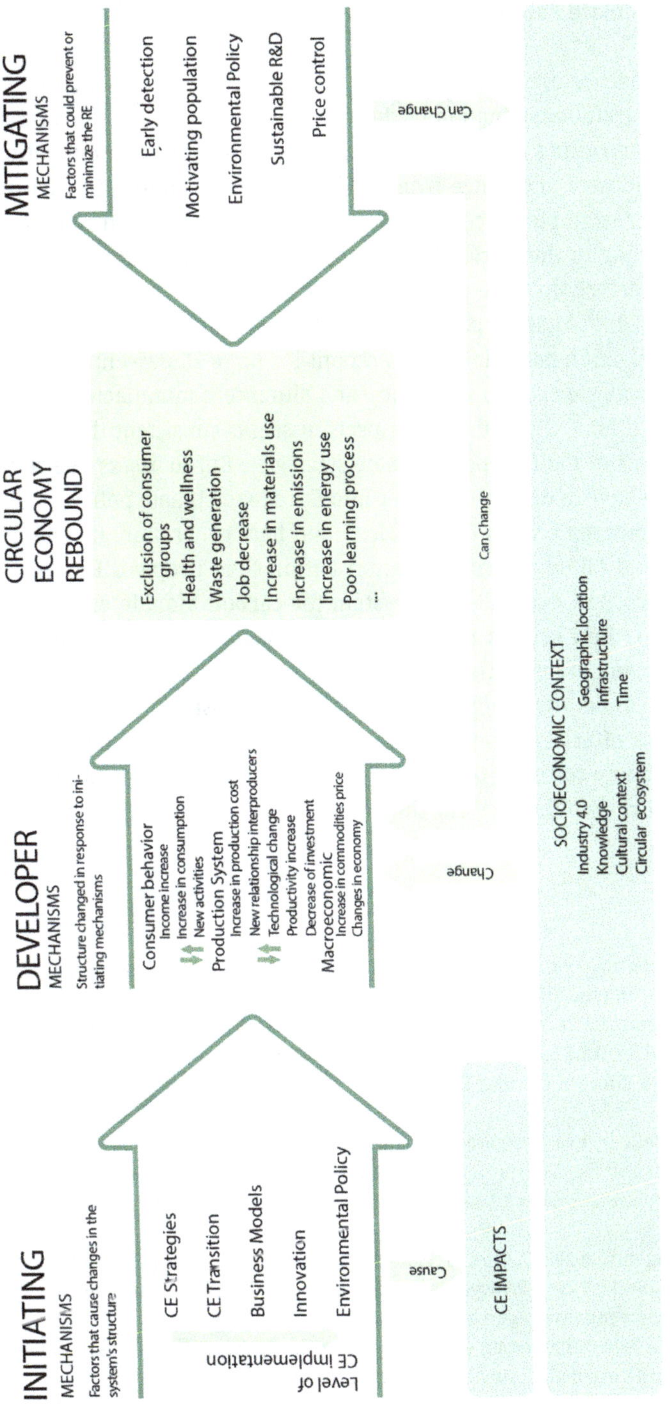

Figure 24.5: Circular Economy Rebound Effect, source: Castro et al. (2022).

Circular Economy and Climate Policies

Circular economy approaches are viewed as climate-friendly alternatives. Cantzler et al. (2020) provides a careful examination of the studies that investigate this probable interrelationship.[33] According to the authors, studies suggest that the mitigating effects depend on the industry and range from a 5 percent (for example, in agriculture) to a 90 percent (for example, in construction-related businesses) carbon emission reduction.[34] According to the European Environmental Agency, using circular economy ideas might decreese up to 60 percent of construction emissions (European Environment Agency 2020).[35] Another paper by the Elle McArthur Foundation indicates that worldwide mitigation potential might account for up to 20 percent of global emissions if applied to food, steel, cement, plastics, and aluminum manufacturing (ElleMcArthur Foundation 2019).[36] Given these prospects, it seems consistent that circular economy goals are part of the European climate strategies. But to assess, the relevance for businesses, we have to dive a bit deeper into European climate policy.

It should be noted, that the sectors (1) electricity and heat generation, (2) energy intensive industries,[37] and (3) commercial aviation within the European Economic Area are part of the European cap-and-trade system for carbon dioxide emissions. The overall amount of CO_2 these sectors are allowed to emit is fixed. Circular economy approaches might help companies in these sectors to reduce carbon emission cost. However, the circular economy will not additional reduce emissions, because they are already capped. This effect is known in the literature as the waterbed effect: Whatever pressure we apply to the bed to lower water in one area, the water will simply flow to another corner of the bed. The total amount of water remains constant.

---

33 Cantzler, Jasmin, Felix Creutzig, Eva Ayargarnchanakul, Aneeque Javaid, Liwah Wong, and Willi Haas. 2020. "Saving Resources and the Climate? A Systematic Review of the Circular Economy and Its Mitigation Potential." *Environmental Research Letters* 15 (12): 123001. https://doi.org/10.1088/1748-9326/abbeb7. Castro, Camila Gonçalves, Adriana Hofmann Trevisan, Daniela C. A. Pigosso, and Janaina Mascarenhas. 2022. "The Rebound Effect of Circular Economy: Definitions, Mechanisms and a Research Agenda."

34 Cantzler, Jasmin, Felix Creutzig, Eva Ayargarnchanakul, Aneeque Javaid, Liwah Wong, and Willi Haas. 2020. "Saving Resources and the Climate? A Systematic Review of the Circular Economy and Its Mitigation Potential." *Environmental Research Letters* 15 (12): 123001. https://doi.org/10.1088/1748-9326/abbeb7.

35 European Environment Agency. 2020. "Cutting Greenhouse Gas Emissions Through Circular Economy Actions in the Buildings Sector." https://www.eea.europa.eu/themes/climate/cutting-greenhouse-gas-emissions-through/cutting-greenhouse-gas-emissions-through.

36 ElleMcArthur Foundation. 2019. "Completing the Picture: How the Circula Economy Tackles Climate Change." https://emf.thirdlight.com/link/rl0yth77pffc-jqkp5d/@/preview/1?o.

37 Including oil refineries, steel works, and production of iron, aluminium, metals, cement, lime, glass, ceramics, pulp, paper, cardboard, acids and bulk organic chemicals.

Edenhofer et al. (2017) provide an intuitive introduction into this topic.[38] On the other hand, for companies covered by the European Emisions Trading System (EU-ETS), adapting circular economy approaches could be a way to save cost. This scenario occurs when it is cheaper for businesses to adopt circular economy techniques than to buy EU-ETS certificates and to stick to the established production method.

Setting up legislative frameworks to encourage the circular transformation in sectors except from the European Emissions Trading System (EU-ETS) on the other hand can have a direct carbon emission reduction effect, in principle. The lack of fixed emissions on the European level is the reason. But, the requirements companies must follow depend more on the legislation of the member states in which they operate. Some countries have their own cap-and-trade systems, while others rely on carbon taxes, and still others use a combination of the two. In essence, it is possible that the waterbed effect can occur here on a national level as well.

However, even if emissions are fixed, regulations or incentives for applying circular economy ideas do not have to be useless. Carbon-prices are a necessary condition for climate protection. However, they are insufficient on their own. Carbon prices should be supplemented by subsidies or other incentives in order to encourage environmental innovation and technological adaptation in this field. Hence, carbon pricing should be supported by subsidies or other incentive schemes to boost the development of green technologies [Acemoglu et al. (2012);[39] Koch et al. (2022)].[40] The reason for this is one of the features of any successful invention: the society benefits from successful innovation, while the company that produced the innovation bears the majority of the development weight. Such favorable external impacts require governmental backing, or else an economy's innovation activities will be inefficiently low.

The EU-Circular Economy Action Plans

With these conditions in mind, we may now investigate the European circular economy approach. The European Commission published the first Circular Economy Ac-

38 Edenhofer, Ottmar, Flachsland Christian, Christoph Wolff, Lisa Kahtarina Schmid, Anna Leipprand, Nicolas Koch, Ulrike Kornek, and Michael Pahle. 2017. "Decarbonization and EU ETS Reform: Introducing a Price Floor to Drive Low-Carbon Investments." Policy Paper. Mercator Research Institute on Global Commons and Climate Change. https://www.mcc-berlin.net/fileadmin/data/C18_MCC_Publications/Decarbonization_EU_ETS_Reform_Policy_Paper.pdf.

39 Acemoglu, Daron, Philippe Aghion, Leonardo Bursztyn, and David Hemous. 2012. "The Environment and Directed Technical Change." *The American Economic Review* 102 (1): 131–66. https://doi.org/10.1257/ aer.102.1.131.

40 Koch, Nicolas, Lennard Naumann, Felix Pretis, Nolan Ritter, and Moritz Schwarz. 2022. "Attributing Agnostically Detected Large Reductions in Road Co2 Emissions to Policy Mixes." *Nature Energy*.

tion Plan in 2015. Zink and Geyer (2017) describes the European Commission's (EC) goal as follows:[41]

> The EC's position is that the circular economy transition will be driven by private firms and consumers, with regulatory agencies actively promoting the concept by creating regulatory frameworks, sending economic signals such as recovery targets and quotas, and providing economic incentives and assistance, including preferential governmental procurement programs.

The second Circular Economy Action Plan was released in 2020 (European Commission 2020a), with the same goal in mind.[42] It should be noted that such action plans are not legislation. They are more of a form of agenda that the European Commission establishes for a specific time period. Following that, the appropriate rules and regulations are outlined in Directives that must be adopted by the European Parliament and the European Council. National implementation acts follow in each member states before a proposal becomes legally binding.

Both programs included numerous proposed measures and targets. I will concentrate on the second action plan. The European Commission defines a policy framework based on three areas. The first is the development of environmentally friendly products. In this case, the European Commission mostly depends on product standards, such as those for specific chemicals. The second component is consumer and public buying empowerment. Customers should be able to make more circular product choices as a result of updated consumer information regulations. The third element is the control of production processes to promote increased circularity, such as recycling quotas for specific materials. At its core, Brussels wishes to improve the usage of the Rs in the European economies. Durability, upgradeability, reusability, reparability, remanufacturing, and high-quality recycling are all important considerations. Furthermore, the text advocates for reduced carbon and environmental footprints, restrictions on the use of dangerous chemicals, and advances in energy and resource efficiency.

The proposed rules address seven significant product value chains: (1) electronics and information and communication technology, (2) batteries and cars, (3) packaging, (4) plastics, (5) textiles, (6) construction and buildings, and (7) food, water, and nutrients. As we will see, the European Commission is addressing industries that generate a lot of (recyclable) waste (building) or a lot of junk (packaging) and where the recycling rates are low (electronics and textiles).

The plan identifies research, innovation, and digitization as the primary drivers of the circular economy transition. The European Commission itself reports funding

---

41 Zink, Trevor, and Roland Geyer. 2017. "Circular Economy Rebound." *Journal of Industrial Ecology* 21 (3): 593–602. https://doi.org/10.1111/jiec.12545.

42 European Commission. 2020a. *Circular Economy Action Plan: For a Cleaner and More Competitive Europe*. Luxembourg: Publications Office of the European Union.

of 10 billion Euros for the period between 2016 and 2020. The sources of finance in place today remain predominantly the same, but the European Union communicates much bigger funds on its European Circular EConomy Stakeholder Platform:[43]

- European Investment Bank through its program "EU Finance for Innovators" where circular innovators can apply for funding and for advise
- European Investment Fund who supports SMEs in general
- Horizon Europe: EU Research and Innovation program, that offers innovation funding
- Regional policy when investments can contribute to the circular transition
- LIFE programme: funding instrument for the environment and climate action created
- Single Market Programme
- Specific programs of the member states
- EU sustainable finance taxonomy also possibly guide private funds towards sustainable investments. This is especially relevant for companies, that want to establish a sustainable ways for critical raw material supply.
- Furthermore, there is a new energy lending policy, that is meant to support "projects relating to the supply of critical raw materials needed for low-carbon technologies" (European Commission 2020b).[44]

Summing up without going too much into detail, the selected policy approach seems to be a mixture out of command-and-control regulation, improved consumer information and support for innovation. The issue with this method is that it does not create new price mechanisms. Prices provide information concerning scarcity. This is also true when it comes to pricing the usage of natural capital. The higher the price, the less natural capital is available for utilization. As a result, beyond a certain extent, a working price mechanism makes it unappealing for firms to employ nature as capital. Prices can act as a deterrent to natural fatigue which is the entire point of carbon pricing. Section three will go into more detail on this. Before, I examine the existing situation of the circular economy in Europe.

43 https://circulareconomy.europa.eu/platform/en/financing-circular-economy.

44 European Commission 2020b. "Critical Raw Materials Resilience: Charting a Path Towards Greater Security and Sus- tainability: COMMUNICATION FROM THE COMMISSION TO THE EUROPEAN PARLIAMENT, THE COUNCIL, THE EUROPEAN ECONOMIC AND SOCIAL COMMITTEE AND THE COMMITTEE OF THE REGIONS." Edited by European Commission. https://eur-lex.europa.eu/legal-content/EN/TXT/?uri=CELEX:52020DC0474.

24.3 The current State of the Circular Economy in Europe

The European Union put in place a monitoring system offered by Eurostat, the European statistics office.[45] Its target is to provide statistics on the development of circular economies in Europe and other big economies. In this section, we will take a look into the main indicators. The time series cover around 15 to 20 years of circular developments, from the early 2000s to the end of the 2010s. So, what is the current state of the European Union's circular economy? The resource productivity is a good place to start. It is calculated by dividing the gross domestic product (GDP) by the material flow. The resource productivity can be interpreted as measure of the importance of the inner circles of the circular economy graphic model relative to the input arrows that stand for the extraction of natural resources. The greater this number, the fewer material flows are required to generate the given amount of GDP. Material flows in the European Union are declining while GDP is increasing on average. As a result, resource productivity rises.

However, this aggregate picture on the left hand side of Figure 24.6 is incomplete. First, resource productivity appears to have slowed in recent years. Second, the circular economy is gaining traction in the European Union generally, but not in all member countries and not at the same rate. While the largest European economies, Germany and France, are gradually but steadily increasing their resource productivity, Italy and Spain experienced a significant increase in resource productivity beginning in 2010. Poland, on the other hand, has yet to make the transition to more immaterial production.

Although significant, resource productivity is simply one facet of the circular economy. Eurostat gives a plethora of additional indicators gathered to assess the state of a circular economy. The information comes from Eurostat's corporate surveys of European firms. There are statistics for four areas: production and consumption, waste management, secondary raw materials, and competitiveness and innovation. It would take far too much time to analyze all of them. However, having a perspective on some is enlightening. Readers who are not interested in the technical details can skip this section towards section three which applies these findings to the business rationales companies face.

24.3.1 Production and Consumption

Progress in this field is measured in four main categories: 1. EU-self sufficiency for raw materials, 2. green public procurement, 3. waste generation, 4. food waste. Be-

45 https://ec.europa.eu/eurostat/web/circular-economy/.

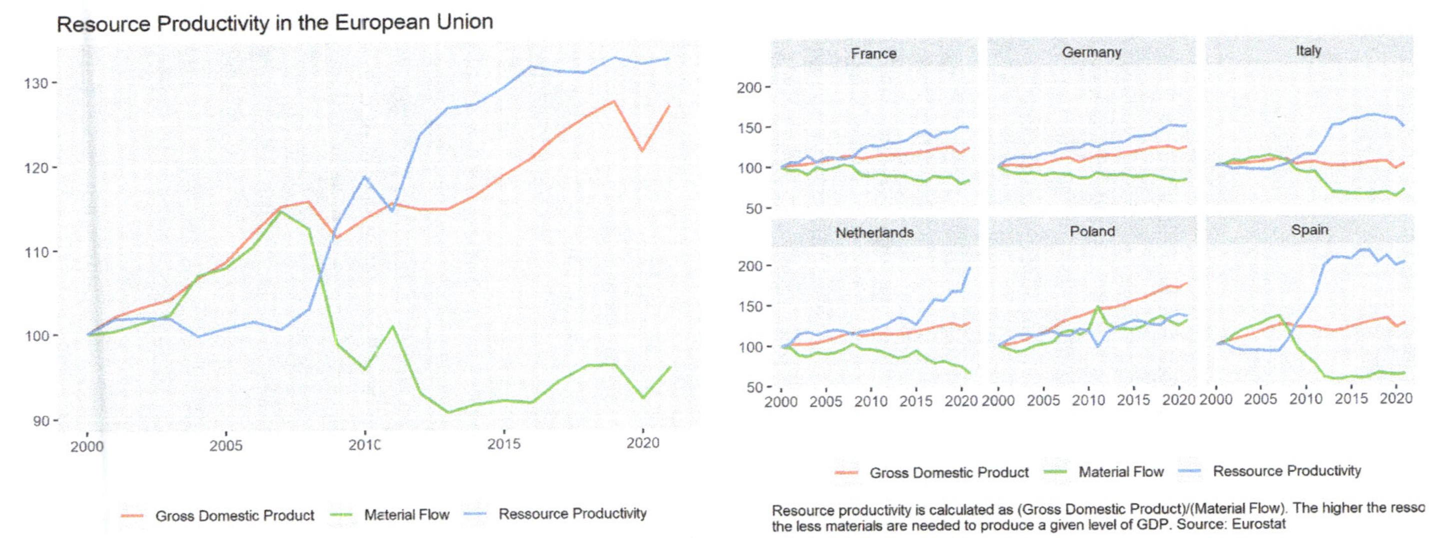

Figure 24.6: Resource Productivity in Europe; figure created by author.

cause there is currently no data on green public procurement and food waste, the analysis focuses on waste generation and self sufficiency.

Waste Generation

When we talk about garbage in our everyday lives, we usually mean municipal waste. Between 2004 and 2018, the average municipal garbage per inhabitant in the European Union was roughly 500 kilograms. That is a lot of trash, especially when viewed globally. Municipal garbage production per person in Asia and Africa is less than 300 kg annually, and in certain places, it is considerably less than 200 kilogram. However, municipal garbage has decreased in many, but not all, European countries over the last 20 years. Municipal waste levels in Germany, for example, have risen on average since 2005. The Netherlands and Spain have reduced their trash production, whereas Poland appears to be catching up from a low base. This is a rather common pattern. There are countries that have advanced in the circular economy, countries where the transition has stalled, such as Germany, and those that are less circularly transformative but economically coming up, such as Poland. Although this should not be considered a general rule because Eastern European countries vary greatly. Some have developed in the similar way as Poland (e.g., the Baltic nations), while others, such as Romania, have significantly decreased municipal garbage levels.

Municipal garbage levels in the European Union as a whole have fallen from more than 515 kilograms per capita in 2010 to less than 480 in 2014. However, municipal garbage has begun to rise again after 2015. This can be explained by the increasing volumes of packaging waste in general, and specifically plastic packaging waste, which is showing the same trend. One probable cause is that packaging-intensive online commerce has gained increasing market share since the 2000s (see Figure 24.7).

Intuitively, one could assume that households are the primary generators of garbage because consumption and waste from packaging and discarding old items are closely associated. However, that is untrue. The largest contribution to overall waste levels is not "our" junk. While 500 kilograms of waste per capita may appear to be a lot, it accounts for only 10 percent of the total waste produced per capita in the European Union. It is useful to gain an understanding of the sectors of the economy that generate the most garbage in order to identify where the most waste is generated. In the European Union, households are the fourth largest producers of garbage, after industries like manufacturing, mining, and construction. Figure 24.8 shows the general trend that trash levels are not declining quickly. There is some waste reduction in the manufacturing, energy, mining and quarrying, and wholesale trash and scrap markets. However, it is rising in the fields of building, water supply, and waste management. Because it must otherwise be disposed of in a landfill or burned, these absolute levels are useful in determining the necessity for effective recycling techniques. For

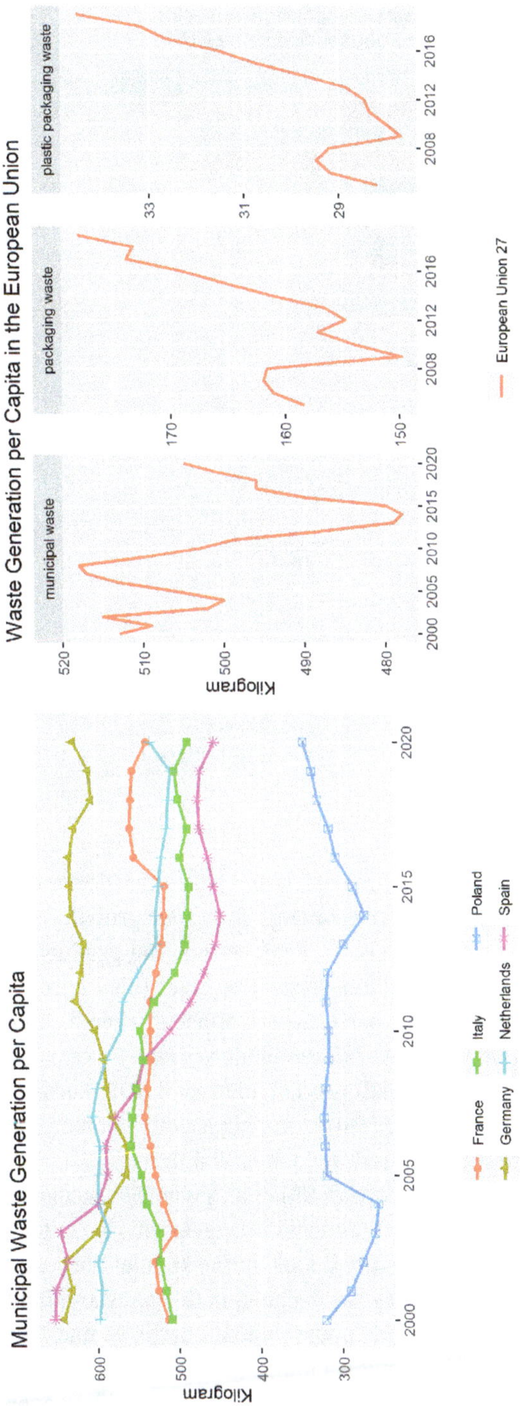

Figure 24.7: Municipal Waste Levels and Composition in Europe; figure created by author.

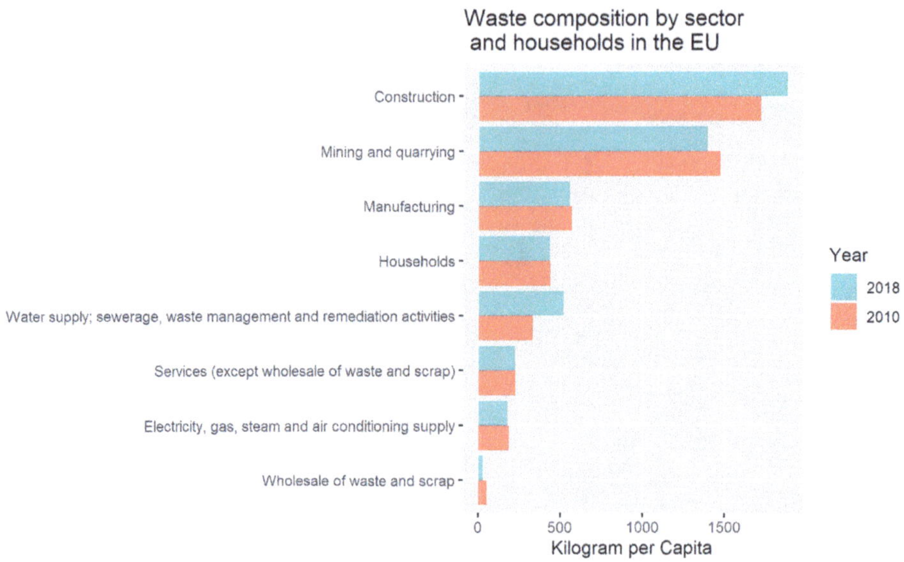

Figure 24.8: Waste Production by Sector; figure created by author.

an economy to grow more circular, recycling must become more productive as trash levels rise.

Waste Intensity and Eco-efficiency

Decoupling economic growth from the use of raw materials and waste creation is one of the main objectives of implementing a circular economy. Economic growth must not lead to proportionally increasing waste levels. To achieve this, the production practices of European economies should become less waste-intensive. This improves in turn an economy's eco-efficiency. An economy performs eco-efficient when as little waste as feasible is produced in relation to the overall economic activity. By calculating the kilogram of trash (mineral wastes excluded) per 1,000 Euros of GDP, Eurostat provides a number that measures this kind of efficiency. The higher this number, the lower the eco-efficiency, because more waste is produced per GDP-unit.

Looking at Figure 24.9, it is clear that the European Union as a whole is becoming more environmentally friendly. The waste intensity reduces. However, this is not true for all member countries. To see this, we have to take a look on the scale of the y-axis of the graphs on the right hand side. Eco-efficiency has declined in Germany and Italy as waste intensity rises for many years. But a plateau was attained in the second half of the 2010s. Since than in Germany waste intensity is falling again. France, the Netherlands, and Spain have reduced waste intensity while increasing eco-efficiency over the last decade. Based on these figures, we may conclude that aggregated eco-efficiency is growing. Since waste intensity is falling. Nevertheless, this result does not apply to all European economies. In Europe, there is a disparity in circular speed.

The waste intensity of consumption is the final indicator in this dimension. Eurostat divides the amount of all waste generated in a country by the amount of materials used for domestic consumption. Because both waste and consumed resources are measured in tonnes, the index displays percentages. The greater the number, the greater the waste produced in relation to the materials consumed. That is, the lower the value, the greater an economy's circular performance. The conclusion is simple for the European Union as a whole and its six largest economies: The performance was poor, and levels in all countries have been growing. With every kilogram of materials consumed, European economies produced more garbage. The trend has shifted in Poland and Spain. A plateau might have been reached in France, Germany, and Italy. We should be cautious in assessing the circular economy based on this indicator. Because of methodological difficulties the interpretation of this indication may require additional context indicators as recommended by Eurostat (2021).[46]

To summarize, the European economy appears to be decoupling in some parts, as waste levels remain constant or fall despite rising GDP. However, trash levels are rising in various industries, most notably construction, which is the largest generator of waste. Furthermore, waste intensities are dropping across Europe as a whole. However, this development is not unique to all member countries. Since 2008, consumption has gotten more wasteful – with the restriction that this indicator is difficult to interpret.

Anyway, rising waste levels, in and of themselves, do not reveal the whole story about the state of the circular economy. It also depends on the economy's ability to utilize used materials as an input for manufacturing. Before we look at the European Union's recycling record, a look at self-sufficiency in raw material utilization provides an explanation for why it may be politically and economically advantageous to focus on the construction of a more circular European economy.

---

46 This performance could be attributed to the index's design: (waste generated in 1000 kg)/(domestic material consumption in 1000 kg). The domestic material consumption (DMC) in the denominator has been decreasing faster than the waste generated in the nominator has. Eurostat uses the same data sources to calculate resource productivity as for domestic material consumption, which is decreasing in many European Union member countries. However, when waste levels remain more or less constant, lowering DMC leads to an increase in waste intensity of consumption. Furthermore, Eurostat reports that the non-metallic mineral component of DMC has a considerable influence on this ratio. Eurostat. 2020. "Metadata - Private Investments, Jobs and Gross Value Added Related to Circular Economy Sectors." https://ec.europa.eu/eurostat/cache/metadata/de/cei_cie010_esmsip2.htm.

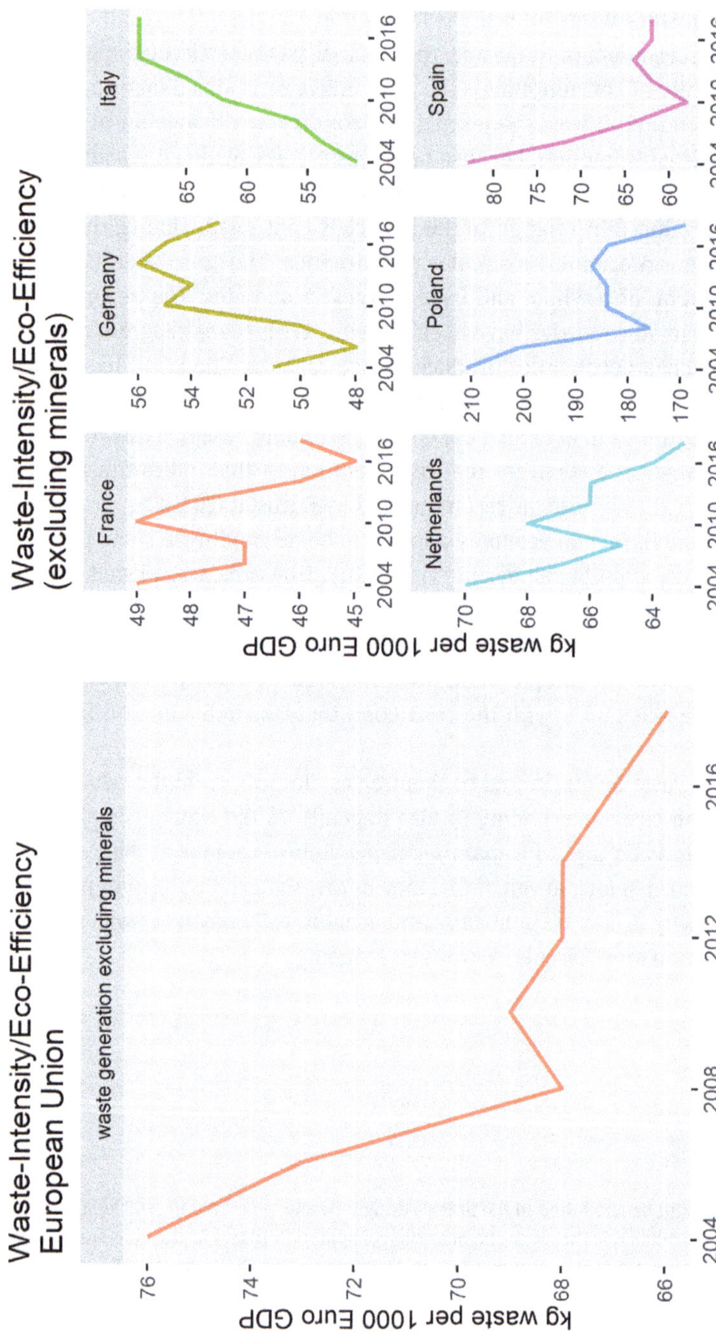

Figure 24.9: Waste Intensity in Europe; figure created by author.

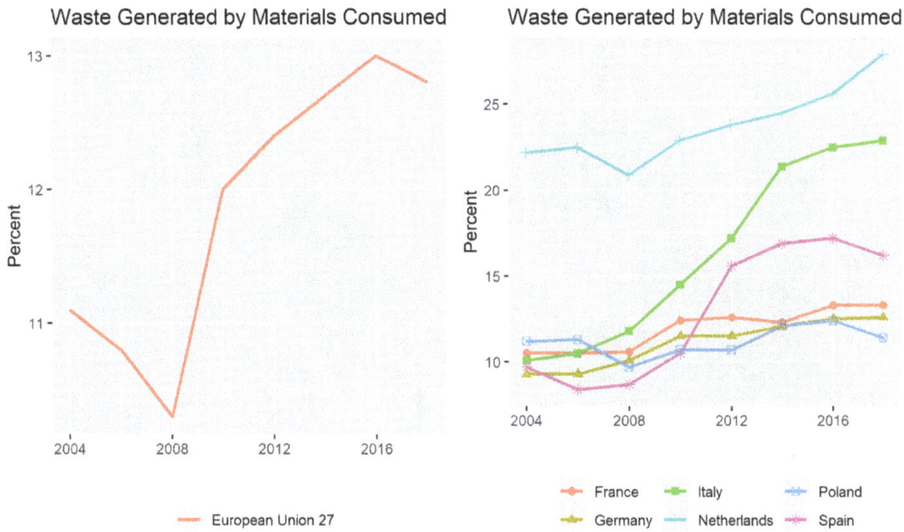

Figure 24.10: Waste generated by Consumption; figure created by author.

Self sufficiency in critical Raw Materials

European economies could be fragile, because they significantly rely on global raw material supply chains. Critical raw materials are defined "raw materials of high importance to the economy of the EU and whose supply is associated with high risk" (Blengini and Grohol 2017).[47] Critical raw materials are widely employed in European industrial manufacturing. Phones (Tungsten), light emitting diodes (LED), semiconductors (Silicon), hydrogen fuel cells, and electrolyzers are among examples (Platinum). They come from all around the world. In any case, even though the market is globally, the crucial raw material supply is not diverse. On the contrary, numerous critical raw material supplier countries have a market share of more than two-thirds. Among these are democratic countries with established trade agreements with the European Union (Brazil and Chile), allied democracies without EU-trade agreements (the United States), autocratic countries (the People's Republic of China), and potentially unstable states such as the Democratic Republic of the Congo. Furthermore, the EU relies on single businesses for specific elements, especially hafnium and strontium (see Figure 24.11).

The level of self-sufficiency varies depending on the materials. The proportion of critical raw materials required in Europe supplied by producers within the European

47 Blengini, Gian Andrea, and Milan Grohol. 2017. Methodology for Establishing the EU List of Critical Raw Materials: Guidelines. Luxembourg: Europäische Kommission; Publications Office of the European Union. https://doi.org/10.2873/769526.

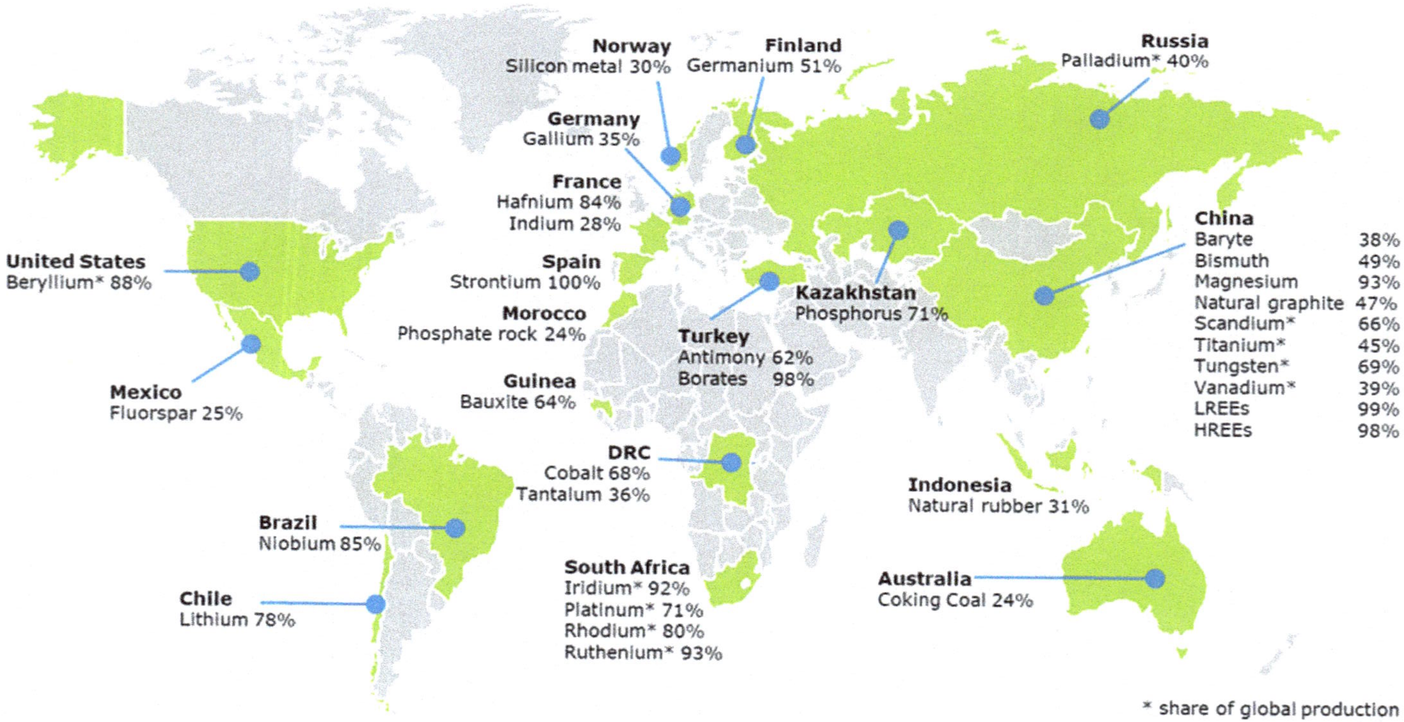

Figure 24.11: European Union Supply of critical Raw Materials, source: European Commission (2020b).

Union is measured as self-sufficiency. Critical raw materials are classified into two types. To begin, there are materials extracted (EXR) in the European Union. Most of them have a very low or no level of self reliance. On average, material extraction from all critical raw materials in the EU is roughly 16.5 percent. Second, there are materials that are processed within the European Union (PCS). The percentage of self reliance for processed materials is substantially greater, with an average of 32.7 percent. Processing means preparing extracted resources for industrial application. For some resources, Europe lacks both processing and extraction capacity. In the European Union, other vital raw materials are extracted and processed (Aluminum, Cobalt, Copper, Iron and Flerovium). There are also resources extracted but not processed in the European Union. This may seem unusual, but certain elements, such as lithium, are mined in Europe and processed somewhere else in the world (European Commission 2020b).[48] The Table 24.1 below provides an overview.

Table 24.1: Materials with some and with no self-sufficiency.

Materials with some self-sufficiency	Materials with no self-sufficiency
Aluminium (EXR & PCS)	Borate
Cobalt (EXR & PCS)	Dysprosium
Gallium (PCS)	Europium
Germanium (PCS)	Magnesium
Iron (EXR & PCS)	Neodymium
Indium (PCS)	Phosphorus
Flerovium (EXR & PCS)	Tantalum
Lithium (EXR)	Yttrium
Limestone (EXR)	
Molybdenum (PCS)	
Natural Graphite (EXR)	
Platinum (PCS)	
Silicium (PCS)	
Vanadium (PCS)	

In the European Union, just a few resources are extracted and processed at a high level. The share of self sufficiency in total material demand in Europe is also modest, with some resources declining while others increasing. As the following graphs in Figure 24.12 demonstrates, the overall picture is mixed. There is certainly no clear direction in which the time series are heading. In terms of circular economy measures, cer-

48 European Commission 2020b. "Critical Raw Materials Resilience: Charting a Path Towards Greater Security and Sus- tainability: COMMUNICATION FROM THE COMMISSION TO THE EUROPEAN PARLIAMENT, THE COUNCIL, THE EUROPEAN ECONOMIC AND SOCIAL COMMITTEE AND THE COMMITTEE OF THE REGIONS." Edited by European Commission. https://eur-lex.europa.eu/legal- content/EN/TXT/?uri=CELEX:52020DC0474.

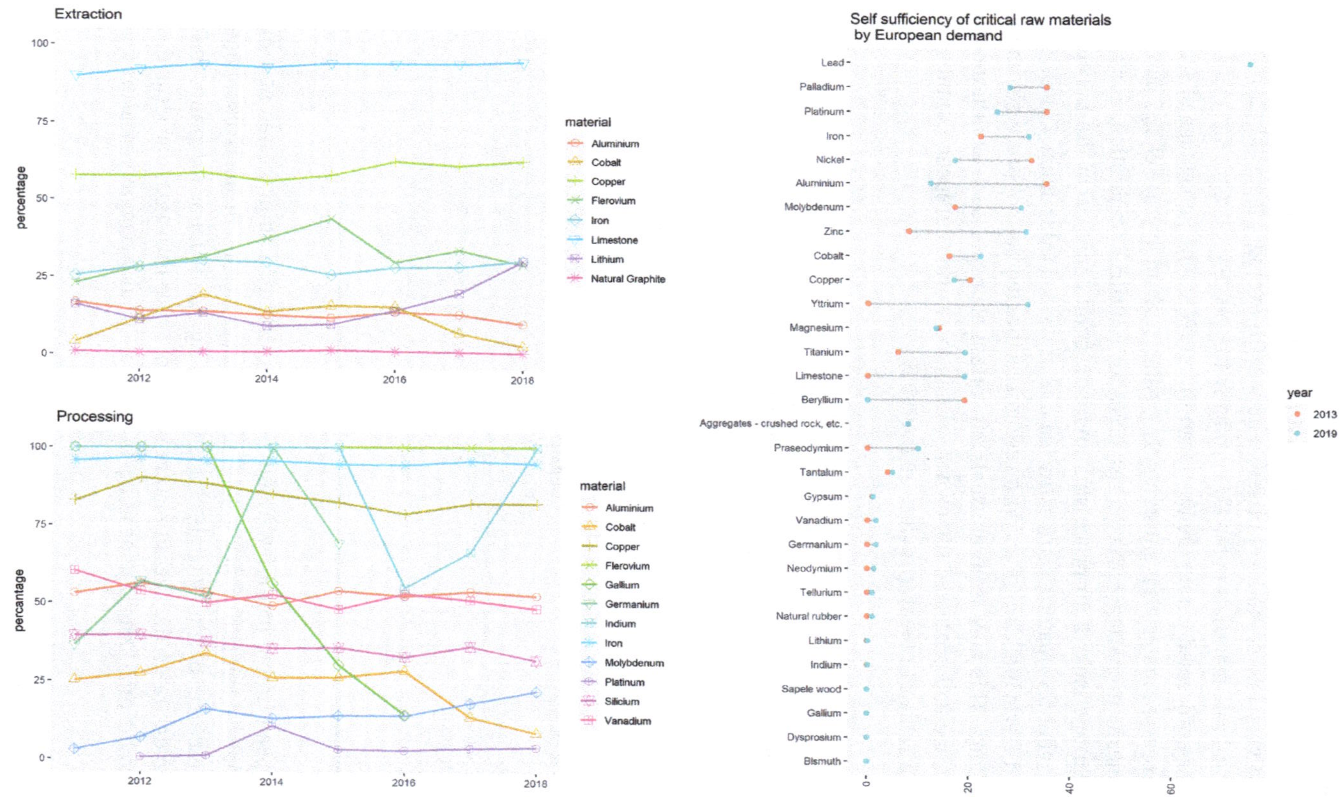

Figure 24.12: Self-sufficiency for Materials that are extracted or processed in the European Union; figure created by author.

tain areas have made growth, while others have stagnated or even shrunk. However, self-sufficiency can be improved. Critical raw materials can be mined and processed within the European Union. For further reading, the European Commission's report on critical raw material resilience is recommended (European Commission 2020b).[49]

While the picture of material sufficiency remains diversified, the European Union's material footprint in Europe has shrunk since 2008. The material footprint is a metric that measures according to Eurostat (2022b) the global demand for material extractions generated by consumption and investment by households, governments and enterprises in the EU.[50] The index includes biomass, metal ores, non-metallic minerals, and fossil energy materials/carriers. The numbers provide information on the quantity and type of materials needed to meet the EU's product demand.

Material footprints like in Figure 24.13 provide useful information. Much of the manufacturing of commodities purchased in Europe is done in Asia. As a result, the EU accounts for a higher share of worldwide consumption and investment than it does of global output. Therefore, the Circular Economy Action Plan asks for material footprint indicators that highlight the EU's environmental responsibilities as a result of products exported to the EU (Eurostat 2022b).[51] A lower material footprint means that the economy relies less on raw materials.

In the 2000s, there were considerable decreases on a European level. However, development has stalled in the last ten years. Furthermore, the result that European countries' development is highly different is proven once more. Once again, we witness a growth in the material footprint in Poland, the largest Eastern European country. However, as in Germany, which is also above the European average in terms of material footprint, the local maximum occurred around 2017. It will be interesting to observe if this trend continues in the future years. However, we can also say that the circular economy lost steam in the 2010s in terms of this statistic, too.

Waste Management

One side of the coin is waste generation and the use of virgin raw materials. The other side of the coin is the management of waste and used materials. The better the waste management, the more circular is an economy and the more sustainable are

49 ———. 2020b. "Critical Raw Materials Resilience: Charting a Path Towards Greater Security and Sustainability: Communication from the commission to the european parliament, the council, the european economic and social committee and the committee of the regions." Edited by European Commission. https://eur-lex.europa.eu/legal- content/EN/TXT/?uri=CELEX:52020DC0474.

50 Eurostat 2022b. "What Are the Main Destinations of EU Export of Waste?" https://ec.europa.eu/eurostat/de/web/products-eurostat-news/-/ddn-20220525-1.

51 Eurostat. 2020. "Metadata - Private Investments, Jobs and Gross Value Added Related to Circular Economy Sectors." https://ec.europa.eu/eurostat/cache/metadata/de/cei_cie010_esmsip2.htm.

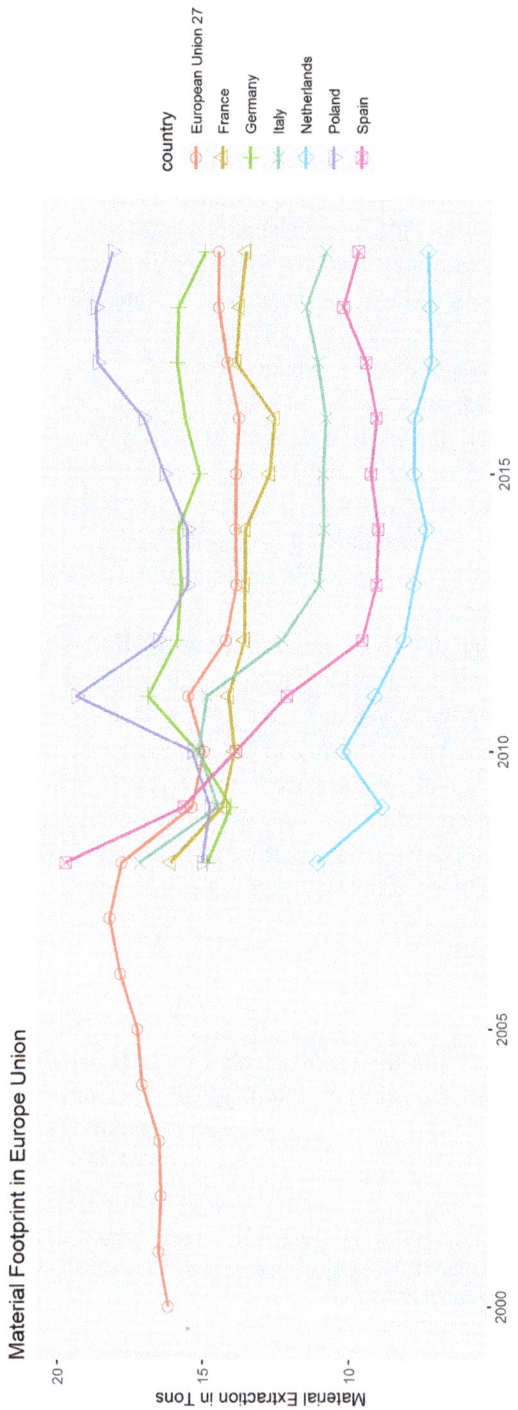

Figure 24.13: Material Footprint in Europe; figure created by author.

high waste levels. Simply put, circular economy is about both reducing and recycling. Looking at the graph in Figure 24.14, we can observe that recycling rates are particularly high for construction, biowaste, paper, metal, and glass. They all recycle at rates greater than 75%. These figures are quite impressive. However, we must consider trade-offs. A high recycling rate may result in increased CO_2-emissions. Municipal waste is burnt to a high degree throughout Europe, as shown by the right side of the figure, primarily for energy generation. This process is frequently referred to as "thermal recycling," which is a euphemism for the fact that it is not a cycle. When garbage is burned, its materials can no longer be utilized, and carbon emissions are not reduced. As a result, incineration does not constitute a component of the circular economy in the sense indicated above.

A look at the time series is interesting since it informs us if there are trends toward higher recycling rates (see Figure 24.15). Because there are still numerous materials with recycling rates that are significantly lower than what we would expect from a circular economy. The picture is mixed again. Recycling of electronic trash, metal, packaging, paper, plastics, and notably wood has taken a negative turn after the rates have increased for many years. It is too early to draw definite conclusions, especially because overall the recycling rate is increasing and around 55 percent. But maybe recycling has reached its maximum capacity for certain kinds of wastes. Rising waste levels raise the question of whether it is necessary to enhance other method for an economy of becoming more circular. Designing things to be easy to recycle, and repairing, remanufacturing, or reusing as additional possibilities, to name a few.

At this point, it is clear why the European Commission has prioritized not only recycling but all of the "Rs." However, the picture is not full. We must also consider the utilization of recycled resources in the manufacturing process. If recycling rates rise, companies must increase their demand for recycled resources. And if they do so in the future, the incentives to recycle more will likely increase. So, let's have a look at how recycled materials have been utilized in manufacturing to see whether we have the same developments as the ones observed so far.

24.3.2 Secondary Raw Materials

Again, recycling used materials is simply one side of the market. Companies in the circular economy sector recycle old materials to create new resources for production. As a result, having instruments that measure the utilization of secondary raw materials is critical. Eurostat provides three perspectives: (1) Contribution of recycled materials to raw materials demand, (2) Circular economy use rate, (3) Trade in recycled raw materials.

The contribution of recycled materials to raw materials demand measures end-of-life recycling input rates (see Figure 24.16). It is an indicator for the employment of recycled scrap and waste in the production of new goods. Thereby the indicator represents the treatment of "end-of-life" products. These products contain certain raw

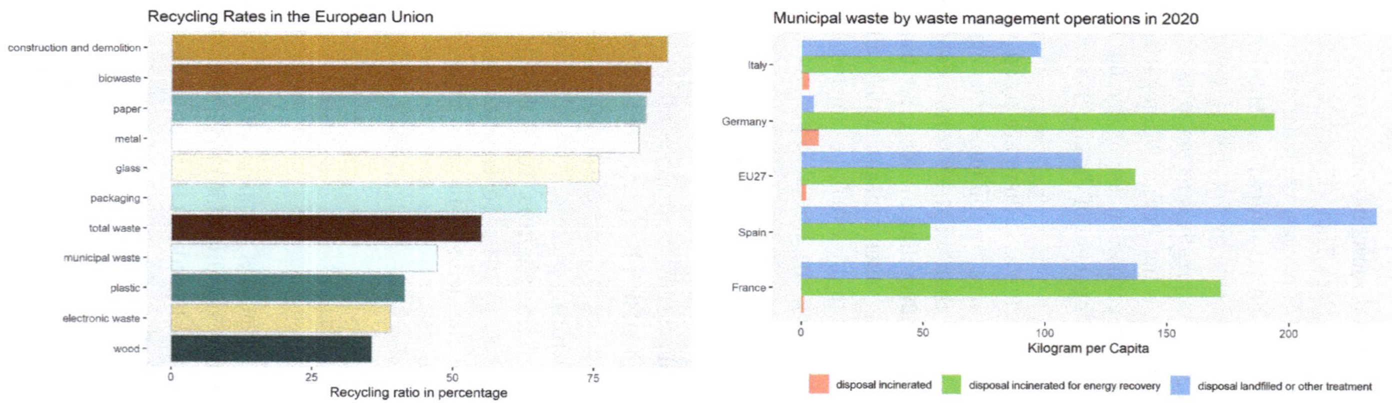

Figure 24.14: Waste Treatment in Europe; figure created by author.

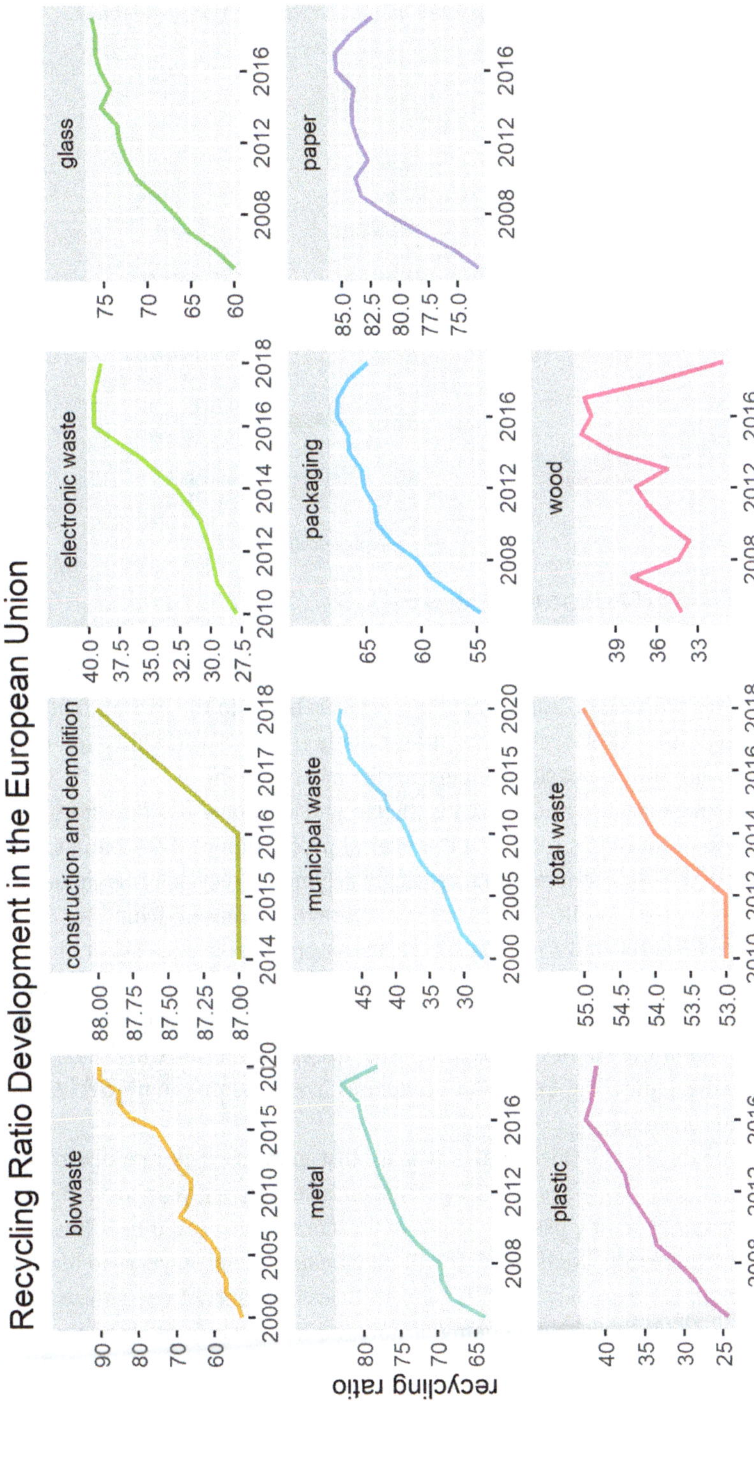

Figure 24.15: Development of Recycling Rates by Waste Type; figure created by author.

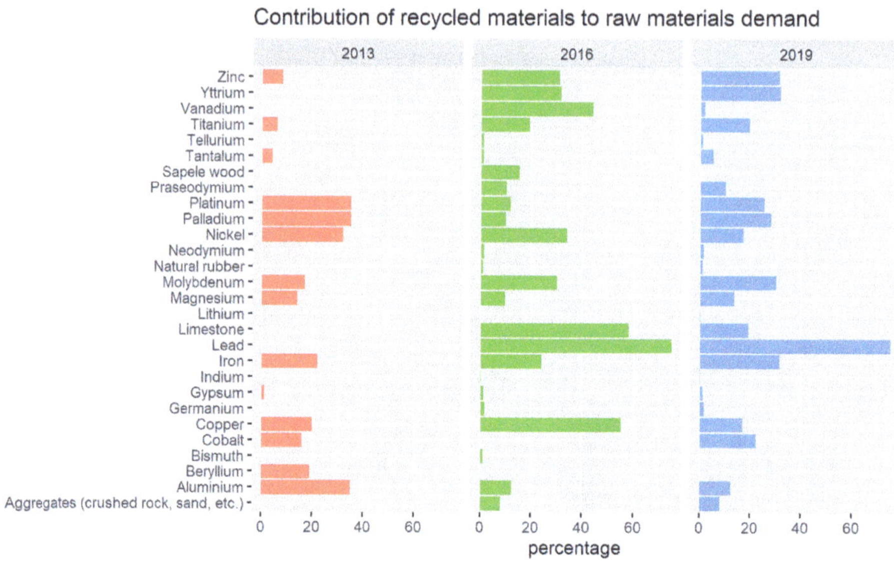

Figure 24.16: Use of Recycled Raw Materials in the European Union.

materials. Through recycling these materials become useable again. The indicator measures the share of these secondary raw materials relative to the overall material input. This contribution of recycled raw materials is increasing steadily for most critical raw materials, but it is below 30 percent for most of them.

The most famous measure in this area is the circular economy use rate. It quantifies the proportion of materials reused in the overall production of the economy. Rising statistics imply that an economy is increasingly relying on recycled raw materials rather than virgin ones. While the European average increased from less than eight percent in 2004 to around 13 percent in 2020, the progress is once again diverse across the six largest economies in Figure 24.17. With the exception of Poland, which continues to employ the circular economy at a rate lower than the European average, the index rises in all of the major economies. Fastest recently in France and Italy, and highest in the Netherlands. The European economies are becoming more circular in terms of raw material reuse.

However, growth rates in Europe as a whole must increase. The rate of circular material use has increased by around 2.5 percent on average per year over the last decade. If the circular economy use rate grows at the same rate as it did in 2019 and if a logarithmic growth process is assumed, it will take an additional 56 years to reach a 50 percent criterion, 72 years to reach a 75 percent threshold, and 84 years to reach a hypothetical 100 percent threshold. Looking at how the Netherlands attained its high level, or France and Italy achieved their excellent growth rates, could be a good place to start. Of course, this computation is hardly a serious empirical inquiry. Multiplying factors does not provide many insights. These numbers are more of an illustration of

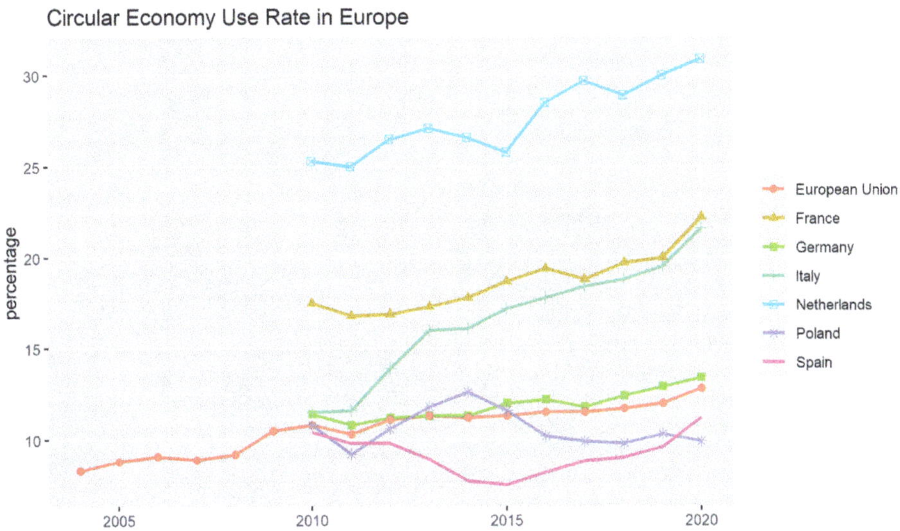

Figure 24.17: Development of the Circular Economy Use Rate; figure created by author.

the intensity of the struggle to become circular than a statistic on which decision makers should rely. The point is that while the current trend is encouraging, the usage of secondary raw materials must continue to rise. However, there are some encouraging signals. The utilization of the circular materials increased by more than 6.5 percent between 2019 and 2020, which is the latest year with the respective data available at the time I am writing this article. This is a promising step in the right direction.

With a wink, an optimist could say, "Only 22 years to get to 50 percent, and 33 years to get to 100 percent!" That would be in line with the time frame of the European Green Deal.

The indicator trade in recycled raw materials measures the recyclable waste, scrap and secondary raw materials shipped between the European Union economies (intra-EU) and with the rest of the world (extra-EU). This indicator has an important background. In a European market not every country must have a fully functional recycling industry. When countries trade, there is space for specialization. What reveals the trade statistics for recycled materials? An increase in absolute numbers for commerce in recycled raw materials throughout the European Union during the last two decades.

Imports into the European Union have been less dynamic since 2010 (see Figure 24.18). However, recycled material exports to non-EU destinations have steadily increased since 2005. The trade balance with the rest of the world has improved as a result of increased export of recyclable raw materials. Remarkably, the primary trade partners for imports and exports differ significantly. Brazil, Argentina, Russia, the United Kingdom, and the United States were the primary importers of recyclable raw materials. Turkey is the most important export destination, followed by the United Kingdom, India, Switzerland, and Indonesia. By the way, the latter countries are also

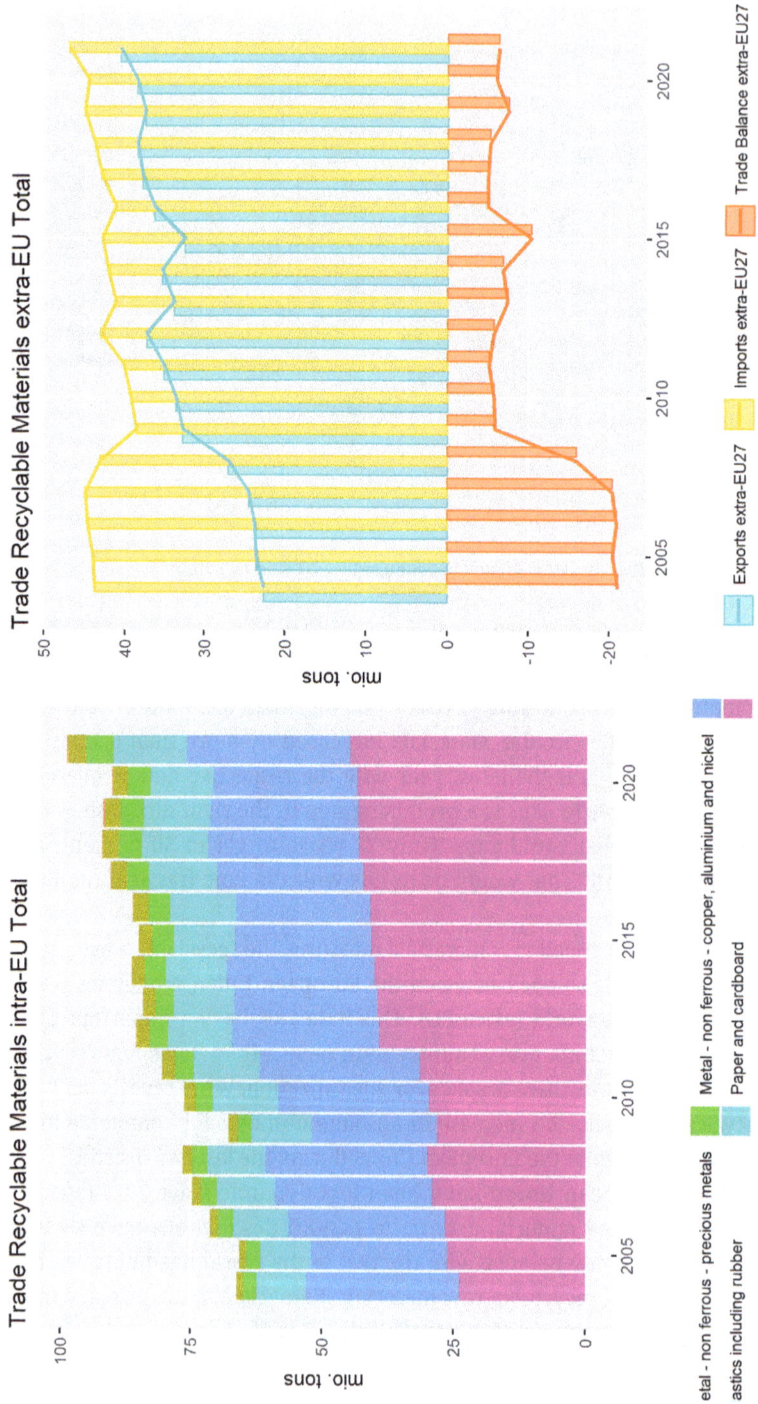

Figure 24.18: Trade with Recyclable Materials inside the European Union; figure created by author.

the largest importers of European waste in general. Overall, trade volumes in recyclable raw materials far outnumber waste exports and imports. In general, the signals are pointing in the correct direction toward a more circular economy (Eurostat 2022b, 2022a).[52,53]

24.3.3 Competitiveness and Innovation

So far, we've observed that while many indicators point to a more circular future, the transformation has stalled in at least some areas. What can be done to accelerate things? Innovation and investment are essential drivers of the circular economy. As a result, as a final feature, the European Union monitoring framework includes measurements for competition and innovation. In the circular economy sector, competitiveness and innovation are measured by value added, employment, investments, and patents (Figure 24.19). But what are the branches of the circular economy? Eurostat classifies all commercial activities including recycling, repair, and reuse as part of the circular economy in terms of value added, employment, and investments (Eurostat 2022b).[54]

In the late 2010s, value added, employment, and investment all increased in absolute and relative terms to GDP. All of these data are trending in the right direction, but it remains to be seen if there will be any future setbacks. A trend cannot be predicted based on the very small time intervals observed. When it comes to recycling of secondary raw materials, patents are counted for the statistic. However, not all waste management or other parts of the circular economy are covered by the statistic (Eurostat 2020).[55] This means that circular economy activities and innovation from enterprises or service providers that are not classified using Eurostat's methodology are not included in these figures.

As a result, time series and international comparisons are more telling. The data presented in Figure 24.20 show a decrease or, at the very least, a slowdown in innovation related to "climate change mitigation technologies related to wastewater treatment or waste management," as the indicator is defined precisely. In this regard, the European Union's six largest economies have become less inventive. However, the

52 Eurostat 2022b. "What Are the Main Destinations of EU Export of Waste?" https://ec.europa.eu/eurostat/de/web/products-eurostat-news/-/ddn-20220525-1.
53 Eurostat 2022a. "Metadata - Material Footprint." https://ec.europa.eu/eurostat/cache/metadata/en/cei_pc0 20_esmsip2.htm.
54 Eurostat 2022b. "What Are the Main Destinations of EU Export of Waste?" https://ec.europa.eu/eurostat/de/web/products-eurostat-news/-/ddn-20220525-1.
55 Eurostat. 2020. "Metadata - Private Investments, Jobs and Gross Value Added Related to Circular Economy Sectors." https://ec.europa.eu/eurostat/cache/metadata/de/cei_cie010_esmsip2.htm.

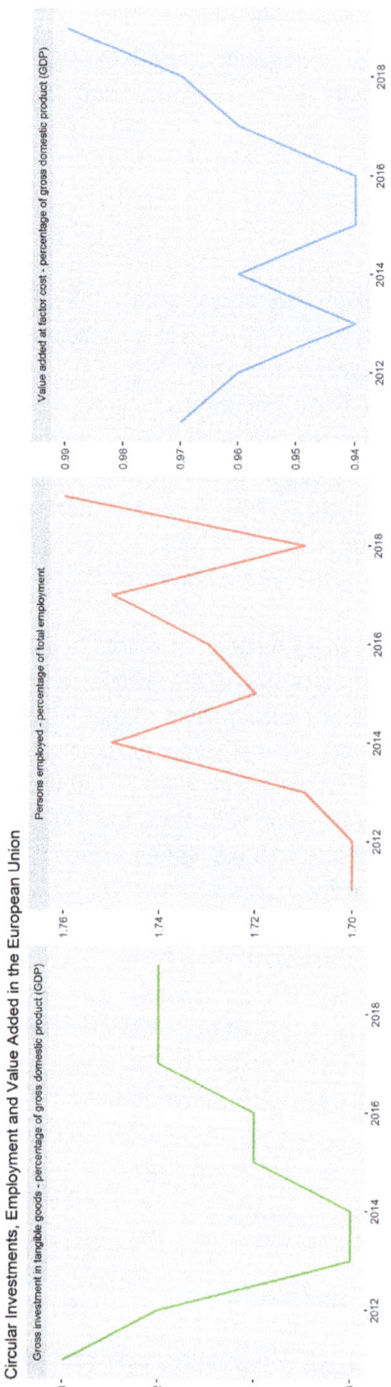

Figure 24.19: Competition Measures for the Circular Economy in Europe; figure created by author.

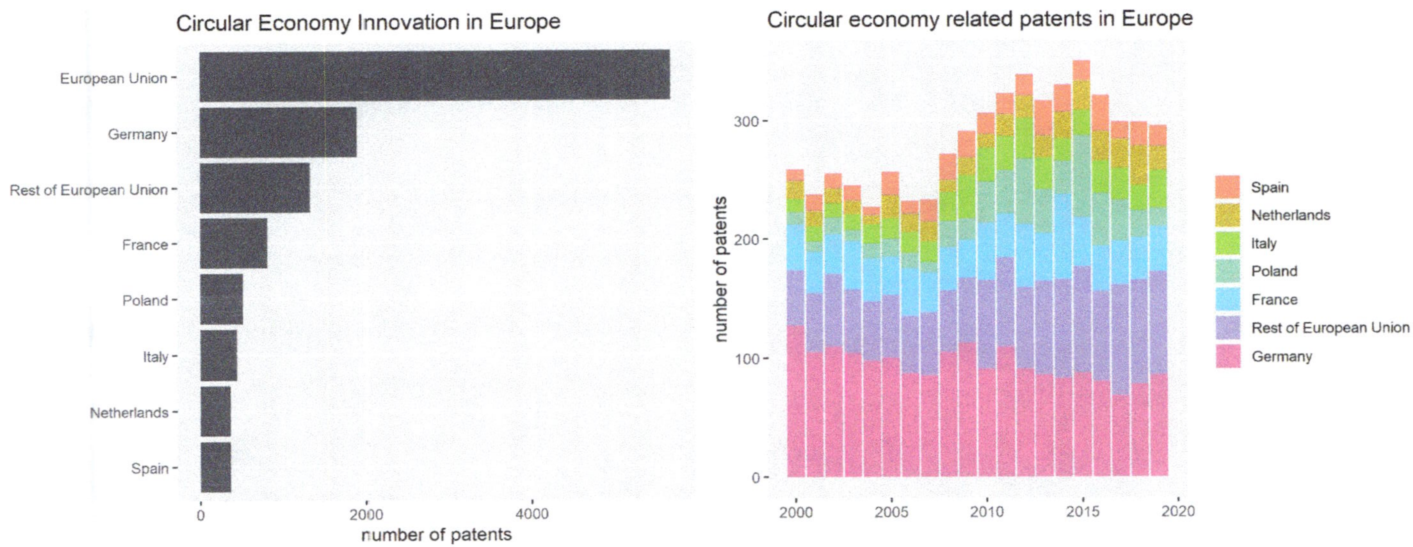

Figure 24.20: Circular Economy related Patents in Europe; figure created by author.

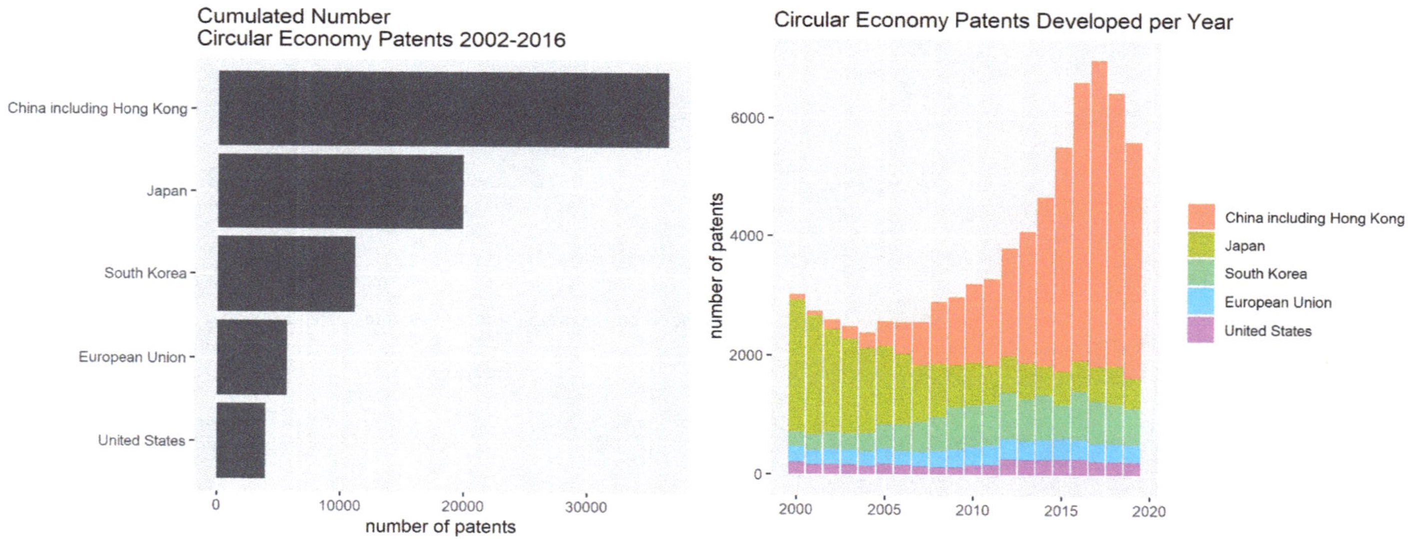

Figure 24.21: Circular Economy related Patents Worldwide; figure created by author.

Netherlands and Poland were highly dynamic in this field during the 2010s, whereas Germany is declining at a rather high rate.

Absolute levels of innovations are critical for an economy's or the common European market's innovativeness. The sheer volume of breakthroughs pushes the technological edge. So, how does Europe fare in contrast to other major economies? Figure 24.21 gives a clue. East Asia has had great success in developing green technologies. One could argue that these figures reflect a catching-up process, given European countries such as Germany have been more inventive in this field for a longer period of time. This would also explain why Japan and South Korea have seen a decrease in the quantity of newly developed patents. But looking objective at these numbers, it becomes obvious that the European Union is falling behind China and South Korea in circular innovation.

24.4 Business Rationales and Steps forwards

The examination of these indicators reveals the following: Because the circular economy is neither established nor on the verge of being formed, the European Commission felt the need to develop a second Circular Economy Action Plan. As we've seen, the European Commission takes a regulatory approach rather than relying solely on market processes like price. For the time being, I'll focus on the rationales and incentives that exist to encourage firms to take the essential steps toward a circular economy. I will start with a discussion on the innovation needed to form a circular economy.

24.4.1 Circular Innovation

There is no doubt that it is beneficial to fund research and development or innovation in general. Especially because the innovation dynamic has slowed down in Europe, support for research and development can be a leverage for more circular innovation. The European Union has stated that it will support creative ventures, which makes perfect sense. Innovation is costly for businesses, yet an extended technological frontier benefits not only the innovative firm but the economy as a whole. An innovation in the sense of the Schumpetarian paradigm has according to Aghion, Antonin, and Bunel (2021) three characteristics.[56] First, innovation lies at the heart of economic growth. Second, if we want the circular economy to grow, innovation is therefore a

56 Aghion, Philippe, Céline Antonin, and Simon Bunel. 2021. *The Power of Creative Destruction: Economic Upheaval and the Wealth of Nations*. Cambridge, Massachusetts: The Belknap Press of Harvard University Press.

key component. Furthermore, incentives are essential for innovation. And third, innovation causes creative destruction. That means new innovations replace former ones.

Non-Radical Innovation

Not all innovations possess these qualities. What I refer to as "escape innovations" are one example of an innovative activity that results from incentives, replaces prior methods, but has long-term very negative repercussions. I refer to creative strategies to get around tougher regulations. One instance was the "Diesel- Skandal" in Germany, where inventive testing techniques were created to forcibly circumvent the European standards for vehicle emissions. Another possibility for such an innovative action would be to develop some fresh lobbying methods to avoid the introduction of more stringent regulations in the first place. Innovation of this nature hinders the circular economy rather than helping it. Setting incentives for such innovation should be avoided in an effective green industrial policy (Rodrik 2014).[57]

Then there are innovations that improve technologies, but do not replace former innovations. They increase an economy's productivity, but their impact on growth is viewed as being less significant than that of creatively destructive innovation (Aghion, Antonin, and Bunel 2021).[58] Companies create such incremental improvements to enhance their processes, techniques of production, or products themselves (Swann 2009).[59] Smartphones get new features all the time. These updates represent incremental improvements. It is by far the most typical sort of innovation. It is so popular among businesses because it improves on current technology rather than replacing earlier innovations. Companies do not destroy their own technologies when making incremental improvements to them. Kasmire, Korhonen, and Nikolic (2012) describes incremental innovation this way:[60]

---

57 Rodrik, Dani. 2014. "Green Industrial Policy." *Oxford Review of Economic Policy* 30 (3): 469–91. https://doi.org/10.1093/oxrep/gru025.

58 Aghion, Philippe, Céline Antonin, and Simon Bunel. 2021. *The Power of Creative Destruction: Economic Upheaval and the Wealth of Nations*. Cambridge, Massachusetts: The Belknap Press of Harvard University Press.

59 Swann, G. M. Peter. 2009. *The Economics of Innovation: An Introduction*. Cheltenham: Edward Elgar. van Leeuwen, George, and Pierre Mohnen. 2017. "Revisiting the Porter Hypothesis: An Empirical Analysis of Green Innovation for the Netherlands." Economics of Innovation and New Technology 26 (1–2): 63–77. https://doi.org/10.1080/10438599.2016.1202521.

60 Kasmire, J., Janne M. Korhonen, and Igor Nikolic. 2012. "How Radical Is a Radical Innovation? An Outline for a Computational Approach." *Energy Procedia* 20: 346–53. https://doi.org/10.1016/j.egypro.2012.03.034. Kirchherr, Julian. 2022. "Bullshit in the Sustainability and Transitions Literature: A Provocation." *Circular Economy and Sustainability*. https://doi.org/10.1007/s43615-022-00175-9.

> Incremental innovations may advance research but inspire no new fields, may increase the efficiency or capabilities of current technologies, but not make competitors obsolete, and uphold the status quo.

Incremental innovation is likely to play a role in the transition to a circular economy in the near future. However, it is necessary to implement new technology. Here is where creative destruction enters the picture. The idea was created by Schumpeter, who portrayed it as a force "that incessantly revolutionizes the economic structure from within, incessantly destroying the old one, incessantly creating a new one. This process of Creative Destruction is the essential fact about capitalism."(Schumpeter and Swedberg 2013).[61] A radical invention is one that completely replaces an existing technology. Disruption results from radical innovations since they render the preceding innovation worthless. A business that develops radical innovation has the potential to destroy existing markets and construct new ones. In essence, the European Commission requests more radical innovation when it requests new forms of circular production. The smartphone itself is an illustration of a radical innovation. The mobile phone market underwent creative destruction as a result. New players entered the phone market, established companies lost their market shares.

Radical Innovation and Creative Destruction

Radical innovation carries a significant risk but promises great profits. Radical innovations are brand-new, so there is no prior knowledge of them. Markets have not yet tried them. The radical, creative product or service might not be popular with customers. It can be harder to produce than anticipated. Or rival businesses engage in predatory behavior in response to the threat of radical innovation. However, if a company's radical invention results in creative destruction, it will provide a monopolistic profit as long as the underlying technology is shielded, such as by intellectual property rights. The surviving established businesses and new market entrants have an incentive to adapt by replicating the radical innovation of the first mover once a successful radical innovation has occurred. No wonder, companies developing radical innovation are often lead by younger CEOs (Acemoglu, Akcigit, and Celik 2022).[62] If the innovation fails they have more time to recover than older executives. Kolev et al. (2022) argue that start-ups have a significant advantage in developing such radical innova-

61 Schumpeter, Joseph A., and Richard Swedberg. 2013. *Capitalism, Socialism and Democracy*. London; New York: Routledge.

62 Acemoglu, Daron, Ufuk Akcigit, and Murat Alp Celik. 2022. "Radical and Incremental Innovation: The Roles of Firms, Managers, and Innovators." *American Economic Journal: Macroeconomics* 14 (3): 199–249. https://doi.org/10.1257/mac.20170410.

tions.[63] It may be a good approach for existing businesses to wait and observe which innovations have the most radical potential before attempting to work with the creative entrepreneurs. A pattern known as David-Goliath symbiosis by Baumol (2002).[64] But given the potential for creative devastation, this might be dangerous. A radical breakthrough with a significant impact may render established businesses obsolete.

Business Model Innovation

An innovative business model is often another cause for disruption. According to Guzzo (2019),[65] business models are "conceptual frameworks" that illustrate the reasoning behind how a certain company or solution works. A business model innovation is a novel method of generating customer values. One way to redesign business models is to incorporate organizational innovation, for instance, through the development of new circular supply chains or new varieties of it. For instance, finding new markets for raw materials that have been recycled calls for organizational innovation – as well on the demand as on the supply side. This way new circular business models can form new markets. Especially for the circular economy this is important, because every "R" – reuse, remanufacture, recycle, etc. – needs a new market (Zink and Geyer 2017).[66] For example, firms might be able in a circular economy to price a good twice. Once they deliver a certain material and once again when they take it back to recycle it.

New business models could also arise, because the circular economy might change the allocation of property rights. It may make sense for businesses to sell the usage of the product rather than the product itself when there is a shortage of a given material or when they are required by law to maintain their items after customers have used them. The asset would continue to belong to the business. Customers would be charged for use. Companies who early adopt such business models could be more attractive to costumers since they do not have to care how the items purchased are fed into the circuit again. As an illustration, consider Rolls Royce, which leases its turbines to the aviation industry.

---

63 Kolev, Julian, Alexis Haughey, Fiona Murray, and Scott Stern. 2022. "Of Academics and Creative Destruction: Startup Advantage in the Process of Innovation." Cambridge, MA: National Bureau of Economic Research. https://doi.org/10.3386/w30362.

64 Baumol, William J. 2002. "Entrepreneurship, Innovation and Growth: The David-Goliath Symbiosis." *Journal of Entrepreneurial Finance* 7 (2): 1–10.

65 Guzzo, Daniel, Adriana Hofmann Trevisan, Marcia Echeveste, and Janaina Mascarenhas Hornos Costa. 2019. "Circular Innovation Framework: Verifying Conceptual to Practical Decisions in Sustainability-Oriented Product-Service System Cases." *Sustainability* 11 (12): 3248. https://doi.org/10.3390/su11123248.

66 Zink, Trevor, and Roland Geyer. 2017. "Circular Economy Rebound." *Journal of Industrial Ecology* 21 (3): 593–602. https://doi.org/10.1111/jiec.12545.

From an economic standpoint, taking into consideration all of these factors, the shift to a circular economy will necessitate a lot of innovation from both outsiders and from insiders in the market, radical as well as incremental innovations. But it's possible that the level of innovation in the technologies used by businesses has peaked now. Further innovation is hard to achieve, new ideas might be getting harder to find, as Bloom et al. (2020) puts it.[67] The demands for circularity and carbon neutrality, however, are increasing quickly. In theory, this circumstance requires businesses to invest in new technologies. Due to the destruction effect brought on by radical innovation to a company's own assets, it might be questioned whether established companies would put forth enough effort in innovation. Therefore creative destruction is often the result from radical innovation by market entrants, being it start-ups or established companies from other branches. Aghion, Antonin, and Bunel (2021) is heavily recommended for further reading on this subject.[68]

Innovating the Circular Economy

For the circular economy to work, we need a lot of creative destruction and because it is unlikely that established companies have enough incentives for radical innovation, the circular economy depends on entrepreneurship. People in established firms or start ups that have the courage and the means to try something new, which are often comparatively young (Acemoglu, Akcigit, and Celik 2022).[69] Why? Given that innovation efforts often fail, it is likely that many circular innovations will as well. So, we require a lot more innovation in new technology, business models, and organizational structures to maximize the probability of creative destruction to cover up for the fails. The European Union's falling circular innovation activity is an indication that such entrepreneurial activity is lacking. Finding incentives for entrepreneurs to enter new markets with innovative circular concepts is vital for policymakers.

Of course, raising entrepreneurship and creative destruction is not everything needed to form a circular economy. Governmental investments in innovative technologies are another requirement for the circular economy to succeed. Infrastructure is crucial to circular production. An economy needs a lot of processing steps that start with waste collection and end with new material or a recycled product in order to

67 Bloom, Nicholas, Charles I. Jones, John van Reenen, and Michael Webb. 2020. "Are Ideas Getting Harder to Find?" *American Economic Review* 110 (4): 1104–44. https://doi.org/10.1257/aer.20180338.
68 Aghion, Philippe, Céline Antonin, and Simon Bunel. 2021. *The Power of Creative Destruction: Economic Upheaval and the Wealth of Nations*. Cambridge, Massachusetts: The Belknap Press of Harvard University Press.
69 Acemoglu, Daron, Ufuk Akcigit, and Murat Alp Celik. 2022. "Radical and Incremental Innovation: The Roles of Firms, Managers, and Innovators." *American Economic Journal: Macroeconomics* 14 (3): 199–249. https://doi.org/10.1257/mac.20170410.

recycle materials. When considering hydrogen, the issue is clear. With the aid of hydrogen, steel production can become circular in nature. However, a transportation network is required for hydrogen. A network of some form must transport hydrogen from hydrogen generating sites to consumers, comparable to a pipeline system or a network of gas stations. For market newcomers, building such a network is difficult or nearly impossible. Since these networks have extremely high fix costs, their returns on scale are extremely high. Massive hurdles to entrance into the market are created by these networks. The screws that economic policy has to tighten are investing in innovative infrastructure, creating incentives for innovation, allowing creative destruction, and accepting the results of this competition. This is not always easy because in the short run it can mean companies to fail resulting in the temporary loss of employment (Aghion, Antonin, and Bunel 2021;[70] Aghion et al. 2016).[71]

24.4.2 Circular Incentives

In principle, the interaction of carbon price, regulation, and innovation support can provide positive incentives for enterprises to innovate, while other instruments may be more successful or efficient than those proposed by the European Union. Regulation in general may trigger innovation. And not only in the country regulating but also in foreign companies who want to be active on European markets. The mechanism is described as "California-Effect". From a European perspective, there may be a reason to aspire for a California-effect via circular economy regulation (Vogel 1997;[72] Perkins and Neumayer 2012).[73] This impact suggests that regulation can result in a race to the top in terms of regulatory norms. When applied to the European scenario, this means that European standards will become binding, and other economies will alter their standards to match the European level. The European Union would be the leading light for circular economy. As a result, these standards would incentivize enterprises both inside and outside the European Union to innovation, because they would otherwise be excluded from the European market.

70 Aghion, Philippe, Céline Antonin, and Simon Bunel. 2021. *The Power of Creative Destruction: Economic Upheaval and the Wealth of Nations*. Cambridge, Massachusetts: The Belknap Press of Harvard University Press.

71 Aghion, Philippe, Ufuk Akcigit, Angus Deaton, and Alexandra Roulet. 2016. "Creative Destruction and Subjective Well-Being." *The American Economic Review* 106 (12): 3869–97. https://doi.org/10.1257/aer.20150338.

72 Vogel, David. 1997. *Trading up: Consumer and Environmental Regulation in a Global Economy*. Paperback ed. Cambridge, Mass.: Harvard University Press.

73 Perkins, Richard, and Eric Neumayer. 2012. "Does the 'California Effect' Operate Across Borders? Trading- and Investing-up in Automobile Emission Standards." *Journal of European Public Policy* 19 (2): 217–37. https://doi.org/10.1080/13501763.2011.609725.

Regulation: Incentive to innovate or Market Barrier

That regulation can lead to innovation is known for a while in economics and management science. The Porter Hypothesis in its strong and weak form postulates, that regulation trigger innovation (van Leeuwen and Mohnen 2017).[74] In this view, firms fail to innovate ecologically, because of incomplete information, organizational and coordination problems. Therefore, they are blind to the advantages of eco-innovation such as lower energy usage or fewer material inputs. Thereby firms are unwillingly ignoring cost saving potentials (Horbach, Rammer, and Rennings 2012;[75] Porter and van der Linde 1995).[76] According to the Porter hypothesis, firms could realize first mover advantage over companies that are located in countries with lower standards. In this view, regulation is helpful for companies to be innovative. The Porter hypothesis comes in a weak and in a strong form. In the weak form companies become more innovative, in the strong version, their overall performance increases – they benefit economically from regulation in this case. Furthermore, it is a long standing insight that governments should encourage the "private innovation matching" in order to support the adoption of clean technologies (Veugelers 2012).[77] Companies might have information deficits regarding which technologies they should innovate upon.

But which factors determine eco-innovation? Horbach, Rammer, and Rennings (2012) give an answer.[78] They use four factors to identify determinants of eco-innovations. The first is regulation as a push factor: Governments force firms to eco-innovate by setting the rules. This is one way the European Commission wants to go. The second factor is a market pull factor: Customers reward eco-innovations in certain areas, "such as food or baby cloth", where they are willing to pay a premium, but not in the field of electricity. The Circular Economy Action Plan sees for this reasons better consumer information as a core instrument. Third companies become more innovative because of a technology push. The improvement and availability of knowl-

74 Swann, G. M. Peter. 2009. The Economics of Innovation: An Introduction. Cheltenham: Edward Elgar. van Leeuwen, George, and Pierre Mohnen. 2017. "Revisiting the Porter Hypothesis: An Empirical Analysis of Green Innovation for the Netherlands." *Economics of Innovation and New Technology* 26 (1–2): 63–77. https://doi.org/10.1080/10438599.2016.1202521.

75 Horbach, Jens, Christian Rammer, and Klaus Rennings. 2012. "Determinants of Eco-Innovations by Type of Environmental Impact — the Role of Regulatory Push/Pull, Technology Push and Market Pull." *Ecological Economics* 78: 112–22. https://doi.org/10.1016/j.ecolecon.2012.04.005.

76 Porter, Michael E., and Claas van der Linde. 1995. "Toward a New Conception of the Environment-Competitiveness Relationship." *Journal of Economic Perspectives* 9 (4): 97–118. https://doi.org/10.1257/jep.9.4.97.

77 Veugelers, Reinhilde. 2012. "Which Policy Instruments to Induce Clean Innovating?" *Research Policy* 41 (10): 1770–78. https://doi.org/10.1016/j.respol.2012.06.012.

78 Horbach, Jens, Christian Rammer, and Klaus Rennings. 2012. "Determinants of Eco-Innovations by Type of Environmental Impact — the Role of Regulatory Push/Pull, Technology Push and Market Pull." *Ecological Economics* 78: 112–22. https://doi.org/10.1016/j.ecolecon.2012.04.005.

edge capital "triggers eco-innovations". Therefore, innovation support is essential, also from this point of view. Organizational innovation – such as the availability of an environmental management system –plays an important part, too. And the fourth driver are firm specific factors such as knowledge transfer mechanisms, access and participation in networks and green capabilities.

On the other hand, regulation can also inhibit innovation, when it makes doing business complicated. Especially start-ups suffer from over-regulation. For them, regulation is a cost burden, not every entrepreneur is ready to bear. Imagine an entrepreneur who needs to hire compliance officers before her start-up can enter a market, because regulation is complex and high. As a consequence, entrepreneurs could decide not to enter the market. Circular innovation does not take place and some ideas do not come into effect. The regulation reduces creative destruction. Especially harmful is regulation, when "regulators are in bed with business" as Rodrik (2014) has put it.[79] Large corporations may even be effective in convincing government representatives to tighten rules in order to raise the costs of their smaller competitors (Bailey and Thomas 2017).[80] When established businesses make the rules, they will probably not favor competition from the outside. On the contrary, established companies have an incentive to build market entry barriers, because this way, they can protect their market shares from outsiders.

Prices and Cost Incentives

On the other side, using cost or price incentives prevents this form of business-biased regulation making. Saving money is a major motivation for both consumers and businesses. Companies are enticed to select cheaper inputs since, ceteris paribus, doing so maximizes earnings. From an economic standpoint, increasing the price of these linear items, for example by a tax that directly raises the cost of utilizing raw materials, is the obvious course of action when linear production results in cheaper products and greater environmental impacts. If employing circular inputs is at least not more expensive than using linear inputs, businesses would select circular production. Due to their proportional decrease in price, more households would buy circular products more frequently. A positive circular supply and demand side effect would result from setting the prices "correct," as Schlosser, Chenavaz, and Dimitrov (2021) convincingly points out.[81]

---

79 Rodrik, Dani. 2014. "Green Industrial Policy." *Oxford Review of Economic Policy* 30 (3): 469–91. https://doi.org/10.1093/oxrep/gru025.

80 Bailey, James B., and Diana W. Thomas. 2017. "Regulating Away Competition: The Effect of Regulation on Entrepreneurship and Employment." *Journal of Regulatory Economics* 52 (3): 237–54. https://doi.org/10.1007/s11149-017-9343-9.

81 Schlosser, Rainer, Régis Y. Chenavaz, and Stanko Dimitrov. 2021. "Circular Economy: Joint Dynamic Pricing and Recycling Investments." *International Journal of Production Economics* 236: 108117. https://doi.org/10.1016/j.ijpe.2021.108117.

Such pricing signals are lacking, according to Europe's status of the circular economy. When waste levels are not shrinking, it is obviously still relatively cheap to dispose. It appears that using "virgin" resources is advantageous for producers if the use of recyclable raw materials is still typically below 30 percent. The circular economy's core tenet is that waste, pollution, and landfills must be decreased because their generation incurs costs, at the very least in the long term. A Pigouvian tax on nonrecycled materials has been suggested by Sørensen (2018) as one potential method to internalize these costs.[82] Each unit of virgin raw materials used by businesses would be subject to an additional charge. An optimal degree of circularity may be ensured by this process, which would correct the flawed price mechanism that promotes the usage of virgin raw materials.

Pricing has the main advantage that neither producers nor consumers must believe that a circular economy is required. Price and financial constraints alone are enough to change their behavior. Innovation is also fueled by a workable price structure that internalizes the negative externalities of linear production. While monitoring and enforcing current restrictions had no impact in the United States during the 1980s and early 1990s, research by Brunnermeier and Cohen (2003) shows that green innovation reacts to rising pollution abatement costs.[83] Although the authors acknowledge that the effect is relatively modest. These findings teach us how crucial it is to internalize the damaging external environmental effects in order to increase circular economy activities.

The burning of waste is a prime example of what occurs when price signals are absent. As we've seen, a sizable portion of municipal waste is burned and used, for instance, to produce energy. Thus, it produces short-term economic benefit (energy, heat), but long-term environmental costs (global warming). Since energy recycling is not included by the EU-ETS, it is not subject to European carbon price which fluctuates in 2022 around 84 Euros. National carbon emission price regimes may apply, however the Tax Foundation reports that the average carbon tax in Europe will be about 42 Euros in 2022 (Bray 2022).[84] If this procedure were more expensive, there would be less incentive to burn waste, especially plastic waste, and other uses would be taken into consideration. Alternative uses, such chemical recycling, are more wide-

82 Sørensen, Peter Birch. 2018. "From the Linear Economy to the Circular Economy: A Basic Model."

83 Bray, Sean. 2022. "Carbon Taxes in Europe." https://taxfoundation.org/carbon-taxes-in-europe-2022/. Brunnermeier, Smita B., and Mark A. Cohen. 2003. "Determinants of Environmental Innovation in US Manufacturing Industries." *Journal of Environmental Economics and Management* 45 (2): 278–93. https://doi.org/10.1016/S0095-0696(02)00058-X.

84 Bray, Sean. 2022. "Carbon Taxes in Europe." https://taxfoundation.org/carbon-taxes-in-europe-2022/. Brunnermeier, Smita B., and Mark A. Cohen. 2003. "Determinants of Environmental Innovation in US Manufacturing Industries." *Journal of Environmental Economics and Management* 45 (2): 278–93. https://doi.org/10.1016/S0095-0696(02)00058-X.

spread but emit fewer greenhouse gases. Therefore, increasing the cost of energetic recycling would benefit less detrimental environmental methods.

Regulation combined with a lack of price signals can result in very ludicrous circumstances. Quotas for recycled resources in the manufacture of items might create bizarre scenarios. Although their quality could be worse, recycled materials might command greater market pricing than raw resources. While that would provide incentives for companies to invest in recycling and enter the market, it can also have major negative consequences. When more recycling companies require recyclable materials, the incentive to reduce waste may dwindle because it is required for recycling. A price on trash production that includes the environmental cost of the waste, in conjunction with the possibility of recycling, would lead to more efficient waste generation.

Effective pricing could help to balance the circular economy. To achieve the same result, regulation must be both sophisticated and dynamic with the risk of turning into detailed and difficult. Internalization of external externalities can be accomplished in theory by legislation or through a direct cost effect on businesses. However, in general, regulation is an imperfect substitute for pricing. Veugelers (2012) puts it this way:[85]

> If governments want to leverage the needed private innovations for clean energy technologies, they will have to provide a well-designed, time consistent policy, by a combination of consistent carbon pricing, performance based regulations and public funding.

24.4.3 Circular Strategies

For a company, to develop an innovation is a decision made in an uncertain environment. As long as the regulatory framework is not binding but in process, it adds to the uncertainty. Because there is uncertainty, it is important to construct scenarios (Martelli 2014) and prepare for their realization.[86] Companies can design their strategies based on this knowledge. An example for this type of strategic decision making is described by the aphorism "hope for the best, be prepared for the worst."

Waiting and seeing is an intuitive coping method in times of uncertainty. However, this could be a disastrous strategy. Not merely because competitors may come up with startling new ideas. The major Asian economies are far more inventive in terms of circular economy technologies than European economies. If European enterprises do not keep up with this trend, they may lose competitive advantages in the future. The circular economy promotes greater material efficiency. And this efficiency is being created at a faster rate in other parts of the world. Waiting and seeing could

85 Veugelers, Reinhilde. 2012. "Which Policy Instruments to Induce Clean Innovating?" *Research Policy* 41 (10): 1770–78. https://doi.org/10.1016/j.respol.2012.06.012.

86 Martelli, Antonio. 2014. *Models of Scenario Building and Planning: Facing Uncertainty and Complexity. Bocconi on Management*. New York: Palgrave Macmillan.

result in the circular spaceship leaving. Especially when other countries try to create a California-effect and make their norms applicable to European exports.

Circular Timing and Growth

But what precisely should businesses be bracing for? According to the Circular Gap Reporting Initiative, just 8.6 percent of the globe currently functions in a circular economy. The World Economic Forum anticipates 4.5 trillion dollars in economic gains from the circular economy until 2030 (World Economic Forum 2022).[87] That is substantially more than the the German GDP – a pretty big cake. Estimates from 2015 place the worth of the European Union's transition to a circular economy at roughly 600 billion Euros per year (Grafström and Aasma 2021).[88]

When we consider the world's current scarcity, the usefulness of circular economy appears fairly obvious: Raw materials are becoming more expensive, energy prices are skyrocketing, and supply chains are massively vulnerable – all of this suggests that there are opportunities for businesses that produce less dependent on "virgin" materials, are not reliant on fossil fuels, and have resilient and diverse supply chains. Many of these trends are likely to continue. According to the European Commission (2020b),[89] the demand for raw materials will increase significantly compared to today. In comparison to the current supply to the entire EU economy, the EU would require up to 18 times more lithium and 5 times more cobalt in 2030 and over 60 times more lithium and 15 times more cobalt in 2050 for energy storage and batteries for electric vehicles. By 2050, there may be a tenfold rise in demand for rare earths, which are utilized in permanent magnets, for things like digital electronics, wind turbines, and electric automobiles. According to the World Bank, if the Paris climate agreement is fulfilled, the demand for metals and minerals will increase by more than 1,000% by 2050 (World Bank 2017).[90] According to the OECD's predictions, through 2060, the world's total material consumption will double (OECD 2018).[91]

---

87 World Economic Forum. 2022. "Circular Economy and Material Value Chains." https://www.weforum.org/projects/circular-economy.

88 Grafström, Jonas, and Siri Aasma. 2021. "Breaking Circular Economy Barriers." *Journal of Cleaner Production* 292: 126002. https://doi.org/10.1016/j.jclepro.2021.126002.

89 European Commission 2020b. "Critical Raw Materials Resilience: Charting a Path Towards Greater Security and Sus- tainability: Communication from the commission to the European parliament, the council, the european economic and social committee and the committee of the regions." Edited by European Commission. https://eur-lex.europa.eu/legal-content/EN/TXT/?uri=CELEX:52020DC0474.

90 World Bank. 2017. *The Growing Role of Minerals and Metals for a Low Carbon Future*. World Bank, Washington, DC. https://doi.org/10.1596/28312.

91 OECD. 2018. "Global Material Global Material Resources Outlook to 2060: Economic Drivers and Environ- mental Consequences." Edited by OECD. https://www.oecd.org/environment/waste/highlights-global-material-resources-outlook-to-2060.pdf.

These figures demonstrate the enormous potential market for circular solutions that maintain material flow in a circuit. In these situations, businesses will benefit from the efficient use of their resources. Companies that contribute to meeting this demand, such as those that offer recycled raw materials, will grow in systemic importance. The earlier discussed restrictions make it highly unlikely that the entire economy will operate in a closed loop. However, if these projections are accurate, the predicted material shortage that the global economy will soon suffer provides a strong incentive to reconsider the corporate circular strategy.

Little time remains for preparation. The finish line has been determined. Germany aims to achieve carbon neutrality by 2045, and the European Union by 2050. Carbon emissions must decrease more quickly than they do now in order to achieve this goal. The Commission suggests raising the EU's commitment to reduce greenhouse gas emissions to at least 55% below 1990 levels by 2030 with the 2030 Climate Target Plan. Considering that the European Commission has tied its Circular Economy Action Plan to the European Green Deal, it is likely that regulation will be related to this timescale. Circular economy concept supports both climate protection and climate change adoption. When the weather becomes dryer, for example, it is beneficial to consider circular solutions for water-intensive manufacturing processes. Given that circular economy production does result in carbon reductions, it will also enable enterprises to significantly reduce carbon costs. Thinking, planning, developing, and acting on the circular economy will be highly relevant for all sectors and many business units inside firms for at least the next two decades: manufacturing, supply, sales, and research and development, to mention a few.

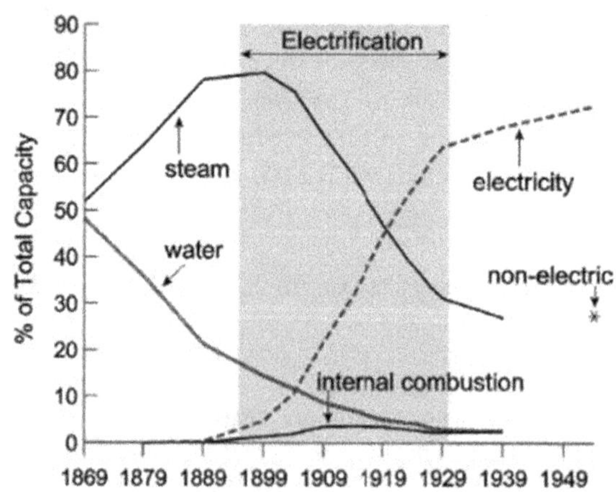

Shares of total horsepower generated by the main sources in U.S. manufacturing, 1869–1954.

Figure 24.22: Diffusion of Electricity in the United States, source: Jovanovic and Rousseau (2005).

Furthermore, even if the development and diffusion of circular technologies are on a comparative level right now, things can speed up quite fast as Jovanovic and Rousseau (2005) famously demonstrated.[92] Old methods of production can decline increasingly, while new ones start to grow very fast. Figure 24.22 shows it for energy generation. Electricity needed a while until it had penetrated the US-American manufacturing sector. But once electricity accelerated it forced a rapid transition among manufacturers. The same could be true for circular economy technologies.

Circular Strategy Building

The first section of this article demonstrated the power of the circular economy narrative. It's possible that the circular economy dynamic is currently in crisis, but the massive amounts of popular and scholarly writing on the subject, along with growing global interest in it, suggest that there is a growing need for circular solutions. Where should businesses begin when they want to increase the circularity of their production? The circular economy focuses on lowering the intake of raw materials while also lowering non-recyclable trash or emissions. The Rs – reuse, repair, redistribute, refurbish, and reduce or recycle – might serve as a beginning point for developing a strategy. More circular economy will result from thinking about how to include these Rs into a company's business model or production system. This circle's use leads to consideration of alternative product designs and the development of methods for establishing a recycling channel from customers to manufacturing. Atasu, Dumas, and van Wassenhove (2021) distills this thinking into two straightforward inquiries that aid in strategy development:[93] 1. What happens to our product once it is used? 2. Where do the materials we used in the production end up?

Additionally, Atasu, Dumas, and van Wassenhove (2021) create three models of a circular economy strategy, notably for manufacturers.[94] The first strategy is known as "Retain Product Ownership (RPO)". Instead of selling their items, businesses rent or lease them. The product's makers are still in charge. Castro et al. (2022) describe this

---

92 Jovanovic, Boyan, and Peter L. Rousseau. 2005. "General Purpose Technologies." In *Handbook of Economic Growth*, edited by Philippe Aghion and Peter L. Rousseau, 1:1181–1224. Handbooks in Economics. Amsterdam; Heidelberg: Elsevier. https://doi.org/10.1016/S1574-0684(05)01018-X.

93 Atasu, Atalay, Céline Dumas, and Luk N. van Wassenhove. 2021. "The Circular Business Model: Pick a Strategy That Fits Your Resources and Capabilities." *Harvard Business Review*, no. July-August. https://hbr.org/2021/07/the-circular-business-model.

94 Atasu, Atalay, Céline Dumas, and Luk N. van Wassenhove. 2021. "The Circular Business Model: Pick a Strategy That Fits Your Resources and Capabilities." *Harvard Business Review*, no. July-August. https://hbr.org/2021/07/the-circular-business-model.

kind of strategy.[95] The circular economy "encourages strategies that remove the sense of ownership, and customers are stimulated to rent products", satisfying the demand to use the function without owning the object, shifting the "focus from product to function, like sharing economy and the Product-service-system." As a result of this strategy, businesses are incentivized to reconsider the quality, durability, and reusability of their products. Instead of selling copper pipes, construction businesses may lend them. This way, they can protect themselves from potential copper scarcity. In general, the company can compute a stream of secondary raw materials, decreasing supply and pricing concerns. As the authors point out, such a business model necessitates corporations investing "heavily in after-sales and maintenance capabilities." The businesses are not only producers, but also, to a wider extent, service providers.

"Product Life Extension (PLE)" is the second strategy. Companies that manufacture things with a long product life expectancy may be able to create a secondary market in which they sell used or refurbished goods. This strategy is becoming more common among electronics manufacturers who sell refurbished products such as smartphones and tablets. This method enables corporations to establish an additional market for their products, allowing less rich buyers to purchase this expensive type of product in a subsequent round. Companies can obtain these clients using PLE. This gives them the opportunity to develop a long-term consumer relationship. As a result, they may compete with companies who provide lesser-quality products at lower prices.

The third strategy is "Design for Recycling (DFR)". The strategy's name accurately expresses its purpose: Redesigning manufacturing methods, products, and materials to maximize the recyclability of the product or materials utilized. Because the competence to implement this plan is sometimes unavailable within a corporation, Atasu, Dumas, and van Wassenhove (2021) emphasizes that this technique is frequently used in collaboration with strategic partners who have "particular technological knowledge."[96]

Aside from formulating plans based on the "Rs," identifying ways to assess a company's circularity performance is beneficial. For this, various approaches have been developed: Life Cycle Assessment (LCA), Data Envelopment Analysis (DEA), Simulation, and Material Flow Analysis are only a few examples (Castro et al. 2022).[97] The most well-known is the LCA, which is defined as "a strategy for assessing the environ-

95 Castro, Camila Gonçalves, Adriana Hofmann Trevisan, Daniela C. A. Pigosso, and Janaina Mascarenhas. 2022. "The Rebound Effect of Circular Economy: Definitions, Mechanisms and a Research Agenda." Journal of Cleaner Production 345: 131136. https://doi.org/10.1016/j.jclepro.2022.131136.

96 Atasu, Atalay, Céline Dumas, and Luk N. van Wassenhove. 2021. "The Circular Business Model: Pick a Strategy That Fits Your Resources and Capabilities." *Harvard Business Review*, no. July-August. https://hbr.org/2021/07/the-circular-business-model.

97 Castro, Camila Gonçalves, Adriana Hofmann Trevisan, Daniela C. A. Pigosso, and Janaina Mascarenhas. 2022. "The Rebound Effect of Circular Economy: Definitions, Mechanisms and a Research Agenda." *Journal of Cleaner Production* 345: 131136. https://doi.org/10.1016/j.jclepro.2022.131136.

mental factors connected with a product over its life cycle" by Muralikrishna and Manickam (2017).[98] According to Muralikrishna and Manickam (2017), an LCA typically consists of four steps:[99]

1. Definition of scope and goals that a company wants to achieve
2. Inventory Analysis that describes the material and energy use, flows and how both interacts with environment
3. Impact Assessment based on the inventory analysis
4. Interpretation of the life cycle

This methodology can be used for both strategy development and product development. Many more ideas have been discussed and explored in the literature, including some highly thorough ways for constructing a circular strategy (Guzzo et al. 2019).[100] These ideas can serve as a starting point for firms looking to develop a circular strategy for existing products. There are compelling reasons for corporate plans to include more circular economy approaches. Different rules govern the circular economy; a strategy review should take these rules into account. As a result, firms should aim to foresee the influence of European law on their operations in the future. Furthermore, plans should take into account the fact that prices do not reflect genuine scarcity or the true cost of production inputs. Building strategies around low-cost inputs today could lead to existential concerns in the future when the actual cost becomes apparent. Such examples are rapidly growing prices for European ETS-certificates since 2021 and Europe's skyrocketing energy prices in 2022.

24.5 Conclusions

So, what exactly is the circular economy? The circular economy is a technology idea with significant political implications that is being studied in management schools. The circular economy is a compelling story. The current understanding of the circular economy concept is built upon three principles – according to Castro et al. (2022):[101]

---

98 Muralikrishna, Iyyanki V., and Valli Manickam. 2017. *Environmental Management.* [Lieu de publication non identifié]: Butterworth-Heinemann.

99 Muralikrishna, Iyyanki V., and Valli Manickam. 2017. *Environmental Management.* [Lieu de publication non identifié]: Butterworth-Heinemann.

100 Guzzo, Daniel, Adriana Hofmann Trevisan, Marcia Echeveste, and Janaina Mascarenhas Hornos Costa. 2019. "Circular Innovation Framework: Verifying Conceptual to Practical Decisions in Sustainability-Oriented Product-Service System Cases." *Sustainability* 11 (12): 3248. https://doi.org/10.3390/su11123248.

101 Castro, Camila Gonçalves, Adriana Hofmann Trevisan, Daniela C. A. Pigosso, and Janaina Mascarenhas. 2022. "The Rebound Effect of Circular Economy: Definitions, Mechanisms and a Research Agenda." Journal of Cleaner Production 345: 131136. https://doi.org/10.1016/j.jclepro.2022.131136.

Figure 24.23: A Space Cowboy looking at a Spaceship that already departed, figure created by author with DALL.E).

"(1) Designing waste and pollution, (2) keeping products and materials in use, and (3) regenerating natural systems". As such, the circular economy is an alternative approach of production that is opposed to the so called "linear" production model of "take, make, waste". However, it is economically in a niche – both practically and scientifically: neither are circular economy production methods prominent in most sectors, markets, or economies, nor is the topic well covered in mainstream academic economic study. Kirchherr (2022) gives an interesting analysis on the current state of circular economy related research.[102] However, the circular economy has a significant political impact because it presents an enticing promise: on the one hand, a developing industry-based economy that generates wealth for European citizens, and on the other, a remedy to the exploitation of natural capital. In other words, sustainable wealth. Circular economy in this sense is a political program used to achieve climate goal on the one hand and strategic independence from raw material suppliers.

But the circular economy is not yet on a fast track to substitute the linear production method. Progress as well as regress and stagnation is visible in the indicators. The European Commission's Circular Economy Action Plan thrives to give the circular economy a new push with a variety of proposed regulation. The proposed measures by the European Commission will likely change the business environment into more

102 Kirchherr, Julian, Laura Piscicelli, Ruben Bour, Erica Kostense-Smit, Jennifer Muller, Anne Huibrechtse- Truijens, and Marko Hekkert. 2018. "Barriers to the Circular Economy: Evidence from the European Union (EU)." Ecological Economics 150: 264–72. https://doi.org/10.1016/j.ecolecon.2018.04.028.

circular requirements. Although this might trigger innovation, it might also erect market barriers thereby reducing incentives for market entrants, radical innovation and creative destruction that could help to fasten the circular economy transformation. The transition towards a circular economy requires a lot of innovation activities with uncertain outcome.

So, what is the business case for investing in circular economies? In this uncertain situation, a "wait and see" strategy may be the most promising. However, this could be a mistake since it would force businesses to board a departing spaceship. Of course, there are risks associated with circular innovation due to uncertainty. The most important question is how dedicated national governments and the European Union are to establishing a circular economy? When meeting climate objectives is a severe constraint for European economies, firms have a strong incentive to plan for the next steps in their circular strategy development. Especially when projections indicate that circular economy activities will become increasingly relevant as raw material scarcity grows. Boulding had the vision in the 1960s that we will transition from the irresponsible cowboy economy to a spaceship economy based on circular economy principles. In such a world, what constitutes the biggest risk for the cowboy? To miss boarding (see Figure 24.23).

Bibliography

Acemoglu, Daron, Philippe Aghion, Leonardo Bursztyn, and David Hemous. 2012. "The Environment and Directed Technical Change." *The American Economic Review* 102 (1): 131–66. https://doi.org/10.1257/aer.102.1.131.

Acemoglu, Daron, Ufuk Akcigit, and Murat Alp Celik. 2022. "Radical and Incremental Innovation: The Roles of Firms, Managers, and Innovators." *American Economic Journal: Macroeconomics* 14 (3): 199–249. https://doi.org/10.1257/mac.20170410.

Aghion, Philippe, Ufuk Akcigit, Angus Deaton, and Alexandra Roulet. 2016. "Creative Destruction and Subjective Well-Being." *The American Economic Review* 106 (12): 3869–97. https://doi.org/10.1257/aer.20150338.

Aghion, Philippe, Céline Antonin, and Simon Bunel. 2021. *The Power of Creative Destruction: Economic Upheaval and the Wealth of Nations*. Cambridge, Massachusetts: The Belknap Press of Harvard University Press.

Atasu, Atalay, Céline Dumas, and Luk N. van Wassenhove. 2021. "The Circular Business Model: Pick a Strategy That Fits Your Resources and Capabilities." *Harvard Business Review*, no. July-August. https://hbr.org/2021/07/the-circular-business-model.

Ayres, Robert U. 1998. "Eco-Thermodynamics: Economics and the Second Law." *Ecological Economics* 26 (2): 189–209. https://doi.org/10.1016/S0921-8009(97)00101-8.

Bailey, James B., and Diana W. Thomas. 2017. "Regulating Away Competition: The Effect of Regulation on Entrepreneurship and Employment." *Journal of Regulatory Economics* 52 (3): 237–54. https://doi.org/10.1007/s11149-017-9343-9.

Baumol, William J. 2002. "Entrepreneurship, Innovation and Growth: The David-Goliath Symbiosis." *Journal of Entrepreneurial Finance* 7 (2): 1–10

Blengini, Gian Andrea, and Milan Grohol. 2017. *Methodology for Establishing the EU List of Critical Raw Materials: Guidelines*. Luxembourg: Europäische Kommission; Publications Office of the European Union. https://doi.org/10.2873/769526.

Bloom, Nicholas, Charles I. Jones, John van Reenen, and Michael Webb. 2020. "Are Ideas Getting Harder to Find?" *American Economic Review* 110 (4): 1104–44. https://doi.org/10.1257/aer.20180338.

Boulding, Kenneth E. 2013. "The Economics of the Coming Spaceship Earth." In *Interdisciplinary Economics*, edited by Kenneth E. Boulding, Wilfred Dolfsma, and Stefan Kesting, 3–14. Routledge Studies in the History of Economics. Abingdon, Oxon: Routledge.

Bray, Sean. 2022. "Carbon Taxes in Europe." https://taxfoundation.org/carbon-taxes-in-europe-2022/.

Brunnermeier, Smita B., and Mark A. Cohen. 2003. "Determinants of Environmental Innovation in US Manufacturing Industries." *Journal of Environmental Economics and Management* 45 (2): 278–93. https://doi.org/10.1016/S0095-0696(02)00058-X.

Cantzler, Jasmin, Felix Creutzig, Eva Ayargarnchanakul, Aneeque Javaid, Liwah Wong, and Willi Haas. 2020. "Saving Resources and the Climate? A Systematic Review of the Circular Economy and Its Mitigation Potential." *Environmental Research Letters* 15 (12): 123001. https://doi.org/10.1088/1748-9326/abbeb7.

Castro, Camila Gonçalves, Adriana Hofmann Trevisan, Daniela C. A. Pigosso, and Janaina Mascarenhas. 2022. "The Rebound Effect of Circular Economy: Definitions, Mechanisms and a Research Agenda." *Journal of Cleaner Production* 345: 131136. https://doi.org/10.1016/j.jclepro.2022.131136.

Dasgupta, Partha, ed. 2021. *The Economics of Biodiversity: The Dasgupta Review: Abridged Version*. Updated: 2 February 2021. London: HM Treasury.

Drabe, Viktoria. 2022. *Innovating in a Circular Economy*. Nomos Verlagsgesellschaft mbH & Co. KG. https://doi.org/10.5771/9783828878426.

Drew Schmidt, and Christian Heckendorf. 2013. "Ngram: Fast n-Gram 'Tokenization'." https://cran.r-project.org/web/packages/ngram/ngram.pdf.

ECPI Group. 2019. "ECPI Circular Economy Leaders Quantity Index." https://www.ecpigroup.com/wp-content/uploads/rules/ECPI_Circular_Economy_Leaders_Equity_INDEX_RULES.pdf.

Edenhofer, Ottmar, Flachsland Christian, Christoph Wolff, Lisa Kahtarina Schmid, Anna Leipprand, Nicolas Koch, Ulrike Kornek, and Michael Pahle. 2017. "Decarbonization and EU ETS Reform: Introducing a Price Floor to Drive Low-Carbon Investments." Policy Paper. Mercator Research Institute on Global Commons and Climate Change. https://www.mcc-berlin.net/fileadmin/data/C18_MCC_Publications/Decarbonization_EU_ETS_Reform_Policy_Paper.pdf.

ElleMcArthur Foundation. 2019. "Completing the Picture: How the Circula Economy Tackles Climate Change." https://emf.thirdlight.com/link/rl0yth77pffc-jqkp5d/@/preview/1?o.

European Commission. 2020a. *Circular Economy Action Plan: For a Cleaner and More Competitive Europe*. Luxembourg: Publications Office of the European Union.

———. 2020b. "Critical Raw Materials Resilience: Charting a Path Towards Greater Security and Sustainability: COMMUNICATION FROM THE COMMISSION TO THE EUROPEAN PARLIAMENT, THE COUNCIL, THE EUROPEAN ECONOMIC AND SOCIAL COMMITTEE AND THE COM-MITTEE OF THE REGIONS." Edited by European Commission. https://eur-lex.europa.eu/legal-content/EN/TXT/?uri=CELEX:52020DC0474.

European Environment Agency. 2020. "Cutting Greenhouse Gas Emissions Through Circular Economy Actions in the Buildings Sector." https://www.eea.europa.eu/themes/climate/cutting-greenhouse-gas-emissions-through/cutting-greenhouse-gas-emissions-through.

Eurostat. 2020. "Metadata – Private Investments, Jobs and Gross Value Added Related to Circular Economy Sectors." https://ec.europa.eu/eurostat/cache/metadata/de/cei_cie010_esmsip2.htm.

———. 2021. "Metadata – Generation of Waste Excluding Major Mineral Wastes Per Domestic Material Consumption." https://ec.europa.eu/eurostat/cache/metadata/de/cei_pc033_esmsip2.htm.

———. 2022a. "Metadata – Material Footprint." https://ec.europa.eu/eurostat/cache/metadata/en/cei_pc020_esmsip2.htm.

———. 2022b. "What Are the Main Destinations of EU Export of Waste?" https://ec.europa.eu/eurostat/de/web/products-eurostat-news/-/ddn-20220525-1.

Geissdoerfer, Martin, Paulo Savaget, Nancy M. P. Bocken, and Erik Jan Hultink. 2017. "The Circular Economy – a New Sustainability Paradigm?" *Journal of Cleaner Production* 143: 757–68. https://doi.org/10.1016/j.jclepro.2016.12.048.

Georgescu-Roegen, Nicholas. 2013. *The Entropy Law and the Economic Process*. Harvard University Press. https://doi.org/10.4159/harvard.9780674281653.

Grafström, Jonas, and Siri Aasma. 2021. "Breaking Circular Economy Barriers." *Journal of Cleaner Production* 292: 126002. https://doi.org/10.1016/j.jclepro.2021.126002.

Guzzo, Daniel, Adriana Hofmann Trevisan, Marcia Echeveste, and Janaina Mascarenhas Hornos Costa. 2019. "Circular Innovation Framework: Verifying Conceptual to Practical Decisions in Sustainability-Oriented Product-Service System Cases." *Sustainability* 11 (12): 3248. https://doi.org/10.3390/su11123248.

Horbach, Jens, Christian Rammer, and Klaus Rennings. 2012. "Determinants of Eco-Innovations by Type of Environmental Impact – the Role of Regulatory Push/Pull, Technology Push and Market Pull." *Ecological Economics* 78: 112–22. https://doi.org/10.1016/j.ecolecon.2012.04.005.

Jovanovic, Boyan, and Peter L. Rousseau. 2005. "General Purpose Technologies." In *Handbook of Economic Growth*, edited by Philippe Aghion and Peter L. Rousseau, 1:1181–1224. Handbooks in Economics. Amsterdam; Heidelberg: Elsevier. https://doi.org/10.1016/S1574-0684(05)01018-X.

Kasmire, J., Janne M. Korhonen, and Igor Nikolic. 2012. "How Radical Is a Radical Innovation? An Outline for a Computational Approach." *Energy Procedia* 20: 346–53. https://doi.org/10.1016/j.egypro.2012.03.034. Kirchherr, Julian. 2022. "Bullshit in the Sustainability and Transitions Literature: A Provocation." *Circular Economy and Sustainability*. https://doi.org/10.1007/s43615-022-00175-9.

Kirchherr, Julian, Laura Piscicelli, Ruben Bour, Erica Kostense-Smit, Jennifer Muller, Anne Huibrechtse-Truijens, and Marko Hekkert. 2018. "Barriers to the Circular Economy: Evidence from the European Union (EU)." *Ecological Economics* 150: 264–72. https://doi.org/10.1016/j.ecolecon.2018.04.028.

Koch, Nicolas, Lennard Naumann, Felix Pretis, Nolan Ritter, and Moritz Schwarz. 2022. "Attributing Agnostically Detected Large Reductions in Road Co2 Emissions to Policy Mixes." *Nature Energy*.

Kolev, Julian, Alexis Haughey, Fiona Murray, and Scott Stern. 2022. "Of Academics and Creative Destruction: Startup Advantage in the Process of Innovation." Cambridge, MA: National Bureau of Economic Research. https://doi.org/10.3386/w30362.

Korhonen, Jouni, Antero Honkasalo, and Jyri Seppälä. 2018. "Circular Economy: The Concept and Its Limitations." *Ecological Economics* 143: 37–46. https://doi.org/10.1016/j.ecolecon.2017.06.041.

Martelli, Antonio. 2014. *Models of Scenario Building and Planning: Facing Uncertainty and Complexity*. Bocconi on Management. New York: Palgrave Macmillan.

Muralikrishna, Iyyanki V., and Valli Manickam. 2017. *Environmental Management*. [Lieu de publication non identifié]: Butterworth-Heinemann.

OECD. 2018. "Global Material Global Material Resources Outlook to 2060: Economic Drivers and Environmental Consequences." Edited by OECD. https://www.oecd.org/envlronment/waste/highlights-global-material-resources-outlook-to-2060.pdf.

Pearce, David W., and R. Kerry Turner. 1990. *Economics of Natural Resources and the Environment*. Baltimore: Johns Hopkins University Press.

Perkins, Richard, and Eric Neumayer. 2012. "Does the 'California Effect' Operate Across Borders? Trading- and Investing-up in Automobile Emission Standards." *Journal of European Public Policy* 19 (2): 217–37. https://doi.org/10.1080/13501763.2011.609725.

Porter, Michael E., and Claas van der Linde. 1995. "Toward a New Conception of the Environment-Competitiveness Relationship." *Journal of Economic Perspectives* 9 (4): 97–118. https://doi.org/10.1257/jep.9.4.97.

Rodrik, Dani. 2014. "Green Industrial Policy." *Oxford Review of Economic Policy* 30 (3): 469–91. https://doi.org/10.1093/oxrep/gru025.

Schiller, Georg, Felix Müller, and Regine Ortlepp. 2017. "Mapping the Anthropogenic Stock in Germany: Metabolic Evidence for a Circular Economy." *Resources, Conservation and Recycling* 123: 93–107. https://doi.org/10.1016/j.resconrec.2016.08.007.

Schlosser, Rainer, Régis Y. Chenavaz, and Stanko Dimitrov. 2021. "Circular Economy: Joint Dynamic Pricing and Recycling Investments." *International Journal of Production Economics* 236: 108117. https://doi.org/10.1016/j.ijpe.2021.108117.

Schumpeter, Joseph A., and Richard Swedberg. 2013. *Capitalism, Socialism and Democracy*. London; New York: Routledge.

Shiller, Robert. 2017. "Narrative Economics." *American Economic Review*, no. 107 (4): 967–1004. https://www.aeaweb.org/articles?id=10.1257/aer.107.4.967.

Sørensen, Peter Birch. 2018. "From the Linear Economy to the Circular Economy: A Basic Model." *FinanzArchiv* 74 (1): 71. https://doi.org/10.1628/001522118X15097191506475.

Swann, G. M. Peter. 2009. *The Economics of Innovation: An Introduction*. Cheltenham: Edward Elgar. van Leeuwen, George, and Pierre Mohnen. 2017. "Revisiting the Porter Hypothesis: An Empirical Analysis of Green Innovation for the Netherlands." *Economics of Innovation and New Technology* 26 (1–2): 63–77. https://doi.org/10.1080/10438599.2016.1202521.

Veugelers, Reinhilde. 2012. "Which Policy Instruments to Induce Clean Innovating?" *Research Policy* 41 (10): 1770–78. https://doi.org/10.1016/j.respol.2012.06.012.

Vogel, David. 1997. *Trading up: Consumer and Environmental Regulation in a Global Economy*. Paperback ed. Cambridge, Mass.: Harvard University Press.

Wickham, Hadley. 2016. *Ggplot2: Elegrant Graphics for Data Analysis*. Second edition. Use r! Switzerland: Springer.

World Bank. 2017. *The Growing Role of Minerals and Metals for a Low Carbon Future*. World Bank, Washington, DC. https://doi.org/10.1596/28312.

World Economic Forum. 2022. "Circular Economy and Material Value Chains." https://www.weforum.org/projects/circular-economy.

Zink, Trevor, and Roland Geyer. 2017. "Circular Economy Rebound." *Journal of Industrial Ecology* 21 (3): 593–602. https://doi.org/10.1111/jiec.12545.

Stefan Wenzel, Adrien Pagano, Clemens Weiß

25 Nuclear Power

25.1 Introduction

25.1.1 On the Historical Development of the Use of Nuclear Energy

The term radioactivity was first coined in 1898 by the married couple Marie and Pierre Curie for the phenomenon discovered two years earlier by Antoine Henri Becquerel. Marie Curie was also particularly interested in using it for medical purposes.

Shortly before the start of the Second World War, the discovery of nuclear fission led to initial speculation about military possibilities for its use. The development of research and explicit hints from Germany moved some scientists to write the so-called Einstein-Szilard letter to the American government, recommending their own research and development work. The reason for that was the concern that the Nazi regime would be the first to develop an atomic weapon with great explosive power and range. To this day, the state of military development in Germany has been the subject of historical research. However, it is considered certain that the Nazi regime worked on the development of nuclear weapons and carrier rockets in several research groups.[1]

The letter from Albert Einstein and Leo Szilard ultimately led to the Manhattan Project, in which nuclear weapons were developed at great financial and human expense and used in Hiroshima and Nagasaki in August 1945. The consequences of the explosion and the radiation damage killed far more than 200,000 people in Japan. The use of the new weapon violated the rules of martial law in force at the time,[2] led to Japan's capitulation and triggered a global nuclear arms race that was accompanied by more than 2,000 nuclear tests. Above- and underground tests led to a massive increase in radioactive contamination through caesium-137 incorporation at the beginning of the 1960s.

The tests were accompanied by national programs to expand a nuclear arsenal that at the time comprised more than 30,000 weapon systems. Most of them were located close to the German-German border because the range of the carrier rockets was limited. Today, the Stockholm International Peace Research Institute (SIPRI) estimates that there are about 13,000 weapon systems. Most of them were built by the USA and the Soviet Union. A smaller number by China, France, and the UK. India and

1 ZDF (2021).

2 International Committee of the Red Cross (1907).

https://doi.org/10.1515/9783110784299-025

Pakistan also have nuclear weapons today. Israel and North Korea are assumed to possess them. Iran seeks ownership of nuclear weapons.[3]

In addition to the military purpose, US President Dwight D. Eisenhower presented a strategy for the civilian use of nuclear energy under the title "Atoms for peace" to the UN General Assembly in New York City in 1953.

In an interview on 5 April 1957, Federal Chancellor Konrad Adenauer expressed the view that he saw nuclear weapons merely as a "further development of artillery". In response, 18 German nuclear scientists led by Otto Hahn and Carl Friedrich von Weizsäcker published the Göttingen Appeal on April 12, 1957, which pointed out the destructive power of these weapons and warned of the military and political consequences of nuclear armament.

The Nuclear Non-Proliferation Treaty was signed by the USA, the UK, and the Soviet Union in 1968. Germany ratified the treaty only in 1975 after long domestic political debates. In the run-up to the signing of the treaty, there had been heated debates in Germany since the 1950s about demands to abandon reprocessing and uranium enrichment. Finally, the Safeguards System was introduced to monitor the whereabouts of weapons-grade material in each nuclear facility to prevent the proliferation of nuclear weapons.

From the 1960s to the 1980s, there was increased research worldwide on breeder reactors to extend the range of limited uranium supplies. The technology failed to gain acceptance due to increased risks, failures, and accidents. The fuel continues to pose intractable problems for final disposal today.

After the oil crisis, nuclear energy was increasingly expanded in Germany to reduce dependence on crude oil-producing countries. The German government launched an extensive campaign to promote the expansion of nuclear energy. In a brochure on nuclear energy, which was distributed in large numbers, the risk of a serious accident was estimated at a probability of one in a million per year.[4] There was a serious warning sign with the Three Miles Island reactor accident in 1979 in the USA. With the Chernobyl nuclear accident, these predictions were reduced to absurdity. After the Chernobyl reactor disaster in 1986, no further orders were placed in Germany for the construction of a reactor.

The handling of nuclear waste is full of errors and misguided developments. In 1969, Carl Friedrich von Weizsäcker believed that "the entire nuclear waste that will exist in the Federal Republic of Germany by the year 2000 will fit into a box, a cube some 20 meters long. If the box gets closed and sealed well and stuck into a mine shaft, then one may hope that the problem is solved."[5]

---

3 Sipri Institute (2021).

4 Bundesminister für Forschung und Technologie (1976), p. 82.

5 Eckert (2019).

Until the 1960s, German nuclear waste was also dumped in the sea off the coast of Portugal. In the 60s and 70s, a supposedly more cost-effective solution was found. Radioactive waste was dumped in the Asse near Wolfenbüttel in an abandoned salt mine according to mining law. At the time, this "final repository" was considered "safe for all time". This quickly proved to be an illusion because water from the overburden leaked into the mine just 10 years after the end of storage. In 2012, the Bundestag passed an amendment to the **Atomgesetz** (the German Atomic Energy Act), which now provides for the retrieval of nuclear waste before the Asse mine is decommissioned.

On 30 August 1976 with the fourth law amending the **Atomgesetz**, legal regulations on the handling of radioactive waste were established for the first time. These included the so-called disposal foresight certificate in § 9a, which tied the licensing of nuclear power plants (NPP) to precautionary measures taken by the operators for the storage of nuclear fuel. For a long time, the arbitrarily selected site Gorleben had to serve as a supposed "foresight certificate". After only four decades and tough sociopolitical conflicts, the site was abandoned. Today, the **Standortauswahlgesetz** (the German Site Selection Act) regulates the search for a safe site for the permanent storage of radioactive waste. However, the implementation of these plans will take decades.

In 2002, an amendment to the Atomgesetz was made in Germany on the proposal of the red-green federal government, which provided for the phase-out of nuclear energy. In 2010, the Merkel government decided on a lifetime extension, but this was only valid for a few months. After the devastating reactor disaster of Fukushima in Japan, the Bundestag decided on 30 June 2011 with a very broad majority to immediately shut down eight nuclear power plants. For a further nine reactors, staggered shutdown dates were set until 31 December 2022.

The new World Nuclear Industry Status Report (WNISR) shows that nuclear power has peaked decades ago, e.g. when considering the highest number of operating nuclear power plants was reached in 2002, the highest share of nuclear power in the electricity mix was reached in 1996, and the largest number first time start-ups were reached in the mid-1980s. "Nothing works without government help," said Mycle Schneider (Publisher). Also, "nearly 90 percent of the nuclear plants under construction are either built in states possessing nuclear weapons or erected by those states in third states."[6]

Today, only 32 countries in the world have operational nuclear power plants.[7] 161 countries and UN members do not use nuclear energy. Yet there are countries which, despite large fossil and nuclear power plants and a state electricity supplier,

6 Kersting (2022).

7 Verbruggen & Schneider (2022), p. 18. See https://www.worldnuclearreport.org/IMG/pdf/wnisr2023-v5.pdf.

only provide a good 75% of the population with an electricity grid connection. Especially in rural regions, decentralized solutions or island power supplies with local grids are clearly more economical than a centralized supply with large power plants. Given the degressive cost development of wind power, photovoltaics and solar thermal energy, this trend of recent years will increase significantly.[8]

This article delves into some of the above-mentioned developments and clarifies the technical foundations as well as the historical and socio-political context. Many topics can only be touched upon and await a more in-depth scientific and historical reappraisal.

25.2 In Search of Nuclear Waste Solutions

25.2.1 Squaring the Nuclear Fuel Cycle

During the operation of a nuclear power plant, the fuel elements generate nuclear fission products, such as caesium-137, as well as transuranics, such as plutonium-239. Because of these extremely radioactive isotopes, spent fuel either has to be stored in a deep geological repository or has to be reprocessed. The reprocessing of nuclear fuel allows the extraction of plutonium from the spent fuel in order to manufacture the so-called Mixed Oxide Fuel (MOX), which contains both enriched uranium and plutonium. This procedure saves uranium, which, however, is currently not a scarce resource. It also produces weapon-grade plutonium and high-level vitrified waste that needs to be stored.

The spent MOX fuel cannot be reprocessed a second time and also needs to be stored. In addition, the reprocessing of nuclear fuel generates significant water discharges (essentially tritium), air discharges (for example krypton-85) and numerous high-level-waste transports, which inherently represent a risk to the environment. Overall, the commonly agreed term "nuclear fuel cycle" rather conceals a straight line from mining to final storage, with one single loop along the line – reprocessing – at a high environmental but also economic cost. Just to indicate the scale of reprocessing-expenses: Out of the expected 124 billion pounds to dismantle nuclear facilities in the UK, 75.6% will be needed for the Sellafield reprocessing plant.[9]

Germany used to send nuclear spent fuel to France and the United Kingdom for reprocessing purposes. In 2005, the red-green alliance decided to focus on direct final storage and prohibited nuclear waste reprocessing abroad. This decision has never been called into question since then, but alternatives are regularly mentioned in the political debate, most notably partitioning and transmutation (P&T). In comparison to

8 BloombergNEF (2021).
9 GOV.UK (2019).

the regular reprocessing process, partitioning would theoretically allow to efficiently separate radioactive waste in different streams, also isolating minor actinides, a special and problematic kind of highly radioactive transuranics (e.g. americium). In the next step, the targeted use of nuclear fission on all extracted radiotoxic nuclides, called transmutation, would allow them to be converted into harmless ones.

As the German Federal Office for the Safety of Nuclear Waste Management stated in a study published in 2021, P&T is not available today anywhere in the world and decades of R&D would be necessary to try and implement it.[10] First, the separation process (partitioning) has proven to be uneconomical and is either associated with the production of high amounts of secondary waste (hydrochemical process) or could not be scaled up yet (pyrochemical process). Moreover, as it is meant to separate plutonium from the fuel, partitioning genuinely poses a great risk for proliferation.

Second, there has been no breakthrough in most transmutation technologies in the last decades, for example in the field of molten salt reactors (using molten salt as a coolant instead of water) and accelerator-driven systems (using a particle accelerator outside the reactor). But even the use of the only available transmutation technology, i.e. fast metal-cooled reactors, would not make sense, as the German Federal Office for the Safety of Nuclear Waste Management showed: 23 fast metal-cooled reactors would have to be built and operated over 300 years in order to exhaust the possibilities for transuranium reduction. 300,000 cubic meters of medium and low-level waste (MAW and LAW) would thereby be generated and would need to be stored anyway. In conclusion, this very expensive endeavour would only displace the problem.[11]

Indeed, P&T could only be used to try and transform high-level waste (HAW), i.e. spent fuel. MAW and LAW, mainly originating from the operation and decommissioning of nuclear power plants, will need final storage anyways. Those account for the biggest share of nuclear waste. Germany, for example, expects to store more than 600,000 cubic meters of MAW/LAW. In comparison, 27,000 cubic meters of HAW are to be disposed of. The same goes for depleted uranium, a residue from uranium enrichment, which is stored all over the world in tanks and cannot be processed with P&T. In fact, P&T does not even work for every HAW type. The vitrified waste mentioned above, a residue of present-day spent fuel reprocessing, is also not suitable for P&T. This means that every country, which already generated that kind of waste, such as France, Germany and the UK, will need a HAW-Repository anyway.

10 Institut für Sicherheits- und Risikowissenschaften Universität für Bodenkultur (BOKU) Wien (2021)., p. 26 et. seq.

11 Ibid, p. 34.

25.2.2 Radioactive Waste – What do we do with it?

The history of nuclear power is also a history of ignoring its consequences. The accidents, but also the radiating legacy was hardly considered for decades, and the enormous challenges of highly toxic and highly radioactive waste were not recognized, or rather grossly negligently downplayed. It is true that the Federal Ministry for Nuclear Affairs was already established in 1955 (called the Federal Ministry of Research from 1962 onwards). But broader research, legal requirements for nuclear power plant operators or mandatory waste standards, however, failed to materialize. Although the Kahl experimental nuclear power plant in Bavaria, Germany's first commercial nuclear reactor, went into operation in 1960, the Atomic Energy Act, which came into force in the same year, did not address the final storage of radioactive waste.

Today we know that there will be at least 600,000 cubic meters of low- and intermediate-level radioactive waste and 27,000 cubic meters of high-level radioactive waste left for final disposal at the end of commercial nuclear power use in Germany. While the high-level waste, which mainly consists of spent fuel elements from the nuclear power plants in 1,900 containers, accounts for only a few percent of the volume, it contains 99% of the total radioactivity of all waste. The long-term safety of the deep geological repository must be certified. According to German law, the final repository for highly radioactive waste, which has yet to be found, must guarantee the permanent protection of people and the environment from the harmful effects of this waste for no less than one million years. Three generations have used nuclear power and 30,000 have to deal with its legacy.

The example of Denmark shows that things can be done differently. Although the country could rely on the experience of the leading nuclear researcher Niels Bohr and had invested in civilian nuclear research at an early stage, the Danish government decided in 1974 to hold an open social debate on the advantages and disadvantages before starting to use nuclear power. Considering the strong anti-nuclear movement, it was not the government, but the people's representatives in the Danish parliament who were to decide whether to go nuclear or not. Intensive debates in society and within the parties, especially on how to deal with radioactive waste, led to the Danish Parliament's decision from 1985 to exclude nuclear power from future energy planning.

In Germany, the enquiry committee called Future Nuclear Energy Policy (Zukünftige Kernenergie-Politik) presented a recommendation with four energy paths at the beginning of the 1980s, including a clear phase-out option. But it was not until 2002 that the red-green federal government first decided to phase out the use of nuclear power. Instead of dealing with the problems of safe storage openly, transparently and on a scientific basis, the industry and several public authorities played them down for a long time, covered up mistakes and took decisions primarily based on short-term profit rather than on safety considerations. Between the 1960s and the 1980s, eight European countries dumped low- and intermediate-level radioactive waste in the Atlantic. Germany also participated, but the majority came from Great Britain, Switzerland and Belgium. In other parts of the world, this practice extended until 1993.

25.2.3 Asse – the Failed German Repository

One of the most momentous projects in Germany was the storage of low and intermediate-level radioactive waste in the former Asse 2 salt mine in Remlingen, Lower Saxony. Between 1967 and 1978, about 47,000 cubic meters of low and intermediate-level radioactive waste were stored there.

In 1969, the operator and the Federal Ministry of Research promised that Asse would be "safe for all times". This promise did not even last for 20 years. It took an inquiry committee from the Lower Saxony state parliament, 70 meetings, 50 witnesses and countless file reviews between 2009 and 2012, to reveal the attempted cover-up from operators and authorities. According to the report, the former mine was operated under the guise of research but was de facto used as final storage since the beginning of the seventies as it was both available and cheap. Critical voices about the suitability of the mine were systematically ignored. Those responsible assumed that Asse would be stable for at least "about 100 years". The common "safety philosophy" of operators and authorities was: Even if something happens in Asse, in reality, nothing will happen. Moreover, the storage criteria turned out to be flexible and adapted to the delivered nuclear waste: After the permissible limits had been clearly exceeded in 1969/70, they were simply increased fivefold for up to 10% of the containers. The amount of plutonium stored is also significantly higher than officially stated. There was talk of "collective deception and cover-up in an advanced stage".

After the extent of the carelessness in Asse became known, in particular regarding safety-related questions, it was placed under nuclear law. With the so-called Lex Asse, the Bundestag decided in 2013 to retrieve the waste and decommission the mine. Only through retrieval and subsequent safe final storage in a new, still-to-be-found repository, can the legal protection goals for humans and the environment be met. Furthermore, the Bundestag promised transparency and comprehensive participation of the citizens.

Consequent water ingress in Asse also raised serious questions about the safety of salt formations as host rock for repositories, questions that had long been ignored by the industry and authorities alike. As early as 1959, the Federal Institute for Soil Research (now the Federal Institute for Geosciences and Natural Resources) committed itself to the storage of radioactive waste in rock salt formations at an international nuclear conference, but without any reliable research. In Gorleben, the conflict between nuclear power plant operators and authorities on the one hand and social groups and the critical public on the other escalated.

25.2.4 From Gorleben to the New Search for a Final Repository

As early as 1977, Lower Saxony's then Prime Minister Ernst Albrecht identified the Gorleben salt dome on the border with East Germany as a provisional site for a disposal center for radioactive waste. In 1983, the Federal Government decided to explore the salt dome for its suitability as a final repository. This was a highly controversial political decision without any reliable research, the exact circumstances of which could not be clarified by a parliamentary investigation committee in the Bundestag even after three years of work. It became clear, however, that the Gorleben salt dome was not selected as a site for a repository for highly radioactive waste by a scientifically verifiable selection process. Instead, it was the result of a political, arbitrary decision that was also based on economic pressure from the nuclear power plant operators. Safety was a secondary concern.

It took decades to come to terms with the failed repository tests and to pacify the social conflict. A serious solution to the permanent conflict over the use of nuclear power and the question of final storage was first attempted by the red-green federal government in 2000. In an agreement with the nuclear power plant operators, it limited the operating lives of German nuclear power plants until the early 2020s, thus ending the commercial use of nuclear power in Germany.

In 2002, the independent Working Group on the Selection Procedure for Repository Sites (Arbeitskreis Auswahlverfahren Endlagerstandorte – AkEnd) set up by the government presented for the first time a recommendation for a comprehensible and transparent procedure for the search and selection of repository sites. When the black-yellow federal government abandoned the nuclear phase-out in 2010, the conflict was once again fueled. Mass demonstrations and the nuclear disaster on 11 March 2011 in Fukushima, Japan, ended this brief relaunch of nuclear power in Germany. Three months later, the Bundestag decided on the second nuclear phase-out at the end of 2022. Limiting the quantities of highly radioactive waste opened the possibility of finding a consensus for the final storage issue as well. Based on an initiative of Baden-Württemberg, the Federal Government and the German Bundesländer agreed to restart the search for a site based on a "white map", i.e. without any predefined criteria. The previous commitment to the Gorleben site was abandoned and exploration stopped, but the salt dome was not excluded from the new search.

The Site Selection Act (Standortauswahlgesetz – StandAG), accepted by the Bundestag and the Bundesrat in 2013, provided for the establishment of a repository commission to develop the criteria for the new search for a repository and to evaluate the law. The commission, consisting of politicians from the Federal Government and the German Bundesländer, scientists, representatives of civil society and companies, spent two years discussing all the main issues relating to the search for a final repository in Germany, working through the mistakes of the past and presenting criteria for a new, open-ended search for a site that "guarantees the best possible safety for a million years".

Based on its final report, the Bundestag decided in 2017 to restart the search for a final repository. According to the amended Site Selection Act, the process is to be participatory, science-based, transparent, self-questioning and learning. A new organizational structure with a supervisory authority (Bundesamt für die Sicherheit der kerntechnischen Entsorgung – BASE), a federally owned company as the project sponsor and operator (Bundesgesellschaft für Endlagerung – BGE), and an independent National Monitoring Committee with comprehensive review rights is to play a part in ensuring that past mistakes are not repeated (Nationales Begleitgremium – NBG). Extensive opportunities for citizens to participate are intended to ensure quality, transparency, and comprehensibility in all phases of the search and even before interim decisions are made. It is the most extensive procedure for the search for a repository in the world. Whether it will be successful in the end, whether the old conflicts can actually be overcome and whether the complex search can be comprehensible for the residents of the repository in a transparent process, still has to be proven.

With the first step of the new search for a final repository, the so-called partial report on areas (Zwischenbericht Teilgebiete) in autumn 2020, the Gorleben salt dome was eliminated from the procedure. From a scientific point of view, it is not a suitable site for a final repository for highly radioactive waste. The "Gorleben method", i.e. arbitrary political and not science-based decisions, has also had its day. According to official plans, the site for a final repository for highly radioactive waste should be found by 2031. But public statements made by the responsible authorities in 2023 indicate that the time period between 2046 and 2068 appears more realistic for a decision on the location.

25.2.5 Schacht Konrad – The only Final Repository Approved to Date

A final repository has already been found in Germany, but for low- and intermediate-level radioactive waste only: The Konrad mine in Salzgitter, Lower Saxony. The former iron ore mine in clay rock received planning approval in 2002 and is scheduled to go into operation in 2027. After several lawsuits, this approval was confirmed by the Federal Administrative Court in 2007. The site was not determined in a comparative procedure as it is foreseen for the final repository site for high-level radioactive waste. Especially because the licensing procedure started in 1982, critics question whether the repository still meets today's safety criteria. For this reason, the federally owned operator BGE is currently carrying out an extensive review to determine whether the safety-relevant requirements still correspond to the state of the art in science and technology.

The Konrad repository only has space for up to 303,000 cubic meters of low- and intermediate-level waste, about half of what will be produced in the coming decades, e.g. in the wake of power plant decommissioning. Germany will therefore have to find another repository for low and intermediate-level waste in the foreseeable future.

25.2.6 Covering the Costs – The German Repository Fund

Not only the radiation is almost eternal, but also the follow-up costs of the use of nuclear power will occupy Germany for many decades to come. Nuclear power is the most expensive way of generating electricity and a not inconsiderable part of these costs arises from the safe storage of hazardous waste. Part of the nuclear consensus was for the state to assume responsibility for the radiating legacy, also to limit the influence of the nuclear power plant operators. The latter paid their reserves of 24 billion euro into a public fund that is to bear the costs of interim and final storage. Through a clever investment strategy, this 24 billion is to be increased to 140 billion over the next few decades. That's how high the costs of nuclear waste storage will be, at least.

25.2.7 Interim Storage

Closely linked to the search for repositories is the question of where, for how long and how safely the radioactive waste will be stored above ground. All 16 existing interim storage facilities for highly radioactive nuclear waste in Germany have been licensed for a maximum of 40 years. The first interim storage facilities are approaching the end of their licensed operating phase in 2034 and an assessment must be made in good time, i.e. many years in advance, as to whether the facilities and their containers still meet current safety standards and whether a new license can be issued. This will be a challenge in the next decades, as concerns regarding the safety of these temporary installations have risen in the last decade, e.g. leading the interim storage of Brunsbüttel to lose its operating license in a lawsuit. As there is no other available temporary storage, Brunsbüttel continues to operate.

In order to foster acceptance by the local population, interim storage facilities must not turn into permanent repositories. But their existence is directly related to the success of the repository search.

25.3 The Role of Nuclear Power in Achieving Climate Neutrality

25.3.1 Greenhouse Gas Emissions of Nuclear Power

As of 2018, the installed nuclear capacity worldwide started to decrease. In the same year, Fridays for Future and the climate strikes gained a whole new dimension, becoming a global phenomenon. This was when the nuclear industry came up with a new argument for nuclear power: it would be a carbon-neutral source of power and therefore a key element of the energy transition. Electricity suppliers like the French Electricité de France (EDF) and operators all around the world launched advertisement campaigns claiming that nuclear power would be clean energy.

It can be stated that nuclear power is not climate neutral. But its CO_2 emissions vary depending on the part of the value chain that is considered. As demonstrated by Benjamin K. Sovacool, studies on greenhouse gas emissions from nuclear power plants allocate the highest share of emissions to the front-end (uranium mining, milling, conversion, enrichment and fabrication) with an average of 25.09 CO_2-equivalents per kilowatt-hour (gCO_2e/kWh). The decommissioning of power plants and mines comes second with 12,01 gCO_2e/kWh, the operation of power plants third with 11.58 gCO_2e/kWh and the backend (fuel conditioning, reprocessing and storage) last with 9.20 gCO_2e/kWh in average. All steps leading to the fuel fabrication require a great amount of power over the whole lifespan of a plant and are therefore the most carbon-intensive ones. Of course, these processes also go along with a direct environmental impact, as for instance soil pollution.

Such calculations depend on a large variety of factors, some of which were looked at more closely only a short time ago. This is the case for sulphur hexafluoride (SF6), which is used for the electrical insulation of high-voltage lines and is known to be the most climate-damaging gas in the world (1 kg = 22,800 kg CO_2e). In September 2021, the French media took interest for the first time in a report from EDF, stating that the plants Flamanville 1 and 2 had already exceeded the authorized limit of 100 kg of SF6 emissions for the year. But this was neither a new nor an isolated occurrence, as other reports from EDF show, for example for the Penly power plant in 2020. In a scientific contribution published in 2020, the engineer Bernard Laponche took the example of the Belleville NPP and showed that between 2003 and 2004, solely the SF6-emissions – not even taking into account the refrigerant gases – of the plant had reached 1.78 g CO_2e/kWh. This shows the wide range of parameters that needs to be taken into account and also how fast the relevant ones can be set aside.

Relying on the existing literature, Benjamin Sovacool obtains a total mean value of 66 gCO_2e/kWh for nuclear power. In a more recent publication, the German Öko-Institut estimated the global emissions of nuclear power to reach 104 gCO_2e/kWh. In a more conservative approach, the International Panel on Climate Change (IPCC) adopted a median value of 12 gCO_2e/kWh, recognizing however, that emissions could reach a maximum of 110 g CO_2e/kWh. By way of comparison, the IPCC sets a median value of 11 g CO_2e/kWh for onshore wind and 820 g CO_2e/kWh for coal. The huge discrepancies in the figures show how difficult nuclear CO_2 emissions are to evaluate. But they also let us draw the conclusion that the lower estimates the nuclear industry likes to rely on, are probably not the ones closest to reality. Indeed, it is interesting to note that the IPCC refers to the work of two authors, but only the lowest figures from Manfred Lenzen appear in the IPCC chart. According to the results of the other author, Garvin A. Heath, emissions of nuclear power plants could even reach a maximum of 220 g CO_2e/kWh. In the end, such calculations do vary strongly depending on the data included or left aside. Further research including the whole supply chain in different countries would be helpful.

25.3.2 The "Partnership" with Renewables

Because of their fluctuant nature, it is often argued that renewable energies need conventional power plants as reliable partners providing baseload. But as the Deutsches Institut für Wirtschaftsforschung (DIW) showed in a publication from 2021 relying on data from the International Atomic Energy Agency (IAEA), the reliability of nuclear power plants is significantly overrated. The load factor of all nuclear power plants worldwide since 1970 reaches 66%. This means contrariwise that one-third of the capacity is not used due to outages, whether planned or not. Planned outages include for example fuel replacement and maintenance while unplanned outages can be caused by technical problems, human errors, or external factors such as droughts reducing the availability of cooling water.

In countries relying strongly on nuclear power, the spontaneous unavailability of capacities can even destabilize the whole electricity grid. France, which has a 69% nuclear share in electricity production, meanwhile has a tradition of seasonal electricity shortages. Electricity production barely meets the demand in the winter when the use of electric heating is high or in the summer when many reactors are in maintenance. When malfunctions ad up to the already tight capacity management, a shortage situation occurs and the electricity prices soar. After several summers with high NPP-unavailability rates in the summer (DIW S.113), France reached a peak of more than 30 unavailable plants in the summer of 2022 because of a stress corrosion issue affecting the welds of several reactors. Such unplanned outages will increase, not only in France, as the worldwide nuclear park ages. It already reached a mean age of 31.9 years.

In sum, nuclear power plants provide for constant electricity production, but their failure rate is high and outages are both unforeseeable and long-lasting. Therefore, they are not the reliable and available partner renewables are looking for.

But most of all, renewables are lacking flexible partners, which NPPs are not. In Germany, the feed-in priority for renewables is anchored in law, in the well-known Erneuerbaren Energien Gesetz – EEG. But even so, when it comes to bottleneck management, renewable energies are often required to reduce their output to a greater degree than nuclear power plants. In 2021, the Federal Network Agency estimated that 3% of the renewable production in Germany had been subject to curtailment. In contrast, in the same year, only 1.44% of the nuclear energy produced was subject to curtailment.

For some of them, this can be explained by their physical location. Considering that the rollout of renewables is less advanced in southern Germany, the southern NPP tended to be more relevant for power supply. But it is also caused by technical and economic limitations specific to NPP. As the technological impact assessment centre of the Bundestag showed (TAB), a load-following operation is much more demanding for NPP-components, which are affected by radical temperature and pressure changes. NPPs in operation today need a few days to start up and are not meant to be running long with a load under 60%. In order to spare components, operators even seek to keep slight load adjustments to a minimum. Consequently, a load-following operation of NPP goes along with more outages and higher costs for maintenance. As NPP are a very capital-intensive infrastructure, operators also have a strong incentive to keep them running constantly to cover the capital costs. The TAB concludes that the operation of nuclear power plants becomes unprofitable when the workload sinks too much. This is one of the reasons, which led the nuclear industry to develop small modular reactors with great determination. The modular approach is meant, among others, to overcome the blatant lack of compatibility between nuclear power and renewables.

25.3.3 Nuclear Power and Climate Change

In the process of producing electricity, nuclear reactors need to be cooled down constantly. To do so, they rely on external coolants such as rivers and seas. Water is withdrawn and returned later on in a heated-up state. In order to diminish the amount of water withdrawn and the thermal discharge, most nuclear power plants use cooling towers, but not all of them. For instance, the Fessenheim plant in Alsace did not. Other reactors use smaller towers as a means of ensuring landscape protection, such as in Neckarwestheim (Germany). The capacity of the cooling tower is an important factor of the plant resilience to heat waves. Because this equipment relies on the cooling action of the air, its efficiency decreases with rising outdoor temperatures. And in the process of cooling down, part of the river water evaporates. In 2019 for example, only 82 million cubic meters out of 105 extracted to cool down Isar 2 in Germany were given back, representing a 23% loss of cooling water.

In order not to unreasonably warm up rivers and to protect fish stocks, local authorities generally set limits for the temperature of rejected water, for the downstream river temperature and the maximal river warming. In France, the operation permits of nuclear power plants generally prohibit a difference greater than two degrees between upstream and downstream water. However, against the background of climate change, complying with these limits becomes increasingly difficult and exemptions are getting more and more common. In the summer of 2022, in the wake of corrosion problems affecting several NPPs and to avoid shortages, the French nuclear regulator Autorité de sûreté nucléaire (ASN) authorized 16 nuclear power plants located in the south of France to exceed the maximal thermal discharge.

As heat waves and droughts are becoming more frequent, so will heat-induced outages and exemptions. The former member of the German parliament Sylvia Kotting-Uhl illustrated that by collecting a wide range of data regarding French NPP. It shows that between 2003 and 2019, 22 French reactors out of 58 had to reduce their output or to stop electricity production because of heat waves.

In 2003, the cumulative total of all French outages reached 6,308 GWh, which is about the production of one 900 MW unit over one year. Some reactors in the southern part of the country are particularly affected, such as the Bugey nuclear power plant. In 2018, the third Bugey unit had to reduce its output or to be stopped for a total period of 52 days. In such situations, France strongly relies on electricity imports from neighbouring countries such as Germany to fill in the gap.

Periods of cold weather can also affect the functioning of a nuclear power plant. This was the case in February 2021 in Texas, when temperatures unexpectedly reached up to −26 degrees Celsius. The massive failure of the energy infrastructure was mainly caused by fuel shortages and the equipment freeze of gas power plants. But unit one of the south Texas project nuclear power plant also faced an unplanned complete outage. It was caused by low steam generator levels due to the loss of two water pumps. According to the NRC, the reason for this event is unknown. But considering one particularity of this NPP, a probable reason appears very clearly. Indeed, because of the favourable weather in Texas and in order to cut expenses, the steam turbine has been placed directly on the roof of the power plant, instead of being placed in an independent building. This means that the cooling circuit partly runs outside. As a consequence, it is highly sensitive to extreme outdoor conditions and it most certainly did not withstand the negative temperatures.

This means that, in addition to coolant shortages and use restrictions, some nuclear power plants will have to face coolant freeze in the future. But just like in Texas, these phenomena tend to happen unexpectedly. NPPs relying on external water tanks as well as on seawater could be surprised by external events, like unforeseeable weather conditions. As NPPs show a vulnerability to climate change, they are also unfit as a means of climate change adaptation. This crucial element should not be forgotten while designing the future carbon-free electricity system.

25.3.4 Nuclear Power and Carbon Neutrality by 2050

Even if nuclear power is said to be instrumental in reaching climate neutrality by 2050, new builds do not seem to gain momentum in a lot of countries. The only countries in the world to effectively build several new NPP are China, India, and the United Arab Emirates. Even in those with a traditionally high share of nuclear power such as France and the USA, the actual implementation of new NPP projects remain rare.

In order to make the technology more attractive, new types of reactors were to be developed, offering more safety and efficiency. And above all, they would solve the nuclear waste issue. But since the 1960s, unsuccessful research has been carried out on the so-called generation IV. In Germany, research on the high-temperature reactor (HTR) in Jülich and at the Hamm-Uentrop high-temperature thorium reactor has been stopped due to insurmountable problems, similarly in other countries. The HTR operation was characterized by the accumulation of technical failures, has proven to be unprofitable, and therefore does not interest private sector operators. As the German Öko-Institute sums it up, the very high degree of energy efficiency it aimed for was never reached and the resulting theoretical benefits for electricity generation and for the provision of process heat (e.g. to produce hydrogen) have never materialized.

HTRs also do not solve the final repository issue, as they produce waste fuel for which no conditioning and disposal method exists. Finally, HTRs are less proliferation-resistant than light-water reactors. Approximately the same goes for other types of "Gen IV reactors". Fast breeders, which could use nuclear fuel more efficiently, have been subject to in-depth research and development, but the complex proliferation issue they also pose, and their great operating costs do not make a convincing option out of it. And the molten salt reactor (MSR), which uses liquid fuel, never even went beyond the development phase.

Small modular reactors (SMR), conversely, are still considered a potential game changer. They do not rely on a specific technology but on the mere concept of downsizing and standardizing other reactor technologies to avoid construction problems. But as the German Federal Office for the Safety of Nuclear Waste Management (BASE) puts it in a study published in 2021, the modular approach corresponds to the mere spreading of risks (nuclear accidents but also fuel transportation) and of proliferation issues. Moreover, as no SMR has yet been constructed in the world, a long R&D and licensing phase still lie ahead and commercial operation cannot be expected to begin before 15 years at least. It appears dangerous to let the decarbonisation schemes we need to design today rely on the potential availability of this technology only some years before the 2050 deadline.

The same goes for nuclear fusion. Until now, prototypes of fusion reactors such as the Wendelstein 7-X in Greifswald (Germany) managed to maintain plasma for several seconds. But the real challenge lies within the continued operation of fusion reactors. In a study published in 2020, Michael Dittmar, a former researcher at the European Organization for Nuclear Research (CERN), illustrates it in the case of the biggest fusion project currently implemented, the International thermonuclear Experimental Reactor (ITER) in Cadarache (France). According to Dittmar, no material capable of lastingly withstanding the expected temperature of 150 million degrees has been found yet to build the reactor's inner wall. Also, tritium resources needed to power such a reactor in the long term do not exist in the quantity required and are particularly difficult to produce.

But even if all these issues were to be ultimately solved, commercial fusion reactors would come too late to help achieve the 2050 target. ITER is not designed to produce electricity but only heat in 2042 at the earliest, the deadline was already moved back several times. After this, a demonstration reactor called DEMO is to be built and to produce electricity for the first time. Even if all technical problems are solved and if this plan is affected by no further delay, the potential licensing and commercialisation of fusion reactors would not start before several decades either. In sum, it is already too late for new nuclear technologies to help us fight climate change.

25.4 Nuclear Power as a means of Energy Transition – Economical and Geopolitical Costs

25.4.1 The Resources bound by Nuclear Power

ITER is also a good example of how money could have been better invested to fight climate change. In the framework of the EU research program Horizon 2020 (for the period 2014–2020), 5.08 billion euro have been spent on nuclear research, out of which 2.707 billion euro have gone to ITER. Comparatively, in the same period, all other EU energy research areas taken together have been funded with 5.93 billion euro, creating a serious imbalance between nuclear and non-nuclear research.[12] In the framework of the following program (for the period 2021–2027), the budget of ITER alone was more than doubled to reach 5.6 billion euro.

This disproportionate investment in nuclear research would already have contributed to fighting climate change had it for example been directed at solar panels and heat pump efficiency. In addition, even in countries phasing out nuclear energy, such as Germany, follow-up costs of nuclear power will burden the national budget in the long term. At least 3.7 billion have been budgeted until 2033 only for pre-retrieval preparation in the failed Asse final storage.[13] For 2022 alone, the costs of decommissioning old power plants and of interim and final storage amount to more than 1.2 billion euro in the budget of the German federal environment, finance, and re-

12 Deutscher Bundestag (2019), p. 46 f.
13 Deutscher Bundestag (2022).

search ministries. If it was not for the nuclear heritage, this money would instead be spent on nature and climate protection. A study commissioned by the German federal economy ministry in 2015 calculated that the total costs of all kinds of decommissioning, transportation and disposal in Germany could reach about 170 billion euro.[14] Seven years later, however, this can already be considered a low estimate.

The costs of nuclear power are also increasingly showing in new builds. Since 2005, the French electricity supplier EDF is building a third reactor at the Flamanville plant. Instead of going online in 2012, as originally planned, the European Pressurized Water Reactor (EPR) is to be commissioned in 2024. The expected costs of 3.2 billion euro have skyrocketed and will reach an expected 19.1 billion euro.[15]

This is not an isolated case. All over Europe, the nuclear industry struggles with meeting the necessary high safety standards of new NPP. As of 2018, one megawatt hour of nuclear power was expected to cost 100 euro in Paks II in Hungary, 119 euro in Hinkley Point C in the UK and up to 125 euro in Flamanville.[16] This is a gigantic cost for electricity generation as compared for example to the levelized cost for onshore wind generation (56 Euro/MWh) and for solar photovoltaics (85 Euro/MWh) in the same year.[17] Since 2018, the levelized costs for renewable generation have sunk dramatically, but those for nuclear power did not; the Lazard investment bank even considered the average levelized costs of nuclear to reach 167 dollar in October 2021.[18]

25.4.2 Tracing Back the Lack of Investment

By mid-2023, 58 reactors worldwide were listed as under construction, as opposed to 234 in 1979. In the same year, only three reactors were being built in the European Union and one in the USA.[19] These figures are somewhat puzzling when considering both the intensive lobbying from the nuclear industry and the nuclear history it could theoretically build on in countries such as France and the United Kingdom. Even eastern European countries showed an overall pro-nuclear stance in the last decades. But despite of that, the number of new projects remains ridiculously low, especially in Europe.

This gap between pro-nuclear stances and a very low implementation rate of specific projects can be explained by the costs of new builds. In fact, such expenses require for example an enormous capacity to incur excessive debt on the part of the purchasing company, as in the case of the French EDF. Then again, investing in new

14 Warth & Klein Grant Thornton (2015), p. 57.
15 Cour de comptes (2020), p. 68.
16 Huneke & Heidinger (2018).
17 IRENA (2019), p. 18 and 20.
18 Lazard (2021).
19 https://www.worldnuclearreport.org/IMG/pdf/wnisr2023-v5.pdf.

builds put the company in a very difficult position, forcing the French government to renationalise it in 2022, mostly to save its nuclear branch. Hungary chose another financing method and contracted a Russian loan to build two new reactors in Paks – a Trojan horse from a geopolitical point of view. As for Slovakia, it chose to complete two soviet-time NPP ruins in Mochovce to save costs. But even so, the Italian company Enel actively tries to get rid of its 66% stake in the purchasing company Slovenske Elektrarne.[20] Given the constant overruns and delays, NPP new-builds are not profitable and a real burden for purchasing companies.

Overall, decision-makers in democratic countries find it difficult to justify the gigantic amounts of taxpayer money new builds require. A closer look at the countries building the biggest share of new NPP in the world in 2022 gives a hint as to the significant steering capacity and the low accountability a state needs to implement nuclear projects nowadays. With 23 reactors under construction, China is the number one country with the most new builds in the world in 2023. The Bertelsmann Stiftung transformation index considers China a "hard-line autocracy".[21] With eight reactors under construction, India is the second country in the world in terms of new builds in 2022. The Bertelsmann Stiftung transformation index considers it a "defective democracy". Out of the five countries with the most new builds in the world in 2022, only one, South Korea, is considered by the Bertelsmann Stiftung to be a working democracy. The other two, Russia and Turkey, are not.

In order for democratic countries to build new NPP without excessively burdening taxpayers, they need private investors. However, several German banks have already excluded nuclear power from their portfolios, such as the Deutsche Kreditbank (DKB) or GLS. Banks, which do not exclude nuclear power from their portfolio, tend to have precise risk and profitability criteria for such investment. The Lazard investment bank for example does not consider new builds in Europe as a promising market, most of all because of the significant upfront costs.[22] This results in a lack of capital for the nuclear industry.

In order to address the financing issue, several countries in the European Union, led by France, strongly advocated for nuclear power to be included in the taxonomy for sustainable activities. This EU regulation sets a list of green activities meant at guiding investors and customers.[23] Banks are allowed to label a fund or investment as "green" only if it contains assets in line with the EU taxonomy. In February 2022, after a long-lasting dispute mostly between France and Germany, the European Commission decided to include nuclear power in the taxonomy as a transitional activity until at least 2045. This apparently insignificant regulatory issue escalated and reached the highest political level for a good reason: It was identified by pro-nuclear countries as

---

20 Enel S.p.A. (2020).

21 BTI Transformation Index (2022).

22 Smith (2022).

23 European Union (2020).

the last available means to flush fresh money into the nuclear industry in Europe. Its effect will have to be measured in the upcoming years. Austria decided to challenge this decision in court.

25.4.3 Between Dependency and Path Dependency

In 2021, about 70% of French electricity generation was covered by nuclear energy. France heats predominantly with electricity and its NPPs are consequently very in demand during the winter. As the ageing French power plant fleet requires continuous maintenance and upgrades, overhauls are closely timed in the summer – otherwise, maximum availability in winter would be jeopardized. This leaves little room for unplanned outages. In the summer of 2022, however, several French nuclear power plants had to curtail their output due to heat: The rivers from which they draw cooling water were overheating. In addition, eleven French nuclear power plants were shut down due to both proven and suspected corrosion damage to their emergency cooling systems. As a consequence, in 2022, up to 32 out of 56 reactors had to be disconnected from the grid simultaneously. Electricity prices skyrocketed.

The French corrosion issue is a generic failure, meaning that it affects all nuclear power plants from at least one series (in this case, the four reactors of the N4 series).[24] As France is strongly dependent on nuclear power, generic issues can jeopardize the country's security of supply in the very short term. Repairs are generally both long and expensive. Therefore, in the worst case, a balance must be struck between nuclear safety and security of supply. The dependency on one single complex and thus vulnerable technology turns out to be a weakness.

In all countries using nuclear energy, the costs of such repairs add up to the costs of decommissioning and in some cases to the costs of NPP upgrades, if the long-term operation (LTO) is decided on. This means that nuclear power does not only have exorbitant upfront costs, but also tremendous final costs, that have to be dealt with towards the end of NPP-lifetime. Even when leaving the costs of final storage asidc (as this burden is often taken over by tax-payers). In order to cover those costs, NPP operators need to generate profit by producing as much electricity as possible from existing plants and, if possible, to build new ones. It is a real lock-in and a headlong rush for nuclear production, always delaying the final payment.

The nuclear path dependency does not only originate from the infrastructure and its costs, but also from the lobby and institutions that grew with it. Of course, nuclear power generation goes along with public agencies such as nuclear safety authorities or bodies in charge of nuclear waste. These can become very influential over time, as is the case with the French Atomic Energy Commission (CEA – Commissariat à l'Éner-

24 Autorité de sureté nucléaire (Asn) (2022).

gie Atomique), which has been in charge of nuclear research since 1945. But as France once again exemplifies, the interplay between public authorities and industry – materialized by the historic state company EDF – can be key in giving rise to a lastingly pro-nuclear mindset of the administration. Students from French elite engineering schools with a strong pro-nuclear stance eventually become both civil servants and top-level managers of the nuclear industry and goad each other and the state into perpetuating nuclear power. This is called "nucleocracy". The word even found its way into the French dictionary in 2021.[25]

25.4.4 The Nuclear Fuel – Europe's Dependency on Russia

A distinction must be made between the markets for uranium, enriched uranium and fuel assemblies. In 2021, 19.69% of natural uranium imported into the EU came from Russia and 22.99% from Kazakhstan.[26] Since the beginning of the Russian aggression war on Ukraine, this balance further shifted in favour of Kazakhstan. However, not only is Kazakhstan part of the Russian sphere of influence, but the Russian uranium mining holding ARMZ also owns shares in uranium mining facilities in Kazakhstan. Uranium originating from Kazakhstan can therefore be considered to be controlled by Russia to a large extent.

Since the beginning of the war, the European Union has tried to stop its reliance on Russian raw materials as quickly as possible. But to replace these uranium sources, Western mining companies would have to expand their production. Australia accounted for 8.7% of global uranium production in 2021 and Canada for 9.7%.[27] However, production in these countries has been almost halved in the last decade, not least due to decreasing demand following the Fukushima disaster. Uranium companies will therefore slowly ramp up production to avoid bad investments. Under these circumstances, Russian natural uranium can at best be replaced in the medium term.

The uranium enrichment sector is even more dependent. In 2021 and 2022, Russia owned 46% of the world's enrichment capacity and 31% of the enriched uranium imported into the EU came from Russia.[28] Western uranium enrichment capacities can only be ramped up to a limited extent. The German-Dutch-British company Urenco, which has terminated its contracts with Russia, was already very busy in recent years. At the Gronau site, only 18% of the licensed capacity was not used in 2020, which is also partly due to maintenance activities.

Therefore, western uranium enrichment facilities will not be able to fully compensate for the lost Russian capacities. In addition, not all Western companies have

25 Wettach (2021).

26 ESA (2022), p. 19.

27 World Nuclear Association (2022).

28 ESA (2022), p. 24 and 68.

terminated contracts with Russia. Because of the close partnership between the French Framatome and the Russian Rosatom, switching to enrichment capacities in France would mostly mean carrying out the dependency on Russia. As the Center on Global Energy Policy of Columbia University explains, even the USA uses Russian enrichment capacities.[29]

The creation of new enrichment capacities is associated with complex licensing processes and huge investments, which do not make it possible to replace Rosatom in the short term. Capacity building is bound by the requirements of the Nuclear Non-Proliferation Treaty and is subject to the IAEA's safeguards regime to prevent the proliferation of weapons-grade material.

However, the most direct reliance on Russia is to be found in the sector of fuel assembly production. 18 nuclear reactors currently operating in the EU are of Russian design (VVER reactors) and depend on special hexagonal fuel assemblies from Rosatom. Those reactors are located in Finland, the Czech Republic, Hungary, Slovakia and Bulgaria.

The Euratom Supply Agency assesses this dependency as "a matter of concern" and a "significant vulnerability".[30] Western countries largely lack the technological capability to produce such hexagonal fuel assemblies. Only Westinghouse partly masters this technology.

The reliance on Russia in the nuclear fuel sector explains why, in November 2022, sanctions had still not been imposed on the Russian nuclear industry for its aggression war in Ukraine. The European dependency on Russian nuclear fuel makes it even more vulnerable than the gas market and a key weakness of the European energy sector.

25.5 Conclusion

The ambivalence of civilian and military use has pervaded the history of nuclear energy from the beginning. For decades, the close connection between the two options was denied in public speeches. Instead, nuclear energy was promoted as a cheap source of energy. In the process, however, a wide range of indirect subsidies were concealed – from necessary investments in research to the costs of dismantling and final storage to the lack of liability insurance for serious accidents and disasters.

During a visit to the Framatome plant in Le Creusot, French President Emmanuel Macron displayed a new openness in the early 20s. Here, Macron mentioned the nuclear industry and the "coherence between strategic autonomy and energy independence". He made it clear that "without civil nuclear power" there could be no nuclear

29 Bowen & Dabbar (2022).
30 ESA (2021), p. 26.

weapons, "without military nuclear power no civil nuclear power," and "both in research and production."[31]

Macron's openness is probably related to the enormous economic challenges associated with the construction, permanent readiness, and operational capability of nuclear weapons. Only through decades of cross-subsidization and synergies in terms of production, research and personnel could the effort be economically managed. This also explains why nearly 90% of the nuclear power plants under construction are either built in nuclear weapon states or are built by those in non-nuclear weapon states.[32] Economies of scale are used to reduce production costs. Russia has even developed a business model from this that forces third countries into decades-long geopolitical dependencies. Given the very high capital costs, very long construction times and years of construction delays, nuclear energy for the production of electricity is no longer competitive worldwide. Heat extraction is not possible anyway. In 2021, more than two-thirds of global investments for the production of electricity already flowed into renewable energies. Only a fraction was invested in nuclear technology.[33]

Another argument has pervaded the debates on nuclear energy for many years and has been increasingly put forward by the industry in recent years. Supposedly, the production of nuclear power is free of greenhouse gases and therefore helps to protect the climate. However, this argument does not stand up to closer scrutiny, as the above chapter shows. The finite nature of uranium resources and the dependence on a few countries with uranium deposits also argue against this. Moreover, an expansion of production to all countries on earth would massively increase the proliferation risk. Costs will also continue to rise against the background of the war-related attacks on nuclear power plants in Ukraine because protection against so-called "third-party interference" presents operators with further challenges that are difficult to solve. So far, all security requirements have assumed the absence of war and hybrid third-party attacks.

Against this background, there is much to be said for the assessment that maintaining or building up weapons programs is the real driver of nuclear energy use. But even countries with nuclear weapons programs will rely on lower-cost renewables for most of their energy supply in the future because they cannot otherwise ensure the competitiveness of their industry.

Most tragically, Russia has invaded Ukraine, which once voluntarily gave up its nuclear weapons. Future disarmament negotiations with nuclear-weapon states will be made considerably more difficult as a result. It is to be hoped that Russia's threats with the nuclear option will not be crowned with success. Otherwise, other states could feel compelled to invest in nuclear weapons, either openly or covertly.

---

31 Bezat (2020).

32 Kersting (2022) /Verbruggen & Schneider (2022).

33 Bloomberg NEF (2021).

Bibliography

Standortauswahlgesetz – StandAG: Gesetz zur Suche und Auswahl eines Standortes für ein Endlager für hochradioaktive Abfälle. Fundstelle: §1, Abs. 2. Online verfügbar unter https://www.gesetze-im-internet.de/standag_2017/BJNR107410017.html, zuletzt geprüft am 13.10.2022.

Autorité de sureté nucléaire (asn) (2022): L'ASN modifie temporairement ses prescriptions encadrant les rejets thermiques des centrales nucléaires de Blayais, Bugey, Golfech, Saint-Alban et Tricastin. Hg. v. Autorité de sureté nucléaire (asn). Online verfügbar unter https://www.asn.fr/l-asn-informe/actualites/modification-temporaire-des-prescriptions-encadrant-les-rejets-thermiques-de-5-centrales-nucleaires, zuletzt aktualisiert am 13.10.2022.

Autorité de sureté nucléaire (asn) (2022): Stress corrosion phenomenon affecting nuclear power reactors: ASN considers that EDF's inspection strategy is appropriate. Online verfügbar unter https://www.french-nuclear-safety.fr/asn-informs/news-releases/stress-corrosion-phenomenon-asn-considers-that-edf-s-inspection-strategy-is-appropriate, zuletzt aktualisiert am 07.11.2022, zuletzt geprüft am 13.10.2022.

Bezat, Jean-Michel (2020): Nucléaire : « Pour Emmanuel Macron, c'est la filière militaire qui prime ». Hg. v. Le Monde. Online verfügbar unter https://www.lemonde.fr/idees/article/2020/12/21/nucleaire-pour-emmanuel-macron-c-est-la-filiere-militaire-qui-prime_6064052_3232.html, zuletzt geprüft am 13.10.2022.

Bloomberg NEF (2021): New Energy Outlook 2021. Executive Summary. Online verfügbar unter https://about.bnef.com/new-energy-outlook/, zuletzt geprüft am 13.10.2022.

Bowen, Matt; Dabbar, Paul (2022): Reducing Russian Involvement in Western Nuclear Power Markets. Online verfügbar unter https://www.energypolicy.columbia.edu/research/commentary/reducing-russian-involvement-western-nuclear-power-markets, zuletzt geprüft am 13.10.2022.

Breteau, Par Pierre; Parienté, Jonathan et al (2019): Comment la canicule affecte le fonctionnement des centrales nucléaires. Online verfügbar unter https://www.lemonde.fr/les-decodeurs/article/2019/07/26/comment-la-canicule-affecte-le-fonctionnement-des-centrales-nucleaires_5493867_4355770.html, zuletzt aktualisiert am 13.10.2022.

Bruckner, Thomas; Fulton, Lew et al: Annex III: Technology-specific Cost and Performance Parameters. In: Climate Change 2014: Mitigation of Climate. Online verfügbar unter https://www.ipcc.ch/site/assets/uploads/2018/02/ipcc_wg3_ar5_annex-iii.pdf, zuletzt geprüft am 13.10.2022.

BTI Transformation Index (2022): China. Bertelsmann Stiftung. Online verfügbar unter https://bti-project.org/en/reports/country-dashboard/CHN, zuletzt geprüft am 13.10.2022.

Bundesamt für die Sicherheit der nuklearen Entsorgung: Abfallarten und ihre Entstehung. Hg. v. Bundesamt für die Sicherheit der nuklearen Entsorgung. Online verfügbar unter https://www.endlagersuche-infoplattform.de/webs/Endlagersuche/DE/Radioaktiver-Abfall/Abfallarten/abfallarten_node.html., zuletzt geprüft am 24.08.2022.

Bundesanzeiger (1959): Bundesgesetzblatt. Teil 1. Bonn, 31.12.1959. Online verfügbar unter http://www.bgbl.de/xaver/bgbl/start.xav?startbk=Bundesanzeiger_BGBl&jumpTo=bgbl159056.pdf, zuletzt geprüft am 24.08.2022.

Bundesministerium für Forschung und Technologie (1976): Kernenergie. Eine Bürgerinformation. Bonn, S. 82.

Bundesnetzagentur (Hg.) (2021): Bericht Netzengpassmanagement Gesamtes Jahr 2021. Bundesnetzagentur. Online verfügbar unter https://www.bundesnetzagentur.de/SharedDocs/Downloads/DE/Sachgebiete/Energie/Unternehmen_Institutionen/Versorgungssicherheit/Engpassmanagement/Zahlen%20Ganzes%20Jahr2021.pdf;jsessionid=FD314F6C7FD934954F1409B35F1EB566?__blob=publicationFile&v=4, zuletzt geprüft am 13.10.2022.

Cour des comptes (2020): RPT La filière EPR. Rapport public thématique. Online verfügbar unter https://www.ccomptes.fr/system/files/2020-08/20200709-rapport-filiere-EPR.pdf.

Detlev Möller (2012): Es geschah am helllichten Tag. In: *Bundesamt für Strahlenschutz: Asse Einblicke* (Nr. 19), S.3. Online verfügbar unter https://archiv.bge.de/archiv/www.asse.bund.de/SharedDocs/Downloads/Asse/DE/asse-einblicke/asse-einblicke-19c2ae.pdf?__blob=publicationFile&v=3, zuletzt geprüft am 13.10.2022.

Deutscher Bundestag (2011): Gesetzentwurf der Fraktionen der CDU/CSU und FDP. Entwurf eines Dreizehnten Gesetzes zur Änderung des Atomgesetzes, 06.06.2011 (Drucksache 17 60/70). Online verfügbar unter https://dserver.bundestag.de/btd/17/060/1706070.pdf, zuletzt geprüft am 13.10.2022.

Deutscher Bundestag (2012): Gesetzentwurf der Bundesregierung. Entwurf eines Gesetzes über die Feststellung des Bundeshaushaltsplans für das Haushaltsjahr 2022. (Haushaltsgesetz 2022) (20/1000). Online verfügbar unter https://dserver.bundestag.de/btd/20/010/2001000.pdf, zuletzt geprüft am 13.10.2022.

Deutscher Bundestag (2012): Entwurf eines Gesetzes zur Beschleunigung der Rückholung radioaktiver Abfälle und der Stilllegung der Schachtanlage Asse II. Bundestagsdrucksache 17/11822, 11.12.2012. Online verfügbar unter https://dserver.bundestag.de/btd/17/118/1711822.pdf, zuletzt geprüft am 13.10.2022.

Deutscher Bundestag (2013): Beschlussempfehlung und Bericht. des 1. Untersuchungsausschusses nach Artikel 44 des Grundgesetzes. Drucksache 17/13700, 13.05.2013. Online verfügbar unter https://dserver.bundestag.de/btd/17/137/1713700.pdf.

Deutscher Bundestag (2013): Beschlussempfehlung und Bericht des 1. Untersuchungsausschusses nach Artikel 44 des Grundgesetzes in der 17. Wahlperiode des Deutschen Bundestags (Parlamentarischer Untersuchungsausschuss Gorleben). Drucksache 17/13700, 23.05.2013. Online verfügbar unter https://dserver.bundestag.de/btd/17/137/1713700.pdf.

Deutscher Bundestag (2019): Schriftliche Fragen. mit den in der Woche vom 20. Mai 2019 eingegangenen Antworten der Bundesregierung, 2019 (Drucksache 19/10441). Online verfügbar unter https://dserver.bundestag.de/btd/19/104/1910441.pdf, zuletzt geprüft am 13.10.2022.

Deutscher Bundestag (18.03.2022): Entwurf eines Gesetzes über die Feststellung des Bundeshaushaltsplans für das Haushaltsjahr 2022. Haushaltsgesetz 2022 2022 (Drucksache 20/1000). Online verfügbar unter https://www.bundesrat.de/SharedDocs/drucksachen/2022/0101-0200/115-22.pdf?__blob=publicationFile&v=1, zuletzt geprüft am 30.11.2022.

Dittmar, Michael (2019): Status and Prospects of the ITER Plasma Physics Experiment. Is it time to terminate the project? (Part 1). Online verfügbar unter https://www.gruene-bundestag.de/fileadmin/media/gruenebundestag_de/themen_az/atomausstieg/PDF/iter-project-2019-part-1.pdf, zuletzt geprüft am 13.10.2022.

Dr. Detlef Appel, Dr. Bruno Baltes, Dr. Volkmar Bräuer et al (2002): Auswahlverfahren für Endlagerstandorte. Empfehlungen des AkEnd – Arbeitskreis Auswahlverfahren Endlagerstandorte. Köln: W & S Druck GmbH. Online verfügbar unter https://www.bundestag.de/endlager-archiv/blob/281906/c1fb3860506631de51b9f1f689b7664c/kmat_01_akend-data.pdf, zuletzt geprüft am 13.10.2022.

Eckert, Astrid M.: Closing the Nuclear Fuel Cycle at Gorleben? West Germanys Energy Future in the Borderlands. In: West Germany and the Iron Curtain, 201–244. Online verfügbar unter https://academic.oup.com/book/36494/chapter-abstract/321192508, zuletzt geprüft am 07.12.2022.

Électricité de France (edf) (2021): Informations réglementaires. Cumul des émissions de gaz SF6 supérieur à 100kg depuis le 1er janvier. Online verfügbar unter https://www.edf.fr/la-centrale-nucleaire-de-flamanville-1-2/les-actualites-de-la-centrale-nucleaire-de-flamanville-1-2/cumul-des-emissions-de-gaz-sf6-superieur-a-100kg-depuis-le-1er-janvier, zuletzt aktualisiert am 13.10.2022.

Électricité de France (edf) (19.05.2022): Update on the stress corrosion phenomenon and adjustment of 2022 French nuclear output estimate. Online verfügbar unter https://www.edf.fr/en/the-edf-group/dedicated-sections/journalists/all-press-releases/update-on-the-stress-corrosion-phenomenon-and-adjustment-of-2022-french-nuclear-output-estimate, zuletzt geprüft am 13.10.2022.

Enel S.p.A. (2020): Enel updates agreement with EPH for sale of stake in Slovenské elektrárne. Hg. v. Enel Spa All Rights Reserved Enel Spa VAT. Online verfügbar unter https://www.enel.com/media/explore/search-press-releases/press/2020/12/enel-updates-agreement-with-eph-for-sale-of-stake-in-slovensk-elektrrne, zuletzt aktualisiert am 13.10.2022.

Euratom Supply Agency (ESA) (2021): 60 years of ESA. Annual Report of 2020. Hg. v. Publications office of the European Union. European Union. Online verfügbar unter https://euratom-supply.ec.europa.eu/system/files/2021-10/MJAA21001ENN_002.pdf, zuletzt geprüft am 13.10.2022.

Euratom Supply Agency (ESA) (2022): Euratom Supply Agency. Annual Report 2021. Hg. v. Publications office of the European Union. European Union. Luxembourg. Online verfügbar unter https://euratom-supply.ec.europa.eu/system/files/2022-10/20222911_PDF_MJAA22001ENN_002.pdf, zuletzt geprüft am 13.10.2022.

European Union (2020): REGULATION (EU) 2020/852 OF THE EUROPEAN PARLIAMENT AND OF THE COUNCIL. on the establishment of a framework to facilitate sustainable investment, and amending Regulation (EU) 2019/2088. EUR-Lex -Acess to European Union Law. Brüssel (Document 32020R0852). Online verfügbar unter https://eur-lex.europa.eu/legal-content/EN/TXT/?uri=celex:32020R0852, zuletzt geprüft am 13.10.2022.

Geufroi, Stéphane (2021): La centrale nucléaire de Flamanville a dépassé le seuil de rejet d'un puissant gaz à effet de serre. En neuf mois, la quantité de SF6, le plus puissant des gaz à effet de serre, rejetée par la centrale nucléaire de Flamanville, dans la Manche, est déjà supérieure au seuil de déclaration annuelle. Ce seuil a été atteint le 27 septembre 2021, comme le confirme EDF ce jeudi 14 octobre 2021. Hg. v. ouest france. Online verfügbar unter https://www.ouest-france.fr/normandie/flamanville-50340/la-centrale-nucleaire-de-flamanville-a-depasse-le-seuil-de-rejet-d-un-puissant-gaz-a-effet-de-serre-4ef83064-2cdc-11ec-9285-f388b2ea32b0, zuletzt geprüft am 30.11.2022.

GOV. UK (2019): Nuclear Provision: the cost of cleaning up Britain's historic nuclear sites. Online verfügbar unter https://www.gov.uk/government/publications/nuclear-provision-explaining-the-cost-of-cleaning-up-britains-nuclear-legacy/nuclear-provision-explaining-the-cost-of-cleaning-up-britains-nuclear-legacy, zuletzt geprüft am 13.10.2022.

Grünwald, Reinhard (2016): Wie flexibel können Kernkraftwerke? (TAB-Brief, Nr. 47). Online verfügbar unter https://www.itas.kit.edu/downloads/tab-brief/tb047_grue16a.pdf, zuletzt geprüft am 13.10.2022.

Grünwald, Reinhard; Caviezel, Claudio (2017): Lastfolgefähigkeit deutscher Kernkraftwerke. Monitoring (Hintergrundpapier, Nr. 21).

Hamm, Horst (2022): Urban Atlas. Fakten und Daten über den Rohstoff des Atomzeitalters 2022. Online verfügbar unter https://www.bund.net/fileadmin/user_upload_bund/publikationen/atomkraft/atomkraft_bund_uranatlas_2022.pdf, zuletzt geprüft am 13.10.2022.

Huneke, F.; Heidinger, P. (2018): Steuerbare erneuerbare Energien als Alternative zur Kernkraft Steuerbare erneuerbare Energien als Alternative zur Kernkraft Kostenvergleich für Polen, Slowakei, Tschechien und Ungarn. Kurzstudie für Greenpeace Energy eG. Hg. v. Energy Brainpool. Online verfügbar unter https://www.energybrainpool.com/fileadmin/download/Studien/Kurzstudie_2018-04-20_GPE_Steuerbare_erneuerbare_Energien_als_Alternative_zur_Kernkraft.pdf, zuletzt geprüft am 13.10.2022.

Institut für Sicherheits-und Risikowissenschaften Universität für Bodenkultur (BOKU) Wien (Hg.) (2021): Sicherheitstechnische Analyse und Risikobewertung von Konzepten zu P&T für hochradioaktive Abfälle. *BASE- Forschungsbericht zur Sicherheit der nuklearen Entsorgung*. Berlin.

Institut de radioprotection et sureté nucléaire (IRSN) (2022): Impact de la sécheresse sur le fonctionnement des centrales nucléaires. Online verfügbar unter https://www.irsn.fr/FR/connaissances/Installations_nucleaires/Les-centrales-nucleaires/impact-secheresse-centrale/Pages/Impact-secheresse-fonctionnement-centrales.aspx#.YwaXuXZBxPY, zuletzt aktualisiert am 13.10.2022.

International Atomic Energy Agency (1999): Inventory of Radioactive Waste Disposals at Sea. In: *International Atomic Energy Agency*. Online verfügbar unter https://www-pub.iaea.org/MTCD/Publications/PDF/te_1105_prn.pdf, zuletzt geprüft am 13.10.2022.

International Atomic Energy Agency (IAEA) (2022): Nuclear Power Capacity Trend. Sum of Reference Unit Power (net capacity) reactors in operation within and at the end of reported years. Hg. v. International Atomic Energy Agency (IAEA). Online verfügbar unter https://pris.iaea.org/pris/worldstatistics/worldtrendnuclearpowercapacity.aspx, zuletzt aktualisiert am 03.11.2022, zuletzt geprüft am 13.10.2022.

International committee of the red cross (1907): Convention (IV) respecting the Laws and Customs of War on Land and its annex. Regulations concerning the Laws and Customs of War on Land. Hg. v. International committee of the red cross (ICRC). International Peace Conference. Den Haag. Online verfügbar unter https://ihl-databases.icrc.org/ihl/INTRO/195, zuletzt geprüft am 07.12.2022.

International Renewable Energy Agency (IRENA) (2019): RENEWABLE POWER RENEWABLE POWER GENERATION COSTS IN 2018. Hg. v. International Renewable Energy Agency. Abu Dhabi. Online verfügbar unter file:///C:/Users/marie/Downloads/IRENA_Renewable%20Power%20Generations%20Costs%20in%202018.pdf, zuletzt geprüft am 13.10.2022.

Kauder, Volker; Hasselfeldt, Gerda und Fraktion; Brüderle, Rainer und Fraktion (2011): Gesetzentwurf der Fraktionen CDU/CSU und FDP: Entwurf eines Dreizehnten Gesetzes zur Änderung des Atomgesetzes. 17/6070 (Bundesdrucksache), 06.06.2011. Online verfügbar unter https://dserver.bundestag.de/btd/17/060/1706070.pdf, zuletzt geprüft am 13.10.2022.

Kenfo-Fonds zur Finanzierung der kerntechnischen Entsorgung. Online verfügbar unter https://www.kenfo.de/der-fonds, zuletzt geprüft am 09.09.2022.

Kersting, Silke (2022): Anteil der Kernenergie am Strommix sinkt weltweit unter zehn Prozent. Die Bundesregierung streitet weiter über die Verlängerung der Atomkraft in Deutschland. International fallen beim Bau neuer Reaktoren besonders zwei Länder auf. Hg. v. Handelsblatt. Handelsblatt Online. Online verfügbar unter https://www.handelsblatt.com/politik/deutschland/atomkraft-anteil-der-kernenergie-am-strommix-sinkt-weltweit-unter-zehn-prozent/28723946.html, zuletzt geprüft am 30.11.2022.

Klaus Reinhold et al (Hg.) (2014): Verbreitung, Zusammensetzung und geologische Lagerungsverhältnisse flach lagernder Steinsalzfolgen in Deutschland – Zwischenbericht (9Y2013030000). Hannover: Bundesanstalt für Geowissenschaften und Rohstoffe.

Kolhoff, Werner (2019): Frankreichs Atomkraftwerke haben in der zurückliegenden Hitzeperiode nicht zuverlässig Strom geliefert, so dass das Land zeitweise auf Energielieferungen aus Deutschland angewiesen war. Hg. v. SAARBRÜCKER ZEITUNG. Berlin. Online verfügbar unter https://www.saarbruecker-zeitung.de/nachrichten/politik/wie-hitze-den-franzoesischen-atomkraftwerken-zusetzt_aid-44682175, zuletzt aktualisiert am 13.10.2022.

Kommission Lagerung hoch radioaktiver Abfallstoffe (2016): Verantwortung für die Zukunft. Ein faires und transparentes Verfahren für die Auswahl eines nationalen Endlagerstandortes. ABSCHLUSSBERICHT (K.-Drs.268), 24.05.2016, S.68ff. Online verfügbar unter https://www.bundestag.de/resource/blob/434430/35fc29d72bc9a98ee71162337b94c909/drs_268-data.pdf, zuletzt geprüft am 13.10.2022.

Krey V., O. Masera, G. Blanford, T. Bruckner, R. Cooke, K. Fisher-Vanden, H. Haberl, E. Hertwich, E. Kriegler, D. Mueller, S. Paltsev, L. Price, S. Schlömer, D. Ürge-Vorsatz, D. van Vuuren, and T. Zwickel (2014): Annex II: Metrics & Methodology. In: Climate Change 2014: Mitigation of Climate Change. Contribution of Working Group III to the Fifth Assessment Report of the Intergovernmental Panel on Climate Change [Edenhofer, O., R. Pichs-Madruga, Y. Sokona, E. Farahani, S. Kadner, K. Seyboth, A. Adler, I. Baum, S. Brunner, P. Eickemeier, B. Kriemann, J. Savolainen, S. Schlömer, C. von Stechow, T. Zwickel and J.C. Minx (eds.)]. Cambridge University Press, Cambridge, United Kingdom and New York, NY, USA. Online verfügbar unter: https://www.ipcc.ch/site/assets/uploads/2018/02/ipcc_wg3_ar5_annex-ii.pdf, zuletzt geprüft am 25.01.2023.

Laponche, Bernard (2020): FLUIDES FRIGORIGENES ET HEXAFLUORURE DE SOUFRE. GAZ A EFFET DE SERRE DES CENTRALES NUCLEAIRES, 20.10.2020. Online verfügbar unter https://www.global-chance.org/IMG/pdf/bl_final_ges_fluides_frigorige_nes_et_sf6_20201020.pdf, zuletzt geprüft am 13.10.2022.

Lazard (2021): Levelized Cost Of Energy, Levelized Cost Of Storage, and Levelized Cost Of Hydrogen. Lazard. Online verfügbar unter https://www.lazard.com/perspective/levelized-cost-of-energy-levelized-cost-of-storage-and-levelized-cost-of-hydrogen/, zuletzt aktualisiert am 13.10.2022.

Leininger, Elodie (2020): Les Novelles de Penly. LE MENSUEL D'INFORMATION DE LA CENTRALE DE PENLY (259), April 2020. Online verfügbar unter https://www.edf.fr/sites/default/files/contrib/groupe-edf/producteur-industriel/carte-des-implantations/centrale-penly/actualites/2020-04_les_nouvelles-de-penly_avril-259.pdf, zuletzt geprüft am 13.10.2022.

Meyer, Jan-Henrik (2019): Denmark Short Country Report. Unter Mitarbeit von Jan-Henrik Meyer. Copenhagen: History of Nuclear Energy and Society. Online verfügbar unter http://www.honest2020.eu/sites/default/files/deliverables_24/DK.pdf, zuletzt geprüft am 13.10.2022.

Pistner, Christoph; Englert, Matthias (2017): Neue Reaktorkonzepte. Eine Analyse des aktuellen Forschungsstands. Online verfügbar unter https://www.oeko.de/fileadmin/oekodoc/Neue-Reaktorkonzepte.pdf, zuletzt geprüft am 13.10.2022.

PreussenElektra GmbH (2020): Umwelterklärung 2020. Kernkraftwerk Isar. Hg. v. PreussenElektra GmbH. Online verfügbar unter https://www.preussenelektra.de/content/dam/revu-global/preussenelektra/documents/UnsereKraftwerke/Isar_1_und_2/Umwelterkl%C3%A4rung_2020_Kernkraftwerk_Isar.pdf, zuletzt geprüft am 13.10.2022

Röhlig, Klaus-Jürgen; Küppers, Christian; Schäfer, Thorsten; Walther, Clemens (2019): Überprüfung der sicherheits-technischen Anforderungen des Endlagers Konrad nach dem Stand von Wissenschaft und Technik (ÜsiKo). Review der Phase 1 „Ermittlung des Überprüfungsbedarfs", Januar 2019, S.44. Online verfügbar unter https://www.bge.de/fileadmin/user_upload/Konrad/Wesentliche_Unterlagen/UEsiKo/UEsiKo_Review-Bericht_final.pdf.

Schneider; Mycele; Froggatt, Antony (2021): The World Nuclear Industry. Status Report 2021. Online verfügbar unter https://www.worldnuclearreport.org/IMG/pdf/wnisr2021-lr.pdf.

Sipr Institute (2021): SIPRI Yearbook 2021. Arments, Disarments and International Security: Oxford Univ Pr.

Small Modular Reactors – Was ist von den neuen Reaktorkonzepten zu erwarten? (2021): Small Modular Reactors – Was ist von den neuen Reaktorkonzepten zu erwarten? Online verfügbar unter https://www.base.bund.de/DE/themen/kt/kta-deutschland/neue_reaktoren/neue-reaktoren_node.html#:~:text=Um%20weltweit%20dieselbe%20elektrische%20Leistung,bis%20zehntausend%20SMR%2DAnlagen%20bedeuten, zuletzt geprüft am 13.10.2022.

Smith, Thomas (2022): Nuclear Power's Role in Promoting Energy Security. Hg. v. Lazard Asset Management LLC. Online verfügbar unter https://www.lazardassetmanagement.com/at/de_de/references/fundamental-focus/nuclear-powers-role-in-promoting-energy-security, zuletzt aktualisiert am 13.10.2022.

Sovacool, Benjamin K. (2008): Valuing the greenhouse gas emissions from nuclear power: A critical survey. In: *Energy Policy* 36 (8), S. 2950–2963. DOI: 10.1016/j.enpol.2008.04.017.

U.S.NRC (United States Nuclear Regulatory Commission (2021): Event Notification Report for February 16, 2021. Hg. v. United States Nuclear Regulatory Commission (U.S.NRC). Online verfügbar unter https://www.nrc.gov/reading-rm/doc-collections/event-status/event/2021/20210216en.html, zuletzt aktualisiert am 13.10.2022.

Verbruggen, Aviel; Schneider, Mycle (2022): The World Nuclear Industry. Status Report 2022. Hg. v. A Mycle Schneider Consulting Project. Paris. Online verfügbar unter https://www.worldnuclearreport.org/IMG/pdf/wnisr2022-lr.pdf, zuletzt geprüft am 13.10.2022.

Warth & Klein Grant Thornton (2015): Gutachtliche Stellungnahme. zur Bewertung der Rückstellungen im Kernenergiebereich, 09.10.2015. Online verfügbar unter https://www.atommuellreport.de/fileadmin/Dateien/pdf/Kosten/20151009_WKGT_Stresstest_Rueckst.pdf, zuletzt geprüft am 30.11.2022.

Wealer, Ben; Hirschhausen, Christian von; Kemfert, Claudia; Präger, Fabian; Steigerwald, Björn (2021): Zehn Jahre nach Fukushima – Kernkraft bleibt gefährlich und unzuverlässig. Online verfügbar unter https://www.diw.de/documents/publikationen/73/diw_01.c.811435.de/21-8-1.pdf, zuletzt geprüft am 13.10.2022.

Wettach, Silke (2021): Der Coup der französischen Atom-Lobbyisten. Atomkraft soll in der EU künftig als nachhaltig gelten – ein Verdienst von gut vernetzten Ingenieuren in Frankreichs Staatsapparat. Berlin ist über diesen Lobbyerfolg gar nicht glücklich. In: *WirtschaftsWoche*, 2021. Online verfügbar unter https://www.wiwo.de/my/politik/europa/atomkraft-der-coup-der-franzoesischen-atom-lobbyisten/27767102.html?ticket=ST-1829265-dGqRsasSfbFWQfsXN7Va-cas01.example.org, zuletzt geprüft am 13.10.2022.

World Nuclear Association (2022): South Texas Project 2, United States Of America. Online verfügbar unter https://www.world-nuclear.org/reactor/default.aspx/SOUTH%20TEXAS-2, zuletzt aktualisiert am 13.10.2022.

World Nuclear Association (2022): World Uranium Mining Production. Hg. v. World Nuclear Association. Online verfügbar unter https://world-nuclear.org/information-library/nuclear-fuel-cycle/mining-of-uranium/world-uranium-mining-production.aspx, zuletzt geprüft am 13.10.2022.

ZDF (2021): Die geheimste Unterwelt der SS. Hg. v. ZDF. Online ZDF-History. Online verfügbar unter https://www.zdf.de/dokumentation/zdf-history/die-geheimste-unterwelt-der-ss-100.html, zuletzt geprüft am 13.10.2022.

Part 4: **Industry Specific Solutions**

Jörg Böttcher

26 Market Failure and Consequences for the Financing of the Energy Transition

26.1 Significance of Market Failures in Project Financing

The energy transition can only succeed if the transition to new markets is successful. We have referred to the challenging transition phase from Technology Readiness Level 7 to a broad-based, proven and economical application technology (see section 16) and also to the key steps of a market introduction strategy (see section 17). In this section, we look at market failures where legislators must also take appropriate countermeasures.

The preferred financing method in this phase will be project financin.[1] In order for it to be used, a stable cash flow of the project is necessary, which also requires that a stable regulatory environment exists and, in particular, that "surprises" due to possible market failures do not occur. Of course, the considerations regarding market failures also apply to all other forms of financing.

A particular challenge in this context is to identify the criteria according to which incentive-compatible contracts are concluded and the particular importance of state institutions in shaping the legal and regulatory environment. This applies in particular to areas for which the application of the project financing method is new and for which no established standards have yet been established, as is currently the case for geothermal energy projects, for example. As the uncertainties are greater here, there is also a need for further involvement of project participants in the project structure.

In economics, the tools used to analyze these issues are treated under the generic term of market failure. Market failure is generally defined as an economic constellation in which the market mechanism is incapable of efficiently coordinating supply and demand. Possible causes are assumed to be asymmetric information, external effects, natural monopolies and public goods. The analysis of these issues usually results in recommendations to the state to intervene to correct the situation. Firstly, this is of course important, as it provides indications as to which framework data in particular the state should take into account when shaping the legal and regulatory environment. In addition, the analysis is important insofar as it provides information on the form in which project financing can be organized in a contractually efficient manner.

Before we look at the relevance of various market failures, the reference model of a market in equilibrium is presented below (see Figure 26.1). In neoclassical theory,

1 Some reasons for this development are outlined in section 4.2.1.

https://doi.org/10.1515/9783110784299-026

the price mechanism in perfect markets reaches a market equilibrium that reflects an efficient allocation of resources. This equilibrium is Pareto-optimal, i.e. no market participant can be made better off by redistributing resources without making another worse off.

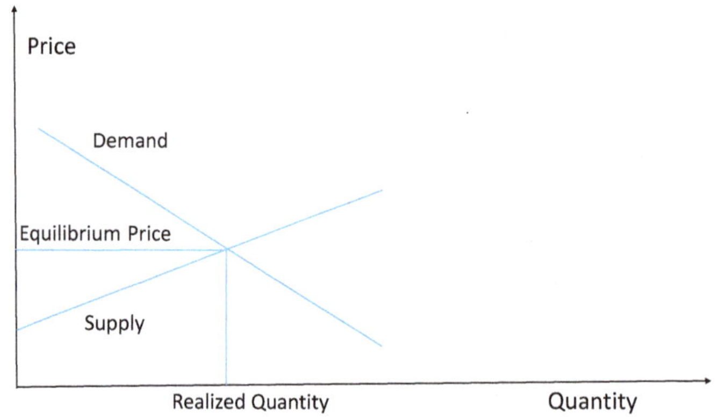

Figure 26.1: Emergence of the Equilibrium Price; figure created by author.

In the event of market failure, resources are wasted. Nevertheless, deriving real recommendations for action on the basis of these model results seems questionable:

- On the one hand, the underlying model of full competition is a model that reflects the welfare-optimizing model assumptions – such as the absence of transaction costs – but these assumptions are not very realistic.
- On the other hand, it is questionable whether the Paretian welfare optimum describes a permanently desirable state from an economic policy perspective, as a fixation on a static state of equilibrium ignores dynamic competitive elements, which are, however, indispensable with a view to long-term welfare increases.

Therefore, in the following analysis, the concept of market failure is based on the market competition process. The main function of the market is seen as the coordination of the individual plans of the market participants. If no market exchange takes place within the framework of coordination, this is referred to as total market failure. If the market coordinates supply and demand, but without a tendency towards a market-clearing equilibrium, the situation is referred to as partial market failure.

The consideration of market failures is first and foremost an economic exercise that is interested in the efficient allocation of resources. A market failure can justify political action to avoid this situation. The existence of external effects, for example, can be cited as a reason for special sector regulation of the energy markets. However, individual market failures also make it possible to show how efficient contract design and partner selection can take place.

In the following, we will take up the first aspect and show the particular importance of efficient allocation in project finance markets and the conclusions that can be drawn from this for the structuring of project finance.

Things without all remedy
Should be without regard: what's done is done.
WILLIAM SHAKESPEARE: MACBETH, ACT III, SCENE II, FIG. 11F.

26.2 Transactional Market Failure – Why Markets can Stall

Neoclassical theory has dealt extensively with the issue of the circumstances under which non-market-clearing prices can occur. A distinction must be made between two levels – the microeconomic view and the aggregated view on the markets.

In addition to the actual market price, market participants also take into account transaction costs that arise before or during the market transaction. Rationally acting market participants will refrain from a market exchange due to the level of transaction costs if they either value the transaction costs higher than the benefits of the transaction or than those of a non-market alternative service relationship. In these cases, there is no market coordination. However, observing transaction costs is just as difficult as forecasting them, as it reflects the knowledge and expectations of all transaction partners. Nevertheless, it is conceded that transaction costs become lower the more frequently similar transactions are carried out. This aspect may be relevant for certain transactions, as the total number of transactions already carried out is low.

The facts of a transactional market failure can be visualized as follows (Figure 26.2):

In this example, supply and demand cannot be brought together and no market result is achieved. This entails the risk of a transactional market failure in an early market phase of a technology, which makes further development of the industry impossible. At this point, state intervention could be justified in order to help a technology that can only be competitive in the future to achieve a breakthrough. We have discussed this risk with regard to the drilling risk of geothermal projects (see section 12).

However, it seems questionable whether state intervention can be justified on the basis of transaction costs that cannot be determined in practice. Nevertheless, the fact that coordination between supply and demand is not possible is important to us for another reason.

As we have shown in section 5.3.2, the risk of discoverability means that investors and banks will refrain from financing unless this risk is largely socialized. In contrast to other transactional market failures, both the economic cost of shearing and the economic benefit can be roughly quantified here. In this respect, this case is not about

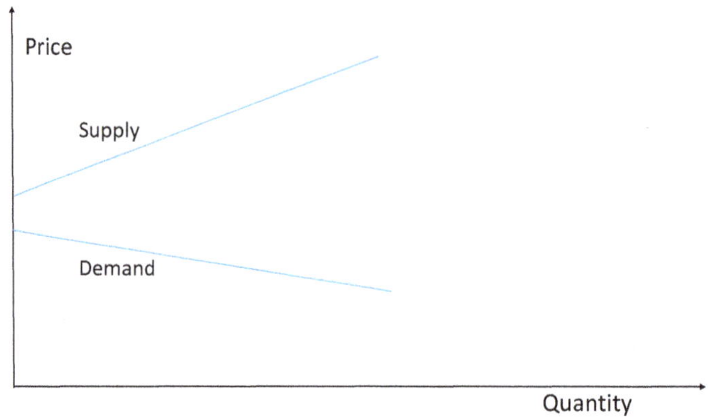

Figure 26.2: Example of a Transactional Market Failure; figure created by author.

initiating a process in which a market has to be created through economies of scale or competition, but about overcoming an initial obstacle to investment.

The constellation considered above represents a situation in which supply and demand cannot be brought together within the framework of a market coordination process. In this case, the literature speaks of a transactional market failure that can justify state intervention. In these cases, the stability of the regulatory environment and the associated remuneration rates are of paramount importance for the realization of project financing.

Having shown in this chapter that markets react very sensitively to the setting of political framework data, in the following section we look at the market failure situation of information asymmetries, which is of outstanding importance for the bilateral structuring of project contracts.

26.3 Information Asymmetries – Adverse Selection and Moral Hazard

The unequal distribution of knowledge is usually referred to as asymmetric information distribution. An asymmetric distribution of information occurs when the contractual partners in a market do not have the same information about the quality of a product or service on offer.

The question of what consequences this has for the contractual relationship between contracting parties should be usefully split: 1. what consequences arise if a contracting party has an information advantage before the contract is concluded? In the literature, this question is discussed under the generic term of adverse selection. 2. what are the consequences for the conclusion of the contract and the content of the contract of the knowledge that the contractual partner gains room for maneuver during the execution of the contract, which he can exercise in his own interest? The situation described as moral hazard is based on hidden action or hidden information that the client cannot observe free of charge. Hold-up is a hidden intention on the part of the contractor before the contract is concluded, which only becomes apparent after the contract has been concluded.

The basic model for information asymmetry was presented by George A. Akerlof in 1970 using the example of the used car market. The seller of a used car knows more about this experience than the potential buyer – he knows the previous history, the frequency of maintenance and his own driving behavior, the buyer only knows the average quality of vehicles. If a certain price is achieved on the used car market for cars of a brand with comparable equipment, a car that is prone to repairs tends to be offered. The true value of the vehicle is below this price, but the average quality of the used cars offered at a given price will be poor compared to the asking price. Buyers cannot expect to get a good deal. The buyer will form an expected value for the quality, which corresponds to the average quality of all used cars of a type, and derive a reservation price from this, i.e. the maximum price he is prepared to pay. Conversely, the sellers will demand a minimum selling price for the good cars – i.e. their reservation price – which is higher than the maximum purchase price of the buyers. No contract is therefore concluded for these good cars; only the poorer quality cars are sold, as their minimum selling price is below the buyer's maximum. In a dynamic view, the suppliers of good cars are forced out of the market, the expectations of the buyers and their reservation price fall, so that the market price tends to fall and in the end only bad cars are offered.

In principle, there are several ways to prevent a suboptimal market result:

- In **signaling**, the better-informed market participants try to reduce the information asymmetry. It is not the case that all suppliers have an advantage if they sell poorer quality to consumers. Suppliers of better quality products certainly have an interest in credibly differentiating themselves from suppliers of poor quality products. To do this, they incur the cost of producing a signal, such as ISO certification. However, they will only do this if the benefits of producing a signal are greater than the costs involved.
- During **screening**, less well-informed market participants try to better assess the quality of products by obtaining additional information. In the case of a used car purchase, for example, this would be an appraisal of the vehicle by an expert. Here too, poorly informed market participants will only do this if the resulting benefits are greater than the costs.
- With **self-selection**, the better-informed parties are offered different contract menus so that the good and bad providers each choose different contracts. The providers of good products can, for example, provide a guarantee or make their remuneration dependent on performance. This would be too expensive for bad

providers, so that any provider who accepts a contract with the appropriate conditions will be recognized as a good provider.

All of the possible solutions mentioned incur costs, so that the market equilibrium does not correspond to the market equilibrium under perfect competition and is therefore only second best. However, the dynamic effects must not be ignored.

The phenomenon of asymmetric information is highly significant for project financing. Various aspects of risk assumption and the design of an incentive system will be presented in the following chapters.

26.4 Existence of Natural Monopolies

A classic market failure is assumed to exist in the case of a natural monopoly. A natural monopoly exists if the supplier has a subadditive cost function. The supply of additional consumers is possible at ever lower additional costs, so that the entire demand can be met most efficiently by one supplier. Each additional supplier would only increase the economic costs of supplying this good, which would represent a misallocation of resources. The situation can be illustrated graphically as shown in Figure 26.3:

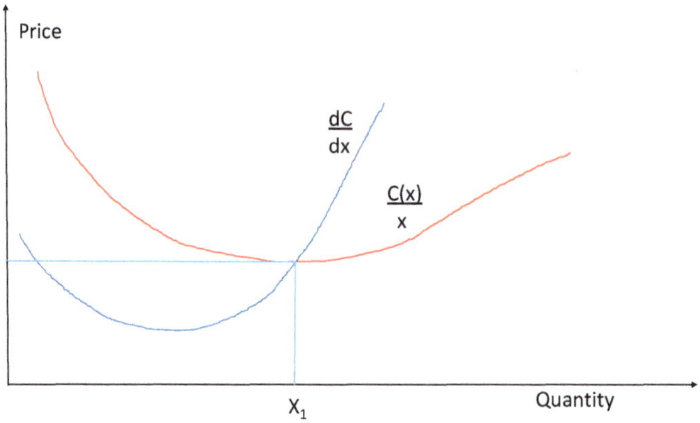

Figure 26.3: Cost Structure for a Natural Monopoly; figure created by author.

Figure 26.3 shows the course of a – normal – demand function, the average costs C(X)/X and the marginal costs dC/dX. The average costs C(X)/X fall up to the output quantity x_1. At this production quantity, the marginal costs are equal to the average costs. Increasing returns to scale therefore exist in the range between 0 and x_1. In this range, the marginal costs are below the average costs.

How would the market function under a natural monopoly? Initially, the natural monopolist would behave like a normal monopolist, producing at the quantity of marginal revenue equal to the marginal price and realizing a price that is above the price under competitive conditions. In the long term, however, it is questionable whether this monopoly price can be maintained. As a signal of scarcity, the price would attract further competitors, who would ideally push the price down to the level that would also result under competitive conditions. The special feature of a natural monopoly is that, in a dynamic view, only one supplier remains, but it cannot enforce the monopoly price. However, price competition will hardly be frictionless, as the respective supplier has made irreversible investments. There are sunk costs that can no longer be earned in the event of a market exit. In the competitive process, only the marginal costs are relevant to the incumbent's decision. The incumbent company can threaten to reduce prices to marginal costs if a new competitor enters the market. For the potential competitor, however, the entire costs are relevant to the decision. This asymmetry can deter potential competitors, so that established companies can ultimately impose a higher price than the competitive price.

The existence of natural monopolies is assumed, particularly in the case of rail and pipeline networks. A second supply from a network that runs parallel to the existing network cannot be Pareto-optimal, provided that the capacity of the existing network is not yet exhausted. Competition between different providers for feeding electricity into the grid need not therefore be ruled out, but a competitive organization does not appear to make economic sense for the actual transport service. If competing lines were established, each provider would try to offer at minimum average costs, and since only one provider can succeed in this due to the limited demand, this would result in ruinous competition, which would ultimately leave only one provider, so that competition would have to be restricted. On the other hand, the monopoly company protected in this way should not exploit its monopoly position to the detriment of consumers and would therefore have to be regulated by the state. Monopolistically excessive prices should be prevented by price regulation.

From a project financing perspective, it is imperative that the overall project is able to receive remuneration for the electricity produced, which requires the grid to be in place at the latest when the first production unit is ready for operation. This leads to practical problems.

How can it be ensured that the network is also able to make its capacity available for other projects? As I said, if it is a natural monopole, it would be economically desirable if the grid could be used by all providers. From a project-specific perspective, the aim is to isolate the projects as far as possible from external risks. This means that it is unacceptable if the project has to bear the costs of investments that it considers to be oversized, but for which the financing has not yet been secured. In a worst case scenario, the project would bear the costs of a – possibly oversized – pipeline network in addition to the actual project costs, which would massively impair economic efficiency and lead to considerable problems as early as the planning phase.

If there is no planning certainty, the lenders would assume the worst case scenario for the project – complete coverage of the pipeline network costs by a project –

and use this as the basis for dimensioning the borrowed funds. This approach can lead to a situation where the debt capitalization is so low that the internal rate of return of the project is too low from the investor's point of view and the realization of the project fails due to this partial aspect. This is a classic prisoner's dilemma.

Recognizing that grid-bound networks can lead to a natural monopoly with the associated market failure, the legislator has often unbundled in these cases. Unbundling means that there is a separation between network infrastructure operation and the upstream and downstream value creation stages. As there is usually no tendency towards a natural monopoly in the latter areas, these areas can also be organized on a market basis, while the actual network infrastructure is subject to more far-reaching regulation.

From the grid operator's point of view, it is important that the grid is sufficiently dimensioned so that all relevant generation units can be connected, but also to ensure that all planned generation units are realized so that the refinancing of the grid is secured. In practice, this is problematic insofar as not all projects that are to be connected to the grid are at a comparable stage of project development. While an initial project is so far advanced that construction could already begin without further ado, other projects that are to be connected to the grid are still in an early project phase. If, in this situation, the pioneer were to be slowed down until the last project has reached the appropriate project maturity, this could, in extreme cases, result in not a single project being realized if even one project is delayed for a long time – for whatever reason. Such a market failure can be avoided, for example, if the legislator obliges the responsible transmission system operators to construct the grid connection on their own responsibility and – for example – to refinance it via the grid fees.

Even if electricity generation and transmission are separated from each other through unbundling, there are considerable coordination requirements between the project company and the grid operator. The project company requires an existing grid connection or one that has been contractually agreed at a certain point in time. The project will only be able to generate income if it is clear that the electricity generated can also be fed into the grid. On the other hand, the grid operator also needs a certain degree of planning certainty in order to actually take on the risks and costs associated with laying the cables. For the grid operator, it is important to rule out sunk investments from the outset.

The question in each case will therefore be to what extent the grid operator's need for security should be taken into account. The grid operators will set the requirements correspondingly high. It is known from a similar issue relating to the connection of offshore wind farms that the grid operators had demanded that binding commitments for the financing of the project were already in place before a grid connection was approved. In turn, the banks responsible for external financing required a grid connection commitment to ensure that the project was also able to generate cash flows.

26.5 External Effects

An external effect exists when the economic decisions of an economic entity have an impact on third parties that is not compensated for.

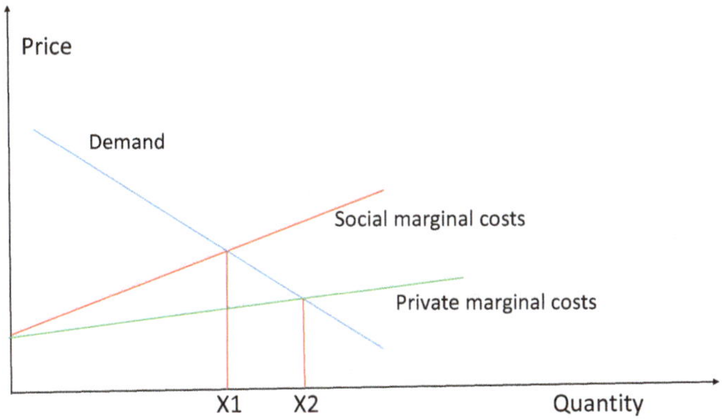

Figure 26.4: External Effect; figure created by author.

The starting point for the graph above (Figure 26.4) is the consideration that consumers only have to bear the private marginal costs, but not the social marginal costs. Their demand is therefore higher than it would be if the marginal social costs were taken into account and they pay less than they would if the marginal social costs were taken into account.

Since the external effects lie outside the cost accounting of the market participants, the competitive process between the suppliers is disturbed and – in the case of negative external effects – leads to a higher utilization of a resource than would be economically desirable; in the example of production at x_2. In order to achieve a welfare-economic optimum, the external effects should be internalized, i.e. assigned to a party that can take them into account in its cost accounting. The various theoretical methods that have been developed in this context will not be discussed in detail here. The various approaches can be subsumed under two concepts:

- Either the causes of the cost distortions are addressed by confronting the providers who burden the general public with additional social costs with corresponding bans and prohibitions or with financial burdens.
- Or, in accordance with the second-best criterion, an attempt is made to compensate for allocation distortions in one energy source through corrective measures, such as subsidies for competing energy sources.

However, when assessing whether external effects exist and how they should be evaluated, there are considerable value judgments for both conventional and renewable energies. The consideration of external effects has been closely linked to a subjective value judgment, which already makes a qualitative assessment appear difficult. This is why it is so important to recognise and quantify external costs.

Despite all the difficulties in operationally qualifying and quantifying external effects, it is precisely their existence that is cited as the main justification for the creation of competitive exceptions and state intervention. The Stern Report in particular has made the connection between climate change and economic consequential costs clear.

26.6 Public Goods in Project Financing

Markets can fail with regard to the pareto-efficient provision of public goods. Public goods are characterized by non-rivalry in consumption and non-excludability from consumption.

The consumption of the good in question by an additional demander does not reduce the opportunities for consumption of this good by other demanders and, conversely, is not impaired by their consumption (non-rivalry in consumption). The exclusion principle is not applicable to public goods, i.e. individuals who are unwilling to pay cannot be excluded from consuming the good. For example, national defense is a public good – everyone living in the country benefits from it and individual participation does not impair the protection of all other individuals.

The private provision of public goods of this kind suffers from free-rider behavior, which consists of having the good provided by others in order to then enjoy the good free of charge. Even if there were a sufficient willingness to pay overall, there would still be no demand for the good due to an individual-rational calculation and non-excludability.

In the case of project financing, this problem regularly arises when the project financing method is used for the first time in a new field of application. This new field of application can be, for example, the use of a new technology, but also the use in a country that is using a regulatory system for a specific technology for the first time. Project financing requires a calculable and reliable basis, so that influencing factors that increase the uncertainty of the expected cash flows are viewed particularly critically and significantly limit the use of project financing. This means that the use of project financing in the aforementioned cases is not a matter of course, as there is not yet any positive experience of its use in the specific application.

If a project with innovative technology is successful, the project participants have created a public good that benefits not only the project participants but the entire industry. The public good consists of the fact that it has been proven that the technology in question can be used on a large industrial scale and that the project financing method is fundamentally suitable. This means that there is a strong incentive for interested parties not to participate in pioneering projects and only to get involved in this technology once its success has been proven by other participants. The project participants who do not participate in the first phase, but hope that others will join in earlier and pave the way, win.

As a consequence, this can lead to a market failure and no such project being realized. In this case, the state is often called upon to provide the public good, which could, for example, consist of the public financing of pilot plants in the region. In this context, it is difficult to resolve the question of what quantity of the public good is efficient, as this requires information on individual willingness to pay.

However, experience shows that project financing is actually realized even though the application of the methodology in this area is new. There are two possible explanations:

- Assume that there is an innovative technology which, however, offers the opportunity to be duplicated if successful and generates additional income for the project participants. The additional earnings result from the fact that the participants acquire knowledge with this project that the competition can only catch up with after a time delay, so that follow-up projects are also realized within the framework of the previous participants.
- A second approach is aimed at the fact that the project participants involved in the project are often active in the very business area that is now up for financing and that project financing represents a new financing option that also provides additional remuneration opportunities for the participants. If remuneration systems are linked to business transactions without the need to prove long-term viability, there tends to be an adverse incentive to behave more risk-averse.

The development in the use of many new areas of renewable energies illustrates this statement. The successful commissioning of the first projects will significantly accelerate further expansion, as the risks are now much more tangible and therefore quantifiable. The further entry of new developers and investors will also lead to more competition, which will then also help to reduce costs.

In summary, it can be stated that market failures in RE projects must be assessed in a differentiated manner and the results will be industry-specific:

- In some cases, there are alleged market failures that can only claim a small degree of validity.
- Other possible market failures can be avoided by the political design of the regulatory environment.
- Finally, the issue of information asymmetries is of paramount importance for the understanding of project financing and the design of project contracts.

In any case, the discussion about possible market failures has important consequences for the design of project structures and project contracts.

The discussion about a transactional market failure has made it clear that energy markets must be assessed on a national basis and require country-specific solutions.

The discussion about the natural monopoly has shown how complex the issue of pipeline connections is and what obstacles there are to finding a practicable solution.

In this section, we have described the significance that market failures can have for project financing. The next step is to show the consequences of a market failure, namely the asymmetric distribution of information, on the design of the project structure and the project contracts.

> Denn der Wille
> Und nicht die Gabe macht den Geber.
> [For it is the will that counts, not the gift.]
> LESSING, NATHAN THE WISE, ZFN. 539 F.

26.7 Economic Analysis of Contractual Relationships and Partner Selection

26.7.1 Risk Premium and Risk-Bearing Capacity

Neoclassical capital market theory has dealt intensively with the price of risk. A key finding is that in a world of perfect capital markets, complete diversification of investments is possible. Risks that are only related to a single security have a price of zero, as they can be completely eliminated through diversification. In this model world, there is only a price for systematic risk, i.e. the risk of a fully diversified portfolio that is driven by the general market trend.

The findings of this theory form an important basis for today's understanding of how the capital markets work. However, the assumptions under which they are derived are hardly realistic. Markets do not function without friction, and perfect diversification is neither possible nor sensibly achievable. However, the degree of approximation to this ideal varies. A capital market investor who wants to invest extensive free assets can very well achieve extensive diversification. A company whose success depends on the success of a project, on the other hand, is poorly diversified. Accordingly, you would have to pay a correspondingly high price to this company to take on an additional share of risk from this project. There is another effect behind this, which is not foreseen in the model world of perfect capital markets. The company would also like to avoid the additional project risk because if very poor results are realized – e.g. the project fails – further costs are incurred in addition to the loss from the project: In the worst case, the company goes bankrupt and follow-up costs are incurred. These problems do not arise for a capital market investor who invests his own money.

In economics, the aversion to risk is referred to as risk aversion. In order to analyze the effects of risk-averse behavior, economists construct individual utility functions. A utility function is a function that assigns a certain utility value to each pay-off that an individual receives. Individuals strive to maximize their utility value. However, they often do not know at the time of a decision which pay-off they will receive later. If they can estimate the probability with which they will receive the various pay-offs, they can maximize the expected value of the utility instead of the utility value. Obviously, individuals reject risks if they can receive the expected value of this cash flow as a secure payment instead of a risky cash flow. The utility value of this expected value is therefore higher for a risk-averse individual than the expected utility of the risky cash flow. For this to be guaranteed for all conceivable risky cash flows, an individual's utility function must be concave.

The idea can be illustrated using a simple example (see Figure 26.5). A project can either succeed or fail with the same probability of 0.5. If the project is successful, the profit is 40 GE. If it fails, the profit is 0, so the project has an expected profit of 20 MU and is therefore worthwhile. However, a risk-averse investor would rather have the expected value of 20 MU than a project with the same expected value, which can also go wrong. The concave utility function illustrates this situation:

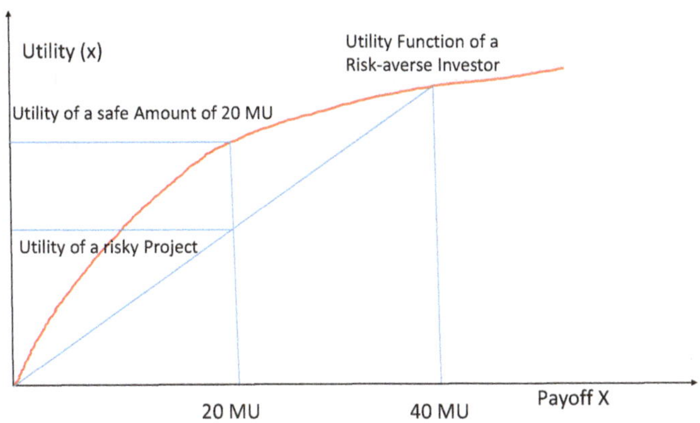

Figure 26.5: Benefit Function of a risk-averse Investor; figure created by author.

Let's take a closer look at the parties involved in project financing from the point of view of who can best assume risks in terms of diversification and possible bankruptcy costs (see Table 26.1):

In the case of project financing, the integration of project participants into the project structure is of particular importance. A distinction must be made below between contractors who provide a one-off trade or service for the project company and those who are involved in multiple transactions with the project company. The working hypothesis is that in the case of repeated transactions, the risk for both parties to the contract is reduced due to lower sunk costs and therefore other incentive structures are required. A distinction should therefore be made between the contractor, who is responsible for completion, and the project participants, who are responsible

Table 26.1: Risk-Bearing Capacity for Project Financing (compiled by author) .

	Characteristics	Risk Bearing Capability
Contractors (during completion)	Smaller Contractors are dependent on the success of a single project, insolvency costs are highly relevant. Bigger contractors benefit from a better diversification	little to medium
Contractors (during operation)	Contractors can be supposed to act diversified, since they can offer their services to a number of companies	medium to high
Lender	Diversification is one core principle of the business model of a bank. The real degree of diversification depends on the business model and the size of the project. Regulation requirements should restrict the relevance of insolvency costs.	medium to high
Insurance Companies	Risk Assessment and risk diversifikation are core business of insurers. Cumulative Riks may be covered by re-insureres.	High

for operational management during the operating phase, for example. We will structure the presentation in such a way that we first run through the case of a one-off transaction and then consider what changes in the case of multiple transactions.

What are the consequences of these considerations for risk allocation in project financing? As an example, let's imagine a contractor who is to supply a precisely specified system worth €4 million for a project, which also corresponds to his initial assets. As the project is clearly very large for him, he is risk-averse, while the other project participants are much better diversified and behave in a risk-neutral manner. How should the contract with the contractor be structured?

The preferences of the risk-averse contractor are to be described by the utility function $X^{0.5}$. Let X be the final wealth of the contractor. In order for the contractor to deliver the system, he must be at least as well off as if he did not deliver. The easiest way would be to pay him the € 4 million that the system costs. However, as the system is essential for the project, he will regularly be asked to bear part of the risk. One option would be to offer him a share in the company's profits. For this to be attractive to the contractor, however, the expected benefit must be just as attractive to him as the € 4 million he has to spend on the component. In the above example, we expect the project to generate a profit of € 0 or € 30 million with a probability of 20% (failure) or 80% (success). In the first case, the investment is worthless; in the second case, the contractor receives an amount of €30 million* a from his investment share a. His expected benefit 0.8 * (a * € 30 million)^{0,5} must therefore be exactly as high as the benefit from his previous assets – € 4 million^{0,5} = € 2 million. If this system of equations is solved, the required shareholding is 20.84%. The expected value of this participation, 0.2084 * 80% * 30 M€ = 5.00 M€, is significantly greater than the initial assets or the final assets in the case of a simple purchase agreement with a fixed price. The difference 5.0 M€ – 4.0 M€ = 1.0 M€ is the premium that the contractor demands for his participation in the risk of the project. This is at the expense of the other participants, for whom the project becomes correspondingly less valuable. This means that in this example, the contractor is better off as a result of a participation agreement, but the other participants are worse off. Depending on the above assumptions, the contractor's preference may change, but the other project participants will be worse off in any case.

These results can be generalized. It does not make sense to burden risk-averse project participants with high risks simply because the principle of risk sharing postulates this. Risks do not disappear by spreading them over as many parties as possible. Rather, the burden should be placed on those parties that can bear these risks well. Ideally, the sponsors should realize from prohibitively high premiums for further risk assumption that they are unduly burdening the risk-bearing capacity of a project participant. However, there is also the danger that the consistent pursuit of the risk-sharing principle could lead to project participants who are already so badly off that they have no choice but to accept any risk. In any case, risk sharing requires caution, even if it is advisable from an incentive point of view to allow risk-averse contractual partners to share in the risk. We look at these aspects in the following sections.

26.7.2 Risk Assumption with Asymmetric Information

As we showed above, information asymmetries in a market can systematically lead to non-optimal or sub-optimal results.

In the case of project financing, the hidden information prior to the conclusion of the contract is of particular importance for at least three reasons: 1. firstly, the sponsors are regularly released from liability at the time of completion, which underlines the particular importance of the timely and scheduled construction of the project. 2) Furthermore, many aspects of project financing are one-off transactions that make it difficult to correct contracts at a later date. 3. the consequences of wrong decisions are also considerable because the economic viability of project financing is determined by the bilateral relationships between the project company and the project participants and is therefore fixed at the time the respective contracts are signed.

If we take a closer look at the project financing process, a distinction can be made between the completion and operation phases: While the focus in the completion phase is on the correct selection of the general contractor who is responsible for the project work, the focus in the operating phase is on ensuring ongoing operations by involving suitable contractors. The main difference is that a wrong decision in the selection of the contractor has a considerable impact on completion that is difficult to correct, whereas the operating phase requires the control of several parties, who can usually be replaced relatively quickly. The sunk costs will generally be greater in the first case, so that the possibilities for replacement are also limited in practice.

Quality information is only available via markets if the quality can be determined directly. Many aspects of product quality cannot be determined before purchase, even after careful analysis, but the products only reveal their true characteristics after prolonged use; they are so-called experience goods. This is a critical point for project financing in particular, as its feasibility depends to a large extent on the quality and prices of the contracts relevant to the projects being predictable and reliable.

The issue of asymmetric information plays a particularly important role in a large number of questions in the area of project financing. In addition to the optimal design of risk assumption and incentive systems, this also includes the question of who is eligible as a risk carrier and who is not. We will look at the first question in the following chapters and the second question immediately afterwards. It is the starting point for the problems relating to hidden information prior to the conclusion of the contract.

The allocation problem described by Akerlof is also relevant for project financing. The following section examines the question of which group of capital providers is suitable for assuming certain project risks.

Risks that cannot be influenced by transparent projects are best placed on the capital market. They can be widely spread there and capital market investors cannot demand more than the market risk premium for assuming the risks. If all such risks of such projects are placed on the capital market, the project has the lowest possible financing costs and, conversely, achieves the highest possible profit for the sponsors. However, the formulation also shows its limited application: Which project is so transparent that it can be offered to the broad capital market? Which risks can actually be influenced by the project participants?

First of all, the question of transparency. If capital market participants are not sufficiently informed about the value of a project, they will demand a risk premium. Let us imagine two different projects that are offered to the capital market. The first project is equally likely to have a cash flow of € 0 million or € 12 million after deducting the purchase price for the investment. Let us assume that the second project generates a cash flow of € 2 million or € 18 million. For the sake of simplicity, let us assume that both projects are equally likely to occur and that capital market investors cannot distinguish between the two projects. What would be the price for the projects if the risk-neutral capital market investors want to maximize their expected value?

At first glance, this appears to be a simple calculation: the first project has an expected value of € 6 million, the second € 10 million. Since both are equally likely, risk-neutral investors would pay a maximum of € 8 million to win this type of lottery. The sponsors of the bad project would earn € 2 million more than their project is worth. However, the sponsors of the good project lose € 2 million, which they have to hand over to the capital market investors as a risk premium for their willingness to buy a non-transparent risk. However, the average information premium would be 0, as the projects are fairly priced due to the level of information on the capital market and the sponsors, although they know the actual characteristics of their project, do not draw any conclusions. This is obviously an unrealistic assumption.

Let's also assume that for some projects there are other players, e.g. certain sponsors or banks and insurance companies, whose risk aversion is positive but not very pronounced, and who are also informed as to whether the project is more valuable or less valuable. Let's imagine, for example, an insurance company that is so informed that it could also be sold the project risk. Nothing changes for bad projects. For good projects, however, we could consider transferring the risk to this insurance company and not to the capital markets. Let us assume, for example, that the preferences of this particularly risk-averse insurance company are represented by a utility function $x^{0,9}$ and that it has initial assets of € 10 million. What is the maximum price the insurance company would pay for the project with the higher value? Without this purchase, its utility from its assets would be exactly $10^{0.9}$. If it buys the project, it must pay the purchase price P from its assets and receives the full project cash flow, i.e. either € 6 million or € 10 million. The resulting expected benefit can be represented as

Equation 22: $0.5^* \ (10-P+6)^{0,9}+0.5^* \ (10-P+10)^{0,9}$.

This value must be at least as high as the benefit without the purchase of the risky project, i.e. $10^{0.9}$. If we solve the corresponding equation, we find that the insurance company is prepared to pay up to € 9.38 million for the good project. This is significantly more than the € 8 million that would be paid on the capital market. These projects would therefore not be offered on the capital market.

In fact, all good projects for which not too risk-averse inside investors can be found would no longer appear on the capital market. If the utility function is formulated more generally with X^a with a as the inverse measure of risk aversion, the good projects that can find an informed investor with a value for a greater than 0.35 would no longer be offered to uninformed capital market participants at a market price of € 8 million. However, the effect of so-called adverse selection now occurs. Although capital market investors cannot distinguish between good and bad projects, they know that at a market price of € 8 million, many good projects will no longer be offered on the capital market. Let us assume, for example, that the good projects are equally distributed in the interval from 0 to 1 with regard to the lowest risk aversion (i.e. the highest a) of an inside investor. In this case, with a market price of € 8 million and thus a critical value for a of 0.35, just 25.9% of the good projects would remain on the capital market. The capital market investors now know that only 35% / (100% + 35%) = 25.9% of the projects on the capital market have the higher value. They will adjust their willingness to pay accordingly. This results in a new, lower price of

Equation 23: $0.259^* \, 10 \, \text{M€} \ + (1-0.259)^* \, 6 \, \text{M€} \ = 7.04 \, \text{M€}$.

What is particularly serious is that this process continues. Due to the lower price, the threshold value for a, above which the good projects no longer appear on the capital market, also falls, in our example to 0.07. However, this means that only 6.5% of the projects on the market are still of high value, and the market value continues to fall in line with the above consideration. This process continues until only bad projects are offered on the capital market and these are offered at the right price of € 6 million. Figure 26.6 once again illustrates the competitive failure in the case of adverse selection:

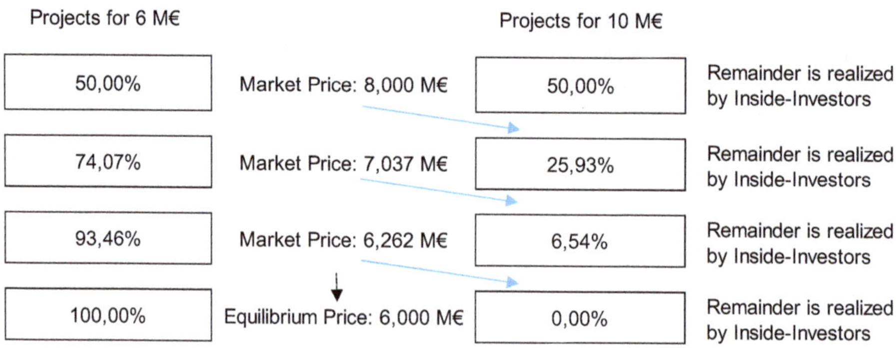

Figure 26.6: Competitive Failure due to Adverse Selection; figure created by author.

The capital market financing of good projects thus fails completely. This market failure expands even further if there are many projects of very different quality. Of these, too, only the very worst would be offered on the capital market. All others would be financed by inside investors. However, the share of these projects in the market for project financing would be negligible.

Even if risk allocation on the capital market is preferable from a risk theory perspective, it will generally fail in project financing due to information problems and the associated problems of adverse selection. Conversely, projects can only be financed via the public capital market if they are sufficiently transparent for capital market participants.

The considerations on adverse selection are also helpful for the practice of project financing for two further reasons:

- The person to whom a certain offer is made can draw negative conclusions about the quality of the offer from the mere fact of the offer against the background of adverse selection. This problem arises on an ongoing basis with project financing. The invitation to participate in a syndication of project financing is not unproblematic. Why are you invited? Is the syndicating bank interested in a long-term relationship, possibly also in several syndication transactions, and is it therefore acting seriously? Similar questions may arise when searching for suitable suppliers, manufacturers, operators or buyers. It is anything but indifferent who takes the initiative.
- Although the majority of project financing is provided by banks, there has been a market for project bonds since the 1990s. From the sponsors' point of view, the question arises with every project financing whether and when this route is feasi-

ble. In the case of project bonds, it is mostly international investors who expect their funds to be serviced from the cash flows of the specific project. Projects that are eligible for the use of project bonds must be largely transparent, relatively simply structured and relatively secure. In addition, the ability to react flexibly to necessary adjustments to project, loan or collateral agreements is limited in the case of project bonds, as only the bondholders as a whole can regularly decide on contractual amendments, which is often only possible to a very limited extent due to free rider phenomena.

Projects in which these conditions may be met can be found in the renewable energy sector in particular, with solar energy, wind energy and biomass being the priority areas. Once the projects have been successfully launched, there is rarely a need for adjustment and the risks shift to risk aspects that the capital market participants can assess just as well as the sponsors, such as the development of energy prices or the assessment of resource risk. In this case, the capital markets actually take on a relevant risk package, but not with a poorer level of information than the risk seller.

How critical the capital markets are in terms of meeting these conditions will always depend on the general market situation. While it was possible to finance many things via the public capital markets before the financial crisis – including many things that are not suitable for this – such aspects are viewed much more critically today, making it much more difficult to place project bonds.

Jörg Böttcher

27 Markets and Markets Behaviour – Some Practical Aspects

27.1 Introduction

In addition to understanding market failures, it is really important to understand how markets work in the interaction between project developers, investors and banks.

I would like to present the following aspects: How have Levelized Costs of Energy developed for different renewable energies? What were the drivers of this development and can you draw conclusions for new markets? To what extent is there a relationship between investor's interest claims and the purchase price of a project? Are there any conclusions to be drawn here about the design of a regulatory design? And finally: Overall, are there any recommendations to the regulator regarding possible changes of the regulatory regime?

27.2 The Development of Levelized Cost of Energy

The **Levelized Cost of Energy** (LCOE) is the net present value of the electricity costs over the lifetime of a generation plant. They include a number of costs over the lifetime – in particular the initial investment and all operating costs as the numerator and the total electricity production of the plant as the denominator. All these values are calculated in present value terms so that the capital costs are also taken into account. The LCOE is often used as a proxy for the average price that the generation plant must achieve on a market in order to cover its costs over its lifetime. It is also suitable and appropriate for calculating these costs as long as it also takes into account the integration costs and the system costs of the energy under consideration.

The development of the LCOE in this sample is shown in Figure 27.1 (costs are shown in cents/kWh).

The project contracts for the various project financing arrangements show that the LCOE for PV projects has fallen significantly overall.[1] This mainly reflects the decrease in total investment costs and operating costs, which – as mentioned above – is directly related to the general reduction in remuneration. The sample of the last five

1 The peak value in 2010 resulted from some Italian projects, which had an opex ratio that was 50% higher than comparable European projects. The main driver of operating costs here was local and national taxes.

https://doi.org/10.1515/9783110784299-027

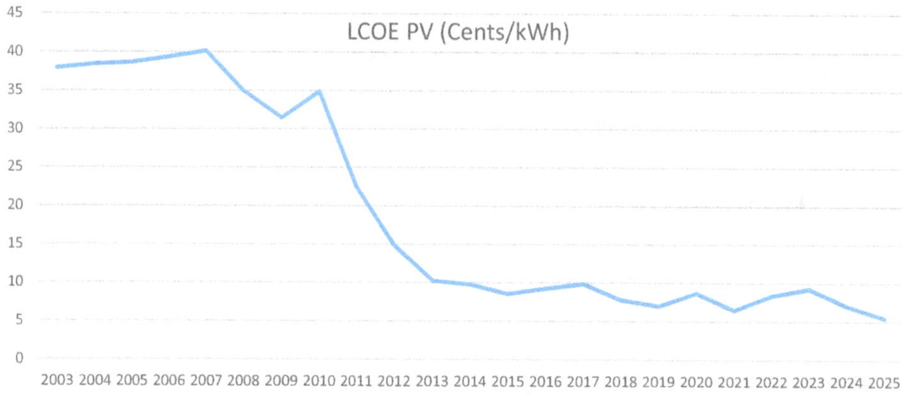

Figure 27.1: LCOE for PV projects; figure created by author.

years consists of projects in Germany and the Netherlands, which means that the costs for ground-mounted PV projects in these countries are close to 5 cents/kWh and continue to fall.

The decline in total investment costs has led to a further trend in the financing of PV projects: While initially only larger banks were willing to finance larger PV projects, the decline in total investment costs and the positive experience with existing PV projects have led to smaller banks also financing PV projects. As the specific investment costs – i.e. the GIK/MW ratio – fell significantly at the same time, the scope of the due diligence review also had to be reduced in order to keep the costs affordable. As a complete due diligence was always a prerequisite for the financing process of the larger banks, some of these banks were no longer in a position to finance PV projects unless they could offer this service themselves.

In the following Figure 27.2 we have compared the remuneration rate with the LCOE value:

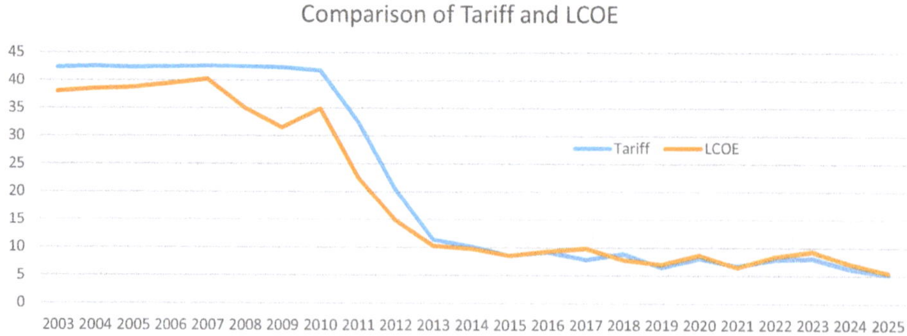

Figure 27.2: Comparison of the LCOE solar and the applicable tariff; figure created by author.

Since 2012, the development of LCOE Solar and the applicable tariff have shown a high degree of consistency, both in terms of the development and the absolute amount. This suggests that – at least since 2012 – a WACC of 4% appears to be a realistic discount rate.

The onshore wind market in Germany is as follows: With one exception in 2015, the remuneration rate has always been higher than the LCOE, with the gap between the two values narrowing significantly from around 2012. The LCOE itself has ranged from around 6 to 8.5 cents/kWh (see Figure 27.3).[2]

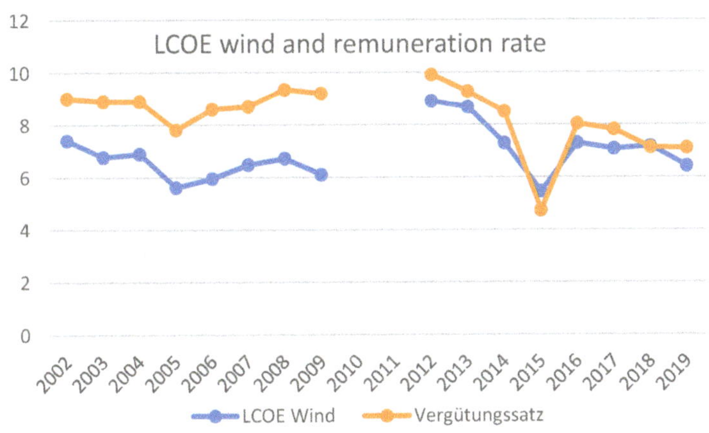

Figure 27.3: LCOE for onshore wind projects and remuneration rate; figure created by author.

In Figure 27.4 we have compared the development of the LCOE of the PV market with the figures for the onshore wind market in Germany. While the LCOE values for the PV markets have fallen significantly to a current level of 5 cents/kWh, the LCOE values for onshore wind projects have remained at a level of Ø 6.9 cents/kWh over 18 years – albeit with fluctuations. This means that onshore wind energy and photovoltaics have now reached a tipping point, as their LCOE is already close to or below that of conventional energy.

2 The German EEG 2017 has two important features in this respect: Firstly, the support regime was changed from a feed-in tariff system to a tendering procedure, which has led to a slight reduction in investment and operating costs. The effect is difficult to quantify, as costs had previously risen significantly due to pull-forward effects. If the average of the last five years is used, this results in a reduction of approx. 10%; higher reductions are also possible in individual areas of operating costs.

In addition, a cash flow stabilizing mechanism was introduced for onshore wind projects by adjusting the project-specific remuneration every five years. This means that both competition and awareness of more or less stable revenues have reduced the WACC requirements.

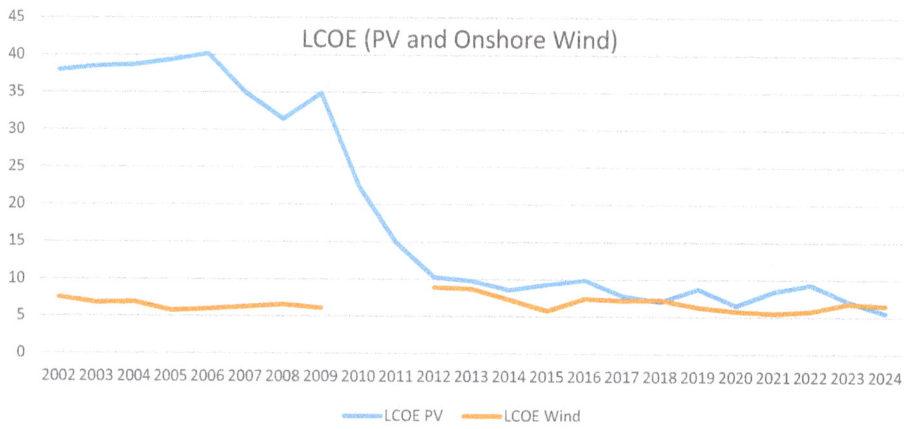

Figure 27.4: Development of LCOE for PV and Onshore Wind Projects; figure created by author.

While the total investment costs for onshore wind energy range between 1.2 M€/MW and 2.0 M€/MW, this ratio has fallen significantly for PV projects from 9.0 M€/MW in the boom years in Spain to less than 0.6 M€/MW in recent years. This development is also triggered by the development of the applicable tariff.

The level of the assumed WACC has a significant influence on the LCOE, as shown in the following Figure 27.5 is shown below:

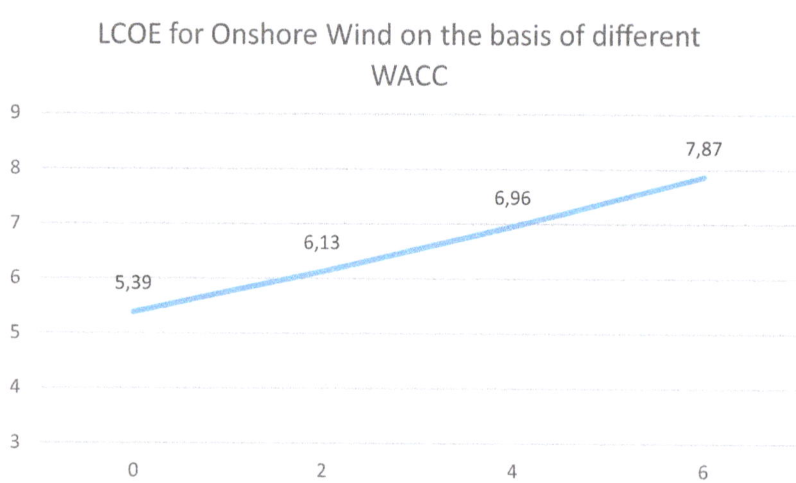

Figure 27.5: LCOE for onshore wind; figure created by author.

Figure 27.5 shows the LCOE (cents/kWh) for onshore wind projects with different WACC values. For example, an expected WACC of 6% p.a. results in an LCOE of 7.87 cents/kWh. This means that a regulatory authority must ensure that the current level of the WACC is appropriately reflected in the tariff in order to enable investments in this asset class. If a WACC of 4% were to function as an adequate proxy for the WACC, a typical onshore wind farm in Germany should receive a tariff of at least 6.96 cents/kWh in 2021.

27.2.1 Financial Learning in the Photovoltaic Sector – the Banks' Perspective

Now that the LCOE development has also revealed something about the expected returns that investors could realize with an investment in a PV project, the view of the banks as lenders is still missing. To this end, we want to look at how certain core parameters of a financing structure have developed, namely the level of **resilience**, the **term of** the project financing loan and the **margin.** Some of these parameters are also examined in the "Financial Learning" research field. Once again, we used the aforementioned set of project financing, whereby we refer the respective values to the average value of all values. This means that the data is appropriately anonymized, but nevertheless allows the desired conclusions to be drawn. The relevant key figures for the photovoltaics sector are shown in Figure 27.6 below:

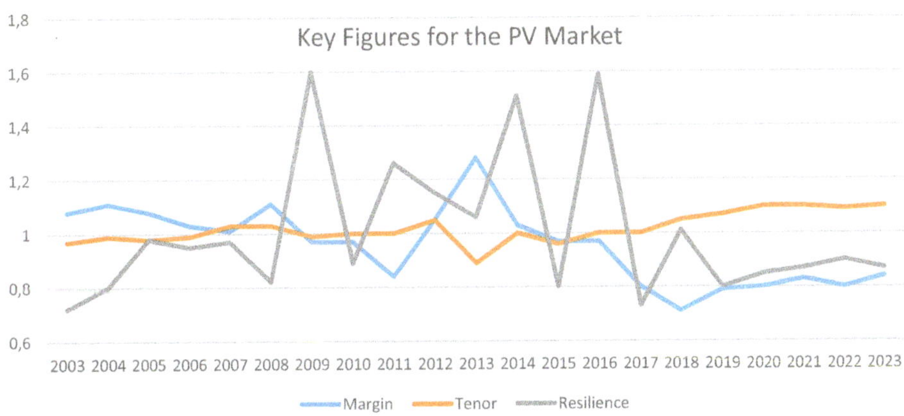

Figure 27.6: Key figures of the PV market; figure created by author.

A slight decrease in the **margin can be seen in** the first eight years, with the absolute extent of the maximum change being around 20 bps. The increase in 2013 is due to the fact that PV projects in France were also financed for the first time, which allowed for a slightly higher margin. However, margins fell again as the range of projects to be financed became increasingly limited and the number of interested banks increased. Overall, however, the changes are within a range of around 40 bps, excluding the one-off outlier in 2013. The **term** of the project financing loans has been highly consistent from the outset. This is also due to the fact that the structure of the typical remuneration systems was only changed slightly and usually covered a uniform period of 20 years. As the experience with the projects was obviously in line with expectations, there was no reason for the banks to make any adjustments. The **resilience** – understood as the maximum possible decline in energy production while maintaining the ability to service the debt – shows quite high fluctuations. These have arisen in cases where a new market was entered for the first time and it was obviously possible for the banks to enforce higher load capacities. This effect disappeared relatively quickly, presumably because the increasing competition between the banks did not allow this. Overall, however, the changes with regard to the three parameters examined are minor – in principle, today's financing structures are easily comparable with those of 17 years ago.

If we look at the same parameters in the onshore wind sector, the results are as follows (Figure 27.7):

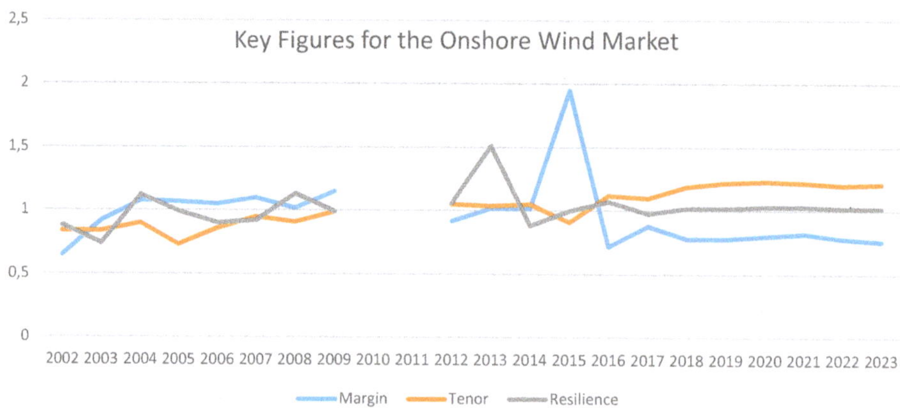

Figure 27.7: Key figures for the onshore wind market; figure created by author.

The **margin shows** a rapid adjustment to the long-term average value, followed by an upward outlier and a significant drop in the last three years. The outlier in 2015 is a wind project in Sweden – the significantly higher volatility of the cash flows here also results in a higher margin requirement. The decrease in the last three years is due to the fact that only EEG 2017 projects are included here, for which the legislator provides for a remuneration adjustment every five years, which results in a significant stabilization of the overall cash flows of the project[3] . In this respect, the default risk of an EEG 2017 project is also significantly lower, which explains the lower risk margin. The slight decrease in the first eight years can be seen, with the absolute extent of

3 J. Böttcher and P. Nagel 2019, p. 402 ff.

the maximum change being around 20 bps. The increase in 2013 is due to the fact that PV projects in France were also financed for the first time, which allowed for a slightly higher margin. However, margins fell again as the range of projects to be financed became increasingly limited and the number of interested banks increased. Overall, the changes are within a range of around 40 bps, excluding the one-off outlier in 2013. In contrast to the PV market, there is a consistent trend towards longer **terms.** This applies in particular to the German market, which had average maximum lifetimes of around 13 years when the EEG was introduced and has now reached more than 20 years. In my opinion, there are two reasons for this development: firstly, there is indeed a process of realization on the part of the banks, which have determined that, as a rule, the technology of a wind turbine has a technical service life of 20 years or more. The second driver, which is particularly responsible for the increase in the last three years, is the regulatory environment: with increasing reliability of cash flows, as a result of the EEG 2017, an extension of the term is also possible. The **resilience** is relatively constant, without any major outliers. However, the increasing experience of lenders, the generally improved measurement results and, again, the influence of the regulatory environment cannot be seen in this indicator. Overall, the changes with regard to the two parameters margin and resilience are small, while increasingly longer financing terms were widely accepted.

If we summarize the findings from the PV and onshore wind markets, there is practically no relation to changes in the LCOE. These were never an issue for the banks either; in principle, they were only interested in whether a project could generate enough cash flow to service the debt.

27.2.2 Closing Words

In particular, the dynamic decline in the LCOE of photovoltaics has shown that the initially generous subsidy regimes, especially in Germany and Spain, allowed considerable income opportunities for developers and module producers. The analysis also shows that it was not the investors who were able to achieve excess returns or particularly high coverage ratios with their projects. In order to assess whether the high subsidy rates were also necessary to organize economies of scale that enabled the PV industry to mature, the annual financial statements of the companies in question would have to be analyzed retrospectively. On a positive note, both onshore wind energy and photovoltaics now have such low LCOE that they have become cheaper than all conventional energy sources and will therefore continue to be a central pillar of the energy transition in the future.

For a long time, the discussion about renewable energies focused on the respective generation classes. This was also justified, as the various renewable energies needed time to prove themselves technically and establish a stable value chain. With the increasing proportion of fluctuating energy in our electricity mix, which is ultimately the result of lower electricity generation costs, the question of a new balance in the energy supply system arises: in my opinion, this requires three sets of measures: Firstly, the technical conditions for storing fluctuating energy must be created. It is also crucial that regulators realize that an incentive mechanism for the storage of renewable energies is needed so that the technical solutions can actually be implemented. And finally, it will also be important to promote decentralized electricity and heat generation so that the entire energy system is neither overburdened nor too supply-centric in its upcoming transformation.

27.3 The Interaction between Developer, Investor and Bank

In this section, we present a typical interaction between the capital providers (investor and banks) and the project developer who sells the project to the capital providers. The range of possibilities is as follows: The investor will expect his investment to generate a certain internal rate of return in line with the market. The bank will expect that the debt service for the loan it has provided can be reliably provided. This means that a certain debt service ratio will be achieved in accordance with its credit standards.

For example, a project could be presented as follows, starting with the central assumptions (Table 27.1):

Table 27.1: Basic Assumptions for Renewable Energy Project (Base Case).

Investment Costs		4200
Equity		1100
Debt		3100
Interest Rate:		4.90%
Repayment in years		19
Income p.a.		580
Sensitivity Income:	100%	580
OPEX-Quota:	26%	150.8
Discount Rate:		6.00%

A simple cash flow statement can be derived from this (Figure 27.8).

Let us assume that the investor wants to achieve an internal rate of return of at least 5.00% p.a. and that the bank wants to achieve a minimum DSCR of 1.10. These plans are well fulfilled in our base case (with an IRR of 7.09% p.a. and a minimum DSCR of 1.36 in the first year of operation).

	Sum		1	2	3	4	5	6	7	8	9	10	11	12	13	14	15	16	17	18	19	20
Income	11.600		580	580	580	580	580	580	580	580	580	580	580	580	580	580	580	580	580	580	580	580
OPEX	3.016		151	151	151	151	151	151	151	151	151	151	151	151	151	151	151	151	151	151	151	151
CFADS	8.584		429	429	429	429	429	429	429	429	429	429	429	429	429	429	429	429	429	429	429	429
Interest	1.519		152	144	136	128	120	112	104	96	88	80	72	64	56	48	40	32	24	16	8	0
Repayment	3.100		163	163	163	163	163	163	163	163	163	163	163	163	163	163	163	163	163	163	163	0
Debt Service	4.619		315	307	299	291	283	275	267	259	251	243	235	227	219	211	203	195	187	179	171	0
Free Cashflow	3.965		114	122	130	138	146	154	162	170	178	186	194	202	210	218	226	234	242	250	258	429
DSCR			1,362	1,398	1,435	1,475	1,516	1,560	1,607	1,657	1,709	1,765	1,826	1,890	1,959	2,033	2,113	2,199	2,293	2,396	2,508	######
Discounted FCF		-1100	107,7	108,7	109,3	109,4	109,2	108,6	107,8	106,7	105,4	103,9	102,2	100,4	98,5	96,5	94,3	92,1	89,9	87,6	85,3	133,8

IRR	7,09%

Figure 27.8: Cashflow-Projection (Base Case) .

At this point, the project developer has the opportunity to adjust the purchase price upwards, as this is permitted by the requirements of both investors. If the project developer knows where the pain threshold lies for each group, he can calculate the maximum loan amount and also the maximum equity contribution. In our example, this results in the following change (Table 27.2):

Table 27.2: Basic Assumptions for Renewable Energy Project (Maximum Sale Price Scenario).

Investment Costs		5340
Equity		710
Debt		4630
Interest Rate:		4.90%
Repayment in years		19
Income p.a.		580
Sensitivity Income:	100%	580
OPEX-Quota:	26%	150.8
Discount Rate:		6.00%

And this translates into the following cash flow calculation (Figure 27.9).

The requirements of the two groups of investors have just been met, which allows him to increase the purchase price from 4,200 monetary units to 5,340 monetary units. The investors' interest claims therefore largely determine the investor's possible selling price.

We now change the picture by assuming – starting from the base case – a 10% decline in revenue. This results in the following cash flow analysis (see Figure 27.10):

This scenario is still acceptable from the point of view of the financing bank, as the minimum DSCR of 1.18 is higher than the minimum required value of 1.10. However, the investor no longer achieves the IRR of 5.0% p.a. required by him, but remains well below this value at 2.23% p.a. He will therefore not participate in the project, meaning that it will not be sold.

However, this does not have to be the end of the transaction, but a negotiation process is likely to begin, which could include the following components, for example: On the one hand, the bank could be prepared to provide more debt to the project. In economic terms, this means that the investor benefits from the leverage effect and his internal rate of return improves because the surpluses generated relate to a smaller equity base. In the example, the bank is prepared to provide a further 200 monetary units of credit. In addition, it is worth talking to the project developer, who is advised to lower the sales price, otherwise the transaction will not go ahead. The economic rationale from the investor's point of view is comparable to an extension of the bank loan. The project developer can consider whether he has alternative – better – sales options so that he has an alternative marketing opportunity or whether he can make

Maximum Sale Price Scenario

	Sum		1	2	3	4	5	6	7	8	9	10	11	12	13	14	15	16	17	18	19	20
Income	11.600		580	580	580	580	580	580	580	580	580	580	580	580	580	580	580	580	580	580	580	580
OPEX	3.016		151	151	151	151	151	151	151	151	151	151	151	151	151	151	151	151	151	151	151	151
CFADS	8.584		429	429	429	429	429	429	429	429	429	429	429	429	429	429	429	429	429	429	429	429
Interest	3.018		227	219	211	203	195	187	179	171	163	155	147	139	131	123	115	107	99	91	83	75
Repayment	3.100		163	163	163	163	163	163	163	163	163	163	163	163	163	163	163	163	163	163	163	0
Debt Service	6.118		390	382	374	366	358	350	342	334	326	318	310	302	294	286	278	270	262	254	246	75
Free Cashflow	2.466		39	47	55	63	71	79	87	95	103	111	119	127	135	143	151	159	167	175	183	354
DSCR			1,100	1,123	1,147	1,173	1,199	1,226	1,255	1,285	1,316	1,349	1,384	1,421	1,459	1,500	1,543	1,589	1,637	1,689	1,744	5,725
Discounted FCF		-710	37,0	42,0	46,3	50,0	53,2	55,8	58,0	59,7	61,0	62,1	62,8	63,2	63,3	63,3	63,0	62,6	62,1	61,3	60,5	110,5

IRR	5,02%

Figure 27.9: Cashflow-Projection (Maximum Sale Price Scenario); figure created by author.

Downside Scenario

	Sum		1	2	3	4	5	6	7	8	9	10	11	12	13	14	15	16	17	18	19	20
Income	10.440		522	522	522	522	522	522	522	522	522	522	522	522	522	522	522	522	522	522	522	522
OPEX	3.016		151	151	151	151	151	151	151	151	151	151	151	151	151	151	151	151	151	151	151	151
CFADS	7.424		371	371	371	371	371	371	371	371	371	371	371	371	371	371	371	371	371	371	371	371
Interest	1.519		152	144	136	128	120	112	104	96	88	80	72	64	56	48	40	32	24	16	8	0
Repayment	3.100		163	163	163	163	163	163	163	163	163	163	163	163	163	163	163	163	163	163	163	0
Debt Service	4.619		315	307	299	291	283	275	267	259	251	243	235	227	219	211	203	195	187	179	171	0
Free Cashflow	2.805		56	64	72	80	88	96	104	112	120	128	136	144	152	160	168	176	184	192	200	371
DSCR			1,178	1,209	1,241	1,275	1,311	1,349	1,390	1,433	1,478	1,527	1,579	1,634	1,694	1,758	1,827	1,902	1,984	2,072	2,169	######
Discounted FCF		-1100	53,0	57,1	60,6	63,5	65,8	67,8	69,2	70,3	71,1	71,5	71,7	71,6	71,3	70,8	70,1	69,3	68,4	67,3	66,1	115,7

IRR	2,23%

Figure 27.10: Cashflow-Projection (Downside Scenario); figure created by author.

a concession on his asking price. Assuming that he is prepared to make the desired concession, the following assumption structure results (see Table 27.3):

Table 27.3: Basic Assumptions for Renewable Energy Project (Compromise) (compiled by author) .

Investment Costs		4000
Equity		700
Debt		3300
Interest Rate:		4.90%
Repayment in years		19
Income p.a.		580
Sensitivity Income:	90%	522
OPEX-Quota:	26%	150.8
Discount Rate:		6.00%

This then translates into the following cash flow statement (see Figure 27.11).

In this case, both investors achieve their minimum requirements: the investor achieves an internal rate of return of 5.17% p.a. and the bank just exceeds its minimum DSCR requirement with 1.107.

We can derive the following findings from this example:

- **Transaction perspective**: All parties involved must understand the objectives and restrictions of the other parties. The requirements of the lenders in terms of their return on capital or ability to service capital largely determine the seller's scope of options. Both sides of capital are interrelated and changes in the cash flow-relevant parameters change the scope of the various parties involved.
- **Regulator perspective**: We have changed one component here, the revenue. If a transaction is no longer attractive from an investor's point of view, the regulator can of course consider taking countermeasures here – e.g. with a higher remuneration tariff. However, as our examples show, this can also mean that it is not the users who benefit, but the sellers. This is a difficult task for the regulator: it is also possible that the sellers are not just free riders, but that the higher tariff is possible, for example, so that the desired expansion targets can also be achieved.

27.4 The Economic Value of Predictability

Investments are usually assessed using the internal rate of return or the net present value method. One of the implicit underlying premises is the mapping of the payment consequences of the investments according to amounts and payment dates. This means, on the one hand, that the investors' plans are implemented in full in accordance with the original investment plan and, on the other, that any deviations from

Compromise

	Sum		1	2	3	4	5	6	7	8	9	10	11	12	13	14	15	16	17	18	19	20
Income	10.440		522	522	522	522	522	522	522	522	522	522	522	522	522	522	522	522	522	522	522	522
OPEX	3.016		151	151	151	151	151	151	151	151	151	151	151	151	151	151	151	151	151	151	151	151
CFADS	7.424		371	371	371	371	371	371	371	371	371	371	371	371	371	371	371	371	371	371	371	371
Interest	1.617		162	153	145	136	128	119	111	102	94	85	77	68	60	51	43	34	26	17	9	0
Repayment	3.300		174	174	174	174	174	174	174	174	174	174	174	174	174	174	174	174	174	174	174	0
Debt Service	4.917		335	327	318	310	301	293	284	276	267	259	250	242	233	225	216	208	199	191	182	0
Free Cashflow	2.507		36	44	53	61	70	78	87	95	104	112	121	129	138	146	155	163	172	180	189	371
DSCR			1,107	1,136	1,166	1,198	1,232	1,268	1,306	1,346	1,389	1,434	1,483	1,535	1,591	1,652	1,717	1,787	1,863	1,946	2,037	######
Discounted FCF		-700	33,8	39,5	44,4	48,6	52,2	55,2	57,8	59,8	61,5	62,8	63,7	64,3	64,7	64,8	64,7	64,4	63,9	63,2	62,5	115,7

IRR	5,17%

Figure 27.11: Cashflow-Projection (Compromise); figure created by author.

the plan do not trigger any countermeasures. Both implicit assumptions do not exist in reality. The concept of options for action deals with the real possibilities of an economic entity and their evaluation. For our purposes, the concept of waiting options is of particular importance, as we will illustrate using two examples.

Assume that an investor can buy an asset, e.g. a ground-mounted solar power plant, today (t_0) at a price of 90 MU ("Money Unit"). In return, he receives a secure, present value return of 105 MU over the project term in the following year (t_1). The discount rate is 10%. In the case of an immediate purchase, the net present value in t_0 is calculated at 5.45 MU, so that the net present value criterion is used for the immediate investment. The example has now been changed to the extent that the investor has the right to wait a further period and can then decide again whether to invest or not. The expected prices for a ground-mounted solar plant after one year are either 60 GE or 120 GE. Both environmental conditions are equally likely. The investor will only invest after one year if the asset has fallen in price to 60 GE, otherwise the net present value of his investment would be negative. The capital value when exercising a waiting option is then calculated as follows:

$$KW_{Wait} = 0.5 * (-60/1.1 + 105/1.1 * 1/1.1) = 16.12\,\text{MU}.$$

The net present value is higher if the investor waits a period instead of investing immediately. Examples of comparable situations can be found quite frequently: The US market for renewable energies has so far provided for the use of certain tax benefits (production tax credits) as a significant consideration, which require that investors also generate profits in order to be able to use the PTCs. However, if profits cannot be made (e.g. due to market distortions), investments in renewable energies are no longer attractive. Such a decline in demand would have to put significant pressure on the achievable prices for economic goods, causing investors to hold back on investments on an individual-rational basis.

A second example: Assume that an investment requires an immediate decision and leads to initial payments of 180 GE. The investment leads to an infinite annuity of 20 MU (e.g. tariff charges), the calculation interest rate is 10 %. This results in a capital value of

$$KW_0 = -180 + 20/0.1 = 20.$$

Let us now assume that the investor waits one period before making the investment. In this case, the entire investment cash flow is postponed by one period, so that the net present value is calculated as follows:

$$KW = KW_0/1.1 = 18.18\,\text{MU}$$

Waiting does not lead to any advantage here, as it is assumed that no new future environmental conditions are expected. This is a general result: waiting options have no value if a development in line with the original investment scenario or even a deterioration is expected, provided that this has no retroactive effect on investments already made.

The picture changes if we make the following modification to the example: Assume that a government plans to change the tariff rates after a certain period of time. After one year, a decision is expected as to whether the tariff rates for new projects will be either 17 or 23 MU. If the investor now waits one period, the following values result:

$$KW_{Wait} = 0.5 * (-180/1.1 + 23/0.1 * 1/1.1) = 22.73 \text{ MU}.$$

If the unfavorable case of 17 MU should arise for the investor, he will not invest, as the capital value would then be negative. He will only take action if the annuity is 23 MU. In this case, waiting results in a higher capital value of 22.73 MU, the value of waiting is 2.73 MU.

The investor behavior described above is individually rational. However, it also means that uncertainties and volatility lead to investment restraint or higher return requirements.

Waiting options are generally of value if it is expected that economically relevant framework data may change significantly. If investors fear that investments will no longer pay off, they will hold back on their investments. However, this also applies vice versa if there are clear indications that general economic data will improve in the future without them being able to benefit from this in their current decision. This may be the case, for example, if there are key date-related regulations that exclude existing users from improved new regulations.

The consequence for the regulator is that it has a clear plan as to how it wants to intervene. Once the regulator has decided on a strategy, it should only deviate from its chosen path for very good reasons. This is where the economic value of predictability comes into play.

27.5 The Concept of a Transformation Funds

27.5.1 Structure of a Transformation Funds

The investment requirements for the energy transition are considerable and can easily overwhelm the financing options of banks, investors and the public sector. In addition, many of the projects are new to investors and the financial world, and until the economic viability or debt service capacity is proven, investors and banks will tend to choose safe investments instead. But if it is in the public interest to develop these environmentally friendly technologies in particular, we should think about alternative financing approaches.

Here it would be worth considering whether suitable financing constructs could be used that could act as a multiplier. I would like to present the idea for such an instrument at the end of the book (see Figure 27.12):[4]

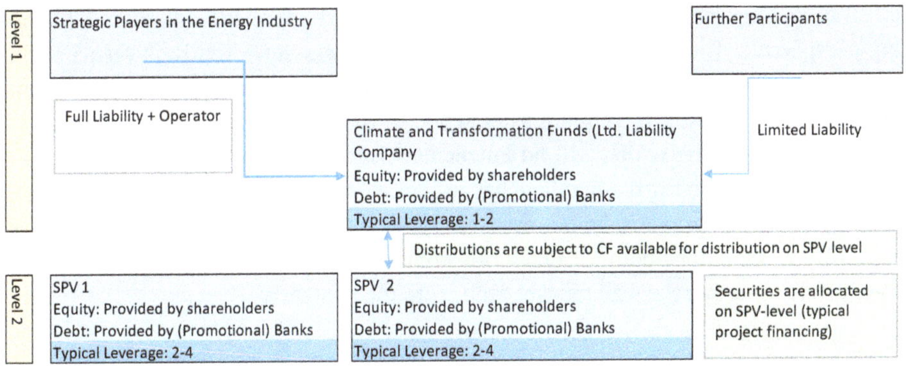

Figure 27.12: Concept of a Transformation Funds; figure created by author.

At the center of the structure is a climate and transformation funds. It is intended to pool funds from strategic and other participants and invest them in individual renewable energy projects. The investment decision lies with one or a few strategic partners, while other participants only contribute equity to the financing of the fund. From the point of view of the projects (Level 2), the funds contributed at fund level are equity capital.[5]

As the collateral is at project level, the level above is structurally subordinate. It only receives the distributions at project level. In order for this to be acceptable to ultimate investors, their risk is reduced by the fact that they have access to a portfolio of projects and the external debt at fund level is lower than at project level.

Experience shows that investors with a lower risk appetite are more likely to participate at the project company level (Level 2), while investors with a somewhat higher risk appetite will participate at the transformation fund level (Level 1).

The construction via a transformation fund helps to implement as many renewable energy projects as possible. This is particularly important in the initial transfor-

4 The idea is based on a paper of the Hans-Boeckler-Foundation: https://www.boeckler.de/pdf/p_imk_study_71_2021.pdf. However, this concept is amended in some aspects here, especially the concept of a holding-level-structure.

5 It is also conceivable that the capital is contributed as a subordinated loan or junior loan. A subordinated loan or a junior loan is subordinated in its servicing to the servicing of a senior loan (these are the loans at project level), which generally qualifies them as functional equity from a project perspective. Nevertheless, the structure as equity seems to make more sense: here there is complete flexibility with regard to interest and repayment. A subordinated loan is accompanied by a repayment schedule that does not have the flexibility of a profit-related distribution.

mation phase, as the investment requirements for infrastructure projects are very high at the beginning, while they become increasingly smaller later on and result in a return on investment. A second effect is that investors at fund level can provide more equity at project level, especially in the start-up phase of a technology. This has the effect of making the projects more secure and therefore more attractive from a banking perspective. The aforementioned returns on investment – firstly in terms of the output of the individual system, but then in particular in terms of the number of implementations – result in economies of scale, so that more economical projects will emerge in the future, which will no longer necessarily require the start-up aid mentioned here. And the third advantage lies in the fact that the state does not participate in projects with a lost subsidy, but provides funds that generally flow back with interest. From a regulatory point of view, this approach is preferable, provided it fulfills the purpose and there are no market failures to the contrary.

27.5.2 Clarification of State Aid Issues

The granting of state aid is regulated at EU level in order to ensure the fairest possible competition within the EU. Article 107 of the Treaty on the Functioning of the European Union prohibits state aid that is granted selectively, (potentially) distorts competition and affects trade between member states. As the investments of the Transformation Fund constitute state funds selectively granted to companies, it must be regularly examined whether this constitutes unlawful aid.

However, the Treaty on the Functioning of the European Union ("TFEU") and supplementary EU legislation provide for extensive exemptions, particularly in areas where there is a common European interest or where the development of certain economic sectors or areas is promoted without harming common interests or disrupting trade.

Also relevant for the transformation fund is the admissibility of risk financing, which is intended to eliminate market failure in the financing of small and medium-sized enterprises (SMEs). The targeted promotion of start-ups and SMEs through equity investments is also explicitly addressed in the Commission's communication on the European Green Deal. The guidelines on aid for risk finance explicitly exclude risk finance for listed companies because the listing is proof of their ability to raise private funds. However, this is irrelevant for the investments in question here, as they typically involve many decentralized energy and heat transition projects that are organized as special purpose vehicles but will not be listed on the stock exchange.

27.5.3 Assessment on Funds-Level

In this section, we will show the possibility of leveraging investments through the use of a higher-level fund (Level 1) using two typical examples, one from the geothermal energy sector and one from the ground-mounted PV sector. The example of geothermal energy shows relatively high planned DSCR values (at least 1.25) at project level (Level 2), which on the one hand correspond to a higher degree of uncertainty with regard to the heat yield realized in the end, but on the other hand show good economic efficiency. PV projects must be assessed differently from a risk and return perspective: Here we have a low DSCR requirement (at least 1.05), which results from the low uncertainty of performance, but has only a scarce economic efficiency. These characteristics already allow us to draw a few conclusions for the operation of the fund, as we will show in a moment.

First of all, the assumptions and structures of the geothermal project at fund level (see Table 27.4).

Table 27.4: Basic Assumptions of a geothermal project at funds level (compiled by author) .

	in T€	Interest Rate p.a.	Tenor in years	Repayment p.a	Discount Rate
Equity	7,540				0.00%
Debt (1st level)	7,540	5.00%	25	3,01,600	
Sum	15,080				
Debt-to-equity-ratio	1.00				

This then translates into the following cash flows at holding level (see Figure 27.13).

First of all, the cash flow that could be distributed at project level is transferred to the holding level. There it follows the waterfall principle, as we know it from other projects: First, the banks at the holding level are serviced and then the cash flow amount that can be distributed to investors at the holding level is obtained.

The DSCR values at holding level are significantly higher than at project level. This is the result of a longer loan term than would be usual for a project and the lower leverage: in the example, we have assumed a debt-to-equity ratio of 1.0, which means that equity and debt each finance half of the fund. The internal rate of return at holding level is very high at 36.7% and is certainly acceptable for an investor.

We have presented the example of the PV project below (see Table 27.5).

The cash flow distribution mechanism described above also applies here. The distribution of cash flows is as follows (see Figure 27.14).

In this case, the same split between debt and equity means that the debt service at holding level cannot be covered in 2026 and 2027. A modified debt-to-equity ratio could of course be considered here. It should be borne in mind that the example calculation is based on one project that is financed at holding level. If several projects are implemented – and that is the actual idea – these projects will take place at differ-

		Geothermal Project	Funds Level									
		1	2	3	4	5	6	7	8	9	10	11
		2025	**2026**	**2027**	**2028**	**2029**	**2030**	**2031**	**2032**	**2033**	**2034**	**2035**
CF (for distrbution)		2.669.995	3.569.781	3.607.740	3.645.226	3.682.229	3.718.739	3.754.746	3.790.241	3.825.213	3.859.652	3.893.547
Sum Income (1st level)		2.669.995	3.569.781	3.607.740	3.645.226	3.682.229	3.718.739	3.754.746	3.790.241	3.825.213	3.859.652	3.893.547
Interest		377.000	361.920	346.840	331.760	316.680	301.600	286.520	271.440	256.360	241.280	226.200
Reapayment		301.600	301.600	301.600	301.600	301.600	301.600	301.600	301.600	301.600	301.600	301.600
CF after debt service (1st level)		1.991.395	2.906.261	2.959.300	3.011.866	3.063.949	3.115.539	3.166.626	3.217.201	3.267.253	3.316.772	3.365.747
Taxes on Holding-Level		0	0	0	0	0	0	0	0	0	0	0
Distributable CF (1st level)		1.991.395	2.906.261	2.959.300	3.011.866	3.063.949	3.115.539	3.166.626	3.217.201	3.267.253	3.316.772	3.365.747
Discounted CF	###	1.991.395	2.906.261	2.959.300	3.011.866	3.063.949	3.115.539	3.166.626	3.217.201	3.267.253	3.316.772	3.365.747

	2025	**2026**	**2027**	**2028**	**2029**	**2030**	**2031**	**2032**	**2033**	**2034**	**2035**
DSCR	3,93	5,38	5,56	5,76	5,96	6,17	6,38	6,61	6,86	7,11	7,38
Loan (in T€)	7.238.400	6.936.800	6.635.200	6.333.600	6.032.000	5.730.400	5.428.800	5.127.200	4.825.600	4.524.000	4.222.400
IRR	36,67%										

Figure 27.13: Cashflow Projection of Geothermal Project at holding level; figure created by author.

Table 27.5: Basic Assumptions for a PV Project (Fund Level) (compiled by author).

	in T€	Interest Rate p.a.	Tenor in years	Repayment p.a	Discount Rate
Equity	1,383				0.00%
Debt (1st level)	691	5.00%	35	19,755	
Sum	2,074				
Debt-to-equity-ratio	0.50				

ent times, so that a stabilization of cash flows at holding level can be expected. And even in this case, it is a temporary phenomenon from the bank's point of view, as the ability to service debt is good over the total period.

The internal rate of return is in the double-digit range, which can certainly be considered acceptable for a PV project from an investor's perspective.

The instrument of a holding structure is generally well suited to financing long-term infrastructure investments, as it is easier for stakeholders at holding level to implement a long-term financing structure. This allows the projects at project level to receive more "equity" from the fund level, which makes the overall financing at project level much easier. In principle, structural subordination at holding level can also be implemented for banks, provided that the aim is to achieve a lower level of debt than is otherwise usual at project level.

The interim result is as follows:

The fund structure described above is particularly helpful for long-term infrastructure projects in order to make such investments economically viable. It also allows reliable projects to be generated at project level in the start-up phase of a market, thus enabling a market to ramp up. It allows the implementation of long-term infrastructure projects without leading to a permanent outflow of public funds.

The fund develops its potential in the integration of several projects, which should also take place at different times.

Projects from different asset classes are certainly desirable, as this can also help to stabilize cash flows at holding level.

	Solar Project		Funds Level							
	1	2	3	4	5	6	7	8	9	10
	2025	**2026**	**2027**	**2028**	**2029**	**2030**	**2031**	**2032**	**2033**	**2034**
CF (for distrbution)	315.109	61.740	62.985	81.044	99.010	116.879	134.651	152.322	169.892	187.356
Sum Income (1st level)	315.109	61.740	62.985	81.044	99.010	116.879	134.651	152.322	169.892	187.356
Interest	51.858	50.376	48.894	47.413	45.931	44.449	42.968	41.486	40.004	38.523
Repayment	29.633	29.633	29.633	29.633	29.633	29.633	29.633	29.633	29.633	29.633
CF after debt service (1st level)	233.618	-18.268	-15.542	3.999	23.446	42.797	62.050	81.203	100.254	119.201
Taxes on Holding-Level	0	0	0	0	0	0	0	0	0	0
Distributable CF (1st level)	233.618	-18.268	-15.542	3.999	23.446	42.797	62.050	81.203	100.254	119.201
Discounted CF ###	233.618	-18.268	-15.542	3.999	23.446	42.797	62.050	81.203	100.254	119.201

	2025	**2026**	**2027**	**2028**	**2029**	**2030**	**2031**	**2032**	**2033**	**2034**
DSCR	3,87	0,77	0,80	1,05	1,31	1,58	1,85	2,14	2,44	2,75
Loan (in T€)	1.007.517	977.884	948.251	918.619	888.986	859.353	829.720	800.087	770.454	740.821
IRR	11,62%									

Figure 27.14: Cashflow Projection of PV Project at holding level; figure created by author.

List of Figures

https://doi.org/10.1515/9783110784299-028

List of Tables

https://doi.org/10.1515/9783110784299-029

List of Contributors

Kwesi Annan-Takyi is a graduate of the Berlin Professional School, and with additional background as Green Energy Finance Expert from the Renewables Academy, in Berlin. The author has over 15 years' field experience in the energy efficiency sector in Ghana's green energy activities, having been part of various national scale pioneering projects that have reduced Ghana's energy efficiency among industry and domestic consumers over these years. As a sustainable energy projects developer and solutions provider for industry, commercial buildings and other large-scale consumers, he employs an entrepreneurial approach to decarbonization and the low carbon economy. He yearns to share the experiences and insights gained as he continues to learn how the stars would align for a green energy financing utopia in his part of the world.

Jim Armbruster is member of the building technology group of ITE – Institute for the Transformation of the Energy system. He is working on a project covering potential use of large-scale heat pumps in heating networks. He holds a professional bachelors degree in heating technology and a Bachelor of Engineering in Green Building Systems (FH Westkueste, Heide, Germany).

Jörg Böttcher, a graduate in economics and a qualified bank management assistant, has been working with IB.SH (Promotional Bank in Schleswig-Holstein) since 2021. Before that time, he worked for 25 years with Hamburg Commercial Bank (formerly HSH Nordbank AG) and has been specializing in renewable energies projects. Jörg has produced a series of publications on project finance and renewable energies since 2004. He completed his doctorate at Justus-Liebig University in Giessen in 2011 with a thesis entitled "Possibilities for Financing CSP Projects."

Roman Buss holds a doctorate in political science. From 2022 to April 2024, he worked as a senior manager and consultant at the World Energy Council – Germany, where he supervised the German-Colombian industrial dialog on green hydrogen and a study on sustainable ammonia, among other things. As RENAC project manager in the field of energy policy, he implemented numerous international training projects for GIZ and on behalf of federal ministries between 2010 and 2022. In the IASS-led project "CO-BENEFITS", which was funded by the International Climate Initiative (Internationale Klimaschutzinitiative) of the BMU, he organized the work packages on capacity building. Between 2015 and 2024, he supported the "International Communication of the Energy Transition" on behalf of the Federal Foreign Office (Auswärtiges Amt).

Darius Bonk is a member of the building technology group of ITE – Institute for the Transformation of the Energy system. He is working on a project covering potential use of large-scale heat pumps in heating networks. Previously, he studied green building systems (B.Eng.) and automation technology (M.Eng.) at FH Westkueste in Heide, Germany.

Heinz-J. Bontrup, Prof. Dr. rer. pol., graduate economist, graduate business economist, studied business administration in Bielefeld and economics in Bremen. He holds a doctorate from the University of Bremen. He has practical experience in the aerospace industry and as a board member at Stahlwerke Bochum AG. From 1996-2019 he was Professor of Economics at the Westphalian University of Applied Sciences, and since 2018 he is Visiting Professor at the University of Siegen, Faculty III, Economics. He has numerous book publications, including with Prof. Dr. Ralf-M. Marquardt (*Energiewende, Verteilungskonflikte, Kosten Folgen*), and *Die Zukunft der großen Energieversorger*. An expert in the German Bundestag and various state parliaments and board member of the Westphalian Energy Institute ("Westfälisches Energieinstitut"), on March 21, 2018, he was awarded the Federal Cross of Merit on Ribbon by the Federal President for his diverse economic education work against the neoliberal mainstream.

https://doi.org/10.1515/9783110784299-030

Torsten Cziesla studied mechanical engeering (Dipl.-Ing.) and completed his PhD (Dr.-Ing.) at Ruhr-University Bochum. After a longtime career in the energy industry he accepted the position of Professor and Head of Studies Energy Engineering and Resource Optimazation at the newly founded University of Applied Sciences Hamm-Lippstadt (HSHL) in 2009. His research topic is integration of renewables into the energy supply system. As part of a research group concerning energy transition and transformation design he has also contributed the strategic research direction of the university. Currently, he is founding member of the university´s new institute for sector coupling (Institut für Sektorenkopplung in der Energiewende – IfSK).

Andreas Gabler is a lawyer and partner at the law firm Hoffmann Liebs Partnerschaft von Rechtsanwälten mbB in Düsseldorf. He advises domestic and foreign clients in all areas of German and European energy law. Dr. Gabler has extensive experience in energy contract law as well as in the planning and contractual structuring of energy projects (power plant construction, contracting). A major focus of his work is in the field of renewable energies, advising on wind, photovoltaic and biomass power plants. He advises on all project contracts (construction, operational management, grid connection) as well as on the regulatory framework conditions (remuneration claims, direct marketing, handling of load balancing).

Thies Goldner is a corporate lawyer based in the Hamburg office of Osborne Clarke. He has a focus on the Energy & Utilities Sector and advises banks, institutional and private-equity investors as well as developers and construction firms primarily in the areas of renewable energy generation, particularly on- and offshore wind and photovoltaics. Since 2007 he has worked for two other international law firms before he joined Osborne Clarke in 2016. He has a doctor's degree from the Georg-August University in Göttingen. During his professional practice, he has published several publications mostly concerning banking and real estate aspects of renewable energy projects.

Tibor Fischer is Director Renewable Energies at the German Energy Agency (dena). His work focuses on analysing framework conditions for renewable energies and new business models in the electricity and heating market. He has 18 years of experience in the renewable sector. Mr Fischer is the author of numerous studies and analyses on current legal and economic issues and trends in the energy sector in Germany and abroad. He is also a member of the UNECE Group of Experts on Renewable Energies (GERE) since 2019. His current work focuses on issues relating to the financing of renewable energies and new financial instruments and business models as well as new approaches to decarbonising individual sectors via integrated supply concepts. Tibor Fischer studied political science at the University of Leipzig (Germany) and the Universidad de Granada (Spain).

Tomas Haug is a Senior Managing Director at NERA Economic Consulting based in Berlin, where he leads NERA's energy and infrastructure practice. He provides advisory services to private companies, investors, and regulatory authorities on a wide range of economic and financial matters, including financial and tariff modelling, access pricing, competition policy matters, tariff structure, efficiency studies, and wider questions of regulatory reform. He is a Chartered Financial Analyst and holds academic degrees from the University of Quebec in Montreal and the University of Magdeburg.

Sebastian Helgenberger graduated in Environmental Sciences at Leuphana University in Germany and ETH Zurich. He also holds a PhD in socio-economics, acquired from BOKU University Vienna for his work on organizational resilience and the relevance of global warming on investment decisions in small and medium enterprises. In 2014 he joined the Research Institute for Sustainability as research team leader to support authorities and industries in identifying and realizing country-specific opportunities of renewable energy and energy efficiency solutions for planet and people.

Thoralf Herbold is an attorney and Partner in the Cologne and Hamburg office of GÖRG Partnerschaft von Rechtsanwälten mbB. He specializes in the area of energy law, in particular renewable energies. Thoralf advises national and international investors, project developers and power plant constructors in the project development and M&A-transactions in the energy sector (onshore and offshore wind farms, solar plants, energy storages).

He is also an expert in the field of energy regulatory law, energy trading and the development of decentralized energy concepts. Thoralf is further specialized in the area of public commercial law (permits, grid connection). Prior to joining GÖRG, Thoralf had an engagement with Hengeler Mueller in Düsseldorf from 2007 through 2009. He joined GÖRG in 2009 and was announced Partner in 2016.

Nicolai Herrmann is a managing partner for enervis energy advisors GmbH in Berlin. He has vast experience in energy economics, market analysis and modelling, investment appraisal for energy assets, PPAs, energy market design and scenario studies. He holds a degree as an industrial engineer and a PhD in economics from Flensburg University (GER). He attended academic programs at Griffith University in Brisbane, Australia and was a visiting scholar to the University of Maryland, USA.

Philipp Hiemann is a doctoral student in economics and assistant to the Board of Directors at RWI – Leibniz Institute for Economic Research in Essen. He studied Business Administration and Economics at Bielefeld University and holds a master's degree in economics from St Catharine's College, University of Cambridge. Before joining RWI, he worked for several years as an economist at NERA Economic Consulting in Berlin. At NERA, he worked on topics in economic regulation with a particular focus on energy networks, supporting clients in regulatory and arbitration proceedings as well as transactions of energy assets. At RWI, where he works in the Environment and Resources department, he specialises in energy and transport economics as well as energy and infrastructure regulation.

Jerrit Hilgedieck studied Energy and Environmental Engineering as well as Renewable Energy at the Hamburg University of Technology (TUHH). He has been working on his PhD in renewable energy system technology since 2016 with a focus on the integration of renewable energies in the electricity system at the Hamburg University of Technology.

Attakan Janpidok, a seasoned professional in the energy sector, brings two decades of knowledge and expertise across oil & gas upstream, deep geothermal drilling operations, and green energy finance. In his previous roles, Attakan has worked on multiple geothermal projects in Indonesia and served as the regional focal point for geothermal project development. His understanding of Indonesia's renewable energy policy and the relationships with private sectors have been key to his achievement. Currently pursuing a master's degree in Business Management with a focus on Green Energy and Climate Finance, Attakan continues to deepen his knowledge and reflects a profound commitment to sustainable energy.

Martin Kaltschmitt obtained his PhD as well as his Habilitation in the field of renewable energies at Stuttgart University/Germany. After a research stay at King's College in London and at the University of California at Berkeley he became the managing director of the Leipzig Institute for Energy non-profit research company. In 2006 he was promoted to a full professor at Hamburg University of Technology where he is heading the Institute of Environmental Technology and Energy Economics. Between 2008 and 2010 he was in parallel the scientific managing director of the German Biomass Research Centre (DBFZ).

Jennifer Kowallik works at the Agency for the Environment, Climate, Energy and Agriculture as a consultant for Green Economy and Sustainable Finance. Previously, she worked at the Energy Agency of Investitionsbank Schleswig-Holstein as a project manager for the Energy and Climate Initiative (EKI). She holds a doctorate from the Institute for Banking, Finance and Start-up Management at the Leuphana

University of Lüneburg. Prior to this, she held various professional positions in the finance and energy industry.

Jens Kühne works at AGFW ("Energy Efficiency Association for Heating, Cooling and CHP"). He is responsible for the department's staff and all topics in the field of heat generation (conventional or renewable), sector coupling (heat pumps or synthetic fuels) and heat accumulation in all variations. He holds a doctorate from the Technical University of Darmstadt (TU Darmstadt). Prior to this, he worked as Scientific Researcher with TU Darmstadt (Institute for Energy and Power Plant Technology).

Jelto Lange studied General Engineering Science (B.Sc.) and Energy Engineering (M.Sc.) at Hamburg University of Technology (TUHH). In December 2017, he started his work as a research assistant and PhD student (Dr.-Ing.) at the institute of environmental technology and energy economics at TUHH. His research is focused on the improved integration of fluctuating renewable energies into electricity systems by means of sector coupling.

Kathrin Langewald gained practical experience and deepened her knowledge, particularly in the field of international relations and development policy during her academic career from 2016 to 2024. Formative stages of her education were an internship at the German Institute of Development and Sustainability (IDOS) in Bonn and an exchange semester exchange semester at SciencesPo in Paris. There she dealt intensively with relations between the EU and Africa in the context of sustainable development development goals and renewable energies. As part of her master's degree in International Public Affairs at the Central European University (CEU) in Vienna (2020-2024), she wrote her master's thesis entitled "The Strategic Role of Green Hydrogen in German Foreign Policy: A Case Analysis of German-South African Energy Relations" with the strategic importance of green hydrogen in German foreign policy. Through her work as a junior consultant at the International PtX Hub of the German Gesellschaft für Internationale Zusammenarbeit (GIZ) GmbH in Berlin (2023) and as a master's student in the Global-H2-Upscaling project at the Institute for Future Energy and Material Flow Systems GmBH (IZES) in Saarbrücken (2024), she was able to substantiate her research question both empirically and theoretically. This experience enabled her to analyze the role of states in the development of global energy markets and challenges in the field of sustainable development in the field of green hydrogen.

Andreas Luczak is Professor at Kiel University of Applied Sciences focussing on Energy Transition and Energy Storage. Having a PhD in Physics he worked 16 years at Siemens in several positions in the Industrial and Power Generation area. Being responsible for the solar inverter business between 2008 and 2012 he experienced the challenging time of the solar boom in Germany/Europe and supported the first solar projects in the United States in the Megawatt range. After leading a corporate wide energy storage strategy project he left Siemens to become in 2013 the Managing Director of Vanadis Power, a German subsidiary of a Chinese Vanadium Flow Battery company deploying some of Europe´s largest Flow Batteries. In 2016 he accepted the appointment as Professor for Renewable Energy.

Jan Lüken studied Economics and Political Science at the University of Heidelberg. As a researcher at the Chair of Public Finance at the Catholic University of Eichstätt-Ingolstadt, he worked on the relationship between tax systems and on the relationship between tax systems and innovation. He then worked for five years in the representation of interests and as office manager of the board at the Bavarian Cooperative Association ("Genossenschaftsverband Bayern"). In 2019, Jan Lüken was appointed to the professorship for Business Economics at the Rosenheim University of Applied Sciences. He has been head of the "Circular Economy" Master's degree program there since 2021. His research interests include the areas of circular economy, innovation economics and economic policy.

Farid Mohamadi has been working in renewables since 2009 as business leader in markets such as Europe, North Africa and Latin America. Now he advises project promoters to raise capital: equity (through M&A transactions or with mezzanine/bridge loans from dedicated energy funds) and senior long-term debt. He has a degree in industrial engineering from the business and technology school ESTA in Belfort, France, a Masters in Business Administration and Engineering from the university of technology and economy HTW in Berlin, Germany and a Masters in Business Management – Green Energy and Climate Finance from the University of Economy and Law HWR in Berlin. He is a certified Financial Modelling and Valuation Analyst (FMVA) from the Toronto-based Corporate Finance Institute (CFI) and Green Energy Finance Specialist (GEFS) from Green Banking and RENAC in Berlin, among other certifications he holds in the area of climate and sustainability finance. Furthermore, he is author of *Introduction to Project Finance in Renewable Energy Infrastructure* published by Springer Nature and works on further publications of relevance in his field. He lectures at the National University of Colombia. He is interested in topics related to climate finance, blended finance for developing countries, ethical banking, public private partnerships, digital innovation and strategic transformation.

Alexandra Oboda holds an MBA in Sustainability Management from Leuphana University Lüneburg and has been working as a sustainability manager for different banks for six years. She currently works for Investitions- und Förderbank Niedersachsen, where she coordinates all ESG measures from strategy to reporting.

Oliver Opel is Professor for energetic optimization of buildings. He is scientific director of the ITE – Institute for the Transformation of the Energy system at FH Westkueste and scientific coordinator of the HY.SH competence centre of Schleswig-Holstein. He holds a doctorate in environmental science from the Institute for Environmental Chemistry at Leuphana University of Lueneburg.

Jane O'Sullivan is an Honorary Senior Research Fellow at the University of Queensland, based in Brisbane, Australia. With an overarching interest in human ecology and planetary health, she worked on tropical semi-subsistence farming systems before focusing on the threats to security posed by population growth and the taboos and misrepresentations that prevent effective responses. She is a co-convenor of The Overpopulation Project and a board member of Sustainable Population Australia.

Adrien Pagano studied EU politics and international relations in France, Germany and Belgium. He started working at the Bundestag in 2019 as a scientific assistant to Sylvia Kotting-Uhl, Chairperson of the Environment Committee, where he was in charge of nuclear safety in Germany and international nuclear monitoring. Since 2022, he is the head of Stefan Wenzels office in the Bundestag. Also relying on two years in the French administration, he particularly enjoys fostering the Franco-German discussion and cooperation in the energy sector.

Jasmine Ramsebner studied business administration at Vienna University of Economics and Business. In her doctoral thesis at the Vienna University of Technology, she dealt with the topic "The role of sector coupling in the energy transition". She initially worked at the Austrian Federal Association for Electric Mobility (BEÖ) and has been with the KEBA Group since 2021, where she is Portfolio Director and leads a team that develops charging solutions, digital solutions and service products for emobility.

Michael Schulthoff studied Mechanical Engineering with a focus on Chemical Engineering and Energy Science (M. Sc.) at Ruhr-University Bochum (RUB). After an exchange at Virginia Polytechnic Institute and State University in 2018 he worked at Massachusetts Institute of Technology (MIT) in 2019 for his master´s thesis. He pursued working at the MIT Energy Initiative as a Researcher in 2020 until he started as a research assistant and PhD student (Dr.-Ing.) at the Institute of Environmental Technology and Energy Economics at TUHH.

Alexander Stuckenholz studied Electrical Engineering with a focus on Software Engineering (Dipl.-Ing.) at the Ruhr-University Bochum (RUB). In his doctoral thesis at the Fernuniversität Hagen, he dealt with the unplanned evolution of component-based software systems. During his research activities, he spent several months at San Diego State University. After his research, Dr. Stuckenholz worked for several years in industry, e.g. as a software architect and IT manager, mainly for large energy companies. In 2012, Dr. Stuckenholz was announced as a Professor of Practical Information at Hamm-Lippstadt University of Applied Sciences. He is head of the Intelligent Systems Design course there. He is also the author of the book *Basiswissen Energieinformatik*.

Jens Thorn studied economics at the Ruhr University Bochum (Dipl.-Ök.). He continued his university education with his doctoral studies, where he also worked as a research assistant. As a research assistant, he carried out several research projects with companies. Subsequently, he worked for a technology-oriented industrial company for measurement and control technology. There he accompanied an IPO at the New York Stock Exchange. At Hamm-Lippstadt University of Applied Sciences, Jens is active in the research field of energy transition and transformation design with several research projects. Currently, he is founding member of the university´s new institute for sector coupling (Institut für Sektorenkopplung in der Energiewende – IfSK).

Christoph Torwegge is corporate and M&A partner and focuses his advice on matters of M&A transactions and selected areas of national and international commercial law in the Energy & Utilities and Infrastructure Sectors.

His clients are benefitting from his long-term experience gained in numerous international M&A transaction work and as General Counsel of an international group. Being focused on Energy & Utilities and Infrastructure Christoph advises regularly investors, banks, project developers and operators in planning, construction, financing and on transactions of renewable energy and infrastructure projects. Besides his experience in drafting and negotiating supply, maintenance and service agreements as well as engineering, procurement and construction agreements he has achieved numerous successful closings of transactions driven by his special expertise on transaction work in projects.

Christoph studied in Bielefeld, Bristol, and Leipzig. He worked for an international law firm and as head of legal before becoming a partner in Osborne Clarke in 2012. He is a member in various associations and author of numerous publications.

Jochen Weilepp holds a doctorate in physics and an MBA from INSEAD. He is a professor of energy economics at Biberach University of Applied Sciences in southern Germany. He teaches in particular finance and investment theory as well as strategic management and entrepreneurship with a focus on energy projects and start-ups in the energy sector.

Previously, Jochen was Managing Director of Voith Hydro Ocean Current Technologies and Division Head for New Technologies at Voith Hydro. He started his career as a management consultant and project manager at the Boston Consulting Group.

In the course of his career, Jochen has held several advisory board mandates in the marine energy sector. He was also Vice President and Co-President of Ocean Energy Europe for many years.

Clemens Weiß has a background in sustainability geography, social sciences and worked for several years in the renewable energy sector. In order to have more influence on politics, he joined the office of the Chairperson of the Bundestag Committee on the Environment, Nature Conservation and Nuclear Safety, in particular dealing with the issue of radioactive waste management and the search for a final repository. Since 2022, he has been working with Stefan Wenzel in the Federal Ministry for Economic Affairs and Climate Action.

Stefan Wenzel is a member of the party Bündnis 90/Die Grünen since 1986 and is currently working as a Parliamentary State Secretary in the Federal Ministry for Economic Affairs and Climate Action. He studied Agronomy and has experience as a distributor in the ecological winery sector. From 1998 to 2021, he was a member of the Landtag of Lower Saxony and from 2013 to 2017 the Minister for Environment, Energy and Climate Protection of Lower Saxony. In this position, he also headed the nuclear inspectorate of Lower Saxony.

Marlies Wiegand is the team leader of the group "building technology" at the ITE – Institute for the Transformation of the Energy system since 2019. Before, she worked at Leuphana University of Lueneburg as a scientific associate. She studied Environmental Sciences (B.Sc.) and Sustainability Science (M.Sc.) at Leuphana University of Lueneburg, Germany and at the Hamburg University of Technology.

Lorenz Wieshammer is a Director at NERA Economic Consulting based in Berlin. He advises utilities, regulatory authorities, investors, and lawyers in the fields of regulatory, financial, and energy economics. He specializes in energy network regulation and energy transition topics. He is a Chartered Financial Analyst (CFA). He holds an M.Sc. in Economics from the London School of Economics and a B.Sc. in Economics from the University of Mannheim.

Index

https://doi.org/10.1515/9783110784299-031

www.ingramcontent.com/pod-product-compliance
Lightning Source LLC
Chambersburg PA
CBHW080936240926
51849CB00004BA/470
9783110784176